ENVIRONMENTAL ENGINEERING
FE Review Manual

Brightwood
ENGINEERING EDUCATION

This publication is designed to provide accurate and authoritative information in regard to the subject matter covered. It is sold with the understanding that the publisher is not engaged in rendering legal, accounting, or other professional service. If legal advice or other expert assistance is required, the services of a competent professional person should be sought.

Executive Director of Engineering Education: Brian S. Reitzel, PE

ENVIRONMENTAL ENGINEERING: FE REVIEW MANUAL

© 2016 Brightwood College

Published by Brightwood Engineering Education

2800 E. River Road

Dayton, OH 45439

1-800-420-1432

www.brightwoodengineering.com

All rights reserved. The text of this publication, or any part thereof, may not be reproduced in any manner whatsoever without permission in writing from the publisher.

Printed in the United States of America.

16 17 18   10 9 8 7 6 5 4 3 2 1

ISBN: 978-1-68338-014-6

CONTENTS

CHAPTER 1 — **Introduction** 1
- HOW TO USE THIS BOOK 1
- BECOMING A PROFESSIONAL ENGINEER 1
- FUNDAMENTALS OF ENGINEERING/ENGINEER-IN-TRAINING EXAMINATION 2

CHAPTER 2 — **Mathematics** 7
- ALGEBRA 8
- COMPLEX QUANTITIES 12
- TRIGONOMETRY 12
- GEOMETRY AND GEOMETRIC PROPERTIES 16
- PLANE ANALYTIC GEOMETRY 20
- VECTORS 28
- LINEAR ALGEBRA 31
- NUMERICAL METHODS 40
- NUMERICAL INTEGRATION 41
- NUMERICAL SOLUTIONS OF DIFFERENTIAL EQUATIONS 43
- DIFFERENTIAL CALCULUS 44
- INTEGRAL CALCULUS 52
- DIFFERENTIAL EQUATIONS 56
- FOURIER SERIES AND FOURIER TRANSFORM 64
- DIFFERENCE EQUATIONS AND Z-TRANSFORMS 67
- PROBLEMS 68
- SOLUTIONS 81

CHAPTER 3 — **Probability, Statistics, and Sampling** 93
- COUNTING SETS 93
- PROBABILITY 96
- STATISTICAL TREATMENT OF DATA 100
- STANDARD DISTRIBUTION FUNCTIONS 103
- CONFIDENCE BOUNDS 110
- HYPOTHESIS TESTING 112
- SAMPLING 114
- PROBLEMS 116
- SOLUTIONS 120

CHAPTER 4 — Engineering Economics 125

- CASH FLOW 126
- TIME VALUE OF MONEY 128
- EQUIVALENCE 128
- COMPOUND INTEREST 129
- NOMINAL AND EFFECTIVE INTEREST 136
- SOLVING ENGINEERING ECONOMICS PROBLEMS 138
- PRESENT WORTH 138
- FUTURE WORTH OR VALUE 141
- ANNUAL COST 141
- RATE OF RETURN ANALYSIS 143
- BENEFIT-COST ANALYSIS 148
- BREAKEVEN ANALYSIS 149
- OPTIMIZATION 149
- VALUATION AND DEPRECIATION 151
- TAX CONSEQUENCES 156
- INFLATION 157
- RISK 159
- REFERENCE 160
- PROBLEMS 161
- SOLUTIONS 170

CHAPTER 5 — Ethics and Professional Practices 179

- MORALS, PERSONAL ETHICS, AND PROFESSIONAL ETHICS 180
- CODES OF ETHICS 180
- AGREEMENTS AND CONTRACTS 183
- ETHICAL VERSUS LEGAL BEHAVIOR 185
- PROFESSIONAL LIABILITY 186
- PUBLIC PROTECTION ISSUES 186
- REFERENCES 189
- PROBLEMS 190
- SOLUTIONS 199

CHAPTER 6 — Environmental Management Systems 205

- SUSTAINABILITY 206
- WASTE MANAGEMENT HIERARCHY 207
- POLLUTION PREVENTION AND WASTE MINIMIZATION 208
- LIFE CYCLE ASSESSMENT 209
- INDUSTRIAL AND OCCUPATIONAL HEALTH AND SAFETY 216
- PROBLEMS 221
- SOLUTIONS 223

CHAPTER 7 — Environmental Science and Ecology 225
MASS BALANCE APPROACH TO PROBLEM SOLVING 226
INTRODUCTION TO MICROBIOLOGY 227
BIOLOGICAL SYSTEMS 234
BIOPROCESSING 238
DO SAG MODELING 241
LIMNOLOGY 248
WETLANDS 250
PROBLEMS 254
SOLUTIONS 257

CHAPTER 8 — Environmental Chemistry 261
PRINCIPLES OF MATTER 262
PERIODICITY 266
NOMENCLATURE 267
METALS AND NONMETALS 268
ORGANIC CHEMISTRY 271
MATERIAL PHYSICAL PROPERTIES 274
CONCENTRATION UNITS 274
EQUATIONS AND STOICHIOMETRY 277
REACTION ORDER AND KINETICS 279
EQUILIBRIUM 287
OXIDATION AND REDUCTION 291
ELECTROCHEMISTRY 293
COMBUSTION 295
PROBLEMS 298
SOLUTIONS 306

CHAPTER 9 — Material Science 319
ATOMIC ORDER IN ENGINEERING MATERIALS 320
ATOMIC DISORDER IN SOLIDS 329
MICROSTRUCTURES OF SOLID MATERIALS 332
MATERIALS PROCESSING 339
MATERIALS IN SERVICE 345
COMPOSITES 353
SELECTED SYMBOLS AND ABBREVIATIONS 357
REFERENCE 357
PROBLEMS 358
SOLUTIONS 369

CHAPTER 10: Thermodynamics and Phase Equilibrium 375

- THERMODYNAMIC SYSTEMS 376
- FIRST LAW OF THERMODYNAMICS 377
- SECOND LAW OF THERMODYNAMICS 379
- PROPERTIES OF PURE SUBSTANCES 382
- IDEAL GASES 389
- PHASE EQUILIBRIA 395
- PROCESSES 402
- CYCLES 404
- HEAT TRANSFER 415
- SELECTED SYMBOLS AND ABBREVIATIONS 419
- PROBLEMS 420
- SOLUTIONS 429

CHAPTER 11: Fluid Mechanics 439

- FLUID PROPERTIES 440
- FLUID STATICS 443
- CONSERVATION LAWS 449
- VELOCITY AND FLOW MEASURING DEVICES 458
- PUMP SELECTION 461
- TURBINES 469
- OPEN CHANNEL FLOW 470
- INCOMPRESSIBLE FLOW OF GASES 477
- FANS AND BLOWERS 478
- AIR AND GAS COMPRESSORS 483
- PROBLEMS 492
- SOLUTIONS 500

CHAPTER 12: Water Resources Engineering 509

- WATER RESOURCES PLANNING 509
- HYDROLOGIC ELEMENTS 509
- WATERSHED HYDROGRAPHS 512
- PEAK DISCHARGE ESTIMATION 518
- HYDROLOGIC ROUTING 521
- WATER DEMAND 525
- WATER DISTRIBUTION NETWORKS 529
- WASTEWATER FLOWS 531
- SEWER SYSTEM DESIGN 532
- PROBLEMS 536
- SOLUTIONS 539

CHAPTER 13

Soils and Groundwater 543

SOIL CLASSIFICATION 543
GROUNDWATER 549
CONSOLIDATION 556
GROUNDWATER FLOW 557
GROUNDWATER CONTAMINATION 566
PROBLEMS 574
SOLUTIONS 576

CHAPTER 14

Water and Wastewater 581

WATER QUALITY INDICATORS 582
HYDRAULIC CHARACTERISTICS OF REACTORS 594
WATER TREATMENT 601
ADVANCED WATER TREATMENT 612
WATER TREATMENT RESIDUALS MANAGEMENT 617
WASTEWATER TREATMENT 627
WASTEWATER SLUDGE TREATMENT 649
PROBLEMS 658
SOLUTIONS 663

CHAPTER 15

Air Quality and Atmospheric Pollution Control 675

AIR QUALITY 675
METEOROLOGY 677
ATMOSPHERIC DISPERSION MODELING 678
VENTILATING 681
INTRODUCTION TO ATMOSPHERIC POLLUTION CONTROL 684
REMOVAL OF PARTICULATE MATTER 684
ABSORPTION 689
ADSORPTION 690
THERMAL CONTROLS AND DESTRUCTION 694
PROBLEMS 696
SOLUTIONS 698

CHAPTER 16

Solid and Hazardous Waste 703

MUNICIPAL SOLID WASTE 704
THERMAL TREATMENT OF MSW 705
MSW COMPOSTING 714
LANDFILLS 717
HAZARDOUS WASTE 730
SITE REMEDIATION 740
RADIOACTIVE WASTE 748
PROBLEMS 750
SOLUTIONS 752

PERMISSIONS

"Rules of Professional Conduct," Chapter 5, reprinted by permission of NCEES.
Source: Model Rules, National Council of Examiners for Engineering and Surveying, 2007. www.ncees.org

Figure 11.7 reprinted by permission of ASME.
Source: Moody, L.F, Transactions of the ASME, Volume 66: pp. 671–684. 1944.

The figure on page 390 courtesy of DuPont.
Source: Thermodynamic Properties of HFC-134a, DuPont Company.

Chapter 9 Exhibit 2 and Figures 9.4a-b, 9.5a-b, 9.6a, 9.9, 9.10, 9.18, 9.21, 9.22, and 9.23 reprinted with permission of John Wiley & Sons, Inc.
Source: Callister, William D., Jr. Materials Science and Engineering: An Introduction, 6/e. J. Wiley & Sons. 2003.

Figures 9.4c, 9.5c, and 9.6b reprinted by permission of the estate of William G. Moffatt.
Source: Moffatt, William G. The Structure and Property of Materials, Volume 1. J. Wiley & Sons. 1964.

Tables 9.4 and 9.5 reprinted by permission of McGraw-Hill Companies.
Source: Fontana, M., Corrosion Engineering. McGraw-Hill Companies.

Figure 9.13 used by permission of ASM International.
Source: Mason, Clyde W., Introductory Physical Metallurgy: p. 33. 1947.

Figure 9.26 used by permission of ASM International.
Source: Rinebolt, J.A., and W. J. Harris, Jr., "Effect of Alloying Elements on Notch Toughness of Pearlitic Steels." Transactions of ASM, Volume 43: pp. 1175–1201. 1951

CHAPTER AUTHORS

Philip J. Parker, PhD, PE, Professor of Civil Engineering at University of Wisconsin–Platteville

Ben J. Stuart, PhD, PE, Professor of Civil Engineering at Ohio University

CHAPTER 1

Introduction

OUTLINE

HOW TO USE THIS BOOK 1

BECOMING A PROFESSIONAL ENGINEER 1
Education ■ Fundamentals of Engineering/Engineer-in-Training Examination ■ Experience ■ Professional Engineer Examination

FUNDAMENTALS OF ENGINEERING/ENGINEER-IN-TRAINING EXAMINATION 2
Examination Development ■ Examination Structure ■ Examination Dates ■ Examination Procedure ■ Examination-Taking Suggestions ■ License Review Books ■ Textbooks ■ Examination Day Preparations ■ Items to Take to the Examination ■ Special Medical Condition ■ Examination Scoring and Results ■ Errata

HOW TO USE THIS BOOK

Fundamentals of Engineering FE/EIT Exam Preparation is designed to help you prepare for the Fundamentals of Engineering/Engineer-in-Training exam. The book covers the full breadth and depth of topics covered by the new Fundamentals of Engineering exams.

Each chapter of this book covers a major topic on the exam, reviewing important terms, equations, concepts, analysis methods, and typical problems. Solved examples are provided throughout each chapter to help you apply the concepts and to model problems you may see on the exam. After reviewing the topic, you can work the end-of-chapter problems to test your understanding. The problems are typical of what you will see on the exam, and complete solutions are provided so that you can check your work and further refine your solution methodology.

The following sections provide you with additional details on the process of becoming a licensed professional engineer and on what to expect at the exam.

BECOMING A PROFESSIONAL ENGINEER

To achieve registration as a Professional Engineer, there are four distinct steps: (1) education, (2) the Fundamentals of Engineering/Engineer-in-Training (FE/EIT) exam, (3) professional experience, and (4) the professional engineer (PE) exam. These steps are described in the following sections.

Education

Generally, no college degree is required to be eligible to take the FE/EIT exam. The exact rules vary, but all states allow engineering students to take the FE/EIT exam before they graduate, usually in their senior year. Some states, in fact, have no education requirement at all. One merely need apply and pay the application fee. Perhaps the best time to take the exam is immediately following completion of related coursework. For most engineering students, this will be the end of the senior year.

Fundamentals of Engineering/Engineer-in-Training Examination

This six-hour, multiple-choice examination is known by a variety of names—Fundamentals of Engineering, Engineer-in-Training (EIT), and Intern Engineer—but no matter what it is called, the exam is the same in all states. It is prepared and graded by the National Council of Examiners for Engineering and Surveying (NCEES).

Experience

States that allow engineering seniors to take the FE/EIT exam have no experience requirement. These same states, however, generally will allow other applicants to substitute acceptable experience for coursework. Still other states may allow a candidate to take the FE/EIT exam without any education or experience requirements.

Typically, several years of acceptable experience is required before you can take the Professional Engineer exam—the duration varies by state, and you should check with your state licensing board for details.

Professional Engineer Examination

The second national exam is called Principles and Practice of Engineering by NCEES, but many refer to it as the Professional Engineer exam or PE exam. All states, plus Guam, the District of Columbia, and Puerto Rico, use the same NCEES exam. Review materials for this exam are found in other engineering license review books.

FUNDAMENTALS OF ENGINEERING/ENGINEER-IN-TRAINING EXAMINATION

Laws have been passed that regulate the practice of engineering in order to protect the public from incompetent practitioners. Beginning in 1907 the individual states began passing *title* acts regulating who could call themselves engineers and offer services to the public. As the laws were strengthened, the practice of engineering was limited to those who were registered engineers, or to those working under the supervision of a registered engineer. Originally the laws were limited to civil engineering, but over time they have evolved so that the titles, and sometimes the practice, of most branches of engineering are included.

There is no national licensure law; licensure is based on individual state laws and is administered by boards of registration in each state. You can find a list of contact information for and links to the various state boards of registration at the Brightwood Engineering Web site: *www.brightwoodengineering.com*. This list also shows the exam registration deadline for each state.

Examination Development

Initially, the states wrote their own examinations, but beginning in 1966 NCEES took over the task for some of the states. Now the NCEES exams are used by all states. Thus it is easy for engineers who move from one state to another to achieve licensure in the new state. About 50,000 engineers take the FE/EIT exam annually. This represents about 65% of the engineers graduated in the United States each year.

The development of the FE/EIT exam is the responsibility of the NCEES Committee on Examination for Professional Engineers. The committee is composed of people from industry, consulting, and education, all of whom are subject-matter experts. The test is intended to evaluate an individual's understanding of mathematics, basic sciences, and engineering sciences obtained in an accredited bachelor degree of engineering. Every five years or so, NCEES conducts an engineering task analysis survey. People in education are surveyed periodically to ensure the FE/EIT exam specifications reflect what is being taught.

The exam questions are prepared by the NCEES committee members, subject matter experts, and other volunteers. All people participating must hold professional licensure. When the questions have been written, they are circulated for review in workshop meetings and by mail. You will see mostly metric units (SI) on the exam. Some problems are posed in U.S. customary units (USCS) because the topics typically are taught that way. All problems are four-way multiple choice.

Examination Structure

The FE/EIT exam will be six hours in length, which includes a tutorial, a break, the exam, and a brief survey at the conclusion of the exam. There are 110 questions total on the exam.

The exam will be divided into two sections with a 25-minute break in the middle. Examinees will be given 5 hours and 20 minutes to complete approximately 55 questions prior to the scheduled break and the remaining questions afterward.

Seven different exams are in the test booklet, including one for each of the following six branches: civil, mechanical, electrical, chemical, industrial, environmental. An Other Disciplines exam is included for those examinees not covered by the six engineering branches. If you are taking the FE/EIT as a graduation requirement, your school may compel you to take the exam that matches the engineering discipline in which you are obtaining your degree. Otherwise, you can choose the exam you wish to take.

Examination Dates

The FE is administered year-round at NCEES-approved Pearson VUE test centers. Registration will be open year-round. Those wishing to take the exam must apply to their state board several months before the exam date.

Examination Procedure

You will register and schedule your appointment through your My NCEES account on the NCEES Web site. You will first select your exam location, and then you will be presented with a list of available exam dates for your appointment. If you are not happy with the choices, you can browse through the available dates at another NCEES-approved testing center.

The examination is closed book. You may not bring any reference materials with you to the exam. To replace your own materials, NCEES has prepared a *FE Reference Handbook*. The handbook contains engineering, scientific, and mathematical formulas and tables for use in the examination. Examinees will receive the handbook from their state registration board prior to the examination. The *FE Reference Handbook* is also included in the exam materials distributed at the beginning of each exam period.

Examination-Taking Suggestions

Those familiar with the psychology of examinations have several suggestions for examinees:

1. There are really two skills that examinees can develop and sharpen. One is the skill of illustrating one's knowledge. The other is the skill of familiarization with examination structure and procedure. The first can be enhanced by a systematic review of the subject matter. The second, exam-taking skills, can be improved by practice with sample problems—that is, problems that are presented in the exam format with similar content and level of difficulty.

2. Examinees should answer every problem, even if it is necessary to guess. There is no penalty for guessing. The best approach to guessing is to try to eliminate one or two of the four alternatives. If this can be done, the chance of selecting a correct answer obviously improves from 1 in 4 to 1 in 2 or 3.

3. Plan ahead with a strategy and a time allocation. Compute how much time you will allow for each question. You might allocate a little less time per problem for the areas in which you are most proficient, leaving a little more time for those that may be more difficult. Your time plan should include a reserve block for especially difficult problems, for checking your scoring sheet, and finally for making last-minute guesses on problems you did not work. Your strategy might also include time allotments for two passes through the exam—the first to work all problems for which answers are obvious to you, the second to return to the more complex, time-consuming problems and the ones at which you might need to guess.

4. Read all four multiple-choice answer options before making a selection. All distractors (wrong answers) are designed to be plausible. Only one option will be the best answer.

5. Do not change an answer unless you are absolutely certain you have made a mistake. Your first reaction is likely to be correct.

6. If time permits, check your work.

7. Do not sit next to a friend, a window, or other potential distraction.

License Review Books

To prepare for the FE/EIT exam you need one or two review books.

1. This book, to provide a review of the discipline specific examination.
2. *FE Reference Handbook*. At some point this NCEES-prepared book will be provided to applicants by their state registration board. You may want to obtain a copy sooner so you will have ample time to study it before the exam. Pay close attention to the *FE Reference Handbook* and the notation used in it. You will have two computer screens, one with a searchable PDF version of the *FE Reference Handbook* and one with the exam.

Textbooks

If you still have your university textbooks, they can be useful in preparing for the exam, unless they are out of date. To a great extent the books will be like old friends with familiar notation. You probably need both textbooks and license review books for efficient study and review.

Examination Day Preparations

The exam day will be a stressful and tiring one. You should take steps to eliminate the possibility of unpleasant surprises. If at all possible, visit the examination site ahead of time to determine the following:

1. How much time should you allow for travel to the exam on that day? Plan to arrive about 15 minutes early. That way you will have ample time, but not too much time. Arriving too early, and mingling with others who are also anxious, can increase your anxiety and nervousness.
2. Where will you park?
3. How does the exam site look? Will you have ample workspace? Will it be overly bright (sunglasses), or cold (sweater), or noisy (earplugs)? Would a cushion make the chair more comfortable?
4. Where are the drinking fountain and lavatory facilities?
5. What about food? Most states do not allow food in the test room (exceptions for ADA). Should you take something along for energy in the exam? A light bag lunch during the break makes sense.

Items to Take to the Examination

Although you may not bring books to the exam, you should bring the following:

- *Calculator*—NCEES has implemented a more stringent policy regarding permitted calculators. For a list of permitted models, see the NCEES Web site *(www.ncees.org)*. You also need to determine whether your state permits pre-programmed calculators. Bring extra batteries for your calculator just in case, and many people feel that bringing a second calculator is also a very good idea.
- *Clock*—You must have a time plan and a clock or wristwatch.

- *Exam Assignment Paperwork*—Take along the letter assigning you to the exam at the specified location to prove that you are the registered person. Also bring something with your name and picture (driver's license or identification card).

- *Items Suggested by Your Advance Visit*—If you visit the exam site, it will probably suggest an item or two that you need to add to your list.

- *Clothes*—Plan to wear comfortable clothes. You probably will do better if you are slightly cool, so it is wise to wear layered clothing.

Special Medical Condition

If you have a medical situation that may require special accommodation, notify the licensing board well in advance of exam day.

Examination Scoring and Results

Examinees will be notified via e-mail when their results are available for viewing in My NCEES. The process is still being finalized, but most examinees should receive their results within 7 to 10 business days.

Errata

The authors and publisher of this book have been careful to avoid errors, employing technical reviewers, copyeditors, and proofreaders to ensure the material is as flawless as possible. Any known errata and corrections are posted on the product page at our Web site, *www.brightwoodengineering.com*. If you believe you have discovered an inaccuracy, please notify Customer Service at *enginfo@brightwood.edu*.

CHAPTER 2

Mathematics

OUTLINE

ALGEBRA 8
Factorials ■ Exponents ■ Logarithms ■ The Solution of Algebraic Equations ■ Progressions

COMPLEX QUANTITIES 12
Definition and Representation of a Complex Quantity ■ Properties of Complex Quantities

TRIGONOMETRY 12
Definition of an Angle ■ Measure of an Angle ■ Trigonometric Functions of an Angle ■ Fundamental Relations among the Functions ■ Functions of Multiple Angles ■ Functions of Half Angles ■ Functions of Sum or Difference of Two Angles ■ Sums, Differences, and Products of Two Functions ■ Properties of Plane Triangles

GEOMETRY AND GEOMETRIC PROPERTIES 16
Right Triangle ■ Oblique Triangle ■ Equilateral Triangle ■ Square ■ Rectangle ■ Parallelogram ■ Regular Polygon of n Sides ■ Circle ■ Ellipse ■ Parabola ■ Cube ■ Prism or Cylinder ■ Pyramid or Cone ■ Sphere

PLANE ANALYTIC GEOMETRY 20
Rectangular Coordinates ■ Polar Coordinates ■ Relations Connecting Rectangular and Polar Coordinates ■ Points and Slopes ■ Locus and Equation ■ Straight Line ■ Circle ■ Conic ■ Parabola ■ Ellipse ■ Hyperbola

VECTORS 28
Definition and Graphical Representation of a Vector ■ Graphical Summation of Vectors ■ Analytic Representation of Vector Components ■ Properties of Vectors ■ Vector Sum V of any Number of Vectors, $V_1, V_2, V_3, \ldots$ ■ Product of a Vector V and a Scalar s ■ Scalar Product or Dot Product of Two Vectors: $V_1 \bullet V_2$ ■ Vector Product or Cross Product of Two Vectors: $V_1 \times V_2$

LINEAR ALGEBRA 31
Matrix Operations ■ Types of Matrices ■ Elementary Row and Column Operations ■ Determinants

NUMERICAL METHODS 40
Root Extraction ■ Newton's Method

NUMERICAL INTEGRATION 41
Euler's Method ■ Trapezoidal Rule

NUMERICAL SOLUTIONS OF DIFFERENTIAL EQUATIONS 43
Reduction of Differential Equation Order

DIFFERENTIAL CALCULUS 44
Definition of a Function ▪ Definition of a Derivative ▪ Some Relations among Derivatives ▪ Table of Derivatives ▪ Slope of a Curve: Tangent and Normal ▪ Maximum and Minimum Values of a Function ▪ Points of Inflection of a Curve ▪ Taylor and Maclaurin Series ▪ Evaluation of Indeterminate Forms ▪ Differential of a Function ▪ Functions of Several Variables, Partial Derivatives, and Differentials

INTEGRAL CALCULUS 52
Definition of an Integral ▪ Fundamental Theorems on Integrals

DIFFERENTIAL EQUATIONS 56
Definitions ▪ Notation ▪ Equations of First Order and First Degree: $M\,dx + N\,dy = 0$ ▪ Constant Coefficients ▪ Variation of Parameters ▪ Undetermined Coefficients ▪ Euler Equations ▪ Laplace Transform

FOURIER SERIES AND FOURIER TRANSFORM 64
Fourier Series ▪ Fourier Transform

DIFFERENCE EQUATIONS AND Z-TRANSFORMS 67
Z-Transforms

PROBLEMS 68

SOLUTIONS 81

ALGEBRA

Factorials

Definition. The factorial of a non-negative integer, n, is defined as $n!$ $n! = n(n-1)(n-2)(n-3)\ldots$ and so forth.

For example, $6! = 6(5)(4)(3)(2)(1) = 720$.

Also, $1! = 1$ and $0! = 1$.

Factorials can be written as multiples of other factorials:

$6! = 6(5!) = 6(5)(4!) = 6(5)(4)(3!) = 6(5)(4)(3)(2!) = 6(5)(4)(3)(2)(1!) = 720$

Factorials can be multiplied and divided.

$$\frac{n!}{(n-1)!} = \frac{n(n-1)!}{(n-1)!} = n$$

Exponents

Definition. Any number defined as base, b, can be multiplied by itself x number of times, which is denoted as b^x.

For example, $b^4 = b(b)(b)(b)$.

Properties of Exponents

$$b^0 = 1; \quad b^1 = b; \quad b^{-1} = \frac{1}{b}; \quad b^{-x} = \frac{1}{b^x}$$

$$b^{x+y} = (b^x)(b^y); \quad b^{x-y} = \frac{b^x}{b^y}; \quad b^{x \times y} = (b^x)^y$$

$$b^{\frac{1}{2}} = \sqrt{b}; \quad b^{\frac{1}{x}} \sqrt[x]{b}; \quad b^{\frac{x}{y}} = \left(\sqrt[y]{b}\right)^x$$

Logarithms

Definition. If b is a finite positive number, other than 1, and $b^x = N$, then x is the logarithm of N to the base b, or $\log_b N = x$. If $\log_b N = x$, then $b^x = N$.

Properties of Logarithms

$$\log_b b = 1; \quad \log_b 1 = 0; \quad \log_b 0 = \begin{cases} +\infty, \text{ when } b \text{ lies between 0 and 1} \\ +\infty, \text{ when } b \text{ lies between 1 and } \infty \end{cases}$$

$$\log_b (M \circ N) = \log_b M + \log_b N \quad \log_b M/N = \log_b M - \log_b N$$

$$\log_b N^P = p \log_b N \quad \log_b \sqrt[r]{N^P} = \frac{p}{r} \log_b N$$

$$\log_b N = \log_a N / \log_a b; \quad \log_b b^N = N; \quad b^{\log_b N} = N$$

Systems of Logarithms

Common (Briggsian)—base 10.

Natural (Napierian or hyperbolic)—base 2.7183 (designated by e or ε).

The abbreviation of *common logarithm* is log, and the abbreviation of *natural logarithm* is ln.

Example 2.1

(i) Solve for a if $\log_a 10 = 0.25$ and (ii) find $\log(\frac{1}{x})$, if $\log x = 0.3332$.

Solution

(i) If $\log_a N = x$, then $N = a^x$ or $a = N^{(1/x)}$

Here, $N = 10$ and $x = 0.25$; then $a = 10^{\frac{1}{0.25}} = 10,000$

(ii) Since $\log \frac{M}{N} = \log M - \log N$, $\log(\frac{1}{x}) = \log 1 - \log x = -0.3332$

Example 2.2

Solve the equation $\log x + \log (x - 3) - \log 4 = 0$.

Solution

Since $(\log M + \log N - \log P) = \log (\frac{MN}{P})$, $\log x + \log (x - 3) - \log 4$

$= \log \left[\frac{x(x-3)}{4}\right] = 0$

$\left[\frac{x(x-3)}{4}\right] = 10^0 = 1$; simplifying, $x^2 - 3x - 4 = 0$. Finding the roots,

$x = 4$ and -1; then $x = 4$ (-1 is not an answer because the log of a negative number is undefined).

The Solution of Algebraic Equations

Definition. A root, x, is any value such that $f(x) = 0$.

The Quadratic Equation

If $ax^2 + bx + c = 0$, then

$$x = \frac{-b \pm \sqrt{b^2 - 4ac}}{2a}$$

If $b^2 - 4ac > 0$, the two roots are real and unequal; if $b^2 - 4ac = 0$, the two roots are real and equal; if $b^2 - 4ac < 0$, the two roots are imaginary.

Example 2.3

Find the root(s) of the following equations (i) $x + 4 = 0$, (ii) $x^2 + 4x + 3 = 0$, (iii) $x^2 - 4x + 4 = 0$, (iv) $x^2 + 4 = 0$, and (v) $3x^3 + 3x^2 - 18x = 0$.

Solution

For the quadratic equations in (ii), (iii), and (iv), use either the equation or simply a scientific calculator to find the roots. The results yield:

(i) -4, (ii) -1 and -3 (real and distinct roots), (iii) 2 and 2 (real and equal roots),

(iv) $+i2$ and $-i2$ (complex roots; always occur in pairs called *conjugates*)

For (v), factoring, $3x(x^2 + x - 6) = 0$. Roots are 0, -3, and 2.

Example 2.4

Find the equation whose roots are 3 and -2.

Solution

If the roots are x_1, x_2, x_3, etc., the equation is $(x - x_1)(x - x_2)(x - x_3)\ldots = 0$; here, $(x - 3)(x - (-2)) = 0$; $(x - 3)(x + 2) = 0$; $x^2 - x - 6 = 0$.

Progressions

Arithmetic Progression

An arithmetic progression is $a, a + d, a + 2d, a + 3d, \ldots$, where d = common difference.

The nth term is $t_n = a + (n - 1)d$

The sum of n terms is $S_n = \frac{n}{2}[2a+(n-1)d] = \frac{n}{2}(a+t_n)$

Geometric Progression

A geometric progression is $a, ar, ar^2, ar^3, \ldots$, where r = common ratio.

The nth term is $t_n = ar^{n-1}$

The sum of n terms is $S_n = a\left(\frac{1-r^n}{1-r}\right)$

If $r^2 < 1$, S_n approaches a definite limit as n increases indefinitely, and

$$S_\infty = \frac{a}{1-r}$$

Example 2.5

Consider the arithmetic progression $1, 3, 5, 7, 9, 11, 13, \ldots$ (i) Find the sum of the first seven terms and (ii) the 18th term of the progression.

Solution

(i) First term $a = 1$, number of terms $n = 7$, and the common difference $d = 2$

sum $S_n = \frac{n}{2}[2a+(n-1)d] = \frac{7}{2}[2+6(2)] = 49$

(ii) Number of terms $n = 18$, first term $a = 1$, and the common difference $d = 2$

The last term or the 18th term is $= a + (n - 1)d = 1 + (18 - 1)2 = 35$.

Example 2.6

Find the sum of the series $1, 0.5, 0.25, 0.125, 0.0625, \ldots$

Solution

This geometric series is convergent. First term $a = 1$ and the common ratio $r = 0.5$.

As the number of terms tend to infinity, sum $S = \frac{a}{1-r} = \frac{1}{1-0.5} = 2$

Example 2.7

Consider the geometric progression $2, 4, 8, 16, 32, 64, 128, \ldots$ (i) Find the sum of the first seven terms and (ii) the 20th term of the series.

Solution

(i) First term $a = 2$, common ratio $r = 2$, and the number of terms $n = 7$

sum $S = a\frac{(1-r^n)}{(1-r)} = \frac{2(1-2^7)}{(1-2)} = 256$

(ii) The 20th term of the series is $= ar^{(n-1)} = 2(2)^{(20-1)} = 1{,}048{,}576$.

COMPLEX QUANTITIES

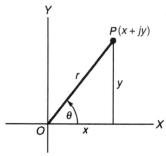

Figure 2.1

Definition and Representation of a Complex Quantity

If $z = x + jy$, where $j = \sqrt{-1}$ and x and y are real, z is called a complex quantity and is completely determined by x and y.

If $P(x, y)$ is a point in the plane (Figure 2.1), then the segment OP in magnitude and direction is said to represent the complex quantity $z = x + jy$.

If θ is the angle from OX to OP and r is the length of OP, then $z = x + jy = r(\cos\theta + j\sin\theta) = re^{j\theta}$, where $\theta = \tan^{-1} y/x$, $r = +\sqrt{x^2 + y^2}$ and e is the base of natural logarithms. The pair $x + jy$ and $x - jy$ are called complex conjugate quantities.

Properties of Complex Quantities

Let z, z_1, and z_2 represent complex quantities; then

Sum or difference: $z_1 \pm z_2 = (x_1 \pm x_2) + j(y_1 \pm y_2)$

Equation: If $z_1 = z_2$, then $x_1 = x_2$ and $y_1 = y_2$

Periodicity: $z = r(\cos\theta + j\sin\theta) = r[\cos(\theta + 2k\pi) + j\sin(\theta + 2k\pi)]$, or $z = re^{j\theta} = re^{j(\theta + 2k\pi)}$ and $e^{j2k\pi} = 1$, where k is any integer.

Exponential-trigonometric relations: $e^{jz} = \cos z + j\sin z$, $e^{-jz} = \cos z - j\sin z$,

$$\cos z = \frac{1}{2}\left(e^{jz} + e^{-jz}\right), \quad \sin z = \frac{1}{2j}\left(e^{jz} - e^{-jz}\right)$$

TRIGONOMETRY

Definition of an Angle

An angle is the amount of rotation (in a fixed plane) by which a straight line may be changed from one direction to any other direction. If the rotation is counterclockwise, the angle is said to be positive; if clockwise, negative.

Measure of an Angle

A degree is $\frac{1}{360}$ of the plane angle about a point, and a radian is the angle subtended at the center of a circle by an arc equal in length to the radius. One complete circle contains 180 degrees or 2π radians; 1 radian = $\pi/180$ degrees.

Trigonometric Functions of an Angle

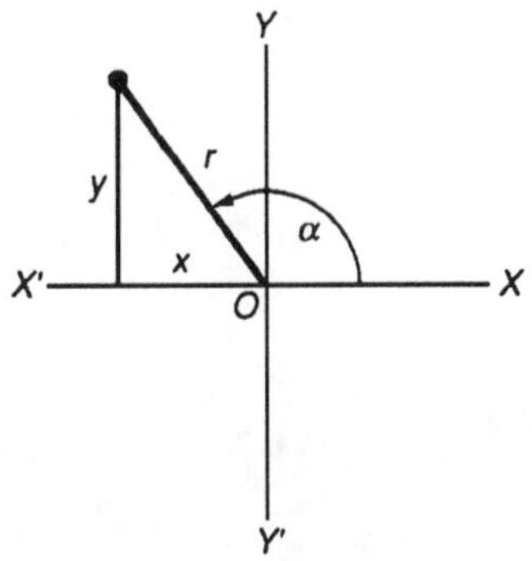

sine $(\sin)\alpha = y/r$ cosecant $(\csc)\alpha = r/y$

cosine $(\cos)\alpha = x/r$ secant $(\sec)\alpha = r/x$

tangent $(\tan)\alpha = y/x$ cotangent $(\cot)\alpha = x/y$

The variable x is positive when measured along OX and negative along OX'. Similarly, y is positive when measured parallel to OY, and negative parallel to OY'.

$\sin 0° = 0$; $\sin 90° = 1$; $\sin 180° = 0$; $\sin 270° = -1$

$\cos 0° = 1$; $\cos 90° = 0$; $\cos 180° = -1$; $\cos 270° = 0$

Figure 2.2

Fundamental Relations among the Functions

$$\sin\alpha = \frac{1}{\csc\alpha}; \quad \cos\alpha = \frac{1}{\sec\alpha}; \quad \tan\alpha = \frac{1}{\cot\alpha} = \frac{\sin\alpha}{\cos\alpha}$$

$$\csc\alpha = \frac{1}{\sin\alpha}; \quad \sec\alpha = \frac{1}{\cos\alpha}; \quad \cot\alpha = \frac{1}{\tan\alpha} = \frac{\cos\alpha}{\sin\alpha}$$

$$\sin^2\alpha + \cos^2\alpha = 1; \quad \sec^2\alpha - \tan^2\alpha = 1; \quad \csc^2\alpha - \cot^2\alpha = 1$$

Functions of Multiple Angles

$\sin 2\alpha = 2\sin\alpha\cos\alpha$

$\cos 2\alpha = 2\cos^2\alpha - 1 = 1 - 2\sin^2\alpha = \cos^2\alpha - \sin^2\alpha$

$\tan 2\alpha = (2\tan\alpha)/(1 - \tan^2\alpha)$

$\cot 2\alpha = (\cot^2\alpha - 1)/(2\cot\alpha)$

Functions of Half Angles

$$\sin\tfrac{1}{2}\alpha = \sqrt{\frac{1-\cos\alpha}{2}}; \quad \cos\tfrac{1}{2}\alpha = \sqrt{\frac{1+\cos\alpha}{2}}$$

$$\tan\tfrac{1}{2}\alpha = \frac{1-\cos\alpha}{\sin\alpha} = \frac{\sin\alpha}{1+\cos\alpha} = \sqrt{\frac{1-\cos\alpha}{1+\cos\alpha}}$$

Functions of Sum or Difference of Two Angles

$\sin(\alpha \pm \beta) = \sin\alpha\cos\beta \pm \cos\alpha\sin\beta$

$\cos(\alpha \pm \beta) = \cos\alpha\cos\beta \mp \sin\alpha\sin\beta$

$$\tan(\alpha \pm \beta) = \frac{\tan\alpha \pm \tan\beta}{1 \mp \tan\alpha\tan\beta}$$

Sums, Differences, and Products of Two Functions

$$\sin\alpha + \sin\beta = 2\sin\frac{1}{2}(\alpha+\beta)\cos\frac{1}{2}(\alpha-\beta)$$

$$\sin\alpha - \sin\beta = 2\cos\frac{1}{2}(\alpha+\beta)\sin\frac{1}{2}(\alpha-\beta)$$

$$\cos\alpha + \cos\beta = 2\cos\frac{1}{2}(\alpha+\beta)\cos\frac{1}{2}(\alpha-\beta)$$

$$\cos\alpha - \cos\beta = 2\sin\frac{1}{2}(\alpha+\beta)\sin\frac{1}{2}(\alpha-\beta)$$

$$\tan\alpha \pm \tan\beta = \frac{\sin(\alpha+\beta)}{\cos\alpha\cos\beta}$$

$$\sin^2\alpha - \sin^2\beta = \sin(\alpha+\beta)\sin(\alpha-\beta)$$

$$\cos^2\alpha - \cos^2\beta = \sin(\alpha+\beta)\sin(\alpha-\beta)$$

$$\cos^2\alpha - \sin^2\beta = \cos(\alpha+\beta)\cos(\alpha-\beta)$$

$$\sin\alpha\sin\beta = \frac{1}{2}\cos(\alpha-\beta) - \frac{1}{2}\cos(\alpha+\beta)$$

$$\cos\alpha\cos\beta = \frac{1}{2}\cos(\alpha-\beta) + \frac{1}{2}\cos(\alpha+\beta)$$

$$\sin\alpha\cos\beta = \frac{1}{2}\sin(\alpha+\beta) + \frac{1}{2}\sin(\alpha-\beta)$$

Example 2.8

Simplify the following expressions to trigonometric functions of angles less than 90 degrees. Note: There is more than one correct answer for each.

i) cos 370°
ii) sin 120°

Solution

Typical strategies for simplifying the functions are shown below:

i) cos 370° = cos (360° + 10°) = (cos 360°)(cos 10°) − (sin 360°)(sin 10°) = (1)(cos 10°) − (0)(sin 10°) = cos 10°

ii) sin 120° = sin (90° + 30°) = (sin 90°)(cos 30°) + (cos 90°)(sin 30°) = (1)(cos 30°) + (0)(sin 30°) = cos 30°

or

sin 120° = sin (180° − 60°) = (sin 180°)(cos 60°) − (cos 180°)(sin 60°) = (0)(cos 60°) − (−1)(sin 60°) = sin 60°

Properties of Plane Triangles

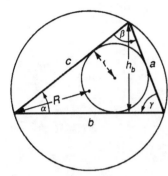

Figure 2.3

Notation. α, β, γ = angles; a, b, c = sides; A = area; h_b = altitude on b;
$s = \frac{1}{2}(a+b+c)$; r = radius of inscribed circle; R = radius of circumscribed circle

$$\alpha + \beta + \gamma = 180° = \pi \text{ radians}$$

$$\frac{a}{\sin \alpha} = \frac{b}{\sin \beta} = \frac{c}{\sin \gamma}$$

$$\frac{a+b}{a-b} = \frac{\tan \frac{1}{2}(\alpha+\beta)}{\tan \frac{1}{2}(\alpha-\beta)}$$

$$a^2 = b^2 + c^2 - 2bc \cos \alpha \qquad a = b \cos \gamma + c \cos \beta$$

$$\cos \alpha = \frac{b^2 + c^2 - a^2}{2bc} \qquad \sin \alpha = \frac{2}{bc}\sqrt{s(s-a)(s-b)(s-c)}$$

$$\sin \frac{\alpha}{2} = \sqrt{\frac{(s-b)(s-c)}{bc}} \qquad \cos \frac{\alpha}{2} = \sqrt{\frac{s(s-a)}{bc}}$$

$$\tan \frac{\alpha}{2} = \sqrt{\frac{(s-b)(s-c)}{s(s-a)}} = \frac{r}{s-a}$$

$$h_b = c \sin \alpha = a \sin \gamma = \frac{2}{b}\sqrt{s(s-a)(s-b)(s-c)}$$

$$r = \sqrt{\frac{(s-a)(s-b)(s-c)}{s}} = (s-a)\tan \frac{\alpha}{2}$$

$$R = \frac{a}{2 \sin \alpha} = \frac{abc}{4A}$$

$$A = \frac{1}{2}bh_b = \frac{1}{2}ab \sin \gamma = \frac{a^2 \sin \beta \sin \gamma}{2 \sin \alpha} = \sqrt{s(s-a)(s-b)(s-c)} = rs$$

Example 2.9

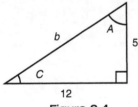

Figure 2.4

Find the side b and the angles A and C for the triangle in Figure 2.4.

Solution

$\tan(A) = 12/5$; then, Angle $A = \tan^{-1}(12/5) = 67.38°$

Since the sum $(A + C + 90°) = 180°$; $C = 90° - A = 22.62°$

Now, $\cos(A) = \frac{5}{b}$; then, $b = \frac{5}{\cos(A)} = 13$

Example 2.10

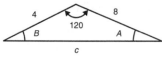

Figure 2.5

Find the side c and the angles A and B for the triangle in Figure 2.5.

Solution

Since two sides and an included angle are given, use the law of cosines.

$c^2 = 4^2 + 8^2 - 2(4)(8) \cos 120$; solving $c = 10.583$

Now use law of sines to find the remaining angles.

$$\frac{10.583}{\sin 120} = \frac{4}{\sin A} = \frac{8}{\sin B};$$ solving, $A = 19.1°$ and $B = 40.89°$

(Check: sum of the angles = 180°)

Example 2.11

Simplify: (i) $(\sec^2 \theta)(\sin^2 \theta)$ (ii) $\sin (A + B) + \sin (A - B)$
(iii) $2\sin^2 \theta + 1 + \cos^2 \theta$

Solution

(i) $(1/\cos^2 \theta) \sin^2 \theta = \tan^2 \theta$ (ii) $2\sin A \cos B$

(iii) $2\sin^2 \theta + 1 + (2\cos^2 \theta - 1) = 2(\sin^2 \theta + \cos^2 \theta) = 2$

GEOMETRY AND GEOMETRIC PROPERTIES

Notation. $a, b, c, d,$ and s denote lengths, A denotes area, V denotes volume

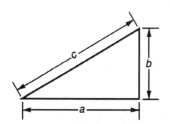

Figure 2.6

Right Triangle

$$A = \frac{1}{2} ab$$

$$c = \sqrt{a^2 + b^2}, \quad a = \sqrt{c^2 - b^2}, \quad b = \sqrt{c^2 - a^2}$$

Figure 2.7

Oblique Triangle

$$A = \frac{1}{2} bh$$

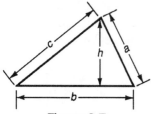

Figure 2.8

Equilateral Triangle

All sides are equal and all angles are 60°.

$$A = \frac{1}{2} ah = \frac{1}{4} a^2 \sqrt{3}, \qquad h = \frac{1}{2} a\sqrt{3}, \qquad r_1 = \frac{a}{2\sqrt{3}}, \qquad r_2 = \frac{a}{\sqrt{3}}$$

Geometry and Geometric Properties

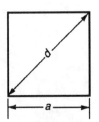

Figure 2.9

Square

All sides are equal, and all angles are 90°.

$$A = a^2, \qquad d = a\sqrt{2}$$

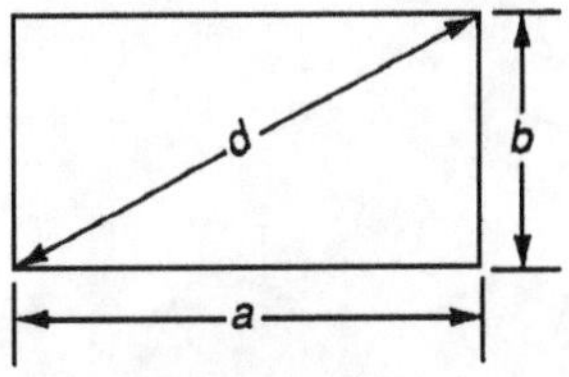

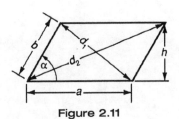

Figure 2.10

Rectangle

Opposite sides are equal and parallel, and all angles are 90°.

$$A = ab, \qquad d = \sqrt{a^2 + b^2}$$

Parallelogram

Opposite sides are equal and parallel, and opposite angles are equal.

$$A = ah = ab \sin \alpha, \quad d_1 = \sqrt{a^2 + b^2 - 2ab \cos \alpha}, \quad d_2 = \sqrt{a^2 + b^2 + 2ab \cos \alpha}$$

Figure 2.11

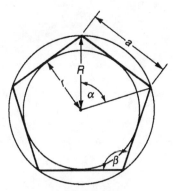

Regular Polygon of n Sides

All sides and all angles are equal.

$$\beta = \frac{n-2}{n} 180° = \frac{n-2}{n} \pi \text{ radians}, \qquad \alpha \frac{360°}{n} = \frac{2\pi}{n} \text{ radians}, \qquad A = \frac{nar}{2}$$

Figure 2.12

Circle

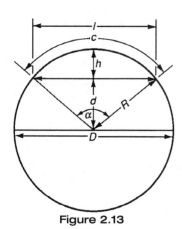
Figure 2.13

Notation. C = circumference, α = central angle in radians

$$C = \pi D = 2\pi R$$

$$c = R\alpha = \frac{1}{2}D\alpha = D\cos^{-1}\frac{d}{R} = D\tan^{-1}\frac{1}{2d}$$

$$l = 2\sqrt{R^2 - d^2} = 2R\sin\frac{\alpha}{2} = 2d\tan\frac{\alpha}{2} = 2d\tan\frac{c}{D}$$

$$d = \frac{1}{2}\sqrt{4R^2 - l^2} = \frac{1}{2}\sqrt{D^2 - l^2} = R\cos\frac{\alpha}{2}$$

$$h = R - d$$

$$\alpha = \frac{c}{R} = \frac{2c}{D} = 2\cos^{-1}\frac{d}{R}$$

$$A_{(circle)} = \pi R^2 = \frac{1}{4}\pi D^2 = \frac{1}{2}RC = \frac{1}{4}DC$$

$$A_{(sector)} = \frac{1}{2}Rc = \frac{1}{2}R^2\alpha = \frac{1}{8}D^2\alpha$$

Ellipse

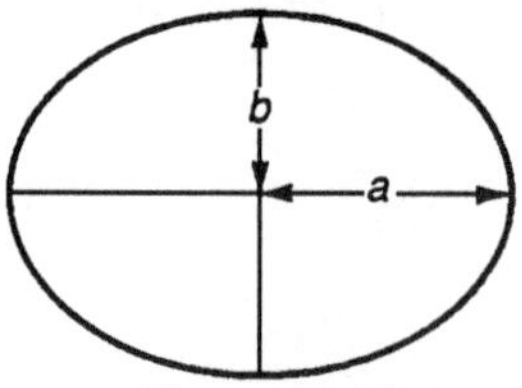

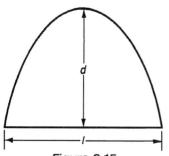
Figure 2.14

$$A = \pi ab$$

$$\text{Perimeter }(s) = \pi(a+b)\left[1 + \frac{1}{4}\left(\frac{a-b}{a+b}\right)^2 + \frac{1}{64}\left(\frac{a-b}{a+b}\right)^4 + \frac{1}{256}\left(\frac{a-b}{a+b}\right)^6 + \cdots\right]$$

$$\text{Perimeter }(s) \approx \pi\frac{a+b}{4}\left[3(1+\lambda) + \frac{1}{1-\lambda}\right], \quad \text{where } \lambda = \left[\frac{a-b}{2(a+b)}\right]^2$$

Parabola

Figure 2.15

$$A = \frac{2}{3}ld$$

Cube

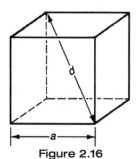
Figure 2.16

$$V = a^3 \qquad d = a\sqrt{3}$$

Total surface area = $6a^2$

Prism or Cylinder

$V =$ (area of base) (altitude, h)

Lateral area = (perimeter of right section)(lateral edge, e)

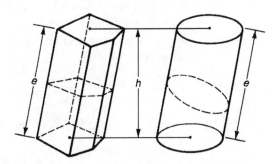

Figure 2.17

Pyramid or Cone

$V = \dfrac{1}{3}$ (area of base) (altitude, h)

Lateral area of regular figure = $\dfrac{1}{2}$ (perimeter of base)(slant height, s)

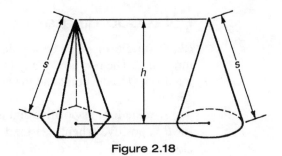

Figure 2.18

Sphere

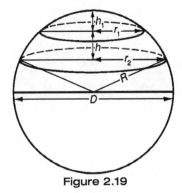

Figure 2.19

$$A_{(sphere)} = 4\pi R^2 = \pi D^2$$

$$A_{(zone)} = 2\pi Rh = \pi Dh$$

$$V_{(sphere)} = \dfrac{4}{3}\pi R^3 = \dfrac{1}{6}\pi D^3$$

$$V_{(spherical\ sector)} = \dfrac{2}{3}\pi R^2 h = \dfrac{1}{6}\pi D^2 h$$

PLANE ANALYTIC GEOMETRY

Rectangular Coordinates

Let two perpendicular lines, $X'X$ (x-axis) and $Y'Y$ (y-axis) meet at a point O (origin). The position of any point $P(x, y)$ is fixed by the distances x (abscissa) and y (ordinate) from $Y'Y$ and $X'X$, respectively, to P. Values of x are positive to the right and negative to the left of $Y'Y$; values of y are positive above and negative below $X'X$.

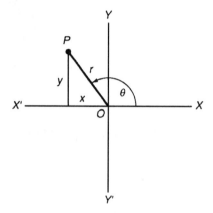

Figure 2.20

Polar Coordinates

Let O (origin or pole) be a point in the plane and OX (initial line) be any line through O. The position of any point $P(r, \theta)$ is fixed by the distance r (radius vector) from O to the point and the angle θ (vectorial angle) measured from OX to OP (Figure 2.20).

A value for r is positive and is measured along the terminal side of θ; a value for θ is positive when measured counterclockwise and negative when measured clockwise.

Relations Connecting Rectangular and Polar Coordinates

$$x = r\cos\theta, \quad y = r\sin\theta$$

$$r = \sqrt{x^2 + y^2}, \quad \theta = \tan^{-1}\frac{y}{x}, \quad \sin\theta = \frac{y}{\sqrt{x^2 + y^2}},$$

$$\cos\theta = \frac{x}{\sqrt{x^2 + y^2}}, \quad \tan\theta = \frac{y}{x}$$

Points and Slopes

Let $P_1(x_1, y_1)$ and $P_2(x_2, y_2)$ be any two points, and let α_1 be the angle from the x axis to P_1P_2, measured counterclockwise.

The length P_1P_2 is $d = \sqrt{(x_2 - x_1)^2 + (y_2 - y_1)^2}$.

The midpoint of P_1P_2 is $\left(\dfrac{x_1 + x_2}{2}, \dfrac{y_1 + y_2}{2}\right)$.

The point that divides P_1P_2 in the ratio $n_1:n_2$ is $\left(\dfrac{n_1 x_2 + n_2 x_1}{n_1 + n_2}, \dfrac{n_1 y_2 + n_2 y_1}{n_1 + n_2}\right)$.

The slope of P_1P_2 is $\tan \alpha = m = \dfrac{y_2 - y_1}{x_2 - x_1}$.

The angle between two lines of slopes m_1 and m_2 is $\beta = \tan^{-1} \dfrac{m_2 - m_1}{1 + m_1 m_2}$.

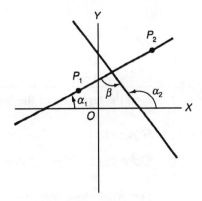

Figure 2.21

Two lines of slopes m_1 and m_2 are perpendicular if $m_2 = -\dfrac{1}{m_1}$.

Example 2.12

Find the distance between the points $(1, -2)$ and $(-4, 2)$.

Solution

The distance is $d = \sqrt{(x_2 - x_1)^2 + (y_2 - y_1)^2} = \sqrt{[1 - (-4)]^2 + [-2 - 2]^2}$
$= \sqrt{5^2 + 4^2} = 6.4$

Locus and Equation

The collection of all points that satisfy a given condition is called the **locus** of that condition; the condition expressed by means of the variable coordinates of any point on the locus is called the **equation of the locus**.

The locus may be represented by equations of three kinds: (1) a rectangular equation involves the rectangular coordinates (x, y); (2) a polar equation involves the polar coordinates (r, θ); and (3) parametric equations express x and y or r and θ in terms of a third independent variable called a parameter.

The following equations are generally given in the system in which they are most simply expressed; sometimes several forms of the equation in one or more systems are given.

Straight Line

$Ax + By + C = 0$ $\quad$ [$-A/B$ = slope]

$y = mx + b$ $\quad$ [m = slope, b = intercept on OY]

$y - y_1 = m(x - x_1)$ $\quad$ [m = slope, $P_1(x_1, y_1)$ is a known point on the line]

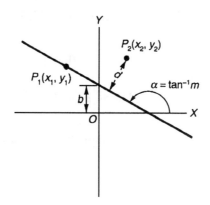

Figure 2.22

Example 2.13

Find the equation of the line passing through the points (2, 1) and (3, −3).

Solution

Slope can be found as $m = \dfrac{y_2 - y_1}{x_2 - x_1} = \dfrac{1-(-3)}{2-3} = -4$.

The point-slope form is $(y - 1) = -4(x - 2)$ or $(y + 3) = -4(x - 3)$.

Simplifying, either equation yields $y = -4x + 9$ or equivalently $4x + y - 9 = 0$.

Example 2.14

Find the equation of the straight line passing through the point (3, 1) and perpendicular to the line passing through the points (3, −2) and (−3, 7).

Solution

Slope of the line passing through (3, −2) and (−3, 7) is

$m_1 = \dfrac{y_2 - y_1}{x_2 - x_1} = \dfrac{7-(-2)}{-3-3} = -\dfrac{3}{2}$

Slope of the line passing through (3, 1) is $m_2 = -\dfrac{1}{m_1} = \dfrac{2}{3}$, since the two lines are perpendicular to each other.

Equation is $(y-1) = \dfrac{2}{3}(x - 3)$ or $2x - 3y - 3 = 0$.

Circle

The locus of a point at a constant distance (radius) from a fixed point C (center) is a circle.

$(x-h)^2 + (y-k)^2 = a^2$ $C(h, k)$, radius $= a$
$r^2 + b^2 \pm 2\,br\cos(\theta - \beta) = a^2$ $C(b, \beta)$, radius $= a$ [Figure 2.23(a)]

$x^2 + y^2 = 2ax$ $C(a, 0)$, radius $= a$
$r = 2a\cos\theta$ $C(a, 0)$, radius $= a$ [Figure 2.23(b)]

$x^2 + y^2 = 2ay$ $C(0, a)$, *radius* $= a$
$r = 2a\sin\theta$ $C(0, a)$, *radius* $= a$ [Figure 2.23(c)]

$x^2 + y^2 = a^2$ $C(0, 0)$, radius $= a$
$r = a$ $C(0, 0)$, radius $= a$ [Figure 2.23(d)]

$x = a\cos\phi,\ y = a\sin\phi$ ϕ = angle from OX to radius

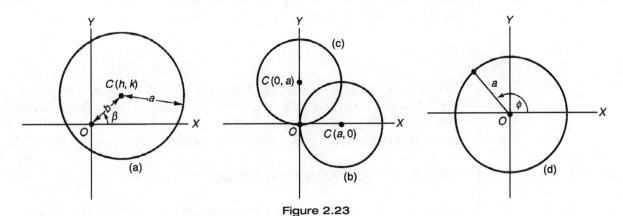

Figure 2.23

Example 2.15

Find the equation of the circle (i) with center at (0, 0) and radius 3, and (ii) with center at (1, 2) and radius 4.

Solution

Equation of a circle with center at (h, k) and radius r is $(x - h)^2 + (y - k)^2 = r^2$.

i) Here, $h = 0$, $k = 0$, and $r = 3$; equation of the circle is $x^2 + y^2 = 3^2$ or $x^2 + y^2 - 9 = 0$
ii) Here, $h = 1$, $k = 2$, and $r = 4$; equation of the circle is $(x - 1)^2 + (y - 2)^2 = 4^2$
Simplifying, $x^2 - 2x + y^2 - 4y - 11 = 0$

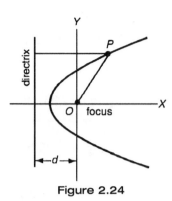

Figure 2.24

Conic

A **conic** is the locus of a point whose distance from a fixed point (focus) is in a constant ratio e, called the eccentricity, to its distance from a fixed straight line (directrix).

$$x^2 + y^2 = e^2(d + x)^2 \qquad d = \text{distance from focus to directrix}$$

$$r = \frac{de}{1 - e\cos\theta}$$

The conic is called a parabola when $e = 1$, an ellipse when $e < 1$, and a hyperbola when $e > 1$.

Example 2.16

Which conic section is represented by each of the following equations: (i) $x^2 + 4xy + 4y^2 + 2x = 10$ and (ii) $x^2 + y^2 - 2x - 4y - 11 = 0$.

Solution

The general equation of a conic section is $Ax^2 + 2Bxy + Cy^2 + 2Dx + 2Ey + F = 0$, where both A and C are not zeros. If $B^2 - AC > 0$, a *hyperbola* is defined; if $B^2 - AC = 0$, a *parabola* is defined; if $B^2 - AC < 0$, an *ellipse* is defined. (Note: If B is zero and A = C, a *circle* is defined.)

If $A = B = C = 0$, a *straight line* is defined.

If $B = 0$, $A = C$, a *circle* is defined with equation $x^2 + y^2 + 2ax + 2by + c = 0$.

Center is at $(-a, -b)$ and radius = $\sqrt{a^2 + b^2 - c}$ provided $a^2 + b^2 - c > 0$.

(i) Here, $A = 1$, $B = 2$, and $C = 4$. Then, $B^2 - AC = (2)^2 - (1)(4) = 0$. The equation represents a parabola.

(ii) Here, $A = 1$, $B = 0$, $C = 1$. Because A = C and B = 0, the equation represents a circle.

[Note that the center is at (1, 2), and the radius is $\sqrt{(-1)^2 + (-2)^2 - (-11)} = 4$.]

Parabola

A parabola is a special case of a conic where $e = 1$.

$(y - k)^2 = a(x - h)$ Vertex (h, k), axis $\parallel OX$
$y^2 = ax$ Vertex $(0, 0)$, axis along OX [Figure 2.25(a)]
$(x - h)^2 = a(y - k)$ Vertex (h, k), axis $\parallel OY$
$x^2 = ay$ Vertex $(0, 0)$, axis along OY [Figure 2.25(b)]

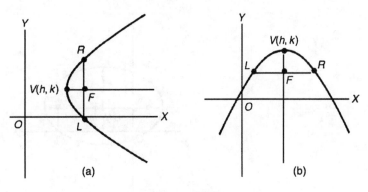

Figure 2.25

Distance from vertex to focus = $VF = \dfrac{1}{4}a$. Latus rectum = $LR = a$.

Example 2.17

Find the equation of a parabola (i) with center at $(0, 0)$ and focus at $(4, 0)$ and (ii) with center at $(4, 2)$ and focus at $(8, 2)$.

Solution

The equation of a parabola with center at (h, k) and focus at $(h + p/2, k)$ is given as

$(y - k)^2 = 2p(x - h)$

(i) $h = 0$; $k = 0$; $h + p/2 = 4$; then, $p = 8$ and the equation is $y^2 = 16x$.

(ii) $h = 4$; $k = 2$; $h + p/2 = 8$; then, $p = 8$ and the equation is $(y - 2)^2 = 16(x - 4)$.

Ellipse

This is a special case of a conic where $e < 1$.

$$\frac{(x-h)^2}{a^2} + \frac{(y-k)^2}{b^2} = 1 \quad \text{Center } (h, k), \text{ axes } \| \ OX, OY$$

$$\frac{x^2}{a^2} + \frac{y^2}{b^2} = 1 \quad \text{Center } (0, 0), \text{ axes along } OX, OY$$

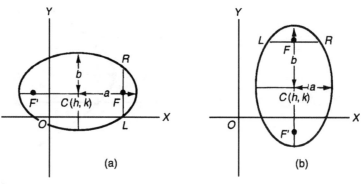

Figure 2.26

	$a > b$, Figure 2.26(a)	$b > a$, Figure 2.26(b)
Major axis	$2a$	$2b$
Minor axis	$2b$	$2a$
Distance from center to either focus	$\sqrt{a^2 - b^2}$	$\sqrt{b^2 - a^2}$
Latus rectum	$\dfrac{2b^2}{a}$	$\dfrac{2a^2}{b}$
Eccentricity, e	$\sqrt{\dfrac{a^2 - b^2}{a}}$	$\sqrt{\dfrac{b^2 - a^2}{b}}$
Sum of distances of any point P from the foci, $PF' + PF$	$2a$	$2b$

Example 2.18

Find the equation of an ellipse with center at origin, x-axis intercept $(4, 0)$, and y-axis intercept $(0, 2)$.

Solution

Equation of an ellipse with center at origin and x-axis intercept of $(a, 0)$ and y-axis intercept of $(0, b)$ is $\dfrac{x^2}{a^2} + \dfrac{y^2}{b^2} = 1$. Here, $a = 4$ and $b = 2$; then, the equation is $\dfrac{x^2}{4^2} + \dfrac{y^2}{2^2} = 1$.

Hyperbola

This is a special case of a conic where $e > 1$.

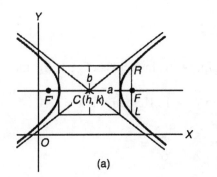
(a)

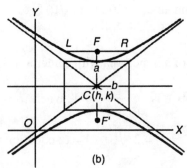
(b)

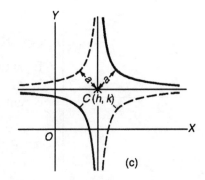
(c)

Figure 2.27

$$\frac{(x-h)^2}{a^2} - \frac{(y-k)^2}{b^2} = 1 \qquad C(h, k), \text{ transverse axis} \parallel OX$$

$$\frac{x^2}{a^2} - \frac{y^2}{b^2} = 1 \qquad C(0, 0), \text{ transverse axis along } OX$$

$$\frac{(y-k)^2}{a^2} - \frac{(x-h)^2}{b^2} = 1 \qquad C(h, k), \text{ transverse axis} \parallel OY$$

$$\frac{y^2}{a^2} - \frac{x^2}{b^2} = 1 \qquad C(0, 0), \text{ transverse along } OY$$

Transverse axis = $2a$; conjugate axis = $2b$

Distance from center to either focus = $\sqrt{a^2 + b^2}$

Latus rectum = $\dfrac{2b^2}{a}$

Eccentricity, $e = \dfrac{\sqrt{a^2 + b^2}}{a}$

Difference of distances of any point from the foci = $2a$.

The asymptotes are two lines through the center to which the branches of the hyperbola approach arbitrarily closely; their slopes are $\pm b/a$ [Figure 2.27(a)] or $\pm a/b$ [Figure 2.27(b)].

The rectangular (equilateral) hyperbola has $b = a$. The asymptotes are perpendicular to each other.

$$(x - h)(y - k) = \pm e = \sqrt{2} \qquad \text{Center } (h, k), \text{ asymptotes} \parallel OX, OY$$

$$xy = \pm e = \sqrt{2} \qquad \text{Center } (0, 0), \text{ asymptotes along } OX, OY$$

The + sign gives the solid curves in Figure 2.27(c); the − sign gives the dotted curves in Figure 2.27(c).

Example 2.19

What is the equation of the hyperbola with center at origin, passing through (±2, 0), and an eccentricity of $\sqrt{10}$?

Solution

Equation of a hyperbola with center at origin, x-axis intercepts of (± a, 0), and eccentricity e is $\dfrac{x^2}{a^2} - \dfrac{y^2}{b^2} = 1$, where $b = a\sqrt{e^2 - 1}$. Here, $a = 2$; $e = \sqrt{10}$; then, $b = a\sqrt{e^2 - 1} = 2\sqrt{10 - 1} = 6$; equation is $\dfrac{x^2}{2^2} - \dfrac{y^2}{6^2} = 1$.

VECTORS

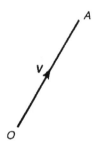

Figure 2.28

Definition and Graphical Representation of a Vector

A vector (**V**) is a quantity that is completely specified by magnitude *and* a direction. A scalar (*s*) is a quantity that is completely specified by a magnitude *only*.

The vector (**V**) may be represented geometrically by the segment $\overrightarrow{OA}$, the length of *OA* signifying the magnitude of **V** and the arrow carried by *OA* signifying the direction of **V**. The segment $\overrightarrow{AO}$ represents the vector –**V**.

Graphical Summation of Vectors

If $\mathbf{V}_1$ and $\mathbf{V}_2$ are two vectors, their graphical sum $\mathbf{V} = \mathbf{V}_1 + \mathbf{V}_2$ is formed by drawing the vector $\mathbf{V}_1 = \overrightarrow{OA}$, from any point *O*, and the vector $\mathbf{V}_2 = \overrightarrow{AB}$ from the end of $\mathbf{V}_1$ and joining *O* and *B*; then $\mathbf{V} = \overrightarrow{OB}$. Also, $\mathbf{V}_1 + \mathbf{V}_2 = \mathbf{V}_2 + \mathbf{V}_1$ and $\mathbf{V}_1 + \mathbf{V}_2 - \mathbf{V} = 0$ (Figure 2.29(a)).

Similarly, if $\mathbf{V}_1, \mathbf{V}_2, \mathbf{V}_3, \ldots, \mathbf{V}_n$ are any number of vectors drawn so that the initial point of one is the end point of the preceding one, then their graphical sum $\mathbf{V} = \mathbf{V}_1 + \mathbf{V}_2 + \ldots + \mathbf{V}_n$ is the vector joining the initial point of $\mathbf{V}_1$ with the end point of $\mathbf{V}_n$ (Figure 2.29(b)).

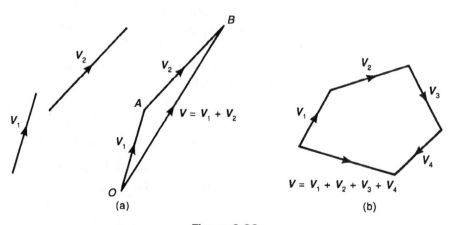

Figure 2.29

Analytic Representation of Vector Components

A vector **V** that is considered as lying in the *x-y* coordinate plane (Figure 2.30(a)) is completely determined by its horizontal and vertical components *x* and *y*. If **i** and **j** represent vectors of unit magnitude along *OX* and *OY*, respectively, and *a* and *b* are the magnitude of *x* and *y*, then **V** may be represented by $\mathbf{V} = a\mathbf{i} + b\mathbf{j}$, its magnitude by $|\mathbf{V}| = +\sqrt{a^2 + b^2}$, and its direction by $\alpha = \tan^{-1} b/a$.

A vector **V** in three-dimensional in space is completely determined by its components *x*, *y*, and *z* along three mutually perpendicular lines *OX*, *OY*, and *OZ*, directed as shown in Figure 2.30(b). If **i**, **j**, and **k** represent vectors of unit magnitude along *OX*, *OY*, *OZ*, respectively, and *a*, *b*, and *c* are the magnitudes of the components *x*, *y*, and *z*, respectively, then **V** may be represented by $\mathbf{V} = a\mathbf{i} + b\mathbf{j} + c\mathbf{k}$, its magnitude by, $|\mathbf{V}| = +\sqrt{a^2 + b^2 + c^2}$, and its direction by $\cos \alpha : \cos \beta : \cos \gamma = a : b : c$.

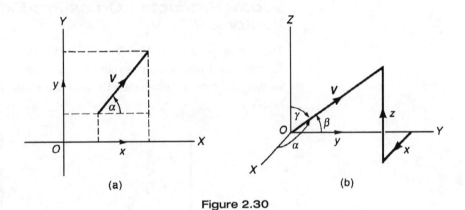

Figure 2.30

Properties of Vectors

$\mathbf{V} = a\mathbf{i} + b\mathbf{j}$ or $\mathbf{V} = a\mathbf{i} + b\mathbf{j} + c\mathbf{k}$

Vector Sum V of any Number of Vectors, $\mathbf{V}_1, \mathbf{V}_2, \mathbf{V}_3, \ldots$

$$\mathbf{V} = \mathbf{V}_1 + \mathbf{V}_2 + \mathbf{V}_3 + \ldots = (a_1 + a_2 + a_3 + \ldots)\mathbf{i} + (b_1 + b_2 + b_3 + \ldots)\mathbf{j} + (c_1 + c_2 + c_3 + \ldots)\mathbf{k}$$

Product of a Vector V and a Scalar s

The product *s***V** has the same direction as **V**, and its magnitude is *s* times the magnitude of **V**.

$$s\mathbf{V} = (sa)\mathbf{i} + (sb)\mathbf{j} + (sc)\mathbf{k}$$
$$(s_1 + s_2)\mathbf{V} = s_1\mathbf{V} + s_2\mathbf{V} \qquad (\mathbf{V}_1 + \mathbf{V}_2)s = \mathbf{V}_1 s + \mathbf{V}_2 s$$

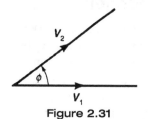

Figure 2.31

Scalar Product or Dot Product of Two Vectors: $V_1 \bullet V_2$

$V_1 \bullet V_2 = |V_1||V_2|\cos\phi$, where ϕ is the angle between V_1 and V_2

$V_1 \bullet V_2 = V_2 \bullet V_1$; $\quad V_1 \bullet V_1 = |V_1|^2$; $\quad (V_1 + V_2) \bullet V_3 = V_1 \bullet V_3 + V_2 \bullet V_3$

$(V_1 + V_2) \bullet (V_3 + V_4) = V_1 \bullet V_3 + V_1 \bullet V_4 + V_2 \bullet V_3 + V_2 \bullet V_4$

$\mathbf{i} \bullet \mathbf{i} = \mathbf{j} \bullet \mathbf{j} = \mathbf{k} \bullet \mathbf{k} = 1$; $\quad \mathbf{i} \bullet \mathbf{j} = \mathbf{j} \bullet \mathbf{k} = \mathbf{k} \bullet \mathbf{i} = 0$

In a plane, $V_1 \bullet V_2 = a_1 a_2 + b_1 b_2$; in space, $V_1 \bullet V_2 = a_1 a_2 + b_1 b_2 + c_1 c_2$.

The scalar product of two vectors $V_1 \bullet V_2$ is a scalar quantity and may physically represent the work done by a constant force of magnitude $|V_1|$ on a particle moving through a distance $|V_2|$, where ϕ is the angle between the direction of the force and the direction of motion.

Vector Product or Cross Product of Two Vectors: $V_1 \times V_2$

The vector product is $V_1 \times V_2 = \mathbf{l}\,|V_1||V_2|\sin\phi$, where ϕ is the angle from V_1 to V_2 and $\mathbf{l}$ is a unit vector perpendicular to the plane of the vectors V_1 to V_2 and so directed that a right-handed screw driven in the direction of $\mathbf{l}$ would carry V_1 into V_2.

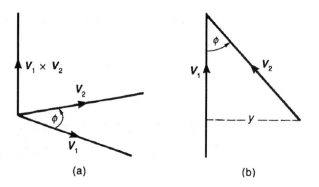

(a)   (b)

Figure 2.32

$V_1 \times V_2 = -V_2 \times V_1$; $\quad V_1 \times V_1 = 0$

$(V_1 + V_2) \times V_3 = V_1 \times V_3 + V_2 \times V_3$

$V_1 \bullet (V_2 \times V_3) = V_2 \bullet (V_3 \times V_1) = V_3 \bullet (V_1 \times V_2)$

$\mathbf{i} \times \mathbf{i} = \mathbf{j} \times \mathbf{j} = \mathbf{k} \times \mathbf{k} = 0$; $\quad \mathbf{i} \times \mathbf{j} = \mathbf{k}$; $\quad \mathbf{j} \times \mathbf{k} = \mathbf{i}$; $\quad \mathbf{k} \times \mathbf{i} = \mathbf{j}$

In the x-y plane, $V_1 \times V_2 = (a_1 b_2 - a_2 b_1)\mathbf{k}$.

In space, $V_1 \times V_2 = (b_2 c_3 - b_3 c_2)\mathbf{i} + (c_3 a_1 - c_1 a_3)\mathbf{j} + (a_1 b_2 - a_2 b_1)\mathbf{k}$.

The vector product of two vectors is a vector quantity and may physically represent the moment of a force V_1 about a point O placed so that the moment arm is $y = |V_2|\sin\phi$ (see Figure 2.32(b)).

Example 2.20

Vectors **A** and **B** are defined as: $\mathbf{A} = i - 2j + 3k$, $\mathbf{B} = 2i + j - 2k$

Find (i) $\mathbf{A} + \mathbf{B}$, (ii) $\mathbf{A} - \mathbf{B}$, (iii) $2\mathbf{A}$, (iv) $|\mathbf{A}|$, (v) dot product $\mathbf{A} \cdot \mathbf{B}$, and (vi) the cross product $\mathbf{A} \times \mathbf{B}$.

Solution

(i) $\mathbf{A} + \mathbf{B} = 3i - j + k$ (ii) $\mathbf{A} - \mathbf{B} = -i - 3j + 5k$ (iii) $2\mathbf{A} = 2i - 4j + 6k$

(iv) $|\mathbf{A}| = \sqrt{1^2 + (-2)^2 + 3^2} = \sqrt{14}$

(v) Dot product $\mathbf{A} \cdot \mathbf{B} = (2) + (-2) + (-6) = -6$

(vi) Cross product $\mathbf{A} \times \mathbf{B} = \begin{vmatrix} i & j & k \\ 1 & -2 & 3 \\ 2 & 1 & -2 \end{vmatrix} = i + 8j + 5k$

LINEAR ALGEBRA

Matrix Operations

Matrices are rectangular arrays of real or complex numbers. Their great importance arises from the variety of operations that may be performed on them. Using the standard convention, the across-the-page lines are called **rows** and the up-and-down-the-page lines are **columns**. Entries in a matrix are **addressed** with double subscripts (always row first, then column). Thus the matrix

$$\mathbf{A} = \begin{bmatrix} 1 & 2 & 3 \\ 0 & 9 & -3 \end{bmatrix}$$

is 2×3, and the "9" is a_{22}. The "2" is a_{12} and the "0" is a_{21}. One also can refer to entries with square brackets, the "9" being $[A]_{22}$ and the "3" $[A]_{13}$.

If two matrices are the same size, they may be added: $[A + B]_{ij} = [A]_{ij} + [B]_{ij}$. Thus,

$$\begin{bmatrix} 1 & 2 & 3 \\ 0 & 9 & -3 \end{bmatrix} + \begin{bmatrix} 1 & 3 \\ 2 & 4 \end{bmatrix}$$

is not defined, but

$$\begin{bmatrix} 1 & 2 & 3 \\ 0 & 9 & -3 \end{bmatrix} + \begin{bmatrix} 1 & 3 & 5 \\ 2 & 4 & 6 \end{bmatrix} = \begin{bmatrix} 2 & 5 & 8 \\ 2 & 13 & 3 \end{bmatrix}$$

is proper.

Any matrix may be multiplied by a **scalar** (a number): $[c\mathbf{A}]_{ij} = c[A]_{ij}$, so that

$$5 \begin{bmatrix} 1 & 5 \\ 0 & 6 \end{bmatrix} = \begin{bmatrix} 5 & 25 \\ 0 & 30 \end{bmatrix}$$

The most peculiar matrix operation (and the most useful) is matrix multiplication. If **A** is $m \times n$ and **B** is $n \times p$, then $\mathbf{A} \bullet \mathbf{B}$ (or **AB**) is of size $m \times p$, and

$$[AB]_{ij} = \sum_{k=1}^{n} a_{jk} \bullet b_{kj}$$

The **dot product** (scalar product) of the ith row of **A** with the jth column of **B**, as in

$$\begin{bmatrix} 1 & 2 & 3 \\ 0 & 9 & -3 \end{bmatrix} \bullet \begin{bmatrix} 1 & 5 \\ 0 & 6 \end{bmatrix}$$

is not defined (owing to the mismatch of row and column lengths), but

$$\begin{bmatrix} 1 & 2 & 3 \\ 0 & 9 & -3 \end{bmatrix} \bullet \begin{bmatrix} 1 & 5 \\ 0 & 6 \\ 7 & 8 \end{bmatrix} = \begin{bmatrix} 1(1)+2(0)+3(7) & 1(5)+2(6)+3(8) \\ 0(1)+9(0)-3(7) & 0(5)+9(6)-3(8) \end{bmatrix} = \begin{bmatrix} 22 & 41 \\ -21 & 30 \end{bmatrix}$$

is correct.

A matrix with only one row or one column is called a vector, so a matrix times a vector is a vector (if defined). Thus $\mathbf{A}\,(m \times n) \bullet \mathbf{X}\,(n \times 1) = \mathbf{Y}(m \times 1)$, so a matrix can be thought of as an **operator** that takes vectors to vectors.

Another useful way of working with matrices is transposition: If **A** is $m \times n$, $\mathbf{A}^t$ is $n \times m$ and is the result of interchanging rows and columns. Hence

$$\begin{bmatrix} 1 & 2 & 3 \\ 0 & 9 & -3 \end{bmatrix}^t = \begin{bmatrix} 1 & 0 \\ 2 & 9 \\ 3 & -3 \end{bmatrix}$$

These various operations interact in the usual pleasant ways (and one decidedly unpleasant way); the standard convention is that all of the following combinations are defined:

$$\mathbf{A} + \mathbf{B} = \mathbf{B} + \mathbf{A}$$
$$\mathbf{A} + (\mathbf{B} + \mathbf{C}) = (\mathbf{A} + \mathbf{B}) + \mathbf{C}$$
$$c(\mathbf{A} + \mathbf{B}) = c\mathbf{A} + c\mathbf{B}$$
$$(c + d)\mathbf{A} = c\mathbf{A} + d\mathbf{A}$$
$$(-1)\mathbf{A} + \mathbf{A} = (0\mathbf{A})$$
$$\mathbf{A} \bullet \mathbf{B} \neq \mathbf{B} \bullet \mathbf{A} \text{ (in general)}$$
$$\mathbf{A} \bullet (\mathbf{B} \bullet \mathbf{C}) = (\mathbf{A} \bullet \mathbf{B}) \bullet \mathbf{C}$$
$$\mathbf{A} \bullet (\mathbf{B} + \mathbf{C}) = \mathbf{A} \bullet \mathbf{B} + \mathbf{A} \bullet \mathbf{C}$$
$$(\mathbf{A} + \mathbf{B}) \bullet \mathbf{C} = \mathbf{A} \bullet \mathbf{C} + \mathbf{B} \bullet \mathbf{C}$$
$$(\mathbf{A} + \mathbf{B})^t = \mathbf{A}^t + \mathbf{B}^t$$
$$(\mathbf{A} \bullet \mathbf{B})^t = \mathbf{B}^t \bullet \mathbf{A}^t$$

In addition, matrices **I**, which are $n \times n$ and whose entries are 1 on the diagonal $i = j$ and 0 elsewhere, are multiplicative identities: $\mathbf{A} \bullet \mathbf{I} = \mathbf{A}$ and $\mathbf{I} \bullet \mathbf{A} = \mathbf{A}$. Here the two **I** matrices may be different sizes; for example,

$$\begin{bmatrix} 1 & 2 & 3 \\ 0 & 9 & -3 \end{bmatrix} \bullet \begin{bmatrix} 1 & 0 & 0 \\ 0 & 1 & 0 \\ 0 & 0 & 1 \end{bmatrix} = \begin{bmatrix} 1 & 2 & 3 \\ 0 & 9 & -3 \end{bmatrix}$$

but

$$\begin{bmatrix} 1 & 0 \\ 0 & 1 \end{bmatrix} \bullet \begin{bmatrix} 1 & 2 & 3 \\ 0 & 9 & -3 \end{bmatrix} = \begin{bmatrix} 1 & 2 & 3 \\ 0 & 9 & -3 \end{bmatrix}$$

I is called the identity matrix, and the size is understood from context.

Example 2.21

Verify that the transpose of $\mathbf{A} + \mathbf{BC}$ is $\mathbf{C}^t\mathbf{B}^t + \mathbf{A}^t$ if

$$\mathbf{A} = \begin{bmatrix} 1 & 1 \\ 2 & 3 \end{bmatrix} \qquad \mathbf{B} = \begin{bmatrix} 1 & 2 & 3 \\ 0 & 9 & -3 \end{bmatrix} \qquad \mathbf{C} = \begin{bmatrix} 1 & 5 \\ 0 & 6 \\ 7 & 8 \end{bmatrix}$$

Solution

$$\mathbf{BC} \begin{bmatrix} 22 & 41 \\ -21 & 30 \end{bmatrix}, \text{ so } \mathbf{A} + \mathbf{BC} = \begin{bmatrix} 23 & 42 \\ -19 & 33 \end{bmatrix} \text{ and } [\mathbf{A} + \mathbf{BC}]^t = \begin{bmatrix} 23 & -19 \\ 42 & 33 \end{bmatrix}$$

On the other hand, $\mathbf{C}^t\mathbf{B}^t = \begin{bmatrix} 1 & 0 & 7 \\ 5 & 6 & 8 \end{bmatrix} \bullet \begin{bmatrix} 1 & 0 \\ 2 & 9 \\ 3 & -3 \end{bmatrix} = \begin{bmatrix} 22 & -21 \\ 41 & 30 \end{bmatrix}$ and

$$\mathbf{A}^t = \begin{bmatrix} 1 & 2 \\ 1 & 3 \end{bmatrix}, \text{ so } \mathbf{A}^t + \mathbf{C}^t\mathbf{B}^t = \mathbf{C}^t\mathbf{B}^t + \mathbf{A}^t = \begin{bmatrix} 23 & -19 \\ 42 & 33 \end{bmatrix}$$

Types of Matrices

Matrices are classified according to their appearance or the way they act. If **A** is square and $\mathbf{A}^t = \mathbf{A}$, then **A** is called symmetric. If $\mathbf{A}' = -\mathbf{A}$, then it is skew-symmetric.

If **A** has complex entries, **A*** then is called the Hermitian adjoint of **A**. If $\mathbf{A}^* = \overline{\mathbf{A}^t}$ (complex conjugate), then

$$\begin{bmatrix} 1+i & i \\ 3 & 4-i \end{bmatrix}^* = \overline{\begin{bmatrix} 1-i & -i \\ 3 & 4+i \end{bmatrix}^t} = \begin{bmatrix} 1-i & 3 \\ -i & 4+i \end{bmatrix}$$

If $\mathbf{A} = \mathbf{A}^*$, then A is called Hermitian. If $\mathbf{A}^* = -\mathbf{A}$, the name is skew-Hermitian.

If **A** is square and $a_{ij} = 0$ unless $i = j$, **A** is called diagonal. If **A** is square and zero below the diagonal ($[A]_{ij} = 0$ if $i > j$), **A** is called upper triangular. The transpose of such a matrix is called lower triangular.

If **A** is square and there is a matrix $\mathbf{A}^{-1}$ such that $\mathbf{A}^{-1} \bullet \mathbf{A} = \mathbf{A} \bullet \mathbf{A}^{-1} = \mathbf{I}$, **A** is nonsingular. Otherwise, it is singular. If **A** and **B** are both nonsingular $n \times n$ matrices, then **AB** is nonsingular and $(\mathbf{AB})^{-1} = \mathbf{B}^{-1}\mathbf{A}^{-1}$, because $(\mathbf{AB})(\mathbf{B}^{-1}\mathbf{A}^{-1}) = \mathbf{A}(\mathbf{B}\,\mathbf{B}^{-1})\mathbf{A}^{-1} = \mathbf{A}\mathbf{I}\mathbf{A}^{-1} = \mathbf{A}\mathbf{A}^{-1} = \mathbf{I}$, as does $(\mathbf{B}^{-1}\mathbf{A}^{-1}) \bullet (\mathbf{AB})$.

If $\mathbf{A}'\mathbf{A} = \mathbf{A}\mathbf{A}' = \mathbf{I}$ and **A** is real, it is called orthogonal (the reason will appear below). If $\mathbf{A}^*\mathbf{A} = \mathbf{A}\mathbf{A}^* = \mathbf{I}$ (**A** complex), **A** is called unitary. If **A** commutes with $\mathbf{A}^*$, so that $\mathbf{A}\mathbf{A}^* = \mathbf{A}^*\mathbf{A}$, then **A** is called normal.

Elementary Row and Column Operations

The most important tools used in dealing with matrices are the elementary operations: R for row, C for column. If **A** is given matrix, performing $R(i \leftrightarrow j)$ on **A** means interchanging Row i and Row j. $R_i(c)$ means multiplying Row i by the number c (except $c = 0$). $R_j + cR_i$ means multiply Row i by c and add this result into Row j ($i \neq j$). Thus, if

$$\mathbf{A} = \begin{bmatrix} 1 & 2 & 3 \\ 4 & 5 & 6 \\ 7 & 8 & 0 \end{bmatrix}$$

then

$$R(2 \leftrightarrow 3)(\mathbf{A}) = \begin{bmatrix} 1 & 2 & 3 \\ 7 & 8 & 0 \\ 4 & 5 & 6 \end{bmatrix} \quad C_1(2)(\mathbf{A}) = \begin{bmatrix} 2 & 2 & 3 \\ 8 & 5 & 6 \\ 14 & 8 & 0 \end{bmatrix}$$

$$R_1 - R_2(\mathbf{A}) = \begin{bmatrix} -3 & -3 & -3 \\ 4 & 5 & 6 \\ 7 & 8 & 0 \end{bmatrix}$$

These operations are used in reducing matrix problems to simpler ones.

Example 2.22

Solve $\mathbf{AX} = \mathbf{B}$ where

$$\mathbf{A} = \begin{bmatrix} 1 & 2 & 3 \\ 4 & 5 & 6 \\ 7 & 8 & 9 \end{bmatrix}, \quad \mathbf{X} = \begin{bmatrix} x \\ y \\ z \end{bmatrix}, \quad \mathbf{B} = \begin{bmatrix} 1 \\ 1 \\ 1 \end{bmatrix}$$

Solution

Form the "augmented" matrix

$$[\mathbf{A}|\mathbf{B}] = \begin{bmatrix} 1 & 2 & 3 & 1 \\ 4 & 5 & 6 & 1 \\ 7 & 8 & 9 & 1 \end{bmatrix}$$

and perform elementary row operations on this matrix until the solution is apparent:

$$\begin{bmatrix} 1 & 2 & 3 & 1 \\ 4 & 5 & 6 & 1 \\ 7 & 8 & 9 & 1 \end{bmatrix} \begin{array}{c} R_2 - 4R_1 \\ R_3 - 7R_1 \end{array} \begin{bmatrix} 1 & 2 & 3 & 1 \\ 0 & -3 & -6 & -3 \\ 7 & -6 & -12 & -6 \end{bmatrix} \begin{array}{c} R_2\left(-\dfrac{1}{3}\right) \\ R_3\left(-\dfrac{1}{6}\right) \end{array} \begin{bmatrix} 1 & 2 & 3 & 1 \\ 0 & 1 & 2 & 1 \\ 0 & 1 & 2 & 1 \end{bmatrix}$$

$$R_3 - R_2 \begin{bmatrix} 1 & 2 & 3 & 1 \\ 0 & 1 & 2 & 1 \\ 0 & 0 & 0 & 0 \end{bmatrix}$$

The answer is now apparent: $y + 2z = 1$ and $x + 2y + 3z = 1$, or, z arbitrary, $y = 1 - 2z$, $x = 1 - 2(1 - 2z) - 3z = -1 + z$. This system of equations has an infinite number of solutions.

Example 2.23

Solve the system of equations

$$\begin{aligned} x + y - z &= a \\ 2x - y + 3z &= 2 \\ 3x + 2y + z &= 1 \end{aligned}$$

for x, y, and z in terms of a.

Solution

Strip off the variables x, y, and z:

$$\begin{bmatrix} 1 & 1 & -1 & a \\ 2 & -1 & 3 & 2 \\ 3 & 2 & 1 & 1 \end{bmatrix} \begin{array}{c} R_2 - 2R_1 \\ R_3 - 3R_1 \end{array} \begin{bmatrix} 1 & 1 & -1 & a \\ 0 & -3 & 5 & 2 - 2a \\ 0 & -1 & 4 & 1 - 3a \end{bmatrix}$$

$$\begin{array}{c} R_2 (2 \leftrightarrow 3) \\ R_2(-1) \end{array} \begin{bmatrix} 1 & 1 & -1 & a \\ 0 & 1 & -4 & 3a - 1 \\ 0 & -3 & 5 & 2 - 2a \end{bmatrix} \begin{array}{c} R_1 - R_2 \\ R_3 + 3R_2 \end{array} \begin{bmatrix} 1 & 0 & 3 & 1 - 2a \\ 0 & 1 & -4 & 3a - 1 \\ 0 & 0 & -7 & 7a - 1 \end{bmatrix}$$

The solution is now clear:

$$z = \frac{7a - 1}{-7} = -a + \frac{1}{7}$$

$$y = 3a - 1 + 4z = 3a - 1 - 4a + \frac{4}{7} = -a - \frac{3}{7}$$

$$x = 1 - 2a - 3z = 1 - 2a + 3a - \frac{3}{7} = a + \frac{4}{7}$$

Example 2.24

Find $\mathbf{A}^{-1}$ if

$$\mathbf{A} = \begin{bmatrix} 1 & 1 & -1 \\ 1 & 2 & 3 \\ 3 & 2 & 1 \end{bmatrix}$$

Solution

Since this amounts to solving $\mathbf{AX} = \mathbf{B}$ three times, with

$$\mathbf{B} = \begin{bmatrix} 1 \\ 0 \\ 0 \end{bmatrix} \quad \mathbf{B} = \begin{bmatrix} 0 \\ 1 \\ 0 \end{bmatrix} \quad \mathbf{B} = \begin{bmatrix} 0 \\ 0 \\ 1 \end{bmatrix}$$

form

$$[\mathbf{A}|\mathbf{I}] = \begin{bmatrix} 1 & 1 & -1 & 1 & 0 & 0 \\ 1 & 2 & 3 & 0 & 1 & 0 \\ 3 & 2 & 1 & 0 & 0 & 1 \end{bmatrix}$$

and perform row operations until a solution emerges.

$$[\mathbf{A}|\mathbf{I}] = \begin{matrix} \\ R_2 - R_1 \\ R_3 - 3R_1 \end{matrix} \begin{bmatrix} 1 & 1 & -1 & 1 & 0 & 0 \\ 0 & 1 & 4 & -1 & 1 & 0 \\ 0 & -1 & 4 & -3 & 0 & 1 \end{bmatrix}$$

$$\begin{matrix} R_1 - R_2 \\ R_3 + R_2 \end{matrix} z \begin{bmatrix} 1 & 0 & -5 & 2 & -1 & 0 \\ 0 & 1 & 4 & -1 & 1 & 0 \\ 0 & 0 & 8 & -4 & 1 & 1 \end{bmatrix}$$

$$\begin{matrix} R_3\left(\frac{1}{8}\right) \\ R_2 - 4R_3 \\ R_1 + 5R_3 \end{matrix} \begin{bmatrix} 1 & 0 & 0 & -\dfrac{1}{2} & -\dfrac{3}{8} & \dfrac{5}{8} \\ 0 & 1 & 0 & 1 & \dfrac{1}{2} & -\dfrac{1}{2} \\ 0 & 0 & 1 & -\dfrac{1}{2} & \dfrac{1}{8} & \dfrac{1}{8} \end{bmatrix}$$

Thus,

$$\mathbf{A}^{-1} = \begin{bmatrix} -\dfrac{1}{2} & -\dfrac{3}{8} & \dfrac{5}{8} \\ 1 & \dfrac{1}{2} & -\dfrac{1}{2} \\ -\dfrac{1}{2} & \dfrac{1}{8} & \dfrac{1}{8} \end{bmatrix}$$

Example 2.25

Verify that $\mathbf{A}^{-1}\mathbf{A} = \mathbf{I}$ in Example 2.24.

Solution

$$8\mathbf{A}^{-1}\mathbf{A} = \begin{bmatrix} -4 & -3 & 5 \\ 8 & 4 & -4 \\ -4 & 1 & 1 \end{bmatrix} \begin{bmatrix} 1 & 1 & -1 \\ 1 & 2 & 3 \\ 3 & 2 & 1 \end{bmatrix}$$

$$= \begin{bmatrix} -4-3+15 & -4-6+10 & 4-9+5 \\ 8+4-12 & 8+8-8 & -8+12-4 \\ -4+1+3 & -4+2+2 & 4+3+1 \end{bmatrix} = 8 \begin{bmatrix} 1 & 0 & 0 \\ 0 & 1 & 0 \\ 0 & 0 & 1 \end{bmatrix} = 8\mathbf{I}$$

Example 2.26

Describe the set of solutions of $\mathbf{AX} = \mathbf{B}$.

Solution

If $\mathbf{AX}_0 = \mathbf{B}$ is one solution, and $\mathbf{AY} = 0$, then $\mathbf{A}(\mathbf{X}_0 + \mathbf{Y})$ is a solution, so all solutions are of the form $\mathbf{X} = \mathbf{X}_0 + \mathbf{Y}$ where $\mathbf{AY} = 0$. Thus, if $\mathbf{N} = \{\mathbf{Y} : \mathbf{AY} = 0\}$ is the null space of $\mathbf{A}$, the set of solutions to $\mathbf{AX} = \mathbf{B}$ is $\mathbf{X}_0 + \mathbf{N} = \{\mathbf{X}_0 + \mathbf{Y} : \mathbf{Y} \in \mathbf{N}\}$.

Determinants

The determinant of a square matrix is a scalar representing the *volume* of the matrix in some sense. Matrices that are not square do not have determinants.

The determinant is frequently indicated by vertical lines, viz. $|A|$. It is a complicated formula, and one way to find it is by induction. The determinant of a 1×1 matrix is $|a| = a$. The determinant of a 2×2 matrix is

$$\begin{vmatrix} a & b \\ c & d \end{vmatrix} = ad - bc$$

The determinant of an $n \times n$ matrix is given in terms of n determinants, each of size $(n-1) \times (n-1)$. If $\mathbf{A}$ is $n \times n$ and $\mathbf{M}_{ij}$ is the matrix obtained by removing the ith row and the jth column from $\mathbf{A}$, then

$$|A| = \sum_{j=1}^{n} (-1)^{1+j} a_{1j} |M_{1j}|$$

Example 2.27

Find the determinant

$$\begin{vmatrix} 1 & 2 & 3 \\ 4 & 0 & 6 \\ 7 & 8 & 9 \end{vmatrix}$$

Solution

$$|A| = (-1)^{1+1} a_{11} |M_{11}| + (-1)^{1+2} a_{12} |M_{12}| + (-1)^{1+3} a_{13} |M_{13}|$$
$$= 1 \begin{vmatrix} 0 & 6 \\ 8 & 9 \end{vmatrix} - 2 \begin{vmatrix} 4 & 6 \\ 7 & 9 \end{vmatrix} + 3 \begin{vmatrix} 4 & 0 \\ 7 & 8 \end{vmatrix}$$
$$= -48 - 2(36 - 42) + 3(32) = 60$$

Example 2.28

Find the determinant

$$\begin{vmatrix} 0 & 0 & 2 & 0 \\ 1 & 2 & 7 & 3 \\ 4 & 0 & 3 & 6 \\ 7 & 8 & -6 & 9 \end{vmatrix}$$

Solution

$$|A| = a_{11} |M_{11}| - a_{12} |M_{12}| + a_{13} |M_{13}| - a_{14} |M_{14}|$$
$$= 0 |M_{11}| - 0 |M_{12}| + 2 |M_{13}| - 0 |M_{14}| = 2 (60) = 120$$

The last example provides a clue to the evaluation of large determinants, but the use of the first row of **A** in the definition of a determinant was arbitrary. For any row or column (fix i or j),

$$|A| = \sum_{j=1}^{n} (-1)^{1+j} a_{i1j} |M_{ij}|$$

The interaction of the determinant with elementary row or column operations is simple: Interchanging two rows changes the sign of the determinant; multiplying a row by a constant multiplies the determinant by that constant.

Example 2.29

Evaluate the determinant

$$\begin{vmatrix} 1 & 2 & 3 & 4 \\ 1 & 1 & 1 & 0 \\ 4 & 0 & 3 & 2 \\ 0 & 3 & 0 & 1 \end{vmatrix}$$

Solution

Choose a row or column with many zeroes and introduce still more:

$$|A| \underset{C_2 - 3C_4}{=} |A| = \begin{vmatrix} 1 & -10 & 3 & 4 \\ 1 & 1 & 1 & 0 \\ 4 & -6 & 3 & 2 \\ 0 & 0 & 0 & 1 \end{vmatrix} = (-1)^{4+4} a_{44} \begin{vmatrix} 1 & -10 & 3 \\ 1 & 1 & 1 \\ 4 & -6 & 3 \end{vmatrix}$$

$$\underset{\substack{R_2 - R_1 \\ R_3 - 4R_1}}{=} \begin{vmatrix} 1 & -10 & 3 \\ 0 & 11 & -2 \\ 0 & 34 & -9 \end{vmatrix} = (-1)^{1+1} a_{11} \begin{vmatrix} 11 & -2 \\ 34 & -9 \end{vmatrix} = -99 + 68 = -31$$

Example 2.30

Find which values, if any, of the number c make $\mathbf{A}$ singular if

$$\mathbf{A} = \begin{vmatrix} 1 & 2 & c \\ 4 & 5 & 6 \\ 1 & 1 & 1 \end{vmatrix}$$

Solution

$|\mathbf{A}| = (-1)^2(5-6) + (-1)^3(2)(4-6) + (-1)^4 c(4-5) = -1 + 4 - c = 0$. Hence $\mathbf{A}$ is singular for only one value of c, $c = 3$.

Cramer's Rule is a consequence of adj(A): If $\mathbf{A}$ is nonsingular, the ith component of the solution of $\mathbf{AX} = \mathbf{B}$ is $x_i = \dfrac{|A_i|}{|A|}$, where A_i is the result of replacing the ith column of $\mathbf{A}$ by $\mathbf{B}$.

Example 2.31

Find x_2 in $\mathbf{AX} = \mathbf{B}$ by Cramer's Rule if

$$\mathbf{A} = \begin{bmatrix} 1 & 2 & 1 & 1 \\ 3 & 4 & 5 & -2 \\ 6 & 7 & 1 & 5 \\ -1 & 0 & 2 & 0 \end{bmatrix} \quad \text{and} \quad \mathbf{B} = \begin{bmatrix} 1 \\ 2 \\ 3 \\ 4 \end{bmatrix}$$

Solution

First,

$$|A| = \begin{vmatrix} 1 & 2 & 3 & 1 \\ 3 & 4 & 11 & -2 \\ 6 & 7 & 13 & 5 \\ -1 & 0 & 0 & 0 \end{vmatrix} = (-1)^{4+1}(-1) \begin{vmatrix} 2 & 3 & 1 \\ 4 & 11 & -2 \\ 7 & 13 & 5 \end{vmatrix}$$

$$= \begin{vmatrix} 0 & 0 & 1 \\ 8 & 17 & -2 \\ -3 & -2 & 5 \end{vmatrix} = (-1)^{1+3}(1) \begin{vmatrix} 8 & 17 \\ -3 & -2 \end{vmatrix} = -16 + 51 = 35$$

Next, the numerator of x_2 is

$$\begin{vmatrix} 1 & 1 & 1 & 1 \\ 3 & 2 & 5 & -2 \\ 6 & 3 & 1 & 5 \\ -1 & 4 & 2 & 0 \end{vmatrix} = \begin{vmatrix} 1 & 0 & 0 & 0 \\ 3 & -1 & 2 & -5 \\ 6 & -3 & -5 & -1 \\ -1 & 5 & 3 & 1 \end{vmatrix} = \begin{vmatrix} -1 & 2 & -5 \\ -3 & -5 & -1 \\ 5 & 3 & 1 \end{vmatrix} = \begin{vmatrix} -1 & 2 & -5 \\ 0 & -11 & 14 \\ 0 & 13 & -24 \end{vmatrix}$$

$$= -\begin{vmatrix} -11 & 14 \\ 13 & -24 \end{vmatrix} = -\begin{vmatrix} -11 & 14 \\ 2 & -10 \end{vmatrix} = -(110 - 28) = -82, \; x_2 = -\frac{82}{35}$$

NUMERICAL METHODS

This portion of numerical methods includes techniques of finding roots of polynomials by the Routh-Hurwitz criterion and Newton methods, Euler's techniques of numerical integration and the trapezoidal methods, and techniques of numerical solutions of differential equations.

Root Extraction

Routh-Hurwitz Method (without Actual Numerical Results)

Root extraction, even for simple roots (i.e., without imaginary parts), can become quite tedious. Before attempting to find roots, one should first ascertain whether they are really needed or whether just knowing the area of location of these roots will suffice. If all that is needed is knowing whether the roots are all in the left half-plane of the variable (such as is in the s-plane when using Laplace transforms—as is frequently the case in determining system stability in control systems), then one may use the Routh-Hurwitz criterion. This method is fast and easy even for higher-ordered equations. As an example, consider the following polynomial:

$$p_n(x) = \prod_{m=1}^{n}(x - x_m) = x^n + a_1 x^{n-1} + a_2 x^{n-2} + \cdots + a_{n-1} \tag{2.1}$$

Here, finding the roots, x_m, for $n > 3$ can become quite tedious without a computer; however, if one only needs to know if any of the roots have positive real parts, one can use the Routh-Hurwitz method. Here, an array is formed listing the coefficients of every other term starting with the highest power, n, on a line, followed by a line listing the coefficients of the terms left out of the first row. Following rows are constructed using Routh-Hurwitz techniques, and after completion of the array, one merely checks to see if all the signs are the same (unless there is a zero coefficient—then something else needs to be done) in the first column; if none, no roots will exist in the right half-plane. In case of zero coefficient, a simple technique is used; for details, see almost any text dealing with stability of control systems. A short example follows.

$F(s) = s^3 + 3s^2 + 10$ Array:

	s^3	1	2
	s^2	3	10
$= (s + ?)(s + ?)(s + ?)$	s^1	$-\frac{4}{3}$	0
	s^0	10	0

Where the s^1 term is formed as

$(3 \times 2 - 10 \times 1)/3 = -\frac{4}{3}$. For details, refer to any text on control systems or numerical methods.

Here, there are two sign changes: one from 3 to $-\frac{4}{3}$, and one from $-\frac{4}{3}$ to 10. This means there will be two roots in the right half-plane of the s-plane, which yield an unstable system. This technique represents a great savings in time without having to factor the polynomial.

Newton's Method

The use of Newton's method of solving a polynomial and the use of iterative methods can greatly simplify a problem. This method utilizes synthetic division and is based upon the remainder theorem. This synthetic division requires estimating a root at the start, and, of course, the best estimate is the actual root. The root is the correct one when the remainder is zero. (There are several ways of estimating this root, including a slight modification of the Routh-Hurwitz criterion.)

If a $P_n(x)$ polynomial (see Equation (2.1)) is divided by an estimated factor $(x - x_1)$, the result is a reduced polynomial of degree $n - 1$, $Q_{n-1}(x)$, plus a constant remainder of b_{n-1}. Thus, another way of describing Equation (2.1) is

$$P_n(x)/(x - x_1) = Q_{n-1}(x) + b_{n-1}/(x - x_1) \quad \text{or} \quad P_n(x) = (x - x_1)Q_{n-1}(x) + b_{n-1} \quad (2.2)$$

If one lets $x = x_1$, Equation (2.2) becomes

$$P_n(x = x_1) = (0)Q_{n-1}(x) + b_{n-1} = b_{n-1} \quad (2.3)$$

Equation 2.3 leads directly to the remainder theorem: "The remainder on division by $(x - x_1)$ is the value of the polynomial at $x = x_1$, $P_n(x_1)$."[1]

Newton's method (actually, the Newton-Raphson method) for finding the roots for an nth-order polynomial is an iterative process involving obtaining an estimated value of a root (leading to a simple computer program). The key to the process is getting the first estimate of a possible root. Without getting too involved, recall that the coefficient of x^{n-1} represents the sum of all of the roots and the last term represents the product of all n roots; then the first estimate can be "guessed" within a reasonable magnitude. After a first root is chosen, find the rate of change of the polynomial at the chosen value of the root to get the next, closer value of the root x_{n+1}. Thus the new root estimate is based on the last value chosen:

$$x_{n+1} = x_n - P_n(x_n)/P_n'(x_n), \quad (2.4)$$

where $P_n'(x_n) = dP_n(x)/dx$ evaluated at $x = x_n$

NUMERICAL INTEGRATION

Numerical integration routines are extremely useful in almost all simulation-type programs, design of digital filters, theory of z-transforms, and almost any problem solution involving differential equations. And because digital computers have essentially replaced analog computers (which were almost true integration devices), the techniques of approximating integration are well developed. Several of the techniques are briefly reviewed below.

---

[1] Gerald & Wheatley, *Applied Numerical Analysis*, 3rd ed., Addison-Wesley, 1985.

Euler's Method

For a simple first-order differential equation, say $dx/dt + ax = af$, one could write the solution as a continuous integral or as an interval type one:

$$x(t) = \int^{t} [-ax(\tau) + af(\tau)] d\tau \qquad (2.5a)$$

$$x(kT) = \int^{kT-T} [-ax + af] d\tau + \int_{kT-T}^{kT} [-ax + af] d\tau = x(kT - T) + A_{rect} \qquad (2.5b)$$

Here, A_{rect} is the area of $(-ax + af)$ over the interval $(kT - T) < \tau < kT$. One now has a choice looking back over the rectangular area or looking forward. The rectangular width is, of course, T. For the forward-looking case, a first approximation for x_1 is[2]

$$\begin{aligned} x_1(kT) &= x_1(kT - T) + T[ax_1(kT - T) + af(kT - T)] \\ &= (1 - aT)x_1(kT - T) + aTf(kT - T) \end{aligned} \qquad (2.5c)$$

Or, in general, for Euler's forward rectangle method, the integral may be approximated in its simplest form (using the notation $t_{k+1} - t_k$ for the width, instead of T, which is $kT-T$) as

$$\int_{t_k}^{t_{k+1}} x(\tau) d\tau \approx (t_{k+1} - t_k) x(t_k) \qquad (2.6)$$

Trapezoidal Rule

This trapezoidal rule is based upon a straight-line approximation between the values of a function, $f(t)$, at t_0 and t_1. To find the area under the function, say a curve, is to evaluate the integral of the function between points a and b. The interval between these points is subdivided into subintervals; the area of each subinterval is approximated by a trapezoid between the end points. It will be necessary only to sum these individual trapezoids to get the whole area; by making the intervals all the same size, the solution will be simpler. For each interval of delta t (i.e., $t_{k+1} - t_k$), the area is then given by

$$\int_{t_k}^{t_{k+1}} x(\tau) d\tau \approx (1/2)(t_{k+1} - t_k)[x(t_{k+1}) + x(t_k)] \qquad (2.7)$$

This equation gives good results if the delta t's are small, but it is for only one interval and is called the "local error." This error may be shown to be $-(1/12)$ (delta $t)^3 f''(t = \xi_1)$, where ξ_1 is between t_0 and t_1. For a larger "global error" it may be shown that

$$\text{Global error} = -(1/12)(\text{delta } t)^3 [f''(\xi_1) + f''(\xi_2) + \cdots + f''(\xi_n)] \qquad (2.8)$$

---

[2] This method is as presented in Franklin & Powell, *Digital Control of Dynamic Systems*, Addison-Wesley, 1980, page 55.

Following through on Equation 2.8 allows one to predict the error for the trapezoidal integration. This technique is beyond the scope of this review or probably the examination; however, for those interested, please refer to pages 249–250 of the previously mentioned reference to Gerald & Wheatley.

NUMERICAL SOLUTIONS OF DIFFERENTIAL EQUATIONS

This solution will be based upon first-order ordinary differential equations. However, the method may be extended to higher-ordered equations by converting them to a matrix of first-ordered ones.

Integration routines produce values of system variables at specific points in time and update this information at each interval of delta time as T (delta $t = T = t_{k+1} - t_k$). Instead of a continuous function of time, $x(t)$, the variable x will be represented with discrete values such that $x(t)$ is represented by $x_0, x_1, x_2, ..., x_n$. Consider a simple differential equation as before as based upon Euler's method,

$$dx/dt + ax = f(t)$$

Now assume the delta time periods, T, are fixed (not all routines use fixed step sizes); then one writes the continuous equations as a difference equation where $dx/dt \approx (x_{k+1} - x_k)/T = -ax_k + f_k$ or, solving for the updated value, x_{k+1},

$$x_{k+1} = x_k - Tax_k + Tf_k \qquad (2.9a)$$

For fixed increments by knowing the first value of $x_{k=0}$ (or the initial condition), one may calculate the solution for as many "next values" of x_{k+1} as desired for some value of T. The difference equation may be programmed in almost any high-level language on a digital computer; however, T must be small as compared to the shortest time constant of the equation (here, $1/a$).

The following equation—with the "f" term meaning "a function of" rather than as a "forcing function" term as used in Equation (2.5a)—is a more general form of Equation (2.9a). This equation is obtained by letting the notation x_{k+1} become $y[k+1 \, \Delta t]$ and is written (perhaps somewhat more confusingly) as

$$y[(k+1)\Delta t] = y(k\Delta t) + \Delta t f[y(k\Delta t), k\Delta t] \qquad (2.9b)$$

Reduction of Differential Equation Order

To reduce the order of a linear time-dependent differential equation, the following technique is used. For example, assume a second-order equation: $x'' + ax' + bx = f(t)$. If we define $x = x_1$ and $x' = x_1' = x_2$, then

$$x_2' + ax_2 + bx_1 = f(t)$$
$$x_1' = x_2 \text{ (by definition)}$$
$$x_2' = -bx_1 - ax_2 + f(t)$$

This technique can be extended to higher-order systems and, of course, be put into a matrix form (called the state variable form). And it can easily be set up as a matrix of first-order difference equations for solving digitally.

DIFFERENTIAL CALCULUS

Definition of a Function

Notation. A variable y is said to be a function of another variable x if, when x is given, y is determined. The symbols $f(x)$, $F(x)$, etc., represent various functions of x. The symbol $f(a)$ represents the value of $f(x)$ when $x = a$.

Definition of a Derivative

Let $y = f(x)$. If Δx is any increment (increase or decrease) given to x, and Δy is the corresponding increment in y, then the derivative of y with respect to x is the limit of the ratio of Δy to Δx as Δx approaches zero; that is,

$$\frac{dy}{dx} = \lim_{\Delta x \to 0} \frac{\Delta y}{\Delta x} = \lim_{\Delta x \to 0} \frac{f(x + \Delta x) - f(x)}{\Delta x} = f'(x)$$

Some Relations among Derivatives

If $x = f(y)$, then $\dfrac{dy}{dx} = 1 \div \dfrac{dx}{dy}$

If $x = f(t)$ and $y = F(t)$, then $\dfrac{dy}{dx} = \dfrac{dy}{dt} \div \dfrac{dx}{dt}$

If $y = f(u)$ and $u = F(x)$, then $\dfrac{dy}{dx} = \dfrac{dy}{du} \times \dfrac{du}{dx}$

If $x = f(t)$, then $f''(t) = \dfrac{d^2 x}{dt^2} = \dfrac{d}{dt}\left(\dfrac{dx}{dt}\right)$

Table of Derivatives

Functions of x are represented by u and v, and constants are represented by a, n, and e.

$$\frac{d}{dx}(x) = 1 \qquad \frac{d}{dx}(a) = 0$$

$$\frac{d}{dx}(u \pm v \pm \ldots) = \frac{du}{dx} \pm \frac{dv}{dx} \pm \ldots \qquad \frac{d}{dx}(au) = a\frac{du}{dx}$$

$$\frac{d}{dx}(uv) = u\frac{dv}{dx} + v\frac{du}{dx} \qquad \frac{d}{dx}\left(\frac{u}{v}\right) = \frac{v\dfrac{du}{dx} - u\dfrac{dv}{dx}}{v^2}$$

$$\frac{d}{dx}(u^n) = nu^{n-1}\frac{du}{dx} \qquad \frac{d}{dx}\log_a u = \frac{\log_a e}{u}\frac{du}{dx}$$

$$\frac{d}{dx}a^u = a^u \ln a \frac{du}{dx}$$

$$\frac{d}{dx}e^u = e^u \frac{du}{dx} \qquad \frac{d}{dx}u^v = vu^{v-1}\frac{du}{dx} + u^v \ln u \frac{dv}{dx}$$

$$\frac{d}{dx}\sin u = \cos u \frac{du}{dx} \qquad \frac{d}{dx}\cot u = -\csc^2 u \frac{du}{dx}$$

$$\frac{d}{dx}\cos u = -\sin u \frac{du}{dx} \qquad \frac{d}{dx}\sec u = \sec u \tan u \frac{du}{dx}$$

$$\frac{d}{dx}\tan u = \sec^2 u \frac{du}{dx} \qquad \frac{d}{dx}\csc u = -\csc u \cot u \frac{du}{dx}$$

$$\frac{d}{dx}\sin^{-1} u = \frac{1}{\sqrt{1-u^2}}\frac{du}{dx} \quad \text{where} \quad -\pi/2 \leq \sin^{-1} u \geq \pi/2$$

$$\frac{d}{dx}\cos^{-1} u = -\frac{1}{\sqrt{1-u^2}}\frac{du}{dx} \quad \text{where} \quad 0 \leq \cos^{-1} u \geq \pi$$

$$\frac{d}{dx}\tan^{-1} u = \frac{1}{1+u^2}\frac{du}{dx}$$

$$\frac{d}{dx}\cot^{-1} u = -\frac{1}{1+u^2}\frac{du}{dx}$$

$$\frac{d}{dx}\sec^{-1} u = \frac{1}{u\sqrt{u^2-1}}\frac{du}{dx} \quad \text{where} \quad 0 \leq \sec^{-1} u \leq \pi/2 \text{ and } -\pi \leq \sec^{-1} u \leq -\pi/2$$

$$\frac{d}{dx}\csc^{-1} u = -\frac{1}{u\sqrt{u^2-1}}\frac{du}{dx} \quad \text{where} \quad -\pi < \csc^{-1} u \leq -\pi/2 \text{ and } 0 < \csc^{-1} u \leq \pi/2$$

Example 2.32

Find the derivatives of the following functions with respect to x.

(i) $x^3 + 4e^{-2x}$, (ii) $x^2 \sin x$, (iii) $2\sin^2 x$, (iv) $(\dfrac{e^{-x}}{x})$

Solution

(i) $\dfrac{d}{dx}(x^3 + 4e^{-2x}) = \dfrac{d}{dx}(x^3) + \dfrac{d}{dx}(4e^{-2x}) = 3x^2 + 4(-2)e^{-2x} = 3x^2 - 8e^{-2x}$

(ii) $\dfrac{d}{dx}(x^2 \sin x) = x^2 \dfrac{d}{dx}(\sin x) + (\sin x)\dfrac{d}{dx}(x^2) = x^2 \cos x + 2x \sin x$

(iii) $\dfrac{d}{dx}(2\sin^2 x) = 2(2)(\sin^{2-1} x)\dfrac{d}{dx}(\sin x) = (4 \sin x)\cos x$

(iv) $\dfrac{d}{dx}(\dfrac{e^{-x}}{x}) = \dfrac{x \dfrac{d}{dx}(e^{-x}) - e^{-x} \dfrac{d}{dx}(x)}{x^2} = \dfrac{x(-e^{-x}) - (e^{-x})1}{x^2} = \dfrac{-e^{-x}(x+1)}{x^2}$

Slope of a Curve: Tangent and Normal

The slope of the curve (slope of the tangent line to the curve) whose equation is $y = f(x)$ is

$$\text{Slope} = m = \tan \phi = \frac{dy}{dx} = f'(x)$$

Slope at x_1 is $m_1 = f'(x_1)$

The equation of a tangent line at $P_1(x_1, y_1)$ is $y - y_1 = m_1(x - x_1)$. The equation of a normal at $P_1(x_1, y_1)$ is

$$y - y_1 = -\frac{1}{m_1}(x - x_1)$$

The angle β of the intersection of two curves whose slopes at a common point are m_1 and m_2 is

$$\beta = \tan^{-1} \frac{m_2 - m_1}{1 + m_1 m_2}$$

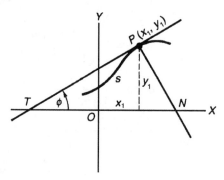

Figure 2.33

Example 2.33

Find the slope of the curve $y = (4x^2 - 8)$ at $x = 1$.

Solution

Slope $\dfrac{dy}{dx} = \dfrac{d}{dx}(4x^2 - 8) = \dfrac{d}{dx}(4x^2) - \dfrac{d}{dx}(8) = 4\dfrac{d}{dx}(x^2) - 0 = 8x;$

at $x = 1$, slope $= 8$

Example 2.34

Find the equation of the tangent to the curve $y = (x^2 - x - 4)$ at $(1, -4)$.

Solution

Slope of the curve $\dfrac{dy}{dx} = \dfrac{d}{dx}(x^2 - x - 4) = \dfrac{d}{dx}(x^2) - \dfrac{d}{dx}(x) - \dfrac{d}{dx}(4) = 2x - 1;$

at $(1, -4)$, slope $= 2(1) - 1 = 1$

Equation of the tangent is $(y + 4) = 1(x - 1)$ or $y = x - 5$.

Maximum and Minimum Values of a Function

The maximum or minimum value of a function $f(x)$ in an interval from $x = a$ to $x = b$ is the value of the function that is larger or smaller, respectively, than the values of the function in its immediate vicinity. Thus, the values of the function at M_1 and M_2 in Figure 2.34 are maxima, and its values at m_1 and m_2 are minima.

Test for a maximum at $x = x_1$: $f'(x_1) = 0$ or ∞, and $f''(x_1) < 0$

Test for minimum at $x = x_1$: $f'(x_1) = 0$ or ∞, and $f''(x_1) > 0$

If $f''(x_1) = 0$ or ∞, then for a maximum, $f'''(x_1) = 0$ or ∞ and $f^{IV}(x_1) < 0$; for a minimum, $f'''(x_1) = 0$ or ∞ and $f^{IV}(x_1) > 0$, and similarly if $f^{IV}(x_1) = 0$ or ∞, and so on, where f^{IV} represents the fourth derivative.

In a practical problem that suggests that the function $f(x)$ has a maximum or has a minimum in an interval from $x = a$ to $x = b$, simply equate $f'(x)$ to 0 and solve for the required value of x. To find the largest or smallest values of a function $f(x)$ in an interval from $x = a$ to $x = b$, find also the values $f(a)$ and $f(b)$. L and S may be the largest and smallest values, although they are not maximum or minimum values (see Figure 2.34).

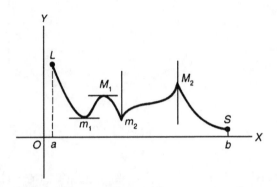

Figure 2.34

Points of Inflection of a Curve

Wherever $f''(x) < 0$, the curve is concave down.

Wherever $f''(x) > 0$, the curve is concave up.

The curve is said to have a point of inflection at $x = x_1$ if $f''(x_1) = 0$ or ∞, and the curve is concave up on one side of $x = x_1$ and concave down on the other (see points I_1 and I_2 in Figure 2.35).

Example 2.35

For the function $y = f(x) = (x^3 - 3x)$, find the maximum, minimum, and the point of inflection.

Solution

The derivative of $f(x)$, $f'(x) = 3x^2 - 3$.

Equating $f'(x)$ to 0, $3x^2 - 3 = 0$, which has roots at $x = 1$ and -1.

The derivative of $f'(x)$, $f''(x) = 6x$. At $x = 1$, $f''(1) = 6$ and at $x = -1$, $f''(-1) = -6$.

Minimum occurs at $x = 1$, since $f''(1) > 0$ and $f''(1) = 0$. Value of minimum $= f(1) = -2$.

Maximum occurs at $x = -1$, since $f'(-1) < 0$ and $f'(-1) = 0$. Value of maximum $= f(-1) = 2$.

$f''(x) = 0$ at $x = 0$; $f''(x)$ changes sign as x passes through 0; then $x = 0$ is the inflection point (it is between 1 and -1).

Taylor and Maclaurin Series

In general, any $f(x)$ may be expanded into a **Taylor series**:

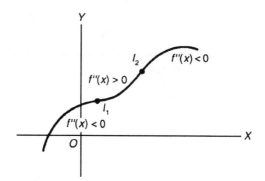

Figure 2.35

$$f(x) = f(a) + f'(a)\frac{x-a}{1} + f''(a)\frac{(x-a)^2}{2!} + f'''(a)\frac{(x-a)^3}{3!} + \ldots$$

$$+ f^{(n-1)}(a)\frac{(x-a)^{n-1}}{(n-1)!} + R_n$$

where a is any quantity whatever, so chosen that none of the expressions $f(a)$, $f'(a)$, $f''(a)$,... become infinite. If the series is to be used for the purpose of computing the approximate value of $f(x)$ for a given value of x, a should be chosen such that $(x - a)$ is numerically very small, and thus only a few terms of the series need be used. If $a = 0$, this series is called a Maclaurin series.

Example 2.36

Find the power series expansion of sin x about the point 0.

Solution

In the Taylor's series expansion, $f(x) = \sin(x)$ and $a = 0$. Derivatives of $f(x)$ are: $f'(x) = \cos(x), f''(x) = -\sin(x), f'''(x) = -\cos(x)$, and so on.

Substituting $f(x)$ and its derivatives at a in the expansion,

$$\sin(x) = \sin(0) + x\cos(0) + \frac{x^2(-\sin(0))}{2!} + \frac{x^3(-\cos(0))}{3!} + \ldots = x - \frac{x^3}{3!} + \frac{x^5}{5!} - \ldots$$

Evaluation of Indeterminate Forms

Let $f(x)$ and $F(x)$ be two functions of x, and let a be a value of x.

1. If $\dfrac{f(a)}{F(a)} = \dfrac{0}{0}$ or $\dfrac{\infty}{\infty}$, use $\dfrac{f'(a)}{F'(a)}$ for the value of this fraction.

   If $\dfrac{f'(a)}{F'(a)} = \dfrac{0}{0}$ or $\dfrac{\infty}{\infty}$, use $\dfrac{f''(a)}{F''(a)}$ for the value of this fraction, and so on.

2. If $f(a) \bullet F(a) = 0 \bullet \infty$ or if $f(a) - F(a) = \infty - \infty$, evaluate the expression by changing the product or difference to the form $\frac{0}{0}$ or $\frac{\infty}{\infty}$ and use the previous rule.

3. If $f(a)^{F(a)} = 0^0$ or ∞^0 or 1^∞, then form $e^{F(a) \bullet \ln f(a)}$, and the exponent, being of the form $0 \bullet \infty$, may be evaluated by rule 2.

Example 2.37

Find the following limits: (i) $\lim_{x \to 0} \frac{\sin x}{x}$ (ii) $\lim_{x \to 0} \frac{1 - \cos x}{x^2}$

Solution

(i) Since $\frac{\sin(0)}{0} = \frac{0}{0}$, we apply the limiting theorem:

$$\lim_{x \to 0} \frac{\sin x}{x} = \lim_{x \to 0} \frac{d(\sin x)}{d(x)} = \lim_{x \to 0} \frac{\cos x}{1} = 1$$

(ii) Since $\frac{1 - \cos(0)}{0^2} = \frac{0}{0}$, we apply the limiting theorem:

$$\lim_{x \to 0} \frac{1 - \cos x}{x^2} = \lim_{x \to 0} \frac{\sin x}{2x} = \lim_{x \to 0} \frac{\cos x}{2} = \frac{1}{2}$$

Example 2.38

The cubic $y = x^3 + x^2 - 3$ has one point of inflection. Where does it occur?

Solution

The answer requires knowing where y' changes sign. Now $y' = 3x^2 + 2x$ and $y'' = 6x + 2$, which is 0 where $x = -\frac{1}{3}$. Thus the only inflection point is at $x = -\frac{1}{3}$.

Example 2.39

The function of Example 2.38 has one local maximum and one local minimum. Where are they?

Solution

Setting $y' = 0$ ($3x^2 + 2x = 0$) yields $x = 0$ or $x = -\frac{2}{3}$. Since the second derivative is 2 at $x = 0$, this is the local minimum. At $x = -\frac{2}{3}$, $y'' = -2$, so $x = -\frac{2}{3}$ is the local maximum.

Differential of a Function

If $y = f(x)$ and Δx is an increment in x, then the differential of x equals the increment of x, or $dx = \Delta x$; and the differential of y is the derivative of y multiplied by the differential of x; thus

$$dy = \frac{dy}{dx} dx = \frac{df(x)}{dx} dx = f'(x) dx \quad \text{and} \quad \frac{dy}{dx} = dy \div dx$$

If $x = f_1(t)$ and $y = f_2(t)$, then $dx = f_1'(t) dt$, and $dy = f_2'(t) dt$.

Every derivative formula has a corresponding differential formula; thus, from the Table of Derivatives subsection, we have, for example,

$$d(uv) = u\,dv + v\,du; \quad d(\sin u) = \cos u\,du; \quad d(\tan^{-1} u) = \frac{du}{1+u^2}$$

Functions of Several Variables, Partial Derivatives, and Differentials

Let z be a function of two variables, $z = f(x, y)$; then its partial derivatives are

$$\frac{\partial z}{\partial x} = \frac{dz}{dx} \text{ when } y \text{ is kept constant} \qquad \frac{\partial z}{\partial y} = \frac{dz}{dy} \text{ when } x \text{ is kept constant}$$

Example 2.40

Find the partial derivatives of $f(x, y) = 4x^2 y - 2y$ (i) with respect to x, and (ii) with respect to y.

Solution

(i) $\dfrac{\partial f}{\partial x} = \dfrac{\partial}{\partial x}(4x^2 y - 2y) = 4y\dfrac{\partial(x^2)}{\partial x} - 0 = (4y)(2x) = 8xy$

(ii) $\dfrac{\partial f}{\partial y} = \dfrac{\partial}{\partial y}(4x^2 y - 2y) = 4x^2 \dfrac{\partial(y)}{\partial y} - \dfrac{\partial(2y)}{\partial y} = 4x^2 - 2$

Example 2.41

Two automobiles are approaching the origin. The first one is traveling from the left on the x-axis at 30 mph. The second is traveling from the top on the y-axis at 45 mph. How fast is the distance between them changing when the first is at $(-5, 0)$ and the second is at $(0, 10)$? (Both coordinates are in miles.)

Solution

If $x(t)$ is taken as the position of the first auto at time t and $y(t)$ as the position of the second auto at time t, then the distance between them at time t is $s(t) = \sqrt{[x(t)]^2 + [y(t)]^2}$. Using the chain rule,

$$s'(t) = \frac{ds}{dt} = \frac{1}{d_2 s(t)} \frac{d}{dt}\{[x(t)]^2 + [y(t)]^2\} = \frac{1}{2s(t)}[2x(t)x'(t) + 2y(t)y'(t)]$$

Now $x'(t) = 30$ and $y'(t) = -45$ for all t; and when $t = t_0$, $x(t_0) = -5$ and $y(t_0) = 10$.

Therefore,

$$s'(t_0) = \frac{-2(5)(30) - 2(10)(45)}{2\sqrt{(-5)^2 + (10)^2}} = \frac{-1200}{2\sqrt{125}} = \frac{-120}{\sqrt{5}} = 24\sqrt{5} \approx 54$$

Thus, the two automobiles are "closing" at about 54 mph.

Example 2.42

How close do the two automobiles in Example 2.41 get?

Solution

One wants to minimize $s(t)$ in Example 2.41, so set $s'(t) = 0$. Thus,

$$\frac{xx' + yy'}{s} = 0 \quad \text{or} \quad xx' + yy' = 0$$

Since $x' = 30$ and $y' = -45$, $30x = 45y$. However, since $x'(t) = 30$, $x(t) = 30t + x_0$, and similarly $y(t) = -45t + y_0$. If one takes $t_0 = 0$ when the problem starts, $x_0 = -5$ and $y_0 = 10$, so $30(30t - 5) = 45(-45t + 10)$ gives time of minimum distance. Solving for t, factor out 75 from both sides to get $2(6t - 1) = 3(-9t + 2)$, or $39t = 8$.

Thus the minimum distance occurs at 8/39 of an hour after the initial conditions of Example 2.41. At this time $x = -5 + 240/39$ and $y = 10 - 360/39$, so $x = 45/39$ and $y = 30/39$. The minimum distance is

$$s\left(\frac{8}{39}\right) = \frac{\sqrt{(45)^2 + (30)^2}}{39} = \frac{15}{39}\sqrt{9 + 4} = \frac{15\sqrt{13}}{39} \approx 1.4 \text{ miles}$$

Example 2.43

In Example 2.41, which reaches the origin car first, Car 1 or 2?

Solution

This is obvious if, in Example 2.42, one notices that x is positive and y is (still) positive. Alternatively, notice that Car 1 takes 5/30 of an hour to reach the origin and Car 2 takes 10/45 of an hour. The time 5/30 < 10/45, so Car 1 gets there first.

INTEGRAL CALCULUS

Definition of an Integral

The function $F(x)$ is said to be the integral of $f(x)$ if the derivative of $F(x)$ is $f(x)$, or if the differential of $F(x)$ is $f(x)\,dx$. In symbols,

$$F(x) = \int f(x)\,dx \quad \text{if} \quad \frac{dF(x)}{dx} = f(x), \quad \text{or} \quad dF(x) = f(x)\,dx$$

In general, $\int f(x)\,dx = F(x) + C$, where C is an arbitrary constant.

Fundamental Theorems on Integrals

$$\int df(x) = f(x) + C$$

$$\int df(x)\,dx = f(x)\,dx$$

$$\int [f_1(x) \pm f_2(x) \pm \cdots]\,dx = \int f_1(x)\,dx \pm \int f_2(x)\,dx \pm \cdots$$

$$\int af(x)\,dx = a\int f(x)\,dx, \text{ where } a \text{ is any constant}$$

$$\int u^n\,du = \frac{u^{n+1}}{n+1} + C \quad (n \neq -1), \text{ where } u \text{ is any function of } x$$

$$\int \frac{du}{u} = \ln u + C, \text{ where } u \text{ is any function of } x$$

$$\int u\,dv = uv - \int v\,du, \text{ where } u \text{ and } v \text{ are any functions of } x$$

$$\int [u(x) \pm v(x)]\,dx = \int u(x)\,dx \pm \int v(x)\,dx$$

$$\int \frac{dx}{ax+b} = \frac{1}{a}\ln|ax+b|$$

$$\int \frac{dx}{\sqrt{x}} = 2\sqrt{x}$$

$$\int a^x\,dx = \frac{a^x}{\ln a}$$

$$\int \sin x \, dx = -\cos x$$

$$\int \sin^2 x \, dx = \frac{x}{2} - \frac{\sin 2x}{4}$$

$$\int x \sin x \, dx = \sin x - x \cos x$$

$$\int \cos x \, dx = \sin x$$

$$\int \cos^2 x \, dx = \frac{x}{2} + \frac{\sin 2x}{4}$$

$$\int x \cos x \, dx = \cos x + \sin x$$

$$\int \sin x \cos x \, dx = (\sin^2 x)/2$$

$$\int \tan x \, dx = -\ln|\cos x| = \ln|\sec x|$$

$$\int \tan^2 x \, dx = \tan x - x$$

$$\int \cot x \, dx = -\ln|\csc x| = \ln|\sin x|$$

$$\int \cot^2 x \, dx = -\cot x - x$$

$$\int e^{ax} \, dx = (1/a)e^{ax}$$

$$\int \ln x \, dx = x[\ln(x) - 1] \qquad (x > 0)$$

Example 2.44

Evaluate the following integrals:

(i) $\int (x^3 + 4) dx$ (ii) $\int \sin 3x \, dx$ (iii) $\int \sqrt{(1-x)} \, dx$

Solution

(i) $\int (x^3 + 4) dx = \left(\frac{x^4}{4} + 4x \right) + \text{constant}$

(ii) $\int \sin 3x \, dx$; let $t = 3x$; then $dt = 3dx$ and

$$\int \frac{1}{3} \sin t \, dt = -\frac{\cos t}{3} = \frac{-\cos 3x}{3} + \text{constant}$$

(iii) $\int \sqrt{(1-x)} \, dx$; let $t = 1 - x$; then $dt = -dx$;

$$\int -\sqrt{t} \, dt = -\frac{2}{3} t^{\frac{3}{2}} = -\frac{2}{3}(1-x)^{\frac{3}{2}} + \text{constant}$$

Moment of Inertia

Moment of inertia J of a mass m:

About $OX: I_x = \int y^2 \, dm = \int r^2 \sin^2 \theta \, dm$

About $OY: I_y = \int x^2 \, dm = \int r^2 \cos^2 \theta \, dm$

About $O: J_0 = \int (x^2 + y^2) \, dm = \int r^2 \, dm$

Center of Gravity

Coordinates $(\bar{x}, \bar{y})$ of the center of gravity of a mass m:

$$\bar{x} = \frac{\int x \, dm}{\int dm}, \qquad \bar{y} = \frac{\int y \, dm}{\int dm}$$

The center of gravity of the differential element of area may be taken at its midpoint. In the above equations, x and y are the coordinates of the center of gravity of the element.

Work

The work W done in moving a particle from $s = a$ to $s = b$ against a force whose component in the direction of motion is F_s is

$$dW = F_s \, ds, \qquad W = \int_a^b F_s \, ds$$

where F_s must be expressed as a function of s.

Example 2.45

Consider the function $y = x^2 + 1$ between $x = 0$ and $x = 2$. What is the area between the curve and the x-axis?

Solution

$$A = \int_0^2 (x^2 + 1) \, dx = \left(\frac{x^3}{3} + x\right)\Bigg|_0^2 = \frac{8}{3} + 2 = \frac{14}{3}$$

Example 2.46

Consider the area bounded by $x = y^2$, the x-axis, and the line $x = 4$. Find (1) the area; (2) the first moment of area with respect to the x-axis; (3) the first moment of area with respect to the y-axis; (4) the centroid; (5) the second moment of area with respect to the x-axis; (6) the second moment of area with respect to the y-axis; (7) the moment of inertia about the line $x = -2$; and (8) the moment of inertia about the line $y = 4$.

Solution
The area is shown shaded in Exhibit 1.

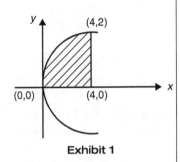

Exhibit 1

For the vertical strip shown in Exhibit 2, $dA = y\, dx = \sqrt{x}\, dx$.

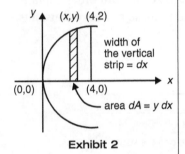

Exhibit 2

(1) area $A = \int dA = \int y\, dx = \int_0^4 \sqrt{x}\, dx = \dfrac{16}{3}$ cm^2

(3) first moment with respect to y-axis $M_y = \int x\, dA = \int_0^4 x\sqrt{x}\, dx = 12.8$ cm^3

(6) second moment with respect to y-axis

$$I_y = \int x^2\, dA = \int_0^4 x^2 \sqrt{x}\, dx = 36.57 \text{ cm}^4$$

For the horizontal strip shown in Exhibit 3, $dA_1 = (4 - x)\, dy = (4 - y^2)\, dy$

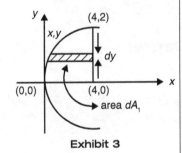

Exhibit 3

Note: area $A = \int_0^2 (4 - y^2)\, dy$

(2) first moment with respect to x-axis $M_x = \int y\, dA_1 = \int_0^2 y(4 - y^2)\, dy = 4$ cm^3

(4) x coordinate of centroid $x_c = \dfrac{M_y}{A} = \dfrac{12.8}{5.33} = 2.4$ cm

y coordinate of the centroid $y_c = \dfrac{M_x}{A} = \dfrac{4}{5.33} = 0.75$ cm

(5) second moment with respect to the x-axis

$$I_x = \int y^2\, dA_1 = \int_0^2 y^2(4 - y^2)\, dy = 4.27 \text{ cm}^4$$

Now, using parallel-axis theorem,

(7) moment of inertia about the line $x = -2$ is $I_y + Ad^2 = 36.57 + (5.333)(2 + 2.4)^2 = 139.8$ cm^4

(8) moment of inertia about the line $y = 4$ is $I_x + Ad^2 = 4.27 + (5.333)(4 - 0.75)^2 = 60.6$ cm^4

DIFFERENTIAL EQUATIONS

Definitions

A **differential equation** is an equation involving differentials or derivatives.

The **order** of a differential equation is the order of the derivative of highest order that it contains.

The **degree** of a differential equation is the power to which the derivative of highest order in the equation is raised, that derivative entering the equation free from radicals.

The **solution** of a differential equation is the relation involving only the variables (but not their derivatives) and arbitrary constants, consistent with the given differential equation.

The most **general solution** of a differential equation of the nth order contains n arbitrary constants. If particular values are assigned to these arbitrary constants, the solution is called a particular solution.

Notation

Symbol or Abbreviation	Definition
M, N	Functions of x and y
X	Function of x alone or a constant
Y	Function of y alone or a constant
C, c	Arbitrary constants of integration
$a, b, k, 1, m, n$	Given constants

Equations of First Order and First Degree: $M\,dx + N\,dy = 0$

Variables Separable: $X_1 Y_1\,dx + X_2 Y_2\,dy = 0$

Solution

$$\int \frac{X_1}{X_2}dx + \int \frac{Y_2}{Y_1}dy = 0$$

Linear Equation: $dy + (X_1 Y - X_2)\,dx = 0$

Solution

$$y = e^{-\int X_1 dx}\left(\int X_2 e^{\int X_1 dx}\,dx + C\right)$$

Second-Order Differential Equations

A second-order differential expression, $L(x, y, y', y'')$, is linear if

$$L(x, ay_1 + by_2, ay_1' + by_2', ay_1'' + by_2'') = aL(x, y_1, y_1', y_1'') + bL(x, y_2, y_2', y_2'')$$

or, if it has the form

$$L(x, y_1, y_1', y_1'') = f(x)y + g(x)y' + h(x)y''$$

A second-order linear differential equation is
$$L(x, y, y', y'') = F(x)$$
If $F(x) \equiv 0$, it is homogeneous; if $F(x)$ is nonzero, it is inhomogeneous.

Constant Coefficients

If $L = ay'' + by' + cy$ where a, b, and c are constants with $a \neq 0$, the first step is to solve the associated homogeneous equation $ay'' + by' + cy = 0$. By replacing y by 1, y' by r, and y'' by r^2, one obtains the characteristic equation $ar^2 + bc + c = 0$ with roots r_1 and r_2 obtained from factoring or from the quadratic formula. There are three cases to consider:

Case 1: $r_1 \neq r_2$, both real; $y = c_1 e^{r_1 x} + c_2 e^{r_2 x}$, where c_1 and c_2 are arbitrary constants.

Case 2: $r_1 = r_2$; $y = c_1 e^{r_1 x} + c_2 x e^{r_2 x}$.

Case 3: $r_1 = \alpha + j\beta$, $r_2 = \alpha - j\beta$, where α and β are real and $j^2 = -1$;
$y = d_1 e^{r_1 x} + d_2 e^{r_2 x} = e^{\alpha x}(c_1 \sin \beta x + c_2 \cos \beta x)$. In particular, if $\alpha = 0$,
$y = c_1 \sin \beta x + c_2 \cos \beta x$.

After finding the two solutions to the associated homogeneous equation (y_1 is the result of setting $c_1 = 1$ and $c_2 = 0$, whereas y_2 has $c_1 = 0$ and $c_2 = 1$), one proceeds in either of the following two ways.

Variation of Parameters

If $L(y) = F(x)$ in which the coefficient of y'' is 1, and $W(x) = y_1(x)y_2'(x) - y_1'(x)y_2(x)$ in which y_1 and y_2 are those solutions found above, and if

$$u_1' = \frac{-F(x)y_2(x)}{W(x)} \quad \text{and} \quad u_2' = \frac{F(x)y_1(x)}{W(x)}$$

one solution to the inhomogeneous equation is $y_p = u_1(x)y_1(x) + u_2(x)y_2(x)$.

Undetermined Coefficients

In this technique, one guesses y_p by using the following patterns.

One guesses that the solution may be of the same form as the $F(x)$ function but with coefficients to be determined. This method requires modification if $F(x)$ and $c_1 y_1 + c_2 y_2$ from the associated homogeneous equation interfere, but it is frequently easier than the integration required in the Variation of Parameters technique to construct u_1 and u_2 from their derivatives.

To guess, classify $F(x)$. If it is a polynomial of degree k, the guess will be a polynomial of degree k. However, if $r = 0$ occurs in the homogeneous equation, increase the degree of the polynomial by one. If $F(x)$ is a polynomial times e^{Ax}, so will be the guess. Once again, the degree may have to be increased by one. If $F(x)$ contains sines and cosines, so should the guess.

After making the guess, differentiate it twice and put it into the equation. The coefficients may be determined at this time.

When mixed forms of functions are present in $F(x)$—for example, $x^2 + 2 + 3 \sin 2x$—treat the terms $x^2 + 2$ and $3 \sin 2x$ independently. The principle of *superposition* then permits you to add the results.

Now, after finding the solution $c_1 y_1 + c_2 y_2$ to the associated homogeneous equation, and the particular solution y_p to the inhomogeneous equation, form the general solution $y = c_1 y_1 + c_2 y_2 + y_p$. If initial values are required, such as $y(1) = 2$ and $y'(1) = 3$, the final step is to determine the values of c_1 and c_2 that fit the initial conditions.

Example 2.47

Specify the order of each of the following differential equations and also identify each as linear/nonlinear and homogeneous/nonhomogeneous.

Note: $y' = \dfrac{dy}{dt}; \; y'' = \dfrac{d^2 y}{dt^2}$

(i) $y'' + 4y' + 4y = 0$ (ii) $3y' + 2y = 0$ (iii) $y'' + 2y' + 2y = e^{-t}$

(iv) $2y' + y + 2 = 0$ (v) $y' + y^2 + 4 = 0$

Solution

(i) second order, linear, homogeneous

(ii) first order, linear, homogeneous

(iii) second order, linear, nonhomogeneous

(iv) first order, linear, nonhomogeneous

(v) first order, nonlinear, nonhomogeneous

Example 2.48

Find the homogeneous solutions for each of the following differential equations and specify whether it belongs to either overdamped, underdamped, or critically damped case.

(i) $y'' + 3y' + 2y = 0$ (ii) $y'' + 2y' + y = 4$ (iii) $y'' + 2y' + 2y = 0$

Solution

(i) Characteristic equation is $s^2 + 3s + 2 = 0$, and the characteristic roots are -1 and -2. Homogeneous solution is $y_h(t) = C_1 e^{-1t} + C_2 e^{-2t}$. The roots are real and distinct, so the function is overdamped.

(ii) Characteristic equation is $s^2 + 2s + 1 = 0$, and the characteristic roots are -1 and -1. Homogeneous solution is $y_h(t) = (C_1 + C_2 t)e^{-1t}$. The roots are real and equal, so the function is critically damped.

(iii) Characteristic root is $s^2 + 2s + 2 = 0$ and the characteristic roots are $-1 + i1$ and $-1 - i1$. Homogeneous solution is $y_h(t) = e^{-1t}(C_1 \cos t + C_2 \sin t)$. As the roots are complex, the function is underdamped.

Example 2.49

Solve the differential equation $y' + 2y = 0$ with initial condition $y(0) = 3$.

Solution

This is a first-order, linear, homogeneous differential equation with constant coefficient.

Characteristic equation is $s + 2 = 0$, and the root is -2; solution is $y(t) = Ce^{-2t}$.

To find C, $y(0) = C = 3$; then, $y(t) = 3e^{-2t}$.

Example 2.50

Solve the differential equation $y'' + 2y' + y = 4$ with initial conditions $y(0) = 0$, $y'(0) = 1$.

Solution

Characteristic equation is $s^2 + 2s + 1 = 0$.

Characteristic roots are: -1 and -1 (roots are real and equal; the solution is critically damped).

Natural or homogeneous solution $y_h(t) = (C_1 + C_2 t)e^{-t}$

Particular solution $y_p(t) = B$ due to forcing function $f(t) = 4$, a constant

Substituting $y_p(t)$ in the differential equation, $y_p'' + 2y_p' + y_p = 4$; $0 + 0 + B = 4$; or $B = 4$

Complete solution $y(t) = y_h(t) + y_p(t) = C_1 e^{-t} + C_2 t e^{-t} + 4$

To find C_1 and C_2, use the initial conditions in $y(t)$ and $y'(t)$; $y'(t) = -C_1 e^{-t} - C_2 t e^{-t} + C_2 e^{-t}$

$y(0) = C_1 + B = 0$ and $y'(0) = -C_1 + C_2 = 1$; solving $C_1 = -4$; $C_2 = -3$

Complete solution is $y(t) = -4e^{-t} - 3te^{-t} + 4$.

Example 2.51

Find the solution of $y'' + 2y' = x^2 + 2 + 3\sin 2x$ subject to $y(0) = 1$, $y'(0) = 0$.

Solution

Begin with $y'' + 2y' = 0$. The characteristic equation is $r^2 + 2r = 0$, which has roots 0 and -2. Thus, the two solutions to the associated homogeneous equation are $y_1 = 1$ and $y_2 = e^{-2x}$. To use the method of Undetermined Coefficients, guess $(ax^2 + bx + c) \bullet x$ for the $x^2 + 2$ term (the x is needed because $y_1 = 1$). Differentiate twice and insert in the equation: $(6ax + 2b) + 2(3ax^2 + 2bx + c)$ should be the same as $x^2 + 2$. Thus, $6a = 1$, $4b + 6a = 0$, and $2b + 2c = 2$. Consequently, $a = \frac{1}{6}$, $b = -\frac{1}{4}$, and $c = \frac{5}{4}$. Next, guess $c \sin 2x + d \cos 2x$ for the other term. Then $y'' + 2y' = -4c \sin 2x - 4d \cos 2x + 2(2c \cos 2x - 2d \sin 2x)$ should match $3 \sin 2x$, so $-4c - 4d = 3$ and $-4d + 4c = 0$. Thus $c = d = -\frac{3}{8}$.

Putting all this together, one has the general solution

$$y = A + Be^{-2x} + \frac{1}{6}x^3 - \frac{1}{4}x^2 + \frac{5}{4}x - \frac{3}{8}\sin 2x - \frac{3}{8}\cos 2x$$

Now to fit the initial conditions,

$$y(0) = 1 = A + B - \frac{3}{8} \quad \text{and} \quad y'(0) = 0 = -2B + \frac{5}{4} - \frac{3}{4}$$

Hence, $B = \frac{1}{4}$ and $A = \frac{9}{8}$, so

$$y = \frac{9}{8} + \frac{1}{4}e^{-2x} + \frac{1}{6}x^3 - \frac{1}{4}x^2 + \frac{5}{4}x - \frac{3}{8}\sin 2x - \frac{3}{8}\cos 2x$$

Euler Equations

An equation of the form $x^2 y'' + axy' + by = F(x)$, with a and b constants, may be solved as readily as the constant coefficient case. These are called **Euler equations**. Upon substituting $y = x^m$, one obtains the *indicial equation* $m(m-1) + am + b = 0$. This quadratic equation has two roots, m_1 and m_2.

If $m_1 \neq m_2$, both real, then $y = c_1 |x|^{m_1} + c_2 |x|^{m_2}$.

If $m_1 = m_2$, then $y = |x|^{m_1} (c_1 + c_2 \ln |x|)$.

If $m_1 = p + jq$ and $m_2 = p - jq$, then $y = |x|^p [c_1 \cos(q \ln |x|) + c_2 \sin(q \ln |x|)]$.

Once y is determined, y_p for the inhomogeneous equation may be found by Variation of Parameters. The method of Undetermined Coefficients is not recommended for Euler equations.

Higher-order linear equations with constant coefficients or of Euler form may be solved analogously.

Laplace Transform

The Laplace transform is an operation that converts functions of x on the half-line $[0, \infty]$ into functions of p on some half-line (a, ∞). The damping power of e^{-xp} is the basis for this useful technique. If $f(x)$ is a piecewise continuous function on $[0, \infty]$ that does not grow too fast, $L(f)$ is the function of p defined by

$$L[f(p)] = \int_0^\infty e^{-xp} f(x)\, dx$$

for the values of p for which the integral converges. For example, if $f(x) \equiv 1$,

$$L(f) = L(1) = \int_0^\infty e^{-xp} dx = \frac{1}{p} \quad (\text{for } p > 0)$$

As a further example,

$$L(f) = L(e^{ax}) = \int_0^\infty e^{-xp} e^{ax} dx = \int_0^\infty e^{-x(p-a)} dx = \frac{1}{p-a} \quad (\text{for } p > a)$$

The basic connection between the Laplace transform and differential equations is the following result achieved by integration by parts:

$$L[y'(p)] = \int_0^\infty e^{-xp} y'(x)\,dx = y(x)e^{-xp}\Big|_0^\infty + p\int_0^\infty e^{-xp} y(x)\,dx = -y(0) + pL[y(p)]$$

Consequently, the solution to $ay'' + by' + cy = F(x)$ may be obtained by transforming $L[ay'' + by' + cy] = L(F)$, so $a[p^2 L(y) - py(0) - y'(0)] + b[pL(y) - y(0)] + cL(y) = L(F)$. Solving for $L(y)$,

$$L(y) = \frac{L(F) + apy(0) + ay'(0) + by(0)}{ap^2 + bp + c}$$

If one were able to "invert" this result,

$$y = L^{-1}\left[\frac{L(F) + apy(0) + ay'(0) + by(0)}{ap^2 + bp + c}\right]$$

the solution would appear, complete with initial values. The Laplace transform is invertible, and the process of finding $L^{-1}[f(p)]$ as a function of x is much like the process of integration.

Table 2.1 presents a tabulation of selected transforms, where the transform of $f(x)$ is called $F(p)$. Line 3 in Table 2.1 reveals that the operation of multiplying by x corresponds to the negative of the operation of differentiating with respect to p. Lines 2 and 8 have been previously discussed. The δ in Line 1 is a *pseudo-function* with great utility defined by $\delta(x) = 0$ for all x except $x = 0$, and $\int_0^\infty \delta(x)\,dx = 1$, so $\delta(0) = +\infty$. The u in Line 7 is called the *Heaviside function*, and it is *zero* until $x - c > 0$. Thus $u(x-c)f(x-c)$ is $f(x)$ shifted right to the point $x = c$. For example, if $f(x) = x$ and $c = 1$, $u(x-1)f(x-1)$ has the graph shown in Figure 2.36, whereas f has the graph shown in Figure 2.37. The operation in the left column of Line 9 is a new way to multiply functions, called **convolution**.

Table 2.1 Selected transforms

$f(x)$	$F(p)$
1. δ	1
2. 1	p^{-1}
3. $xf(x)$	$\dfrac{-dF}{dp}$
4. $e^{ax} f(x)$	$F(p-a)$
5. $\sin ax$	$\dfrac{a}{p^2 + a^2}$
6. $\cos ax$	$\dfrac{p}{p^2 + a^2}$
7. $u(x-c)f(x-c)$	$e^{-cp} F(p)$
8. $f'(x)$	$pF(p) - f(0)$

	(continued)
9. $\int_0^x f(u)g(x-u)du$	$F(p) \bullet G(p)$
10. $f(x)$, if f is periodic, of period L	$\dfrac{\int_0^L e^{-px} f(x)dx}{1-e^{-pL}}$

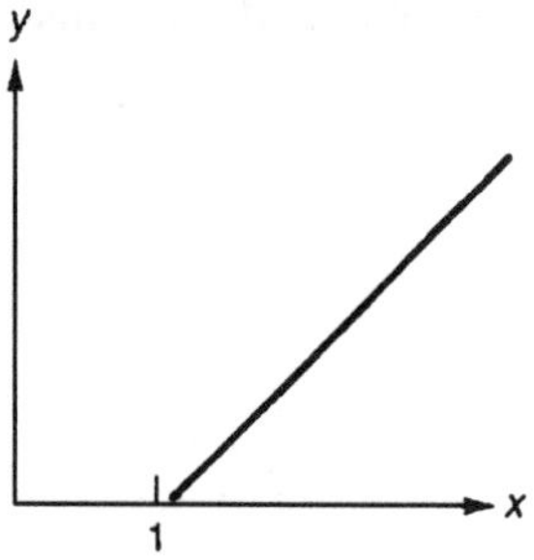

For an example of inversion of a transform, consider

$$F(p) = \frac{2p+3}{(p^2+1)(p-2)}$$

Begin by using partial fractions to write

$$F(p) = \frac{A}{p-2} + \frac{Bp+C}{p^2+1}$$

from which

$$F(p) = \frac{7/5}{p-2} + \frac{(-7/5)p - 4/5}{p^2+1}$$

Table 2.1 shows that $p/(p^2+1)$ is $L(\cos x)$, so

$$L^{-1}\left[\frac{(-7/5)p}{p^2+1}\right]$$

is $-(7/5) \cos x$, and similarly

$$L^{-1}\left(-\frac{4}{5} \bullet \frac{1}{p^2+1}\right)$$

is $-(4/5) \sin x$. Since $L(1) = 1/p$, $L(e^{2x})$ is $1/(p-2)$ by Line 4, and combining these three terms yields

$$f(x) = \frac{7}{5}e^{2x} - \frac{7}{5}\cos x - \frac{4}{5}\sin x$$

The *Heaviside* operation in Line 7 leads to an easy solution of differential equations whose right-hand side is not continuous. For example, consider the response of $y'' + y$ to a driving function $f(x)$ that is 1 for x between 0 and 2 and then becomes 0. Suppose $y(0) = 1$ and $y'(0) = 3$. By applying Line 8 in Table 2.1 twice,

$$L(y) = \frac{L(f) + p + 3}{p^2 + 1}$$

and since $f(x) = 1 - u(x-2)$,

$$L(f) = \frac{1}{p} - \frac{e^{-2p}}{p}$$

so

$$L(y) = \frac{p + 3 + (1 - e^{-2p})/p}{p^2 + 1}$$

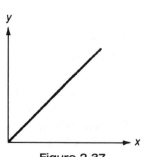

Figure 2.36

Figure 2.37

The e^{-2p} portion represents delay, so write

$$L(y) = \frac{p^2 + 3p + 1}{p(p^2+1)} - \frac{e^{-2p}}{p(p^2+1)} = \frac{1}{p} + \frac{3}{p^2+1} - e^{-2p}\left(\frac{1}{p} - \frac{p}{p^2+1}\right)$$

from which $y = 1 + 3\sin x - u(x-2) + u(x-2)\cos(x-2)$.

Example 2.52

Find the Laplace transform of each of the following functions:

(i) $2u(t) + e^{-3t}$ (ii) $2te^{-t}$ (iii) $e^{-t}\sin 4t$ (iv) $\sin 4t$ (v) $\cos 2t$

Solution

(i) $\dfrac{2}{s} + \dfrac{1}{s+3} = \dfrac{2s+7}{s(s+3)}$ (ii) $\dfrac{2}{(s+1)^2}$ (iii) $\dfrac{4}{(s+1)^2 + 4^2}$

(iv) $\dfrac{4}{s^2 + 4^2}$ (v) $\dfrac{s}{s^2 + 2^2}$

Example 2.53

Find an expression for $Y(s) = £\, y(t)$ for each of the following differential equations.

(i) $\dfrac{dy}{dt} + 6y(t) = 4u(t);\ y(0) = 0$

(ii) $\dfrac{dy}{dt} + 2y(t) = 0;\ y(0) = -2$

(iii) $\dfrac{d^2y}{dt^2} + 2\dfrac{dy}{dt} + 3y(t) = \sin 2t$ with zero initial conditions

(iv) $\dfrac{d^2y}{dt^2} + 2\dfrac{dy}{dt} + y(t) = e^{-2t}$ with initial conditions $y(0) = -1$ and $\dfrac{dy(0)}{dt} = 1$

Solution

(i) $[sY(s) - 0] + 6Y(s) = \dfrac{4}{s};\ Y(s) = \dfrac{4}{s(s+6)}$

(ii) $[sY(s) - (-2)] + 2Y(s) = 0;\ Y(s) = -\dfrac{2}{s+2}$

(iii) $[s^2 Y(s) - 0 - 0] + 2[sY(s) - 0] + 3Y(s) = \dfrac{2}{s^2+4}$;

$Y(s) = \dfrac{2}{(s^2+4)(s^2+2s+3)}$

(iv) $\left[s^2Y(s)-(-s)-1\right]+2\left[sY(s)-(-1)\right]+Y(s)=\dfrac{1}{s+2}$

simplifying $Y(s)=\dfrac{(-s^2-3s-1)}{(s+2)(s^2+2s+1)}$

Example 2.54

Find the initial and final values of the functions whose Laplace transforms are given:

(i) $F(s)=\dfrac{2(s+1)}{s(s+4)(s+6)}$ (ii) $F(s)=\dfrac{4s}{s^2+2s+2}$

Solution

(i) Initial value is: $\lim\limits_{t\to 0}f(t)=\lim\limits_{s\to\infty}sF(s)=\lim\limits_{s\to\infty}s\,\dfrac{2(s+1)}{s(s+4)(s+6)}=0$

Final value is: $\lim\limits_{t\to\infty}f(t)=\lim\limits_{s\to 0}sF(s)=\lim\limits_{s\to 0}s\,\dfrac{2(s+1)}{s(s+4)(s+6)}=\dfrac{1}{12}$

(ii) Initial value is: $\lim\limits_{t\to 0}f(t)=\lim\limits_{s\to\infty}sF(s)=\lim\limits_{s\to\infty}s\,\dfrac{4s}{s^2+2s+2}=4$

Final value is: $\lim\limits_{t\to\infty}f(t)=\lim\limits_{s\to 0}sF(s)=\lim\limits_{s\to 0}s\,\dfrac{4s}{s^2+2s+2}=0$

FOURIER SERIES AND FOURIER TRANSFORM

Fourier Series

A periodic function $F(t)$ with period T can be expanded into Fourier series as

$$F(t)=a_0+\sum_{n=1}^{\infty}(a_n\cos n\omega_0 t+b_n\sin n\omega_0 t) \qquad (2.10)$$

where $\omega_0=\dfrac{2\pi}{T}$ and the Fourier coefficients are defined as

$$a_0=\left(\dfrac{1}{T}\right)\int_0^T F(t)\,dt \qquad (2.10a)$$

$$a_n=\left(\dfrac{2}{T}\right)\int_0^T F(t)\cos(n\omega_0 t)\,dt \qquad (2.10b)$$

$$b_n=\left(\dfrac{2}{T}\right)\int_0^T F(t)\sin(n\omega_0 t)\,dt \qquad (2.10c)$$

For a truncated series, the root mean square (RMS) value F_N is defined as

$$F_N^2 = a_0^2 + \left(\frac{1}{2}\right)\sum_{n=1}^{N}(a_n^2 + b_n^2) \qquad (2.11)$$

Example 2.55

Find the Fourier coefficients of the periodic waveform shown in Exhibit 4 and the RMS value of the truncated Fourier series including five harmonics.

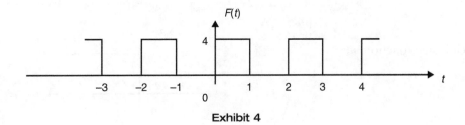

Exhibit 4

Solution

Period $T = 2\ s$; $w_0 = \dfrac{2\pi}{T} = \pi\ rad/s$

From Equation (2.10a) $a_0 = \dfrac{1}{2}\left[\displaystyle\int_0^1 4\,dt + \int_1^2 0\,dt\right] = 2$

From Equation (2.10b) $a_n = \dfrac{2}{2}\left[\displaystyle\int_0^1 4\cos n\omega_0 t\,dt + \int_1^2 0\,dt\right] = 0$, when $n \neq 0$

From Equation (2.10c) $b_n = \dfrac{2}{2}\displaystyle\int_0^1 4\sin n\omega_0 t\,dt + \int_1^2 0\,dt = \dfrac{4}{n\pi}(1-\cos n\pi)$

$b_n = 0$ for all even n; $b_n = \dfrac{8}{n\pi}$ for all odd n

Fourier series is $F(t) = 2 + \dfrac{8}{\pi}\sin \omega_0 t + \dfrac{8}{3\pi}\sin 3\omega_0 t + \dfrac{8}{5\pi}\sin 5\omega_0 t + \ldots$

From Equation (2.11), for $N = 5$, $F_N^2 = 2^2 + \dfrac{1}{2}\left[\left(\dfrac{8}{\pi}\right)^2 + \left(\dfrac{8}{3\pi}\right)^2 + \left(\dfrac{8}{5\pi}\right)^2\right] = 7.732$;

$F_N = 2.7807$

Fourier Transform

The Fourier transform of a function $x(t)$ and its inverse relation are:

$$X(f) = \int_{-\infty}^{+\infty} x(t)\exp(-j2\pi ft)\,dt \qquad (2.12a)$$

$$X(t) = \int_{-\infty}^{+\infty} x(f)\exp(j2\pi ft)\,dt \qquad (2.12b)$$

Table 2.2 lists the Fourier transforms for a few commonly used functions in communication systems. The NCEES *Fundamentals of Engineering Supplied-Reference Handbook* includes a more complete table of Fourier transform pairs.

Table 2.2 Fourier Transform Pairs

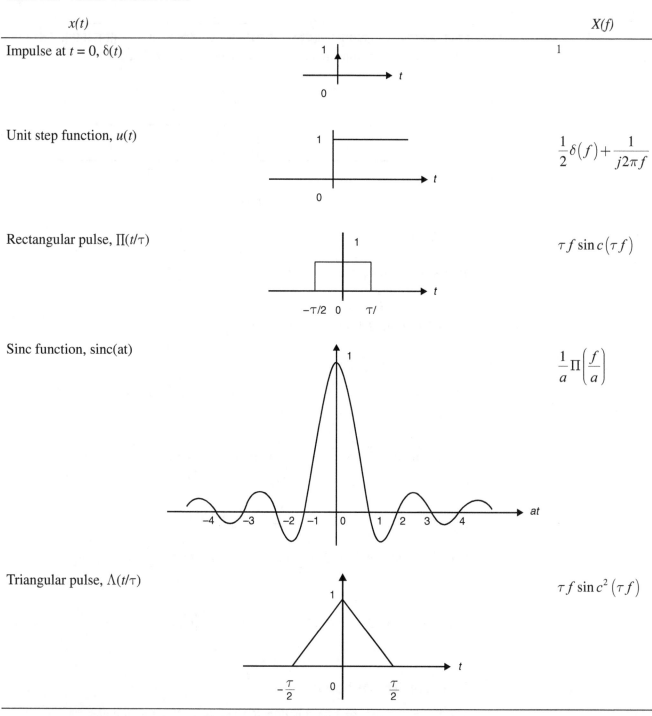

$x(t)$		$X(f)$
Impulse at $t = 0$, $\delta(t)$		1
Unit step function, $u(t)$		$\dfrac{1}{2}\delta(f) + \dfrac{1}{j2\pi f}$
Rectangular pulse, $\Pi(t/\tau)$		$\tau f \, \text{sinc}(\tau f)$
Sinc function, sinc(at)		$\dfrac{1}{a}\Pi\left(\dfrac{f}{a}\right)$
Triangular pulse, $\Lambda(t/\tau)$		$\tau f \, \text{sinc}^2(\tau f)$

Example 2.56

Find the Fourier transform of (i) $4\,\delta(t)$, (ii) $2\,u(t)$, (iii) $\prod(t/2)$, and (iv) $6\cos(100\pi t)$.

Solution

(i) 4   (ii) $\delta(f) + 1/j\pi f$ (iii) $2f\,\text{sinc}(2f)$ (iv) $3[\delta(f-50) + \delta(f+50)]$

DIFFERENCE EQUATIONS AND Z-TRANSFORMS

Difference equations are used to model discrete systems; they are analogous to differential equations that describe continuous systems. The equations $y(k) - y(k-1) = 10$ and $y(k+1) - y(k) = 5$ are some examples of first order linear difference equations; $y(k) = y(k-1) + y(k-2)$ is an example of second order difference equation.

Example 2.57

Find the values of $y(1)$, $y(2)$, and $y(3)$ of the equation $y(k) - 1.01y(k-1) = -50$ with the initial condition $y(0) = 1000$.

Solution

$y(1) - 1.01\,y(0) = -50;\quad y(1) = \$960$

$y(2) - 1.01\,y(1) = -50;\quad y(2) = \$919.60$

$y(3) - 1.01\,y(2) = -50;\quad y(3) = \$878.80$ and so on

Z-Transforms

Z-transform of a discrete sequence is defined as

$$F(z) = \sum_{k=0}^{\infty} f[k]z^{-k} \qquad (2.13)$$

For example, if the discrete sequence is $f[k] = 0, 1, 4$, then its z-transform is $(F_z) = 0 + z^{-1} + 4z^{-2}$.

Example 2.58

Find the z-transform of the function $f(k) = 3\,u(k) + 2^k$ for $k \geq 0$.

Solution

Using the Z-transform table from the *Fundamentals of Engineering Supplied-Reference Handbook*, $F(z) =$

$$\frac{3}{1-z^{-1}} + \frac{1}{1-2z^{-1}}$$

Simplifying, $F(z) = \dfrac{3(1-2z^{-1}) + (1-z^{-1})}{(1-z^{-1})(1-2z^{-1})} = \dfrac{4-7z^{-1}}{(1-z^{-1})(1-2z^{-1})}$

Multiplying by z^2, $F(z) = \dfrac{4z^2 - 7z}{z^2 - 3z + 2}$

PROBLEMS

2.1 The simplest value of $\dfrac{[(n+1)!]^2}{n!(n-1)!}$ is:
- a. n^2
- b. $n(n+1)$
- c. $n+1$
- d. $n(n+1)^2$

2.2 If $x^{3/4} = 8$, x equals:
- a. 6
- b. 9
- c. −9
- d. 16

2.3 If $\log_a 10 = 0.250$, $\log_{10} a$ equals:
- a. 4
- b. 0.50
- c. 2
- d. 0.25

2.4 If $\log_5 x = -1.8$, $\log_x 5$ is:
- a. 0.35
- b. 0.79
- c. −0.56
- d. undefined

2.5 If $\log x + \log (x - 10) - \log 2 = 1$, x is:
- a. −1.708
- b. 5.213
- c. 7.824
- d. 11.708

2.6 A right circular cone, cut parallel with the axis of symmetry, reveals a(n):
- a. circle
- b. hyperbola
- c. eclipse
- d. parabola

2.7 The expression $\dfrac{6!}{3!0!}$ is equal to:
- a. ∞
- b. 120
- c. 2!
- d. 0

2.8 To find the angles of a triangle, given only the lengths of the sides, one would use:
- a. the law of cosines
- b. the law of tangents
- c. the law of sines
- d. the inverse-square law

2.9 If $\sin \alpha = \dfrac{a}{\sqrt{a^2+b^2}}$, which of the following equations is true?

   a. $\tan^{-1} \dfrac{b}{a} = \dfrac{\pi}{2} - \alpha$

   b. $\tan^{-1} \dfrac{b}{a} = -\alpha$

   c. $\cos^{-1} \dfrac{b}{\sqrt{a^2+b^2}} = \dfrac{\pi}{2} - \alpha$

   d. $\cos^{-1} \dfrac{a}{\sqrt{a^2+b^2}} = \alpha$

2.10 The sine of 840° equals:
   a. $-\cos 30°$
   b. $-\cos 60°$
   c. $\sin 30°$
   d. $\sin 60°$

2.11 One root of $x^3 - 8x - 3 = 0$ is:
   a. 2
   b. 3
   c. 4
   d. 5

2.12 Roots of the equation $3x^3 - 3x^2 - 18x = 0$ are:
   a. −2, 3
   b. 0, −2, 3
   c. $2 + 1i, 2 - 1i$
   d. 0, 2, −3

2.13 The equation whose roots are $-1 + i1$ and $-1 - i1$ is given as:
   a. $x^2 + 2x + 2 = 0$
   b. $x^2 - 2x - 2 = 0$
   c. $x^2 + 2 = 0$
   d. $x^2 - 2 = 0$

2.14 Natural logarithms have a base of:
   a. 3.1416
   b. 2.171828
   c. 10
   d. 2.71828

2.15 $(5.743)^{1/30}$ equals:
- a. 1.03
- b. 1.04
- c. 1.05
- d. 1.06

2.16 The value of $\tan(A + B)$, where $\tan A = 1/3$ and $\tan B = 1/4$ (A and B are acute angles) is:
- a. 7/12
- b. 1/11
- c. 7/11
- d. 7/13

2.17 To cut a right circular cone in such a way as to reveal a parabola, it must be cut:
- a. perpendicular to the axis of symmetry
- b. at any acute angle to the axis of symmetry
- c. at any obtuse angle to the axis of symmetry
- d. none of these

2.18 The equation of the line perpendicular to $3y + 2x = 5$ and passing through $(-2, 5)$ is:
- a. $2x = 3y$
- b. $2y = 3x$
- c. $2y = 3x + 16$
- d. $3x = 2y + 8$

2.19 Equation of a line that has a slope of -2 and passes through $(2, 0)$ is:
- a. $y = -2x$
- b. $y = 2x + 4$
- c. $y = -2x + 4$
- d. $y = 2x - 4$

2.20 Equation of a line that intercepts the x-axis at $x = 4$ and the y-axis at $y = -6$ is:
- a. $3x - 2y = 12$
- b. $2x - 3y = 12$
- c. $x + y = 6$
- d. $x - y = 4$

2.21 The x-axis intercept and the y-axis intercept of the line $x + 3y + 9 = 0$ are:
- a. 0 and 3
- b. -9 and -3
- c. -3 and -9
- d. 9 and 0

2.22 The distance between the points $(1, 0, -2)$ and $(0, 2, 3)$ is:
- a. 3.45
- b. 5.39
- c. 6.71
- d. 7.48

2.23 The equation of a parabola with center at $(0, 0)$ and directrix at $x = -2$ is:
- a. $y^2 = 4x$
- b. $y = 2x^2$
- c. $y^2 = 8x$
- d. $y^2 = 2x$

2.24 The equation of the directrix of the parabola $y^2 = -4x$ is:
- a. $y = 2$
- b. $x = 1$
- c. $x + y = 0$
- d. $x = -2$

2.25 The equation of an ellipse with foci at (± 2, 0) and directrix at $x = 6$ is:

a. $\dfrac{x^2}{12} + \dfrac{y^2}{8} = 1$

b. $\dfrac{x^2}{8} + \dfrac{y^2}{8} = 1$

c. $\dfrac{x^2}{12} + \dfrac{y^2}{12} = 1$

d. $\dfrac{x^2}{4} + \dfrac{y^2}{4} = 1$

2.26 The foci of the ellipse $\left(\dfrac{x}{3}\right)^2 + \left(\dfrac{y}{2}\right)^2 = 1$ are at:

a. ($\pm\sqrt{5}$, 0)   c. (0, $\pm\sqrt{2}$)
b. ($\pm\sqrt{2}$, 0)   d. (0, $\pm\sqrt{5}$)

2.27 The equation of a hyperbola with center at (0, 0), foci at (± 4, 0), and eccentricity of 3 is:

a. $\dfrac{x^2}{16} - \dfrac{y^2}{16} = 1$

b. $\dfrac{x^2}{16} - \dfrac{y^2}{128} = 1$

c. $\dfrac{x^2}{64} - \dfrac{y^2}{48} = 1$

d. $\dfrac{x^2}{128} - \dfrac{y^2}{128} = 1$

2.28 The equation of a circle with center at (1, 2) and passing through the point (4, 6) is:

a. $x^2 + y^2 = 25$
b. $(x + 1)^2 + (y - 2)^2 = 25$
c. $x^2 + (y - 2)^2 = 25$
d. $(x - 1)^2 + (y - 2)^2 = 25$

2.29 The length of the tangent from (4, 8) to the circle $x^2 + (y - 1)^2 = 3^2$ is:

a. 3.81   c. 5.66
b. 4.14   d. 7.48

2.30 The conic section described by the equation $x^2 - 10xy + y^2 + x + y + 1 = 0$ is:

a. circle   c. hyperbola
b. parabola   d. ellipse

2.31 A triangle has sides of length 2, 3, and 4. The angle subtended by the sides of length 2 and 4 is:

a. 21.2° c. 46.6°
b. 35.0° d. 61.2°

2.32 Length a of one side of the triangle below is:

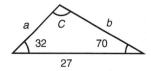

a. 25.9 c. 12.7
b. 19.1 d. 4.8

2.33 The relation $\sec\theta - (\sec\theta)(\sin^2\theta)$ can be simplified as:

a. $\sin\theta$ c. $\cot\theta$
b. $\tan\theta$ d. $\cos\theta$

2.34 If $\sin\theta = m$, $\cot\theta$ is:

a. $\dfrac{\sqrt{1-m^2}}{m}$

b. $\dfrac{m}{\sqrt{1-m^2}}$

c. $\sqrt{1-m^2}$

d. m

2.35 If vectors $\mathbf{A} = 3i - 6j + 2k$ and $\mathbf{B} = 10i + 4j - 6k$, their cross product $\mathbf{A} \times \mathbf{B}$ is:

a. $-12i + 38j$ c. $-12i + 18j + 24k$
b. $12i + 24j + 36k$ d. $28i + 38j + 72k$

2.36 The sum of all integers from 10 to 50 (both inclusive) is:

a. 990 c. 1230
b. 1110 d. 1420

2.37 The 50th term of the series 10, 16, 22, 28, 34, 40. . . is:

a. 272 c. 428
b. 304 d. 584

2.38 Sum of the infinite series 4, 2, 1, 0.5, 0.25. . . is:

a. 8 c. 10,400
b. 128 d. ∞

2.39 If $\mathbf{A} = \begin{bmatrix} 1 & 2 & 3 \\ 1 & 2 & 9 \end{bmatrix}$ and $\mathbf{B} = \begin{bmatrix} 5 & 1 \\ 6 & 0 \\ 4 & 7 \end{bmatrix}$, the (2,1) entry of AB is:

a. 29 c. 33
b. 53 d. 64

2.40 The inverse of the matrix $\begin{bmatrix} 1 & 1 \\ 3 & 2 \end{bmatrix}$ is:

a. $\begin{bmatrix} 2 & -1 \\ -3 & 1 \end{bmatrix}$ b. $\begin{bmatrix} 2 & 3 \\ 1 & 1 \end{bmatrix}$ c. $\begin{bmatrix} 1 & 3 \\ 1 & 2 \end{bmatrix}$ d. $\begin{bmatrix} -2 & 1 \\ 3 & -1 \end{bmatrix}$

2.41 The determinant of the matrix $\begin{bmatrix} 1 & 2 & -1 \\ 3 & 0 & 2 \\ 2 & -2 & -1 \end{bmatrix}$ is:

a. 4 c. 24
b. 16 d. −16

2.42 In the system of equations
$$3x_1 + 2x_2 - x_3 = 5$$
$$x_2 - x_3 = 2$$
$$x_1 + 2x_2 - 3x_3 = -1$$

the value of $x_2 = $ is:
a. 2 c. 4
b. −1 d. 6

2.43 What is the determinant of M?

$$M = \begin{bmatrix} 0 & 1 & 1 & 1 \\ 1 & 1 & 1 & 1 \\ 1 & 1 & 3 & 1 \\ 2 & 1 & 3 & 4 \end{bmatrix}$$

a. −6 c. 0
b. 6 d. 7

2.44 $\int_{\pi/2}^{\pi} \sin 2x \, dx =$

a. 2
b. 1
c. 0
d. −1

2.45 $\int_{0}^{2} x^2 \sqrt{1+x^3} \, dx =$

a. 52/9
b. 0
c. 52/3
d. 26/3

2.46 $\int_{1}^{e} x(\ln x) dx =$

a. $\frac{1}{2}e^2 + 1$
b. $\frac{1}{2}e^2 - e + \frac{1}{2}$
c. $\frac{1}{4}e^2 + \frac{1}{4}$
d. $\frac{1}{4}e^2 - \frac{1}{2}e + \frac{1}{4}$

2.47 If the first derivative of the equation of a curve is constant, the curve is a:
a. circle
b. hyperbola
c. parabola
d. straight line

2.48 Which of the following is a characteristic of all trigonometric functions?
a. The values of all functions repeat themselves every 45 degrees.
b. All functions have units of length or angular measure.
c. The graphs of all functions are continuous.
d. All functions have dimensionless units.

2.49 For a given curve $y = f(x)$ that is continuous between $x = a$ and $x = b$, the average value of the curve between the ordinates at $x = a$ and $x = b$ is represented by:

a. $\dfrac{\int_a^b x^2 \, dy}{b-a}$
b. $\dfrac{\int_a^b y^2 \, dx}{b-a}$
c. $\dfrac{\int_a^b x \, dy}{a-b}$
d. $\dfrac{\int_a^b y \, dx}{b-a}$

2.50 If $y = \cos x$, $\dfrac{dy}{dx}$ is:

a. sin x
b. −tan x cos x
c. $\dfrac{1}{\sec x}$
d. sec x sin x

2.51 The derivative of $\cos^3 5x$ is:

a. $3 \sin^2 5x$
b. $15 \sin^2 5x$
c. $\cos^2 5x \sin x$
d. $-15 \cos^2 5x \sin 5x$

2.52 The slope of the curve $y = 2x^3 - 3x$ at $x = 1$ is:

    a. -1 c. 3
    b. 0   d. 5

2.53 A stone is dropped from the top of a building at $t = 0$. The position of the stone is given by the equation $s(t) = 16\,t^2$ m. Acceleration of the stone (in m/s²) 2 seconds after it is dropped is:

    a. 64   c. 24
    b. 32   d. 16

2.54 Maximum value of the function $f(x) = x^3 - 5x - 4$ occurs at:

    a. 0   c. -0.30
    b. 1.29   d. -1.29

2.55 The partial derivative with respect to x of the function $xy^2 - 5y + 6$ is:

    a. xy c. y^2
    b. $2y$ d. $-5y$

2.56 The power series expansion of $\cos(x)$ about the point $x = 0$ is:

    a. $1 - \dfrac{x^2}{2!} + \dfrac{x^4}{4!} - \dfrac{x^6}{6!} + \ldots$ c. $1 - \dfrac{x}{2!} + \dfrac{x}{4!} - \dfrac{x}{6!} + \ldots$

    b. $1 + \dfrac{x^2}{2!} + \dfrac{x^4}{4!} + \dfrac{x^6}{6!} + \ldots$ d. $1 + \dfrac{x}{2!} + \dfrac{x}{4!} + \dfrac{x}{6!} + \ldots$

2.57 The value of $\lim\limits_{x \to 2} \dfrac{x^2 - 4}{x - 2}$ is:

    a. 0   c. 1
    b. ∞ d. 4

2.58 If $A = \int_0^{\frac{\pi}{4}} \sin^2\theta\, d\theta$, the value of A is:

    a. 0.29   c. 1.75
    b. 0.58   d. 3.14

2.59 If $x^3 + 3x^2y + y^3 = 4$ defines y implicitly, $dy/dx =$

    a. $-\dfrac{x^2 + 2xy}{x^2 + y^2}$ c. $-\dfrac{x^2 + y^2}{x^2 + 2xy}$

    b. $3x^2 + 3y^2$ d. $-\dfrac{x^2 + 2xy}{x^2 + y^2}$

2.60 Estimate $\sqrt{34}$ using differentials. The answer is closest to:

a. $6 + \dfrac{1}{6}$ c. 6

b. $6 - \dfrac{1}{6}$ d. $6 - \dfrac{1}{3}$

2.61 The only relative maximum of $f(x) = x^4 - \dfrac{4}{3}x^3 - 12x^2 + 1$ is:

a. -1 c. 0

b. 1 d. -1

2.62 The area between $y = x^2$ and $y = 2x + 3$ is:

a. 9 c. $6\dfrac{1}{3}$

b. 20 d. $10\dfrac{2}{3}$

2.63 The area enclosed by the curve $r = 2(\sin\theta + \cos\theta)$ is:

a. π c. 2π

b. $\dfrac{\pi}{2}$ d. $\pi\sqrt{2}$

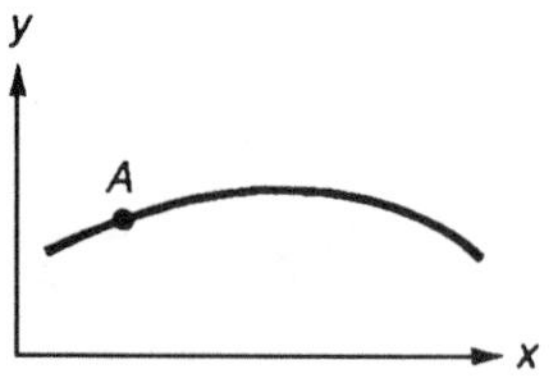

2.64 The curve in Exhibit 2.64 has the equation $y = f(x)$. At point A, what are the values of $\dfrac{dy}{dx}$ and $\dfrac{d^2y}{dx^2}$?

a. $\dfrac{dy}{dx} < 0$, $\dfrac{d^2y}{dx^2} < 0$ c. $\dfrac{dy}{dx} = 0$, $\dfrac{d^2y}{dx^2} = 0$

b. $\dfrac{dy}{dx} < 0$, $\dfrac{d^2y}{dx^2} > 0$ d. $\dfrac{dy}{dx} > 0$, $\dfrac{d^2y}{dx^2} < 0$

Exhibit 2.64

2.65 The area of the shaded region in Exhibit 2.65 is:

a. 1.37   c. 5.33

b. 3.82   d. 6.80

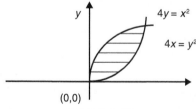

Exhibit 2.65

Refer to Exhibit 2.66 for problems 2.66 through 2.71.

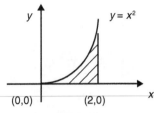

Exhibit 2.66

2.66 Area of the shaded region is:

a. 1.10
b. 2.67
c. 4.02
d. 6.80

2.67 The first moment of the shaded area with respect to the x-axis is:

a. 1.20
b. 2.30
c. 3.20
d. 4.60

2.68 The first moment of the shaded area with respect to the y-axis is:

a. 0.85
b. 1.90
c. 3.07
d. 4.00

2.69 The centroid of the shaded area is:

a. (1.5, 1.2)
b. (1.5, 0)
c. (0, 1.2)
d. (1.0, 1.0)

2.70 The second moment of the shaded area with respect to the x-axis is:

a. 6.10
b. 9.05
c. 12.11
d. 18.32

2.71 The second moment of the shaded area with respect to the y-axis is:

a. 1.98
b. 3.12
c. 4.63
d. 6.40

2.72 If the characteristic roots of a differential equation are $-4 - i4$ and $-4 + i4$, the homogeneous solution is:

a. $C_1 \cos 4x + C_2 \sin 4x$
b. $e^{-4x}(C_1 \cos 4x + C_2 \sin 4x)$
c. $C_1 e^{-i4x} + C_2 e^{i4x}$
d. $C_1 \cos(4x + \theta)$

2.73 Characteristic roots of a differential equation are -2 and -2. If the forcing function is e^{-2x}, the particular solution, $y_p(x)$ is:

a. Ax^2
b. Ae^{-2x}
c. Axe^{-2x}
d. $Ax^2 e^{-2x}$

2.74 The solution of the differential equation $y'' + 5y' + 6y = 2e^{-2x}$ with zero initial conditions is:

a. $y = -2e^{-2x} + 2e^{-3x} + 2xe^{-2x}$
b. $y = x - 2e^{-2x} + 2e^{-3x}$
c. $y = 2e^{-3x} + 2xe^{-2x}$
d. $y = -2e^{-2x} + 2xe^{-2x}$

2.75 $\lim_{x \to 1} \dfrac{x^2 - 1}{x - 1} =$

    a. 2   c. 0
    b. ∞ d. 1

2.76 The solution to $xy' + 2y = e^{3x}$ is:

    a. $y = e^{3x} - \dfrac{e^{3x}}{x} + \dfrac{c}{x}$

    b. $y = \dfrac{xe^{3x} - 3e^{3x} + 3c}{3x^2}$

    c. $y = xe^{3x} - 3e^{3x} + c$

    d. $y + x = e^{3x} + c$

2.77 Solve $xy'' - 2(x+1)y' + (x+2)y = 0$.

    a. $y = Ae^x + Bx^3 e^x$ c. $y = A\sin(x+1) + B\cos(x+2)$
    b. $y = Ae^x + Be^{2x}$ d. $y = Ae^x + Be^{-x}$

2.78 Solve $y'' + 4y = 8\sin x$.
    a. $y = Ae^{2x} + Be^{-2x}$
    b. $y = A\sin 2x + B\cos 2x$
    c. $y = A\sin 2x + B\cos 2x + \sin x$
    d. $y = A\sin 2x + B\cos 2x + \dfrac{8}{3}\sin x$

2.79 The family of trajectories orthogonal to the family $x^2 + y^2 = 2cy$ is:

    a. $x - y = c$ c. $x^2 + y^2 = c$
    b. $x^2 - y^2 = cx$ d. $x^2 + y^2 = 2cx$

2.80 The Laplace transform of the function $e^{-t}\cos(t)$ is:

    a. $\dfrac{s}{s^2 + 1^2}$ c. $\dfrac{(s+1)}{s^2 + 1^2}$

    b. $\dfrac{(s+1)}{(s+1)^2 + 1^2}$ d. $\dfrac{(s-1)}{(s-1)^2 + 1^2}$

2.81 For the differential equation $\dfrac{d^2 y}{dt^2} + 2\dfrac{dy}{dt} + y(t) = \cos 2t$ with zero initial conditions, the Laplace transform $Y(s)$ of $y(t)$ is:

    a. $\dfrac{s}{(s^2 + 2s + 1)}$ c. $\dfrac{s}{(s^2 + 4)(s^2 + 2s + 1)}$

    b. $\dfrac{2}{(s^2 + 2)(s^2 + 2s + 1)}$ d. $\dfrac{s}{(s^2 + 4)(s^2 + 1)}$

2.82 The initial value of the function whose Laplace transform $F(s) = \dfrac{s+5}{s(s+1)(s+10)}$ is:

a. 2.0 c. 0.5
b. 1.0 d. 0

2.83 For the difference equation $y(k) = 2y(k-1) + 3y(k-2)$ with initial conditions $y(-1) = 1$; $y(-2) = 1$, the value of $y(1)$ is:

a. 1 c. 13
b. 5 d. 41

2.84 For the difference equation $y(k+1) + y(k) = u(k)$ with $y(0) = 0$, find $Y(z)$, the z-transform of the function $y(k)$ is:

a. $\dfrac{1}{(z-1)(z+2)}$ c. $\dfrac{z}{(z+1)(z-2)}$

b. $\dfrac{z}{(z-1)(z+2)}$ d. $\dfrac{2}{(z+1)(z+2)}$

2.85 The initial value of the function whose Z-transform $F(z) = \dfrac{z+2}{z-4}$ is:

a. 0   c. 0.5
b. −0.5 d. 1.0

2.86 Find the Fourier coefficient b_3 of the waveform shown in Exhibit 2.86.

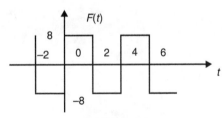

Exhibit 2.86

a. 1.17 c. 3.40
b. 2.23 d. 4.78

2.87 Find the Fourier transform, $F(f)$, of a triangular pulse of width 2.

a. sinc(f) c. sinc$(2f)$
b. sinc$^2(f)$ d. f sinc$^2(f)$

2.88 Find the Fourier transform of $f(t) = \delta(t-1) + \delta(t+1)$.

a. $\exp(-j2\pi f)$ c. $2\cos(2\pi f)$
b. $\exp(j2\pi f)$ d. $j2\sin(2\pi f)$

2.89 Find $y(1)$ if $y(k) = 2y(k-1) + 3y(k-2)$ with $y(-1) = 1; y(-2) = 1$.

    a. 1   c. 13
    b. 5   d. 41

2.90 Find the final value of the function $f(k)$ whose Z-transform is $F(z) = \dfrac{2(z+1)}{(z-1)}$.

    a. 1   c. 4
    b. 2   d. ∞

SOLUTIONS

2.1 d. The value $(n+1)!$ may be written as $(n+1)(n)[(n-1)!]$. It may be written also as $n!(n+1)$. Hence the given expression may be written as follows:

$$\frac{\{(n+1)(n)[(n-1)!]\}\{n!(n+1)\}}{n!(n-1)!} = (n+1)^2 n$$

2.2 d. Raise both sides of the equation to the 4/3 power:

$$[x^{3/4}]^{4/3} = 8^{4/3}$$
$$x = \sqrt[3]{8^4} = \sqrt[3]{(2^3)^4} = 2^{\frac{3 \cdot 4}{3}} = 2^4 = 16$$

2.3 a. $\log_a 10 = 0.250$ can be written as $10 = a^{0.250}$. Taking $\log_{10}$,

$$\log_{10} 10 = \log_{10} a^{0.250}$$
$$1 = 0.250 \log_{10} a$$

and

$$\log_{10} a = \frac{1}{0.250} = 4$$

2.4 c. Since $(\log_5 x)(\log_x 5) = 1$, $\log_x 5 = \dfrac{1}{-1.8} = -0.556$

2.5 d. Since $\log(a) + \log(b) - \log(c) = \log \dfrac{ab}{c}$, the given equation simplifies to $\log \dfrac{x(x-10)}{2} = 1$. Equivalently, $\dfrac{x(x-10)}{2} = 10$ or $x^2 - 10x - 20 = 0$. The roots are 11.708 and −1.708. Since log is not defined for negative values, $x = 11.708$.

2.6 b.

2.7 b. $\dfrac{6!}{3!\,0!} = \dfrac{6(5)(4)(3)!}{3!(1)} = 120$

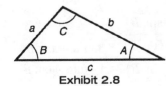
Exhibit 2.8

2.8 a. The law of cosines is $a^2 = b^2 + c^2 - 2bc \cos A$ for any plane triangle with angles A, B, C and sides a, b, c, respectively.
    This law can be applied to solve for the angles, given three sides in a plane triangle (Exhibit 2.8).

2.9 a. The triangle appears in Exhibit 2.9.

$$\tan\left(\frac{\pi}{2} - \alpha\right) = \frac{b}{a}$$
$$\tan^{-1}\frac{b}{a} = \frac{\pi}{2} - \alpha$$

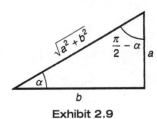
Exhibit 2.9

2.10 d.

$$840° = 2(360) + 120 = 2(2\pi) \text{ rad} + 120°$$
$$\sin [2(2\pi) \text{ rad} + 120°] = \sin 120° = \sin 60°$$

2.11 b. The solution is obtained by seeing which of the five answers satisfies the equation.

x	$x^3 - 8x - 3$
2	−11
3	0
4	29
5	82
6	165

2.12 b. Any scientific calculator can be used to find the roots as 0, 3, and −2.

2.13 a. If x_1 and x_2 are the roots, the equation is $(x - x_1)(x - x_2) = 0$. Since the roots are $-1 + i$ and $-1 - i$, the equation is $(x - (-1 + i))(x - (-1 - i)) = 0$ or $(x + 1 - i)(x + 1 + i) = 0$ or $x^2 + 2x + 2 = 0$.

2.14 d. Common logarithms have base 10. Natural, or napierian, logarithms have base $e = 2.71828$.

2.15 d.

$$\log (5.743)^{1/30} = \frac{1}{30} \log 5.743 = \frac{1}{30}(0.7592) = 0.0253$$

The antilogarithm of 0.0253 is 1.06.

2.16 c.

$$\sin (A + B) = (\sin A \cos B) + (\cos A \sin B)$$

$$\cos (A + B) = (\cos A \cos B) - (\sin A \sin B)$$

$$\tan (A + B) = \frac{\sin(A + B)}{\cos(A + B)} = \frac{(\sin A \cos B) + (\cos A \sin B)}{(\cos A \cos B) - (\sin A \sin B)}$$

Dividing by $\cos A \cos B$,

$$\tan(A + B) = \frac{\dfrac{(\sin A \cos B)}{(\cos A \cos B)} + \dfrac{(\cos A \sin B)}{(\cos A \cos B)}}{\dfrac{(\cos A \cos B)}{(\cos A \cos B)} - \dfrac{(\sin A \sin B)}{(\cos A \cos B)}} = \frac{\tan A + \tan B}{1 - \tan A \tan B}$$

$$= \frac{\dfrac{1}{3} + \dfrac{1}{4}}{1 - \dfrac{1}{3} \times \dfrac{1}{4}} = \frac{\dfrac{4}{12} + \dfrac{3}{12}}{1 - \dfrac{1}{12}} = \frac{\dfrac{7}{12}}{\dfrac{11}{12}} = \frac{7}{11}$$

The problem could also be solved by determining angle A (whose tangent is 1/3) and angle B (whose tangent is 1/4). Then we could find the tangent of $(A + B)$.

$$\tan^{-1}\frac{1}{3} = 18.435° \quad \tan^{-1}\frac{1}{4} = 14.036°$$

$$\tan(18.435 + 14.036)° = \tan(32.471°) = 0.6364 = \frac{7}{11}$$

2.17 d. To reveal a parabola, a right circular cone must be cut parallel to an element of the cone and intersecting the axis of symmetry.

2.18 c. Rewriting the given line, $y = \frac{5}{3} - \frac{2}{3}x$. This line has slope $-\frac{2}{3}$, so a perpendicular line must have slope $\frac{3}{2}$. Using the point-slope form, $\frac{y-5}{x+3} = \frac{3}{2}$. Simplifying, $y = \frac{3}{2}x + 8$.

2.19 d. The equation of a straight line is $y = mx + b$ where the slope is m, (x, y) is any point on the line, and b is the y-axis intercept. Here, $m = -2$ and $(2, 0)$ is a point. Substituting these values, $0 = -2(2) + b$ or $b = 4$. Then, the equation is $y = -2x + 4$ or $y + 2x = 4$.

2.20 a. $(4,0)$ and $(0, -6)$ are two points on the straight line. Then, the slope $m = \frac{y_2 - y_1}{x_2 - x_1} = \frac{0 - (-6)}{4 - 0} = \frac{3}{2}$. Substituting one of the points, say, $(0, -6)$ in the general equation $y = mx + b$; $-6 = 0 + b$. Then, $b = -6$ and the equation is $y = \left(\frac{3}{2}\right)x - 6$ or $3x - 2y = 12$.

2.21 b. For y-axis intercept, $x = 0$. Substituting this in the equation, $0 + 3y + 9 = 0$ or $y = -3$ is the y-axis intercept. For x-axis intercept, $y = 0$. Substituting this in the equation, $x + 0 + 9 = 0$ or $x = -9$ is the x-axis intercept.

2.22 b. Distance d between any two points (x_1, y_1, z_1) and (x_2, y_2, z_2) can be determined as $d^2 = (x_1 - x_2)^2 + (y_1 - y_2)^2 + (z_1 - z_2)^2$. Here, $d^2 = (1 - 0)^2 + (0 - 2)^2 + (-2 - 3)^2 = 30$; $d = 5.385$.

2.23 c. Directrix, $x = -p/2 = -2$ or $p = 4$. Equation of a parabola with center at origin is $y^2 = 2px$; as $p = 4$, $y^2 = 8x$.

2.24 b. Equation of a parabola with center at origin is $y^2 = 2px$. Since $y^2 = -4x$, $2p = -4$ or $p = -2$. Equation of a directrix is $x = (-p/2)$; since $p = 2$, the equation is $x = 1$.

2.25 a. Focus $ae = 2$ and directrix $\dfrac{a}{e} = 6$. Solving, $a^2 = 12$ and $e = 1/\sqrt{3}$.
But $e = \sqrt{1 - \dfrac{b^2}{a^2}}$; solving, $b^2 = 8$. The equation of an ellipse is
$$\dfrac{x^2}{a^2} + \dfrac{y^2}{b^2} = 1 \text{ or } \dfrac{x^2}{12} + \dfrac{y^2}{8} = 1.$$

2.26 a. The equation of an ellipse is $\dfrac{x^2}{a^2} + \dfrac{y^2}{b^2} = 1$. Here, $a = 3$ and $b = 2$.
The eccentricity $e = \sqrt{1 - \dfrac{b^2}{a^2}} = \sqrt{\dfrac{5}{9}}$, and the foci $= (\pm ae, 0) = (\pm\sqrt{5}, 0)$.

2.27 b. Focus $ae = 4$ and eccentricity $e = 3$. Solving, $a = 4/3$.
Also, $b = a\sqrt{e^2 - 1} = \dfrac{4\sqrt{8}}{3}$. Equation for a hyperbola is
$$\dfrac{x^2}{a^2} - \dfrac{y^2}{b^2} = 1 \Rightarrow \dfrac{x^2}{16} - \dfrac{y^2}{128} = 1.$$

2.28 d. Radius of a circle with center at (h, k) and passing through a point (x, y) is $r^2 = (x - h)^2 + (y - k)^2$. Here, $r^2 = (4 - 1)^2 + (6 - 2)^2 = 25$. Equation of a circle with center at (h, k) is $(x - h)^2 + (y - k)^2 = r^2$; here, $(x - 1)^2 + (y - 2)^2 = 25$.

2.29 d. Length of the tangent from any point (x', y') outside a circle with center at (h, k) and radius r is given as $t^2 = (x' - h)^2 + (y' - k)^2 - r^2$. Here, $(x', y') = (4, 8)$, center $(h, k) = (0, 1)$, and $r^2 = 3^2$. Then, $t^2 = (4 - 0)^2 + (8 - 1)^2 - 3^2 = 56$ or $t = 7.483$.

2.30 c. General equation of a conic section is $Ax^2 + 2Bxy + Cy^2 + 2Dx + 2Ey + F = 0$, where both A and C are not zeros. Here, A = 1; B = –5; C = 1; then, $(B^2 - AC) > 0 \Rightarrow$ hyperbola.

2.31 c. Since three sides of a triangle are given, use the law of cosines; $3^2 = 2^2 + 4^2 - 2(2)(4)\cos(\theta)$ where θ is the angle opposite to side 3. Solving, $\theta = 46.6°$.

2.32 a. Angle C = 180 – (70 + 32) = 78°. Using the law of sines,
$$\dfrac{a}{\sin 70} = \dfrac{b}{\sin 32} = \dfrac{27}{\sin 78}.$$ Solving, $a = 25.94$.

2.33 d. $(\sec \theta)(1 - \sin^2 \theta) = \sec \theta \cos^2 \theta = \dfrac{1}{\cos \theta} \cos^2 \theta = \cos \theta$

2.34 a. $\sin\theta = \dfrac{m}{1} = \dfrac{\text{opposite side}}{\text{hypotenuse}}$; then, adjacent side $= \sqrt{1-m^2}$

and $\cos\theta = \dfrac{\text{adjacent side}}{\text{hypotenuse}} = \dfrac{\sqrt{1-m^2}}{1}$; then, $\cot\theta = \dfrac{\cos\theta}{\sin\theta} = \dfrac{\sqrt{1-m^2}}{m}$

2.35 d. Cross product, $\mathbf{A} \times \mathbf{B} = \begin{vmatrix} i & j & k \\ 3 & -6 & 2 \\ 10 & 4 & -6 \end{vmatrix}$

Expanding, $i[(-6)(-6) - (2)(4)] - j[(3)(-6) - (2)(10)] + k[(3)(4) - (-6)(10)] = 28\mathbf{i} + 38\mathbf{j} + 72\mathbf{k}$

2.36 c. This is an arithmetic series; first term $a = 10$, last term $l = 50$, and common difference $d = 1$. The number of terms, n, can be calculated as, $l = a + (n-1)d$; $50 = 10 + (n-1)1$; solving, $n = 41$.

sum $S = \dfrac{n(a+l)}{2} = \dfrac{41}{2}(10+50) = 1230$

2.37 b. This is an arithmetic series; first term $a = 10$ and the common difference $d = 6$. Taking the number of terms n as 50, the nth term (last term) can be calculated as, $l = a + (n-1)d$. In this case, $l = 10 + (50-1)6 = 304$.

2.38 a. This is a geometric series; first term $a = 4$ and common ratio $r = 0.5$. Since $r < 1$, the series is convergent and the sum as the number of terms n tend to infinity is $S = \dfrac{a}{1-r} = \dfrac{4}{1-0.5} = 8$.

2.39 b. To compute the (2,1) entry, take $[1\ 2\ 9] \bullet [5\ 6\ 4] = 5 + 12 + 36 = 53$.

2.40 d. To invert a 2×2 matrix,

$$\begin{bmatrix} a & b \\ c & d \end{bmatrix}^{-1} = \dfrac{1}{ad-bc}\begin{bmatrix} d & -b \\ -c & a \end{bmatrix} = \dfrac{1}{2-3}\begin{bmatrix} 2 & -1 \\ -3 & 1 \end{bmatrix}$$

2.41 c. This 3×3 determinant can be computed quickly be expanding it in minors, especially around the second column:

$$\begin{vmatrix} 1 & 2 & -1 \\ 3 & 0 & 2 \\ 2 & -2 & 1 \end{vmatrix} = (-1)^{1+2}(2)\begin{vmatrix} 3 & 2 \\ 2 & -1 \end{vmatrix} + (-1)^{2+2}(0)\begin{vmatrix} 1 & -1 \\ 2 & -1 \end{vmatrix} + (-1)^{3+2}(-2)\begin{vmatrix} 1 & -1 \\ 3 & 2 \end{vmatrix}$$

$$= -2(-3-4) + 0 + 2(2+3) = 24$$

2.42 d. By Cramer's Rule,

$$x_2 = \frac{\text{Det}\begin{bmatrix} 3 & 5 & -1 \\ 0 & 2 & -1 \\ 1 & -1 & -3 \end{bmatrix}}{\text{Det}\begin{bmatrix} 3 & 2 & -1 \\ 0 & 1 & -1 \\ 1 & 2 & -3 \end{bmatrix}} = \frac{3\begin{vmatrix} 2 & -1 \\ -1 & -3 \end{vmatrix} + 1\begin{vmatrix} 5 & -1 \\ 2 & -1 \end{vmatrix}}{3\begin{vmatrix} 1 & -1 \\ 2 & -3 \end{vmatrix} + 1\begin{vmatrix} 2 & -1 \\ 1 & -1 \end{vmatrix}} = \frac{3(-7) + (-3) = -24}{3(-1) + (-1) = -4} = 6$$

2.43 a. To evaluate a 4×4 matrix, one must do some row or column operations and expand by minors:

$$\begin{bmatrix} 0 & 1 & 1 & 1 \\ 1 & 1 & 1 & 1 \\ 1 & 1 & 3 & 1 \\ 2 & 1 & 3 & 4 \end{bmatrix} \sim \begin{bmatrix} 0 & 1 & 1 & 1 \\ 1 & 1 & 1 & 1 \\ 0 & 0 & 2 & 0 \\ 0 & -1 & 1 & 2 \end{bmatrix}$$

Taking minors of column 1,

$$\text{Det}(M) = (1)(-1)^{2+1} \text{Det}\begin{bmatrix} 1 & 1 & 1 \\ 0 & 2 & 0 \\ -1 & 1 & 2 \end{bmatrix} = -(4+2) = -6$$

2.44 d.

$$\int_{\pi/2}^{\pi} \sin 2x \, dx = -\frac{1}{2}\cos 2x = -\frac{1}{2}\cos 2\pi + \frac{1}{2}\cos \pi = -\frac{1}{2} - \frac{1}{2} = -1$$

2.45 a. Let $u = 1 + x^3$, so $du = 3x^2 dx$. The integral becomes $\frac{1}{3}\int_1^9 \sqrt{u}\, du$

$$= \frac{2}{9}u^{3/2} = \frac{2}{9}(27 - 1).$$

2.46 c. Prepare to solve using $\int u\, dv = uv - \int v\, du$.

Let $u = \ln x$ and $dv = x\, dx$.

Therefore, $du = \frac{1}{x} dx$ and $v = \frac{1}{2}x^2$

$$\int_1^e x(\ln x) dx = \ln x \left(\frac{1}{2}x^2\right)\Big|_1^e - \int_1^e \frac{1}{2}x^2 \left(\frac{1}{x}\right) dx$$

$$= \ln x \left(\frac{1}{2}x^2\right)\Big|_1^e - \frac{1}{4}x^2 \Big|_1^e = \ln e\left(\frac{1}{2}e^2\right) - \ln 1\left(\frac{1}{2}1^2\right) - \left[\frac{1}{4}e^2 - \frac{1}{4}1^2\right]$$

$$= \frac{1}{4}e^2 + \frac{1}{4}$$

2.47 d. If $\frac{dy}{dx} = m$, $y = \int m\, dx = m \int dx = mx + b$, so $y = mx + b$ is a straight line.

2.48 d. All trigonometric functions are ratios of lengths, with the result that they are dimensionless.

2.49 d.
$$\text{Area} = \int_a^b y\, dx$$

$$\text{Average value} = \frac{\text{Area}}{\text{Base width}} = \frac{\int_a^b y\, dx}{b - a}$$

2.50 b. Since $\frac{dy}{dx} = -\sin x$ and $\tan x = \frac{\sin x}{\cos x}$, then $\sin x = \tan x \cos x$. Thus, the derivative is
$$\frac{dy}{dx} = -\tan x \cos x$$

2.51 d. Apply the chain rule. The "outside" function is u^3, so $y' = 3 \cos^2 5x$ $(\cos 5x)' = 3 \cos^2 5x\, (-\sin 5x)\, (5)$.

2.52 c. Derivative of y, $y' = 6x^2 - 3$. Since the slope is the derivative, at $x = 1$ the slope is $y'|_{x=1} = 3$.

2.53 b. Position $s(t) = 16t^2$; then, velocity $v(t) = s'(t) = 32t$ m/s and acceleration $a(t) = s''(t) = 32$ m/s^2; at any time the acceleration is 32 m/s^2.

2.54 d. Derivative of y is $y' = 3x^2 - 5$. Equating y' to 0, $3x^2 - 5 = 0$ has roots at $x = -1.29, +1.29$.

Second derivative of y is $y'' = 6x$.

$y''|_{x=-1.29} < 0$; then, the maximum of y occurs at $x = -1.29$.
(Note: value of $y_{\max} = (-1.29)^3 - 5(-1.29) - 4 = 0.30$

$y''|_{x=1.29} > 0$; then, the minimum of y occurs at $x = 1.29$; $y_{\min} = (1.29)^3 - 5(1.29) - 4 = -8.30$.

At the inflection point $y'' = 0$: $y'' = 6x = 0$; then, $x = 0$. Also, y'' changes sign at $x = 0$.)

2.55 c. y^2

2.56 a. $f(x) = \cos x; f'(x) = -\sin x; f''(x) = -\cos x; f'''(x) = \sin x; f^{iv}(x) = \cos x$
Substituting these values in the Taylor's series expansion,

$$f(x) = \cos(0) + \frac{-\sin 0}{1!}x + \frac{-\cos 0}{2!}x^2 + \frac{\sin 0}{3!}x^3 + \frac{\cos 0}{4!}x^4$$

$$= 1 - \frac{x^2}{2!} + \frac{x^4}{4!} - \frac{x^6}{6!} + \ldots$$

2.57 d. Using L'Hopital's rule, $\lim_{x \to 2} \frac{x^2 - 4}{x - 2} = \lim_{x \to 2} \frac{2x}{1} = 4$.

2.58 a. Using the integral table, the integral =

$$2\left[\frac{\theta}{2} - \frac{\sin 2\theta}{4}\right]_0^{\pi/4} = \left[(\frac{\pi}{4} - \frac{1}{2}) - 0\right] = 0.285$$

2.59 a. Taking the derivative with respect to x,

$$3x^2 + 6xy + 3x^2 y' + 3y^2 y' = 0, \text{ so } y' = -\frac{3x^2 + 6xy}{3x^2 + 3y^2}$$

2.60 b. Since $\sqrt{36} = 6$, take $x_0 = 36$ and $f(x) = \sqrt{x}$. In general,

$$\Delta y = f(x) - f(x_0) = f'(x_0)(x - x_0)$$

$$\sqrt{34} - \sqrt{36} = \frac{1}{2}\frac{1}{\sqrt{36}}(-2) = -\frac{1}{6}$$

2.61 b. Here,
$f'(x) = 4x^3 - 4x^2 - 24x = x(4x^2 - 4x - 24) = 4x(x - 3)(x + 2) = 0$,
so possible extrema are at 0, 3, and –2. Since $f''(0) = -24$, it is
the maximum (3 and –2 are minima). Since $f(0) = 1$, the relative
maximum is 1.

2.62 d. The line and the parabola intersect when $x^2 = 2x + 3$, or $x^2 - 2x + 1 = 4$,
or $(x - 1)^2 = 2^2$. The line is above the parabola, so

$$A = \int_{-1}^{3} (2x + 3 - x^2)\, dx = \left[x^2 + 3x - \frac{1}{3}x^3\right]_{-1}^{3} = 9 - \left(-\frac{5}{3}\right)$$

2.63 c. Multiply by r to obtain $x^2 + y^2 = 2y + 2x$,
or $x^2 - 2x + y^2 - 2y = 0$, or $(x - 1)^2 + (y - 1)^2 = 2$ a circle centered
at (1, 1) of radius $\sqrt{2}$. The area is $\pi(\sqrt{2})^2 = 2\pi$.

2.64 d. The first derivative $\frac{dy}{dx}$ is the slope of the curve. At point A the slope
is positive. The second derivative $\frac{d^2 y}{dx^2}$ gives the direction of bending.
A negative value indicates the curve is concave downward.

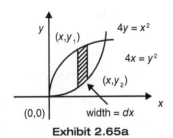

Exhibit 2.65a

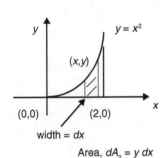

Exhibit 2.66a

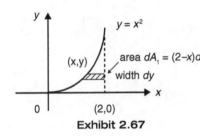

Exhibit 2.67

2.65 c. Point of intersection of the curves is determined as $y^2 = 4x = 4\sqrt{4y}$, $y = 4$; $x = 4$.

Area of the strip in Exhibit 2.65a, $dA = (y_1 - y_2)dx = \left(\sqrt{4x} - \dfrac{x^2}{4}\right)dx$

$$\text{Area} = A = \int_{x=0}^{4}\left(\sqrt{4x} - \dfrac{x^2}{4}\right)dx = \dfrac{16}{3}$$

2.66 b. In Exhibit 2.66a, area $dA_2 = ydx$.

$$\text{Shaded area} = \int dA_2 = \int_{x=0}^{2} y.dx = \int_{x=0}^{2} x^2 dx = \dfrac{8}{3}$$

2.67 c. In Exhibit 2.67, area $dA_1 = (2-x)dy$.

First moment with respect to x-axis of this area $dM_x = y\,dA_1$

First moment with respect to x-axis of shaded area is

$$M_x = \int dM_x = \int_{y=0}^{4} y(2-x)dy = \int_{0}^{4} y(2-\sqrt{y})dy = 3.2$$

2.68 d. From Exhibit 2.66a, first moment with respect to y-axis of the shaded area dA_2 is $dM_y = x\,dA_2 = xy\,dx$; first moment with respect to y-axis of shaded area is

$$M_y = \int dM_y = \int_{x=0}^{2} x.y.dx = \int_{0}^{2} x.x^2 dx = 4$$

2.69 a. x-coordinate $x_c = \dfrac{M_y}{\text{area}} = \dfrac{4}{2.67} = 1.5$; y-coordinate y_c

$$= \dfrac{M_x}{\text{area}} = \dfrac{3.2}{2.67} = 1.2$$

2.70 a. From Exhibit 2.67, moment of inertia with respect to x-axis of area dA_1 is $dMI_x = y^2 dA_1$. Moment of inertia with respect to x-axis of shaded area:

$$MI_x = \int dMI_x = \int y^2 dA_1 = \int_{y=0}^{4} y^2(2-x)dy = \int_{0}^{4} y^2(2-\sqrt{y})dy$$

Integrating, $MI_x = 6.095$.

2.71 d. From Exhibit 2.66a, moment of inertia with respect to y-axis of area dA_2 is $dMI_y = x^2 dA_2$. Moment of inertia with respect to y-axis of shaded area:

$$MI_y = \int dMI_y = \int x^2 dA_2 = \int_{x=0}^{2} x^2.ydx = \int_{x=0}^{2} x^4 dx = 6.4$$

2.72 b. Homogeneous solution, $y_h(x) = e^{-4x}(C_1 \cos 4x + C_2 \sin 4x)$ or $e^{-4x} C_3 \cos(4x + \theta)$

2.73 d. Since the characteristic roots are -2 and -2, the particular solution is $y_p(x) = A\, x^2 e^{-2x}$.

2.74 a. The characteristic equation is $r^2 + 5r + 6 = 0$ and the characteristic roots are $-2, -3$. Then, the homogeneous solution is $y_h(x) = c_1 e^{-2x} + c_2 e^{-3x}$.

The particular solution due to e^{-2x} is $y_p(x) = Bxe^{-2x}$.

[As -2 is a characteristic root, e^{-2x} cannot be a particular solution.]

$y_p' = -2Be^{-2x}.x + Be^{-2x}$ and $y_p'' = 4Be^{-2x}.x - 2Be^{-2x} - 2Be^{-2x}$

$y_p'' + 5y_p' + 6y = 2e^{-2x}$ yields $B = 2$

$y(x) = 2e^{-2x}x + c_1 e^{-2x} + c_2 e^{-3x}$ $y(0) = c_1 + c_2 = 0$ $c_1 = -2$

$y'(x) = 2e^{-2x} - 4xe^{-2x} - 2c_1 e^{-2x} - 3c_2 e^{-3x}$ $y'(0) = 2 - 2c_1 - 3c_2 = 0$ $c_2 = 2$

The solution is $y = -2e^{-2x} + 2e^{-3x} + 2xe^{-2x}$.

2.75 a.

$$\lim_{x \to 1} \frac{x^2 - 1}{x - 1} = \lim_{x \to 1} \frac{(x-1)(x+1)}{x - 1} = \lim_{x \to 1}(x+1) = 2$$

2.76 b. This is linear equation, $y' + \frac{2}{x}y = \frac{1}{x}e^{3x}$. The integrating factor is $e^{\int \frac{2}{x} dx} = x^2$ so the equation becomes $d(x^2 y) = xe^{3x}\, dx$. Integrating,

$x^2 y = \frac{1}{3}xe^{3x} - \frac{1}{9}e^{3x} + c$, or $y = \frac{1}{3x}e^{3x} - \frac{1}{9x^2}e^{3x} + \frac{c}{x^2}$.

2.77 a. By inspection, $y_1 = e^x$ is one solution. Use reduction of order to obtain

$$\left(\frac{y_2}{y_1}\right)' = \frac{e^{\int \frac{-2(x+1)}{x} dx}}{y_1^2} = \frac{e^{2x+2\, \ln x}}{e^{2x}} = x^2$$

Hence $\dfrac{y_2}{y_1} = \dfrac{x^3}{3}$, so $y_2 = \dfrac{x^3}{3} y_1$. Suppressing the $\dfrac{1}{3}$, $y_2 = x^3 e^x$.

2.78 d. The associated homogeneous equation, $y'' + 4y = 0$, has the solution $y_h = A \sin 2x + B \cos 2x$. Using the method of undetermined coefficients,

$y_p = a \sin x + b \cos x$
$y_p'' = -a \sin x - b \cos x$
$y_p'' + 4y_p = (-a + 4a) \sin x + (-b + 4b) \cos x = 8 \sin x$
$3a \sin x = 8 \sin x$
$3b \cos x = 0 \cos x$

Thus, $b = 0$ and $a = \dfrac{8}{3}$.

2.79 d. Begin by eliminating c by solving for it from the derivative of the given equation: $2x + 2yy' = 2cy'$, $c = \dfrac{x + yy'}{y'}$,

$$x^2 + y^2 = 2\dfrac{x + yy'}{y'} y', \quad x^2 y' + y^2 y' = 2xy + 2y^2 y', \text{ and}$$

$$y' = \dfrac{2xy}{x^2 - y^2}.$$

Now, the orthogonal family will have $y'_{\text{new}} = -\dfrac{1}{y'_{\text{old}}}$, so $y'_{\text{old}} = \dfrac{y^2 - x^2}{2xy}$.

Letting $u = \dfrac{y}{x}$, $xu' + u = \dfrac{u^2 - 1}{2u}$, $xu' = \dfrac{u^2 - 1 - 2u^2}{2u} = -\dfrac{1 + u^2}{2u}$,

and $\dfrac{2u\,du}{1 + u^2} = -\dfrac{dx}{x}$. Integrating, $\ln(1 + u^2) = -\ln|x| + c$, and

$$\ln\left\{\left[1 + \left(\dfrac{y}{x}\right)^2\right]|x|\right\} = c$$

$$\left|x + \dfrac{y^2}{x}\right| = e^c = c_1 > 0, \text{ or } x + \dfrac{y^2}{x} = c_2 \,(=\pm c_1)$$

Thus $x^2 + y^2 = c_2 x$, or $x^2 + y^2 = 2cx$.

2.80 b. Using Laplace transform table, the transform is $\dfrac{(s+1)}{(s+1)^2 + 1^2}$.

2.81 c. Taking the Laplace transform of the differential equation and simplifying,

$$s^2 Y(s) + 2sY(s) + Y(s) = \dfrac{s}{s^2 + 2^2} \Rightarrow Y(s) = \dfrac{s}{(s^2 + 4)(s^2 + 2s + 1)}$$

2.82 d. Initial value, $\lim_{t \to 0} f(t) = \lim_{s \to \infty} s \dfrac{(s+5)}{s(s+1)(s+10)} = 0$

2.83 c. $y(k) = 2y(k-1) + 3y(k-2)$ with $y(-1) = 1$ and $y(-2) = 1$
For $k = 0$, $y(0) = 2y(-1) + 3y(-2) = 2(1) + 3(1) = 5$
For $k = 1$, $y(1) = 2y(0) + 3y(-1) = 2(5) + 3(1) = 13$

2.84 a. $y(k+1) + y(k) = u(k)$; taking Z-transform, $[z\,y(z) - 0] + y(z)$

$= \dfrac{1}{1 - z^{-1}}$. Simplifying, $y(z) = \dfrac{z}{(z-1)(z+2)}$.

2.85 d. Using the initial value theorem, the initial value,

$$\lim_{k \to 0} f(k) = \lim_{z \to \infty} \dfrac{z+2}{z-4} = 1$$

2.86 c. Period $T = 4$ s; $\omega_0 = \dfrac{2\pi}{T} = \dfrac{\pi}{2}$ rad/s;

$$b_3 = \dfrac{2}{4}\left[\int_0^2 8\sin\left(\dfrac{3\pi t}{2}\right)dt + \int_2^4 -8\sin\left(\dfrac{3\pi t}{2}\right)dt\right] = \dfrac{16}{3\pi}(1-\cos 3\pi) = 3.40$$

2.87 d. Since the pulse width is 2, $\tau = 1$ and from the Table, $X(f) = f\operatorname{sinc}^2(f)$.

2.88 c. Using time shift theorem for Fourier transforms,

$$X(f) = 1\exp(-j2\pi f) + 1\exp(+j2\pi f) = 2\cos(2\pi f)$$

2.89 c. Substituting $k = 0$, $y(0) = 2y(-1) + 3y(-2) = 2 + 3 = 5$
Substituting $k = 1$, $y(1) = 2y(0) + 3y(-1) = 10 + 3 = 13$

2.90 c. Final value $= \lim_{k\to\infty} f(k) = \lim_{z\to 1}(1-z^{-1})F(z) = \lim_{z\to 1}(1-z^{-1})\dfrac{2(z+1)}{(z-1)} = 4$

CHAPTER 3

Probability, Statistics, and Sampling

OUTLINE

COUNTING SETS 93
Permutations ∎ Combinations

PROBABILITY 96
Definitions ∎ General Character of Probability ∎ Complementary Probabilities ∎ Joint Probability ∎ Conditional Probability

STATISTICAL TREATMENT OF DATA 100
Frequency Distribution ∎ Standard Statistical Measures

STANDARD DISTRIBUTION FUNCTIONS 103
Binomial Distribution ∎ Normal Distribution Function ∎ t-Distribution ∎ X^2 Distribution ∎ The Poisson Distribution

CONFIDENCE BOUNDS 110

HYPOTHESIS TESTING 112

SAMPLING 114

PROBLEMS 116

SOLUTIONS 120

COUNTING SETS

For discrete probability calculations it is important to count the number of elements in sets of possible outcomes. The primary method is simply to write down all the elements in a set and count them. For example, count the number of elements in the set S of all possible outcomes of a six-sided die throw. The set definition is $S = \{1, 2, 3, 4, 5, 6\}$. Simple counting gives six elements.

Most useful sets are large, and simple counting is too time-consuming. There are several methods for simplifying this task. One can use the product set concept introduced in the last section. If A is a set with n elements and B is a set with m elements, then the product set $A \times B$ has the arithmetic $n \times m$ number of elements. To simplify counting then, first count the sets making up the product set (usually containing a much smaller number of elements) and simply multiply these counts.

Example 3.1

Count the number of possible outcomes for tossing five coins.

Solution

The number of outcomes for a single toss defined by the set $R = \{H, T\}$ is 2. The result of five coin tosses is the product set $R \times R \times R \times R \times R$. The total number of possible outcomes is then the arithmetic product of the number of outcomes in each of the individual five tosses. This is $2 \times 2 \times 2 \times 2 \times 2 = 2^5 = 32$.

Permutations

If A is a set with n elements, a **permutation** of A is an ordered arrangement of A. Given the set $A = \{a, b, c\}$, the order a, b, c of the elements is one permutation. Any other order—for example, b, c, a—is another permutation.

The set B of all permutations of the set A is defined as the set of all arrangements of the three elements. These are

$$B = \{\{a, b, c\}, \{a, c, b\}, \{c, b, a\}, \{b, a, c\}, \{b, c, a\}, \{c, a, b\}\}$$

There are six permutations. This number also can be derived as follows. The number of ways an element can be chosen for the first space is three. Then there are two elements left. One of these can go in the second space. Then there is one element left. This must go in the third space. This gives the formula $3 \times 2 \times 1 = 6$. In general, the number of ways n distinct elements can be arranged is given by

$$n! = n \times (n-1) \times (n-2) \times \cdots 1$$

and is called the **factorial** of the number n. The factorial of 0 is 1 ($0! = 1$).

For example, count the number of ways a standard playing deck can be arranged. Since there are 52 distinct cards in a deck, there are 52! different arrangements, or permutations.

Now suppose we have the set of letters L in the word *obtuse* so that $L = \{o, b, t, u, s, e\}$. How many two-letter symbols could be made from this set? Notice that the letters are all distinct. We again count the number of ways the letters can be selected. For the first choice it is six; for the second choice it is five. The two selections are now complete. There are therefore $6 \times 5 = 30$ possibilities.

The general formula for the number of permutations, taking r items from a set of n, is given by

$$P(n, r) = n!/(n-r)!$$

Using this equation, one can express the previous example as $P(6, 2) = 6!/(6-2)! = 30$.

Example 3.2

A jeweler has nine different beads and a bracelet design that requires four beads. To find out which looks the best, he decides to try all the permutations. How many different bracelets will he have to try?

Solution

There are $n = 9$ beads. He selects $r = 4$ at a time. The order is important, because each arrangement of r beads on the bracelet makes a different bracelet. So the number of different bracelets is

$$P(9,4) = 9!/(9-4)! = 9 \times 8 \times 7 \times 6 = 3024$$

If the bracelet is a closed circle, there is no discernible difference when it is rotated. Then one observes four identical states for each unique bracelet. This is called ring permutation and is given by the formula

$$P_{ring}(n,r) = P(n,r)/r$$

There are only $3024/4 = 756$ distinct ring bracelets the jeweler can make.

Example 3.3

(i) In how many ways can four people be asked to form a line of three people?
(ii) In how many ways can the letters of the word BEAUTY be arranged?
(iii) In how many ways can the letters of the word GOOD be arranged?

Solution

(i) $P(4,3) = \dfrac{4!}{(4-3)!} = 24$

(ii) $P(6,6) = \dfrac{6!}{(6-6)!} = 720$

(iii) $P(4;1,1,2) = \dfrac{4!}{1!1!2!} = 12$

Combinations

When the order of the set of r things that are selected from the set of n things does not matter, we talk about combinations.

Again consider the standard playing deck of 52 cards. How many hands of 5 cards can we get from a deck of 52 cards? Count the number of ways the hands can be drawn. The first draw can be any of the 52 cards. The second draw can only be one of the remaining 51 cards. The third draw can only be one of the remaining 50, the next is one of 49, the last one of 48. So the result is

$$52 \times 51 \times 50 \times 49 \times 48 = 52!/(52-5)!$$

This is the formula for permutations discussed in the last section. But the order in which we receive the cards is not important, so many of the hands are the same. In fact there are 5! similar arrangements of cards that make the same hand. The number of distinct hands is

$$(52 \times 51 \times 50 \times 49 \times 48)/(5 \times 4 \times 3 \times 2 \times 1) = 52!/[5! \times (52-5)!]$$

The general form r items taken from a set of n items when order is not important is written as the binomial coefficient $C(n, r)$, also written $\binom{n}{r}$, and is given by the formula

$$C(n, r) = \frac{n!}{r!(n-r)!}$$

Example 3.4

There are six skiers staying in a cabin with four bunks. How many combinations of people will be able to sleep in beds?

Solution

$C(6, 4) = 6!/[4! \times (6 - 4)!] = (6 \times 5 \times 4 \times 3 \times 2 \times 1)/[(4 \times 3 \times 2 \times 1) \times (2 \times 1)]$
$= 15$

PROBABILITY

Definitions

An **experiment**, or **trial**, is an action that can lead to a measurement.

Sampling is the act of taking a measurement. The **sample space** S is the set of all possible outcomes of an experiment (trial). An event e is one of the possible outcomes of the trial.

If an experiment can occur in n mutually exclusive and equally likely ways, and if m of these ways correspond to an event e, then the probability of the event is given by

$$P\{e\} = m/n$$

Example 3.5

A die is a cube of six faces designated as 1 through 6. The set of outcomes R of one die roll is defined as $R = \{1, 2, 3, 4, 5, 6\}$. If two dice are rolled, define trial, sample space, n, m, and the probability of rolling a seven when adding both dice together.

Solution

The trial is the rolling of two dice. The sample space is all possible outcomes of a two-dice roll, and the event is the outcome that the sum is 7.

The number of all possible outcomes, n, is the number of elements in the product set of the outcome of two dice when each is rolled independently. The product set is $R \times R$ and contains 36 elements.

The number of all possible ways, m, that the (7) event can occur is $(1, 6), (2, 5), (3, 4), (4, 3), (5, 2),$ and $(6, 1)$ for a total of six ways. The probability of rolling a 7 is $P\{7\} = \dfrac{6}{36} = \dfrac{1}{6}$.

Example 3.6

What is the probability of (i) a tail showing up when a fair coin is tossed, (ii) number 3 showing up when a fair die is tossed, and (iii) a red king is drawn from a deck of 52 cards.

Solution

(i) 1/2   (ii) 1/6   (iii) 2/52

General Character of Probability

The probability $P\{E\}$ of an event E is a real number in the range 0 through 1. Two theorems identify the range between which all probabilities are defined:

1. If $\emptyset$ is the null set, $P\{\emptyset\} = 0$.
2. If S is the sample space, $P\{S\} = 1$.

The first states that the probability of an impossible event is zero, and the second states that, if an event is certain to occur, the probability is 1.

Complementary Probabilities

If E and E' are complementary events, $P\{E\} = 1 - P\{E'\}$. Complementary events are defined with respect to the sample space. The probability that an event E will happen is complementary to the probability that any of the other possible outcomes will happen.

Example 3.7

If the probability of throwing a 3 on a die is 1/6, what is the probability of not throwing a 3?

Solution

E is the probability of not throwing a 3, so $P\{E\} = 1 - P\{E'\} = 1 - \dfrac{1}{6} = \dfrac{5}{6}$.

Sometimes the complementary property of probabilities can be used to simplify calculations. This will happen when seeking the probability of an event that represents a larger fraction of the sample space than its complement.

Example 3.8

What is the probability $P\{E\}$ of getting at least one head in four coin tosses?

Solution

The complementary event $P\{E'\}$ to getting at least one head is getting no heads (or all tails) in four tosses. So the probability of getting at least one head is

$$P\{E\} = 1 - (0.5)^4 = 1 - 0.0625 = 0.9375$$

Joint Probability

The probability that a combination of events will occur is covered by joint probability rules. If E and F are two events, the joint probability is given by the rule

$$P\{E \cup F\} = P\{E\} + P\{F\} - P\{E \cap F\} \qquad \text{(Rule 1)}$$

A special case of the joint probability rule can be derived by considering two events, E and F, to be mutually exclusive. In this case the last term in Rule 1 is zero since $P\{E \cap F\} = P\{0\} = 0$. Thus, if E and F are mutually exclusive events,

$$P\{E \cup F\} = P\{E\} + P\{F\} \qquad \text{(Rule 2)}$$

Example 3.9

What is the probability of throwing a 7 or a 10 with two dice?

Solution

We will call the event of throwing a 7 A, and of throwing a 10 B. We know from previous examples that $P\{A\} = \frac{1}{6}$, and we can count outcomes to get $P\{B\} = \frac{1}{12}$. Applying the formula,

$$P\{A \cup B\} = P\{A\} + P\{B\} = \frac{1}{6} + \frac{1}{12} = \frac{1}{4}$$

If two events E and F are independent—that is, if they come from different sample spaces—then the probability that both will happen is given by the rule

$$P\{E \cap F\} = P\{E\} \times P\{F\} \qquad \text{(Rule 3)}$$

Example 3.10

What is the probability of throwing two heads in two coin tosses?

Solution

Call the throwing of one head E, the other F. The probability of throwing a single head is $P\{E\} = \frac{1}{2}$, and $P\{F\} = \frac{1}{2}$. The probability of throwing both heads is

$$P\{E \cap F\} = P\{E\} \times P\{F\} = \frac{1}{2} \times \frac{1}{2} = \frac{1}{4}$$

To visualize joint probabilities, we can use a Venn diagram showing two intersecting events, A and B, as shown in Figure 3.1. Let the normalized areas of each event represent the probability that the event will occur. For example, think of a random dart thrown at the Venn diagram: What are the chances of hitting one of the areas? Assume the areas correspond to probabilities and are given by $P\{S\} = 1$, $P\{A\} = 0.3$, $P\{B\} = 0.2$, and $P\{A \cap B\}$ is 0.6. The probability of hitting either area A or area B is calculated as the sum of the areas A and B minus the overlap area so it is not counted twice:

$$P\{A \cup B\} = 0.3 + 0.2 - .06 = .44$$

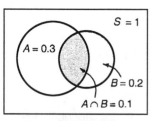

Figure 3.1
Venn diagram of joint probabilities

The result is also equal to the normalized area covered by A and B. The probability of hitting both A and B on one throw is simply the overlap area $P\{A \cap B\}$ = 0.1.

If we throw two darts, the area S is used twice and represents two independent sample spaces. Hence Rule 3 applies.

Example 3.11

A die is tossed. Event A = {an odd number shows up}; event B = {a number > 4 shows up}.

(i) Find the probabilities, P(A) and P(B). (ii) What is the probability that either A or B or both occur?

Solution
(i) P(A) = P(1 or 3 or 5 showing up) = 3 (1/6) = 0.5
    P(B) = P(5 or 6 showing up) = 2 (1/6) = 0.333
(ii) Event (A, B) = {5} and P(A, B) = 1/6;
    then P(A + B) = P(A) + P(B) − P(AB) = 1/2 + 2/6 − 1/6 = 4/6
    Check: Event (A + B) = {1,3,5,6}; then P(A + B) = 4/6

Conditional Probability

The conditional probability of an event E given an event F is denoted by $P\{E \mid F\}$ and is defined as

$$P\{E \mid F\} = P\{E \cap F\}/P\{F\} \quad \text{for } P\{F\} \text{ not zero}$$

Example 3.12

Two six-sided dice, one red and one green, are tossed. What is the probability that the green die shows a 1, given that the sum of numbers on both dice is less than 4?

Solution

Let E be the event "green die shows 1" and let F be the event "sum of numbers shows less than four." Then

$$E = \{(1,1), (1,2), (1,3), (1,4), (1,5), (1,6)\}$$
$$F = \{(1,1), (1,2), (2,1)\}$$
$$E \cap F = \{(1,1), (1,2)\}$$
$$P\{E \mid F\} = P\{E \cap F\}/P\{F\} = (2/36)/(3/36) = 2/3$$

The generalized form of conditional probability is known as Bayes' theorem and is stated as follows: If $E_1, E_2, \ldots, E_n$ are n mutually exclusive events whose union is the sample space S, and E is any arbitrary event such that $P\{E\}$ is not zero, then

$$P\{E_k \mid E\} = \frac{P\{E_k\} \times P\{E \mid E_k\}}{\sum_{j=1}^{n}[P\{E_j\} \times P\{E \mid E_j\}]}$$

STATISTICAL TREATMENT OF DATA

Whether from the outcome of an experiment or trial, or simply the output of a number generator, we are constantly presented with numerical data. A statistical treatment of such data involves ordering, presentation, and analysis. The tools available for such treatment are generally applicable to a set of numbers and can be applied without much knowledge about the source of the data, although such knowledge is often necessary to make sensible use of the statistical results.

In its raw form, numerical data is simply a list of n numbers denoted by x_i, where $i = 1, 2, 3, \ldots, n$. There is no specific significance associated with the order implicit in the i numbers. They are names for the individuals in the list, although they are often associated with the order in which the raw data was recorded. For example, consider a box of 50 resistors. They are to be used in a sensitive circuit, and their resistances must be measured. The results of the 50 measurements are presented in the following table.

Table of Raw Measurements (Ω)				
101	105	110	115	82
86	91	96	117	112
109	103	89	97	98
101	104	99	95	97
85	90	94	112	107
103	94	98	106	98
114	112	108	101	99
93	96	99	104	90
109	106	101	93	92
104	99	109	100	107

Each number is named by the variable x_i, and there are $n = 50$ of them. The numbers range from 82 to 117.

Frequency Distribution

A systematic tool used in ordering data is the frequency distribution. The method requires counting the number of occurrences of raw numbers whose values fall within step intervals. The step intervals (or bins) are usually chosen to (1) be of constant size, (2) cover the range of numbers in the raw data, (3) be small enough in quantity to limit the amount of writing yet not have many empty steps, and (4) be sufficient in quantity so that significant information is not lost.

For example, the aforementioned raw data of measured resistances may be ordered in a frequency distribution table such as Table 3.1. Here the step interval is the event E of a random variable that can be mapped onto the x-axis. The set of eight events is the sample space. If we take a number randomly from the raw measurement set, the probability that it will be in bin 5 is

$$f(E_5) = P\{E_5\} = 10/50 = 0.2$$

Table 3.1 Frequency and cumulative frequency table

Event, E_i	Range, Ω	Frequency	Cumulative Frequency	Probability Density Function, $f(E_i)$
1	80–84	1	1	0.02
2	85–89	3	4	0.06
3	90–94	8	12	0.16
4	95–99	12	24	0.24
5	100–104	10	34	0.20
6	105–109	9	43	0.18
7	110–114	5	48	0.10
8	115–119	2	50	0.04

The last column in Table 3.1 is the probability density function of the distribution. The probability table can be plotted along the *x*-axis in several ways, as shown in Figures 3.2 through 3.4.

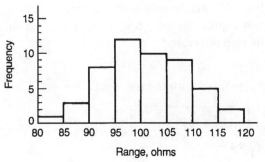

Figure 3.2 Histogram of resistance measurements

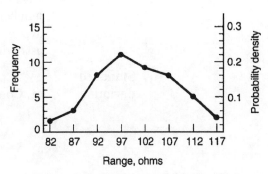

Figure 3.3 Frequency distribution and probability density plot

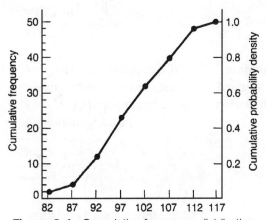

Figure 3.4 Cumulative frequency distribution and cumulative probability density

Standard Statistical Measures

There are several statistical quantities that can be calculated from a set of raw data and its distribution function. Some of the more important ones are listed here, together with the method of their calculation.

Mode The observed value that occurs most frequently; here the mode is bin 4 with a range of 95–99 Ω.

Median The point in the distribution that divides the number of observations such that half of the observations are above and half are below. The median is often the mean of the two middle values; here the median is 4.5 bins, 100 Ω.

Mean The arithmetic mean, or average, is calculated from raw data as

$$\mu = \frac{1}{n}\sum_{i=1}^{n} x_i = 100.6$$

It is calculated from the distribution function as

$$\mu = \sum_{i=1}^{m} b_i \times f(E_i) = 100.4$$

where b_i is the ith event value (for $i = 1$, $b_i = 82$) and m is the number of bins; $f(E_i)$ is the probability density function. (The two averages are not quite the same because of the information lost in assigning the step intervals.)

Standard deviation (a) Computational form for the raw data:

$$\sigma = \sqrt{\frac{1}{n}\left[\left(\sum_{i=1}^{n} x_i^2\right) - n \times \mu^2\right]} = 8.08$$

(b) Computational form for the distribution function:

$$\sigma = \sqrt{\left[\left(\sum_{i=1}^{m} b_i^2 \times f(E_i)\right) - \mu^2\right]} = 8.02$$

Sample standard deviation If the data set is a sample of a larger population, then the sample standard deviation is the best estimate of the standard deviation of the larger population.

The computational form for the raw data set is

$$\sigma = \sqrt{\frac{1}{n-1}\left[\left(\sum_{i=1}^{n} x_i^2\right) - n \times \mu^2\right]} = 8.166$$

Sample standard deviations and the use of $(n-1)$ in the denominator are discussed in the section on sampling.

Skewness This is a measure of the frequency distribution asymmetry and is approximately

$$\text{skewness} \cong 3(\text{mean} - \text{median})/(\text{standard deviation})$$

Example 3.13

Two professors give the following scores to their students. What is the mode and arithmetic mean?

Frequency	1	3	6	11	13	10	2
Score	35	45	55	65	75	85	95

Solution

mode = 75; N = 1 + 3 + 6 + 11 + 13 + 10 + 2 = 46

weighted arithmetic mean = $\overline{X_w} = [35(1) + 45(3) + \ldots 95(2)] / 46 = 70$

STANDARD DISTRIBUTION FUNCTIONS

In the previous section, we calculated several general properties of probability distribution functions.

To know the appropriate probability density function for an actual situation, two general methods are available:

1. The probability density function is actually calculated, as was done in the last section, by analyzing the physical mechanism by which experimental events and outcomes are generated and counting the number of ways an individual event occurs.

2. Recognition of an overall similarity between the present experiment and another for which the probability density function is already known permits the known behavior of the function to be applied to the new experiment. This work-saving method is by far the more popular one. Of course, to apply this method, it is necessary to have a repertoire of known probability functions and to understand the problem characteristics to which they apply.

This section lists several popular probability density functions and their characteristics.

Binomial Distribution

The binomial distribution applies when there is a set of discrete binary alternative outcomes. Deriving this distribution function helps one understand the class of problems to which it applies. For example, given a set of n events, each with a probability p of occurring, what is the probability that r of the events will occur and $(n - r)$ not occur?

The probability of one event occurring is p.

The probability of r events occurring is p^r.

The probability of $(n - r)$ events not occurring is $(1 - p)^{n-r}$.

The probability of exactly r events occurring and $(n - r)$ not occurring in a trial is given by the joint probability Rule 3:

$$P[r \cap (n-r)] = p^r \times (1-p)^{n-r}$$

However, there are many ways of choosing r occurrences out of n events. In fact, the number of different ways of choosing r items from a set of n items when order is not important is given by the binomial coefficient $C(n, r)$. The total probability of r occurrences from n trials, given an individual probability of occurrence as p, is thus given by

$$C(n,r) \times p^r \times (1-p)^{n-r} = f(r)$$

This is the **binomial probability density function**.

The mean of this density function is the first moment of the density function, or expected value, and is calculated as

$$E\{x\} = \sum_{r=0}^{n} r \times f(r) = \sum_{r=0}^{n} r \times \frac{n!}{(r)!(n-r)!} \times p^r \times (1-p)^{n-r}$$

This can be rewritten as

$$\sum_{r=1}^{n} \frac{n!}{(r-1)!(n-r)!} \times p^r \times (1-p)^{n-r}$$

We can now factor out the quantity $n \times p$ and let $r - 1 = y$. This can be rewritten as

$$n \times p \times \sum_{y=0}^{n-1} \frac{(n-1)!}{(y)!(n-1-y)!} \times p^y \times (1-p)^{n-1-y} = n \times p \times [p + (1-p)]^{n-1}$$

Since the sum is merely the expansion of a binomial raised to a power, and the number 1 raised to any power is 1, the mean is

$$\mu = n \times p$$

A similar calculation shows the variance is

$$\text{var} = n \times p \times (1-p)$$

The standard deviation is

$$\sigma = \sqrt{\text{var}} = \sqrt{n \times p \times (1-p)}$$

Example 3.14

A truck carrying dairy products and eggs damages its suspension and 5% of the eggs break.

(i) What is the probability that a carton of 12 eggs will have exactly one broken egg?

(ii) What is the probability that one or more eggs in a carton will be broken?

Solution

(i) Since an egg is either broken or not broken, the binomial distribution applies. The probability p that an egg is broken is 0.05 and that one is not broken is $(1-p) = 0.95$. From the equation for the binomial distribution, with $n = 12$ and $r = 1$,

$$p\{1\} = f(1) = C(12, 1) \times 0.05^1 \times 0.95^{11} = 12 \times 0.05 \times 0.57 = 0.34$$

(ii) The probability that one or more eggs will be broken can be calculated as the sum of each individual probability:

$$p\{x>0\} = p\{1\} + p\{2\} + \cdots + p\{12\}$$

However, this requires 12 calculations. The problem can also be solved using the complementary rule:

$$p\{x>0\} = 1 - p\{0\} = C(12,0) \times 0.05^0 \times 0.95^{12} = 0.95^{12} = 0.54$$

Example 3.15

A biased coin is tossed. Find the probability that a head appears once in three trials.

P(Head) = p = 0.6

Solution

Here q = P(Head not occurring) or P(Tail) = 1 − 0.6 = 0.4.

Then P(1 Head) = $C(3,1)\, 0.6^1\, 0.4^2$ = 0.2880.

Normal Distribution Function

The normal distribution, or Gaussian distribution, is widely used to represent the distribution of outcomes of experiments and measurements. It is popular because it can be derived from a few empirical assumptions about the errors presumed to cause the distribution of results about the mean. One assumption is that the error is the result of a combination of N elementary errors, each of magnitude e and equally likely to be positive or negative. The derivation then assumes $N \to \infty$ and $e \to 0$ in such a way as to leave the standard deviation constant. This error model is universal, since most experiments are analyzed to eliminate systematic errors. What remains is attributable to errors that are too small to explain systematically, so the normal probability distribution is evoked.

The form of the probability density and distribution functions for the **normal distribution** with a mean μ and variance σ^2 is given by

$$f(x) = \frac{e^{-(x-\mu)^2/2\sigma^2}}{\sigma\sqrt{2\pi}} \qquad -\infty < x < \infty$$

$$F(x) = \int_{-\infty}^{x} \frac{e^{-(x-\mu)^2/2\sigma^2}}{\sigma\sqrt{2\pi}}\, dt$$

The normal distribution is the typical bell-shaped curve shown in Figure 3.5. Here we see that the curve is symmetric about the mean μ. Its width and height are determined by the standard deviation σ. As σ increases, the curve becomes wider and lower.

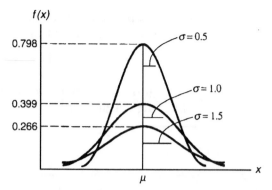

Figure 3.5 Normal distribution curve

Since this function is difficult to integrate, reference tables are used to calculate probabilities in a standard format; then the standard probabilities are converted to the actual variable required by the problem. The relation between the standard variable, z, and a typical problem variable, x, is

$$z = (x - \mu)/\sigma$$

Since μ and σ are constants, the standard probability at a value z is the same as the problem probability for the value at x.

The standard probability density function is

$$f(z) = \frac{1}{\sqrt{2\pi}} \times e^{-z^2/2}$$

Table 3.2 Standard probability table

z	$F(z)$	$f(z)$	z	$F(z)$	$f(z)$
0.0	0.5000	0.3989	2.0	0.9773	0.0540
0.1	0.5398	0.3970	2.1	0.9821	0.0440
0.2	0.5793	0.3910	2.2	0.9861	0.0355
0.3	0.6179	0.3814	2.3	0.9893	0.0283
0.4	0.6554	0.3683	2.4	0.9918	0.0224
0.5	0.6915	0.3521	2.5	0.9938	0.0175
0.6	0.7257	0.3332	2.6	0.9953	0.0136
0.7	0.7580	0.3123	2.7	0.9965	0.0104
0.8	0.7881	0.2897	2.8	0.9974	0.0079
0.9	0.8159	0.2661	2.9	0.9981	0.0060
1.0	0.8413	0.2420	3.0	0.9987	0.0044
1.1	0.8643	0.2179	3.1	0.9990	0.0033
1.2	0.8849	0.1942	3.2	0.9993	0.0024
1.3	0.9032	0.1714	3.3	0.9995	0.0017
1.4	0.9192	0.1497	3.4	0.9997	0.0012
1.5	0.9332	0.1295	3.5	0.9998	0.0009
1.6	0.9452	0.1109	3.6	0.9998	0.0006
1.7	0.9554	0.0940	3.7	0.9999	0.0004
1.8	0.9641	0.0790	3.8	0.9999	0.0003
1.9	0.9713	0.0656	3.9	1.0000	0.0002
			4.0	1.0000	0.0001

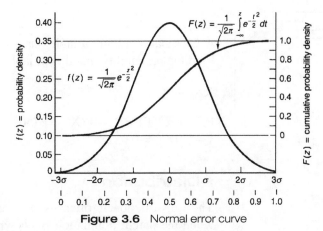

Figure 3.6 Normal error curve

The standard cumulative distribution function is

$$F(z) = \int_{-\infty}^{z} \frac{1}{\sqrt{2\pi}} \times e^{-t^2/2} \times dt$$

The standard probability function is shown graphically in Figure 3.6, and Table 3.2 shows the corresponding numerical values. The standard probability curve is symmetric about the origin and is given in terms of unit *sigma*. To use the table, remember that the function $F(z)$ is the area under the probability curve from minus infinity to the value z. The area under the curve up to $x = 0$ is therefore 0.5. Also, from symmetry,

$$F(-z) = 1 - F(z)$$

Example 3.16

Find the probability that the standard variable z lies within (i) 1σ, (ii) 2σ, and (iii) 3σ of the mean.

Solution

(i) The probability is $P_1 = F(1.0) - F(-1.0)$. From the symmetry of F, $F(-1.0) = 1 - F(1.0)$, so
$$P_1 = 2F(1.0) - 1 = 2(0.8413) - 1$$
$$= 0.6826$$

(ii) In this case, the probability is
$$P_2 = F(2.0) - F(-2.0)$$
$$= F(2.0) - [1 - F(2.0)]$$
$$= 2F(2.0) - 1 = 2(0.9773) - 1$$
$$= 0.9546$$

(iii) In the same way,
$$P_3 = 2F(3.0) - 1$$
$$= 2(0.9987) - 1$$
$$= 0.9974$$

Example 3.17

A Gaussian random variable has a mean of 1830 and standard deviation of 460. Find the probability that the variable will be more than 2750.

Solution

P(X > 2750) = 1 − P(X ≤ 2750) = 1 − F[(2750 − 1830)/460] = 1 − F(2.0) = 1 − 0.9772 = 0.0228

t-Distribution

The *t*-distribution is often used to test an assumption about a population mean when the parent population is known to be normally distributed but its standard deviation is unknown. In this case, the inferences made about the parent mean will depend upon the size of the samples being taken.

It is customary to describe the *t*-distribution in terms of the standard variable *t* and the number of degrees of freedom ν. The number of degrees of freedom is a measure of the number of independent observations in a sample that can be used to estimate the standard deviation of the parent population; the number of degrees of freedom ν is one less than the sample size ($\nu = n - 1$).

The density function of the *t*-distribution is given by

$$f(t) = \frac{\Gamma\left(\frac{\nu+1}{2}\right)}{\sqrt{\nu\pi}\,\Gamma\left(\frac{\nu}{2}\right)\left(1 + t^2/\nu\right)^{(\nu+1)/2}}$$

and is provided in Table 3.3. The mean is $m = 0$, and the standard deviation is

$$\sigma = \sqrt{\frac{\nu}{\nu - 2}}$$

Table 3.3 *t*-Distribution; values of $t_{\alpha,\nu}$

Degrees of Freedom, ν	Area of the Tail				
	$\alpha = 0.10$	$\alpha = 0.05$	$\alpha = 0.025$	$\alpha = 0.01$	$\alpha = 0.005$
1	3.078	6.314	12.706	31.821	63.657
2	1.886	2.920	4.303	6.965	9.925
3	1.638	2.353	3.182	4.541	5.841
4	1.533	2.132	2.776	3.747	4.604
5	1.476	2.015	2.571	3.365	4.032
6	1.440	1.943	2.447	3.143	3.707
7	1.415	1.895	2.365	2.998	3.499
8	1.397	1.860	2.306	2.896	3.355
9	1.383	1.833	2.262	2.821	3.250
10	1.372	1.812	2.228	2.764	3.169
11	1.363	1.796	2.201	2.718	3.106
12	1.356	1.782	2.179	2.681	3.055
13	1.350	1.771	2.160	2.650	3.012
14	1.345	1.761	2.145	2.624	2.977

(continued)

15	1.341	1.753	2.131	2.602	2.947
16	1.337	1.746	2.120	2.583	2.921
17	1.333	1.740	2.110	2.567	2.898
18	1.330	1.734	2.101	2.552	2.878
19	1.328	1.729	2.093	2.539	2.861
20	1.325	1.725	2.086	2.528	2.845
21	1.323	1.721	2.080	2.518	2.831
22	1.321	1.717	2.074	2.508	2.819
23	1.319	1.714	2.069	2.500	2.807
24	1.318	1.711	2.064	2.492	2.797
25	1.316	1.708	2.060	2.485	2.787
26	1.315	1.706	2.056	2.479	2.779
27	1.314	1.703	2.052	2.473	2.771
28	1.313	1.701	2.048	2.467	2.763
29	1.311	1.699	2.045	2.462	2.756
inf.	1.282	1.645	1.960	2.326	2.576

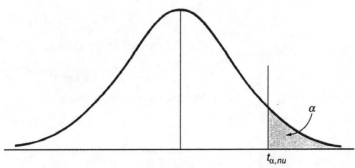

Figure 3.7

Probability questions involving the t-distribution can be answered by using the distribution function $t_{\alpha,\nu}$ shown in Figure 3.7. Table 3.3 gives the value of t as a function of the degrees of freedom ν down the column and the area (α) of the tail across the top. The t-distribution is symmetric. As an example, the probability of t falling within ± 3.0 when a sample size of 8 ($\nu = 7$) is selected is one minus twice the tail ($\alpha = 0.01$):

$$P\{-3.0 < t < 3.0\} = 1 - (2 \times 0.01) = 0.98$$

The t-distribution is a family of distributions that approaches the Gaussian distribution for large n.

X² Distribution

In probability theory and statistics, the chi-square distribution (also chi-squared or x^2-distribution) is one of the most widely used theoretical probability distributions in inferential statistics (e.g., in statistical significance tests). It is useful because, under reasonable assumptions, easily calculated quantities can be proven to have distributions that approximate to the chi-square distribution if the null hypothesis is true.

The best known situations in which the chi-square distribution is used are the common chi-square tests for goodness of fit of an observed distribution to a theoretical one, and of the independence of two criteria of classification of qualitative data. Many other statistical tests also lead to a use of this distribution.

If $Z_1, Z_2, \ldots, Z_n$ are independent unit normal random variables, then

$$\chi^2 = Z_1^2 + Z_2^2 + \ldots + Z_n^2$$

is said to have a chi-square distribution with n degrees of freedom.

The Poisson Distribution

The Poisson distribution is useful in describing the occurrence of discrete random events. The probability of x events occurring in time t is given by the following expression:

$$P(x) = (e^{-m} m^x) / x! \tag{3.1}$$

where e = base of the natural logarithms, m = mean frequency of occurrence, or number of time periods, and ! = factorial operator (for example, $3! = 3 \times 2 \times 1 = 6$).

The terms of the Poisson distribution may be summed to give the probability of fewer than or more than x events per time periods. If the engineer is interested in the probability of $\leq x$ events occurring, it may be expressed as

$$P(\leq x) = \Sigma \ [(m^i e^{-m})/i!], \text{ for } i = 0, 1 \ldots, x \tag{3.2}$$

For the case of fewer than x,

$$P(< x) = \Sigma \ [(m^i e^{-m})/i!], \text{ for } i = 0, 1 \ldots, x-1 \tag{3.3}$$

For the case of more than x,

$$P(> x) = 1 - \Sigma \ [(m^i e^{-m})/i!], \text{ for } i = 0, 1 \ldots, x \tag{3.4}$$

For the case of x or more,

$$P(\geq x) = 1 - \Sigma \ [(m^i e^{-m})/i!], \text{ for } i = 0, 1 \ldots, x-1 \tag{3.5}$$

or

$$P(\geq x) = \Sigma \ [(m^i e^{-m})/i!], \text{ for } i = x, x+1 \ldots, \infty \tag{3.6}$$

For the case of at least x but not more than y,

$$P(x \leq i \leq y) = \Sigma \ [(m^i e^{-m})/i!], \text{ for } i = x, \ldots, y \tag{3.7}$$

It should be noted that for the Poisson distribution, the mean variances are equal. Therefore, when the ratio of the variance to the mean is markedly different from 1.0, this is an indication that the observed data do not follow a Poisson distribution.

Tabulated values of the Poisson distribution can be found in most standard statistics textbooks.

CONFIDENCE BOUNDS

Referring to Table 3.4, it can be seen that in a normal distribution there is a probability of approximately 0.95 that an outcome will fall within about two standard deviations of the mean. More precisely, there is a probability of 0.95 that an outcome will lie within 1.96 standard deviations of the mean. That is

$$P(-1.96 \leq Z \leq +1.96) = 0.95 \tag{3.8}$$

Table 3.4 Area under the standard normal curve

Boundaries	Area between Boundaries
$\mu \pm 0.5\sigma$	0.383
$\mu \pm 1.0\sigma$	0.683
$\mu \pm 1.5\sigma$	0.866
$\mu \pm 2.0\sigma$	0.954
$\mu \pm 2.5\sigma$	0.977
$\mu \pm 3.0\sigma$	0.988
$\mu \pm 3.5\sigma$	0.999

Recall that any variable x can be expressed in "standard form" by changing from x to $(x - \mu)/\sigma$. For $\bar{x}$ (the sample mean), the corresponding expression is

$$Z = (\bar{x} \pm \mu)/(\sigma\sqrt{n}) \tag{3.9}$$

where $\sigma/\sqrt{n}$ is the standard error of the mean. From Equation 3.7 it follows that

$$P\left[\bar{x} - (1.96\sigma/\sqrt{n}) \leq \mu \leq \bar{x} + (1.96\sigma/\sqrt{n})\right] = 0.95 \tag{3.10}$$

Equation 3.10 can be read as "the probability that the true (population) mean lies in the interval $\bar{x} - (1.96\sigma/\sqrt{n})$ and $\bar{x} + (1.96\sigma/\sqrt{n})$ is 0.95." This is referred to as a *confidence bound*, or confidence interval, on the estimate of the true mean. Other confidence intervals can be constructed by replacing 1.96 with the Z-score corresponding to the desired confidence probability. The general expression for confidence limits on $\bar{x}$ is

$$CL = \bar{x} \pm Z(s/\sqrt{n}) \tag{3.11}$$

The general expression for the probability of a value of x greater or less than the mean is

$$x = \bar{x} \pm Zs \tag{3.12}$$

In the preceding discussion, the Z values of the standard normal distribution are, strictly speaking, applicable only when the sample size is "large" ($n > 30$). For small samples ($n < 30$) values from the *t*-distribution should be used in place of Z. Tabulated t values can be found in most standard statistics textbooks; see also Table 3.5.

If the required confidence bound (or tolerable error) is specified, the sample size required to estimate the mean with this confidence can be estimated by solving the following expression for n:

$$1.96(\sigma/\sqrt{n}) \leq e \tag{3.13}$$

and

$$n = (1.96s)^2/e^2 \tag{3.14}$$

where e = the desired tolerance (error).

Table 3.5 t-Distribution; values of $t_{\alpha,v}$

Degrees of Freedom, v	Area of the Tail				
	$\alpha = 0.10$	$\alpha = 0.05$	$\alpha = 0.025$	$\alpha = 0.01$	$\alpha = 0.005$
1	3.078	6.314	12.706	31.821	63.657
2	1.886	2.920	4.303	6.965	9.925
3	1.638	2.353	3.182	4.541	5.841
4	1.533	2.132	2.776	3.747	4.604
5	1.476	2.015	2.571	3.365	4.032
6	1.440	1.943	2.447	3.143	3.707
7	1.415	1.895	2.365	2.998	3.499
8	1.397	1.860	2.306	2.896	3.355
9	1.383	1.833	2.262	2.821	3.250
10	1.372	1.812	2.228	2.764	3.169
11	1.363	1.796	2.201	2.718	3.106
12	1.356	1.782	2.179	2.681	3.055
13	1.350	1.771	2.160	2.650	3.012
14	1.345	1.761	2.145	2.624	2.977
15	1.341	1.753	2.131	2.602	2.947
16	1.337	1.746	2.120	2.583	2.921
17	1.333	1.740	2.110	2.567	2.898
18	1.330	1.734	2.101	2.552	2.878
19	1.328	1.729	2.093	2.539	2.861
20	1.325	1.725	2.086	2.528	2.845
21	1.323	1.721	2.080	2.518	2.831
22	1.321	1.717	2.074	2.508	2.819
23	1.319	1.714	2.069	2.500	2.807
24	1.318	1.711	2.064	2.492	2.797
25	1.316	1.708	2.060	2.485	1.787
26	1.315	1.706	2.056	2.479	2.779
27	1.314	1.703	2.052	2.473	2.771
28	1.313	1.701	2.048	2.467	2.763
29	1.311	1.699	2.045	2.462	2.756
inf.	1.282	1.645	1.960	2.326	2.576

HYPOTHESIS TESTING

Hypothesis testing belongs to that branch of statistics known as inferential statistics and is the basis for statistical decision making. To test a hypothesis is to make a decision regarding the reasonableness of the results obtained from statistical analyses.

In statistics, there are two hypotheses: the *null hypothesis* (H_0), and the *alternative hypothesis* (H_1). The test procedure uses sample data to make one of two statistical decisions: (1) reject the null hypothesis (as false) or (2) decide *not* to reject the null hypothesis. The test is performed on the null hypothesis. When we reject the null hypothesis, we accept the alternative hypothesis as being true. In making this decision, we incur two possible types of errors: (1) Type I error (concluding the hypothesis is true when it is really false) and (2) Type II error (concluding the hypothesis is false when it is really true). These errors are commonly referred to as α and β, respectively. The basic steps in hypothesis testing are outlined as follows:

Step 1: *State the statistical hypotheses.* If the parameters of interest are the means of two populations, the following hypotheses could be considered:

$$H_0: \mu_1 = \mu_2, \quad H_1: \mu_1 \neq \mu_2$$
$$H_0: \mu_1 \leq \mu_2, \quad H_1: \mu_1 > \mu_2$$
$$H_0: \mu_1 \geq \mu_2, \quad H_1: \mu_1 < \mu_2$$

The first case is an example of a "two-sided" or "two-tailed" hypothesis. In this case, we are asking, "Can we conclude that the two populations have different means?" If the issue is which population has the larger mean, then the second or third statements would be appropriate. These represent "one-sided" or "one-tailed" hypotheses. In hypothesis testing, the alternative hypothesis is the statement of what we expect to be able to conclude. If the questions is whether Population 1 has a larger mean than Population 2, the $H_0: \mu_1 > \mu_2$ would be tested. If this H_0 can be rejected, we accepted the $H_1: \mu_1 > \mu_2$ as true.

Step 2: *Calculate the test statistics.* To test the hypothesis, the analyst selects an appropriate test statistics and specifies its distribution when H_0 is true; that is, the test procedure is based on the underlying distribution of the statistics used to estimate the parameters in question. For example, *testing the significance of the difference between means from two independent samples* may be based on the t statistics

$$t = (\bar{x}_1 - \bar{x}_2) / [s_p^2 (1/n_1 + 1/n_2)]^{1/2} \quad (3.15)$$

where $\bar{x}_1$, $\bar{x}_2$ and n_1, n_2 refer to the means and sample sizes of the two groups, and s_p^2 is obtained by pooling the two sample variances s_1^2 and s_2^2:

$$s_p^2 = [(n_1 - 1)s_1^2 + (n_2 - 1)s_2^2] / (n_1 + n_2 - 2) \quad (3.16)$$

Step 3: *State the "decision rule."* The issue here is to determine whether the magnitude of the test statistic computed from sample data in Step 2 is sufficiently extreme (either too large or too small) to justify rejecting H_0. Two basic approaches are commonly used to formulate the decision rule. In the first, the analyst rejects H_0 if the probability of obtaining a value of the test statistic of a given or more extreme value is equal to or less than some small number α (referred to as the level of significance). Commonly used values for α are 0.10, 0.05, or 0.01. The second approach involves stating the decision rule in terms of critical values of the test statistic. Because critical values are a function of the level of significance, the two approaches are equivalent. Tabulated values for α and the corresponding critical values for commonly used probability distribution can be found in many statistics textbooks. (The decision rule is usually formulated prior to even stating the statistical hypotheses. Its location in the sequence of steps presented here is largely illustrative.)

Step 4: *Apply the decision rule.* If the probability of obtaining the computed or larger value of the test statistic is $\leq \alpha$, reject H_0 and conclude that H_1 is true. Alternatively, if the computed value of the test statistic is greater than the critical value for the stated level of significance, reject H_0 and conclude that H_1 is true.

A summary of test statistics for use in several other important hypothesis testing situations is as follows:

Testing the difference between an observed and a hypothesized mean

$$Z = (\bar{x} - \mu_0) / (\sigma / \sqrt{n}) \tag{3.17}$$

$$t = (\bar{x} - \mu_0) / (s / \sqrt{n}) \tag{3.18}$$

where $\bar{x}$ is the observed mean and μ_0 is the value of the hypothesized mean, such as a know population mean. The test statistic in this case is compared to values in the standard normal distribution or the *t*-distribution, depending upon the size of *n*.

Tests concerning population variances

$$\chi^2 = (n-1)s^2 / \sigma_0^2 \tag{3.19}$$

$$F = s_1^2 / s_2^2 \tag{3.20}$$

Equation 3.19 tests the difference between a hypothesized population variance σ_0^2 and a sample variance (s^2) using the chi-square distribution (χ^2). Equation 3.20 uses values from an *F* distribution to determine whether two populations have equal variances. Values of the chi-square and *F* distribution can be found in standard statistics textbooks.

SAMPLING

The *confidence level* is the probability that a measurement is within a set interval and can be defined as $(1 - \alpha)$, or (1 minus the probability of a type I error), or equivalently, the probability of not making a type I error. A type I error is defined as *the probability of deciding a constituent is present when it is actually absent*, often called a false positive. It is also necessary to define the *power* as $(1 - \beta)$, or (1 minus the probability of a type II error), or equivalently, the probability of not making a type II error. A type II error is defined as *the probability of not detecting a constituent when it actually is present*, often called a false negative. Finally, the *minimum detectable relative difference (MDRD)* is defined as the *relative increase over the background measurement that is detectable with a probability of* $(1 - \beta)$. Mathematically, this may be expressed as

$$\text{MDRD} = \frac{\mu_s - \mu_b}{\mu_b} \times 100 \qquad (3.21)$$

where μ_s and μ_b are the sample and background mean values, respectively. The coefficient of variation (CV) is a measure of the sample variance relative to the sample mean and may be calculated as

$$CV = \frac{s^2}{\bar{x}} \qquad (3.22)$$

Once values for CV, power, confidence level, and MDRD are calculated or established by protocol, the number of samples required to achieve that MDRD is tabulated in the Environmental Engineering section of the *FE Supplied-Reference Handbook*.

Example 3.18

MDRD

A set of samples from an abandoned industrial site detected a suspected groundwater contaminant at a mean concentration of 0.045 µg/L with $s^2 = 0.007$ µg/L. How many samples are required to be collected at a confidence level of 95% and power of 90% if the mean background concentration of the contaminant is 0.035 µg/L?

Solution

The minimum detectable relative difference (MDRD) is calculated from the mean concentration (C_{avg}) and the background concentration (C_{bg}) as follows:

$$\text{MDRD} = \frac{C_{avg} - C_{bg}}{C_{bg}} \times 100 = \frac{0.045 - 0.035}{0.035} \times 100 = 28.6\%$$

Also, the coefficient of variation (CV) may be calculated as

$$CV = \frac{s^2}{\bar{x}} = \frac{0.007}{0.045} = 15.6\%$$

Therefore, from the table for "Number of Samples Required in a One-Sided One-Sample *t*-Test" in the environmental engineering section of the *FE Supplied-Reference Handbook*, at MRDD = 30% and CV = 15% with confidence = 95% and power = 90%,

$$\Rightarrow n = 4$$

PROBLEMS

3.1 What is the probability of drawing a pair of aces in two cards when an ace has been drawn on the first card?
- a. 1/13
- b. 1/26
- c. 3/51
- d. 4/51

3.2 An auto manufacturer has three plants (A, B, C). Four out of 500 cars from Plant A must be recalled, 10 out of 800 from Plant B, and 10 out of 1000 from Plant C. Now a customer purchases a car from a dealer who gets 30% of his stock from Plant A, 40% from Plant B, and 30% from Plant C, and the car is recalled. What is the probability it was manufactured in Plant A?
- a. 0.0008
- b. 0.01
- c. 0.0125
- d. 0.2308

3.3 There are ten defectives per 1000 times of a product. What is the probability that there is one and only one defective in a random lot of 100?
- a. 99×0.01^{99}
- b. 0.01
- c. 0.5
- d. 0.99^{99}

3.4 The probability that both stages of a two-stage missile will function correctly is 0.95. The probability that the first stage will function correctly is 0.98. What is the probability that the second stage will function correctly given that the first one does?
- a. 0.99
- b. 0.98
- c. 0.97
- d. 0.95

3.5 A standard deck of 52 playing cards is thoroughly shuffled. The probability that the first four cards dealt from the deck will be the four aces is closest to:
- a. 2.0×10^{-1}
- b. 8.0×10^{-2}
- c. 4.0×10^{-4}
- d. 4.0×10^{-6}

3.6 In statistics, the standard deviation measures:
- a. a standard distance
- b. a normal distance
- c. central tendency
- d. dispersion

3.7 There are three bins containing integrated circuits (ICs). One bin has two premium ICs, one has two regular ICs, and one has one premium IC and one regular IC. An IC is picked at random. It is found to be a premium IC. What is the probability that the remaining IC in that bin is also a premium IC?
- a. $\frac{1}{5}$
- b. $\frac{1}{4}$
- c. $\frac{1}{3}$
- d. $\frac{2}{3}$

3.8 How many teams of four can be formed from 35 people?
- a. about 25,000
- b. about 2,000,000
- c. about 50,000
- d. about 200,000

3.9 A bin contains 50 bolts, 10 of which are defective. If a worker grabs 5 bolts from the bin in one grab, what is the probability that no more than 2 of the 5 are bad?
  a. about 0.5
  b. about 0.75
  c. about 0.90
  d. about 0.95

3.10 How many three-letter codes may be formed from the English alphabet if no repetitions are allowed?
  a. 26^3
  b. 26/3
  c. $26 \times 25 \times 24$
  d. $26^3/3$

3.11 A widget has three parts, A, B, and C, with probabilities of 0.1, 0.2, and 0.25, respectively, of being defective. What is the probability that exactly one of these parts is defective?
  a. 0.375
  b. 0.55
  c. 0.95
  d. 0.005

3.12 If three students work on a certain math problem, student A has a probability of success of 0.5; student B, 0.4; and student C, 0.3. If they work independently, what is the probability that no one works the problem successfully?
  a. 0.12
  b. 0.25
  c. 0.32
  d. 0.21

3.13 A sample of 50 light bulbs is drawn from a large collection in which each bulb is good with a probability of 0.9. What is the approximate probability of having less than 3 bad bulbs in the 50?
  a. 0.1
  b. 0.2
  c. 0.3
  d. 0.4

3.14 The number of different 3-digit numbers that can be formed from the digits 1, 2, 3, 7, 8, 9 without reusing the digits is:
  a. 10
  b. 20
  c. 30
  d. 40

3.15 The number of different ways that a party of seven councilmen can be seated in a row is:
  a. 1
  b. 560
  c. 2080
  d. 5040

3.16 A student must answer six out of eight questions on an exam. The number of different ways in which he can do the exam is:
  a. 8
  b. 18
  c. 28
  d. 48

3.17 Repeat Problem 3.16 if the first two questions are mandatory.
  a. 4
  b. 8
  c. 12
  d. 15

3.18 A group of five women wishes to form a subcommittee consisting of two of them. The number of possible ways to do so is:
a. 5
b. 10
c. 15
d. 20

3.19 An integer has to be chosen from numbers between 1 and 100 (both inclusive). The probability of choosing a number divisible by 9 (with a remainder of 0) is:
a. 0
b. 0.01
c. 0.11
d. 0.91

3.20 Four fair coins are tossed. The probability of either one head or two heads showing up is:
a. 1/8
b. 2/8
c. 4/8
d. 5/8

3.21 Two identical bags contain ten apples and five oranges each. The probability of selecting an apple from the first bag and an orange from the second bag is:
a. 1/9
b. 2/9
c. 3/9
d. 4/9

3.22 A bag contains 5 red, 10 orange, 15 green, 20 violet, and 25 black cards. The probability that you will get a black card or a red card if you remove a card from the bag is:
a. 5/85
b. 15/85
c. 25/85
d. 35/85

3.23 Two bags each contain two orange balls, five white balls, and three red balls. The probability of selecting an orange ball from the first bag or a white ball from the other bag is:
a. 0
b. 0.2
c. 0.6
d. 1.0

3.24 A bag contains 100 balls numbered 1 to 100. One ball is drawn from the bag. What is the probability that the number on the ball will be even or greater than 72?
a. 0.64
b. 0.50
c. 0.28
d. 0.14

3.25 A circuit has two switches connected in series. For a signal to pass through, both switches must be closed. The probability that the first switch is closed is 0.95, and the probability that a signal passes through is 0.90. The probability that the second switch is closed is:
a. 0.8545
b. 0.9000
c. 0.9474
d. 0.9871

3.26 A coin is weighted so that heads is twice as likely to appear as tails. The probability that two heads occur in four tosses is:
a. 0.15
b. 0.35
c. 0.45
d. 0.75

3.27 It is given that 20% of all employees leave their jobs after one year. A company hired seven new employees. The probability that nobody will leave the company after one year is:
   a. 0.1335   c. 0.3815
   b. 0.2315   d. 0.6510

3.28 If four fair coins are tossed simultaneously, the probability that at least one head appears is:
   a. 0.1335   c. 0.7815
   b. 0.5635   d. 0.9375

3.29 For unit normal distribution, the probability that $(x > 3)$ is:
   a. 0.0013   c. 0.1807
   b. 0.0178   d. 0.5402

3.30 Scores in a particular game have a normal distribution with a mean of 30 and a standard deviation of 5. Contestants must score more than 26 to qualify for the finals. The probability of being disqualified in the qualifying round is:
   a. 0.121   c. 0.304
   b. 0.212   d. 0.540

3.31 The radial distance to the impact points for shells fired by a cannon is approximated by a normal Gaussian random variable with a mean of 2000 m and standard deviation of 40 m. When a target is located at 1980 m distance, the probability that shells will fall within ± 68 m of the target is:
   a. 0.2341   c. 0.5847
   b. 0.3248   d. 0.8710

3.32 The chance of a car being stolen from a residential area is 1 in 120. In one area there are five cars parked in front of the houses. The probability that none will be stolen is:
   a. 0.0131   c. 0.5847
   b. 0.3248   d. 0.9590

3.33 The standard deviation of the sequence 3, 4, 4, 5, 8, 8, 8, 10, 11, 15, 18, 20 is:
   a. 5.36   c. 15.62
   b. 9.35   d. 28.75

3.34 Weighted arithmetic mean of the following 50 data points is:

Frequency	3	8	18	12	9
Score	1.5	2.5	3.5	4.5	5.5

   a. 1.56   c. 5.62
   b. 3.82   d. 8.75

SOLUTIONS

3.1 c. This is a conditional probability problem. Let B be "draw an ace," and let A be "draw a second ace": $P\{B\} = 4/52$ (1/13) and $P\{A\} = 3/51$. Then $P\{A|B\} = P\{A\} \times P\{B\}/P\{B\} = 3/51$.

3.2 d. This is a Bayes' theorem problem application because partitions are involved. The event E is a recall, with E_1 = Plant A, E_2 = Plant B, and E_3 = Plant C. The conditional probabilities of a recall from Plants E_1, E_2, and E_3 are

$$P(E \mid E_1) = 4/500 = 0.008$$
$$P(E \mid E_2) = 10/800 = 0.0125$$
$$P(E \mid E_3) = 10/1000 = 0.01$$

The probabilities that the dealer had a car from E_1, E_2, or E_3 are $P(E_1) = 0.3$, $P(E_2) = 0.4$, and $P(E_3) = 0.3$. Now applying Bayes' formula gives the probability that the recall was built in Plant A (E_1) as

$$P\{E_1 | \text{recall}\} = \frac{P\{E_1\} \times P\{E|E_1\}}{P\{E_1\} \times P\{E \mid E_1\} + P\{E_2\} \times P\{E|E_2\} + P\{E_3\} \times P\{E \mid E_3\}}$$

$$= \frac{0.3 \times 0.008}{0.3 \times 0.008 + 0.4 \times 0.0125 + 0.3 \times 0.01} = 0.2308$$

3.3 d. The problem involves binomial probability. The probability that one item, selected at random, is defective is

$$p_{\text{defective}} = \frac{10}{1000} = 0.01$$

and the probability that one item is good (not defective) is

$$p_{\text{good}} = 1 - p_{\text{defective}} = 0.99$$

The probability that exactly one defective item will be found in a random sample of 100 items is given by the binomial $b(1, 100, 0.01)$, in which

$$b(1, 100, 0.01) = C(100, 1)(0.01)^1 (0.99)^{99}$$

(No. defective, size sample, $p_{\text{defective}}$, No. good, p_{good})

$C(n, r) = \binom{n}{r} = \dfrac{n!}{(n-r)! r!}$ is the number of combinations of n objects taken r at a time without concern for the order of arrangement. $C(100, 1) = \dfrac{100!}{99! 1!} = 100$, so $b(1, 100, 0.01) = 100(0.01)(0.99)^{99} = 0.99^{99} = 0.3697$.

3.4 c. Here, $P(S_1) = 0.98$ and $P(S_2 \cap S_1) = 0.95$ are given. Hence the conditional probability $P(S_2|S_1)$ is

$$P(S_2|S_1) = \frac{P(S_2 \cap S_1)}{P(S_1)} = \frac{0.95}{0.98} = 0.97$$

3.5 d. The probability of drawing an ace on the first card is 4/52. The probability that the second card is an ace is 3/51. The probability that the third card is an ace is 2/50, and probability for the fourth ace is 1/49. The probability that the first four cards will all be aces is

$$P = \frac{4}{52} \cdot \frac{3}{51} \cdot \frac{2}{50} \cdot \frac{1}{49} = 0.00\ 003\ 7 = 3.7 \times 10^{-6}$$

3.6 d.

3.7 d. Since the first IC that is picked is a premium IC, it was drawn from either bin 1 or bin 3. From the distribution of premium ICs, the probability that the premium IC came from bin 1 is $\frac{2}{3}$, and from bin 3 is $\frac{1}{3}$.

In bin 1, the probability that the remaining IC is a premium IC is 1; in bin 3, the probability is 0. Thus, the probability that the remaining IC is a premium IC is

$$\frac{2}{3}(1) + \frac{1}{3}(0) = \frac{2}{3}$$

An alternative solution using Bayes' theorem for conditional probability is

$$P(\text{bin 1}|\text{drew premium}) = \frac{P(\text{bin 1 and premium})}{P(\text{premium})}$$

$$= \frac{P(\text{premium}|\text{bin 1}) \cdot P(\text{bin 1})}{\sum_{i=1}^{3} P(\text{premium}|\text{bin 1}) P(\text{bin 1})}$$

$$= \frac{1\left(\frac{1}{3}\right)}{1\left(\frac{1}{3}\right) + 0\left(\frac{1}{3}\right) + \frac{1}{2}\left(\frac{1}{3}\right)} = \frac{2}{3}$$

3.8 c. The answer is the binomial coefficient

$$\binom{35}{4} = \frac{35 \cdot 34 \cdot 33 \cdot 32}{4 \cdot 3 \cdot 2 \cdot 1} = 35 \cdot 34 \cdot 11 \cdot 4 = 52,360$$

3.9 d. The total number of choices of 5 is $\binom{50}{5}$. Of these, $\binom{40}{5}$ have no bad bolts, $\binom{40}{4} \times \binom{10}{1}$ have one bad bolt, and $\binom{40}{3}\binom{10}{2}$ have two bad bolts. Thus,

$$\frac{\binom{40}{5} + \binom{40}{4}\binom{10}{1} + \binom{40}{3}\binom{10}{2}}{\binom{50}{5}}$$

$$= \frac{\frac{40 \cdot 39 \cdot 38 \cdot 37 \cdot 36}{5 \cdot 4 \cdot 3 \cdot 2} + \frac{40 \cdot 39 \cdot 38 \cdot 37}{4 \cdot 3 \cdot 2} \cdot 10 + \frac{40 \cdot 39 \cdot 38}{3 \cdot 2} \cdot \frac{10 \cdot 9}{2}}{\frac{50 \cdot 49 \cdot 48 \cdot 47 \cdot 46}{5 \cdot 4 \cdot 3 \cdot 2}}$$

$$= \frac{658{,}008 + 913{,}900 + 444{,}600}{2{,}118{,}760} = 0.9517$$

3.10 c. There are 26 choices for the first letter; 25 remain for the second, and 24 for the third.

3.11 a. The probability that only A is defective is

$$0.1 \times (1 - 0.2) \times (1 - 0.25) = 0.06$$

The probability that only B is defective is

$$(1 - 0.1) \times (0.2) \times (1 - 0.25) = 0.135$$

The probability that only C is defective is

$$(1 - 0.1) \times (1 - 0.2) \times (0.25) = 0.18$$

Now add to find the final probability, which is

$$0.06 + 0.135 + 0.18 = 0.375$$

3.12 d. Simply multiply the complementary probabilities $(1 - 0.5) \times (1 - 0.4) \times (1 - 0.3) = 0.21$.

3.13 a. Apply the binomial distribution. The probability of 0 bad is $(0.9)^{50}$; of 1 bad, $\binom{50}{1}(0.1)(0.9)^{49}$; and of 2 bad, $\binom{50}{1}(0.1)^2(0.9)^{48}$. Adding these, $(0.9)^{48}\,[(0.9)^2 + 5.0(0.9) + 1225(0.1)^2] = 0.112$.

3.14 b. This is the permutation of arranging 3 objects out of 6:

$$P(6,3) = \frac{6!}{(6-3)!} = 20$$

3.15 d. This is the permutation of arranging 7 persons out of 7:

$$P(7,7) = \frac{7!}{(7-7)!} = 5040$$

3.16 c. This is the selection (or combination) of 6 out of 8:

$$C(8,6) = \frac{8!}{(8-6)6!} = 28$$

3.17 d. Since two questions are mandatory, only four questions have to be selected out of six.

Then, $C(6,4) = \frac{6!}{(6-4)!4!} = 15$.

3.18 b. This is the selection (or combination) of 2 out of 5:

$$C(5,2) = \frac{5!}{3!2!} = 10$$

3.19 c. Since there are 11 integers that are exactly divisible by 9, probability = 11/100 = 0.11.

3.20 d. $P(1 \text{ head}) = \frac{C(4,1)}{2^4} = \frac{4}{16}$; $P(2 \text{ heads}) = \frac{C(4,2)}{2^4} = \frac{6}{16}$

Since P(1 head AND 2 heads) is 0, P(1 head or 2 heads) = (4/16) + (6/16) = 5/8.

3.21 b. Let event A = (Apple from a bag) and event B = (Orange from the other bag).

Then, P(A) = 10/15 and P(B) = 5/15.

Since the events are independent, P(A AND B) = P(A) P(B) = (10/15)(5/15) = 2/9.

3.22 d. P(Black OR Red) = P(Black) + P(Red) − P(Black AND Red) = (5/85) + (30/85) − 0 = 35/85.

3.23 c. Let event A = (Orange from a bag) and event B = (White from the other bag).

P(A) = 2/10 = 0.2 and P(B) = 5/10 = 0.5

P(A OR B) = P(A) + P(B) − P(A AND B) = 0.2 + 0.5 − (0.2)(0.5) = 0.6

Note: P(A AND B) = P(A) P(B) as events A and B are independent.

3.24 a. Let event A = (number is even) and event B = (number > 72).

P(A) = 50/100 = 0.5 and P(B) = 28/100 = 0.28

Event (A and B) = (number is odd and > 72); P(A AND B) = 14/100 = 0.14

P(A OR B) = P(A) + P(B) − P(A AND B) = 0.50 + 0.28 − 0.14 = 0.64

3.25 c. P(both closed) = P(1 is closed) P(2 is closed)
0.90 = (0.95) P(2 is closed); then, P(2 is closed) = 0.90/0.95 = 0.9474

3.26 a. Let p = P(head on the first toss); then, P(tail on the first toss) = $1 - p$
But, $p = 2(1 - p)$; solving, $p = 0.667$.
P(two heads in four tosses) = $C(4,2)(0.667)^2(1 - 0.667)^2 = 0.1481$

3.27 a. P(leaving the job) = 0.25; P(none will leave the job) = $C(7,0)(0.25)^0 (1 - 0.25)^7 = 0.1335$

3.28 d. P(head) = $p = 0.5$; P(at least one head) = 1 − P(no head) = $1 - C(4,0)(0.5)^0(1 - 0.5)^4 = 0.9375$

3.29 a. Using the normal distribution table, P(X > 3) = 1 − F(3) = 0.0013.

3.30 b. $P\{X \le 26\} = F(26) = F\left(\dfrac{26-30}{5}\right) = F(-0.8) = 1 - F(0.8) = 1 - 0.7881 = 0.2119$

3.31 d. $P\{1980 - 68 < x \le 1980 + 68\} = F(2048) - F(1912)$
$= F\left(\dfrac{2048 - 2000}{40}\right) - F\left(\dfrac{1912 - 2000}{40}\right) = F(1.20) - F(-2.2)$
$= 0.8849 - \{1 - 0.9861\} = 0.8710$

3.32 d. $C(5,0)(1/120)^0(119/120)^5 = 0.9590$

3.33 a. mean $= \overline{X} = \dfrac{\sum x}{n} = \dfrac{114}{12} = 9.5$

variance, $\sigma^2 = (1/12)\left[(3-9.5)^2 + (4-9.5)^2 + \ldots\right] = 28.75$

standard deviation $\sigma = 5.36$

3.34 b. $\dfrac{3(1.5) + 8(2.5) + \ldots + 9(5.5)}{3 + 8 + 18 + \ldots + 9} = 3.82$

CHAPTER 4

Engineering Economics

OUTLINE

CASH FLOW 126

TIME VALUE OF MONEY 128
Simple Interest

EQUIVALENCE 128

COMPOUND INTEREST 129
Symbols and Functional Notation ■ Single-Payment Formulas ■ Uniform Payment Series Formulas ■ Uniform Gradient ■ Continuous Compounding

NOMINAL AND EFFECTIVE INTEREST 136
Non-Annual Compounding ■ Continuous Compounding

SOLVING ENGINEERING ECONOMICS PROBLEMS 138
Criteria

PRESENT WORTH 138
Appropriate Problems ■ Infinite Life and Capitalized Cost

FUTURE WORTH OR VALUE 141

ANNUAL COST 141
Criteria ■ Application of Annual Cost Analysis

RATE OF RETURN ANALYSIS 143
Two Alternatives ■ Three or More Alternatives

BENEFIT-COST ANALYSIS 148

BREAKEVEN ANALYSIS 149

OPTIMIZATION 149
Minima-Maxima ■ Economic Problem—Best Alternative ■ Economic Order Quantity

VALUATION AND DEPRECIATION 151
Notation ■ Straight-Line Depreciation ■ Sum-of-Years'-Digits Depreciation ■ Declining-Balance Depreciation ■ Sinking-Fund Depreciation ■ Modified Accelerated Cost Recovery System Depreciation

TAX CONSEQUENCES 156

INFLATION 157
    Effect of Inflation on Rate of Return

RISK 159
    Expected Value

REFERENCE 160

PROBLEMS 161

SOLUTIONS 170

This is a review of the field known variously as *engineering economics, engineering economy*, or *engineering economic analysis*. Since engineering economics is straightforward and logical, even people who have not had a formal course should be able to gain sufficient knowledge from this chapter to successfully solve most engineering economics problems.

There are 33 example problems scattered throughout the chapter. These examples are an integral part of the review and should be examined as you come to them.

The field of engineering economics uses mathematical and economics techniques to systematically analyze situations which pose alternative courses of action. The initial step in engineering economics problems is to resolve a situation, or each alternative in a given situation, into its favorable and unfavorable consequences or factors. These are then measured in some common unit—usually money. Factors which cannot readily be equated to money are called intangible or irreducible factors. Such factors are considered in conjunction with the monetary analysis when making the final decision on proposed courses of action.

CASH FLOW

A cash flow table shows the "money consequences" of a situation and its timing. For example, a simple problem might be to list the year-by-year consequences of purchasing and owning a used car:

Year	Cash Flow	
Beginning of first year 0	$4500	Car purchased "now" for $4500 cash. The minus sign indicates a disbursement.
End of year 1	−350	
End of year 2	−350	Maintenance costs are $350 per year.
End of year 3	−350	
End of year 4	−350 / +2000	This car is sold at the end of the fourth year for $2000. The plus sign represents the receipt of money.

This same cash flow may be represented graphically, as shown in Figure 4.1. The upward arrow represents a receipt of money, and the downward arrows represent disbursements. The horizontal axis represents the passage of time.

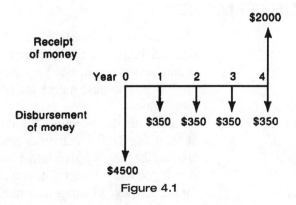

Figure 4.1

Example 4.1

In January 1993 a firm purchased a used typewriter for $500. Repairs cost nothing in 1993 or 1994. Repairs are $85 in 1995, $130 in 1996, and $140 in 1997. The machine is sold in 1997 for $300. Complete the cash flow table.

Solution

Unless otherwise stated, the customary assumption is a beginning-of-year purchase, followed by end-of-year receipts or disbursements, and an end-of-year resale or salvage value. Thus the typewriter repairs and the typewriter sale are assumed to occur at the end of the year. Letting a minus sign represent a disbursement of money and a plus sign a receipt of money, we are able to set up the cash flow table:

Year	Cash Flow
Beginning of 1993	–$500
End of 1993	0
End of 1994	0
End of 1995	–85
End of 1996	–130
End of 1997	+160

Notice that at the end of 1997 the cash flow table shows +160, which is the net sum of –140 and +300. If we define year 0 as the beginning of 1993, the cash flow table becomes

Year	Cash Flow
0	–$500
1	0
2	0
3	–85
4	–130
5	+160

From this cash flow table, the definitions of year 0 and year 1 become clear. Year 0 is defined as the *beginning* of year 1. Year 1 is the *end* of year 1, and so forth.

TIME VALUE OF MONEY

When the money consequences of an alternative occur in a short period of time—say, less than one year—we might simply add up the various sums of money and obtain the net result. But we cannot treat money this way over longer periods of time. This is because money today does not have the same value as money at some future time.

Consider this question: Which would you prefer, $100 today or the assurance of receiving $100 a year from now? Clearly, you would prefer the $100 today. If you had the money today, rather than a year from now, you could use it for the year. And if you had no use for it, you could lend it to someone who would pay interest for the privilege of using your money for the year.

Simple Interest

Simple interest is interest that is computed on the original sum. Thus if one were to lend a present sum P to someone at a simple annual interest rate i, the future amount F due at the end of n years would be

$$F = P + Pin.$$

Example 4.2

How much will you receive back from a $500 loan to a friend for three years at 10% simple annual interest?

Solution

$$F = P + Pin = 500 + 500 \times 0.10 \times 3 = \$650$$

In Example 4.2 one observes that the amount owed, based on 10% simple interest at the end of one year, is $500 + 500 \times 0.10 \times 1 = \$550$. But at simple interest there is no interest charged on the $50 interest, even though it is not paid until the end of the third year. Thus simple interest is not realistic and is seldom used. *Compound interest* charges interest on the principal owed plus the interest earned to date. This produces a charge of interest on interest, or compound interest. Engineering economics uses compound interest computations.

EQUIVALENCE

In the preceding section we saw that money at different points in time (for example, $100 today or $100 one year hence) may be equal in the sense that they both are $100, but $100 a year hence is *not* an acceptable substitute for $100 today. When we have acceptable substitutes, we say they are *equivalent* to each other. Thus at 8% interest, $108 a year hence is equivalent to $100 today.

Example 4.3

At a 10% per year (compound) interest rate, $500 now is equivalent to how much three years hence?

Solution

A value of $500 now will increase by 10% in each of the three years.

$$\text{Now} = \$500.00$$
$$\text{End of 1st year} = 500 + 10\%(500) = 550.00$$
$$\text{End of 2nd year} = 550 + 10\%(550) = 605.00$$
$$\text{End of 3rd year} = 605 + 10\%(605) = 665.50$$

Thus $500 now is *equivalent* to $665.50 at the end of three years. Note that interest is charged each year on the original $500 plus the unpaid interest. This compound interest computation gives an answer that is $15.50 higher than the simple-interest computation in Example 4.2.

Equivalence is an essential factor in engineering economics. Suppose we wish to select the better of two alternatives. First, we must compute their cash flows. For example,

	Alternative	
Year	A	B
0	−$2000	−$2800
1	+800	+1100
2	+800	+1100
3	+800	+1100

The larger investment in alternative B results in larger subsequent benefits, but we have no direct way of knowing whether it is better than alternative A. So we do not know which to select. To make a decision, we must resolve the alternatives into *equivalent* sums so that they may be compared accurately.

COMPOUND INTEREST

To facilitate equivalence computations, a series of compound interest factors will be derived here, and their use will be illustrated in examples.

Symbols and Functional Notation

i = effective interest rate per interest period. In equations, the interest rate is stated as a decimal (that is, 8% interest is 0.08).

n = number of interest periods. Usually the interest period is one year, but it could be something else.

P = a present sum of money.

F = a future sum of money. The future sum F is an amount n interest periods from the present that is equivalent to P at interest rate i.

A = an end-of-period cash receipt or disbursement in a uniform series continuing for n periods. The entire series is equivalent to P or F at interest rate i.

G = uniform period-by-period increase in cash flows; the uniform gradient.

r = nominal annual interest rate.

Table 4.1 Periodic compounding: functional notation and formulas

Factor	Given	To Find	Functional Notation	Formula
Single payment				
Compound amount factor	P	F	$(F/P, i\%, n)$	$F = P(1+i)^n$
Present worth factor	F	P	$(P/F, i\%, n)$	$P = F(1+i)^{-n}$
Uniform payment series				
Sinking fund factor	F	A	$(A/F, i\%, n)$	$A = F\left[\dfrac{i}{(1+i)^n - 1}\right]$
Capital recovery factor	P	A	$(A/P, i\%, n)$	$A = P\left[\dfrac{i(1+i)^n}{(1+i)^n - 1}\right]$
Compound amount factor	A	F	$(F/A, i\%, n)$	$F = A\left[\dfrac{(1+i)^n - 1}{1}\right]$
Present worth factor	A	P	$(P/A, i\%, n)$	$P = A\left[\dfrac{(1+i)^n - 1}{i(1+i)^n}\right]$
Uniform gradient				
Gradient present worth	G	P	$(P/G, i\%, n)$	$P = G\left[\dfrac{(1+i)^n - 1}{i^2(1+i)^n} - \dfrac{n}{i(1+i)^n}\right]$
Gradient future worth	G	F	$(F/G, i\%, n)$	$F = G\left[\dfrac{(1+i)^n - 1}{i^2} - \dfrac{n}{1}\right]$
Gradient uniform series	G	A	$(A/G, i\%, n)$	$A = G\left[\dfrac{1}{i} - \dfrac{n}{(1+i)^n - 1}\right]$

From Table 4.1 we can see that the functional notation scheme is based on writing (to find/given, i, n). Thus, if we wished to find the future sum F, given a uniform series of receipts A, the proper compound interest factor to use would be $(F/A, i, n)$.

Single-Payment Formulas

Suppose a present sum of money P is invested for one year at interest rate i. At the end of the year, the initial investment P is received together with interest equal to Pi, or a total amount $P + Pi$. Factoring P, the sum at the end of one year is $P(1+i)$. If the investment is allowed to remain for subsequent years, the progression is as follows:

Amount at Beginning of the Period	+	Interest for the Period	=	Amount at End of the Period
1st year, P	+	Pi	=	$P(1+i)$
2nd year, $P(1+i)$	+	$Pi(1+i)$	=	$P(1+i)^2$
3rd year, $P(1+i)2$	+	$Pi(1+i)^2$	=	$P(1+i)^3$
nth year, $P(1+i)^{n-1}$	+	$Pi(1+i)^{n-1}$	=	$P(1+i)^n$

The present sum P increases in n periods to $P(1 + i)^n$. This gives a relation between a present sum P and its equivalent future sum F:

$$\text{Future sum} = (\text{Present sum})(1 + i)^n$$
$$F = P(1 + i)^n$$

This is the *single-payment compound amount formula*. In functional notation it is written

$$F = P(F/P, i, n).$$

The relationship may be rewritten as

$$\text{Present sum} = (\text{Future sum})(1 + i)^{-n}$$
$$P = F(1 + i)^{-n}.$$

This is the *single-payment present worth formula*. It is written

$$P = F(P/F, i, n).$$

Example 4.4

At a 10% per year interest rate, $500 now is *equivalent* to how much three years hence?

Solution

This problem was solved in Example 4.3. Now it can be solved using a single-payment formula. $P = \$500$, $n = 3$ years, $i = 10\%$, and $F =$ unknown:

$$F = P(1 + i)^n = 500(1 + 0.10)^3 = \$665.50.$$

This problem also may be solved using a compound interest table:

$$F = P(F/P, i, n) = 500(F/P, 10\%, 3).$$

From the 10% compound interest table, read $(F/P, 10\%, 3) = 1.331$.

$$F = 500(F/P, 10\%, 3) = 500(1.331) = \$665.50$$

Example 4.5

To raise money for a new business, a man asks you to lend him some money. He offers to pay you $3000 at the end of four years. How much should you give him now if you want 12% interest per year?

Solution
$P =$ unknown, $F = \$3000$, $n = 4$ years, and $i = 12\%$:

$$P = F(1 + i)^{-n} = 3000(1 + 0.12)^{-4} = \$1906.55$$

Alternative computation using a compound interest table:

$$P = F(P/F, i, n) = 3000(P/F, 12\%, 4) = 3000(0.6355) = \$1906.50$$

Note that the solution based on the compound interest table is slightly different from the exact solution using a hand-held calculator. In engineering economics the compound interest tables are always considered to be sufficiently accurate.

Uniform Payment Series Formulas

Consider the situation shown in Figure 4.2. Using the single-payment compound amount factor, we can write an equation for F in terms of A:

$$F = A + A(1 + i) + A(1 + i)^2 \qquad (4.1)$$

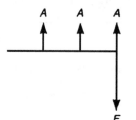

A = End-of-period cash receipt or disbursement in a uniform series continuing for n periods

F = A future sum of money

Figure 4.2

In this situation, with $n = 3$, Equation (4.1) may be written in a more general form:

$$F = A + A(1 + i) + A(1 + i)^{n-1} \qquad (4.2)$$

Multiply Eq. (4.2) by $(1 + i)$ $\quad (1 + i)F = A(1 + i) + A(1 + i)^{n-1} + A(1 + i)^n \qquad (4.3)$

Subtract Eq. (4.2): $\quad -F = A + A(1 + i) + A(1 + i)^{n-1}$

$$iF = -A + A(1 + i)^n$$

This produces the *uniform series compound amount formula*:

$$F = A\left(\frac{(1+i)^n - 1}{i}\right)$$

Solving this equation for A produces the *uniform series sinking fund formula*:

$$A = F\left(\frac{i}{(1+i)^n - 1}\right)$$

Since $F = P(1 + i)^n$, we can substitute this expression for F in the equation and obtain the *uniform series capital recovery formula*:

$$A = P\left(\frac{i(1+i)^n}{(1+i)^n - 1}\right)$$

Solving the equation for P produces the *uniform series present worth formula*:

$$P = A\left(\frac{(1+i)^n - 1}{i(1+i)^n}\right)$$

In functional notation, the uniform series factors are

Compound amount (*F/A*, i, n)
Sinking fund (*A/F*, i, n)
Capital recovery (*A/P*, i, n)
Present worth (*P/A*, i, n)

Example 4.6

If $100 is deposited at the end of each year in a savings account that pays 6% interest per year, how much will be in the account at the end of five years?

Solution

$A = \$100$, F = unknown, $n = 5$ years, and $i = 6\%$:

$$F = A(F/A, i, n) = 100(F/A, 6\%, 5) = 100(5.637) = \$563.70$$

Example 4.7

A fund established to produce a desired amount at the end of a given period, by means of a series of payments throughout the period, is called a *sinking fund*. A sinking fund is to be established to accumulate money to replace a $10,000 machine. If the machine is to be replaced at the end of 12 years, how much should be deposited in the sinking fund each year? Assume the fund earns 10% annual interest.

Solution

Annual sinking fund deposit $A = 10,000(A/F, 10\%, 12)$

$$= 10,000(0.0468) = \$468$$

Example 4.8

An individual is considering the purchase of a used automobile. The total price is $6200. With $1240 as a down payment, and the balance paid in 48 equal monthly payments with interest at 1% per month, compute the monthly payment. The payments are due at the end of each month.

Solution

The amount to be repaid by the 48 monthly payments is the cost of the automobile *minus* the $1240 down payment.

$P = \$4960$, A = unknown, $n = 48$ monthly payments, and $i = 1\%$ per month:

$$A = P(A/P, 1\%, 48) = 4960(0.0263) = \$130.45$$

Example 4.9

A couple sell their home. In addition to cash, they take a mortgage on the house. The mortgage will be paid off by monthly payments of $450 for 50 months. The couple decides to sell the mortgage to a local bank. The bank will buy the mortgage, but it requires a 1% per month interest rate on its investment. How much will the bank pay for the mortgage?

Solution

$A = \$450$, $n = 50$ months, $i = 1\%$ per month, and P = unknown:

$$P = A(P/A, i, n) = 450(P/A, 1\%, 50) = 450(39.196) = \$17,638.20$$

Uniform Gradient

At times one will encounter a situation where the cash flow series is not a constant amount A. Instead, it is an increasing series. The cash flow shown in Figure 4.3 may be resolved into two components (Fig. 4.4). We can compute the value of P^* as equal to P' plus P. And we already have the equation for P': $P' = A(P/A, i, n)$. The value for P in the right-hand diagram is

$$P = G\left[\frac{(1+i)^n - 1}{i^2(1+i)^n} - \frac{n}{i(1+i)^n}\right].$$

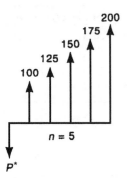

Figure 4.3

This is the *uniform gradient present worth formula*. In functional notation, the relationship is $P = G(P/G, i, n)$.

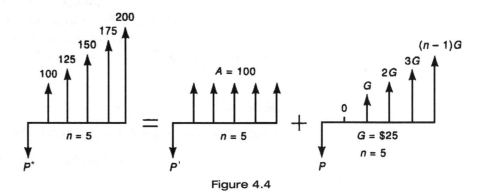

Figure 4.4

Example 4.10

The maintenance on a machine is expected to be $155 at the end of the first year, and it is expected to increase $35 each year for the following seven years (Exhibit 1). What sum of money should be set aside now to pay the maintenance for the eight-year period? Assume 6% interest.

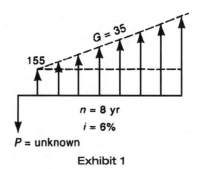

Exhibit 1

Solution

$$P = 155(P/A, 6\%, 8) + 35(P/G, 6\%, 8)$$
$$= 155(6.210) + 35(19.841) = \$1656.99$$

In the gradient series, if—instead of the present sum, P—an equivalent uniform series A is desired, the problem might appear as shown in Figure 4.5. The relationship between A' and G in the right-hand diagram is

$$A' = G\left[\frac{1}{i} - \frac{n}{(1+i)^n - 1}\right].$$

In functional notation, the uniform gradient (to) uniform series factor is: $A' = G(A/G, i, n)$.

The uniform gradient uniform series factor may be read from the compound interest tables directly, or computed as

$$(A/G, i, n) = \frac{1 - n(A/F, i, n)}{i}.$$

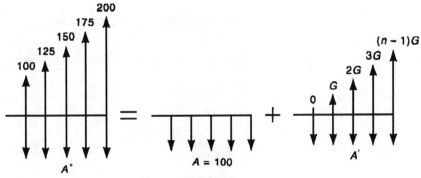

Figure 4.5

Note carefully the diagrams for the uniform gradient factors. The first term in the uniform gradient is zero and the last term is $(n-1)G$. But we use n in the equations and function notation. The derivations (not shown here) were done on this basis, and the uniform gradient compound interest tables are computed this way.

Example 4.11

For the situation in Example 4.10, we wish now to know the uniform annual maintenance cost. Compute an equivalent A for the maintenance costs.

Solution

Refer to Exhibit 2. The equivalent uniform annual maintenance cost is

$$A = 155 + 35(A/G, 6\%, 8) = 155 + 35(3.195) = \$266.83.$$

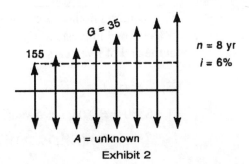

Exhibit 2

Standard compound interest tables give values for eight interest factors: two single payments, four uniform payment series, and two uniform gradients. The tables do *not* give the uniform gradient future worth factor, $(F/G, i, n)$. If it is needed, it may be computed from two tabulated factors:

$$(F/G, i, n) = (P/G, i, n)(F/P, i, n)$$

For example, if $i = 10\%$ and $n = 12$ years, then $(F/G, 10\%, 12) = (P/G, 10\%, 12)(F/P, 10\%, 12) = (29.901)(3.138) = 93.83$.

A second method of computing the uniform gradient future worth factor is

$$(F/G, i, n) = \frac{(F/A, i, n) - n}{i}.$$

Using this equation for $i = 10\%$ and $n = 12$ years, $(F/G, 10\%, 12) = [(F/A, 10\%, 12) - 12]/0.10 = (21.384 - 12)/0.10 = 93.84$.

Continuous Compounding

Table 4.2 Continuous compounding: functional notation and formulas

Factor	Given	To Find	Functional Notation	Formula
Single payment				
Compound amount factor	P	F	$(F/P, r\%, n)$	$F = P[e^{rn}]$
Present worth factor	F	P	$(P/F, r\%, n)$	$P = F[e^{-rn}]$
Uniform payment series				
Sinking fund factor	F	A	$(A/F, r\%, n)$	$A = F\left[\dfrac{e^r - 1}{e^{rn} - 1}\right]$
Capital recovery factor	P	A	$(A/P, r\%, n)$	$A = P\left[\dfrac{e^r - 1}{1 - e^{-rn}}\right]$
Compound amount factor	A	F	$(F/A, r\%, n)$	$A = P\left[\dfrac{e^{rn} - 1}{e^r - 1}\right]$
Present worth factor	A	P	$(P/A, r\%, n)$	$A = P\left[\dfrac{1 - e^{-rn}}{e^r - 1}\right]$

r = nominal annual interest rate, n = number of years.

Example 4.12

Five hundred dollars is deposited each year into a savings bank account that pays 5% nominal interest, compounded continuously. How much will be in the account at the end of five years?

Solution

$A = \$500$, $r = 0.05$, $n = 5$ years.

$$F = A(F/A, r\%, n) = A\left[\frac{e^{rn} - 1}{e^r - 1}\right] = 500\left[\frac{e^{0.05(5)} - 1}{e^{0.05} - 1}\right] = \$2769.84$$

NOMINAL AND EFFECTIVE INTEREST

Nominal interest is the annual interest rate without considering the effect of any compounding. *Effective interest* is the annual interest rate taking into account the effect of any compounding during the year.

Non-Annual Compounding

Frequently an interest rate is described as an annual rate, even though the interest period may be something other than one year. A bank may pay 1% interest on the amount in a savings account every three months. The *nominal* interest rate in this situation is 4 ×1% = 4%. But if you deposited $1000 in such an account, would you have 104%(1000) = $1040 in the account at the end of one year? The answer is no, you would have more. The amount in the account would increase as follows:

Amount in Account
Beginning of year: 1000.00
End of three months: 1000.00 + 1%(1000.00) = 1010.00
End of six months: 1010.00 + 1%(1010.00) = 1020.10
End of nine months: 1020.10 + 1%(1020.10) = 1030.30
End of one year: 1030.30 + 1%(1030.30) = 1040.60

At the end of one year, the interest of $40.60, divided by the original $1000, gives a rate of 4.06%. This is the *effective* interest rate.

$$\text{Effective interest rate per year: } i_{\text{eff}} = (1 + r/m)^m - 1$$

where r = nominal annual interest rate
m = number of compound periods per year
r/m = effective interest rate per period.

Example 4.13

A bank charges 1.5% interest per month on the unpaid balance for purchases made on its credit card. What nominal interest rate is it charging? What is the effective interest rate?

Solution

The nominal interest rate is simply the annual interest ignoring compounding, or 12(1.5%) = 18%.

$$\text{Effective interest rate} = (1 + 0.015)^{12} - 1 = 0.1956 = 19.56\%$$

Continuous Compounding

When m, the number of compound periods per year, becomes very large and approaches infinity, the duration of the interest period decreases from Δt to dt. For this condition of *continuous compounding*, the effective interest rate per year is

$$i_{\text{eff}} = e^r - 1$$

where r = nominal annual interest rate.

Example 4.14

If the bank in Example 4.13 changes its policy and charges 1.5% per month, compounded continuously, what nominal and what effective interest rate is it charging?

Solution

Nominal annual interest rate, $r = 12 \times 1.5\% = 18\%$

$$\text{Effective interest rate per year, } i_{\text{eff}} = e^{0.18} - 1 = 0.1972 = 19.72\%$$

SOLVING ENGINEERING ECONOMICS PROBLEMS

The techniques presented so far illustrate how to convert single amounts of money, and uniform or gradient series of money, into some equivalent sum at another point in time. These compound interest computations are an essential part of engineering economics problems.

The typical situation is that we have a number of alternatives; the question is, which alternative should we select? The customary method of solution is to express each alternative in some common form and then choose the best, taking both the monetary and intangible factors into account. In most computations an interest rate must be used. It is often called the minimum attractive rate of return (MARR), to indicate that this is the smallest interest rate, or rate of return, at which one is willing to invest money.

Criteria

Engineering economics problems inevitably fall into one of three categories:

1. *Fixed input.* The amount of money or other input resources is fixed.
   Example: A project engineer has a budget of $450,000 to overhaul a plant.

2. *Fixed output.* There is a fixed task or other output to be accomplished.
   Example: A mechanical contractor has been awarded a fixed-price contract to air-condition a building.

3. *Neither input nor output fixed.* This is the general situation, where neither the amount of money (or other inputs) nor the amount of benefits (or other outputs) is fixed.
   Example: A consulting engineering firm has more work available than it can handle. It is considering paying the staff to work evenings to increase the amount of design work it can perform.

There are five major methods of comparing alternatives: present worth, future worth, annual cost, rate of return, and benefit-cost analysis. These are presented in the sections that follow.

PRESENT WORTH

Present worth analysis converts all of the money consequences of an alternative into an equivalent present sum. The criteria are

Category	Present Worth Criterion
Fixed input	Maximize the present worth of benefits or other outputs
Fixed output	Minimize the present worth of costs or other inputs
Neither input nor output fixed	Maximize present worth of benefits minus present worth of costs, or maximize net present worth

Appropriate Problems

Present worth analysis is most frequently used to determine the present value of future money receipts and disbursements. We might want to know, for example, the present worth of an income-producing property, such as an oil well. This should provide an estimate of the price at which the property could be bought or sold.

An important restriction in the use of present worth calculation is that there must be a common analysis period for comparing alternatives. It would be incorrect, for example, to compare the present worth (PW) of cost of pump A, expected to last 6 years, with the PW of cost of pump B, expected to last 12 years (Fig. 4.6). In situations like this, the solution is either to use some other analysis technique (generally, the annual cost method is suitable in these situations) or to restructure the problem so that there is a common analysis period.

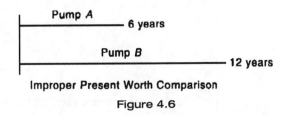

Improper Present Worth Comparison
Figure 4.6

In this example, a customary assumption would be that a pump is needed for 12 years and that pump A will be replaced by an identical pump A at the end of 6 years. This gives a 12-year common analysis period (Fig. 4.7). This approach is easy to use when the different lives of the alternatives have a practical least-common-multiple life. When this is not true (for example, the life of J equals 7 years and the life of K equals 11 years), some assumptions must be made to select a suitable common analysis period, or the present worth method should not be used.

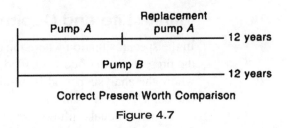

Correct Present Worth Comparison
Figure 4.7

Example 4.15

Machine X has an initial cost of $10,000, an annual maintenance cost of $500 per year, and no salvage value at the end of its 4-year useful life. Machine Y costs $20,000, and the first year there is no maintenance cost. Maintenance is $100 the second year, and it increases $100 per year thereafter. The machine has an anticipated $5000 salvage value at the end of its 12-year useful life. If the minimum attractive rate of return (MARR) is 8%, which machine should be selected?

Solution

The analysis period is not stated in the problem. Therefore, we select the least common multiple of the lives, or 12 years, as the analysis period.
Present worth of cost of 12 years of machine X:

$$PW = 10{,}000 + 10{,}000(P/F, 8\%, 4) + 10{,}000(P/F, 8\%, 8) + 500(P/A, 8\%, 12)$$
$$= 10{,}000 + 10{,}000(0.7350) + 10{,}000(0.5403) + 500(7.536) = \$26{,}521$$

Present worth of cost of 12 years of machine Y:

$$PW = 20{,}000 + 100(P/G, 8\%, 12) - 5000(P/F, 8\%, 12)$$
$$= 20{,}000 + 100(34.634) - 5000(0.3971) = \$21{,}478$$

Choose machine Y, with its smaller PW of cost.

Example 4.16

Two alternatives have the following cash flows:

	Alternative	
Year	A	B
0	−$2000	−$2800
1	+800	+1100
2	+800	+1100
3	+800	+1100

At a 4% interest rate, which alternative should be selected?

Solution

The net present worth of each alternative is computed:

Net present worth (NPW) = PW of benefit − PW of cost
$NPW_A = 800(P/A, 4\%, 3) - 2000 = 800(2.775) - 2000 = \$220.00$
$NPW_B = 1100(P/A, 4\%, 3) - 2800 = 1100(2.775) - 2800 = \$252.50$

To maximize NPW, choose alternative B.

Infinite Life and Capitalized Cost

In the special situation where the analysis period is infinite ($n = \infty$), an analysis of the present worth of cost is called *capitalized cost*. There are a few public projects where the analysis period is infinity. Other examples are permanent endowments and cemetery perpetual care.

When n equals infinity, a present sum P will accrue interest of Pi for every future interest period. For the principal sum P to continue undiminished (an essential requirement for n equal to infinity), the end-of-period sum A that can be disbursed is Pi (Fig. 4.8). When $n = \infty$, the fundamental relationship is

$$A = Pi.$$

Some form of this equation is used whenever there is a problem involving an infinite analysis period.

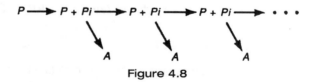

Figure 4.8

Example 4.17

In his will, a man wishes to establish a perpetual trust to provide for the maintenance of a small local park. If the annual maintenance is $7500 per year and the trust account can earn 5% interest, how much money must be set aside in the trust?

Solution

When $n = \infty$, $A = Pi$ or $P = A/i$. The capitalized cost is $P = A/i = \$7500/0.05 = \$150,000$.

FUTURE WORTH OR VALUE

In present worth analysis, the comparison is made in terms of the equivalent present costs and benefits. But the analysis need not be made in terms of the present—it can be made in terms of a past, present, or future time. Although the numerical calculations may look different, the decision is unaffected by the selected point in time. Often we do want to know what the future situation will be if we take some particular couse of action now. An analysis based on some future point in time is called *future worth analysis*.

Category	Future Worth Criterion
Fixed input	Maximize the future worth of benefits or other outputs
Fixed output	Minimize the future worth of costs or other inputs
Neither input nor output fixed	Maximize future worth of benefits minus future worth of costs, or maximize net future worth

Example 4.18

Two alternatives have the following cash flows:

	Alternative	
Year	A	B
0	−$2000	−$2800
1	+800	+1100
2	+800	+1100
3	+800	+1100

At a 4% interest rate, which alternative should be selected?

Solution

In Example 4.16, this problem was solved by present worth analysis at year 0. Here it will be solved by future worth analysis at the end of year 3.
Net future worth (NFW) = FW of benefits − FW of cost

$$NFW_A = 800(F/A, 4\%, 3) - 2000(F/P, 4\%, 3)$$
$$= 800(3.122) - 2000(1.125) = +\$247.60$$

$$NFW_B = 1100(F/A, 4\%, 3) - 2800(F/P, 4\%, 3)$$
$$= 1100(3.122) - 2800(1.125) = +\$284.20$$

To maximize NFW, choose alternative B.

ANNUAL COST

The annual cost method is more accurately described as the method of equivalent uniform annual cost (EUAC). Where the computation is of benefits, it is called the method of equivalent uniform annual benefits (EUAB).

Criteria

For each of the three possible categories of problems, there is an annual cost criterion for economic efficiency.

Category	Annual Cost Criterion
Fixed input	Maximize the equivalent uniform annual benefits (EUAB)
Fixed output	Minimize the equivalent uniform annual cost (EUAC)
Neither input nor output fixed	Maximize EUAB − EUAC

Application of Annual Cost Analysis

In the section on present worth, we pointed out that the present worth method requires a common analysis period for all alternatives. This restriction does not apply in all annual cost calculations, but it is important to understand the circumstances that justify comparing alternatives with different service lives.

Frequently, an analysis is done to provide for a more-or-less continuing requirement. For example, one might need to pump water from a well on a continuing basis. Regardless of whether each of two pumps has a useful service life of 6 years or 12 years, we would select the alternative whose annual cost is a minimum. And this still would be the case if the pumps' useful lives were the more troublesome 7 and 11 years. Thus, if we can assume a continuing need for an item, an annual cost comparison among alternatives of differing service lives is valid. The underlying assumption in these situations is that the shorter-lived alternative can be replaced with an identical item with identical costs, when it has reached the end of its useful life. This means that the EUAC of the initial alternative is equal to the EUAC for the continuing series of replacements.

On the other hand, if there is a specific requirement to pump water for 10 years, then each pump must be evaluated to see what costs will be incurred during the analysis period and what salvage value, if any, may be recovered at the end of the analysis period. The annual cost comparison needs to consider the actual circumstances of the situation.

Examination problems are often readily solved using the annual cost method. And the underlying "continuing requirement" is usually present, so an annual cost comparison of unequal-lived alternatives is an appropriate method of analysis.

Example 4.19

Consider the following alternatives:

	A	B
First cost	$5000	$10,000
Annual maintenance	500	200
End-of-useful-life salvage value	600	1000
Useful life	5 years	15 years

Based on an 8% interest rate, which alternative should be selected?

Solution

Assuming both alternatives perform the same task and there is a continuing requirement, the goal is to minimize EUAC.

Alternative A:

$$EUAC = 5000(A/P, 8\%, 5) + 500 - 600(A/F, 8\%, 5)$$
$$= 5000(0.2505) + 500 - 600(0.1705) = \$1650$$

Alternative B:

$$EUAC = 10{,}000(A/P, 8\%, 15) + 200 - 1000(A/F, 8\%, 15)$$
$$= 10{,}000(0.1168) + 200 - 1000(0.0368) = \$1331$$

To minimize EUAC, select alternative B.

RATE OF RETURN ANALYSIS

A typical situation is a cash flow representing the costs and benefits. The rate of return may be defined as the interest rate where PW of cost = PW of benefits, EUAC = EUAB, or PW of cost − PW of benefits = 0.

Example 4.20

Compute the rate of return for the investment represented by the following cash flow table.

Year:	0	1	2	3	4	5
Cash flow:	−$595	+250	+200	+150	+100	+50

Solution

This declining uniform gradient series may be separated into two cash flows (Exhibit 3) for which compound interest factors are available.

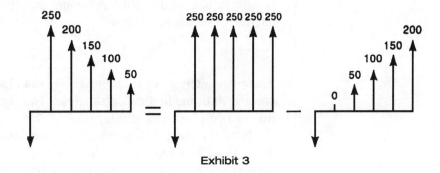

Exhibit 3

Note that the gradient series factors are based on an increasing gradient. Here the declining cash flow is solved by subtracting an increasing uniform gradient, as indicated in the figure.

PW of cost − PW of benefits = 0

$$595 - [250(P/A, i, 5) - 50(P/G, i, 5)] = 0$$

Try $i = 10\%$:
$$595 - [250(3.791) - 50(6.862)] = -9.65$$
Try $i = 12\%$:
$$595 - [250(3.605) - 50(6.397)] = +13.60$$

The rate of return is between 10% and 12%. It may be computed more accurately by linear interpolation:

$$\text{Rate of return} = 10\% + (2\%)\left(\frac{9.65 - 0}{13.60 + 9.65}\right) = 10.83\%$$

Two Alternatives

Compute the incremental rate of return on the cash flow representing the difference between the two alternatives. Since we want to look at increments of investment, the cash flow for the difference between the alternatives is computed by taking the higher initial-cost alternative minus the lower initial-cost alternative. If the incremental rate of return is greater than or equal to the predetermined minimum attractive rate of return (MARR), choose the higher-cost alternative; otherwise, choose the lower-cost alternative.

Example 4.21

Two alternatives have the following cash flows:

Year	Alternative A	Alternative B
0	−$2000	−$2800
1	+800	+1100
2	+800	+1100
3	+800	+1100

If 4% is considered the minimum attractive rate of return (MARR), which alternative should be selected?

Solution

These two alternatives were previously examined in Examples 4.16 and 4.18 by present worth and future worth analysis. This time, the alternatives will be resolved using a rate-of-return analysis.

Note that the problem statement specifies a 4% MARR, whereas Examples 4.16 and 4.18 referred to a 4% interest rate. These are really two different ways of saying the same thing: The minimum acceptable time value of money is 4%.

First, tabulate the cash flow that represents the increment of investment between the alternatives. This is done by taking the higher initial-cost alternative minus the lower initial-cost alternative:

	Alternative		Difference between Alternatives
Year	A	B	B − A
0	−$2000	−$2800	−$800
1	+800	+1100	+300
2	+800	+1100	+300
3	+800	+1100	+300

Then compute the rate of return on the increment of investment represented by the difference between the alternatives:

$$\text{PW of cost} = \text{PW of benefits}$$
$$800 = 300(P/A, i, 3)$$

$$(P/A, i, 3) = 800/300 = 2.67$$
$$i = 6.1\%$$

Since the incremental rate of return exceeds the 4% MARR, the increment of investment is desirable. Choose the higher-cost alternative B.

Before leaving this example, one should note something that relates to the rates of return on alternative A and on alternative B. These rates of return, if calculated, are

	Rate of Return
Alternative A	9.7%
Alternative B	8.7%

The correct answer to this problem has been shown to be alternative B, even though alternative A has a higher rate of return. The higher-cost alternative may be thought of as the lower-cost alternative plus the increment of investment between them. Viewed this way, the higher-cost alternative B is equal to the desirable lower-cost alternative A plus the difference between the alternatives.

The important conclusion is that computing the rate of return for each alternative does not provide the basis for choosing between alternatives. Instead, incremental analysis is required.

Example 4.22

Consider the following:

| | Alternative | |
Year	A	B
0	−$200.0	−$131.0
1	+77.6	+48.1
2	+77.6	+48.1
3	+77.6	+48.1

If the MARR is 10%, which alternative should be selected?

Solution

To examine the increment of investment between the alternatives, we will examine the higher initial-cost alternative minus the lower initial-cost alternative, or $A - B$.

| | Alternative | | Increment |
Year	A	B	A − B
0	−$200.0	−$131.0	−$69.0
1	+77.6	+48.1	+29.5
2	+77.6	+48.1	+29.5
3	+77.6	+48.1	+29.5

Solve for the incremental rate of return:

$$\text{PW of cost} = \text{PW of benefits}$$
$$69.0 = 29.5(P/A, i, 3)$$
$$(P/A, i, 3) = 69.0/29.5 = 2.339$$

From compound interest tables, the incremental rate of return is between 12% and 18%. This is a desirable increment of investment; hence we select the higher-initial-cost alternative A.

Three or More Alternatives

When there are three or more mutually exclusive alternatives, proceed with the same logic presented for two alternatives. The components of incremental analysis are listed below.

Step 1. Compute the rate of return for each alternative. Reject any alternative where the rate of return is less than the desired MARR. (This step is not essential, but helps to immediately identify unacceptable alternatives.)

Step 2. Rank the remaining alternatives in order of increasing initial cost.

Step 3. Examine the increment of investment between the two lowest-cost alternatives as described for the two-alternative problem. Select the better of the two alternatives and reject the other one.

Step 4. Take the preferred alternative from step 3. Consider the next higher initial-cost alternative and proceed with another two-alternative comparison.

Step 5. Continue until all alternatives have been examined and the best of the multiple alternatives has been identified.

Example 4.23

Consider the following:

	Alternative	
Year	A	B
0	−$200.0	−$131.0
1	+77.6	+48.1
2	+77.6	+48.1
3	+77.6	+48.1

If the MARR is 10%, which alternative, if any, should be selected?

Solution

One should carefully note that this is a *three-alternative* problem, where the alternatives are A, B, and *Do nothing*. In this solution we will skip step 1. Reorganize the problem by placing the alternatives in order of increasing initial cost:

		Alternative	
Year	Do Nothing	B	A
0	0	−$131.0	−$200.0
1	0	+48.1	+77.6
2	0	+48.1	+77.6
3	0	+48.1	+77.6

Examine the B − *Do nothing* increment of investment:

Year	B − Do Nothing
0	−$131.0 − 0 = −$131.0
1	+48.1 − 0 = +48.1
2	+48.1 − 0 = +48.1
3	+48.1 − 0 = +48.1

Solve for the incremental rate of return:

$$\text{PW of cost} = \text{PW of benefits}$$
$$131.0 = 48.1(P/A, i, 3)$$
$$(P/A, i, 3) = 131.0/48.1 = 2.723$$

From compound interest tables, the incremental rate of return is about 5%. Since the incremental rate of return is less than 10%, the *B – Do nothing* increment is not desirable. Reject alternative *B*.

Year	A – Do Nothing
0	–$200.0 – 0 = -$200.0
1	+77.6 – 0 = +77.6
2	+77.6 – 0 = +77.6
3	+77.6 – 0 = +77.6

Next, consider the increment of investment between the two remaining alternatives.

Solve for the incremental rate of return:

$$PW \text{ of cost} = PW \text{ of benefits}$$
$$200.0 = 77.6(P/A, i, 3)$$
$$(P/A, i, 3) = 200.0/77.6 = 2.577$$

The incremental rate of return is 8%, less than the desired 10%. Reject the increment and select the remaining alternative: *Do nothing*.

If you have not already done so, you should go back to Example 4.22 and see how the slightly changed wording of the problem has radically altered it. Example 4.22 required a choice between two undesirable alternatives. This example adds the *Do nothing* alternative, which is superior to *A* and *B*.

BENEFIT-COST ANALYSIS

Generally, in public works and governmental economic analyses, the dominant method of analysis is the *benefit-cost ratio*. It is simply the ratio of benefits divided by costs, taking into account the time value of money.

$$B/C = \frac{PW \text{ of benefits}}{PW \text{ of cost}} = \frac{\text{Equivalent uniform annual benefits}}{\text{Equivalent uniform annual cost}}$$

For a given interest rate, a B/C ratio ≥ 1 reflects an acceptable project. The B/C analysis method is parallel to rate-of-return analysis. The same kind of incremental analysis is required.

Example 4.24

Solve Example 4.22 by benefit-cost analysis.

Solution

	Alternative		Increment
Year	A	B	A – B
0	–$200.0	–$131.0	–$69.0
1	+77.6	+48.1	+29.5
2	+77.6	+48.1	+29.5
3	+77.6	+48.1	+29.5

The benefit-cost ratio for the $A - B$ increment is

$$B/C = \frac{\text{PW of benefits}}{\text{PW of cost}} = \frac{29.5(P/A, 10\%, 3)}{69.0} = \frac{73.37}{69.0} = 1.06.$$

Since the B/C ratio exceeds 1, the increment of investment is desirable. Select the higher-cost alternative A.

BREAKEVEN ANALYSIS

In business, "breakeven" is defined as the point where income just covers costs. In engineering economics, the breakeven point is defined as the point where two alternatives are equivalent.

Example 4.25

A city is considering a new $50,000 snowplow. The new machine will operate at a savings of $600 per day compared with the present equipment. Assume that the MARR is 12%, and the machine's life is 10 years with zero resale value at that time. How many days per year must the machine be used to justify the investment?

Solution

This breakeven problem may be readily solved by annual cost computations. We will set the equivalent uniform annual cost (EUAC) of the snowplow equal to its annual benefit and solve for the required annual utilization. Let X = breakeven point = days of operation per year.

$$\text{EUAC} = \text{EUAB}$$
$$50{,}000(A/P, 12\%, 10) = 600X$$
$$X = 50{,}000(0.1770)/600 = 14.8 \text{ days/year}$$

OPTIMIZATION

Optimization is the determination of the best or most favorable situation.

Minima-Maxima

In problems where the situation can be represented by a function, the customary approach is to set the first derivative of the function to zero and solve for the root(s) of this equation. If the second derivative is *positive*, the function is a minimum for the critical value; if it is *negative*, the function is a maximum.

Example 4.26

A consulting engineering firm estimates that their net profit is given by the equation

$$P(x) = -0.03x^3 + 36x + 500 \quad x \geq 0$$

where x = number of employees and P(x) = net profit. What is the optimal number of employees?

Solution

$$P'(x) = -0.09x^2 + 36 = 0 \qquad P''(x) = -0.18x$$
$$x^2 = 36/0.09 = 400$$
$$x = 20 \text{ employees.}$$
$$P''(20) = -0.18(20) = -3.6$$

Since $P''(20) < 0$, the net profit is maximized for 20 employees.

Economic Problem—Best Alternative

Since engineering economics problems seek to identify the best or most favorable situation, they are by definition optimization problems. Most use compound interest computations in their solution, but some do not. Consider the following example.

Example 4.27

A firm must decide which of three alternatives to adopt to expand its capacity. It wants a minimum annual profit of 20% of the initial cost of each increment of investment. Any money not invested in capacity expansion can be invested elsewhere for an annual yield of 20% of the initial cost.

Alternative	Initial Cost	Annual Profit	Profit Rate
A	$100,000	$30,000	30%
B	300,000	66,00	22
C	500,000	80,000	16

Which alternative should be selected?

Solution

Since alternative C fails to produce the 20% minimum annual profit, it is rejected. To decide between alternatives A and B, examine the profit rate for the $B - A$ increment.

Alternative	Initial Cost	Annual Profit	Incremental Cost	Incremental Profit	Incremental Profit Rate
A	$100,000	$30,000			
			$200,000	$36,000	18%
B	300,000	66,000			

The $B - A$ incremental profit rate is less than the minimum 20%, so alternative B should be rejected. Thus the best investment of $300,000, for example, would be alternative A (annual profit = $30,000) plus $200,000 invested elsewhere at 20% (annual profit = $40,000). This combination would yield a $70,000 annual profit, which is better than the alternative B profit of $66,000. Select A.

Economic Order Quantity

One special case of optimization occurs when an item is used continuously and is periodically purchased. Thus the inventory of the item fluctuates from zero (just prior to the receipt of the purchased quantity) to the purchased quantity (just after receipt). The simplest model for the economic order quantity (EOQ) is

$$EOQ = \sqrt{\frac{2BD}{E}}$$

where
- B = ordering cost, \$/order
- D = demand per period, units
- E = inventory holding cost, \$/unit/period
- EOC = economic order quantity, units.

Example 4.28

A company uses 8000 wheels per year in its manufacture of golf carts. The wheels cost \$15 each and are purchased from an outside supplier. The money invested in the inventory costs 10% per year, and the warehousing cost amounts to an additional 2% per year. It costs \$150 to process each purchase order. When an order is placed, how many wheels should be ordered?

Solution

$$EOQ = \sqrt{\frac{2 \times \$150 \times 8000}{(10\% + 2\%)(15.00)}} = 1155 \text{ wheels}$$

VALUATION AND DEPRECIATION

Depreciation of capital equipment is an important component of many after-tax economic analyses. For this reason, one must understand the fundamentals of depreciation accounting.

Notation

BV = book value
C = cost of the property (basis)
D_j = depreciation in year j
S_n = salvage value in year n

Depreciation is the systematic allocation of the cost of a capital asset over its useful life. *Book value* is the original cost of an asset, minus the accumulated depreciation of the asset.

$$BV = C - \Sigma(D_j)$$

In computing a schedule of depreciation charges, four items are considered.

1. Cost of the property, C (called the *basis* in tax law).

2. Type of property. Property is classified as either *tangible* (such as machinery) or *intangible* (such as a franchise or a copyright), and as either *real property* (real estate) or *personal property* (everything that is not real property).

3. Depreciable life in years, n.

4. Salvage value of the property at the end of its depreciable (useful) life, S_n.

Straight-Line Depreciation

The depreciation charge in any year is

$$D_j = \frac{C - S_n}{n}.$$

An alternative computation is

Depreciation charge in any year, $D_j = \dfrac{C - \text{Depreciation taken to beginning of year } j - S_n}{\text{Remaining useful life at beginning of year } j}.$

Sum-of-Years'-Digits Depreciation

Depreciation charge in any year, $D_j = \dfrac{\text{Remaining depreciable life at beginning of year}}{\text{Sum of years' digits for total useful life}} \times (C - S_n)$

Declining-Balance Depreciation

Double declining-balance depreciation charge in any year, $D_j = \dfrac{2C}{m}\left(1 - \dfrac{2}{n}\right)^{j-1}$

Total depreciation at the end of n years, $C = \left[1 - \left(1 - \dfrac{2}{n}\right)^n\right]$

Book value at the end of j years, $\text{BV}_j = C\left(1 - \dfrac{2}{n}\right)^j$

For 150% declining-balance depreciation, replace the 2 in the three equations above with 1.5.

Sinking-Fund Depreciation

Depreciation charge in any year, $D_j = (C - S_n)(A/F, i\%, n)(F/P, i\%, j - 1)$.

Modified Accelerated Cost Recovery System Depreciation

The modified accelerated cost recovery system (MACRS) depreciation method generally applies to property placed in service after 1986. To compute the MACRS depreciation for an item, one must know

1. Cost (basis) of the item.

2. Property class. All tangible property is classified in one of six classes (3, 5, 7, 10, 15, and 20 years), which is the life over which it is depreciated (see Table A.3). Residential real estate and nonresidential real estate are in two separate real property classes of 27.5 years and 39 years, respectively.

3. Depreciation computation.
   - Use double-declining-balance depreciation for 3-, 5-, 7-, and 10-year property classes with conversion to straight-line depreciation in the year that increases the deduction.
   - Use 150%-declining-balance depreciation for 15- and 20-year property classes with conversion to straight-line depreciation in the year that increases the deduction.
   - In MACRS, the salvage value is assumed to be zero.

Table 4.3 MACRS classes of depreciable property

Property Class	Personal Property (All Property Except Real Estate)
3-year property	Special handling devices for food and beverage manufacture Special tools for the manufacture of finished plastic products, fabricated metal products, and motor vehicles Property with an asset depreciation range (ADR) midpoint life of 4 years or less
5-year property	Automobiles* and trucks Aircraft (of nonair-transport companies) Equipment used in research and experimentation Computers Petroleum drilling equipment Property with an ADR midpoint life of more than 4 years and less than 10 years
7-year property	All other property not assigned to another class Office furniture, fixtures, and equipment Property with an ADR midpoint life of 10 years or more, and less than 16 years
10-year property	Assets used in petroleum refining and preparation of certain food products Vessels and water transportation equipment Property with an ADR midpoint life of 16 years or more, and less than 20 years
15-year property	Telephone distribution plants Municipal sewage treatment plants Property with an ADR midpoint life of 20 years or more, and less than 25 years
20-year property	Municipal sewers Property with an ADR midpoint life of 25 years or more

Property Class	Real Property (Real Estate)
27.5 years	Residential rental property (does not include hotels and motels)
39 years	Nonresidential real property

* The depreciation deduction for automobiles is limited to $2860 in the first tax year and is reduced in subsequent years.

Half-Year Convention

Except for real property, a half-year convention is used. Under this convention all property is considered to be placed in service in the middle of the tax year, and a half-year of depreciation is allowed in the first year. For each of the remaining years, one is allowed a full year of depreciation. If the property is disposed of prior to the end of the recovery period (property class life), a half-year of depreciation is allowed in that year. If the property is held for the entire recovery period, a half-year of depreciation is allowed for the year following the end of the recovery period (see Table 4.4). Owing to the half-year convention, a general form of the double-declining-balance computation must be used to compute the year-by-year depreciation.

DDB depreciation in any year, $D_j = \dfrac{2}{n}(C - \text{Depreciation in years prior to } j)$

Table 4.4 MACRS* depreciation for personal property half-year convention

If the Recovery Year Is	The Applicable Percentage for the Class of Property Is			
	3-Year Class	5-Year Class	7-Year Class	10-Year Class
1	33.33	20.00	14.29	10.00
2	44.45	32.00	24.49	18.00
3	14.81	19.20	17.49	14.40
4	7.41	11.52	12.49	11.52
5		11.52	8.93	9.22
6		5.76	8.92	7.37
7			8.93	6.55
8			4.46	6.55
9				6.56
10				6.55
11				3.28

* In the *Fundamentals of Engineering Supplied-Reference Handbook*, this table is called "Modified ACRS Factors."

† Use straight-line depreciation for the year marked and all subsequent years.

Example 4.29

A $5000 computer has an anticipated $500 salvage value at the end of its five-year depreciable life. Compute the depreciation schedule for the machinery by (a) sum-of-years'-digits depreciation and (b) MACRS depreciation. Do the MACRS computation by hand, and then compare the results with the values from Table 4.4.

Solution

(a) Sum-of-years'-digits depreciation:

$$D_j = \frac{n-j+1}{\frac{n}{2}(n+1)}(C - S_n)$$

$$D_1 = \frac{5-1+1}{\frac{5}{2}(5+1)}(5000 - 500) = \$1500$$

$$D_2 = \frac{5-2+1}{\frac{5}{2}(5+1)}(5000 - 500) = 1200$$

$$D_3 = \frac{5-3+1}{\frac{5}{2}(5+1)}(5000 - 500) = 900$$

$$D_4 = \frac{5-4+1}{\frac{5}{2}(5+1)}(5000 - 500) = 600$$

$$D_5 = \frac{5-5+1}{\frac{5}{2}(5+1)}(5000 - 500) = 300$$

$$\$4500$$

(b) MACRS depreciation. Double-declining-balance with conversion to straight-line. Five-year property class. Half-year convention. Salvage value S_n is assumed to be zero for MACRS. Using the general DDB computation,

Year
1 (half-year) $\quad D_1 = \dfrac{1}{2} \times \dfrac{2}{5}(5000 - 0) \quad = \quad \$1000$

2 $\quad D_2 = \dfrac{2}{5}(5000 - 1000) \quad = \quad 1600$

3 $\quad D_3 = \dfrac{2}{5}(5000 - 2600) \quad = \quad 960$

4 $\quad D_4 = \dfrac{2}{5}(5000 - 3560) \quad = \quad 576$

5 $\quad D_5 = \dfrac{2}{5}(5000 - 4136) \quad = \quad 346$

6 (half-year) $\quad D_6 = \dfrac{1}{2} \times \dfrac{2}{5}(5000 - 4482) = \quad 104$

$\overline{\quad\quad\$4586}$

The computation must now be modified to convert to straight-line depreciation at the point where the straight-line depreciation will be larger. Using the alternative straight-line computation,

$$D_5 = \frac{5000 - 4136 - 0}{1.5 \text{ years remaining}} = \$576.$$

This is more than the $346 computed using DDB, hence switch to straight-line for year 5 and beyond.

$$D_6 \text{ (half-year)} = \frac{1}{2}(576) = \$288$$

Answers:

	Depreciation	
Year	SOYD	MACRS
1	$1500	$1000
2	1200	1600
3	900	960
4	600	576
5	300	576
6	0	288
	$4500	$5000

The computed MACRS depreciation is identical to the result obtained from Table 4.4.

TAX CONSEQUENCES

Income taxes represent another of the various kinds of disbursements encountered in an economic analysis. The starting point in an after-tax computation is the before-tax cash flow. Generally, the before-tax cash flow contains three types of entries:

1. Disbursements of money to purchase capital assets. These expenditures create no direct tax consequence, for they are the exchange of one asset (money) for another (capital equipment).

2. Periodic receipts and/or disbursements representing operating income and/or expenses. These increase or decrease the year-by-year tax liability of the firm.

3. Receipts of money from the sale of capital assets, usually in the form of a salvage value when the equipment is removed. The tax consequences depend on the relationship between the book value (cost – depreciation taken) of the asset and its salvage value.

Situation	Tax Consequence
Salvage value > Book value	Capital gain on differences
Salvage value = Book value	No tax consequence
Salvage value < Book value	Capital loss on difference

After determining the before-tax cash flow, compute the depreciation schedule for any capital assets. Next, compute taxable income, the taxable component of the before-tax cash flow minus the depreciation. The income tax is the taxable income times the appropriate tax rate. Finally, the after-tax cash flow is the before-tax cash flow adjusted for income taxes.

To organize these data, it is customary to arrange them in the form of a cash flow table, as follows:

Year	Before-Tax Cash Flow	Depreciation	Taxable Income	Income Taxes	After-Tax Cash Flow
0	•				•
1	•	•	•	•	•

Example 4.30

A corporation expects to receive $32,000 each year for 15 years from the sale of a product. There will be an initial investment of $150,000. Manufacturing and sales expenses will be $8067 per year. Assume straight-line depreciation, a 15-year useful life, and no salvage value. Use a 46% income tax rate. Determine the projected after-tax rate of return.

Solution

Straight-line depreciation, $D_j = \dfrac{C - S_n}{n} = \dfrac{\$150,000 - 0}{15} = \$10,000$ per year

Year	Before-Tax Cash Flow	Depreciation	Taxable Income	Income Taxes	After-Tax Cash Flow
0	−150,000				−150,000
1	+23,933	10,000	13,933	−6409	+17,524
2	+23,933	10,000	13,933	−6409	+17,524
.	.	.	.	.	.
.	.	.	.	.	.
.	.	.	.	.	.
15	+23,933	10,000	13,933	−6409	+17,524

Take the after-tax cash flow and compute the rate of return at which the PW of cost equals the PW of benefits.

$$150,000 = 17,524(P/A, i\%, 15)$$

$$(P/A, i\%, 15) = \frac{150,000}{17,524} = 8.559$$

From the compound interest tables, the after-tax rate of return is $i = 8\%$.

INFLATION

Inflation is characterized by rising prices for goods and services, whereas deflation produces a fall in prices. An inflationary trend makes future dollars have less purchasing power than present dollars. This helps long-term borrowers of money, for they may repay a loan of present dollars in the future with dollars of reduced buying power. The help to borrowers is at the expense of lenders. Deflation has the opposite effect. Money borrowed at one point in time, followed by a deflationary period, subjects the borrower to loan repayment with dollars of greater purchasing power than those borrowed. This is to the lenders' advantage at the expense of borrowers.

Price changes occur in a variety of ways. One method of stating a price change is as a uniform rate of price change per year.

f = General inflation rate per interest period
i = Effective interest rate per interest period

The following situation will illustrate the computations. A mortgage is to be repaid in three equal payments of $5000 at the end of years 1, 2, and 3. If the annual inflation rate, f, is 8% during this period, and a 12% annual interest rate (i) is desired, what is the maximum amount the investor would be willing to pay for the mortgage?

The computation is a two-step process. First, the three future payments must be converted to dollars with the same purchasing power as today's (year 0) dollars.

Year	Actual Cash Flow	Multiplied by		Cash Flow Adjusted to Today's (yr. 0) Dollars
0	—	—		—
1	+5000	$(1 + 0.08)^{-1}$	=	+4630
2	+5000	$(1 + 0.08)^{-2}$	=	+4286
3	+5000	$(1 + 0.08)^{-3}$	=	+3969

The general form of the adjusting multiplier is

$$(1 + f)^{-n} = (P/F, f, n).$$

Now that the problem has been converted to dollars of the same purchasing power (today's dollars, in this example), we can proceed to compute the present worth of the future payments.

Year	Adjusted Cash Flow	Multiplied by	Present Worth
0			
1	+4630	$(1 + 0.12)^{-1}$	+4134
2	+4286	$(1 + 0.12)^{-2}$	+3417
3	+3969	$(1 + 0.12)^{-3}$	+2825
			$10,376

The general form of the discounting multiplier is

$$(1 + i)^{-n} = (P/F, i\%, n).$$

Alternative Solution

Instead of doing the inflation and interest rate computations separately, one can compute a combined equivalent interest rate, d.

$$d = (1 + f)(1 + i) - 1 = i + f + i(f)$$

For this cash flow, $d = 0.12 + 0.08 + 0.12(0.08) = 0.2096$. Since we do not have 20.96% interest tables, the problem has to be calculated using present worth equations.

$$\text{PW} = 5000(1 + 0.2096)^{-1} + 5000(1 + 0.2096)^{-2} + 5000(1 + 0.2096)^{-3}$$
$$= 4134 + 3417 + 2825 = \$10,376$$

Example 4.31

One economist has predicted that there will be 7% per year inflation of prices during the next 10 years. If this proves to be correct, an item that presently sells for $10 would sell for what price 10 years hence?

Solution

$$f = 7\%, P = \$10$$
$$F = ?, n = 10 \text{ years}$$

Here the computation is to find the future worth F, rather than the present worth, P.

$$F = P(1 + f)^{10} = 10(1 + 0.07)^{10} = \$19.67$$

Effect of Inflation on Rate of Return

The effect of inflation on the computed rate of return for an investment depends on how future benefits respond to the inflation. If benefits produce constant dollars, which are not increased by inflation, the effect of inflation is to reduce the before-tax rate of return on the investment. If, on the other hand, the dollar benefits increase to keep up with the inflation, the before-tax rate of return will not be adversely affected by the inflation.

Example 4.32

This is not true when an after-tax analysis is made. Even if the future benefits increase to match the inflation rate, the allowable depreciation schedule does not increase. The result will be increased taxable income and income tax payments. This reduces the available after-tax benefits and, therefore, the after-tax rate of return.

A man bought a 5% tax-free municipal bond. It cost $1000 and will pay $50 interest each year for 20 years. The bond will mature at the end of 20 years and return the original $1000. If there is 2% annual inflation during this period, what rate of return will the investor receive after considering the effect of inflation?

Solution

$$d = 0.05, \; i = \text{unknown}, \; j = 0.02$$
$$d = i + j + i(j)$$
$$0.05 = i + 0.02 + 0.02i$$
$$1.02i = 0.03, \; i = 0.294 = 2.94\%$$

RISK

The term risk has a special meaning in statistics. It is defined as a situation where there are two or more possible outcomes and the probability associated with each outcome is known. We cannot know in advance what playing card will be dealt from a deck or what number will be rolled by a pair of dice. However, since the various probabilities could be computed, our definition of risk has been satisfied. Probability and risk are not restricted to gambling games. For example, in a particular engineering course, a student has computed the probability for each of the letter grades he might receive as follows:

Grade	Grade Point	Probability P(Grade)
A	4.0	0.10
B	3.0	0.30
C	2.0	0.25
D	1.0	0.20
F	0	0.15
		1.00

From the table we see that the grade with the highest probability is a B. This, therefore, is the most likely grade. We also see that there is a substantial probability that some grade other than a B will be received. And the probabilities indicate that if a B is not received, the grade will probably be something less than a B. But in saying that the most likely grade is a B, other outcomes are ignored. In the next section we will show that a composite statistic may be computed using all the data.

Expected Value

In the last example the most likely grade of B in an engineering class had a probability of 0.30. That is not a very high probability. In some other course, say a math class, we might estimate a probability of 0.65 of obtaining a B, again making the B the most likely grade. While a B is most likely in both classes, it is more certain in the math class.

We can compute a weighted mean to give a better understanding of the total situation as represented by various possible outcomes. When the probabilities are used as the weighting factors, the result is called the *expected value* and is written

$$\text{Expected value} = \text{Outcome}_A \times P(A) + \text{Outcome}_B \times P(B) + \ldots$$

Example 4.33

An engineer wishes to determine the risk of fire loss for her $200,000 home. From a fire rating bureau she obtains the following data:

Outcome	Probability
No fire loss	0.986 in any year
$10,000 fire loss	0.010
40,000 fire loss	0.003
200,000 fire loss	0.001

Compute the expected fire loss in any year.

Solution

Expected fire loss = 10,000(0.010) + 40,000(0.003) + 200,000(0.001) = $420

REFERENCE

Newnan, Donald G. *Engineering Economic Analysis*, 5th ed. Engineering Press, San Jose, CA, 1995.

PROBLEMS

4.1 A retirement fund earns 8% interest, compounded quarterly. If $400 is deposited every three months for 25 years, the amount in the fund at the end of 25 years is nearest to:
- a. $50,000
- b. $75,000
- c. $100,000
- d. $125,000

4.2 The repair costs for some handheld equipment are estimated to be $120 the first year, increasing by $30 per year in subsequent years. The amount a person needs to deposit into a bank account paying 4% interest to provide for the repair costs for the next five years is nearest to:
- a. $500
- b. $600
- c. $700
- d. $800

4.3 One thousand dollars is borrowed for one year at an interest rate of 1% per month. If this same sum of money were borrowed for the same period at an interest rate of 12% per year, the saving in interest charges would be closest to:
- a. $0
- b. $3
- c. $5
- d. $7

4.4 How much should a person invest in a fund that will pay 9%, compounded continuously, if he wishes to have $10,000 in the fund at the end of ten years?
- a. $4000
- b. $5000
- c. $6000
- d. $7000

4.5 A store charges 1.5% interest per month on credit purchases. This is equivalent to a nominal annual interest rate of:
- a. 1.5%
- b. 15.0%
- c. 18.0%
- d. 19.6%

4.6 A small company borrowed $10,000 to expand its business. The entire principal of $10,000 will be repaid in two years, but quarterly interest of $330 must be paid every three months. The nominal annual interest rate the company is paying is closest to:
- a. 3.3%
- b. 5.0%
- c. 6.6%
- d. 13.2%

4.7 A store's policy is to charge 3% interest every two months on the unpaid balance in charge accounts. The effective interest rate is closest to:
- a. 6%
- b. 12%
- c. 15%
- d. 19%

4.8 The effective interest rate on a loan is 19.56%. If there are 12 compounding periods per year, the nominal interest rate is closest to:
- a. 1.5%
- b. 4.5%
- c. 9.0%
- d. 18.0%

4.9 A deposit of $300 was made one year ago into an account paying monthly interest. If the account now has $320.52, the effective annual interest rate is closest to:
   a. 7%
   b. 10%
   c. 12%
   d. 15%

4.10 If the effective interest rate per year is 12%, based on monthly compounding, the nominal interest rate per year is closest to:
   a. 8.5%
   b. 9.3%
   c. 10.0%
   d. 11.4%

4.11 If 10% nominal annual interest is compounded daily, the effective annual interest rate is nearest to:
   a. 10.00%
   b. 10.38%
   c. 10.50%
   d. 10.75%

4.12 An individual wishes to deposit a certain quantity of money now so that he will have $500 at the end of five years. With interest at 4% per year, compounded semiannually, the amount of the deposit is nearest to:
   a. $340
   b. $400
   c. $410
   d. $416

4.13 A steam boiler is purchased on the basis of guaranteed performance. A test indicates that the operating cost will be $300 more per year than the manufacturer guaranteed. If the expected life of the boiler is 20 years, and the time value of money is 8%, the amount the purchaser should deduct from the purchase price to compensate for the extra operating cost is nearest to:
   a. $2950
   b. $3320
   c. $4100
   d. $5520

4.14 A consulting engineer bought a fax machine with one year's free maintenance. In the second year the maintenance cost is estimated at $20. In subsequent years the maintenance cost will increase $20 per year (that is, third year maintenance will be $40, fourth year maintenance will be $60, and so forth). The amount that must be set aside now at 6% interest to pay the maintenance costs on the fax machine for the first six years of ownership is nearest to:
   a. $101
   b. $164
   c. $229
   d. $284

4.15 An investor is considering buying a 20-year corporate bond. The bond has a face value of $1000 and pays 6% interest per year in two semiannual payments. Thus the purchaser of the bond will receive $30 every six months, and in addition he will receive $1000 at the end of 20 years, along with the last $30 interest payment. If the investor believes he should receive 8% annual interest, compounded semiannually, the amount he is willing to pay for the bond (bond value) is closest to:
   a. $500
   b. $600
   c. $700
   d. $800

4.16 Annual maintenance costs for a particular section of highway pavement are $2000. The placement of a new surface would reduce the annual maintenance cost to $500 per year for the first five years and to $1000 per year for the next five years. The annual maintenance after ten years would again be $2000. If maintenance costs are the only saving, the maximum investment that can be justified for the new surface, with interest at 4%, is closest to:
a. $5500
b. $7170
c. $10,000
d. $10,340

4.17 A project has an initial cost of $10,000, uniform annual benefits of $2400, and a salvage value of $3000 at the end of its ten-year useful life. At 12% interest the net present worth (NPW) of the project is closest to:
a. $2500
b. $3500
c. $4500
d. $5500

4.18 A person borrows $5000 at an interest rate of 18%, compounded monthly. Monthly payments of $167.10 are agreed upon. The length of the loan is closest to:
a. 12 months
b. 20 months
c. 24 months
d. 40 months

4.19 A machine costing $2000 to buy and $300 per year to operate will save labor expenses of $650 per year for eight years. The machine will be purchased if its salvage value at the end of eight years is sufficiently large to make the investment economically attractive. If an interest rate of 10% is used, the minimum salvage value must be closest to:
a. $100
b. $200
c. $300
d. $400

4.20 The amount of money deposited 50 years ago at 8% interest that would now provide a perpetual payment of $10,000 per year is nearest to:
a. $3000
b. $8000
c. $50,000
d. $70,000

4.21 An industrial firm must pay a local jurisdiction the cost to expand its sewage treatment plant. In addition, the firm must pay $12,000 annually toward the plant operating costs. The industrial firm will pay sufficient money into a fund that earns 5% per year to pay its share of the plant operating costs forever. The amount to be paid to the fund is nearest to:
a. $15,000
b. $30,000
c. $60,000
d. $240,000

4.22 At an interest rate of 2% per month, money will double in value in how many months?
a. 20 months
b. 22 months
c. 24 months
d. 35 months

4.23 A woman deposited $10,000 into an account at her credit union. The money was left on deposit for 80 months. During the first 50 months the woman earned 12% interest, compounded monthly. The credit union then changed its interest policy so that the woman earned 8% interest compounded quarterly during the next 30 months. The amount of money in the account at the end of 80 months is nearest to:
a. $10,000
b. $12,500
c. $15,000
d. $20,000

4.24 An engineer deposited $200 quarterly in her savings account for three years at 6% interest, compounded quarterly. Then for five years she made no deposits or withdrawals. The amount in the account after eight years is closest to:
a. $1200
b. $1800
c. $2400
d. $3600

4.25 A sum of money, Q, will be received six years from now. At 6% annual interest the present worth of Q is $60. At this same interest rate the value of Q ten years from now is closest to:
a. $60
b. $77
c. $90
d. $107

4.26 If $200 is deposited in a savings account at the beginning of each year for 15 years and the account earns interest at 6%, compounded annually, the value of the account at the end of 15 years will be most nearly:
a. $4500
b. $4700
c. $4900
d. $5100

4.27 The maintenance expense on a piece of machinery is estimated as follows:

Year	1	2	3	4
Maintenance	$150	$300	$450	$600

If interest is 8%, the equivalent uniform annual maintenance cost is closest to:
a. $250
b. $300
c. $350
d. $400

4.28 A payment of $12,000 six years from now is equivalent, at 10% interest, to an annual payment for eight years starting at the end of this year. The annual payment is closest to:
a. $1000
b. $1200
c. $1400
d. $1600

4.29 A manufacturer purchased $15,000 worth of equipment with a useful life of six years and a $2000 salvage value at the end of the six years. Assuming a 12% interest rate, the equivalent uniform annual cost (EUAC) is nearest to:
a. $1500
b. $2500
c. $3500
d. $4500

4.30 Consider a machine as follows:

> Initial cost: $80,000
> End-of-useful-life salvage value: $20,000
> Annual operating cost: $18,000
> Useful life: 20 years

Based on 10% interest, the equivalent uniform annual cost for the machine is closest to:
a. $21,000
b. $23,000
c. $25,000
d. $27,000

4.31 Consider a machine as follows:

> Initial cost: $80,000
> Annual operating cost: $18,000
> Useful life: 20 years

What must be the salvage value of the machine at the end of 20 years for the machine to have an equivalent uniform annual cost of $27,000? Assume a 10% interest rate. The salvage value S_{20} is closest to:
a. $10,000
b. $20,000
c. $30,000
d. $40,000

4.32 Twenty-five thousand dollars is deposited in a savings account that pays 5% interest, compounded semiannually. Equal annual withdrawals are to be made from the account beginning one year from now and continuing forever. The maximum amount of the equal annual withdrawals is closest to:
a. $625
b. $1000
c. $1250
d. $1265

4.33 An investor is considering the investment of $10,000 in a piece of land. The property taxes are $100 per year. The lowest selling price the investor must receive if she wishes to earn a 10% interest rate after keeping the land for ten years is:
a. $20,000
b. $21,000
c. $23,000
d. $27,000

4.34 The rate of return for a $10,000 investment that will yield $1000 per year for 20 years is closest to:
a. 1%
b. 4%
c. 8%
d. 12%

4.35 An engineer invested $10,000 in a company. In return he received $600 per year for six years and his $10,000 investment back at the end of the six years. His rate of return on the investment was closest to:
a. 6%
b. 10%
c. 12%
d. 15%

4.36 An engineer made ten annual end-of-year purchases of $1000 of common stock. At the end of the tenth year, just after the last purchase, the engineer sold all the stock for $12,000. The rate of return received on the investment is closest to:
a. 2%   c. 8%
b. 4%   d. 10%

4.37 A company is considering buying a new piece of machinery.

> Initial cost: $80,000
> End-of-useful-life salvage value: $20,000
> Annual operating cost: $18,000
> Useful life: 20 years

The machine will produce an annual saving in material of $25,700. What is the before-tax rate of return if the machine is installed? The rate of return is closest to:
a. 6%   c. 10%
b. 8%   d. 15%

4.38 Consider the following situation: Invest $100 now and receive two payments of $102.15—one at the end of year 3 and one at the end of year 6. The rate of return is nearest to:
a. 6%   c. 10%
b. 8%   d. 18%

4.39 Two mutually exclusive alternatives are being considered:

Year	A	B
0	−$2500	−$6000
1	+746	+1664
2	+746	+1664
3	+746	+1664
4	+746	+1664
5	+746	+1664

The rate of return on the difference between the alternatives is closest to:
a. 6%   c. 10%
b. 8%   d. 12%

4.40 A project will cost $50,000. The benefits at the end of the first year are estimated to be $10,000, increasing $1000 per year in subsequent years. Assuming a 12% interest rate, no salvage value, and an eight-year analysis period, the benefit-cost ratio is closest to:
a. 0.78   c. 1.28
b. 1.00   d. 1.45

4.41 Two alternatives are being considered:

	A	B
Initial cost	$500	$800
Uniform annual benefit	$140	$200
Useful life, years	8	8

The benefit-cost ratio of the difference between the alternatives, based on a 12% interest rate, is closest to:
- a. 0.60
- b. 0.80
- c. 1.00
- d. 1.20

4.42 An engineer will invest in a mining project if the benefit-cost ratio is greater than 1.00, based on an 18% interest rate. The project cost is $57,000. The net annual return is estimated at $14,000 for each of the next eight years. At the end of eight years the mining project will be worthless. The benefit-cost ratio is closest to:
- a. 0.60
- b. 0.80
- c. 1.00
- d. 1.20

4.43 A city has retained your firm to do a benefit-cost analysis of the following project:

> Project cost: $60,000,000
> Gross income: $20,000,000 per year
> Operating costs: $5,500,000 per year
> Salvage value after ten years: None

The project life is ten years. Use 8% interest in the analysis. The computed benefit-cost ratio is closest to:
- a. 0.80
- b. 1.00
- c. 1.20
- d. 1.60

4.44 A piece of property is purchased for $10,000 and yields a $1000 yearly profit. If the property is sold after five years, the minimum price to break even, with interest at 6%, is closest to:
- a. $5000
- b. $6500
- c. $7700
- d. $8300

4.45 Given two machines:

	A	B
Initial cost	$55,000	$75,000
Total annual costs	$16,200	$12,450

With interest at 10% per year, at what service life do these two machines have the same equivalent uniform annual cost? The service life is closest to:
- a. four years
- b. five years
- c. six years
- d. eight years

4.46 A machine part that is operating in a corrosive atmosphere is made of low-carbon steel. It costs $350 installed, and lasts six years. If the part is treated for corrosion resistance it will cost $700 installed. How long must the treated part last to be as economical as the untreated part, if money is worth 6%?

   a. 8 years
   b. 11 years
   c. 15 years
   d. 17 years

4.47 A firm has determined that the two best paints for its machinery are Tuff-Coat at $45 per gallon and Quick at $22 per gallon. The Quick paint is expected to prevent rust for five years. Both paints take $40 of labor per gallon to apply, and both cover the same area. If a 12% interest rate is used, how long must the Tuff-Coat paint prevent rust to justify its use?

   a. Five years
   b. Six years
   c. Seven years
   d. Eight years

4.48 Two alternatives are being considered:

	A	B
Cost	$1000	$2000
Useful life in years	10	10
End-of-useful-life salvage value	$100	$400

The net annual benefit of alternative A is $150. If interest is 8%, what must be the net annual benefit of alternative B for the two alternatives to be equally desirable?

   a. $150
   b. $200
   c. $225
   d. $275

4.49 A $5000 municipal bond is offered for sale. It will provide 8% annual interest by paying $200 to the bondholder every six months. At the end of ten years, the $5000 will be paid to the bondholder along with the final $200 interest payment. If you consider 12% nominal annual interest, compounded semiannually, an appropriate bond yield, the amount you would be willing to pay for the bond is closest to:

   a. $2750
   b. $3850
   c. $5000
   d. $7400

4.50 A municipal bond is being offered for sale for $10,000. It is a zero-coupon bond, that is, the bond pays no interest during its 15-year life. At the end of 15 years the owner of the bond will receive a single payment of $26,639. The bond yield is closest to:

   a. 4%
   b. 5%
   c. 6%
   d. 7%

Problems   169

4.51 A firm is considering purchasing $8000 of small hand tools for use on a production line. It is estimated that the tools will reduce the amount of required overtime work by $2000 the first year, with this amount increasing by $1000 per year thereafter. The payback period for the hand tools is closest to:
a. 2.00 years
b. 2.50 years
c. 2.75 years
d. 3.00 years

4.52 Special tools for the manufacture of finished plastic products cost $15,000 and have an estimated $1000 salvage value at the end of an estimated three-year useful life and recovery period. The third-year straight-line depreciation is closest to:
a. $3000
b. $3500
c. $4000
d. $4500

4.53 Refer to the facts of Problem 4.52. The first-year MACRS depreciation is closest to:
a. $3000
b. $3500
c. $4000
d. $5000

4.54 An engineer is considering the purchase of an annuity that will pay $1000 per year for ten years. The engineer feels he should obtain a 5% rate of return on the annuity after considering the effect of an estimated 6% inflation per year. The amount he would be willing to pay to purchase the annuity is closest to:
a. $1500
b. $3000
c. $4500
d. $6000

4.55 An automobile costs $20,000 today. You can earn 12% tax-free on an *auto purchase account*. If you expect the cost of the auto to increase by 10% per year, the amount you would need to deposit in the account to provide for the purchase of the auto five years from now is closest to:
a. $12,000
b. $14,000
c. $16,000
d. $18,000

4.56 An engineer purchases a building lot for $40,000 cash and plans to sell it after five years. If he wants an 18% before-tax rate of return, after taking the 6% annual inflation rate into account, the selling price must be nearest to:
a. $55,000
b. $65,000
c. $75,000
d. $125,000

4.57 A piece of equipment with a list price of $450 can actually be purchased for either $400 cash or $50 immediately plus four additional annual payments of $115.25. All values are in dollars of current purchasing power. If the typical customer considered a 5% interest rate appropriate, the inflation rate at which the two purchase alternatives are equivalent is nearest to:
a. 5%
b. 6%
c. 8%
d. 10%

SOLUTIONS

4.1 d.
$$F = A(F/A,i,n) = 400(F/A,2\%,100)$$
$$= 400(312.23) = \$124{,}890$$

4.2 d.
$$P = A(P/A,i,n) + G(P/G,i,n)$$
$$= 120(P/A,4\%,5) + 30(P/G,4\%,5)$$
$$= 120(4.452) + 30(8.555) = \$791$$

4.3 d.

At $i = 1\%$/month: $F = 1000(1 + 0.01)^{12} = \$1126.83$
At $i = 12\%$/year: $F = 1000(1 + 0.12)^1 = 1120.00$
Saving in interesting charges = $1126.83 - 1120.00 = \$6.83$

4.4 a.
$$P = Fe^{-rn} = 10{,}000e^{-0.09(10)} = 4066$$

4.5 c. The nominal interest rate is the annual interest rate ignoring the effect of any compounding. Nominal interest rate = $1.5\% \times 12 = 18\%$.

4.6 d. The interest paid per year = $330 \times 4 = 1320$. The nominal annual interest rate = $1320/10{,}000 = 0.132 = 13.2\%$.

4.7 d.
$$i_e = (1 + r/m)^m - 1 = (1 + 0.03)6 - 1 = 0.194 = 19.4\%$$

4.8 d.
$$i_e = (1 + r/m)^m - 1$$
$$r/m = (1 + i^e)^{1/m} - 1 = (1 + 0.1956)1/12 - 1 = 0.015$$
$$r = 0.015(m) = 0.015 \times 12 = 0.18 = 18\%$$

4.9 a.
$$i_e = 20.52/300 = 0.0684 = 6.84\%$$

4.10 d.
$$i^e = (1 + r/m)^m - 1$$
$$0.12 = (1 + r/12)12 - 1$$
$$(1.12)1/12 = (1 + r/12)$$
$$1.00949 = (1 + r/12)$$
$$r = 0.00949 \times 12 = 0.1138 = 11.38\%$$

4.11 c.
$$i^e = (1 + r/m)^m - 1 = (1 + 0.10/365)^{365} - 1 = 0.1052 = 10.52\%$$

4.12 c.

$$P = F(P/F,i,n) = 500(P/F,2\%,10) = 500(0.8203) = \$410$$

4.13 a.

$$P = 300(P/A,8\%,20) = 300(9.818) = \$2945$$

4.14 c. Using single payment present worth factors:

$$P = 20(P/F,6\%,2) + 40(P/F,6\%,3) + 60(P/F,6\%,4)$$
$$+ 80(P/F,6\%,5) + 100(P/F,6\%,6) = \$229$$

Alternate solution using the gradient present worth factor:

$$P = 20(P/G,6\%,6) = 20(11.459) = \$229$$

4.15 d.

$$PW = 30\,(P/A,4\%,40) + 1000(P/F,4\%,40)$$
$$= 30(19.793) + 1000(0.2083) = \$802$$

4.16 d. Benefits are $1500 per year for the first five years and $1000 per year for the subsequent five years.

As Exhibit 4.16 indicates, the benefits may be considered as $1000 per year for ten years, plus an additional $500 benefit in each of the first five years.

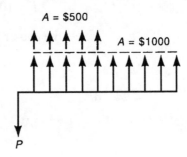

Exhibit 4.16

maximum investment = present worth of benefits
$$= 1000(P/A,4\%,10) + 500(P/A,4\%,5)$$
$$= 1000(8.111) + 500(4.452) = \$10{,}337$$

4.17 c.

NPW = PW of benefits − PW of cost
$$= 2400(P/A,12\%,10) + 3000(P/F,12\%,10) - 10{,}000 = \$4526$$
$$= 2400(5.65) + 3000(.32) - 10{,}000 = \$4526$$

4.18 d.

PW of benefits = PW of cost
$$5000 = 167.10(P/A,1.5\%,n)$$
$$(P/A,1.5\%,n) = 5000/167.10 = 29.92$$

From the $1\frac{1}{2}\%$ interest table, $n = 40$.

4.19 c.

$$\text{NPW} = \text{PW of benefits} - \text{PW of cost} = 0$$
$$= (650 - 300)(P/A, 10\%, 8) + S_8 (P/F, 10\%, 8) - 2000 = 0$$
$$= 350(5.335) + S_8(0.4665) - 2000 = 0$$
$$S_8 = 132.75/0.4665 = \$285$$

4.20 a. The amount of money needed now to begin the perpetual payments is $P' = A/i = 10{,}000/0.08 = 125{,}000$. From this we can compute the amount of money, P, that would need to have been deposited 50 years ago:

$$P = 125{,}000(P/F, 8\%, 50) = 125{,}000(0.0213) = \$2663$$

4.21 d.

$$P = A/i = 12{,}000/0.5 = \$240{,}000$$

4.22 d.

$$2 = 1(F/P, i, n)$$
$$(F/P, 2\%, n) = 2$$

From the 2% interest table, n = about 35 months.

4.23 d. At the end of 50 months

$$F = 10{,}000(F/P, 1\%, 50) = 10{,}000(1.645) = \$16{,}450$$

At the end of 80 months

$$F = 16{,}450(F/P, 2\%, 10) = 16{,}450(1.219) = \$20{,}053$$

4.24 d.

$$FW = 200(F/A, 1.5\%, 12)(F/P, 1.5\%, 20)$$
$$= 200(13.041)(1.347) = \$3513$$

4.25 d. The present amount $P = 60$ is equivalent to Q six years hence at 6% interest. The future sum F may be calculated by either of two methods:

$$F = Q(F/P, 6\%, 4) \text{ and } Q = 60(F/P, 6\%, 6)$$

or

$$F = P(F/P, 6\%, 10)$$

Since P is known, the second equation may be solved directly.

$$F = P(F/P, 6\%, 10) = 60(1.791) = \$107$$

4.26 c.

$$F' = A(F/A,i,n) = 200(F/A,6\%,15) = 200(23.276) = \$4655.20$$
$$F = F'(F/P,i,n) = 4655.20(F/P,6\%,1) = 4655.20(1.06) = \$4935$$

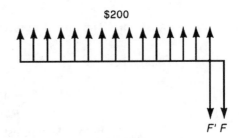

Exhibit 4.26

4.27 c.

$$\text{EUAC} = 150 + 150(A/G,8\%,4) = 150 + 150(1.404) = \$361$$

4.28 b.

$$\text{Annual payment} = 12{,}000(P/F,10\%,6)(A/P,10\%,8)$$
$$= 12{,}000(0.5645)(0.1874) = \$1269$$

4.29 c.

$$\text{EUAC} = 15{,}000(A/P,12\%,6) - 2000(A/F,12\%,6)$$
$$= 15{,}000(0.2432) - 2000(0.1232) = \$3402$$

4.30 d.

$$\text{EUAC} = 80{,}000(A/P,10\%,20) - 20{,}000(A/F,10\%,20)$$
$$+ \text{ annual operating cost}$$
$$= 80{,}000(0.1175) - 20{,}000(0.0175) + 18{,}000$$
$$= 9400 - 350 + 18{,}000 = \$27{,}050$$

4.31 b.

$$\text{EUAC} = \text{EUAB}$$
$$27{,}000 = 80{,}000(A/P,10\%,20) + 18{,}000 - S_{20}(A/F,10\%,20)$$
$$= 80{,}000(0.1175) + 18{,}000 - S_{20}(0.0175)$$
$$S_{20} = (27{,}400 - 27{,}000)/0.0175 = \$22{,}857$$

4.32 d. The general equation for an infinite life, $P = A/i$, must be used to solve the problem.

$$i_e = (1 + 0.025)^2 - 1 = 0.050625$$

The maximum annual withdrawal will be $A = Pi = 25{,}000(0.050625) = \$1266$.

4.33 d.

$$\text{Minimum sale price} = 10{,}000(F/P,10\%,10) + 100(F/A,10\%,10)$$
$$= 10{,}000(2.594) + 100(15.937) = \$27{,}530$$

4.34 c.

$$\text{NPW} = 1000(P/A,i,20) - 10{,}000 = 0$$
$$(P/A,i,20) = 10{,}000/1000 = 10$$

From interest tables: $6\% < i < 8\%$.

4.35 a. The rate of return was $600/10{,}000 = 0.06 = 6\%$.

4.36 b.

$$F = A(F/A,i,n)$$
$$12{,}000 = 1000(F/A,i,10)$$
$$(F/A,i,10) = 12{,}000/1000 = 12$$

In the 4% interest table: $(F/A,4\%,10) = 12.006$, so $i = 4\%$.

4.37 b.

$$\text{PW of cost} = \text{PW of benefits}$$
$$80{,}000 = (25{,}700 - 18{,}000)(P/A,i,20) + 20{,}000(P/F,i,20)$$

Try $i = 8\%$.

$$80{,}000 \stackrel{?}{=} 7700(9.818) + 20{,}000(0.2145) = 79{,}889$$

Therefore, the rate of return is very close to 8%.

4.38 d.

$$\text{PW of cost} = \text{PW of benefits}$$
$$100 = 102.15(P/F,i,3) + 102.15(P/F,i,6)$$

Solve by trial and error. Try $i = 12\%$.

$$100 \stackrel{?}{=} 102.15(0.7118) + 102.15(0.5066) = 124.46$$

The PW of benefits exceeds the PW of cost. This indicates that the interest rate i is too low. Try $i = 18\%$.

$$100 \stackrel{?}{=} 102.15(0.6086) + 102.15(0.3704) = 100.00$$

Therefore, the rate of return is 18%.

4.39 c. The difference between the alternatives:

$$\text{Incremental cost} = 6000 - 2500 = \$3500$$

$$\text{Incremental annual benefit} = 1664 - 746 = \$918$$
$$\text{PW of cost} = \text{PW of benefits}$$
$$3500 = 918(P/A,i,5)$$
$$(P/A,i,5) = 3500/918 = 3.81$$

From the interest tables, i is very close to 10%.

4.40 c.

$$\text{B/C} = \frac{\text{PW of benefits}}{\text{PW of cost}} = \frac{10{,}000\,(P/A,\,12\%,\,8) + 1000\,(P/G,\,12\%,\,8)}{50{,}000}$$

$$= \frac{10{,}000\,(4.968) + 1000\,(14.471)}{50{,}000} = 1.28$$

4.41 c.

$$B/C = \frac{\text{PW of benefits}}{\text{PW of cost}} = \frac{60(P/A,12\%,8)}{300} = \frac{60(4.968)}{300} = 0.99$$

Alternate solution:

$$B/C = \frac{\text{EUAB}}{\text{EUAC}} = \frac{60}{300(A/P,12\%,8)} = \frac{60}{300(0.2013)} = 0.99$$

4.42 a.

$$B/C = \frac{\text{PW of benefits}}{\text{PW of cost}} = \frac{14,000(P/A,18\%,8)}{57,000} = \frac{14,000(4.078)}{57,000} = 1.00$$

4.43 d.

$$B/C = \frac{\text{EUAB}}{\text{EUAC}} = \frac{20,000,000 - 5,500,000}{60,000,000(A/P,8\%,10)} = 1.62$$

4.44 c.

$$F = 10,000(F/P,6\%,5) - 1000(F/A,6\%,5)$$
$$= 10,000(1.338) - 1000(5.637) = \$774$$

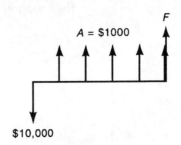

Exhibit 4.44

4.45 d.

$$\text{PW of cost}_A = \text{PW of cost}_B$$
$$55,000 + 16,200(P/A,10\%,n) = 75,000 + 12,450(P/A,10\%,n)$$
$$(P/A,10\%,n) = (75,000 - 55,000)/(16,200 - 12,450)$$
$$= 5.33$$

From the 10% interest tables, $n = 8$ years.

4.46 c.

$$\text{EUAC}_{\text{untreated}} = \text{EUAC}_{\text{treated}}$$
$$350(A/P,6\%,6) = 700(A/P,6\%,n)$$
$$350(0.2034) = 700(A/P,6\%,n)$$
$$(A/P,6\%,n) = 71.19/700 = 0.1017$$

From the 6% interest table, $n = 15+$ years.

4.47 d.

$$\text{EUAC}_{\text{T-C}} = \text{EUAC}_{\text{Quick}}$$
$$(45 + 40)(A/P,12\%,n) = (22 + 40)(A/P,12\%,5)$$
$$(A/P,12\%,n) = 17.20/85 = 0.202$$

From the 12% interest table, $n = 8$.

4.48 d. At breakeven,

$$\text{NPWA} = \text{NPWB}$$
$$150(P/A,8\%,10) + 100(P/F,8\%,10) - 1000 = \text{NAB}(P/A,8\%,10)$$
$$+ 400(P/F,8\%,10) - 2000$$
$$52.82 = 6.71(\text{NAB}) - 1814.72$$

Net annual benefit (NAB) = $(1814.72 + 52.82)/6.71 = \$278$

4.49 b. The number of six-month compounding periods in this problem is 20. So $n = 20$ and $12\%/2 = 6\%$ is the interest rate for the six-month interest period.

$$\text{Bond value} = \text{PW of all future benefits}$$
$$= 200(P/A,6\%,20) + 5000(P/F,6\%,20)$$
$$= 200(11.470) + 5000(0.3118) = \$3853$$

4.50 d. We know $P = 10{,}000$, $F = 26{,}639$, $n = 15$, and i = bond yield. Using the equation for the single payment compound amount:

$$F = P(1 + i)^n$$
$$26{,}639 = 10{,}000(1 + i)^{15}$$
$$2.6639^{1/15} = (1 + i)$$
$$1.0675 = 1 + i$$
$$i = 0.0675 = 6.75\%$$

4.51 c. The annual benefits are $2000, $3000, $4000, $5000, and so on. The payback period is the time when $8000 of benefits are received. This will occur in 2.75 years.

4.52 d.

$$D_3 = (C - S)/n = (15{,}000 - 1000)/3 = \$4666$$

4.53 d. From the modified ACRS table (Table 4.4) read for the first recovery year and three-year recovery the MACRS depreciation is 33.33% × 15,000 = $5000.

4.54 d.

$$d = i + f + (i \times f) = 0.05 + 0.06 + 0.05(0.06) = 0.113 = 11.3\%$$
$$P = A(P/A,11.3\%,10) = 1000\left[\frac{(1+0.113)^{10} - 1}{0.113(1+0.113)^{10}}\right]$$
$$= 1000\left[\frac{1.9171}{0.3296}\right] = \$5816$$

4.55 d.

Cost of auto five years hence $(F) = P(1 + \text{inflation rate})^n$
$$= 20,000(1 + 0.10)5 = 32,210$$

Amount to deposit now to have \$32,210 available five years hence:

$$P = F(P/F,i,n) = 32,210\ (P/F,12\%,5) = 32,210(0.5674) = \$18,276$$

4.56 d.

$$\text{Selling price } (F) = 40,000(F/P,18\%,5)(F/P,6\%,5)$$
$$= 40,000(2.288)(1.338) = \$122,500$$

4.57 b.

PW of cash purchase = PW of installment purchase
$$400 = 50 + 115.25(P/A,d,4)$$
$$(P/A,d,4) = 350/11.25 = 3.037$$

From the interest tables, $d = 12\%$.
$$d = i + f + i\,(f)$$
$$0.12 = 0.05 + f + 0.05f$$
$$f = 0.07/1.05 = 0.0667 = 6.67\%$$

CHAPTER 5

Ethics and Professional Practices

OUTLINE

MORALS, PERSONAL ETHICS, AND PROFESSIONAL ETHICS 180

CODES OF ETHICS 180
NCEES Model Rules of Professional Conduct

AGREEMENTS AND CONTRACTS 183
Elements of a Contract ∎ Contract and Related Legal Terminology

ETHICAL VERSUS LEGAL BEHAVIOR 185
Conflicts of Interest

PROFESSIONAL LIABILITY 186

PUBLIC PROTECTION ISSUES 186
Environmental Regulations

REFERENCES 189

PROBLEMS 190

SOLUTIONS 199

The specifications for the FE exam in environmental engineering indicate that codes, standards, regulations, and guidelines are fair topics for questions on the exam. This chapter discusses the foundations that underlie professional behavior and offers a brief summary of environmental legislation and the ethical mindset useful in minimizing environmental impact and protecting occupational safety and health. It should be noted that additional regulatory and legal issues relevant to specific areas of environmental engineering are also covered in more detail as needed in subsequent chapters.

MORALS, PERSONAL ETHICS, AND PROFESSIONAL ETHICS

To put professional ethics for engineers in perspective, it is helpful to distinguish it from morals and personal ethics. *Morals* are beliefs about right and wrong behaviors that are widely held by significant portions of a given culture. Obviously, morals will vary from culture to culture, and though some are common across different cultures, there seems to be no universal moral code.

Personal ethics are the beliefs that individuals hold that often are more restrictive than and sometimes contradictory to the morals of the culture. An example of personal ethics that might be more restrictive than morals might be the belief of an individual that alcohol should not be consumed in a culture that accepts the use of alcohol.

Professional ethics, on the other hand, is the formally adopted code of behavior by a group of professionals held out to society as that profession's pledge about how the profession will interact with society. Such rules or codes represent the agreed-on basis for a successful relationship between the profession and the society it serves.

The engineering profession has adopted several such codes of ethics, and different practitioners may adhere to or be bound by codes that vary by professional discipline but are similar in their basics. Codes adopted by the state boards of registration are typically codified into law and are legally binding for licensed engineers in the respective state. Codes adopted by professional societies are not legally binding but are voluntarily adhered to by members of those societies. The successful understanding of professional ethics for engineers requires an understanding of various codes; for purposes of examining registration applicants, the NCEES has adopted a "model code" that includes many canons common to most codes adopted nationwide.

CODES OF ETHICS

Codes of ethics are published by professional and technical societies and by licensing boards. Why are codes published and why are they important? These fundamental questions are at the heart of the definition of a "profession." Some important aspects of the definition of a profession might include skills and knowledge vital to society; extensive and intellectual education and training important for proper practice in the profession; an importance of autonomous action by practitioners; a recognition by society of these aspects, leading to a governmentally endorsed monopoly on the practice of the profession; and a reliance on published standards of ethical conduct, usually in the form of a code of ethics (Harris, et al., 2005). Such codes are published and followed to maintain a high standard of confidence in the profession by the public served by the practicing professionals, because without high standards of confidence, the ability of a profession to serve the public need may be seriously impaired.

The FE exam questions on ethics and business practices are based on the NCEES code of ethics, a concise body of model rules designed to guide state boards and practitioners as a model of good practice in the regulation of engineering. These rules do not bind any engineer, but the codes of ethics published by individual state boards and of professional societies will be very similar to these in principle.

NCEES Model Rules of Professional Conduct

A. Licensee's Obligation to Society

1. Licensees, in the performance of their services for clients, employers, and customers, shall be cognizant that their first and foremost responsibility is to the public welfare.

2. Licensees shall approve and seal only those design documents and surveys that conform to accepted engineering and surveying standards and safeguard the life, health, property, and welfare of the public.

3. Licensees shall notify their employer or client and such other authority as may be appropriate when their professional judgment is overruled under circumstances where the life, health, property, or welfare of the public is endangered.

4. Licensees shall be objective and truthful in professional reports, statements, or testimony. They shall include all relevant and pertinent information in such reports, statements, or testimony.

5. Licensees shall express a professional opinion publicly only when it is founded upon an adequate knowledge of the facts and a competent evaluation of the subject matter.

6. Licensees shall issue no statements, criticisms, or arguments on technical matters which are inspired or paid for by interested parties, unless they explicitly identify the interested parties on whose behalf they are speaking and reveal any interest they have in the matters.

7. Licensees shall not permit the use of their name or firm name by, nor associate in the business ventures with, any person or firm which is engaging in fraudulent or dishonest business or professional practices.

8. Licensees having knowledge of possible violations of any of these Rules of Professional Conduct shall provide the board with the information and assistance necessary to make the final determination of such violation. (Section 150, Disciplinary Action, NCEES Model Law)

B. Licensee's Obligation to Employer and Clients

1. Licensees shall undertake assignments only when qualified by education or experience in the specific technical fields of engineering or surveying involved.

2. Licensees shall not affix their signatures or seals to any plans or documents dealing with subject matter in which they lack competence, nor to any such plan or document not prepared under their direct control and personal supervision.

3. Licensees may accept assignments for coordination of an entire project, provided that each design segment is signed and sealed by the licensee responsible for preparation of that design segment.

4. Licensees shall not reveal facts, data, or information obtained in a professional capacity without the prior consent of the client or employer except as authorized or required by law. Licensees shall not solicit or accept gratuities, directly or indirectly, from contractors, their agents, or other parties in connection with work for employers or clients.

5. Licensees shall make full prior disclosures to their employers or clients of potential conflicts of interest or other circumstances which could influence or appear to influence their judgment or the quality of their service.

6. Licensees shall not accept compensation, financial or otherwise, from more than one party for services pertaining to the same project, unless the circumstances are fully disclosed and agreed to by all interested parties.

7. Licensees shall not solicit or accept a professional contract from a governmental body on which a principal or officer of their organization serves as a member. Conversely, licensees serving as members, advisors, or employees of a government body or department, who are the principals or employees of a private concern, shall not participate in decisions with respect to professional services offered or provided by said concern to the governmental body which they serve. (Section 150, Disciplinary Action, NCEES Model Law)

C. Licensee's Obligation to Other Licensees

1. Licensees shall not falsify or permit misrepresentation of their, or their associates', academic or professional qualifications. They shall not misrepresent or exaggerate their degree of responsibility in prior assignments nor the complexity of said assignments. Presentations incident to the solicitation of employment or business shall not misrepresent pertinent facts concerning employers, employees, associates, joint ventures, or past accomplishments.

2. Licensees shall not offer, give, solicit, or receive, either directly or indirectly, any commission, or gift, or other valuable consideration in order to secure work, and shall not make any political contribution with the intent to influence the award of a contract by public authority.

3. Licensees shall not attempt to injure, maliciously or falsely, directly or indirectly, the professional reputation, prospects, practice, or employment of other licensees, nor indiscriminately criticize other licensees' work. (Section 150, Disciplinary Action, NCEES Model Law)

Many ethical questions arise in the formulation of business practices. Professionals should appreciate that expressions like "all is fair in business" and "let the buyer beware" can conflict with fundamental ideas about how a professional engineer should practice. The reputation of the profession, not only the individual professional, is critically important to the ability of all engineers to discharge their duty to protect the public health, safety, and welfare.

The *NCEES Model Rules* addressing a licensee's obligation to other licensees prohibit misrepresentation or exaggeration of academic or professional qualifications, experience, level of responsibility, prior projects, or any other pertinent facts that might be used by a potential client or employer to choose an engineer.

The *Model Rules* also prohibit gifts, commissions, or other valuable consideration to secure work. Political contributions intended to influence public authorities responsible for awarding contracts are also prohibited.

Often, these rules are misunderstood in the arena of foreign practice. Increasingly, engineering is practiced globally, and engineers must deal with foreign clients and foreign governmental officials, many times on foreign soil where laws and especially cultural practices vary greatly. In the United States, the federal Foreign Corrupt Practices Act (FCPA) is a relatively recent recognition and regulation of this problem. Among other purposes, it provides clearer legal boundaries for U.S. engineers involved with international projects.

According to the FCPA, it is not illegal for a U.S. engineer to make petty extortion payments ("grease payments," "expediting payments," and "facilitating payments" are common expressions) to governmental officials when progress of otherwise legitimate projects is delayed by demands for such payments consistent with prevailing practice in that country. It is illegal, however, for U.S. engineers to give valuable gifts or payments to develop contracts for *new business*. In some cultures, reciprocal, expensive gift giving is an important part of business relationships, and the reciprocal nature of this practice can make it acceptable under the FCPA. Most commonly, when the engineer's responsibilities include interactions with foreign clients or partners, the engineer's corporate employer will publish detailed and conservative guidelines intended to guide the engineer in these ethical questions.

The practicing engineer should always be watchful of established and, especially, new business practices to be sure the practices are consistent with the codes of ethics he or she is following.

Example 5.1

The *NCEES Model Rules of Professional Conduct* allow an engineer to do which one of the following?

a). Accept money from contractors in connection with work for an employer or client

b). Compete with other engineers in seeking to provide professional services

c). Accept a professional contract from a governmental body even though a principal or officer of the engineer's firm serves as a member of the governmental body

d). Sign or seal all design segments of the project as the coordinator of an entire project

Solution

Although the other items are not allowed by the *Model Rules*, nowhere does it say that an engineer cannot compete with other engineers in seeking to provide professional services. But, of course, he or she should conduct business in an ethical manner. The correct answer is b).

AGREEMENTS AND CONTRACTS

One aspect of business practice is understanding the concepts and terminology of agreements and contracts.

Elements of a Contract

Contracts may be formed by two or more parties; that is, there must be a party to make an offer and a party to accept.

To be enforceable in a court of law, a contract must contain the following five essential elements:

1. There must be a mutual agreement.

2. The subject matter must be lawful.

3. There must be a valid consideration.
4. The parties must be legally competent.
5. To be a formal contract, the contract must comply with the provisions of the law with regard to form.

A *formal contract* depends on a particular form or mode of expression for legal efficacy. All other contracts are called *informal contracts* since they do not depend on mere formality for their legal existence.

Contract and Related Legal Terminology

Case law—the body of law created by courts interpreting statute law. Judges use precedents, the outcome of similar cases, to construct logically their decision in a given issue.

Changed or concealed conditions—in construction contracting, it is important to specify how changed or concealed conditions will be handled, usually by changes in the contract terms. For example, if an excavation project is slowed by a difficult soil pocket between soil corings, the excavation contractor may be able to support a claim for increased costs due to these unforeseen conditions. When the concealed conditions are such that they should have been foreseen, such claims are more difficult to support.

Common law—the body of rules of action and principles that derive their authority solely from usage and customs.

Damages for delays—in many contracts, completion time is an important concern, and contractual clauses addressing penalties for delays (or rewards for early completion) are often incorporated.

Equal or approved equivalent—terms used in specifications for materials to permit use of alternative but equal material when an original material is not available or an equivalent material can be obtained at lower cost. The engineer is responsible for approving the alternative material.

Equity—system of doctrines supplementing common and statute law, such as the Maxims of Equity.

Errors and omissions—term used to describe the kind of mistakes that can be made by engineers and architects leading to damage to the client. Often, this risk is protected by liability insurance policies.

Force account—a method of work by which the owner elects to do work with his or her own forces instead of employing a construction contractor. Under this method, the owner maintains direct supervision over the work, furnishes all materials and equipment, and employs workers on his or her own payroll.

Hold harmless—clauses are often included requiring one party to agree not to make a claim against the other and sometimes to cooperate in the defense of the other party if a claim is made by a third party.

Incorporate by reference—the act of making a document legally binding by referencing it within a contract, although it is not attached to or reproduced in the contract. This is done to eliminate unnecessary repetition.

Indemnify—to protect another person against loss or damage, as with an insurance policy.

Liquidated damages—a specific sum of money expressly stipulated as the amount of damages to be recovered by either party for a breach of the agreement by the other.

Mechanics' liens—legal mechanism by which unpaid contractors, suppliers, mechanics, or laborers are allowed to claim or repossess construction materials that have been delivered to the worksite in lieu of payment.

Plans—the drawings that show the physical characteristics of the work to be done. The plans and specifications form the guide and standards of performance that will be required.

Punitive damages—a sum of money used to punish the defendant in certain situations involving willful, wanton, malicious, or negligent torts.

Specifications—written instructions that accompany and supplement the plans. The specifications cover the quality of the materials, workmanship, and other technical requirements. The plans and specifications form the guide and standards of performance that will be required.

Statute law—acts or rules established by legislative action.

Statute of limitations—a time limit on claims resulting from design or construction errors, usually beginning with the date the work was performed, but in some cases beginning on the date the deficiency could first have been discovered.

Surety bond—bonds issued by a third party to guarantee the faithful performance of the contractor. Surety bonds are normally used in connection with competitive-bid contracts, namely, bid bonds, performance bonds, and payment bonds.

Workers' compensation—insurance protecting laborers and subcontractors in case of an on-the-job injury; it is often required of contractors.

ETHICAL VERSUS LEGAL BEHAVIOR

Engineers have a clear obligation to adhere to all laws and regulations in their work—what they do must be done legally. But the obligation goes beyond this. Unlike the world of business where cutthroat but legal practices are commonly condoned and frequently rewarded, engineers assume important obligations to the public and to the profession that restrict how they must practice and that often are much more stringent than law or regulation.

When you realize that restricting the practice of engineering to certain licensed professionals by the state is essentially a state-provided monopoly, you may begin to see why there is a difference. Competitive businesses compete in many ways to gain the kind of advantage in their field that engineers and other licensed professionals are given by the state.

Aggressive advertising is one example of a business practice that engineers avoid, even though it is not illegal or prohibited. Before 1978, it was common for professional societies to prohibit or narrowly restrict advertising by their practitioners; however, in 1978 the U.S. Supreme Court ruled such broad restrictions unconstitutional, allowing only reasonable restraints on advertising by professional societies. Since that time, engineering societies have adopted guidelines on advertising. Other professions have been less successful in regulating advertising. For example, the profusion of television advertising by lawyers, and the language of those advertisements, contrasts with the practice of engineering professionals where advertising is more commonly seen in technical journals or trade literature. Many believe the legal profession has suffered a loss of respect as a result, while the profession of engineering still is held in high regard by the public. It is in the interest of the engineering profession to avoid this kind of advertising, even though it is legal, because it can damage the reputation of the profession.

Another example of the importance of self-regulation is the engineer's responsibility to the environment. Although many laws and regulations restrict engineering practices that might damage the environment, there are still many legal ways to accomplish engineering projects that can have adverse environmental effects. Increasingly, codes of ethics are adding requirements for the engineer to consider the environment or the "sustainability" of proposed engineering projects. The engineer's ethical responsibility to work toward sustainable development may go beyond any legal requirements intended to prevent environmental damage.

Conflicts of Interest

A conflict of interest is any situation where the decision of an engineer can have some significant effect on his or her financial situation. It would be a clear conflict of interest for a designing engineer to specify exclusively some component that is only available from a supplier in which that engineer has a significant financial interest, when other components from other suppliers would serve equally well. Engineers must avoid even the *appearance* of a conflict of interest. This is critically important for the reputation of the profession, which the engineer is charged with protecting, in order for engineers to effectively serve the public interest.

An apparent conflict of interest is any situation that might appear to an outside observer to be an actual conflict of interest. For example, if the engineer in the case mentioned above had subsequently divested himself of all interest in the supplier, there is no longer an actual conflict of interest. However, to an outside observer with imperfect information, there might be the appearance of a conflict, resulting in the perception of unethical behavior in the public eye.

The usual remedy for conflicts of interest and apparent conflicts of interest is disclosure and, often, recusal. The engineer's interest must be disclosed in advance, generally to a supervisor, and recusal must at least be discussed. In many cases, recusal may not be necessary, but disclosure is vitally important. In every case, the public perception of the conflict must be considered, with the goal of protecting the reputation of the individuals and the profession.

PROFESSIONAL LIABILITY

Good engineering practice includes numerous checks and conservative principles of design to protect against blunders, but occasional errors and omissions can result in damage or injury. The engineer is responsible for such damage or injury, and it is good practice to carry errors and omissions insurance to provide appropriate compensation to any injured party, whether a client or a member of the public. Such insurance can be a significant cost in some fields of engineering, but it represents a cost of doing business that should be reflected in the fees charged. The most important factor in preventing errors and blunders is to provide adequate time for careful review of all steps in the project by knowledgeable senior licensed engineers. Frantic schedules and unrealistic deadlines can significantly increase the risk.

PUBLIC PROTECTION ISSUES

State boards in all 50 states and the District of Columbia are charged by their states with the responsibility for the licensing of engineers and the regulation of the practice of engineering to protect the health, safety, and welfare of the public. Licensed engineers in each state are legally bound by laws and regulations published by the respective state board. The boards are generally made up of engineers appointed by the state governor; sometimes nonengineering members also are appointed to make sure the public is adequately represented.

State boards commonly issue cease and desist letters to nonengineers who have firms or businesses with names that imply engineering services are being offered to the public or who may actually be offering engineering services without the required state license. These boards also regulate the practice of engineering by their registrants, often sanctioning registrants for inappropriate business practices or engineering design decisions. Many boards require continuing education by registrants for maintenance of proficiency. A weakness of many boards is in the area of discipline for incompetent practices, but this weakness is often offset by tort law whereby incompetent practitioners who cause damage or injury are commonly subject to significant legal damages.

Environmental Regulations

Environmental regulations are usually developed as a direct result of risk assessment data (see Chapter 6) that identifies a potential negative health effect arising from a specific chemical that can be found in the environment. A complete history of environmental legislation is not possible for this review, and the examinee is directed toward any of several texts that provide substantial coverage of the history of environmental law. Therefore, this review will cover the major pieces of legislation in the United States in the areas of water pollution, air pollution, and solid and hazardous waste.

The Federal Water Pollution Control Act (FWPCA) of 1956 serves as the landmark regulatory effort to remove contaminants from the water supply, specifically with respect to subsidies for municipal water treatment plants. The first consumption law was enacted in 1974 as the Safe Drinking Water Act (SDWA), which established drinking water standards for public water supplies. The Clean Water Act (CWA) of 1977 was the first to establish a list of 65 substances or classes of substances, which formed the core for the 127 substances on the current *priority pollutant list*. In 1990, the National Pollution Discharge Elimination System (NPDES) was passed, which requires industrial and municipal water discharges to be permitted and regularly tested for compliance with established contaminant standards.

The Federal Air Pollution Control Act (FAPCA) was the first piece of federal legislation to address air quality, although the act served primarily to establish a research program. It was the Clean Air Act (CAA) of 1963 that allowed enforcement of restrictions on interstate air pollution. Through many amendments, including the establishment of the NAAQS in 1977 and the extensive overhaul in 1990, the CAA has continued to be the primary defense of air quality.

Solid waste was first regulated in 1965 with the passage of the Solid Waste Disposal Act (SWDA), part of which set standards for collection, transport, processing, recovery, and disposal systems. This was followed in 1970 by the Resources Recovery Act, which shifted the mindset away from disposal toward recycling and recovery of energy in power generation facilities. Due to improper handling of wastes, specifically hazardous wastes that negatively impacted the quality of water supplies, Congress passed the Resource Conservation and Recovery Act (RCRA) in 1976. The act regulated solid and hazardous wastes for current and future disposal, arising in part from the recent passage of the CAA and the CWA that required removal of hazardous substances from air and water. These wastes required disposal in a manner that would not have a negative impact on the surrounding environment, and RCRA was intended to assure that was the case. RCRA was amended in 1984 by the Hazardous and Solid Waste Amendments (HSWA), portions of which required increased technology standards for waste landfills and establishing waste minimization as the preferred method of managing hazardous materials.

The passage of RCRA in 1976 addressed current and future disposal practices, however it was clear that past disposal practices had already had an impact on the quality of the environment, and legislation would be required to address the remediation of these sites. The Comprehensive Environmental Response, Compensation, and Liability Act (CERCLA) of 1980 (often referred to as "Superfund") was enacted to give the federal government the authority to complete the remediation of identified waste sites negatively impacting the environment, and to place financial liability on the responsible parties whenever possible. This was followed in 1986 by the Superfund Amendments and Reauthorization Act (SARA), further funding the potential remediation efforts and revising the *national priorities list* and the *hazardous ranking system*, as well as establishing the Right to Know Act, which required public disclosure for discharges.

Table 5.1 Summary of Landmark Environmental Legislation in the United States

Name	Date Enacted (Amended)	Major Provisions
Federal Insecticide, Fungicide, and Rodenticide Act (FIFRA)	1947 (1972)	Regulates the manufacture and use of all pesticides
Federal Water Pollution Control Act (FWPCA)	1956	Attempted to reduce water pollution by funding construction of municipal water treatment plants
Clean Air Act (CAA)	1963 (1970, 1990)	Established National Ambient Air Quality Standards (NAAQSs) for criteria pollutants
Solid Waste Disposal Act (SWDA)	1965	Promoted structured solid waste management through regulation of collection, transport, processing, recovery, and disposal systems
National Environmental Policy Act (NEPA)	1969	Required an environmental impact statement (EIS) for all projects receiving federal funding
Resources Recovery Act (RRA)	1970	Amended SWDA; established reuse and recycle as national priorities for managing solid waste; promoted energy recovery from solid waste
Occupational Safety and Health Act (OSHA)	1970	Established workplace safety standards for potential physical and chemical hazards
Clean Water Act (CWA)	1972 (1977)	Amended FWPCA; established the first EPA list of 65 priority pollutants (now at 127 pollutants); defined total maximum daily loads (TMDLs) and National Pollutant Discharge Elimination System (NPDES) permit program; required use of best available control technology (BACT)
Safe Drinking Water Act (SDWA)	1974 (1986, 1996)	First consumption law; established maximum contaminant levels (MCLs) for potable water

(continued)

Name	Date Enacted (Amended)	Major Provisions
Toxic Substances Control Act (TSCA)	1976	Regulates use of chemical substances; requires toxicity testing for newly introduced compounds
Resource Conservation and Recovery Act (RCRA)	1976	Regulates the generation, handling, storage, and disposal of solid and hazardous wastes
Comprehensive Environmental Response, Compensation, and Liability Act (CERCLA)	1980	Also known as "Superfund"; established the national priorities list (NPL) for remediation of hazardous waste sites using the hazard ranking system (HRS); addressed financial liability
Low-Level Waste Policy Act (LLWPA)	1980 (1985)	Regulates disposal of low-level radioactive wastes
Nuclear Waste Policy Act (NWPA)	1982 (1987)	Regulates storage of high-level radioactive wastes
Hazardous and Solid Waste Amendments (HSWA)	1984	Significantly increased the scope of RCRA; new technology standards for landfills; promotes waste minimization preference for hazardous materials
Superfund Amendments and Reauthorization Act (SARA)	1986	Amended CERCLA; provided cleanup fund and revised NPL and the hazard ranking system
Emergency Planning and Community Right-to-Know Act (EPCRA)	1986	Requires manufacturers to report on-site quantities of hazardous materials and develop emergency response plan in case of an uncontrolled release
Pollution Prevention Act (PPA)	1990	Established source reduction as a national priority

It should be noted that regulations are a direct result of the environmental legislation identified in Table 5.1 and may vary from state to state, as some states pass more stringent laws that exceed the federal requirements, however, they generally follow closely the provisions passed in the federal acts.

REFERENCES

Harris, Charles E., Jr., Michael S. Pritchard, and Michael J. Rabins. *Engineering Ethics: Concepts and Cases*. Thompson Wadsworth, 2005.

National Council of Examiners for Engineering and Surveying. *Model Rules*, September 2006.

PROBLEMS

5.1 Jim is a PE working for an HVAC designer who often must specify compressors and other equipment for his many clients. He reports to Joan, the VP of engineering. Jim specifies compressors from several different manufacturers and suppliers based on the technical specifications and on his experience with those products in past projects. Joan's long-time friend Charlie, who has been working in technical sales of construction materials, takes a new job with one of the compressor suppliers that Jim deals with from time to time. Charlie calls on Joan, inviting her and any of her HVAC designers to lunch to discuss a new line of high-efficiency compressors; Joan invites Jim to come along. Jim should:
a. decline to attend the lunch, citing concerns about conflict of interest
b. agree to attend the lunch but insist on paying for his own meal
c. agree to attend the lunch and learn about the new line of compressors
d. report Joan to the state board and never specify compressors from that supplier again

5.2 Harry C. is an experienced geotechnical engineer who has many years' experience as a PE designing geotechnical projects and who is very familiar with the rules regarding the requirement for trench shoring and trench boxes to protect construction workers during excavations. During a vacation visit to a neighboring state, he observes a city sewer construction project with several workers in an unprotected deep trench, which, to Harry's experienced eye, is probably not safe without a trench box or shoring. Harry should:
a. remember that he is not licensed in the neighboring state and has no authority to interfere
b. approach the contractor's construction foreman and insist that work be halted until the safety of the trench is investigated
c. advise all the workers in the trench that they are in danger and encourage them to go on strike for safer conditions
d. contact the city engineer to report his concerns

5.3 Engineering student Travis is eagerly anticipating his graduation in three months and has interviewed with several firms for entry-level employment as an electrical engineer. He has received two offers to work for firms A and B in a nearby city, and after comparing the jobs, salaries, and benefits and discussing the choice with his faculty advisor, he telephones firm A whose offer is more appealing and advises them he will accept their job offer. Two weeks later he is contacted by firm C in a different city with a job offer that includes a salary more than 15% higher than the offer he has accepted plus a generous relocation allowance. Travis should:
a. decline the offer from firm C, explaining that he has already accepted a position
b. contact firm A and ask if he can reconsider his decision
c. contact firm A to give them a fair chance to match the offer from firm C
d. advise firm C that he can accept their offer if they will contact firm A to inform them of this change

5.4 EIT Jerry works for a small civil engineering firm that provides general civil engineering design services for several municipalities in the region. He has become concerned that his PE supervisor Eddie is not giving careful reviews to Jerry's work before sealing the drawings and approving them for construction. Jerry asks Eddie to review with him the design assumptions from Jerry's latest design, a steel fire exit staircase to be added to an elementary school building, because he has concerns about the appropriate design loadings. But before the design assumptions are reviewed, Jerry notices the drawings have been approved and released to the fabricator. Jerry should:
a. quit his job and find another employer
b. take a review course in live loadings for steel structures
c. in the future mark each drawing he prepares "Not Approved for Construction"
d. None of the above

5.5 Dr. Willis Hemmings, PE, is an engineering professor whose research in fire protection engineering is nationally recognized. He is retained as an expert witness for the defendant, a structural engineering design firm, in a lawsuit filed by a firefighter who was injured while fighting a fire in a steel structure that collapsed during the fire. The plaintiff's lawyer alleges that the original design of certain components of the fire protection system protecting the steel structural members was inadequate. Hemmings reviews the original design documents, which call for a protective coating that is slightly thinner than is required by the local building code. Hemmings testifies that even though the specified coating is thinner than required, he believes that the design was sound because the product used is applied by a new process that is probably more efficient and the thinner coating probably gave the same level of protection. He bases his testimony on his national reputation as an expert in this field. Such expert testimony is:
a. a commonly accepted method of certifying good engineering design in tort law
b. legal only when given by a licensed professional engineer like Hemmings
c. unethical because it contradicts accepted practice without supporting tests or other data
d. effective only because Hemmings is involved in cutting-edge research

5.6 Jackie is a young PE who works for a garden tool manufacturer that has produced about 100,000 shovels, rakes, and other garden implements annually for more than 20 years. The company recently won a contract to manufacture and supply 5000 folding entrenching tools of an existing design to a Central American military client. The vice president of marketing has been working to develop contracts with other military clients and asks Jackie to prepare a statement of qualifications (SOQ). Jackie is asked to describe the design group (consisting of two engineers, one EIT, one student intern, three CAD technicians, and one IT technician) as a "team of eight tool design engineers," and to describe the company as "experienced in the design, testing, and manufacturing of military equipment, with a recent production history of over two million entrenching tools and related hardware." Jackie should:
   a. check the production records to be sure the figures cited are accurate
   b. ask the vice president to sign off on the draft of the SOQ
   c. object to describing the qualifications and experiences of her group in an exaggerated way
   d. be sure to mention that she is a PE and list the states in which she is licensed

5.7 The Ford Motor Company paid millions of dollars to individuals injured and killed in crashes of the Ford Pinto, which had a fuel tank and filler system that sometimes ruptured in rear-end collisions, spilling gasoline and causing fires. While many considered the filler system design deficient because of this tendency, one important factor played a role in the lawsuits. An internal Ford memo was discovered that included the cost-benefit calculations Ford managers used in making the decision not to improve the tank/filler system design. This memo was significant because:
   a. it is unethical to use the cost-benefit method for safety-related decisions
   b. it is illegal to estimate the value of human life in cost-benefit calculations
   c. state law requires estimates of the value of human life be at least $500,000 in such calculations
   d. None of the above

5.8 Charles is tasked to write specifications for electric motors and pumps for a new sanitary sewage treatment plant his employer is designing for a municipal client. Charles is concerned that he doesn't have a very good knowledge about current pump design standards but is willing to learn. Charles's fiancée is an accountant employed by a pump distributor and offers to provide Charles with a binder of specifications for all the pumps her firm distributes. Charles should:
   a. decline to accept the binder, citing concerns about conflict of interest
   b. accept the binder but turn it over to his employer's technical librarian without reviewing it
   c. accept the binder and study the materials to gain a better understanding of pump design and specifications
   d. ask his fiancée if she knows an applications engineer at her firm who would draft specifications for him

5.9 Professor Martinez is a PE who teaches chemical engineering classes at a small engineering school. His student, Erica, recently graduated and took a job with WECHO, a small firm that provides chemicals and support to oil well drilling operations. WECHO has never employed an engineer and has hired Erica, partly on Prof. Martinez's strong recommendation, in hopes that she will one day become their chief engineer. After she has worked at WECHO for about two years, her supervisor Harry calls Prof. Martinez to explain that WECHO has been required to complete an environmental assessment before deploying a new surfactant, and the assessment must be sealed by a PE. Harry explains that Erica has done all the research to collect data and answer questions on the assessment, and everyone at WECHO agrees that she has done a superb job in completing the assessment, but it still requires the seal of a PE before submission. Harry asks Prof. Martinez if he can review Erica's work and seal the report, reminding him that he had given a glowing recommendation of Erica at the time WECHO hired her. Prof. Martinez should:
a. negotiate a consulting contract to allow him sufficient time and funding to review the report before sealing it
b. require Erica to first sign the report as an EIT and graduate engineer before reviewing it
c. require WECHO to purchase a bond against environmental damage before sealing the report
d. decline to review or seal the report, citing responsible charge issues

5.10 Jack Krompten, PE, is an experienced civil engineer working for a land development firm that has completed several successful residential subdivision developments in WoodAcres, a suburban bedroom community of a large, sprawling, and rapidly growing city. The WoodAcres city engineer, who also served half-time as the mayor, has retired, and the city council realizes that with rapid growth ahead it will be important to hire a new city engineer. They approach Krompton with an offer of half-time city engineer, suggesting that he can keep his current job while discharging the responsibilities of the city engineer—primarily reviewing plans for future residential subdivision developments in WoodAcres. Krompton should:
a. recognize that by holding two jobs he is being paid by two parties for the same work
b. insist that he can only accept the offer if his present employer agrees to reduce his responsibilities to half-time
c. recognize that a 60-hour workweek schedule will take time away from his family
d. recognize that this arrangement will probably create a conflict of interest and refuse the offer

5.11 Willis is an aerospace engineering lab test engineer who works for a space systems contractor certifying components for spacecraft service. He is in charge of a team of technicians testing a new circuit breaker design made of lighter weight materials intended for service in unpressurized compartments in rockets and spacecraft. The new design has passed all tests except for some minor overheating during certain rare electrical load conditions. The lead technician notices that this overheating does not occur when a fan is used to cool the test apparatus and proposes to run the test with the fan to complete the certification process. He points out that the load conditions will only occur during thruster operation in space, which is a much colder environment. Willis should:
a. agree to the lead technician's suggestion, since he has many years of experience in testing and certification
b. agree to run the test as suggested but include a footnote explaining the use of the fan
c. insist on running the test as specified without the use of the fan
d. report the technician to the state board for falsifying test reports

5.12 Shamar is a registered PE mechanical engineer assigned as a project manager on a new transmission line project. He is tasked to build a project team to include several engineers and EITs that will be responsible for design and construction of 7.6 miles of high tension transmission lines consisting of steel towers and aluminum conductors in an existing right of way. He realizes that foundation design and soil mechanics will be an important technical area to his project, and he has never studied these subjects. He wonders if he is qualified to supervise such a project. He should:
a. meet with his supervisor to decline the assignment
b. decline the assignment and contact the state board to report that he is being asked to take responsibility for tasks he is not knowledgeable about
c. accept the assignment and check out an introductory soil mechanics textbook from the firm's technical library
d. accept the assignment and be sure his team includes licensed engineers with expertise in these areas

5.13 Julio is a design engineer working for a sheet metal fabricating firm. He is tasked with the design of a portable steel tank for compressed air to be mass produced and sold to consumers for pressurizing automobile tires. He designs a cylindrical tank to be manufactured by rolling sheet metal into a cylinder, closing with a longitudinal weld along the top, and welding on two elliptical heads. His design drawings are approved by his supervisor, Sonja, a licensed engineer, and by the vice president of manufacturing, but when the client reviews the designs, he asks the VP to change the design so that the longitudinal weld along the top is moved to the bottom where it will not be visible to improve the esthetics and marketability of the product. The VP agrees with this change. Julio learns of this change and objects, citing concerns about corrosion at the weld if it is on the bottom. Sonja forwards Julio's objection with a recommendation against the change to the VP, with a copy to the client, but the VP insists, saying esthetics is very important in this product. Julio should:
a. accept the fact that esthetics governs this aspect of the design
b. write a letter to the client stating his objections
c. put a clear disclaimer on the drawing indicating his objections
d. contact the state board to report that his recommendation has been overruled by the VP

5.14 William is a PE who designs industrial incineration systems. He is working on a system to incinerate toxic wastes, and his employer has developed advanced technology using higher temperatures and chemical-specific catalyst systems that minimize the risk to the environment, workers, and the public. A public hearing is scheduled to address questions of safety and environmental risk posed by the project, and William is briefed by the corporate VP for public affairs about how to handle questions from the public. He is told to buy a new suit, project an air of technical competence, point out that his firm is the industry leader with many successful projects around the world, and describe the proposed system as one with "zero risk" to the public. William should:
a. follow his instructions to the letter
b. insist that his old suit is adequate, because he refuses to appear more successful than he really is, but follow the other instructions
c. follow all instructions, except use the term "minimal risk" rather than "zero risk"
d. resign from his position and look for a different employer who won't ask him to face the public

5.15 Darlene is a metallurgical engineering EIT who works for a firm that manufactures automotive body panels. She has been tasked with improvements to the design of inner fender and trunk floor panels to reduce corrosion damage. After several weeks of study and comparison of alternatives, she submits a new trunk floor panel design utilizing a weldable stainless steel that will significantly reduce corrosion compared to the galvanized carbon steel alternatives she has been considering. The new panels will cost more, however, and after much study and debate, the VP of manufacturing rejects her design and approves an alternative made of a cheaper material. Darlene should:
a. accept the decision and work to finalize a workable design
b. resign from her position, since her employer has lost confidence in her
c. contact the state board to advise that her design decision has been overturned by a nonengineering manager
d. None of the above

5.16 Matt is a young PE who has just started his own consulting practice after six years of work with a small consulting firm providing structural engineering design services to architects. He has worked on steel and timber framed churches, prestressed and reinforced concrete parking structures, and many tilt-up strip center buildings. His expertise has been in the area of design of tilt-up concrete construction, where he has developed some innovative details regarding reinforcement at lifting points. His building designs, when constructed by experienced contractors, have reduced construction times and costs. Because of his expertise, he is approached by lawyer Marlene, who tells Matt that she represents a construction worker who is suing a project owner, contractor, and designer over a construction accident in which a tilt-up wall was dropped during construction, seriously injuring several workers. Marlene asks Matt to serve as an expert witness to assess the design and construction practices in the project and testify as to the causes of the accident. Marlene has taken the case on a contingency fee basis, in which she will earn 40% of any settlement, and she asks Matt if he would rather be paid by the hour for his study and testimony or instead accept 5% of any settlement, which she believes could be as high as $25 million. Matt should:
a. compare the 5% contingency fee with an expected fee based on his hourly rates, realizing that there is some chance he will earn nothing
b. be sure to have Marlene put the contingency fee arrangement in the form of a legal contract
c. decline the contingency fee arrangement and bill on an hourly basis
d. accept the contingency fee arrangement but donate the difference over his hourly rate to charity

5.17 Victor is a consulting engineer who is also in charge of a crew providing land surveying and subdivision design services to developers. He has been contracted to provide a survey of a 14-acre tract where a local developer is contemplating a subdivision, and he realizes that his crew had surveyed this same tract last month for another developer who has abandoned the project. He reprints the survey drawings, changing the title block for the new client. With respect to billing for the drawings, Victor should:
  a. bill the new client the same as he billed the original client to be fair to both
  b. bill the new client for half of the amount billed to the original client
  c. bill the new client only for any work he did to change the drawings and reprint them
  d. provide the new client with the drawings without any charge

5.18 Frank is a PE who works for ELEC, an electrical engineering design and construction firm. Frank's job is estimating construction costs, bidding construction projects, and supervision of design of electrical systems for buildings. Frank's bright EIT of five years, Linda, has just received her PE license and has resigned her position to open her own consulting business in a nearby community. Until she is replaced, Frank will have to also do all detail design of electrical systems for their projects. Frank receives a request for proposal (RFP) from a general contractor regarding design of electrical systems for a local independent school district. He realizes that his firm may be in competition with Linda for the engineering design, and knowing that Linda's salary was about half of his, he expects she may have a competitive advantage. Frank should:
  a. ask Linda not to bid on this project
  b. remind his contact with the general contractor that Linda has just left his firm, that she is inexperienced, and hint that she was sometimes slow to complete her design assignments
  c. emphasize his 18 years of experience and subsequent design efficiency in his proposal
  d. promote a CAD technician to a designer position so he can show a lower billing rate for engineering design hours

5.19 It is important to avoid the appearance of a conflict of interest because:
  a. the engineer's judgment might be adversely affected
  b. the engineer's client might suffer financial damages
  c. the appearance of a conflict of interest is a misdemeanor
  d. the appearance of a conflict of interest damages the reputation of the profession

5.20 The code of ethics published by the American Society of Civil Engineers is:
  a. legally binding on all licensed engineers practicing civil engineering
  b. adhered to voluntarily by members of the ASCE as a condition of membership
  c. legally binding on all engineers with a degree in civil engineering
  d. published only as a training guideline for young civil engineers

5.21 Which statement *MOST* accurately describes an engineer's responsibility to the environment?
   a. The engineer has a legal obligation to make sure all development is sustainable.
   b. The engineer has no obligation to the environment beyond protecting public health and safety.
   c. The engineer has a moral obligation to consider the impact of his or her work on the environment.
   d. The engineer's environmental responsibility is primarily governed by specific state laws.

SOLUTIONS

5.1 c. Jim can accept this invitation. We can assume there is no corporate policy prohibiting or restricting lunch invitations since VP Joan has accepted the invitation; therefore, there is no reason for Jim to decline the invitation. The opportunity to learn more about the new product is useful to him, his employer, and his clients; the cost of the lunch presumably would not be considered a "valuable" gift; and the lunch would not create either a conflict of interest or the appearance of a conflict to a reasonable person. If instead of lunch the offer involved a 10-day elk hunting trip or a vacation in the south of France, the solution would be very different because of the obvious "value" of the gift.

5.2 d. Doing nothing (a) is not an option if Harry really believes the trench represents a serious hazard to the workers. His code of ethics requires him to remember that his first and foremost responsibility is the public welfare, which includes the safety of the construction workers. Answers (b) and (c) are not the best way to proceed; his concerns should be reported to an engineer with some authority over the project. Since this is a city-contracted sewer improvement project, the city engineer will have project responsibility and will be the appropriate individual for Harry to take his concerns to.

5.3 a. Travis should decline the offer from firm C, explaining that he has already accepted an offer to work for firm A. While the *NCEES Model Rules* don't specifically address issues of personal integrity, it is clear that the engineer's obligation to employer and clients will not be satisfied by any decision that ignores Travis's verbal agreement to employment with firm A. Furthermore, such actions will tarnish his integrity in the eyes of firm A and by implication will harm the reputation and credibility of other students and the profession.

5.4 d. None of the first three solutions will address the concerns Jerry has raised about the safety of the particular project in question, so (d) is the correct solution. He should instead meet with Eddie to discuss the details of his design and make a determination if the design is completed safely. If it isn't, he will need to take further action to stop fabrication and construction while the design is reviewed and possibly modified to address any deficiencies. After this is done, he might want to consider all three of the other choices for his future. If he is thwarted in these responsibilities by Eddie, he should contact the state board with his concerns.

5.5 **c.** Hemmings cannot offer expert opinion that is contrary to accepted engineering practice without supporting that opinion with computer modeling, lab test results, or study of the literature. He can offer expert opinion that the design is not in line with accepted engineering practice without any supporting calculations, but he can't maintain that a substandard practice is acceptable without rational supporting evidence. Hemmings's credentials and experience may qualify him as an expert, but they do not relieve him of the requirement to base his professional opinion on facts.

5.6 **c.** Jackie should object to the request to exaggerate the size, qualifications, and experience level of her design group. The *Model Rules* require engineers to "be objective and truthful" in all professional matters, and the suggested exaggerations are clearly in conflict with this requirement.

5.7 **d.** It is not unethical or illegal to use the cost-benefit method, nor are certain values for human life prescribed by law; the answer is none of the above. The assumed values and calculations in the memo may have appeared callous or inflammatory to the juries in the resulting lawsuits, but they were not unethical or illegal. They may have been imprudent—an important lesson is that the public (jury) apparently objected to a design decision that increased the risk of a post-crash fire for such a small net benefit. Even though risk of death had been considered by the designers, the mode of death (burning to death in otherwise survivable crashes) was a factor in the strong reactions by the juries.

5.8 **c.** Studying products from a distributor is perfectly acceptable and can be a good way to gain a better understanding of pump equipment on the market. Accepting the binder does not represent a conflict of interest as implied by choices (a) and (b). Option (d) would be clearly setting himself up for an apparent conflict of interest.

5.9 **d.** Since Prof. Martinez has not been in responsible charge of the development of the assessment, he can't seal it, regardless of the level of review or the capabilities of the EIT who has done the work. Engineering students should be cautious in accepting a position as the sole engineering employee in a small firm; they are first encouraged to gain experience as an EIT under the guidance of an experienced PE and qualify for licensure as a PE before taking a position where licensure might be needed.

5.10 **d.** Krompton should recognize that the proposed arrangement will create a conflict of interest by placing him in charge of reviewing and approving plans from developers and potential developers that are in direct competition with his own employer. Any decision by him that might tend to make development less profitable for other developers could be advantageous to his employer by making their services more cost effective. Even if he were able to make all decisions rationally and without bias, the appearance of a conflict of interest would be very real and would cause a loss of credibility in the city engineer's actions and damage the reputation of the engineering profession.

5.11 **c.** Willis should insist on running the test as specified. The technician's suggestion to use a fan to "fudge" the test is technically indefensible, as well as unethical, in any case—the fan simulates convection cooling, which does not occur in the vacuum of space. Even if the technician had suggested a way to simulate an increased radiative heat flux, changes to a specified test procedure are not made casually. Much more study, documentation, and higher level approvals are involved.

5.12 **d.** Shamar should accept the assignment and select team members so that all areas of needed expertise are represented by licensed individuals who can seal appropriate portions of the plans. Option (a) is not recommended for an engineer's career advancement—he is expected to accept assignments of increasing responsibility; (b) is detrimental to his career—he will cause unnecessary concern with the board and with his supervisors; and (c) studying an introductory soil mechanics textbook may give him a better understanding of the problem but probably won't qualify him to seal foundation plans for the project described.

5.13 **d.** Julio is objecting because he knows water accumulates in compressed air storage tanks and causes corrosion, particularly at the bottom where the water will collect. Because of the metallurgy at the weld, corrosion is more aggressive at the weld site, and Julio considers this fundamentally a bad design if the weld is at the bottom. Since his recommendation (based on technical reasons and an increased risk to the public) is overruled under circumstances where the safety of the public is endangered, he is obligated (by the NCEES rules) to "notify his employer or client and other such authority." Having already notified his employer (through Sonja and the VP of manufacturing) and the client, all of whom except Sonja are part of the problem, his next logical step is to contact the state board with his concerns. This is rarely necessary; in most cases, the client and VP will be very interested in Julio's objection as it is based on public safety and will serve to reduce the company's own liability. However, if the employer and VP of manufacturing persist as described here, Julio must take additional action. It is not unreasonable to expect that Julio may face some sort of sanction from a management team that has put him in this position. He may even need a lawyer as this unpleasant situation deteriorates. Whistle-blowing should be considered the solution of last resort, as in this case.

5.14 c. William should avoid the use of the term "zero risk"; he knows that no project has zero risk. He should instead look for ways to quantify the risk that will be informative and meaningful to the public and try to convey the attitude of concern for minimizing the risk consistent with the potential public benefits of the project (increased employment and tax base). He should not do (a), and (b) does not address the problem of misinforming the public about the risk. There is no need to resign; to do so will not help his career or the project.

5.15 a. Darlene should accept the decision and work to make the chosen design successful and profitable for her employer. She does not need to resign—the business decision probably does not reflect a lack of confidence. It was made based on costs and profitability, not public safety, so she should not contact authorities to complain that a manager has overruled her. This kind of decision should be considered a "management decision" because it affects business and profits. Decisions that adversely affect public health, safety, or welfare should be considered "engineering decisions," and when these are overturned by nontechnical managers for other reasons, the engineer may be justified or even obligated to report this to authorities.

5.16 c. Matt should decline the contingency fee arrangement. While lawyers commonly work on contingency fee arrangements, engineers can't do this. Any contingency fee arrangement would put the engineer in a conflict of interest situation, where his engineering judgment can influence his income. In such a situation, his engineering judgment may not be sound or will at least appear to be conflicted to an outside observer.

5.17 c. Victor should be cautious in making sure that no additional surveying or resurveying is needed because of any changes to the tract. If additional fieldwork is not needed, he should bill the new client only for the work required to edit and reprint the drawings. He should not bill the new client the same as the original client, because that would be billing two clients for the same work, which is specifically prohibited by the *Model Rules*. Billing for half of the original amount is the same, just for an arbitrary amount. When two clients require the same survey simultaneously, it may make sense for Victor to facilitate a partnership in the project, but this is not always feasible when one client has already been billed for work done.

5.18 **c.** Option (c) is the only ethical and practical solution listed. Option (a) asking a competitor not to bid is not practical. Option (b) starting rumors about Linda's capabilities is clearly unethical. And option (d) is troublesome—experienced CAD technicians can do some aspects of design if closely supervised by a PE, but it isn't really necessary to make such a promotion just to show a lower billing rate. The billing rate for engineering design could be maintained at the same (competitive) level as when Linda was employed, even if the actual design work is done by Frank at twice the salary until a replacement for Linda can be hired. This in itself is not unethical, but if Frank does not budget sufficient time to do the design work in addition to his other work, it becomes a question of ethics. An engineer must allow sufficient time to do a professional job. The most practical and desirable solution is not listed—Frank should expedite hiring a qualified replacement for Linda, and he might have to consider declining the opportunity to bid on some projects until she is replaced.

5.19 **d.** Even the appearance of a conflict of interest can damage the reputation of the individuals involved and the profession as a whole, and such situations should be avoided. If there is no actual conflict of interest, the engineer's judgment will not be affected and the client will not suffer damages; nor is it criminal.

5.20 **b.** Codes of ethics published by professional societies are voluntarily adhered to by membership. They don't carry the weight of law but are much more than training guidelines—members who do not adhere to the society's code can be sanctioned by the society or forfeit their membership.

5.21 **c.** Many professional societies require the engineer to "consider" the impact on the environment. Some say he or she should consider whether the development is "sustainable." Most legal restrictions only require the engineer to prevent certain kinds of environmental damage and do not require sustainable development. There are, of course, specific state laws that must be followed, but many federal laws and regulations also apply. The engineer's responsibility is broader than laws and regulations in any event, making choice (c) the best answer.

CHAPTER 6

Environmental Management Systems

OUTLINE

SUSTAINABILITY   206

WASTE MANAGEMENT HIERARCHY   207

POLLUTION PREVENTION AND WASTE MINIMIZATION   208

LIFE CYCLE ASSESSMENT   209
Environmental Impact Assessment ■ Risk Assessment

INDUSTRIAL AND OCCUPATIONAL HEALTH AND SAFETY   216
Noise Pollution ■ Radiological Health and Safety

PROBLEMS   221

SOLUTIONS   223

While enacting change in corporate management is always difficult, implementing change specifically to accommodate environmental stewardship may present additional challenges, unless decision makers are enlightened regarding the advantages of the proposed change. The National Environmental Policy Act (NEPA) of 1969 established the requirement of developing an Environmental Impact Statement (EIS) for any federally funded project and is typically used to assess construction projects or industrial processes. This topic is covered in more detail under Environmental Impact Assessments (EIA). Another management approach is the "cradle-to-grave" analysis provided by Life Cycle Assessments (LCA) that evaluates a product throughout its life to evaluate material and energy flows associated with its production, use, and disposal. Design for the Environment (DfE) is a defined, systematic approach to selecting products and their manufacturing processes based on the environmental consequences identified in the LCA and EIA investigations.

Another ideology that is common in the corporate world is Total Quality Management (TQM), which can be described as a process of delivering the highest quality product and services, while attempting to continuously improve in future deliverables. Recently, this has been extended to Total Quality Management Systems and the Environment (TQEM) by applying the TQM philosophy to environmental management and sustainable development. Possibly the most widely used environmental management system is the International Standards Organization's (ISO) 14000 family of international standards, specifically ISO 14001. The ISO 14000 family consists of 26 standards in six classifications to assist corporations in the sound management of economic, social, and environmental issues.

SUSTAINABILITY

The prevailing ethic in the area of environmental concerns has recently focused on the issue of sustainability, which can be defined as the ability to meet society's current resource needs, without inhibiting future generations from doing the same. It has also been described as the nexus or "triple bottom line" between social, economic, and environmental concerns as represented in Figure 6.1. One example of a sustainability mindset is the shift of the LCA model to one that employs a "cradle-to-cradle" concept, which arises when all waste can be assumed to be able to be reprocessed into new raw materials, essentially eliminating the disposal stage of an LCA and greatly minimizing the impact of raw materials acquisition.

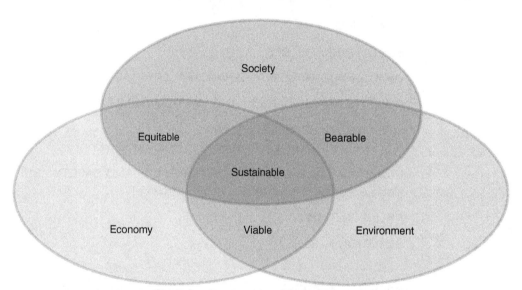

Figure 6.1 Graphical representation of the sustainability concept

It is well documented that those of us living in the U.S. are among the greatest consumers of the earth's resources per capita in the world today. Even more disconcerting is the fact that several societies, specifically China and India, look towards the U.S. as a model for developing and acquiring material possessions. With a combined population of nearly 2.6 billion, 36 percent of the world's inhabitants, those two countries alone have eight times the number of citizens as the U.S. It is a simple fact that the earth's remaining resources will not be able to satisfy the additional demands if such large populations were to consume at the current U.S. rate. In the near future, we will be left with few choices, and making environmentally sustainable choices for development now may be the only hope for future generations. In response, codes of ethics for many engineering disciplines have been recently modified or developed worldwide for protecting environmental integrity. For example, the 1996 update to the American Society of Civil Engineers (ASCE) Code of Ethics included the concept of "sustainable development" in the first of its seven fundamental cannons.

WASTE MANAGEMENT HIERARCHY

As a concept, the waste management hierarchy is simply a prioritized list of possible management choices for any natural or industrial system that must contend with waste materials or energy. In its most basic form, the list would appear as follows: (1) *source reduction*, (2) *recycling*, (3) *waste treatment*, (4) *secure disposal*, and (5) *direct release* to the environment.

While definitions vary slightly by various federal or state agency, in general, *source reduction* can be defined as *any practice that reduces the amount of any hazardous substance entering the waste stream prior to recycling, treatment, or disposal*. The more recent version of this philosophy would be termed *pollution prevention*, as the basic tenets are the same between the two and will be explored in more detail in the following sections of this review. Sometimes, the term *waste minimization* is also used synonymously; however, waste minimization will generally also include some of the recycling options discussed below. Regardless of the term used, source reduction is the idea that we do not have to worry about "what to do with the waste" if it is not generated in the first place.

Once a by-product stream is created, there are choices available on how to manage that material. *Recycling* is a common idea that has many applications, manifested through several subclassifications. Sometimes, the term *reuse* is incorporated as a separate entry in the waste management hierarchy before the recycling options; at other times, it is considered the highest level of recycling. In either case, reuse can be described as *using a discarded item for the original purpose* (for example, used cars or furniture). Often, materials can be *recirculated* within the same process, which is generally called *in-process* recycling. By definition, the U.S. Environmental Protection Agency (U.S. EPA) considers pollution prevention to include source reduction and in-process recycling. When a company chooses to use the waste stream from one process in another process within the plant boundaries, it is called *on-site* recycling. If that material is shipped to another facility to be incorporated into another process, it is called *off-site* recycling. Further, the discussion on life cycle assessments will offer additional possible definitions for recycling. It is interesting to note that many corporations have become engaged in the system of waste trading, where employees identify a waste product stream from another participating corporation that meets their requirements as a process feed stream.

The United States recycles approximately 28% of its waste and has nearly doubled its recycling rate in the past 15 years. Industries are increasingly viewing many "wastes" as potential raw materials. The cost-effectiveness of recycling varies widely, depending on the efficiency of the collection system and the market for the recyclables. The cost obtained for recyclables depends on the supply available to users of recyclables, the type of material available (aluminum, steel, corrugated cardboard, etc.), the geographic locations of seller and buyer, the quality of the materials to be recycled, and so on. Moreover, the market prices vary drastically for some materials within a time frame of years or even months. In many cases, recycling of materials such as aluminum and steel does not require subsidies from the government. Other materials such as mixed glass and low-quality mixed paper are not in great demand and are typically subsidized. Subsidies are in place to take advantage of the many benefits of recycling for those materials for which immediate cost benefits are not seen (for example, the savings in landfill volume).

If it is determined that the generated waste material cannot be used in an economic or appropriate way, several waste treatment technologies are available to transform the material into a less hazardous, or low-volume, high-concentration state. Often, the treatment process still produces a small amount of waste material that requires *secure disposal,* which is generally assumed to be final disposal in a secure landfill. A secure landfill is one that was designed to comply with RCRA regulations for control of the material and will be discussed in Chapter 16. It is unfortunately the case that *direct release* to the environment occurs with a high degree of regularity, both from the permitted (allowed) discharge standpoint and from a fugitive or even illegal discharge perspective. In the case of the permitted discharge, it is usually monitored to determine mass loadings to the environment and generally is followed up with an evaluation of the impact that the discharge has on flora and fauna sustained by the surrounding ecosystem.

POLLUTION PREVENTION AND WASTE MINIMIZATION

The fact that we utilize natural resources to maintain our existence and that natural processes cannot be 100% efficient means that waste streams will always be a part of human activity. In the United States alone, billions of tons of industrial waste are generated each year, and while technology exists to treat much of these streams and reduce the impact, the negative effects on the environment from insufficient treatment and questionable disposal practices are obvious. As defined previously, pollution prevention is any practice that reduces the amount of any hazardous substance entering the waste stream prior to recycling, treatment, or disposal, and waste minimization is pollution prevention with the inclusion of some recycling options. Pollution prevention would encompass the more efficient use of raw materials and energy, not only reducing material and energy waste but also in the process reducing costs and increasing yields. A more recent concept is the idea of *green engineering,* where the practice of pollution prevention becomes part of the design process from the very beginning of a project and is woven throughout in the evaluation of long-term cost and environmental benefit.

For existing manufacturing processes, pollution prevention begins with a *waste audit* (materials accounting in process lines) and *emission inventory* (direct releases through fugitive and secondary sources), where all material and energy flows are identified and evaluated for potential modification for each process unit in the facility. The engineer must also evaluate the interconnectedness of each separate process train in the facility, because a change in one shared material or energy flow could potentially impact several other flows. Special attention is placed on criteria pollutants, toxic chemicals, and other hazardous wastes, and a priority list is generated that provides realistic targets for the amounts of source reduction for those target compounds.

Identification of the sources is generally a simple task, and the real challenge comes in the cost justification to management. Often small changes are quite cost effective, such as valve or seal replacement for reduction of fugitive releases, and can be presented as appropriate maintenance practices. Substantial changes may require a more extensive *total cost accounting* to provide sufficient economic justification. Total cost accounting not only addresses the typical capital, operation and maintenance, and labor costs but also assesses less obvious costs, such as monitoring, permitting, and current or future liability, as well as relationships with employees, customers, and public perception of the corporation as an environmental steward.

LIFE CYCLE ASSESSMENT

In the decision-making process, historically, corporations were nearsighted with respect to their cost-benefit analyses and rarely investigated the impact of production choices over the long term. However, more recently, it has become common practice to evaluate choices from the *cradle-to-grave* perspective. The formal description of this mindset is called *life cycle assessment* (LCA) and can be broken down into three components: (1) an inventory of all material, energy, waste, and emission flows associated with the life cycle of a product; (2) an environmental impact assessment of those flows; and (3) a feedback mechanism to assist in choosing between alternative products or processes or to address identified negative environmental impacts.

A conceptual framework for performing an LCA is provided in Figure 6.2. As seen in the graphical representation, the LCA is comprised of flow inventories during five stages of the product's life, namely, (1) *raw material acquisition* (extraction from nature), (2) *material manufacture* (raw material turned into feedstock), (3) *product manufacture*, (4) *product use*, and (5) *product disposal*. By summing all flows over the five stages, a direct comparison can be made between products or manufacturing processes to determine the one that minimizes the impact to the environment. Classical comparisons that can be made in this way include paper or plastic bags at the supermarket, wax-coated paper or polystyrene cups for take-out hot drinks, or cloth versus disposable diapers, and so on. Sometimes, the choice is clear, but more often it becomes a choice between high flows in one area such as waste compared to high flows in another such as energy, and it is left to the manager to determine the most appropriate choice.

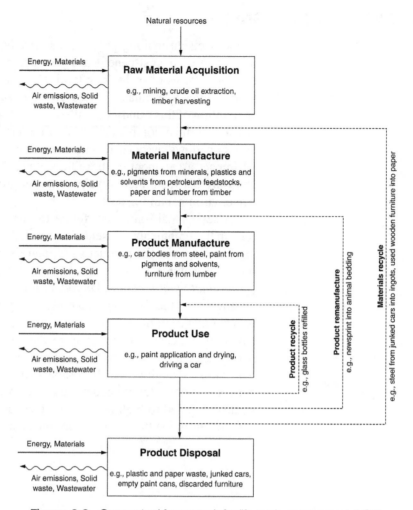

Figure 6.2 Conceptual framework for life cycle assessment (LCA)

Source: Allen and Rosselot, *Pollution Prevention for Chemical Processes*, © 2001, John Wiley & Sons, Inc. Reprinted by permission.

As seen in the figure, recycling has been divided into three categories. The first is termed product recycle and would be equivalent to the reuse term defined previously. The second process is termed product remanufacture and can be defined as conversion of a product for reuse in a different manner. The third term is called material recycle and is most similar to the concept of recycling most people hold.

Environmental Impact Assessment

As the name suggests, an environmental impact assessment (EIA) is an evaluation tool that requires consideration of all potential impacts to an ecosystem for a current or proposed process. The document prepared for external review is called the *environmental impact statement* (EIS), and is required by law in the United States for any federally funded project. The process should be inclusive, not only in technical evaluation, but also in the array of constituents represented in the data gathering and interpretation, impact prioritization, and project management. The four steps in performing an EIA are (1) *screening*, (2) *scoping*, (3) *EIS preparation*, and (4) *review*.

The process of screening can be described as the determination of which projects warrant an EIA. As mentioned previously, some projects are required by law to complete an EIS, specifically when federal dollars are part of the funding source. Another method is the use of plus/minus lists, which often specify types of projects that always require an EIA, or projects that usually require an EIA. Sometimes the project site may be located to an environmentally sensitive area, and impact to that area due to additional human activity should be assessed. Often a project threshold, such as land area used or total energy consumed, can be a good indicator if an assessment is warranted. Finally, an initial environmental evaluation (IEE), sometimes called a mini-EIA, can be used to identify anticipated impacts, without the quantification and evaluation required by a full EIA.

Scoping is used to identify the target (key, critical) impacts that will form the basis of the study. While it would appear important to quantify all potential impacts, the reality is, it is usually a few critical issues that will determine if a project will proceed or be modified. The scoping process should identify all important issues, however the assessment should focus only on those that have a potential to require mitigation, or those that would instigate project modification. This is also a time where public involvement can assist in defining priorities, and expedite the process by not having to face public challenges regarding the selection of critical issues after the submission of the EIS (i.e., during the review process).

EIS preparation needs to consider not only the data to be presented, but *why* the data is important, *to whom* the EIS will be addressed, and *how* the information will be communicated. This requires knowledge of the prospective audience, a clear and defined purpose for the EIS, and a technical framework for the written document. Because the project advocate conducts the assessment and the government agency that requested the EIA usually grants permission for the project to proceed, the EIS must be reviewed by a technically competent but impartial review team. This is done to assure the public that an independent decision was made that upholds the spirit of the EIA, namely, to protect the ecosystem from unnecessary harm. The government agency requesting the EIS is not required to abide by the opinion expressed by the technical review panel, however it is most often the case that a well performed review will carry substantial influence.

Risk Assessment

Risk can be defined as the probability of a negative outcome from a particular activity. The definition implies three separate concepts: (1) risk is a probability, and therefore it is *quantifiable*, (2) negative outcome implies that an *adverse effect* or some *hazard* is present, and (3) this risk is attributable to a singular *causal event* or *pattern of activity*. The scientifically rigorous process to determine risk is called *risk assessment,* while the use of this data to develop regulations that stipulate allowable concentrations of specific compounds in the environment is called *risk management.* Topics covered in this section can be found in the environmental engineering section of the *FE Supplied-Reference Handbook*.

The U.S. EPA has established levels of *acceptable risk* based primarily on the increased chance of the negative outcome occurring as a result of environmental contamination during normal human activity. The acceptable risk of death due to cancer (carcinogenic risk) is usually 10^{-6}, or one in one million, although a range of acceptable risk for different compounds may range from 10^{-4} to 10^{-7}. For non-carcinogenic risk, a hazard index (defined below) greater than 1 is considered unacceptable.

This risk is the *incremental risk* allowed in addition to the *background risk*. The background risk is the risk due to all of the environmental factors to which the average person is exposed. For example, the background risk of contracting cancer is approximately 1 in 4, or $250,000/10^6$. *Toxicology* is the study of the nature, effects, and detection of contaminants in living organisms. A list of helpful definitions is provided in Table 6.1.

Table 6.1 Definitions associated with risk assessment

Acute toxicity	The adverse effect due to a single dose of a contaminant
Carcinogen	A substance capable of inducing cancer in an organism
Chronic toxicity	The adverse effects resulting from repeated doses of or long-term exposure to a substance
Mutagenic	The ability to induce mutations in genes or chromosomes of an organism
Teratogenic	The ability to develop defects in an embryo

The process of performing a risk assessment has five components: (1) data collection and evaluation, (2) toxicity assessment, (3) exposure assessment, (4) risk characterization, or quantitative estimation, and (5) risk management. Data collection and evaluation consists of a thorough examination of a selected site, at which time concerns for human health are identified, in particular, specific chemicals are identified and their concentrations are determined. Often, stages 1 and 2 above are combined and called *hazard assessment*. Potential exposure pathways are suggested based upon the relative soil, water, and air contamination levels.

Toxicity assessment is the process of determining specific negative impacts from quantified doses of the identified chemical. The first part involves *hazard identification*, where specific endpoints (e.g., death, cancer, neurological damage, etc.) are specified. The most important factor that determines the extent of the adverse health effect is the *dose* of the identified chemical, usually expressed as mg of chemical administered per kg body mass per unit time (usually days). While some epidemiological data exists, most toxicity data is obtained from animal studies. These results are often presented in a "dose-response" curve. A typical dose-response curve is shown in Figure 6.3. The results are extrapolated through the use of mathematical models and safety factors to determine acceptable human doses.

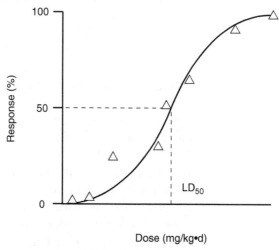

Figure 6.3 Dose-response curve

Data collection attempts to identify target endpoints at discrete doses, specifically the determination of the dose that instigates a negative physiological response. One toxicological measure widely used is the LD_{50}, defined as the dose that is lethal (causes death) in 50 percent of test organisms. Two other measures are the lowest observable adverse effect level (LOAEL) and the no observable adverse effect level (NOAEL). Since the NOAEL establishes the acceptable dose for the test organism, it is used to determine the reference dose (Rf D), which is calculated by multiplying the NOAEL by safety factors, and is generally used for noncarcinogenic endpoints. Another method linearizes the dose response at the lowest administered doses, assuming that there is no threshold dose, and reports the cancer slope factor (CSF or SF).

Exposure assessment identifies all potential pathways for the chemical to enter the body, generally classified as (1) ingestion, (2) inhalation, (3) transdermal, and (4) direct injection. Ingestion occurs through eating and drinking, but also needs to account for the potential of ingestion due to poor hygiene (e.g., not washing hands before eating) and children that place contaminated items in their mouths (e.g., eating dirt, chewing paint off of a windowsill, etc.). Inhalation refers to air contaminants entering the lungs, however some contaminants may be removed by mucous in the nasal passages and subsequently ingested. Transdermal exposure can occur through the skin (e.g., organic solvents) or through the mucous membrane (e.g., around the eyes). Direct injection may be through the obvious use of a hypodermic needle, but may also occur through cuts and abrasions in the skin.

Exposure and intake rates should be measured for site specific cases when possible, however standard rates are generally given for both children and adults, and several are given in the environmental engineering section of the *FE Supplied-Reference Handbook*. The *chronic daily intake* (CDI), called the *daily dose* (DD) in the *FE Supplied-Reference Handbook*, has units of mg of chemical per kg body weight per day, and can be expressed as

$$DD = \frac{(C)(I)(EF)(ED)(AF)}{(AT)(BW)}$$

where C is the species concentration (mg/volume), I is the intake rate (volume/day), EF is the exposure frequency (days/year), ED is the exposure duration (years), AF is the absorption factor (unitless), AT is the averaging time (days), and BW is the body mass (kg). Volume is generally expressed in m^3 for air and L for water.

Risk characterization, or quantitative estimation is the process of determining the value of the risk due to the specified activity that results in the determined exposure to the chemical that has a known dose response. In general, risk of cancer can be expressed as

$$\text{Risk} = (DD)(CSF)$$

while for noncarcinogenic endpoints, the hazard index can be expressed as

$$HI = \frac{DD}{RfD}$$

To account for risk from multiple pathways or from multiple substances, total cancer risk or total hazard index can be calculated as a simple sum of all identified risks. However, care must be taken in the interpretation of these results as synergistic effects may enhance or inhibit true risk.

The final step in risk assessment is often overlooked, but may be the most important. Besides establishing the regulatory limits for environmental exposure, risk management entails public education and risk mitigation. It is not possible to obtain a zero risk for environmental exposure, although it is an admirable goal, and cost-benefit analysis is clouded by the ethical question of what is an acceptable monetary expense for the protection of human health. Unfortunately, public perception of risk is a matter of personal convenience, in that voluntary risk that may have a known negative outcome (e.g., smoking, drinking) is acceptable, while imposed environmental risk (e.g., exposure to a toxic chemical in air or water) that carries a 10^{-6} cancer risk is unacceptable.

Example 6.1

Exposure assessment

A two-year-old child plays on a vacant ground that was once a small industrial site. The child plays on the site five days a week for three years. Soil tests show that the soil contains benzene (C_6H_6) at a concentration of 1.1 µg/kg. What mass of benzene does the child consume?

Solution

Given a soil ingestion rate of 200 mg/day (Table 4.3), the mass of C_6H_6 ingested is

$$\text{mass} = \left(1.1 \times 10^{-6} \frac{g}{kg}\right) \cdot \left(200 \frac{mg}{day}\right) \cdot \left(\frac{1 \text{ kg}}{10^6 \text{ mg}}\right) \cdot \left(\frac{5 \text{ day}}{week}\right) \cdot \left(\frac{52 \text{ week}}{yr}\right) \cdot (3 \text{ yr})$$

$$= 0.18 \text{ µg}$$

This may not seem like a lot of benzene, but consider how many molecules of benzene are ingested by the child. This calculation can be performed relatively easily by knowing the molecular weight of benzene (78 g/mole), recalling Avogadro's number, and taking care of units:

$$(0.18 \text{ µg}) \cdot \left(\frac{1 \text{ g}}{10^6 \text{ µg}}\right) \cdot \left(\frac{1 \text{ mole}}{78 \text{ g}}\right) \cdot \left(\frac{6 \times 10^{23} \text{ molecules}}{mole}\right) = 1.38 \times 10^{15} \text{ molecules}$$

So the child has ingested 1380 trillion molecules of benzene. And given that benzene is "reasonably anticipated to be a human carcinogen," and that theoretically one molecule of a carcinogen can cause cancer, the original answer of 0.18 µg no longer seems insignificant.

Example 6.2

Dose

What is the dose of benzene for the child eating the soil described in Example 6.1, setting the averaging time equal to the time of exposure?

Solution

The dose, averaged over the three years of exposure, is

$$\text{dose} = \frac{0.18 \times 10^{-3} \text{ mg } C_6H_6}{(14 \text{ kg}) \cdot (3 \text{ years}) \cdot \left(\frac{365 \text{ day}}{yr}\right)} = 1.2 \times 10^{-8} \frac{mg}{kg \cdot day}$$

Example 6.3

Cancer risk

What is the risk to the child that has ingested benzene given a potency factor of benzene of 0.029 kg·day/mg?

Solution

The mass of benzene calculated in Example 6.1 can be utilized in this example problem. However, the dose calculated in Example 6.2 cannot be used, as this dose used an averaging time of three years. Rather, the CDI is to be used. The CDI is calculated as

$$\text{CDI} = \frac{0.18 \times 10^{-3} \text{ mg C}_6\text{H}_6}{(14 \text{ kg}) \cdot (70 \text{ years}) \cdot \left(\frac{365 \text{ day}}{\text{yr}}\right)} = 5.0 \times 10^{-10} \frac{\text{mg}}{\text{kg} \cdot \text{day}}$$

Thus, the incremental risk for the child to contract cancer due to the exposure of benzene is:

$$\text{risk} = (\text{CDI})(\text{slope factor}) = \left(5.0 \times 10^{-10} \frac{\text{mg}}{\text{kg} \cdot \text{day}}\right)\left(0.029 \frac{\text{kg} \cdot \text{day}}{\text{mg}}\right) = 1.45 \times 10^{-11}$$

The inverse of this risk is 6.9×10^{10}, so the risk can also be expressed as one in 69 billion.

Example 6.4

Hazard index

Assume that the playground from Example 6.1 also contains 4.2 μg/kg of PCBs. The reference doses for benzene and PCBs are 4.0×10^{-3} mg/kg·day and 7.0×10^{-6} mg/kg·day, respectively.

Solution

The dose of benzene, averaged over the exposure time, was calculated to be

$$1.2 \times 10^{-8} \frac{\text{mg}}{\text{kg} \cdot \text{day}}$$

In a similar manner, the dose of PCBs can be calculated to be

$$2.7 \times 10^{-7} \frac{\text{mg}}{\text{kg} \cdot \text{day}}$$

The hazard index for each contaminant is thus:

$$\text{HI}_{\text{benzene}} = \frac{\text{dose}}{\text{RfD}} = \frac{1.2 \times 10^{-8} \frac{\text{mg}}{\text{kg} \cdot \text{day}}}{4 \times 10^{-3} \frac{\text{mg}}{\text{kg} \cdot \text{day}}} = 3 \times 10^{-6}$$

$$\text{HI}_{\text{PCB}} = \frac{2.7 \times 10^{-7} \frac{\text{mg}}{\text{kg} \cdot \text{day}}}{7.0 \times 10^{-6} \frac{\text{mg}}{\text{kg} \cdot \text{day}}} = 0.039$$

The sum of these hazard indices is clearly controlled by the hazard index of the PCBs. However, since the sum (0.039) is less than unity, the exposure is said to be allowable in terms of noncarcinogenic impacts.

INDUSTRIAL AND OCCUPATIONAL HEALTH AND SAFETY

Nearly all workplace environments can present the labor force with hazards on a regular basis. It is the job of those involved with the study of industrial and occupational health and safety to identify health and safety hazards, estimate the effects these threats pose on the labor force, educate employees and management on the risks associated with the presence of these hazards, and provide safeguards to mitigate the identified risks.

In recent decades, environmental engineers have used these or similar processes to examine the potential risk of human activity on the health and safety of the general population, as well as the world's ecosystems in general. The following sections will look at the specific risks and mitigation techniques involved with noise pollution and radiological health and safety.

Noise Pollution

Noise pollution has been defined as unwanted sound, and can vary in intensity from a mild irritant (nuisance) to a hazard with the potential for permanent hearing loss. In contrast to the material pollution that has been discussed previously and is most commonly addressed in environmental engineering, noise is an energy form, and thus requires a different approach for evaluation and control. Topics covered in this section can be found in the environmental engineering section of the *FE Supplied-Reference Handbook*. Relative levels of sound (measured in dB) are given for a variety of situations in Figure 6.4.

Sound is a result of the vibrations of solids, or the dynamics of fluids as they interact with solid objects. These vibrations cause minute pressure variations which are able to be detected by the ear. It is generally assumed that the pressure fluctuations can be represented by a sinusoidal wave of *period* (*P*) traveling at a velocity commonly called the *speed of sound* (*c*), which may be calculated at atmospheric pressure as

$$c = 20.05 T^{1/2}$$

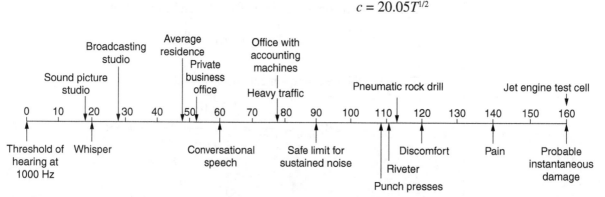

Figure 6.4 Relative scale for various sound pressure levels (Hammer, Willie; and Price, Dennis, *Occupational Safety Management and Engineering*, 5th Edition, © 2001. Adapted by permission of Pearson Education, Inc., Upper Saddle River, NJ.)

where c has units of m/s, and T is the absolute temperature (K). The *frequency* (f) of the wave is the inverse of the period, and the *wavelength* (λ) of the sound wave can be calculated as

$$\lambda = \frac{c}{f} = c\,P$$

The amplitude is defined as the peak height (which is equivalent to the trough depth) measured from an arbitrary zero pressure line (that is, the prevailing pressure is assigned a value of zero). Since the average of the peak and trough would be zero, it is standard convention to describe the average pressure in terms of the *root mean squared pressure* (P_{rms}), which can be expressed as

$$P_{rms} = \left[\frac{1}{T} \int_0^T P^2(t)\,dt \right]^{1/2}$$

where T is the time over which the measurement was taken. Since the variation in pressures can vary greatly depending on the source, sound is measured on a log scale. The most common form of the expression of sound levels is the *sound pressure level* (*SPL* or L_p) in units of *decibels* (*dB*), which can be expressed as

$$SPL = 10 \log \frac{P_{rms}^2}{(P_{rms})_0^2} = 20 \log \frac{P_{rms}}{(P_{rms})_0}$$

where $(P_{rms})_0$ is a reference pressure set at 20 µPa (that is, 2×10^{-5} Pa). Most often, more than one source of noise must be considered simultaneously, in which case the SPL for each source should be determined and combined using the following expression:

$$SPL_{total} = \Sigma\,SPL = 10 \log \left[\Sigma\,10^{(SPL_i/10)} \right]$$

where SPL_i denotes each of the individual sound pressure levels.

Example 6.5

Calculation of total SPL

A factory has four pieces of equipment in the same room with SPLs of 96, 108, 102, and 98 dB measured in the center of the room, which is a distance of 30 ft from each unit. Determine the total SPL from all four units.

Solution

The expression used to calculate the sum of individual noise levels can be found as

$$SPL_{total} = 10 \log \left[\Sigma\,10^{SPL/10} \right]$$

Using the data given in the problem statement, we can solve for SPL_{total} as follows:

$$SPL_{total} = 10 \log \left[10^{\left(\frac{96}{10}\right)} + 10^{\left(\frac{108}{10}\right)} + 10^{\left(\frac{102}{10}\right)} + 10^{\left(\frac{98}{10}\right)} \right] = 109.5\,dB$$

It is often assumed that a wave emanates from a point. In as much, sound that is not reflected back toward the source will radiate in a spherical pattern, thus reducing the intensity inversely proportional to the square of the distance from the source. This change in SPL from a point source is often referred to as attenuation and can be expressed as

$$(\Delta \text{SPL})_{\text{point}} = 10 \log \left(\frac{r_1}{r_2} \right)^2$$

where $(\Delta \text{SPL})_{\text{point}}$ is the change in SPL for a point source (dB), and r_1 and r_2 are the distances between the source and two separate receptors.

Since the nonreflecting assumption is often not valid, another assumption that may be made is that the sound is emanating from a line source in a cylindrical pattern, and dispersion of waves occurs inversely proportional to distance from the source, expressed as

$$(\Delta \text{SPL})_{\text{line}} = 10 \log \left(\frac{r_1}{r_2} \right)$$

where $(\Delta \text{SPL})_{\text{line}}$ is the change in SPL for a line source (dB).

Noise control should always seek to reduce at the source whenever possible. This is usually accomplished by evaluating the source and applying corrective measures, such as machine balancing, friction reduction, or the use of barriers and application of dampening or sound absorbing materials. When potentially damaging sound levels persist after reduction and attenuation measures have been taken, *personal protective equipment* (*PPE*), such as earplugs or cup-style headgear, should be mandatory for operators and other employees.

Radiological Health and Safety

While there are many types of radiation present throughout the electromagnetic spectrum, radiological health generally focuses on *ionizing radiation*, a subset that possesses the potential for cellular damage. *Radioactivity* is the phenomenon whereby an unstable isotope undergoes nuclear disintegration, called *decay*, releasing a particle or electromagnetic radiation to carry off the excess energy. The three major classes of decay products are alpha particles, beta particles, and gamma radiation.

Alpha particles are equivalent to the nucleus of the helium-4 atom (represented as ^4_2He), comprised of 2 protons and 2 neutrons. Loss of an alpha particle must then reduce the mass of the isotope by 4 mass units and reduce the charge by 2. This is usually represented as

$$^A_Z\text{X} \rightarrow\ ^{A-4}_{Z-2}\text{X} + ^4_2\text{He}$$

where Z is the atomic number (number of protons) and A is the atomic mass number, which is equal to the number of neutrons plus the number of protons ($N + Z$). You will note that the original isotope (called a *parent*) yields a different element (called a *daughter*) upon alpha decay, due to the fact that the daughter nucleus contains two less protons than the parent. Alpha decay is common for elements of atomic number greater than 82.

Alpha particles have extremely high energy (exit velocities near 10^4 miles per second) but cannot penetrate dense materials (for example, they cannot penetrate farther than the epidermis) and can be shielded by ordinary clothing or a sheet of paper. Care must be taken for ingested material, where damage is local but intense. Emitted alpha particles, upon striking a surface, cause the substance to release electrons, which the alpha particle uses to form a stable helium atom. The atoms that have lost electrons remain ionized until they can replace their losses.

Another isotope instability results in the decay of a neutron into a proton, which is retained in the atom, and an electron, which is emitted. Because the loss of a neutron is balanced by the gain of a proton, the atomic mass number remains unchanged, but the daughter is a new element due to the change in atomic number. The emitted electron is called a *beta particle* (β), and the transformation is usually represented as

$$^A_Z X \rightarrow ^A_{Z+1} X + \beta^-$$

Beta particles are fast-moving electrons, which have much greater penetrating power than alpha particles due to their small size but are less ionizing. They may be shielded through the use of thin metal plating; however, inappropriate choice of the plating material may result in the generation of X-rays. Appropriate shielding choices include lead, aluminum, and plastics.

Alpha and beta particle decay is often accompanied by *gamma ray* emission, which is a release of electromagnetic radiation energy as the newly formed element settles into a more stable state. Gamma rays are similar in energy and damage potential to X-rays, the latter of which is generated by high energy electrons striking a suitable target material. Both are highly penetrating energies that cause damage throughout a body, not solely on the surface, as is the case with alpha and beta particles. Due to their high penetration potential, gamma rays and X-rays require more substantial (thicker) shielding, with lead as the most common choice.

Radioactive decay is considered constant and is dependent on the original number of nuclei (N) present. The decay of radioactive isotopes can, therefore, be expressed as

$$N = N_0 \, e^{-\lambda t}$$

where N_0 is the original number of nuclei, λ is the radioactive decay constant, and t is the time interval. Note that the use of λ in this expression should not be confused with the use of the same symbol previously for wavelength. Often, it is convenient to express the decay in terms of the species *half-life*, which can be defined as *the amount of time required for 50% of the nuclei to decay*. This can be determined from the above expression by solving for $t_{1/2}$ (sometimes represented as τ) when N is equal to one half of N_0, and is calculated as follows:

$$t_{1/2} = \tau = \frac{\ln 2}{\lambda} = \frac{0.693}{\lambda}$$

Solving for λ in the half-life expression and substituting into the previous one yields

$$N = N_0 \exp\left[\frac{-0.693\, t}{t_{1/2}}\right] = N_0 \exp\left[\frac{-0.693\, t}{\tau}\right]$$

Example 6.6

Radiation Half-Life

How long must you store a 5 μCi/L solution of iodine-131 (half-life of 8.06 days) before safe disposal in a drain if the discharge limit is 10^{-6} μCi/mL?

Solution

The expression for radiation half-life is

$$N = N_0 \exp\left[\frac{-0.693\, t}{\tau}\right]$$

Using the data given in the problem statement, but correcting values for consistent volume units, we can determine the storage time as

$$10^{-6}\frac{\mu Ci}{mL} = 5\times 10^{-3}\frac{\mu Ci}{mL}\exp\left[\frac{(-0.693)(t)}{8.06\text{ day}}\right] \Rightarrow t = 99 \text{ days}$$

The dose for gamma and X-ray exposure is called the *roentgen* (R), while the dose unit that describes the amount of all radiation energies absorbed by the body is called the *rad*. The *relative biological effectiveness* (RBE) factor is a ratio of the absorbed dose of gamma radiation (in rads) to the absorbed dose of another type of radiation required to have an equal biological impact. Since all tissues and organs have unique sensitivity to different radiation types and sources, the *tissue weighting factor* (W_T) has been developed as a metric to assign high values of potential damage to sensitive tissues or organs, and low (or zero) values to other tissues or organs. Finally, the *roentgen equivalent man* (REM) is an indication of the extent of biological injury that is probable from a dose of a specific type of radiation to a specific tissue or organ and can be expressed as

$$\text{dose in REMs} = \text{RBE} \times (\text{dose in rads}) \times W$$

Since the radioactive source is usually considered a point source, gamma and X-ray radiation intensity decreases with the square of the inverse of the distance from the source. Often, separation distance is the most cost-effective form of protection from high dose exposure to a source. The inverse square law may be expressed as

$$I_2 = I_1\left(\frac{r_1}{r_2}\right)^2$$

where I_1 is the radiation intensity at a distance r_1 from the source, and I_2 is the radiation intensity at a distance r_2 from the source. Other means of radiation attenuation generally utilize shielding, and tables and graphs have been created that provide the required thickness of various materials to achieve a desired percentage of radiation intensity reduction.

Finally, there are potential adverse physiological responses to other forms of energy in the electromagnetic spectrum, which are classified as *nonionizing radiation*. *Ultraviolet radiation* can have a mild-to-severe impact on the skin, depending on incident radiation intensity and exposure time, or can cause blindness if not managed properly from industrial sources (for example, electric arc welding). Although *visible light* is the most common, and generally the most safe, it can also have a negative impact on vision if not managed properly. In addition, *infrared radiation* may cause burns to exposed areas, as well as the potential for eye damage. Another source of thermal effects is *microwave radiation*, which causes burns and tissue damage by the creation of heat due to an increase in the kinetic energy of the affected tissue.

PROBLEMS

6.1 Recycling waste streams is preferable to all but which one of the following alternatives?
a. Manufacturing process modification to increase efficiency
b. Secure disposal in a RCRA landfill
c. Incineration
d. Composting

6.2 Which one of the following is a critical entity to quantify in performing a life-cycle assessment?
a. Material flows
b. Energy flows
c. All waste flows
d. All of the above

6.3 Which one of the following is not an example of pollution prevention?
a. Using a less toxic chemical in a manufacturing process
b. Reducing the amount of packaging on a product
c. Using waste steam to preheat a feed stream
d. Upgrading an air pollution control system

6.4 A 170-lb man drinks 2 L of water every day that, unbeknownst to him contain benzene at a concentration of 30 µg/L. If he has been drinking this water for 25 years, estimate his risk of contracting cancer from this exposure. You may assume an absorption factor of 1 and a cancer slope factor of 0.029 (kg·day)/mg.
a. 5.5×10^{-6}
b. 7.1×10^{-6}
c. 9.8×10^{-6}
d. 2.3×10^{-5}

6.5 Which one of the following is not a method used for screening a project to determine if an environmental impact assessment is necessary?
a. Federal government lists
b. Project thresholds
c. Opinion of neighborhood association
d. Project located in sensitive area

6.6 A piece of industrial equipment gives off sound at 105 dB measured at a distance 5 ft in front of the machine. At what distance is the sound level reduced to a safe level (90 dB) assuming the machine acts as a point source?
a. 14 ft
b. 21 ft
c. 28 ft
d. 35 ft

6.7 An accident has caused the release of radioactive materials with an intensity 25,000 times greater than a safe level when measured 10 m from the source. At what distance must the safe perimeter be maintained during the clean-up process?
  a. 0.8 km
  b. 1.6 km
  c. 8.0 km
  d. 16 km

6.8 A rural family has discovered that their well water has a cadmium concentration of 50 µg/L. Determine the hazard index for their 5-year-old daughter if the reference dose is 0.0005 mg/(kg·day).
  a. 10.0   c. 3.3
  b. 7.2   d. 1.6

SOLUTIONS

6.1 a. Secure disposal is one of the least attractive options, even when done responsibly. Incineration and composting are both examples of waste transformation processes and should be implemented when appropriate after recycling efforts have been exhausted. Increased efficiency in manufacturing is an example of reducing waste generation at the source and is therefore highest on the hierarchy.

6.2 d. An extensive life-cycle assessment tracks all material, energy, air emissions, wastewater, and solid waste flows from the manufacture, use, and disposal of a specified product.

6.3 d. Pollution prevention is defined as reducing the amount of hazardous material entering the waste stream prior to any treatment. Therefore choices a, b, and c apply. Any processes added to control byproducts already generated is not considered source reduction.

6.4 d. Use the daily dose (DD) expression and the data given as follows:

$$DD = \frac{C \times I \times EF \times ED \times AF}{AT \times BW} = \frac{\left(30 \frac{\mu g}{L}\right)\left(2 \frac{L}{day}\right)\left(365 \frac{day}{yr}\right)(25 yr)(1)}{\left(365 \frac{day}{yr}\right)(25 yr)\left(\frac{170 \text{ lb}}{2.2 \frac{\text{lb}}{\text{kg}}}\right)}$$

Solving for DD yields

$$DD = 0.776 \frac{\mu g}{kg \cdot day} = 7.76 \times 10^{-4} \frac{mg}{kg \cdot day}$$

Risk can now be estimated as follows:

$$\text{Risk} = DD \times CSF = \left(7.76 \times 10^{-4} \frac{mg}{kg \cdot day}\right)\left(0.29 \frac{kg \cdot day}{mg}\right) = 2.25 \times 10^{-5}$$

6.5 c. Choices a, b, and d are all screening methods to determine EIA necessity. While it is often prudent to respond to public opinion and private organizations may contract their own study, this is not required by law.

6.6 c. The change in SPL is a decrease of 15 dB (–15 dB), and is related to distance for a point source by the equation:

$$\Delta \text{SPL} = 10 \log \left(\frac{r_1}{r_2}\right)^2 = -15 \text{ dB}$$

Using the data given in the problem statement, the safe distance can be calculated as

$$-1.5 = \log\left(\frac{5}{r}\right)^2 \Rightarrow 0.0316 = \left(\frac{5}{r}\right)^2 \Rightarrow (0.0316)^{1/2} = \frac{5}{r}$$

$$\Rightarrow r = 28.13 \text{ ft}$$

6.7 b. Radiation intensity varies with the square of distance as follows:

$$I_2 = I_1 \left(\frac{r_1}{r_2}\right)^2$$

Assuming the safe level at distance r_2 is I_2, and $I_1 = 25{,}000 \times I_2$ at a distance of 10 m, the safe distance may be calculated as

$$I_2 = (25{,}000) I_2 \left(\frac{10 \text{ m}}{r_2}\right)^2 \Rightarrow 0.00004 = \left(\frac{10 \text{ m}}{r_2}\right)^2 \Rightarrow r_2 = 1581 \text{ m}$$

6.8 b. The expression for hazard index is given in the *FE SRH* as

$$HI = \frac{CDI}{RfD}$$

Since the reference dose is given in the problem statement, we only need to determine the *CDI*. Using the data provided in the problem statement, and the information in the *FE SRH*, we may write

$$CDI = \frac{(CW)(IR)(EF)(ED)}{(BW)(AT)} = \frac{\left(0.05 \frac{\text{mg}}{\text{L}}\right)\left(1 \frac{\text{L}}{\text{day}}\right)\left(365 \frac{\text{day}}{\text{yr}}\right)(5 \text{ yr})}{(14 \text{ kg})\left(365 \frac{\text{day}}{\text{yr}}\right)(5 \text{ yr})}$$

$$= 3.6 \times 10^{-3} \frac{\text{mg}}{\text{kg} \cdot \text{day}}$$

Now we can substitute this into the expression for *HI* above as follows:

$$HI = \frac{CDI}{RfD} = \frac{3.6 \times 10^{-3} \frac{\text{mg}}{\text{kg} \cdot \text{day}}}{0.0005 \frac{\text{mg}}{\text{kg} \cdot \text{day}}} = 7.2$$

Since the *HI* > 1, some noncarcinogenic impact exists.

CHAPTER 7

Environmental Science and Ecology

OUTLINE

MASS BALANCE APPROACH TO PROBLEM SOLVING 226

INTRODUCTION TO MICROBIOLOGY 227
Structure, Organization, and Taxonomy ■ Respiration and Metabolism ■ Cell Growth and Reproduction

BIOLOGICAL SYSTEMS 234
Toxicology ■ Bacterial Quantification ■ Microbial Ecology

BIOPROCESSING 238
Water Treatment and Disinfection ■ Wastewater Treatment and DO Sag ■ Solid Waste Treatment ■ Fermentation

DO SAG MODELING 241

LIMNOLOGY 248

WETLANDS 250

PROBLEMS 254

SOLUTIONS 257

Engineering has been defined as using the laws of science and mathematics to design structures or process units that perform a specific, pre-determined function. Applied to issues that impact the air we breathe, the water we drink, and the soils that we obtain food from, environmental engineers prevent the negative impact from, or remediate past practices of, waste discharges from human activities. Prior to the design of treatment processes or pollution control systems, a basic understanding of the science behind the design equations is required.

The next several chapters will serve as a review of the basic science and fundamental engineering principles necessary to solve most environmental engineering problems. Chapter 7 begins with a review of the mass balance approach to problem solving, followed by a comprehensive review of the biology concepts most often applied to environmental systems. Chapter 8 will review a wide array of chemistry principles pertinent to solving environmental problems, while Chapter 9 covers materials science and engineering. Chapter 10 covers principles of thermodynamics and phase equilibrium, and Chapter 11 reviews the basic and applied aspects of fluid mechanics. Water resources are examined in Chapter 12, and their extension into soil/groundwater systems are reviewed in Chapter 13. Taken as a whole, these seven chapters establish the foundation upon which the engineering design (Chapters 14-16) for the protection and treatment of environmental systems will be built.

MASS BALANCE APPROACH TO PROBLEM SOLVING

Many environmental engineering problems can be solved if the engineer has a good command of units and the mass balance approach. Mass balances are applied to a control volume. A *control volume* is an imaginary volume used to identify the system of study. The control volume is bounded by the *control surface*. In environmental engineering problems, the control volume may be a pond, lake, reactor, aquifer, or watershed.

Environmental engineers are often concerned with the concentration of a compound entering or exiting a control volume or the concentration of a compound within the control volume. To solve such problems, it is imperative that all mass flows of the compound *into* and *out of* the control volume be identified, and that any sources or sinks of the compound of interest *within* the control volume are identified.

In simplified terms, the mass balance approach states that all mass of a compound in a control volume must be accounted for. At *steady state* (that is, when the mass of the compound within the control volume is not changing with time), the sum of the mass of compound entering the control volume and the mass of compound being created in the control volume must equal the sum of the mass exiting the control volume and the mass being destroyed within the control volume. In mathematical terms, this can be written as

$$0 = \dot{m}_{in} - \dot{m}_{out} \pm \dot{m}_{rxn} \tag{7.1}$$

where $\dot{m}_{in}$ is the mass flux of compound into the control volume (M/T)

$\dot{m}_{out}$ is the mass flux of compound out of the control volume (M/T)

$\dot{m}_{rxn}$ is the *reaction rate term*, the mass rate at which the compound is either created or destroyed within the control volume (M/T)

The reaction rate term in Equation 7.1 refers to the creation or destruction of mass of the compound of interest. Often, in environmental systems, the compound decays (for example, the decay of biochemical oxygen demand in a biological reactor), and the sign is negative. Alternatively, the compound is formed (for example, the creation of a compound in a reactor from two reactants) in which case the sign on this term is positive. Reaction rates are more thoroughly discussed in Chapter 8.

It is very important to note that the mass balance equation only applies to steady state problems. The equation for non–steady state problems (in which case the mass of compound in the control volume changes with time) is given in Equation 7.2.

$$\left(\frac{dm}{dt}\right)_{cv} = \dot{m}_{in} - \dot{m}_{out} \pm \dot{m}_{rxn} \tag{7.2}$$

where $\left(\frac{dm}{dt}\right)_{cv}$ is the change in mass in the control volume over time.

Note that the terms on the right-hand side are unchanged as compared to the steady state equation (Equation 7.1).

Steady-state problems are more desirable to solve from the standpoint that, unlike non–steady state problems, steady-state problems do not necessitate the use of differential equations. Steady state problems can be identified by such key phrases in the problem statement as "after a long time"; for example, "What is the concentration of mercury exiting the lake after the paper mill has been illicitly discharging mercury at the given concentrations for a long time?"

When solving environmental engineering problems that involve a mass balance approach, the mass fluxes entering and exiting the control volume may be given directly in the problem statement (mass/time). Alternatively, flow rates and associated concentrations may be provided, in which case the mass flux is equal to the product of the flow and concentration. Note that the product of a flow rate and a concentration yields a mass flux.

INTRODUCTION TO MICROBIOLOGY

Discussions of water quality indicators (WQI) typically focus on chemical contaminants or, in the case of suspended or colloidal solids, physical contamination. However, in a global perspective, these are often trivial when compared to the potential negative health effects that are a result of biological contamination. While the United States has not seen large outbreaks of waterborne diseases due to biological contamination, worldwide pathogenic bacteria and other organisms in water can have devastating effects on human health. The ability to identify and quantify the presence of a single strain of organism exists; unfortunately, countless different microorganisms may be present in any ecosystem.

The word *biology* from the original Greek means the study of living organisms. Living things are distinguished from nonliving things by the expression of several life processes, including but not limited to reproduction, growth, respiration, metabolism, adaptation, and response to environmental stimuli. The science of biology is quite broad, and so it is often convenient to divide all life study into subcategories based on common traits. For example, biological systems may be categorized into groups such as the study of animals (*zoology*), the study of plants (*botany*), and the study of organisms that are small enough to be visible only with magnification (*microbiology*).

For environmental engineers, primary application of microbiological activity is in the areas of secondary wastewater treatment and the remediation of contaminated soils and groundwater systems. Microbial populations may also contribute to the depletion of oxygen in WWTP-receiving streams, and the contamination of surface and groundwater and the ability to provide safe drinking water through disinfection are a high priority for all populations. Additional roles for microbes include solid waste decomposition in landfills and composting systems, as well as the novel use of microbes in engineered environments (such as the reduction of the greenhouse gas CO_2, using microalgae in photobioreactors). This section reviews basic microbiological concepts and issues related to microbial ecology, as well as their specific relevance to environmental engineering applications.

Structure, Organization, and Taxonomy

In the biological world, a simple distinction can be made between organisms based on the complexity of cellular structure. The highest order organisms belong to the group known as *eucaryotes* and include animals (both vertebrates and invertebrates), plants, algae, fungi, and protozoa. In addition, eucaryotes may vary in complexity from *multicellular* organisms that display a very wide variety of functionality among individual cells in a single species to *unicellular* species in which each cell possesses identical functionality and cellular subunits. A schematic of two eucaryotic cells, one animal and one plant, is provided in the Biology section of the *FE SRH*. The examinee should be able to discern which schematic represents each type of cell by locating the cellular subunits described below.

Eucaryotes are distinguished by the inclusion of several membrane-delimited *organelles*, which are unique substructures that possess specific functionality. The most notable organelle is the *nucleus*, which contains nearly all of the cell's genetic information and the chromosomes required for cell replication. Additional genetic material may be found in the *mitochondria*, which also serves the cell by supplying the majority of the energy required for cellular activity. Other major organelles common to most eucaryotes include the *Golgi complex* (also called the *Golgi apparatus*) and *endoplasmic reticulum*, both used in the processing of proteins, and *vacuoles*, which serve as primary storage units and clean and maintain the cellular environment. Additional cell structures that are present in plants include a rigid *cell wall* that surrounds the cell's outer membrane and provides structural support and *chloroplasts*, which are organelles that extract energy from sunlight to convert carbon dioxide and water into simple sugars through a process known as *photosynthesis*.

A more physiologically simple form of organisms is called *procaryotes*. The primary difference is the near complete absence of organelles, including a membrane-delimited nucleus. Instead, procaryotes possess a single circle of double-stranded DNA called a *nucleoid* that is free to move within the cell. Of primary interest in this group are the bacteria, due primarily to their ubiquitous nature in natural ecosystems, variation in respiratory and metabolic functionality, and role in water and wastewater systems. A schematic of a typical bacterial cell is provided in the Biology section of the *FE SRH*.

Bacteria shape varies from spherical (*coccus*) to rod (*bacillus*) to curved (*vibrio*) and ranges in size from 0.5 to 2 µm in diameter and up to 5 µm in length. Individual cells may freely roam in the aqueous environment, or they may form clusters (*staphylo*) or chains (*strepto*). Some strains may form *spores* (protective outer covering) when subjected to harsh environmental conditions as a survival tactic, which may affect the selection of the appropriate disinfection technology or parameters employed when attempting to eradicate these particular strains.

Another common historical classification for bacteria is the *Gram reaction*, which identifies chemical (and therefore physical) differences in cell wall construction. The Gram reaction uses a dye-alcohol rinse procedure to determine the characteristics of the *peptidoglycan* (*murein*) cell wall present in a species of bacteria. Cells that retain the purple dye have several layers of peptidoglycan and are termed *Gram positive*. Cells that have very few layers of murein covered by a layer of lipopolysaccharides allow the dye to be rinsed away and appear pink under microscopic observation and are termed *Gram negative*. As a general rule, Gram negative bacteria are more *pathogenic* (disease causing) than Gram positive organisms, and the Gram reaction provides a rapid screening test to assist in bacteria classification. Although Gram reaction is still a common descriptor for bacteria, modern microbiological techniques employ highly specific genetic testing and DNA libraries for species identification.

The procaryotes also include the most primitive life forms, called *archaea* (or equivalently the *archaebacteria*). While archaea and bacteria share the cellular similarities that classify both as procaryotes, the universal phylogenetic tree indicates that archaea are more closely related to the eucaryotes than they are to bacteria, and thus the three-domain classification system was adopted. A summary of organism types and their significance is provided in Table 7.1.

The formal process of classifying organisms into groups (called *taxa*) is called biological *taxonomy*. If the three-domain classification system is accepted, a total of eight levels are used to describe the biological world. The eight levels are as follows:

1. Domain (3)—Eucaryotes, bacteria, archaea
2. Kingdom (6)—Animals, plants, fungi, protista, archaea, procaryotes
3. Phylum (35–40)—classification based on physical (morphological) similarities (*Note:* This taxa level is called Division for (9) plant and (6) fungi classifications.)
4. Class
5. Order
6. Family
7. Genus
8. Species

Table 7.1 Select microorganism classifications and their significance

Domain (Cell Structure)	Organism	Significance
Eucaryotes	Animals	top of food chain predators, animal grazers
	Plants	primary producers (food), carbon cycling
	Algae	primary producers (food), carbon cycling, wastewater treatment (stabilization ponds)
	Fungi	food, food processing (yeast), wastewater treatment, composting, bioremediation
	Protozoa	bacteria grazing, drinking water pathogen
Bacteria (Procaryotic)	Bacteria	nutrient cycling, primary feeders in wastewater treatment plants, bioremediation of contaminated soils and groundwater; drinking water pathogens
Archaea (Procaryotic)	Archaebacteria	thrive in extreme environments; extreme halophiles, hyperthermophiles, acidophiles

Most life forms are identified using the *binomial nomenclature*, where a specific organism is named as *Genus species*. Note the genus name is capitalized, and both genus and species names are italicized. An example might be the difference between *Homo erectus* (meaning *upright man*) and *Homo sapiens* (meaning *wise man*), where both are from the same genus but are two distinct species. For bacteria, common organisms may use abbreviated forms of the binomial nomenclature, such as *E. coli* for the common intestinal bacteria *Escherichia coli*.

Respiration and Metabolism

By definition, *microorganisms* (*microbes*) include bacteria, archaea, algae, fungi, and protista, which all serve as active agents in many engineering applications. This may include the production of food and drinks such as bread and cheese or beer and wine; the development of drugs through fermentations in the pharmaceutical industry; environmental remediation techniques such as secondary wastewater treatment, remediation of contaminated soils, and groundwater systems; composting and solid waste decomposition; air quality management; and ecosystem restoration. Microbial populations may also contribute to the depletion of oxygen in wastewater treatment plants (WWTP) and receiving streams and in the contamination of surface and groundwater. Further, the ability to provide safe drinking water globally through the eradication (disinfection) of microbes is a high priority. For these reasons, it is useful to focus on microbes, and bacteria in particular, in the following discussions on cell growth and activity.

Microbes may be classified by their primary *respiration* process, specifically the entities that serve as the source of mass for growth and replication, and the energy provider to conduct these processes. *Metabolism* refers to two processes: *catabolism*, which describes the breaking down of chemical compounds for energy extraction, and *anabolism*, which is the assimilation and synthesis of new compounds and formation of biomass. *Autotrophs* are characterized as organisms that utilize inorganic carbon (for example, CO_2 and HCO_3) as their sole carbon source, while capturing energy from an external source to synthesize biomass. *Heterotrophs* are characterized by the use of organic matter as a supply of carbon for biomass synthesis and are further classified depending on the energy source. A summary of common terms used to describe several metabolic processes is provided in Table 7.2.

Table 7.2 Definitions for select microbial metabolic processes

Process	Carbon Source	Energy Source
chemoautotrophs	inorganic carbon	organic or inorganic chemicals
photoautotrophs	inorganic carbon	sunlight
photolithotrophs	inorganic carbon	sunlight
chemoheterotrophs	organic matter	organic chemicals
chemoorganotrophs	organic matter	organic chemicals
photoheterotrophs	organic matter	sunlight

The most efficient means of respiration for aqueous microbes is to extract dissolved oxygen from water, a process that is called *aerobic respiration*. The process of converting the substrate into energy requires a *terminal electron acceptor*, which for aerobic processes would be oxygen. Aerobic biological processes require organisms, organic matter, oxygen, and water, and the complete biological transformation can be approximated by the following chemical reaction:

$$C_a H_b O_c N_d + \left(\frac{4a + b - 2c + 3d}{4}\right) O_2 \rightarrow aCO_2 + \left(\frac{b - 3d}{2}\right) H_2O + dNH_3$$

This reaction assumes that the chemical composition given for the organic matter is composed of the biodegradable fraction; that is, plastics and other non-biodegradable compounds should not be included in the chemical formulation. In addition, if sufficient time and oxygen are supplied, ammonia oxidizes to nitrate as follows:

$$NH_3 + 2O_2 \rightarrow H_2O + HNO_3$$

If the generation of biomass and an organic by-product from an organic substrate that does not contain nitrogen is considered, the reaction may be written as

$$CH_m O_n + a\,O_2 + b\,NH_3 \rightarrow c\,CH_\alpha O_\beta N_\delta + d\,CH_x O_y N_z + e\,H_2O + f\,CO_2$$

where $CH_m O_n$ is the substrate, $CH_\alpha O_\beta N_\delta$ is the biomass generated, and $CH_x O_y N_z$ represents any products or by-products generated. From this, a carbon and nitrogen balance may be derived, as well as the ratio of biomass (b—also represented as X in the *FE SRH*) or product (p) generated per unit of substrate (s) consumed (often call the *yield coefficient*—Y) and the ratio of CO_2 produced per unit O_2 consumed (often called the *respiratory quotient*—RQ) as

$$C: 1 = c + d + f$$
$$N: b = c\delta + dz$$
$$Y_{b/s} = c$$
$$Y_{p/s} = d$$
$$RQ = \frac{f}{a}$$

The number of electrons available per unit of carbon consumed is often referred to as the degree of reduction (high reduction is equivalent to little oxidation) and may be represented for the substrate, biomass, and products as

$$\gamma_s = 4 + m - 2n$$
$$\gamma_b = 4 + \alpha - 2\beta - 3\delta$$
$$\gamma_p = 4 + x - 2y - 3z$$

From these expressions, an electron balance may be completed as

$$\gamma_s - 4a = c\gamma_b + d\gamma_p$$

The Biology section of the *FE SRH* has completed an example of aerobic biodegradation using a glucose substrate with no product generation (biomass only) with $RQ = 1.1$. Values for the solved coefficient are provided, and examinees are encouraged to verify the calculations.

Under *anaerobic respiration*, microbes can extract oxygen from oxygen-containing chemicals or use other electron donor compounds in the aqueous environment. Conversion of NO_2^- or NO_3^- to N_2 is called *denitrification*, reduction of SO_4^{2-} to H_2S is called *sulfidogenesis* (meaning sulfide generation), while CO_2 can be reduced to CH_4 in a process called *methanogenesis*, (meaning methane generation). Complete anaerobic transformation (stabilization) may be expressed chemically as

$$C_aH_bO_cN_d + \left(\frac{4a - b - 2c + 3d}{4}\right)H_2O \rightarrow$$

$$\left(\frac{4a + b - 2c - 3d}{8}\right)CH_4 + \left(\frac{4a - b + 2c + 3d}{8}\right)CO_2 + dNH_3$$

For unknown organic compositions, rapidly biodegradable material (for example, food wastes, paper, cardboard, and yard wastes that do not include large stumps or branches) can be estimated to have the composition $C_{70}H_{110}O_{50}N$, while other slower biodegradable matter can be estimated to have the composition $C_{20}H_{30}O_{10}N$. Expressions for the incomplete transformation have been developed, and an example is provided in the Biology section of the *FE SRH*.

Organisms that are capable of only aerobic or anaerobic respiration are termed *obligate* (or *strict*), while those that may use the most energetic pathway available to them (even switching between aerobic and anaerobic as environmental conditions change) are called *facultative*. Methanogenesis and sulfidogenesis are important anaerobic respiration pathways involved in the generation of landfill gases and anaerobic digestion of wastewater sludges.

Cell Growth and Reproduction

A four-phase bacterial growth curve is presented in Figure 7.1, identifying the (1) *lag* or *acclimatization*, (2) *exponential* or *log growth*, (3) *stationary*, and (4) *death* or *decay* phases. Note the use of a logarithmic scale on the vertical axis. Often, the transitions into and out of exponential growth are of particular importance and are identified as *accelerating growth* and *declining growth*, respectively. These transitions are sometimes represented as additional phases on the growth curve and have been added between the appropriate stages in Figure 7.1.

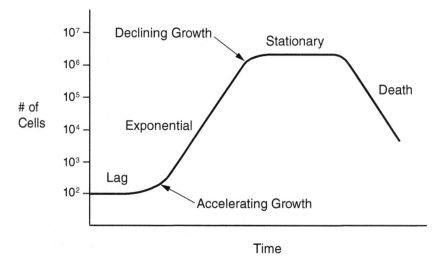

Figure 7.1 Four-phase microbial growth curve

Bacteria reproduce by *binary fission*, each cell dividing every 20–30 minutes under optimum conditions into two cells (often called the *doubling time*), which can be expressed as

$$\frac{dX}{dt} = \mu X$$

which is integrated to yield the cell mass growth curve expressed as

$$X_t = X_0 e^{\mu t}$$

where X_t (mg/L) is the concentration of cells (biomass) at time t, X_0 is the initial cell concentration (mg/L), and μ is the specific growth rate (time^{-1}). In municipal wastewater processes, cell concentrations are generally equal to the measured amount of total volatile suspended solids (TVSS). The form of the equation identifies the second phase of growth due to the exponential dependence of the growth curve.

Example 7.1

Exponential growth

If 100 bacterial cells become 10,000,000 in eight hours, find the doubling period assuming a first-order reaction rate applies.

Solution

Using a first-order reaction rate and the data given in the problem statement we may write

$$N = N_0 e^{\mu t} \Rightarrow \frac{N}{N_0} = \frac{10^7}{10^2} = 10^5 = e^{8\mu} \Rightarrow \frac{\ln(10^5)}{8} = \mu = 1.44 \text{ hr}^{-1}$$

Using this rate constant and the first-order expression, we can now determine the doubling period by selecting a value for N that is twice the value of N_0 as follows:

$$N = N_0 e^{\mu t} = 200 = 100 e^{1.44\tau} \Rightarrow \ln\left(\frac{200}{100}\right) = 1.44\,\tau \Rightarrow$$

$$\tau = 0.48 \text{ hr} \times 60\,\frac{\text{min}}{\text{hr}} = 28.9 \text{ min}$$

While the first-order kinetics model may work well for ideal growth conditions, growth (substrate) limiting conditions may be a more appropriate description for certain environmental systems. In these cases, it may be more appropriate to describe the growth of bacteria by the Monod kinetic model, expressed as

$$\mu = \mu_{max}\left(\frac{S}{K_s + S}\right)$$

where μ_{max} is the maximum specific growth rate (time^{-1}), S is the substrate concentration (mg/L), and K_s is the half-saturation constant (mg/L). Note that Monod kinetics is covered in the Environmental Engineering section of the *FE SRH*. K_s may be defined as the value of S that corresponds to one-half of μ_{max} as shown in Figure 7.2.

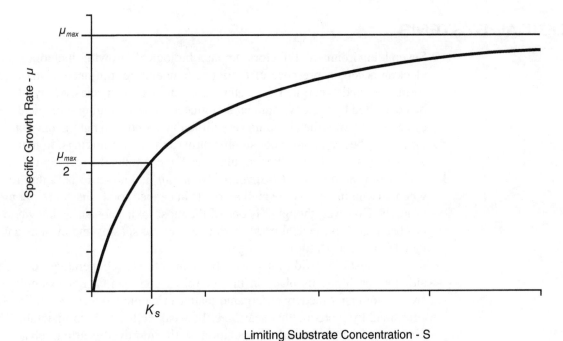

Figure 7.2 Graphical representation of Monod growth equation

Substitution of the Monod kinetic expression into the cell mass growth equation yields

$$\frac{dX}{dt} = \frac{\mu_{max} \, S \, X}{K_s + S}$$

Finally, in order to account for the death of cells during the endogenous phase, we may write

$$\left(\frac{dX}{dt}\right)_{net} = \left(\frac{dX}{dt}\right) - k_d \, X = \left[\frac{\mu_{max} \, S \, X}{K_s + S}\right] - k_d \, X = (\mu - k_d) \, X$$

where $(dX/dt)_{net}$ refers to the net growth rate (mg·L^{-1}·time^{-1}), and k_d is the endogenous decay constant (time^{-1}).

If all of the food in the system was converted to biomass, the rate of substrate utilization would equal the rate of biomass generation, due to the conservation of mass. However, we know that the conversion process is less than perfect, and as such, a ratio of the amount of biomass generated per unit substrate utilized can be evaluated. This quantity is termed the *biological yield* (Y) and can be expressed as

$$Y = \frac{dX/dt}{dS/dt} = \frac{dX}{dS}$$

where Y has units of mg biomass per mg substrate. Finally, substrate utilization can be expressed in terms of the yield as

$$\frac{dS}{dt} = -\frac{\mu_{max} \, S \, X}{Y(K_s + S)}$$

where the endogenous decay constant has been dropped due to the fact that substrate is not being consumed during cell death.

BIOLOGICAL SYSTEMS

Several environmental factors impact biological growth, including substrate, electron acceptor, presence of nutrients, temperature, presence of toxins, biofilm formation, and synergism with other microbiological species. Many of these can be controlled by the environmental engineer when designing biological treatment systems; however, they become very difficult to control in natural environments. In turn, microbes can contribute significantly to their surroundings, in many cases in a way that is harmful to other populations. For example, the secretion of botulinum toxin from the microbe *Clostridium botulinum* is known to be lethal to humans when present in even very small amounts in canned foods that were not processed properly. However, though it is one of the most toxic substances known to man, it has been used in medical practice to treat muscle spasms and even as a cosmetic facial treatment (Botox).

Synergistic relationships can be competitive, cooperative, or predatory, although each has its place in the microbial community. For example, bacteria are the dominant feeders on organic matter in municipal wastewater, but they are consumed by protozoa that scavenge the weak cells and help maintain a healthy, kinetically active bacteria population with improved settling characteristics. Another example is the consumption of organic matter during wastewater processing in treatment ponds by bacteria, which release CO_2 that can be used by algae photosynthetically to produce O_2, which the bacteria use to process more organic matter.

Toxicology

Many indicators of water quality focus on naturally occurring or man-made organic and inorganic chemical contaminants. However, in a global perspective, these are often trivial when compared to the potential negative health effects that are a result of biological contamination. Most often it is direct invasion of tissue (*infection*) or cell destruction through parasitism; however, the biochemical secretions of certain organisms may also adversely affect human health. While the United States has not seen large outbreaks of waterborne diseases due to biological contamination, worldwide pathogenic bacteria and other organisms in water can have devastating effects on human health. *Toxicology* is the study of toxins and potential treatment methodologies. A *toxin* can be defined as a biochemical substance generated (often secreted) by plants, animals, and microorganisms that exhibits characteristics poisonous to humans. Often, these substances are proteins that, in addition to having severe negative impacts in the host, may also induce a reaction such as the production of *antitoxins*. The factors that determine the level of damage these toxins may cause will be examined in the section on risk assessment below.

Countless different microorganisms may be present in, and contribute to the contamination of, any ecosystem. Monitoring all of them is not feasible, typically. Instead, an indicator organism, or class of organisms, is selected to serve the role of identifying the potential for impact to human health. In water intended for human consumption, the most widely used set of indicator organisms is the *coliform* group of bacteria (which includes *E. coli*). They are defined as aerobic or facultative, Gram-negative rods that possess the ability of fermenting lactose within 48 hours at 37°C, producing a gas by-product.

Differentiation is sometimes made between *total coliforms* and *fecal coliforms*, although the majority of total coliforms arise from the feces of warm-blooded animals. Since all waterborne pathogens are transmitted through feces, water that is coliform free is likely to be pathogen free. Since some coliforms may not originate in the intestines of warm-blooded animals, positive tests for total coliforms can be checked for fecal coliforms by transferring a sample to a specific culture medium. Another group of indicator organisms useful in brackish or salt waters is the *fecal streptococci*. Further, the ratio of fecal coliforms to fecal streptococci is a strong indicator as to whether the origin of the fecal contamination is from humans or other animals.

Bacterial Quantification

While the fermentation tubes mentioned previously may indicate the presence or absence of coliforms, it is difficult to quantify organisms present and thereby determine cell concentrations. One method to quantify bacterial coliforms is the most probable number (MPN) approximation, which uses five replicates of several 1:10 dilutions to statistically estimate bacteria concentrations in the original sample. Standard Methods for the Examination of Water and Wastewater dictates that the lowest dilution that results in five positive tubes and the next two dilutions in sequence be used for estimating cell numbers. If no dilution produces five positive tubes, the first three dilutions are used. The number of positive tubes for each dilution are reported as A-B-C and compared to the MPN Index chart shown in Table 7.3. Unfortunately, initial concentrations are usually low, so dilution generates many tubes without growth, and at five samples per dilution the error is quite large and results have little quantitative meaning.

Plate counts have been used, but low concentrations also yield many plates with no colony forming units (CFUs), which can be defined as *one or more organisms that generate a colony*. To overcome the low concentrations, 0.2 or 0.45 μm filters are often employed to strain a known volume of sample (typically 100 mL), and the filters are subsequently incubated in a specific culture medium for 24 hours at 37°C. Filters are then examined for characteristic colonies (reddish or greenish tint), and results are reported as CFU/mL or CFU/100 mL.

Table 7.3 MPN Index and 95% confidence limits for five-tube 1:10 serial dilutions (Standard Methods, APHA, AWWA, WEF)

Combination of Positives	MPN Index per 100 mL	95% Confidence Limits		Combination of Positives	MPN Index per 100 mL	95% Confidence Limits	
		Lower	Upper			Lower	Upper
0-0-0	< 2	—	—	4-3-0	27	12	67
0-0-1	2	1.0	10	4-3-1	33	15	77
0-1-0	2	1.0	10	4-4-0	34	16	80
0-2-0	4	1.0	13	5-0-0	23	9.0	86
0-0-0	< 2	—	—	5-0-1	30	10	110
1-0-0	2	1.0	11	5-0-2	40	20	140
1-0-1	4	1.0	15	5-1-0	30	10	120
1-1-0	4	1.0	15	5-1-1	50	20	150
1-1-1	6	2.0	18	5-1-2	60	30	180
1-2-0	6	2.0	18	5-2-0	50	20	170
2-0-0	4	1.0	17	5-2-1	70	30	210
2-0-1	7	2.0	20	5-2-2	90	40	250
2-1-0	7	2.0	21	5-3-0	80	30	250
2-1-1	9	3.0	24	5-3-1	110	40	300
2-2-0	9	3.0	25	5-3-2	140	60	360
2-3-0	12	5.0	29	5-3-3	170	80	410
3-0-0	8	3.0	24	5-4-0	130	50	390
3-0-1	11	4.0	29	5-4-1	170	70	480
3-1-0	11	4.0	29	5-4-2	220	100	580
3-1-1	14	6.0	35	5-4-3	280	120	690
3-2-0	14	6.0	35	5-4-4	350	160	820
3-2-1	17	7.0	40	5-5-0	240	100	940
4-0-0	13	5.0	38	5-5-1	300	100	1300
4-0-1	17	7.0	45	5-5-2	500	200	2000
4-1-0	17	7.0	46	5-5-3	900	300	2900
4-1-1	21	9.0	55	5-5-4	1600	600	5300
4-1-2	26	12	63	5-5-5	> 1600	—	—
4-2-0	22	9.0	56				
4-2-1	26	12	65				

Microbial Ecology

Ecology is generally considered the study of the interrelationship between living things and the environment in which they exist. The population and its environment together are known as an *ecosystem*. Certain microbiological communities and the conditions under which they interact with other biological and abiotic (that is, chemical or physical in nature) species are of particular interest to environmental engineers. Subspecialties within specific fields of study include aquatic microbiology, soil microbiology, medical sciences, and nutrient cycling or food chains (for example, bioaccumulation and biomagnification).

Within a defined population, *cooperative* or *competitive* influences may have an impact on the health and propagation of the colony. When these influences are extended toward other populations (that is, two different species), it is usually termed *symbiosis*. An example of cooperative symbiosis includes situations where one population benefits while another is neither harmed nor benefited (*commensalism*). In another case, both populations benefit from the interaction, and this relationship may be required (*mutualism*) or optional (*synergism*). Competitive symbiosis situations may include two populations vying for the same limited resources or the extreme case where one population produces an extracellular toxin that inhibits another population (*antagonism*). Finally, many environments exist where one species primarily derives its nutritional needs at the expense of another. This may be accomplished through extracting benefit slowly over relatively long periods of time (*parasitism*), or in a single feeding where the harmed population is used as a primary food source (*predation*).

One important concept in the interactions between living organisms and the world in which they exist is the cycling of all matter. Often called *nutrient cycling* (or more appropriately termed *biogeochemical cycling*), this is the idea that essentially all mass currently providing the sustenance for the populations of today has been in existence from the genesis of the planet. The reuse of carbon, hydrogen, oxygen, nitrogen, sulfur, phosphorus, and all biologically important metals is a fact that impacts every population on a daily basis. The transfer of elements between the different ordered populations is known as the *food web*, as presented by the simplistic schematic of an idealized food web in Figure 7.3. A typical path would identify plant matter as *primary producers* that provide food for grazing animals, which in turn are food for higher predators (that is, humans), possibly along with additional plant matter. Both the *grazers* and *predators* respire CO_2 and return organic matter to the soil in their waste streams, which are used once again by the plants that also produce the O_2 necessary for the animals to live.

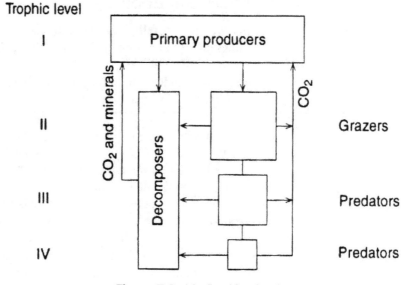

Figure 7.3 Idealized food web

Source: Ronald M. Atlas and Richard Bartha, *Microbial Ecology: Fundamentals and Applications*, 4th ed. © 1998 Benjamin/Cummings Publishing Company, Inc. Reprinted by permission of Pearson Education, Inc.

One role that microbes play in the food chain is the *mineralization* of most organic chemicals, which may be defined as complete biodegradation of large organic molecules and the subsequent release of their elemental or basic ionic species back to the ecosystem. However, several organic compounds are known to be *recalcitrant*, which means they are persistent in the environment due to their resistance to biotransformation. Many of these compounds are man-made (or *xenobiotic*) and inhibit microbial attack or have a toxic effect on the biological community.

Bioconcentration refers to the process whereby contaminants are directly absorbed into an organism from the water surrounding it. *Bioaccumulation* occurs when the contaminants are absorbed from ingested food in addition to the water. The uptake of these chemicals by lower organisms impacts all higher species that incorporate the lower biota in their food cycles. Often, the lower organisms may adapt to the presence of the toxin, or there may be a threshold concentration that triggers a negative response in the higher organism after a period of accumulation. As you travel up the food chain, the concentration of contaminants within the higher-trophic organisms becomes progressively higher, a process known as *biomagnification*.

BIOPROCESSING

With a strong foundational understanding of biological systems and the interaction between these systems and their environment, the engineer has an opportunity to take advantage of the unique capabilities of various life forms. The development of technologies that use living organisms in the transformation of materials or to produce desired products is called *bioprocessing*. For example, the environmental engineer will use knowledge of biological systems in the treatment of water and wastewater or to transform solid and hazardous wastes. The chemical engineer may apply bioprocessing techniques in the pharmaceutical or nutraceutical industries or for a wide variety of products in the processing of foods. Biomedical and biomechanical engineers use biological systems to develop new biocompatible materials or artificial limbs and organs, such as the use of artificial skin in the treatment of burn victims. Recent efforts in the development of bioproducts have led to the use of plant material in the production of biodegradable plastics, which reduces the demand for petroleum imports while offering an alternative to landfill disposal. Several examples are briefly discussed below and many are covered in greater detail in subsequent sections of this review.

Water Treatment and Disinfection

If we assess the value of a substance by the length of time we can survive without it, then water is second only to oxygen as the most valuable substance on Earth. All civilizations arose in locations where there was a ready source, or one able to be developed, of the fresh water necessary for human consumption as well as for farming and animal domestication. While the quality of the water resource varied by location, there is evidence that early civilizations adopted some treatment strategies to purify their water. Water purification today is based on the type of contaminants that are present in a particular source, with water quality standards setting the levels of purity required for human consumption. By far the most important water purification technique is disinfection.

Due to the potential presence of biological agents that are harmful to human health, a primary role of the water treatment plant is the *disinfection* of the water prior to release into the distribution network. The most often used disinfectant is *chlorine* gas. The goal of disinfection is to inactivate any biological agent that could pose a threat to human health, while minimizing the *residual chlorine* levels that may cause a displeasing taste for consumers (usually assumed to be < 0.5 mg/L). This is balanced by the need to ensure complete inactivation of specific persistent pathogens and the desire to have a small (0.1–0.2 mg/L) residual chlorine level that will maintain disinfection throughout the distribution network.

The ability to remove a microbial species is a function of both the *chlorine concentration* and the amount of contact time the organism must experience at that concentration. The combination of these two factors is often referred to as the $C \cdot t$ product, which carries units of (mg/L)·min and is often tabulated for specific microorganisms at specified temperatures and pH values.

Wastewater Treatment and DO Sag

Although not a significant concern in the United States, contaminated drinking water, primarily from biological contamination, is one of the major causes of illness, and potentially death, in the world today. When human population densities were low, natural processes could handle the quantities of human waste discharged to the environment. But as populations increased and towns and cities grew, the mass loadings became too great, and wastewater treatment technologies were necessary. With our increased understanding of causes of illness based on advances in microbiological sciences and medicine, wastewater treatment technologies can be targeted for those contaminants that pose the greatest threats to human health.

The primary consideration for disinfection of WWTP effluents is the removal of pathogens, as quantified through the fecal coliform test, although fecal viruses, protozoal cysts, and helminth eggs may need to be addressed. Two popular choices are chlorination and ultraviolet radiation. Traditional chlorine disinfection similar to that used in water treatment is the most common form of treatment for wastewater effluents. The major difference between chlorination of water supplies and wastewater is that chlorine may react with the organic matter in the wastewater to form *trihalomethanes* and *haloacetic acids* or with the ammonia present in the wastewater to produce *chloramines*. All of these compounds can be toxic at relatively low levels, and as such their creation should be avoided if possible.

Gaining in popularity is the use of *ultraviolet (UV) radiation* for pathogen control in WWTP effluents. Similar to the *C•t product* used in chlorination of drinking water, UV disinfection is a product of the radiation intensity and time of exposure. Treatment times are determined based on the lamp intensity as well as the lamp construction. Low-pressure, low-intensity lamps emit light at 254 nm, which is nearly optimal for disinfection effectiveness. However, lamp efficiency is temperature dependent, with a maximum efficiency at 40°C, and power output is limited to approximately 25 W. Low-pressure, high-intensity lamps provide 2 to 20 times the power output, more emission stability over a greater temperature range, and a 25% life-span increase. Medium-pressure, high-intensity lamps possess a power output 50 to 100 times greater than the low-pressure, low-intensity lamps and are used more commonly in WWTPs that have substantially higher flow rate.

Another potential negative impact of WWTP effluents being discharged to receiving streams is the potential to deplete the *dissolved oxygen (DO)* in the stream that is required by most aquatic species. DO levels below 6 mg/L may inhibit fish reproduction and kill sensitive species; levels below 5 mg/L can render a stream inactive. DO may be consumed by bacteria if a wastewater effluent has a high concentration of organic matter or is discharged to receiving water that does not possess sufficient dilution ability. The decrease of DO in a stream receiving an oxygen consumptive contaminant is called *DO sag*. A classic water-quality model developed by Streeter and Phelps is used to predict the *DO deficit*, which may be defined as the *difference between the DO saturation value and the actual DO concentration of the stream*. In practice, the Streeter-Phelps equation is used to evaluate if a known discharge will have a negative impact on the aquatic life in the receiving stream due to reduced DO levels.

Solid Waste Treatment

Another environmental engineering technology that uses microbes takes advantage of the high organic matter and moisture content of certain fractions of municipal solid waste (MSW) in the process of *composting*. Depending on the feed composition, conditions can be optimized for the biological transformation of MSW into a valuable soil amendment. In general, effective biological activity requires a moisture content in the range of 50% to 60%, a carbon-to-nitrogen ratio of 20 to 25:1, and a pH in the neutral range. The selection of the feed material often dictates the transformation percentage and the usefulness of the end product. Choices include yard waste only, the organic fraction of the segregated MSW (yard waste mixed with food wastes and select paper products), or commingled MSW.

As discussed previously, aerobic and anaerobic processes are both available for the conversion of organic matter. Aerobic processes are simpler to operate; however, anaerobic processes produce methane, which may be collected, purified, and subsequently used for energy production. The more common practice is to use an aerobic process, which has the primary goal of volume reduction (~50%) and a secondary goal of compost production. The source of the microbes and organic matter is the MSW, and, while some moisture is present, more is usually added during the processing. Oxygen is fed to the system by two possible methods. The *windrow* method utilizes large machinery to mix, or turn, a long pile with a triangular cross section to provide aeration throughout the mass. The *aerated static pile* employs an exhaust fan, connected to perforated pipe extending throughout the pile, to draw the necessary oxygen through the pile from the surrounding air. In either case, the processing time is three to four weeks; however, piles are allowed to *cure* for another one to three months without aeration or turning.

As with composting, biological activity within an MSW landfill converts the organic matter through aerobic and anaerobic processes. Available oxygen is depleted quickly, and the majority of the biological activity is anaerobic, primarily through methanogenesis. This produces a landfill gas that is roughly a 50-50 mix by volume of methane and carbon dioxide, with other trace gases such as nonmethane organic compounds (NMOCs) and H_2S (responsible for the *rotten egg* odor). New landfills are designed to remove landfill gases and purify the methane for energy production, while older landfills may be retrofitted with gas extraction technology.

Fermentation

A natural process that has been part of the biogeochemical cycle since life first inhabited the planet, *fermentation* has also been a part of human culture for many millennia. Fermentation is defined as a process where organic compounds serve as both electron donors and electron acceptors during anaerobic conversion by bacteria or yeast. The most notable historical use of fermentation processes is in the production of foods, such as employing yeast cultures for making leavened breads and alcoholic beverages (beer, wine, and distilled spirits), or by the bacterial transformation of milk into yogurt and certain cheeses. Pickling has been a food preservation technique for many centuries and employs fermentation of a brine (salt solution) or the addition of vinegar, itself a fermented product.

Modern manifestations of fermentation include many pharmaceutical and industrial processes, most of which are simply highly engineered versions of natural processes. One example that is enjoying much attention recently is the production of ethanol as a liquid transportation fuel. While it is readily acknowledged that the United States does not have sufficient sugar or starch biomass to completely replace gasoline, it remains to be seen if cellulosic ethanol delivers on the promise to close the gap. Regardless of the stance individuals hold regarding ethanol as a realistic contributor toward energy independence, it is clear to many that fermentation and other biotechnology processes will play a substantial role in the development of energy resources in the future.

DO SAG MODELING

DO is a basic WQI that has a tremendous impact on a surface water's ability to support higher life forms. Typical minimum values for DO in streams are 5 mg/L to support most species of fish, and may be as high as 6 mg/L for more sensitive fish species, such as trout. DO may be consumed by bacteria if a wastewater effluent has a high BOD concentration, or is discharged to receiving water that does not possess sufficient dilution ability. While regulatory limits require average BOD concentrations in a discharge to be less than 30 mg/L, even at this concentration there may be a negative impact to the DO concentration in smaller receiving waters, or in receiving water where several WWTP effluents are discharged in close proximity to each other.

The decrease of DO in a stream receiving an oxygen consumptive contaminant is called DO sag. In order to model the DO sag in a stream, all sources of DO consumption and generation must be identified and quantified. This often proves to be a difficult task, so simplified models can be employed to model select reactions and examine the contributory effect of a target contaminant. For this review, we will assume that a particular stream receives BOD from a single WWTP effluent, the BOD is consumed by bacteria in the stream at a constant rate, and reaeration of the stream occurs only by diffusive flux from the atmosphere at a constant rate.

The classic model developed by Streeter and Phelps is used to predict the DO *deficit* (*D*), defined as the *difference between the DO saturation value and the actual DO concentration*, and can be expressed as

$$D = \frac{k_d L_0}{k_a - k_d}\left[\exp(-k_d\, t) - \exp(-k_a\, t)\right] + D_0 \exp(-k_a\, t)$$

where D is the deficit (mg/L) at time t (days), k_d is the deoxygenation rate constant (day^{-1}), k_a is the reaeration rate constant (day^{-1}), L_0 is the initial ultimate BOD (mg/L), and D_0 is the initial deficit (mg/L).

In this model, initial conditions are determined as the weighted averages of the stream and WWTP discharge after mixing; therefore, initial BOD would be calculated as

$$L_0 = \frac{Q_{ww} L_{ww} + Q_s L_s}{Q_{ww} + Q_s}$$

where Q_{ww} and Q_s are the volumetric flow rates of the WWTP discharge and stream, and L_{ww} and L_s are the ultimate BOD concentrations of the WWTP discharge and stream (mg/L), respectively. The units on volumetric flowrate are not important, as long as both flows have the same units.

The initial deficit is calculated as

$$D_0 = DO_T^{sat} - DO_0$$

where DO_T^{sat} is the saturation DO concentration evaluated at the temperature of the stream after receiving the WWTP discharge (mg/L), and DO_0 is the initial DO in the stream (mg/L). The weighted average temperature of the combined flows may be calculated as

$$T = \frac{Q_{ww} T_{ww} + Q_s T_s}{Q_{ww} + Q_s}$$

where T_{ww} and T_s are the temperatures (°C) of the WWTP discharge and the stream, respectively. This temperature would be located on the DO saturation table to determine DO_T^{sat}, interpolating between temperatures as necessary. Finally, DO_0 is calculated as a weighted average of the DO values in stream and WWTP discharge before mixing, and may be calculated as follows:

Table 7.4 Dissolved oxygen saturation values in river water

Temperature (°C)	Dissolved Oxygen (mg/L)
2	13.84
5	12.80
10	11.33
15	10.15
20	9.17
25	8.38
30	7.63

$$DO_0 = \frac{Q_{ww} DO_{ww} + Q_s DO_s}{Q_{ww} + Q_s}$$

where DO_{ww} and DO_s are the DO concentrations of the WWTP discharge and the stream (mg/L), respectively.

Values for k_d, the deoxygenation rate constant, are usually given or assumed based on values that are determined in the laboratory under controlled conditions. For example, often k_d is determined in the lab at a set temperature, usually 20°C. Typical values of k_d for untreated wastewater range from 0.1 to 0.5 day^{-1}, with an average value of 0.25 day^{-1} used in the absence of system-specific data. Values of k_d for treated effluents are approximately half of the untreated values.

Values for k_a, the reaeration rate constant, are dependent upon the mixing of DO in the stream as oxygen is extracted from the air, and are therefore functions of stream velocity and depth. They may be estimated using the following relationship:

$$\left(k_a\right)_{20°C} = \frac{\left(D_L V_s\right)^{1/2}}{H^{3/2}}$$

where k_a is evaluated at 20°C (s^{-1}), D_L is the oxygen diffusivity in water at 20°C (m²/s), V_s is the velocity of the stream (m/s), and H is the depth of the stream (m). At 20°C, the value of D_L is approximately equal to 2.1×10^{-9} m²/s. However, since most calculations require units for k_a of day^{-1}, the value of k_a as calculated above should be multiplied by 86,400. Combining these factors, a common expression for k_a could be written as

$$\left(k_a\right)_{20°C} = \frac{3.95\, V_s^{1/2}}{H^{3/2}}$$

where V_s has units of m/s, H has units of m, k_a has units of day^{-1}, and the constant 3.95 accounts for the pr oper unit conversions. If insufficient data exists to calculate the reaeration constant, typical values may be used as presented in Table 7.5.

Table 7.5 Typical reaeration coefficients for various water bodies

Water Body	k_a [day^{-1}]
Small ponds	0.15
Sluggish streams, lake	0.3
Large streams, low velocity	0.4
Large streams, high velocity	0.5
Swift streams	0.8

Because both rate constants have assumed or calculated values determined at 20°C, both rate constants need to be corrected to account for reaction at the weighted average temperature using the following relationship:

$$k_T = k_{20}\, \theta^{(T-20)}$$

where T is the weighted average temperature (°C), k_T is the value of the constant at temperature T (day^{-1}), k_{20} is the given or calculated value of the constant at 20°C (day^{-1}), and θ is the temperature coefficient. The temperature coefficient for k_d may range from a value of 1.056 for temperatures between 20°C and 30°C, up to a value of 1.135 for temperatures between 4°C and 20°C. The temperature coefficient for k_a is often assumed to be 1.024.

Example 7.2

Reaeration coefficient

Calculate the reaeration coefficient of a stream 5 ft deep and 10 ft wide if it is flowing at 20 cfs and a temperature of 12°C. You may assume that $D_L = 0.002$ ft²/day and $\theta = 1.024$.

Solution

For the expression given, we will need to determine stream velocity in units of feet per day. This may be accomplished as follows:

$$V_s = \frac{Q}{A} = \frac{20 \frac{\text{ft}^3}{\text{s}}}{(5 \text{ ft})(10 \text{ ft})} = 0.4 \frac{\text{ft}}{\text{s}} \times \frac{86{,}400 \text{ s}}{\text{day}} = 34{,}560 \frac{\text{ft}}{\text{day}}$$

We may now estimate the reaeration coefficient at 20°C as follows:

$$(k_a)_{20°C} = \frac{(0.002 \, V_s)^{1/2}}{H^{3/2}} = \frac{[(0.002)(34{,}560)]^{1/2}}{5^{3/2}} = 0.744 \text{ day}^{-1}$$

This reaeration constant must be corrected to the stated temperature of 12°C. Using the value 1.024 given in the problem statement, we get

$$(k_a)_{12} = (k_a)_{20} (1.024)^{12-20} = (0.744)(1.024)^{-8} = 0.615 \text{ day}^{-1}$$

The Streeter-Phelps model can be solved repeatedly for small increments of t to yield the DO sag curve as shown in Figure 7.4. However, it is often desired to identify the value of the minimum DO concentration (DO_{min}), which corresponds to the maximum (or critical) DO deficit (D_c). This is accomplished by solving the deficit equation using a value of time that corresponds to the critical point (t_c), which can be calculated as

$$t_c = \frac{1}{k_a - k_d} \ln \left[\frac{k_a}{k_d} \left(1 - D_0 \frac{k_a - k_d}{k_d L_0} \right) \right]$$

where t_c is expressed in units of days. Finally, the location (X_c) downstream of the WWTP discharge where the critical deficit occurs can be found using t_c as follows:

$$X_c = t_c V_s$$

where X_c is the critical distance. It is left to the examinee to ensure dimensional homogeneity in the above expression; however, X_c is often reported in units of miles.

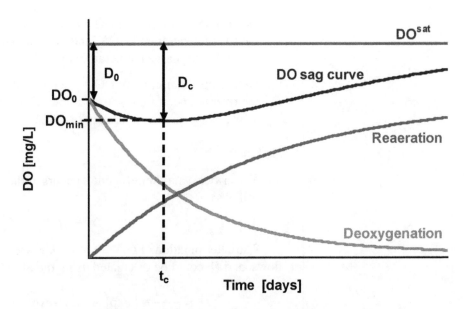

Figure 7.4 Graphical representation of DO sag curve

In practice, the Streeter-Phelps equation is used to evaluate if a known discharge will have a negative impact on the aquatic life in the receiving stream due to reduced DO levels. Because of the form of the deficit equation, one method to determine if a critical deficit will occur due to a particular discharge is to compare the $k_d L_0$ product with the $k_a D_0$ product. If sufficient BOD exists at a high enough rate of degradation, $k_d L_0$ will exceed $k_a D_0$, and a critical deficit will occur at some time and location downstream. Conversely, when a low initial BOD or slow rate of utilization exists, $k_a D_0$ will be greater than $k_d L_0$, and no DO sag will exist downstream.

The system under consideration should be reevaluated in the extreme seasons, because temperature and flow differences can have a substantial impact on the critical deficit for the same discharge. Further, if the critical deficit corresponds to a value of DO that is less than a predetermined, acceptable DO standard, then corrective measures could be attempted. Unfortunately, the only actions available to the engineer that would improve DO in the stream would be to lower the BOD in the discharge or increase the DO of the discharge. Increased DO_{ww} could be obtained through a final aeration step, while lowering BOD_{ww} may require upgrades or additions to the WWTP that would increase the overall plant BOD removal efficiency. In extreme cases, aeration of the receiving stream could be conducted as a last resort.

Example 7.3

Streeter-Phelps modeling

A small river receives a wastewater treatment plant discharge of 5 mgd with a BOD of 20 mg/L and a DO concentration of 1.0 mg/L. The river is 40 feet wide and flows at 1 ft/sec, but depth and temperature varies by season. You may assume the BOD in the river is 0.9 mg/L and the DO concentration is at 90% of the saturation value, which also varies with temperature. You may also assume the biological activity coefficient $[(k_d)_{20}]$ is 0.3 day^{-1}, the stream reaeration coefficient $[(k_a)_{20}]$ is 0.25 day^{-1}, and the temperature correction coefficient for k_d is 1.135, while the value for k_a is 1.024.

a). Determine the minimum DO concentration in the stream in the summer when the river temperature is 22°C and flows at a depth of 3 feet. You may assume a WWTP discharge temperature of 16°C.

b). Determine the minimum DO concentration in the stream in the winter when the river temperature is 4°C and flows at a depth of 4 feet. You may assume a WWTP discharge temperature of 13°C.

Solution

a). The critical (maximum) deficit occurs at the minimum DO concentration as follows:

$$D_c = DO^{sat} - DO_{min}$$

The determination of DO^{sat} requires knowledge of the temperature of the mixed flows, and D_c could be calculated using the Streeter-Phelps equation as follows:

$$D_c = \frac{k_d L_0}{k_a - k_d}\left[\exp(-k_d t_c) - \exp(-k_a t_c)\right] + D_0 \exp(-k_a t_c)$$

where

$$t_c = \frac{1}{k_a - k_d} \ln\left[\frac{k_a}{k_d}\left(1 - D_0 \frac{k_a - k_d}{k_d L_0}\right)\right]$$

First, calculate the biological use coefficient (k_d) and the reaeration coefficient (k_a) at the temperature of the mixed flows (T_m). In order to calculate the flow-weighted temperature, the flow rates of the river and wastewater must be determined in the same units as follows:

$$Q_r = (40 \text{ ft})(3 \text{ ft})\left(1 \frac{\text{ft}}{\text{s}}\right) = 120 \frac{\text{ft}^3}{\text{s}} \quad \text{and} \quad Q_{ww} = (5 \text{ Mgd})\left(1.547 \frac{\text{cfs}}{\text{Mgd}}\right) = 7.735 \frac{\text{ft}^3}{\text{s}}$$

T_m may now be calculated as:

$$T_m = \frac{Q_{ww} T_{ww} + Q_r T_r}{Q_{ww} + Q_r} = \frac{(7.735 \text{ cfs})(16°C) + (120 \text{ cfs})(22°C)}{7.735 \text{ cfs} + 120 \text{ cfs}} = 21.64°C$$

Correction can now be made to k_d and k_a through the temperature correction expression as follows:

$$k_T = k_{20} \theta^{(T-20)}$$

$$(k_d)_{21.64} = (k_d)_{20} 1.135^{(21.64-20)} = (0.3)1.135^{1.64} = 0.37 \text{ day}^{-1}$$

$$(k_a)_{21.64} = (k_a)_{20} 1.024^{(21.64-20)} = (0.25)1.024^{1.64} = 0.26 \text{ day}^{-1}$$

To calculate the deficit, the BOD concentration of the mixed flows is required and may be determined as:

$$L_0 = \frac{Q_{ww} L_{ww} + Q_r L_r}{Q_{ww} + Q_r} = \frac{(7.735 \text{ cfs})\left(20 \frac{\text{mg}}{\text{L}}\right) + (120 \text{ cfs})\left(0.9 \frac{\text{mg}}{\text{L}}\right)}{7.735 \text{ cfs} + 120 \text{ cfs}} = 2.057 \frac{\text{mg}}{\text{L}}$$

The initial deficit (D_0) is also needed, which requires the initial DO in the river. This may be determined by finding 90% of the saturation value at the river temperature as follows:

$$(DO_i)_r = (0.9)(DO^{sat})_{22°C} = (0.9)\left(8.8 \frac{\text{mg}}{\text{L}}\right) = 7.92 \frac{\text{mg}}{\text{L}}$$

This value can be used to determine the flow-weighted DO as follows:

$$DO_0 = \frac{Q_{ww}DO_{ww} + Q_r DO_r}{Q_{ww} + Q_r} = \frac{(7.735 \text{ cfs})\left(1 \frac{\text{mg}}{\text{L}}\right) + (120 \text{ cfs})\left(7.92 \frac{\text{mg}}{\text{L}}\right)}{7.735 \text{ cfs} + 120 \text{ cfs}} = 7.501 \frac{\text{mg}}{\text{L}}$$

From this and the value of DO^{sat} at T_m, the initial deficit may be determined as:

$$D_0 = (DO^{sat})_{T_m} - DO_0 = (DO^{sat})_{21.64°C} - 7.50 \frac{\text{mg}}{\text{L}} = 8.865 - 7.501 = 1.364 \frac{\text{mg}}{\text{L}}$$

Because $k_d > k_a$ and $L_0 > D_0$, we know $k_d L_0 > k_a D_0$, and therefore a deficit will occur.

Plugging the above values in the equation for critical time (t_c) above yields:

$$t_c = \frac{1}{0.26 - 0.37} \ln\left[\frac{0.26}{0.37}\left(1 - (1.364)\frac{0.26 - 0.37}{(0.37)(2.057)}\right)\right] = 1.57 \text{ days}$$

The units were left off of the previous expression for simplicity and it is left to the examinee to verify the homogeneity of units. This value must be put into the deficit equation as follows:

$$D_c = \frac{(0.37)(2.057)}{0.26 - 0.37}\left[\exp(-0.37 \times 1.57) - \exp(-0.26 \times 1.57)\right] + 1.364 \exp(-0.26 \times 1.57)$$

$$D_c = 1.64 \frac{\text{mg}}{\text{L}}$$

Again, the homogeneity of the units is left to the examinee. Finally, DO_{min} may be calculated as follows:

$$DO_{min} = (DO^{sat})_{T_m} - D_c = 8.865 - 1.64 = 7.225 \frac{\text{mg}}{\text{L}}$$

b). Start by calculating k_d, k_a, L_0, and D_0 as before using the cold-weather data:

$$Q_r = (40 \text{ ft})(4 \text{ ft})\left(1 \frac{\text{ft}}{\text{s}}\right) = 160 \frac{\text{ft}^3}{\text{s}} \quad \text{and} \quad Q_{ww} = (5 \text{ Mgd})\left(1.547 \frac{\text{cfs}}{\text{Mgd}}\right) = 7.735 \frac{\text{ft}^3}{\text{s}}$$

$$T_m = \frac{Q_{ww}T_{ww} + Q_r T_r}{Q_{ww} + Q_r} = \frac{(7.735 \text{ cfs})(13°C) + (160 \text{ cfs})(4°C)}{7.735 \text{ cfs} + 160 \text{ cfs}} = 4.415°C$$

$$(k_d)_{4.415} = (k_d)_{20} \, 1.135^{(4.415-20)} = (0.3)\,1.135^{-15.585} = 0.042 \text{ day}^{-1}$$

$$(k_a)_{4.415} = (k_a)_{20} \, 1.024^{(4.415-20)} = (0.25)\,1.024^{-15.585} = 0.173 \text{ day}^{-1}$$

$$L_0 = \frac{Q_{ww}L_{ww} + Q_r L_r}{Q_{ww} + Q_r} = \frac{(7.735 \text{ cfs})\left(20 \frac{\text{mg}}{\text{L}}\right) + (160 \text{ cfs})\left(0.9 \frac{\text{mg}}{\text{L}}\right)}{7.735 \text{ cfs} + 160 \text{ cfs}} = 1.781 \frac{\text{mg}}{\text{L}}$$

$$(DO_i)_r = (0.9)(DO^{sat})_{4°C} = (0.9)\left(13.1 \frac{\text{mg}}{\text{L}}\right) = 11.79 \frac{\text{mg}}{\text{L}}$$

$$DO_0 = \frac{Q_{ww}DO_{ww} + Q_r DO_r}{Q_{ww} + Q_r} = \frac{(7.735 \text{ cfs})\left(1 \frac{\text{mg}}{\text{L}}\right) + (160 \text{ cfs})\left(11.79 \frac{\text{mg}}{\text{L}}\right)}{7.735 \text{ cfs} + 160 \text{ cfs}} = 11.29 \frac{\text{mg}}{\text{L}}$$

$$D_0 = (DO^{sat})_{T_m} - DO_0 = (DO^{sat})_{4.4°C} - 11.29 \frac{\text{mg}}{\text{L}} = 12.98 - 11.29 = 1.69 \frac{\text{mg}}{\text{L}}$$

Since it is not obvious by inspection, we should now calculate the $k_d L_0$ and $k_a D_0$ products to see if a deficit occurs:

$$k_d L_0 = \left(0.042 \text{ day}^{-1}\right)\left(1.781 \frac{\text{mg}}{\text{L}}\right) = 0.075 \frac{\text{mg}}{\text{L} \cdot \text{day}}$$

$$k_a D0 = \left(0.173 \text{ day}^{-1}\right)\left(1.69 \frac{\text{mg}}{\text{L}}\right) = 0.292 \frac{\text{mg}}{\text{L} \cdot \text{day}}$$

Since $k_a D_0 > k_d L_0$, reaeration is greater than demand, and the critical deficit is equal to the initial deficit. This means DO_{min} is equal to DO_0, which was determined to be 11.29 mg/L.

LIMNOLOGY

Limnology is the study of freshwater bodies such as lakes and ponds. Water quality in lakes is characterized by turbidity, secchi disk depth,[1] nutrient concentration (for example, phosphorus and nitrogen), temperature, dissolved oxygen concentration, and so on.

These water quality characteristics vary with the age of a lake. As a lake ages, it transitions from being an *oligotrophic* (literally "few foods") lake, to a *mesotrophic* lake to a *eutrophic* lake ("well fed"). Oligotrophic lakes are generally clear, are low in nutrients, and do not support large fish populations. Eutrophic lakes are high in nutrients and support a large amount of biomass. Eutrophic lakes can support large fish populations but are also susceptible to oxygen depletion. Mesotrophic lakes have characteristics (such as nutrient concentration, fish population, and water clarity) that lie somewhere between those of oligotrophic and eutrophic lakes.

As a lake transitions from an oligotrophic lake to a eutrophic lake, it is said to become more *productive,* in the sense that the eutrophic lake produces more biomass. Eutrophic lakes are considered undesirable from many viewpoints, as the resulting algal growth can severely limit recreational use, decrease aesthetics, cause taste and odor problems if the lake is a source of drinking water, and decrease the dissolved oxygen concentration in the lakes. Algae are plants (and therefore photosynthetic), and thus the decrease in dissolved oxygen typically occurs in the night. During the night no oxygen is produced, but a large oxygen demand is exerted due to the decay of dead algae.

The production of a lake is directly proportional to the concentration of the *limiting nutrient.* A limiting nutrient is the nutrient with the lowest concentration relative to the concentration needed for plant growth. Thus, addition of the limiting nutrient will stimulate additional growth. In nearly all freshwater lakes, phosphorus is the limiting nutrient for algal growth. In many salt water systems, nitrogen is the limiting nutrient.

Naturally, the aging of a lake can take thousands of years, but human impacts (for example, runoff containing excess phosphorus) can rapidly speed up this process. This more rapid process is termed *cultural eutrophication.*

---

1 A secchi disk is a black-and-white patterned disk (usually 12 inches in diameter) that is lowered into the water until the observer can no longer see the disk. The depth at which the disk cannot be seen is directly related to the water clarity.

The water quality of lakes as a function of their trophic state is shown in Table 7.6.

Table 7.6 Lake classification based on productivity

Lake Classification		Chlorophyll *a* Concentration (µg · L⁻¹)	Secchi Depth (m)	Total Phosphorus Concentration (µg · L⁻¹)
Oligotrophic	Average	1.7	9.9	8
	Range	0.3–4.5	5.4–28.3	3.0–17.7
Mesotrophic	Average	4.7	4.2	26.7
	Range	3–11	1.5–8.1	10.9–95.6
Eutrophic	Average	14.3	2.5	84.4
	Range	3–78	0.8–7.0	15–386
Hypereutrophic		> 50	< 0.5	Often > 100

Note: Classification for oligotrophic, mesotrophic, and eutrophic lakes from R. G. Wetzel, *Limnology* (W. B. Saunders, 1983), 767. Classification for hypereutrophic lakes from N. R. Kevern, D. L. King, R. Ring, "Lake classification systems—Part 1," *The Michigan Riparion*, February 1996, last updated December 1999. www.mlswa.org/lkclassill.htm.

Source: M. L. Davis and S. J. Masten, *Principles of Environmental Engineering and Science*, © 2004, McGraw-Hill Education. Reprinted by permission of the McGraw-Hill Companies.

Example 7.4

Algae can be represented as $C_{106}H_{263}O_{110}N_{16}P$. Given this chemical formula, if the concentration of total nitrogen in a lake is 10 ppb and the concentration of total phosphorus is 0.7 ppb, which of the two is the limiting nutrient?

Solution

The stoichiometric ratio of N/P in bacteria is 16:1. Given the atomic weights of N and P (14 and 31, respectively), the *mass ratio* of N:P in the algae is:

$$\left(\frac{N}{P}\right)_{algae} = \frac{16 \text{ moles N} \cdot \frac{14g}{1 \text{ mole N}}}{1 \text{ mole P} \cdot \frac{31g}{1 \text{ mole P}}} = 7.2$$

The mass ratio of N:P in the water is:

$$\left(\frac{N}{P}\right)_{water} = \frac{\frac{10 \mu g}{1 \text{ L}}}{\frac{0.07 \mu g}{1 \text{ L}}} = 14.3$$

Thus, this body of water is phosphorus limited. Adding more nitrogen to the water would not result in greater growth of algae, since there is already more nitrogen in the water than they need to grow. A rule of thumb is that when the mass concentration ratio of nitrogen to phosphorus in the water is greater than 10, the water will be phosphorus limited.

Stratification

Lakes in temperate regions undergo a phenomenon known as *thermal stratification*. Water is most dense at a temperature of 4°C (39°F), and this temperature-density relationship drives the transition of lakes from being stratified to being nonstratified. As a result of thermal stratification, the lake is split into layers:

- An upper layer of warm, lighter water called the *epilimnion*
- A middle, transition zone that prevents mixing, called the *metalimnion*
- A bottom layer of cool, heavier water called the *hypolimnion*

The transition between stratification and complete mixing occurs during a process known as *turnover*. In temperate regions, turnover occurs in the spring and fall. In spring, turnover occurs as the surface water warms to 4°C (and density increases). This water then sinks, bringing colder, deeper water to the surface. As summer progresses, the surface water warms, the deeper waters remain cold, and a *thermocline* sets up between the epilimnion and hypolimnion. This thermocline resists any mixing between the upper and lower layers. Fall turnover occurs as the surface water cools, becomes increasingly dense, and sinks to the bottom of the lake, thus promoting mixing of the lake.

Stratification traps nutrients released from bottom sediments in the hypolimnion. Also, under some circumstances, the hypolimnion can become anoxic as the oxygen is depleted and is not replenished due to the presence of the thermocline. The steep temperature gradient of the metalimnion prevents any surface water with dissolved atmospheric oxygen from reaching the bottom waters.

Water quality in lakes can be modeled using the mass balance approach described at the beginning of this chapter. Many lakes can be modeled as completely mixed reactors (see Chapter 14), especially shallow lakes in which the wind can effectively mix the water. In the case of a stratified lake, the epilimnion may be treated as completely mixed.

WETLANDS

The U.S. Army Corps of Engineers defines a wetland as "those areas that are inundated or saturated by surface or ground water at a frequency and duration sufficient to support, and that under normal circumstances do support, a prevalence of vegetation typically adapted for life in saturated soil conditions. Wetlands generally include swamps, marshes, bogs, and similar areas."

Alternatively, the U.S. EPA Web site states that a wetland is "an area that is regularly saturated by surface water or groundwater and is characterized by a prevalence of vegetation that is adapted for life in saturated soil conditions." These definitions highlight the three traits by which a wetland is characterized: hydrologic characteristics; soil characteristics; and types of vegetation.

1. **Hydrology**. The wetland soil must be saturated at some time during the growing season, but the depth of water must be less than two meters. The seasonal pattern of water levels in a wetland is known as the *hydroperiod*. Examples of hydroperiods for different types of wetlands are shown in Figure 7.5.

2. **Soils**. Wetland soils are classified as *hydric*. A hydric soil is a "soil that formed under conditions of saturation, flooding, or ponding long enough during the growing season to develop anaerobic conditions in the upper part. The concept of hydric soils includes soils developed under sufficiently wet

conditions to support the growth and regeneration of hydrophytic vegetation. Soils that are sufficiently wet because of artificial measures are included in the concept of hydric soils. Also, soils in which the hydrology has been artificially modified are hydric if the soil, in an unaltered state, was hydric. Some soil series, designated as hydric, have phases that are not hydric, depending on water table, flooding, and ponding characteristics."[2] Hydric soils are identified based on their color, permeability, texture, and smell.

3. **Vegetation**. The types of vegetation are characteristics of the vegetation adapted to the soils found in wetlands.

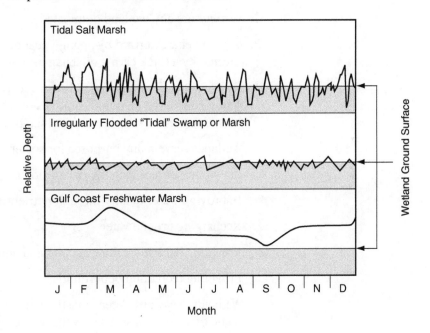

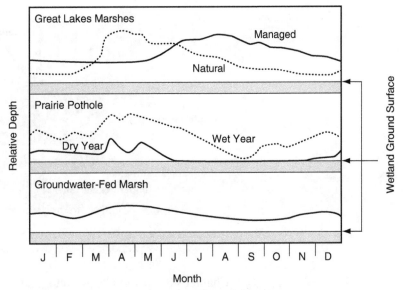

Figure 7.5 Hydroperiods for several different types of wetlands

Source: William J. Mitsch and James G. Gosselink, *Wetlands*, © 2000, John Wiley & Sons, Inc. Reprinted by permission.

---

2   *http://soils.usda.gov/use/hydric/intro.html,* accessed on 6/30/06.

Section 404 of the Clean Water Act regulates the disposal of fill material into waterways, including wetlands. The U.S. Army Corps of Engineers administers the program and is in charge of enforcement. The U.S. EPA develops and interprets policies and criteria used in reviewing permit applications. In general, agricultural and forestry practices are exempt from wetland regulations.

Wetlands can be categorized in several ways, but one common classification scheme is to divide wetlands into four categories: marshes, swamps, bogs, and fens.

1. *Marshes* are frequently or continually inundated with water. They are characterized by soft-stemmed vegetation including reeds, sedges, and cattails.

2. *Swamps* are any wetland dominated by woody plants.

3. *Bogs* are characterized by spongy peat deposits, acidic waters, and a floor covered by a thick carpet of sphagnum moss.

4. *Fens* are peat-forming wetlands that receive nutrients from sources other than precipitation or runoff. Sources of recharge include seeps, springs, and groundwater. Fens are less acidic than bogs.

Wetlands serve many purposes for society. Benefits of wetlands include:

- Preventing or reducing the risk of floods
- Improving water quality through nutrient and suspended solids removal
- Recharging groundwater
- Providing habitat for diverse plant and animal communities
- Providing recreation opportunities

Wetlands have also been constructed for the use of treating wastewater. A schematic of such a constructed wetland is provided in Figure 7.6.

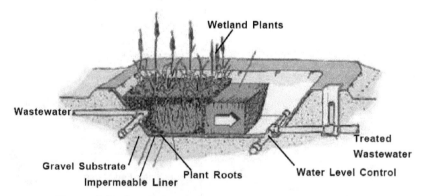

Figure 7.6 Constructed wetland for wastewater treatment

Source: *Constructed Treatment Wetlands*, EPA 843-F-03-013, accessed from www.epa.gov/owow/wetlands/pdf/ConstructedW.pdf.

Chapter 404 of the Clean Water Act also addresses the use of *compensatory mitigation*. Compensatory mitigation is defined by the U.S. Army Corps of Engineers as "the restoration, creation, enhancement, or in exceptional circumstances, preservation of wetlands and/or other aquatic resources for the purpose of compensating for unavoidable adverse impacts which remain after all appropriate and practicable avoidance and minimization has been achieved." In practice, this means that if a developer adversely impacts an existing wetland in the process of developing a portion of land, the developer must replace the affected wetland with another wetland. Of course, this infers that the regulatory agency allows the adverse effect to the wetland in the first place.

The Army Corps of Engineers and the U.S. EPA have established a three-part process, known as *mitigation sequencing*. The process prioritizes the response to the request for mitigation from a developer.

1. Avoid adverse impacts if a reasonable alternative exists

2. Minimize adverse impacts

3. Provide compensatory mitigation if steps 1 and 2 cannot be realized

The Army Corps of Engineers (or approved state authority) is responsible for determining the appropriate amount of mitigation required.

The most common mitigation techniques include the following:

- *Establishment* is the development of a wetland where a wetland did not previously exist.

- *Restoration* is the reestablishment or rehabilitation of an existing wetland with the goal of increasing wetland function or the number of wetland acres.

- *Enhancement* is the improvement of an existing wetland's function (for example, improving water quality, stormwater retention, or habitat quality)

Compensatory mitigation banking allows a third party to establish, restore, or enhance a wetland. A certain number of credits are assigned to this bank, and the third party can then sell these credits to developers needing to provide compensatory mitigation due to adverse effects of their development on an existing wetland.

PROBLEMS

7.1 Which of the following must be *TRUE* about the two organisms *Salmonella bongori* and *Salmonella enterica*?
   a. They are from the same order.
   b. They are from the same family.
   c. They are from the same genus.
   d. All of the above

7.2 A group of microorganisms have been identified as facultative heterotrophs. Which one of the following statements cannot be *TRUE*?
   a. The microbes transform TCE to DCE while consuming acetate.
   b. The microbes transform carbon dioxide and generate oxygen.
   c. The microbes degrade toluene through denitrification.
   d. The microbes degrade benzene aerobically.

7.3 A composting windrow has a triangular cross section and is 10 feet wide at the base, 10 feet high at the peak, and 100 feet long. You may assume the waste has a specific weight of 400 lb/yd³ as placed in the windrow, the waste material is 40% biodegradable, and it has the ultimate composition $C_{60}H_{95}O_{40}N$. Determine the amount of oxygen required to stabilize the waste under aerobic conditions.
   a. 53 tons
   b. 37 tons
   c. 21 tons
   d. 5 tons

7.4 Determine the decay (death) rate constant in the expression below, given the following biological data measured in the laboratory.

$$\left(\frac{dX}{dt}\right)_{net} = 75 \frac{mg}{L \cdot min}; \quad \left(\frac{dS}{dt}\right) = 144 \frac{mg}{L \cdot min}; \quad Y = 0.7; \quad X = 2500 \text{ mg/L}$$

   a. 1.44 hr⁻¹
   b. 0.62 hr⁻¹
   c. 23.7 min⁻¹
   d. 8.7 min⁻¹

7.5 A specific growth rate curve is provided for a laboratory test on a domestic wastewater (see Exhibit 7.5). Determine the substrate utilization rate (dS/dt) if the biomass concentration is 2500 mg/L, the microbial decay coefficient (k_d) is 0.33 hr^{-1}, and the laboratory measured growth yield is 0.58 in a biological growth tank that has a substrate concentration of 165 mg/L.

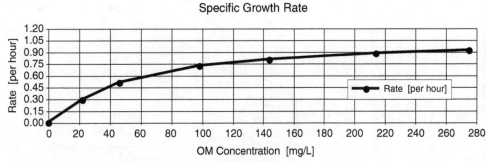

Exhibit 7.5

   a. 56.7 mg/L per min
   b. 31.1 mg/L per min
   c. 15.3 mg/L per min
   d. 2.75 mg/L per min

7.6 The most effective way to prevent the spread of disease is to:
   a. wear a face mask
   b. get a flu shot
   c. gargle daily with an antiseptic mouthwash
   d. wash your hands with an antimicrobial soap

7.7 Personal protective equipment (PPE) performs which of the following functions?
   a. Prevents a diseased person from developing symptoms
   b. Reduces the potential for exposure to an infectious strain
   c. Prevents a virulent strain from entering the body
   d. Kills any microbial strains that contact the body

7.8 Given the following MPN tube results, what is the approximate number of coliforms in the original sample?

Dilution	Results
10^0	+ + + + +
10^{-1}	+ + + + +
10^{-2}	+ + + + +
10^{-3}	+ + + + +
10^{-4}	+ − + + −
10^{-5}	− − + − +
10^{-6}	− + − − −
10^{-7}	− − − − −

a. > 16,000 coliforms/100 mL
b. 90,000 coliforms/100 mL
c. 140,000 coliforms/100 mL
d. 170,000 coliforms/100 mL

7.9 Determine the minimum DO concentration of a stream (DO_{min}) given the following water quality and kinetic parameters: DO^{sat} = 10.78 mg/L, D_0 = 2.73 mg/L, L_0 = 6.57 mg/L, k_a = 1.36 day^{-1}, and k_d = 0.92 day^{-1}.
a. 3.12 mg/L
b. 5.37 mg/L
c. 7.66 mg/L
d. 8.05 mg/L

SOLUTIONS

7.1 d. The two organisms are species from the genus *Salmonella*; therefore, they must also be from the same family and order.

7.2 b. Acetate, toluene, and benzene are all organic compounds, so choices a, c, and d describe heterotrophic metabolism. CO_2 transformation is autotrophic by definition.

7.3 c. First determine the volume of the windrow as follows:

$$V = \frac{1}{2}bhl = (0.5)(10 \text{ ft})(10 \text{ ft})(100 \text{ ft}) = 5000 \text{ ft}^3 \times \left(\frac{1 \text{ yd}}{3 \text{ ft}}\right)^3 = 185.2 \text{ yd}^3$$

The mass of the waste in the windrow may be determined from the specific weight as

$$185.2 \text{ yd}^3 \times 400 \frac{\text{lb}}{\text{yd}^3} = 74{,}074 \text{ lb}$$

To use the aerobic reaction given in the *FE SRH*, mass must be converted to moles. This requires the molecular weight of the biodegradable material, which is determined as follows:

$$MW = (12)(60) + (1)(95) + (16)(40) + (14)(1) = 1469 \frac{\text{lb}}{\text{lbmol}}$$

The number of moles reacted can now be calculated, remembering that only 40% of the waste is biodegradable:

$$n = \frac{74{,}074 \text{ lb}}{1469 \frac{\text{lb}}{\text{lbmol}}} = 50.4 \text{ lbmol} \times (0.4) = 20.17 \text{ lbmol}$$

The moles of oxygen required to complete this transformation is determined by multiplying the moles of waste consumed by the stoichiometric coefficient for oxygen in the reaction given, which may be calculated as follows:

$$n_{O_2} = \left(\frac{(4)(60) + 95 - (2)(40) + (3)(1)}{4}\right) 20.17 = 1301 \text{ lbmol}$$

Finally, the mass of oxygen required is calculated as

$$m_{O_2} = 1301 \text{ lbmol} \times 32 \frac{\text{lb}}{\text{lbmol}} = 41{,}632 \text{ lb } O_2 \times \frac{1 \text{ ton}}{2000 \text{ lb}} = 20.82 \text{ tons } O_2$$

7.4 b. Using the expression for net growth rate and the data provided, we may write

$$\left(\frac{dX}{dt}\right)_{net} = \left(\frac{dX}{dt}\right) - k_d X \quad \Rightarrow \quad k_d = \frac{\left(\frac{dX}{dt}\right) - \left(\frac{dX}{dt}\right)_{net}}{X} = \frac{\left(\frac{dX}{dt}\right) - 75}{2500}$$

From this, we see that we need a value for (dX/dt), which is the rate of biomass generated. Noticing that the yield is given and can be defined as

$$Y = \frac{(dX/dt)}{(dS/dt)}$$

we may rearrange this expression to solve for (dX/dt) as follows:

$$\left(\frac{dX}{dt}\right) = Y\left(\frac{dS}{dt}\right) = (0.7)\left(144 \frac{mg}{L \cdot min}\right) = 100.8 \frac{mg}{L \cdot min}$$

Now, we may substitute this into the first expression to get k_d as follows:

$$k_d = \frac{\left(\frac{dX}{dt}\right) - 75}{2500} = \frac{100.8 - 75}{2500} \quad \Rightarrow \quad k_d = 0.0103 \text{ min}^{-1} \times \frac{60 \text{ min}}{1 \text{ hr}} = 0.62 \text{ hr}^{-1}$$

7.5 a. Substrate utilization can be determined as

$$\frac{dS}{dt} = \frac{\mu_{max} S X}{Y(K_s + S)}$$

Values for S, X, and Y are given. Therefore, only μ_{max} and K_s need to be determined. Extrapolating μ_{max} from the plot provided,

$$\mu_{max} = 1.0 \text{ hr}^{-1}$$

Now, K_s may be evaluated as the substrate concentration at $½\mu_{max}$, which is a rate equal to 0.5 hr^{-1}. From the plot provided, $K_s = 44$ mg/L at a rate of 0.5 hr^{-1}. Now, utilization may be calculated as

$$\frac{dS}{dt} = \frac{(1 \text{ hr}^{-1})(165 \text{ mg/L})(2500 \text{ mg/L})}{(0.58 \text{ mg/mg})(44 + 165 \text{ mg/L})} = 3403 \frac{mg}{L} \text{hr}^{-1} \times \frac{1 \text{ hr}}{60 \text{ min}} = 56.7 \frac{mg}{L} \text{min}^{-1}$$

7.6 d. The CDC states that the most important action that will reduce the spread of disease is the washing of hands with an appropriate sanitizing agent.

7.7 b. PPE does not prevent symptoms from developing once an infection occurs, and it does not possess any properties that would kill bacteria once in contact with the body. It is designed to substantially reduce the potential for exposure but cannot eliminate exposure in all cases.

7.8 c. MPN tables and Standard Methods indicate that the correct selection of tubes is to use the lowest dilution that results in five positive tubes and the following two dilutions. From the data given, the result would be 5-3-2, with a dilution factor of 1000. The MPN tables give a result for 5-3-2 as 140 coliforms per 100 mL. Multiplying by the dilution factor yields 140,000 coliforms/100 mL.

7.9 c. Calculation of critical deficit requires a value for critical time using the equation

$$t_c = \frac{1}{k_a - k_d} \ln\left[\frac{k_a}{k_d}\left(1 - D_0 \frac{k_a - k_d}{k_d L_0}\right)\right]$$

Using data provided in the problem statement, t_c is determined to be

$$t_c = \frac{1}{1.36 - 0.92} \ln\left[\frac{1.36}{0.92}\left(1 - (2.73)\frac{1.36 - 0.92}{(0.92)(6.57)}\right)\right] = 0.385 \text{ days}$$

Next, critical deficit is calculated using t_c in the deficit equation as follows:

$$D_c = \frac{k_d L_0}{k_a - k_d}[\exp(-k_d t_c) - \exp(-k_a t_c)] + D_0 \exp(-k_a t_c)$$

Using the data given with the calculated critical time yields

$$D_c = \frac{(0.92)(6.57)}{1.36 - 0.92}[\exp(-0.92 \times 0.385) - \exp(-1.36 \times 0.385)]$$
$$+ (2.73)\exp(-1.36 \times 0.385)$$

$D_c = 3.12$ mg/L

Now, the DO at the critical time is DO_{min}, and is equal to the difference between DO^{sat} and D_c:

$$DO_{min} = DO^{sat} - D_c = 10.78 - 3.12 \text{ mg/L} = 7.66 \text{ mg/L}$$

CHAPTER 8

Environmental Chemistry

OUTLINE

PRINCIPLES OF MATTER 262

PERIODICITY 266

NOMENCLATURE 267

METALS AND NONMETALS 268

ORGANIC CHEMISTRY 271

MATERIAL PHYSICAL PROPERTIES 274

CONCENTRATION UNITS 274

EQUATIONS AND STOICHIOMETRY 277

REACTION ORDER AND KINETICS 279

EQUILIBRIUM 287

OXIDATION AND REDUCTION 291
Redox Reactions

ELECTROCHEMISTRY 293

COMBUSTION 295

PROBLEMS 298

SOLUTIONS 306

Environmental engineers are often tasked with the design of treatment units or systems that address all potential negative impacts of waste materials that are discharged to the environment. These tasks require a solid understanding of the materials of concern; this includes not only their composition but also how those compounds interact with other substances. It is primarily the ability to manipulate the laws of the basic and life sciences that allows the environmental engineer to design effective and efficient treatment units. To that end, examinees are expected to be familiar with fundamental chemistry definitions and concepts, including but not limited to elemental and molecular weights and structures, inorganic and organic compounds, balanced reactions and stoichiometry, reaction order and kinetics, equilibrium relationships, redox reactions and electrochemistry, and combustion reactions. This section will focus on some of the specific issues in chemistry as they are applied to water, wastewater, atmospheric, solid and hazardous waste environments.

PRINCIPLES OF MATTER

The *atomic number* (Z) is defined as the number of protons present in the nucleus of an element. The *atomic weight* (A) is defined as the weight of one *mole* (6.022×10^{23} atoms per mole) of a substance. While this weight is generally assumed to be contributed by the protons and neutrons in the nucleus, and therefore should be equal to twice the atomic number, most elements may have several isotopes. *Isotopes* are atoms with the same atomic number but that possess different numbers of neutrons, thus yielding a single element with several different atomic weights. In order to report a single atomic weight, a weighted average of all naturally occurring isotopes is used, based upon the mass of a carbon-12 atom. Most chemical species of interest to the environmental engineer contain multiple elements that are bound to form stable molecules. In these cases, the *molecular weight* (MW), also called the *formula weight* (FW), of that compound can be determined by summing the atomic weights of each of the elements that make up the molecule. A copy of the Periodic Table of the Elements is provided in the *Fundamentals of Engineering Supplied-Reference Handbook*.

In acid/base reactions and the treatment of aqueous solutions that contain high quantities of dissolved inorganic species (for example, the removal of hardness in water treatment plants), it is useful to simplify the stoichiometric calculations of reactions occurring in those solutions. In these cases, it is convenient to define the *equivalent weight* (EW) of a compound (often called the *equivalent*) as *the formula weight divided by its equivalency*, where the equivalency can be defined as (1) *the absolute value of the ion charge*, (2) *the number of [H^+] or [OH^-] ions that the species may react with*, or (3) *the absolute value in the change in valency during a redox reaction*. Table 8.1 (elements) and Table 8.2 (molecules) contain several compounds that are commonly encountered in environmental engineering with their corresponding atomic or molecular weight and equivalent weights.

Table 8.1 Common elements in environmental engineering

Name	Symbol	Atomic Weight	EW
Aluminum	Al	27.0	9.0
Calcium	Ca	40.1	20.0
Carbon	C	12.0	
Chlorine	Cl	35.5	35.5
Fluorine	F	19.0	19.0
Hydrogen	H	1.0	1.0
Iodine	I	127	127
Iron	Fe	55.8	27.9
Magnesium	Mg	24.3	12.2
Manganese	Mn	54.9	27.5
Nitrogen	N	14.0	
Oxygen	O	16.0	
Phosphorus	P	31.0	
Potassium	K	39.1	39.1
Sodium	Na	23.0	23.0
Sulfur	S	32.0	16.0

Table 8.2 Common compounds in water treatment processes

Name	Formula	MW	EW
Aluminum hydroxide	$Al(OH)_3$	78.0	26.0
Aluminum sulfate	$Al_2(SO_4)_3 \cdot 14.3H_2O$	600	100
Ammonia	NH_3	17.0	
Ammonium	NH_4^+	18.0	18.0
Ammonium fluorosilicate	$(NH_4)_2SiF_6$	178	
Bicarbonate	HCO_3^-	61.0	61.0
Calcium bicarbonate	$Ca(HCO_3)_2$	162	81.0
Calcium carbonate	$CaCO_3$	100	50.0
Calcium fluoride	CaF_2	78.1	
Calcium hydroxide	$Ca(OH)_2$	74.1	37.0
Calcium hypochlorite	$Ca(ClO)_2 \cdot 2H_2O$	179	
Calcium oxide	CaO	56.1	28.0
Calcium sulfate	$CaSO_4$	136	68.0
Carbonate	CO_3^{2-}	60.0	30.0
Carbon dioxide	CO_2	44.0	22.0
Chlorine	Cl_2	71.0	35.5
Chlorine dioxide	ClO_2	67.0	
Ferric chloride	$FeCl_3$	162	54.1

(continued)

Name	Formula	MW	EW
Ferric hydroxide	$Fe(OH)_3$	107	35.6
Fluorosilicic acid	H_2SiF_6	144	
Hydroxyl ion	OH^-	17.0	17.0
Hypochlorite	OCl^-	51.5	51.5
Magnesium carbonate	$MgCO_3$	84.3	42.1
Magnesium hydroxide	$Mg(OH)_2$	58.3	29.1
Magnesium sulfate	$MgSO_4$	120	60.1
Nitrate	NO_3^-	62.0	62.0
Orthophosphate	PO_4^{3-}	95.0	31.7
Oxygen	O_2	32.0	16.0
Sodium bicarbonate	$NaHCO_3$	84.0	84.0
Sodium carbonate	Na_2CO_3	106	53.0
Sodium hydroxide	$NaOH$	40.0	40.0
Sodium hypochlorite	$NaClO$	74.4	
Sodium fluorosilicate	Na_2SiF_6	188	
Sodium sulfate	Na_2SO_4	142	71.0
Sulfate	SO_4^{2-}	96.0	48.0

Example 8.1

Molecular and equivalent weights

Determine the molecular and equivalent weight of sulfuric acid, H_2SO_4.

Solution

The MW is determined by summing the atomic weights of each of the contributing atoms as

$$MW_{H_2SO_4} = 2\,MW_H + MW_S + 4\,MW_O = (2)(1.0079) + 32.064 + (4)(15.9994)$$
$$= 98.0774$$

Often, solution accuracy does not require six significant figures, and a value of 98.1, or even 98, is acceptable for most calculations. Also note that the units are generally accepted to be g/mol.

Calculation of the equivalent weight requires the determination of the equivalency of sulfuric acid. Using the first definition of equivalency given above, the absolute value of the ion charge of the sulfate ion is 2, and thus 2 is the equivalency of sulfuric acid. However, if the ion charge of hydrogen is used, it might be assumed that the equivalency is equal to 1. The second definition given above helps determine the correct equivalency as it is seen that 2 hydrogen ions are available to react, yielding the true equivalency of 2. The EW may now be calculated as

$$EW_{H_2SO_4} = \frac{MW_{H_2SO_4}}{\text{equivalency}} = \frac{98.0774}{2} = 49.0387 \approx 49.0$$

In its most simple characterization, matter is said to exist in three possible phases: solid, liquid, and gas. While other states are possible, these three phases (along with all of the possible combinations of multiphase systems) are usually sufficient to describe nearly all situations an engineer will encounter in natural environments. The determination of which phase a substance will take is dependent on the specific compound, along with the prevailing temperature and pressure of the environment where the substance is located.

Solids are generally considered the lowest energy state, although all matter at a temperature above absolute zero has molecular energies. Solids have *mass densities* (ρ) [mass per unit volume], or equivalently *specific weights* (γ) [weight per unit volume], that are greater than liquids or gases. The most notable exception to this rule is water, which is slightly less dense as a solid (ice) than it is as a liquid. Solids are usually further defined on a microscopic level by their crystalline structure, or on a macroscopic level by their particle size distribution. Examples of solids that are important to environmental engineers include soils, sand beds for filtration or drying, precipitation reactions and floc management of colloidal or settleable solids in water treatment, and particulate matter in gases from combustion processes or fugitive dusting environments.

Fluids are defined as *substances that deform continuously when acted upon by any shear stress*, or more simply as *substances that take the shape of their container*. Both liquids and gases fall into this category, but material properties are quite different for each type. While liquids possess densities that are similar to that of their solid forms, the gas phase of a substance has a density that is approximately 1000 times smaller. Further, although both types of fluids are able to be poured from their container, the degree of flowability of gases is approximately 1000 greater than the liquid phase. The measure of flowability is called *dynamic* (or *absolute*) *viscosity* (μ) and is most commonly expressed in units of centipoises (*cP*). For liquids, increasing temperature usually reduces viscosity, while reducing temperature increases viscosity until such a point at which the material turns into a solid and no longer flows.

Gases at or near atmospheric pressure are said to behave ideally, referring to the fact that there exists a direct relationship between gas density and the system temperature and pressure. The *ideal gas law* can be expressed as

$$PV = nRT \quad \text{or} \quad P = \rho' RT$$

where P is absolute pressure, V is the gas volume, n is the number of moles of gas, R is the universal gas constant, T is the absolute temperature, and ρ' is the molar density. Several values of the universal gas constant can be found for a variety of convenient units in Table 8.3. Also, note the molar density is reported as moles per unit volume. This must be converted into a mass density (mass per unit volume) by multiplying by the species molecular weight.

Table 8.3 Values for the Universal Gas Constant in various units

Value	Units
8.314	$m^3 \cdot Pa/mol \cdot K$
0.08314	$liter \cdot bar/mol \cdot K$
0.08206	$liter \cdot atm/mol \cdot K$
62.36	$liter \cdot mm\ Hg/mol \cdot K$
0.7302	$ft^3 \cdot atm/lbmole \cdot °R$

	(continued)
Value	Units
10.73	ft³•psia/lbmole•°R
8.314	J/mol•K
1.987	cal/mol•K
1.987	Btu/lbmole•°R

PERIODICITY

The periodicity of the properties of the elements is based on the electronic structure of the atoms and reflected in their order and position in the **periodic table**. Similarities of elemental properties, and their periodicity, were observed long before atomic structure was understood. Definitions of atomic structure, and the understanding thereof, came largely from spectroscopic observations and their mathematical interpretation.

The periodic table now has seven horizontal rows or **periods** and eight, or more, vertical columns or **groups**. The properties of the elements in each period vary from metallic to nonmetallic going from left to right, ending with the nonreactive noble gases. The groups from I to VII (with noble gases as Group VIII), have metals or nonmetals of somewhat similar chemical and physical properties. However, the heavier elements, going down a column in the table, tend to be more metallic in character, as particularly noted on the nonmetal side of the table.

The first period contains only the two elements hydrogen (H) and helium (He), but the second and third periods contain eight elements each, from lithium (Li) through neon (Ne) for the second period and from sodium (Na) through argon (Ar) for the third period. In the fourth period, the number of groups has to be considerably expanded, as ten elements come in order of atomic numbers between calcium (Ca) in the second group and gallium (Ga) in the third group. These are all metals and are called the first transition series.

The fifth period is similar to the fourth, in that ten metallic elements—the second transition series—occur in order of atomic number between strontium (Sr) in the second group and indium (In) in the third group. A third transition series occurs in the sixth period, between barium (Ba) in the second group and thallium (Tl) in the third group.

In order to handle the periods with 18 elements, the modern periodic table divides the elemental groups into A groups and B groups. The eight groups consistent with the elements listed in the second and third periods are IA, IIA, IIIA, IVA, VA, VIA, VIIA and VIIIA. The transition series elements are divided into groups IB, IIB, IIIB, IVB, VB, VIB, VIIB, and VIIIB.

A further complication of the periodic table is the occurrence of 14 elements in the sixth period between lanthanum (La) in Group IIIB and hafnium (Hf) in Group IVB. These are known as the lanthanide series, or **rare earth** elements, and are not listed in the main sequence of the periodic table. Again in the seventh period, 14 elements occur between actinium (Ac) and the unknown element 104, and are known as the actinide series. As in the case of the transition elements, their occurrence at this place in the sequence of atomic numbers can be explained by the theory of atomic structure.

Another comment on the periodic table concerns groups IIIA, IVA, VA, and VIA. These groups all start as nonmetals in the second period, namely with boron (B), carbon (C), nitrogen (N), and oxygen (O), respectively. However, going down the columns corresponding to these groups, the elements change to metallic properties. This occurs in the third, fourth, fifth, and sixth period, respectively, for groups IIIA–VIA; the first metallic elements are aluminum (Al), germanium (Ge), antimony (Sb), and polonium (Po) in each of these groups.

With the overall structure of the periodic table in mind, let us look at some properties that show periodicity. One is atomic radius, which is a maximum for each Group IA element (the alkali metals—Li, Na, K, Rb, and Cs). The atomic radius decreases going across the period to the halogen elements of Group VIIA, complicated by minima in the transition elements and the lanthanide-actinide series. The ionization potential, or first ionization energy, represents the energy needed to remove one electron from an atom. It is a minimum for the Group IA metals and reaches a maximum for the Group VIII rare gases.

NOMENCLATURE

Chemical nomenclature is concerned with the prefixes, suffixes, and other modifications of the root word to denote the state of the elements in a compound. We will be concerned here primarily with inorganic compounds, as the nomenclature of organic compounds is a topic of its own and will be considered in the section on organic chemistry.

When compounds of metals and nonmetals occur, the compound is considered a salt; and the naming of the compound uses the elemental name of the metal first, followed by the root of the nonmetal element name and the ending **-ide**. Sodium brom*ide* and calcium fluor*ide* are examples. The nomenclature becomes slightly more complicated when the metal may have more than one valence state in its compound with the nonmetal. An example is iron oxide, which is known more precisely as either ferrous oxide or ferric oxide, depending on the valence state of the iron. Also, these compounds would be known as oxides, rather than salts, as the compound would not dissociate to oxide ions when dissolved in aqueous solution.

When the metal ion can have both a lower valence and a higher valence, the compounds of the lower valence have the root word with the suffix **-ous**, while the compounds of the higher valence have the root word with the suffix **-ic**. In the case of the iron oxide mentioned above, the elemental root is not iron but **ferr**. The root still designates the same element but comes from a different linguistic source.

The compounds of hydrogen and nonmetals maintain the salt type of nomenclature, with hydrogen fluoride, hydrogen chloride, hydrogen bromide, and hydrogen iodide as designations for the halogen compounds in the unionized gaseous phase. The nomenclature is consistent with other hydrogen compounds such as hydrogen sulfide (H_2S) and hydrogen selenide (H_2Se). However, hydrogen oxide (H_2O) is known universally by its more common name, **water**.

When the above hydrogen compounds are in aqueous solution, they become ionized to inorganic acids. The inorganic acids always contain hydrogen as the cation and the cation root is designated as **hydro**—leading to hydrofluoric (HF), hydrochloric (HCl), hydrobromic (HBr), hydroiodic (HI), hydrosulfuric (H_2S), and hydroselenic (H_2Se) acids. Hydrogen oxide, as the solvent phase, remains designated as water.

Perhaps the most confusing of the designation nomenclature is that of the oxyacids. These are acids containing various amounts of oxygen in the anion.

The compounds with sulfur in the anion are illustrative. Here the designation for hydrogen as the cation is inherent in the name **acid**, so it is not explicitly given. The most common oxyanion of sulfur is the sulfate ion, SO_4^{2-}, where the sulfur has an oxidation number of +6. As the acid, the name becomes sulfuric—with the root **sulfur** and the suffix **-ic**. When the anion has one less oxygen, $\left[SO_3^{2-}\right]$, the name of the acid has the **sulfur** root and the suffix **-ous**. Sulf*ous* acid is H_2SO_3, where the sulfur has an oxidation number of +4.

Example 8.2

The oxyacids of sulfur and phosphorous have the ending **-ic** for the element in the most stable oxidation state and **-ous** for the element in the next lower oxidation state. What are the names of the corresponding salts of these acids?

Solution

The *sulfuric* acid salts have the endings **-ate** for the salt in which both hydrogens are replaced and **hydrogen -ate** when only one hydrogen is replaced—as in sodium sulf*ate* and sodium *hydrogen* sulf*ate*. The latter salt is also known as sodium *bi*sulfate. The *sulfurous* acid salts have similar names, except the *-ate* ending is replaced by *-ite*. The comparable names for the *-ous* acid salts are sodium sulf*ite* and sodium hydrogen sulf*ite* or sodium *bi*sulf*ite*.

The *phosphoric* acid salts also have the ending **-ate** for the salt when all three hydrogens are replaced—as in sodium phosph*ate*. However, the salt is the **dihydrogen -ate** when only one hydrogen is replaced and the **hydrogen -ate** when two of the three hydrogens are replaced, as in sodium dihydrogen phosphate and sodium hydrogen phosphate. The *phosphorous* acid salts have similar names, except that the *-ate* ending is replaced by *-ite*.

The **-ic** and **-ous** suffix designation carries over to the oxidation state of cations in salt-type compounds with nonmetals. Thus iron in the +3 state forms ferr*ic* chloride, $FeCl_3$, and ferr*ic* oxide, Fe_2O_3; iron in the +2 state forms ferr*ous* chloride, $FeCl_2$, and ferr*ous* oxide, FeO. The same designation is used with tin compounds, with the root designated by **stann**. Since the oxidation states of tin are +4 and +2, $SnCl_4$ is stann*ic* chloride and $SnCl_2$ is stann*ous* chloride.

METALS AND NONMETALS

The two general classes of elements are metals and nonmetals. The division between metal and nonmetal is not always sharp, and the nonreactive elements (noble gases) do not fit into either category.

Nevertheless, there is a similarity in the physical aspects and reactivity of the large group known as metals and in the large group known as nonmetals, so that it is convenient and logical to consider the elements as part of each group. Also, the electronic structure of the atoms of the elements can be fundamentally grouped into electron donors, or metals, and electron acceptors, or nonmetals, when participating in chemical reactions. However, there are cases when the metal-nonmetal classification breaks down from this point of view, as, for example, (1) for many elements the loss or gain of electrons in reactions depends on reaction conditions as well as on the other element participating in the reaction, and (2) compound formation may result in the sharing of available electrons rather than an identifiable loss or gain of electrons by the elements participating in the reaction.

Referring to the periodic table of the elements, the period and row in which an element is located is closely related to its metallic or nonmetallic properties.

Generally the elements on the left side of the periodic table are metals and those on the right side are nonmetals. Elements in a particular group become more metallic toward the bottom of the table (heavier elements), and the elements in the middle of the table, including the transition series elements, have metallic properties.

Metals are generally known for their physical properties, which include the characteristic metallic luster, cohesive strength, ductility, and high electrical and thermal conductivity. These properties are well explained by the theory of metals that views metal atoms as points in a crystalline structure surrounded by their **electron cloud**. The outer electrons travel freely through the crystalline structure, bonding the atoms together and providing for flow of electrical charge. The nonmetals, on the other hand, are often gases, or colored solids with no luster and low electrical and thermal conductivity.

The most active metals are those of Group IA of the periodic table, the alkali metals. These are lithium, sodium, potassium, rubidium, and cesium—all with a valence of +1. The next most active metals are the Group IIA alkaline earth metals, beryllium, magnesium, calcium, strontium, barium, and radium—all with a valence of +2. Group IB metals—copper, silver, and gold—are distinctly metallic, with valences from +1 to +3. The Group IIB elements—zinc, cadmium, and mercury—have +1 or +2 valence.

The Group IIIA metals, aluminum, gallium, indium, and thallium, all have a +3 valence. The Group IIIB metallic elements include scandium, yttrium, lanthanum, and actinium. Groups IVB and VB include titanium, zirconium, hafnium, vanadium, niobium, and tantalum, with valences in the +2 to +5 range. The Group VIB and VIIB elements—chromium, molybdenum, tungsten, manganese, technetium, and rhenium—also have variable positive valences. The Group IVB, VB, VIB, and VIIB elements (as well as IB, IIIB, and VIIIB) are **transition metals**.

The nonmetals are on the right side of the periodic table of the elements—or, more specifically, on the upper right side of the table, as elements in Group IIIA through Group VIA become metallic toward the bottom of the table. We include as nonmetals the Group VIIIA elements, also known as the noble gases. These elements, helium (He), neon (Ne), argon (Ar), krypton (Kr), xenon (Xe), and radon (Rn), are certainly nonmetallic, but do not quite fit into the concept of nonmetals because they are almost totally unreactive. However, a limited number of noble gas compounds have been prepared, including the xenon compounds XeF_2, XeF_4, and XeF_6.

The next group of nonmetals to consider are the Group VIIA halogens: fluorine (F), chlorine (Cl), bromine (Br), iodine (I), and astatine (At). These elements are characterized electronically by an unfilled p orbital in the outer electron shell, and a great tendency to fill the orbital and assume the same electron configuration as the rare gases. As elements, this configuration is achieved by the diatomic molecule, x_2, where x stands for any halogen, and the diatomic state shows covalent sharing of an electron pair.

The physical properties of the halogens vary with atomic number and weight. Under most ambient conditions, fluorine is a pale yellow gas, chlorine a greenish-yellow gas, bromine a brownish liquid, and iodine a violet solid. The reactivity of the halogens is outstanding because they react with almost all metals, with hydrogen, with phosphorous, with water, and with each other. Although all halogens are reactive, fluorine and chlorine are more reactive than iodine and bromine. The compounds of the halogens with Group IA and IIA metals are generally ionic, but other metallic compounds, such as $TiCl_4$, are nonionic. Fluorides tend to have the most ionic character and iodides the least ionic character.

The nonmetals of Group VIA are oxygen (O), sulfur (S), selenium (Se), and tellurium (Te). The lightest of these, oxygen, exists in elemental form as the diatomic molecule, O_2, but also can form the triatomic ozone molecule, O_3. Oxygen is very abundant on the earth's surface, not only in the atmosphere, but also as the major weight component of H_2O and in the silicates, oxides, sulfates, and carbonates of the earth's crust.

Elemental oxygen, O_2, is, next to fluorine, the most reactive of the nonmetals; but the reactions may require elevated temperatures to activate and break the O—O bond. All metals, except the noble metals like silver, gold, and platinum, react with O_2, forming oxides or superoxides.

Oxygen also reacts with all nonmetals, except for the halogens and the noble gases. Carbon (C), sulfur (S), and nitrogen (N), all react to form oxides, with the oxide formulation being sensitive to the relative amounts of reactants and the reaction conditions. Oxygen also reacts with compounds, such as metal sulfides and hydrocarbons, to form oxides of each element of the original compound. This is illustrated in the **roasting** of sulfide ores, for example:

$$2\ ZnS + 3\ O_2 = 2\ ZnO + 2\ SO_2$$

S, Se, and Te all form hydrogen compounds of the formula H_2S, where S can also be Se or Te. Many metal sulfides are insoluble, accounting for the use of H_2S as an analytical reagent in tests for metal ions. Metal ions forming insoluble sulfides by reaction with H_2S include bismuth (Bi^{3+}), cadmium (Cd^{2+}), copper (Cu^{2+}), mercury (Hg^{2+}), antimony (Sb^{3+}), iron (Fe^{2+}), manganese (Mn^{2+}), nickel (Ni^{2+}), and zinc (Zn^{2+}).

The Group VA nonmetals are limited to nitrogen (N), phosphorous (P), and arsenic (As), as antimony (Sb) and bismuth (Bi) are considered metals.

Elemental nitrogen is unreactive at ambient temperature, primarily from the strength of the N—N bond in the molecule. At high temperatures nitrogen reacts with a number of metals, forming compounds that are either ionic, interstitial, or covalent. Interstitial means the N atoms are in the interstices of the metal lattice, as in vanadium nitride, VN, and titanium nitride, TiN. These latter materials are high melting, metallic appearing, and relatively inert.

Phosphorous reacts with a number of metals, forming a variety of compounds with Group IA and IIA metals, Group IIIA elements, and the transition metals. Compounds of the Group IA and IIA metals hydrolyze to form phosphine, as in

$$Ca_3P_2 + 6\ H_2O = 3\ Ca(OH)_2 + PH_3$$

The hydrides and halides of the Group VA nonmetals are known. The best known nitrogen hydride is ammonia, NH_3, but hydrazine, N_2H_2, and hydrazoic acid, HN_3, can also be prepared. The trifluoride of nitrogen, NF_3, is a stable gas, but the other trihalides, NX_3, are unstable. The oxides of nitrogen include nitrous oxide, N_2O, which is stable and has anesthetic properties (laughing gas), as well as nitric oxide, NO, which is formed in lightning storms or electric arcs and which reacts with oxygen at ambient temperature. Nitrogen dioxide, NO_2, is a brown gas that exists in equilibrium with colorless dinitrogen tetroxide, N_2O_4: $2\ NO_2 = N_2O_4$.

Carbon (C) and silicon (Si) are the two nonmetals in Group IV, and each has two filled s orbitals and two filled p orbitals in the outer electron shell, so that four electrons are needed to complete an octet. This is generally done by covalent bonding.

The outstanding property of carbon is the ability to form long chains or rings of carbon atoms joined together. Also the carbon atoms may join together in various types of bonds, involving single, double, or triple electron pairs. Elemental carbon occurs in two primary crystalline forms—the three-dimensional tetrahedral structure known as the diamond, and the two-dimensional lattice structure known as graphite. Carbon also joins to other elements, such as O, N, and S, with multiple bonds. This is delineated more fully in the section on organic chemistry. Carbon forms compounds known as **carbides** with a number of metals.

The two most prevalent oxides of carbon are the monoxide, CO, and the dioxide, CO_2. CO is relatively active, as it burns, reacts with halogens and sulfur vapor, and reduces metal oxides,

$$FeO + CO = Fe + CO_2$$

CO_2 is relatively inert and does not burn or support combustion.

Carbon also reacts with other elements to form liquid products that are useful as solvents, such as carbon disulfide, CS_2, and carbon tetrachloride, CCl_4. The latter is one of a series of halocarbon solvents with various commercial names such as **Freon** or **UCON**, followed by a number designating the particular halocarbon. Other important carbon compounds are the carbon-hydrogen compounds, which start with methane, CH_4, and are called **hydrocarbons**. These and compounds containing oxygen, nitrogen, and sulfur are considered further in the section on organic chemistry.

Silicon (Si), the other nonmetallic element of Group IVA, has many similarities to carbon. Elemental silicon is a grey, lustrous solid with a structure similar to the carbon diamond structure. The electronic energy level of elemental silicon is critical to its extensive use in semiconductors. Unlike carbon, the energy gap between the valence band for electrons and the conduction band for electrons is small enough that heating can promote electrons into the conduction band, making the element an intrinsic semiconductor.

The addition of a **dopant** such as arsenic (As) or gallium (Ga) to Si decreases the energy gap and allows conduction at lower temperatures. Arsenic has one more electron in the outer shell than Si, and this electron is promoted into the conduction band to create an n-type semiconductor. Gallium has one less electron than Si in the outer shell and can accept electrons from the occupied level, leaving **holes** and resulting in a p-type semiconductor.

Boron is generally found in the earth's crust as a borate, such as borax, $Na_2B_4O_7 \cdot 8H_2O$. It can be prepared in elemental form (black, lustrous crystals) by the reduction of boron tribromide (BBr_3) with hydrogen (H_2).

ORGANIC CHEMISTRY

Organic compounds are far greater in number than inorganic compounds, even though they are all based on one element, carbon. The chemistry of organic compounds is known as **organic chemistry** and is a distinct discipline. Its importance stems not only from the great number of organic compounds, but also from the fact that all life forms are based on organic compounds. In addition, the importance of petroleum and petroleum products to civilized society has greatly enhanced the interest in organic chemistry.

The unique structure of the carbon atom, with atomic number 6 and atomic weight 12, is the foundation of its ability to form complex compounds. The four bonds are normally directed toward the corners of a tetrahedron, with the carbon nucleus at its center. This allows carbon atoms to form chains of almost unlimited length, as well as side chains, while the remainder of the bonds are directed to hydrogen atoms and other species, such as halogens, oxygen, nitrogen, and sulfur. Another important aspect of carbon bonding is its ability to form double bonds or triple bonds with another carbon and with other elements. The double bonds may alternate between carbon atoms, and this leads to the ability to form rings in which extra bonding strength is provided by **delocalized** electrons around the ring. The most stable ring of this type contains six carbons and is known as the benzene ring. It is the foundation of a large class of ring compounds known as **aromatic compounds**.

We may start in organic chemistry by considering the lower molecular weight compounds, as much of their chemistry is consistent with that of higher molecular weight (longer chain) compounds. The simplest organic compounds are those of only carbon and hydrogen, and are known as **hydrocarbons**. One carbon bonded to four hydrogen atoms is CH_4, or methane. Successive carbons may be added to form the compounds ethane (C_2H_6), propane (C_3H_8), butane (C_4H_{10}), pentane (C_5H_{12}), hexane (C_6H_{14}), heptane (C_7H_{16}), octane (C_8H_{18}), and on to any number—although individual names are lacking above about 30 carbons in a chain. The aforementioned hydrocarbons are known as **alkanes**.

Combustion is a reaction that all organic compounds are subject to, but the hydrocarbons, as mentioned, are often used for that purpose. Carbon dioxide and water are formed, as in the reaction of hexane:

$$2\,C_6H_{14} + 19\,O_2 \rightarrow 12\,CO_2 + 14\,H_2O$$

Another reaction of hydrocarbons is halogenation, such as

$$C_3H_8 + Cl_2 \rightarrow C_3H_7Cl + HCl$$

Example 8.3

A compound has an empirical formula of C_2H_3Br and a molecular weight of 213.9 g/mol (a.w. C = 12.01, H = 1.008, Br = 79.90). What is its molecular formula and a possible structural formula with name?

Solution

The empirical formula weight is $(2 \times 12.01) + (3 \times 1.008) + 79.90 = 106.94$ g/empirical formula.

$$\text{empirical formulas/mol} = \frac{213.9 \text{ g/mol}}{106.94 \text{ g/empirical formula}} = 2.0$$

Therefore, the molecular formula is $C_4H_6Br_2$. A number of structural formulas can be devised that correspond to this molecular formula. Two compounds with this formula are

$$\text{1,4-dibromo-2-butene, or } Br-CH_2-CH=CH-CH_2-Br$$

and

$$\text{3,4-dibromo-1-butene, or } CH_2=CH-CHBr-CH_2Br$$

The carbon compounds with a double bond are called **unsaturated hydrocarbons** or **unsaturates**. They are notably more reactive than the alkanes—one reaction being to hydrogenate (add H_2 with heat, pressure, and possibly catalyst) back to form the **saturated** or alkane compound. They also react readily with halogens, as in

$$H_2C=CH_2 + Br_2 \rightarrow H_2BrC-CBrH_2$$

The product of the reaction is dibromoethane or, more correctly, 1,2-dibromoethane or ethane, 1,2-dibromo. The compound is also known as ethylene dibromide, where, it may be noted, the **-ene** suffix is appropriate because of the reactant material from which the compound was made.

The hydrocarbon compounds may also contain the triple bond, or three pairs of electrons bonding two carbon atoms. The best known example of a compound with this type of bond is acetylene, C_2H_2, which may be written as $HC \equiv CH$. The triple bonded, or acetylene type, hydrocarbons are noted for their reactivity, and the instability of the triple bond. Acetylene is used as a fuel to reach high temperatures with the oxyacetylene torch.

As has been shown, the hydrocarbons may have halogen atoms substituted for the hydrogen, thus giving the generic class of halocarbons. Other groups may also be substituted for one or more hydrogens. These include the OH group, leading to a class of organic compounds called alcohols; the NH_2 group, leading to a class of compounds called amines; the NO_2 group, leading to the nitro compounds; the CN group, leading to a class of compounds called organic cyanides; the O atom in various groups, leading to classes of compounds called aldehydes, ketones, acids, and oxides; and the S atom, leading to classes of compounds called mercaptans and sulfides. There is also a major class of compounds with C—Si bonds, generally known as silanes, as well as compounds of phosphorous, arsenic, and numerous metallo-organics.

The **aromatic** compounds form a large class and can contain substituent groups, just as the alkanes can. When an OH group is substituted on a benzene ring (six carbons in a ring with alternating double and single bonds in resonance), the product is phenol, C_6H_5OH. Unlike the alcohols, phenol is an acid, ionizing in water to yield the proton H^+—or the hydrated proton, hydronium ion, H_3O^+—and the phenylate anion, $C_6H_5O^-$. When two OH groups are on the benzene ring, the compound is resorcinol, and the compound with three OH groups is catechol. Although they have some reactions in common, the aromatic OH compounds, or phenols, are generally considered as a separate chemical group from the alcohols.

The amines, such as methylamine, CH_3NH_2, ethylamine, $C_2H_5NH_2$, and propylamine, $C_3H_7NH_2$, are odoriferous liquids, the NH_2 groups of which are quite reactive.

The nitro compounds, such as nitromethane, NO_2CH_3, nitroethane, $NO_2C_2H_5$, and nitropropane, $NO_2C_3H_7$, are all reactive. Aromatic nitro compounds, such as nitrobenzene ($C_6H_5NO_2$) can be reduced to primary aromatic amines, which are a starting point for many organic synthesis experiments.

Some of the sulfur compounds are analogous to the alcohols, as the S may be substituted for O so that the hydrocarbon is attached to an SH group. These compounds are known as mercaptans, for example, methyl mercaptan, CH_3SH, and ethyl mercaptan, C_2H_5SH. They are odoriferous and reactive.

Sulfur can also be bonded between two carbon atoms, as in the sulfides such as dimethylsulfide, CH_3SCH_3, and diethylsulfide, $C_2H_5SC_2H_5$. The sulfides are reactive.

Besides the alcohols, there are other large classes of organic compounds with oxygen in the substituent group. Among these are aldehydes, ketones, esters, and acids. The characteristic group for aldehydes is HC=O; for ketones, $\overset{|}{\underset{|}{C}}$=O; for ethers, —O—; for esters, —O—$\overset{|}{C}$=O; and for acids, HO—$\overset{|}{\underset{|}{C}}$=O. The double bond is shown to emphasize the fact that in four of these cases the oxygen atom is doubly bonded to the carbon atom. In the other case, ethers, the oxygen atom is singly bonded to each of two carbon atoms.

Aldehydes and ketones are similar in many respects, as both contain the carboxyl group, C=O. In the aldehydes the C=O group is at the end of the carbon chain, with the carboxyl carbon atom also bonded to a hydrogen atom. In ketones, the C=O group is somewhere in the middle of the chain, bonded to carbon atoms on both sides.

MATERIAL PHYSICAL PROPERTIES

An understanding of mass, weight, and volume relationships is crucial to be able to solve environmental engineering problems. The primary quantities used are (SI units are noted in brackets [ ] after equivalent non-SI units):

Density (ρ) = mass/volume, lb/ft^3 [kg/m^3]

Specific volume (v) = 1/density, ft^3/lb [m^3/kg]

Specific weight (γ) = weight/volume = mass × acceleration due to gravity/volume

$= \rho g/g_c$, lbf/ft^3 [N/m^3]

Specific gravity of substance(s) = density of substance/density of water at 4°C

$= \rho(\text{lb/ft}^3)/(62.43 \text{ lb/ft}^3)$
$= \rho(\text{kg/m}^3)/(1000 \text{ kg/m}^3)$

Specific gravity of gas relative to air = molecular weight of gas/28.97

The following constants are used throughout:

g = acceleration due to gravity

= 32.17 ft/s^2 [9.81 m/s^2]; this varies from place to place, and planet to planet

g_c = Newton's-law proportionality factor for the gravitational force unit

= 32.17 lb · ft/s^2 · lbf [1 kg · m/s^2 · N]; this is a constant factor

Specific gravity is dimensionless.

Some approximate densities at standard conditions are the following:
Air: 0.075 lb/ft^3 [1.3 kg/m^3]
Water: 62.4 lb/ft^3 [1000 kg/m^3]
Steel: 462 lb/ft^3 [7400 kg/m^3]
Mercury: 845 lb/ft^3 [13,600 kg/m^3]

CONCENTRATION UNITS

Probably the most common means of expressing the amount of solute in solution is the use of mass concentrations. Many environmental solutions of concern deal with water (aqueous) samples, and concentrations are expressed as mass per unit volume, such as milligrams of solute per liter of solution (mg/L) or micrograms of solute per liter of solution (µg/L). However, it is important to be familiar with the many systems employed to express concentrations of solutions in environmental systems. Molar concentrations are generally employed when dealing with equilibrium chemistry and are expressed in terms of *molarity* (M) as *moles of solute per liter of solution*. Another related means of expressing concentration is *molality* (m), or molal concentrations, which may be defined as *moles of solute per kilogram of solvent*. In solution reactions where it is convenient to use species equivalents (like acid/base and water-softening reactions), it is most common to use the *normality* (N) of the solution, which is expressed as *equivalents of solute per liter of solution*. The use of equivalents is convenient in that various species may be compared, and even combined, using the second definition for equivalency given previously in terms of reactivity, without the need for specific balanced chemical equations.

To convert from units of mg/L (or g/L) to M (moles/L), the molecular weight of the compound must be used. A periodic table listing atomic masses is required to complete these calculations. The conversion from mg/L to moles per liter, or mg/L to mmol/L is given by

$$X \frac{mg}{L} \div \left(MW \frac{g}{mol} \times 1000 \frac{mg}{g} \right) \rightarrow Y \frac{mol}{L}$$

$$X \frac{mg}{L} \div MW \frac{mg}{mmol} \rightarrow Y \frac{mmol}{L}$$

Note again that MW has units of g/mol, which is also equivalent to mg/mmol. It is also useful to note that molarity is related to normality by the equivalency and can be expressed as

$$N = (M)(equivalency)$$

Air pollutants are also usually expressed as mass per unit volume, such as micrograms of contaminant per cubic meter of air ($\mu g/m^3$). If the gas is emitted at temperatures or pressures that are not standard (STP are 0°C and 101.325 kPa) or with high moisture content, concentrations are often corrected to micrograms of contaminant per dry standard cubic meter (µg/dscm). This requires calculations to remove the water mass from the air and use of the ideal gas law for temperature and pressure correction.

Concentrations of species in solids (e.g., soil contamination) and semisolids (such as sludges) are expressed as mass per unit mass, such as milligrams of solute per kilogram of solid (mg/kg) or micrograms of solute per kilogram of solid (µg/kg). Because the ratio of mg/kg is 10^{-6} and µg/kg is 10^{-9}, mass per unit mass concentrations are often expressed as *parts per million (ppm)* or *parts per billion (ppb)*.

In air pollution, when the contaminant species is also present in the gas phase, it is common to see the 10^{-6} or 10^{-9} ratio expressed as ppmv or ppbv. However, the "v" indicates that these are volume ratios, such as milliliters of contaminant per cubic meter of gas, or microliters of contaminant per cubic meter gas. Engineers should be able to convert air concentrations between $\mu g/m^3$ and ppmv using the ideal gas law and molar weights. Further, because the mass of one liter of water is approximately one kilogram, it is not uncommon to see aqueous solution concentrations expressed in ppm and ppb units, although the mass per unit volume units are more appropriate.

Example 8.4

Concentration units I

The secondary water standard for SO_4^{2-} is 250 mg/L. Express this concentration in units of normality.

Solution

Normality has units of equivalents per liter. First, we need to find the equivalent weight of sulfate, which is 48.031 eq/g, which may also be expressed as 48.031 meq/mg. We can now simply convert units as follows:

$$250 \frac{mg}{L} \times \frac{1 \text{ meq}}{48.031 \text{ mg}} = 5.2 \frac{meq}{L} = 5.2 \text{ mN}$$

Example 8.5

Concentration units II

The National Ambient Air Quality Standard (NAAQS) for the criteria pollutant carbon monoxide is 40,000 µg/m³. Express this concentration in units of ppmv if the temperature is 25°C.

Solution

First, determine the number of moles of CO:

$$MW_{CO} = 28 \text{ g therefore } 40 \text{ mg CO} = \frac{0.04 \text{ g CO}}{28 \frac{\text{g CO}}{\text{mol CO}}} = 0.00143 \text{ mol CO}$$

Next, determine the molar volume of a gas at 1 atm and 25°C:

$$\frac{V}{n} = \frac{RT}{P} = \frac{\left(0.08206 \frac{L \cdot atm}{mol \cdot K}\right)(298.15 \text{ K})}{1 \text{ atm}} \Rightarrow 24.47 \frac{L}{mol} = 0.0245 \frac{m^3}{mol}$$

Now, calculate the volume of gas occupied by 0.00143 moles of gas:

$$0.0245 \text{ m}^3/\text{mol} \times 0.00143 \text{ mol} = 0.000035 \text{ m}^3$$

Finally, determine the volume ratio using 1 m³ of air as a basis:

$$[CO] = \frac{0.000035 \text{ m}^3}{1 \text{ m}^3} = 35 \times 10^{-6} \frac{m^3}{m^3} \text{ or equivalently } [CO] = 35 \text{ ppmv}$$

EQUATIONS AND STOICHIOMETRY

The reaction of one chemical species with another is defined by a chemical equation of the form

$$A + B = C + D$$

where A and B represent reactant species and C and D represent product species. Instead of the equal sign to indicate equilibrium, an arrow ($\rightarrow$) is often used to show the direction in which the reaction goes to completion. Also, a double arrow ($\leftrightarrow$ or $\rightleftharpoons$) may be used.

In a **metathetical** reaction, two species react to form two other species without oxidation or reduction. A common example would be an acid-base reaction (in solution) as

$$HCl + NaOH = NaCl + H_2O$$

The simplest form of chemical equation is a unimolecular reaction, such as a dissociation:

$$HgO\ (+\ heat) = Hg + O_2$$

Here we encounter another requirement of chemical equations—that they be **balanced**. A balanced equation has the same number of atoms of each species on the left side of the equation as on the right side of the equation. The above reaction should, therefore, be

$$2\ HgO = 2\ Hg + O_2$$

The balanced equation is very informative in a quantitative manner; the amounts of each species involved in the reaction are given by the relative number of atoms or molecules in the equation. Therefore, the gram-atoms or gram-moles of each species can be calculated. In the case of gases, the volume of the gas at standard conditions can also be calculated, and modified for the volume at nonstandard conditions.

Example 8.6

A reaction that makes gas is that of the metal zinc with hydrochloric acid:

$$Zn + HCl = ZnCl_2 + H_2$$

In a laboratory experiment, 1.00 g Zn is added to a solution of 100 mL of 1.00 M HCl. How many mL of H_2 are produced in the reaction at room temperature (25°C) and one atmosphere pressure? (a.w. Zn = 65.4, Cl = 35.5, H = 1.01; 1.00 mol ideal gas at standard conditions = 22.4 L)

Solution

The coefficients to balance the equation must be added. A causal inspection shows that two molecules of HCl are required to supply two atoms of Cl for $ZnCl_2$ and two atoms of H for H_2:

$$Zn + 2\ HCl = ZnCl_2 + H_2$$

This balanced equation shows that one mole of zinc (65.4 g) reacts with two moles of HCl (73.0 g) to yield one mole of zinc chloride (136.4 g) and one mole (2.0 g) of hydrogen. Any ratio of these amounts can be used to fulfill the reaction predicted by the balanced equation.

In the experiment, a small amount of zinc, 1 g, is added to the solution. The number of moles of Zn is

$$\frac{1.00 \text{ g}}{65.4 \text{ g/mol}} = 0.015 \text{ mol}$$

The number of moles of HCl in solution is

$$0.100 \text{ L} \times \frac{1.00 \text{ mol}}{\text{L}} = 0.100 \text{ mol}$$

The balanced equation shows that 0.015 mol Zn requires 2 × 0.015 = 0.030 mol HCl to completely react, so there is an excess of HCl and the amount of Zn limits the amount of H_2 formed. The amount of H_2 formed will then be the same as the number of moles of Zn, or 0.015 mol. The volume of H_2 can be calculated from the physical constant for an ideal gas at standard conditions (22.4 L/mol):

$$0.015 \text{ mol} \times \frac{22.4 \text{ L}}{\text{mol}} = 0.336 \text{ L at standard conditions}$$

25°C = 298 K, so

$$0.336 \text{ L} \times \frac{298}{273} = 0.367 \text{ L} = 367 \text{ mL}$$

at 25°C and one atmosphere.

The calculation of weights, volumes, and elemental proportions in a chemical reaction can be carried out from basic premises and information. One premise is that the combination of atoms to form a molecule takes place with whole numbers of atoms, and the ratio of the atoms to each other is a ratio of integers and is characteristic of the particular molecular species. Thus, if two atomic species, A and B, join to form a molecule, they are in a fixed ratio and the molecular formula is written with subscript numbers denoting the number of atoms of each atomic species in the molecular formula. A molecular formula may be of the form AB, AB_2, A_2B, A_2B_3 or other whole number subscripts, depending on the valence or bonding capacity of the species A and B. Examples are NaCl, $CaCl_2$, K_2O, and Fe_2O_3.

Chemical compounds can be characterized by both the **molecular formula** and an **empirical formula**. The molecular formula gives the whole numbers of atoms actually joined together by chemical bonding to assemble the molecular species in question. An empirical formula, on the other hand, gives the correct atomic ratio by the subscript numbers, but may be less than the actual number of atoms in the molecule.

For example, the empirical formula for benzene is CH, whereas the molecular formula is C_6H_6, as there are actually six atoms of carbon and six atoms of hydrogen in a molecule of benzene. The empirical formula, or atomic ratio, is the simplest whole number atomic ratio and may be calculated from experimental data, such as a combustion analysis. Either prior knowledge of the chemical species involved or some other experimental data, such as molecular weight determination, is needed to derive the molecular formula.

A second premise is that the equation for the reaction is known and can be presented in balanced form by use of the proper stoichiometric coefficients. The **stoichiometric coefficient** is the number that each molecular species is multiplied by so that the same number of each atom is found on both the left-hand side and the right-hand side of the equation. By convention, the stoichiometric coefficients are usually whole numbers. When fractional values appear, both sides of the equation are multiplied by the lowest number that will make all coefficients whole numbers. Examples of balanced equations are

$$2\ KClO_3 + heat = 2\ KCl + 3\ O_2$$
$$2\ C_2H_6 + 7\ O_2 = 4\ CO_2 + 6\ H_2O$$
$$K_2Cr_2O_7 + 9\ KI + 14\ HCl = 3\ KI(I_2) + 2\ CrCl_3 + 8\ KCl + 7\ H_2O$$

and

$$2\ FeCr_2O_4 + 7\ Na_2O_2 = 2\ NaFeO_2 + 4\ Na_2CrO_4 + 2\ Na_2O$$

Given the whole number ratios of atoms in the molecule and knowledge of a reaction as a balanced equation, the atomic and molecular formulas can be related to the physical world in which quantities of chemicals are measured. This is done by the concept of a quantity called a mole, which is defined in the metric system as the molecular weight of the species expressed in grams, also known as the gram molecular weight. The molecular weight of a molecular species is the sum of the atomic weights, and the atomic weight of an element is the number representing the weight of an atom of the element relative to carbon-12, or ^{12}C. For various reasons the atomic weights are not necessarily whole numbers, primarily because a sampling of a given element in the earth environment may include two or more isotopes of different atomic weight. However, the isotope ratio is, for the purpose of chemical reactions, very constant.

The mole, which has the abbreviation **mol** when used after a number, is a quantity that has, by definition, the same number of **elementary entities** as there are carbon atoms in 12 g of carbon-12. This number is known as the **Avogadro constant** and has been experimentally determined to be 6.022×10^{23}. Since all atomic weights are related to that of carbon-12, a mole of any species has the same number of molecules—for example, there are the same number of molecules in 2.016 g of H_2, 71.0 g of Cl_2, and 58.5 g of NaCl. The great importance of this concept is that the molecular formulas and chemical equations are written for discrete numbers of atoms and molecules, but since all moles have the same number of molecules, the molecular formulas and chemical equations can be related to the measured weights of chemicals in the physical world.

When one of the reactants or products of a chemical reaction is a gas, the usefulness of the mole concept becomes further evident. This is because one mole of gas of any species still stands for a certain number of molecules (6×10^{23}), but in the case of gases this number accounts for a certain volume of gas at a given temperature and pressure. The number that is useful in relating equations to real physical measurements is that one mole of a gas—any gas—occupies 22.4 liters at so-called **standard** conditions. Standard conditions are one atmosphere and 0°C or 273 K. A common ambient temperature is 25°C or 298 K, and the ideal gas law correction indicates that a mole of gas at 25°C occupies 24.4 liters.

REACTION ORDER AND KINETICS

Some reactions may be assumed to be permanent such that the products cannot be further changed to reform the original reactants. These reactions are defined as irreversible, or one-way, and may be represented as

$$aA + bB \rightarrow cC + dD$$

Reversible reactions involve species that may reform the original reactants from the products, depending on changes in the environmental conditions that influence the reaction (such as temperature, pressure, and specific species concentrations) and may be expressed as

$$aA + bB \leftrightarrow cC + dD$$

Notice the two-headed arrow suggests that the reaction may proceed from left to right (forward direction) or from right to left (backwards or reverse direction).

Reaction order is a function of the mechanism of the reaction and is usually determined through experimentation. Most often, reaction rates are written explicitly as a function of reactant consumption and, as such, are always negative, although convention usually reports reaction rates in their positive form. The most simple reaction is not dependent upon the concentration of the reactant; rather, the rate is controlled by other, often physical or environmental, parameters. Such reactions are known as "zero" order reactions and are usually expressed as

$$r_A = \frac{dC_A}{dt} = -k_0$$

where r_A is the rate of reaction for species A, C_A is the concentration of A expressed in molar units, t is time, and k_0 is the zero-order reaction coefficient. Integration of this expression yields

$$dC_A = -k_0\, dt \implies (C_A)_f - (C_A)_i = -k_0(t_f - t_i)$$

where the subscripts f and i indicate the final and initial conditions, respectively. If it is further assumed that the experiment started at $t = 0$, we may express the concentration at any time t as

$$(C_A)_f = (C_A)_i - k_0 t$$

Note that the linearized equation would allow for the plotting of experimental data $(C_A)_f$ vs. time to yield slope of $-k_0$ and a y-intercept of $(C_A)_i$.

First-order reactions are dependent upon the concentration of species A and may be written as

$$r_A = \frac{dC_A}{dt} = -k_1 C_A$$

where k_1 is the first-order reaction rate coefficient. Again, assuming $t_i = 0$, this may be integrated to yield

$$\frac{dC_A}{C_A} = -k_1\, dt \implies \ln\left[\frac{(C_A)_f}{(C_A)_i}\right] = -k_1 t \implies (C_A)_f = (C_A)_i e^{[-k_1 t]}$$

The linearized form of this expression requires plotting the natural log of the concentration vs. time to yield a y-intercept of $\ln[(C_A)_i]$ and a slope of $-k_1$.

Another form of the first-order rate expression found in environmental engineering is used when the concentration of the reacting species has a solubility limit, such as oxygen in aqueous solutions. In these cases, the rate of reaction of species A (oxygen) is dependent upon a concentration gradient that is calculated

Example 8.7

A simple equation involving a gaseous product was previously presented, the decomposition of $KClO_3$ to yield oxygen gas:

$$2\ KClO_3 + heat = 2\ KCl + 3\ O_2$$

How much oxygen could we obtain by heating 100 g of $KClO_3$?

Solution

A mole of $KClO_3$ is 122.6 g (its molecular weight in grams) and the equation shows that 2 moles of $KClO_3$ can yield 3 moles of O_2. However, we only have $100/122.6 = 0.82$ mol $KClO_3$. The equation shows that the number of moles of O_2 yielded will be $3/2 \times 0.82 = 1.23$, and the volume at one atmosphere and 25°C will be $1.23 \times 24.4 = 30.0$ liters.

---

as the difference between the measured value and the value that would occur if the solution was saturated with oxygen under equivalent environmental conditions (same temperature, ion concentration, etc.). In the specific case of oxygen reaeration of water, the reaction rate expression may take the following form

$$r = \frac{dC}{dt} = k_1\left(C - C^S\right) = -k_1\left(C^S - C\right)$$

where C^S is the saturation concentration of oxygen specified at the given environmental conditions. Integration of this equation can be represented as

$$\frac{C^S - C_f}{C^S - C_i} = e^{-k_1 t}$$

The linearized form of this expression requires plotting the natural log of (C^S-C) vs. time to yield a y-intercept of C^S-C_i and a slope of $-k_1$. Additional discussion on the use of the oxygen transfer expression for wastewater treatment systems can be found in Chapter 8.

Second-order reactions may take many different forms, depending on the species dependence. All second-order reactions may be expressed as

$$r = \frac{dC}{dt} = -k_2\ C^2$$

where k_2 is the second-order reaction rate coefficient. However, with the possibility of multiple reactants, this expression may take several forms, such as

$$r = -k_2\ C_A^2 \quad \text{or} \quad r = -k_2\ C_A C_B \quad \text{or} \quad r = -k_2\ C_A^{1.5} C_B^{0.5}$$

As indicated above, in reality the second-order expression may take any number of forms, as long as the total order of the concentration exponents sums to the value 2. For simplicity, if it is assumed that the reaction rate is only dependent upon the concentration of one reactant species, the expression will take the following form

$$r_A = \frac{dC_A}{dt} = -k_2\ C_A^2$$

Again assuming $t_i = 0$, this may be integrated to yield

$$\frac{dC_A}{C_A^2} = -k_2\ dt \quad \Rightarrow \quad \frac{1}{(C_A)_f} = \frac{1}{(C_A)_i} + k_2\ t$$

The linearized form of this expression requires plotting $1/(C_A)_f$ vs. time to yield a y-intercept of $1/(C_A)_i$ and a slope of k_2.

Table 8.4 and Table 8.5 provide a summary comparing zero-, first-, and second-order reactions.

Table 8.4 Reaction rate comparison

Order of Reaction	$n =$	Rate Law	Concentration vs. Time	Dimensions of k	Sample Units of k
Zero	0	$\dfrac{dC_A}{dt} = -k_0$	(linear decrease from C_0)	$[M \cdot T^{-1} \cdot L^{-3}]$	mg/L·sec mol/L·min
First	1	$\dfrac{dC_A}{dt} = -k_1 \cdot C$	(exponential decay from C_0)	$[T^{-1}]$	sec^{-1} min^{-1} hour^{-1}
Second	2	$\dfrac{dC_A}{dt} = -k_2 \cdot C^2$	(steeper decay from C_0)	$[L^3 \cdot M^{-1} \cdot T^{-1}]$	L/mg·sec L/mol·min

Table 8.5 Linearization process for zero-, first-, and second-order reactions

Order of Reaction	Linearized Plot	Value of $k =$
Zero	C vs. t	Negative slope of best-fit line
First	$\ln C$ vs. t	Negative slope of best-fit line
Second	$1/C$ vs. t	Slope of best-fit line

The individual rate law equations are more effectively used in their integrated form (that is, $C = f((C_A)_i, t)$). Moreover, a value often very useful when solving environmental engineering problems is the *half-life,* which is the time required to reduce the initial concentration by 50%. The half-life can be readily deduced from the integrated forms of the rate laws by substituting in $(C_A)_f/(C_A)_i = 0.5$ and solving for t. However, in the interest of ease of use while taking the FE exam, the integrated forms and the equations for half-life have been summarized in Table 8.6.

Table 8.6 Useful equations for zero-, first-, and second-order reactions

Order of Reaction	Integrated Form	Half-Life Equation
Zero	$(C_A)_f = (C_A)_i - k_0 \cdot t$	$(C_A)_i / 2 \cdot k_0$
First	$(C_A)_f = (C_A)_i \cdot e^{-k \cdot t}$	$\ln(2)/k_1$
Second	$(C_A)_f = \left[k_2 \cdot t + \dfrac{1}{(C_A)_i} \right]^{-1}$	$1/k_2 \cdot (C_A)_i$

Some reactions in environmental engineering are dependent upon a reacting species; however, the reaction rate expression may change as that species approaches a certain value. An example of such a reaction includes biological reactions occurring under certain environmental conditions, where the reaction rate is first order until a certain concentration limit is reached, at which point the reaction reverts to a near-zero order rate. Often, these reactions are referred to as saturation reactions and may be expressed in their general form as

$$r_A = \frac{dC_A}{dt} = -\frac{kC_A}{K+C_A}$$

where k is the rate constant and K is the saturation constant. This form of the kinetics expression was covered in more detail in the discussion of Monod growth kinetics for bacteria in wastewater treatment systems in Chapter 7.

While it may be easier to assume that the target species will have only one possible reaction pathway, it is often the case that several reactions are possible with the same set of reactants. One type of complex pathway is called consecutive (or series) reactions and is usually expressed as

$$A + B \xrightarrow{k_C} C \quad \text{and} \quad A + C \xrightarrow{k_D} D$$

where k_C and k_D are the rates of formation of the identified species. The overall reaction rate may be controlled by either k_C or k_D, depending on the relative magnitudes of the two rates. The reaction that possesses the smaller rate coefficient of the two is often called the "rate limiting step" and would be used as the basis for designing the specific treatment unit. However, care must be taken to evaluate the rate coefficients under all anticipated conditions, as changes in the temperature or other environmental conditions may cause the rate limiting reaction to be assigned to different reactions under different conditions.

Another type of complex reaction occurs when there is a competition between two or more reactions that share a common reactant. These reactions are usually called competitive (or parallel) reactions and may be expressed as

$$A + B \xrightarrow{k_C} C \quad \text{and} \quad A + D \xrightarrow{k_E} E$$

The competition for reactive pathway is generally decided by reaction kinetics with the most energetic pathway favored. Further, if species A in the reactions above is in excess, it may be the case that both reactions will proceed to completion, albeit at different rates, upon the exhaustion of either species B or D.

Example 8.8

Zero-order and first-order reaction rates

The data in Exhibit 1 shows how the concentration of BTEX (benzene, toluene, ethylbenzene, and xylene) varies as a function of time.

a). Estimate a rate constant, assuming zero-order decay.

b). Estimate a rate constant, assuming first-order decay.

Time (min)	C (mg/L)
0	400
5	320
10	820
20	200
30	190
40	110
50	50
60	40

Exhibit 1

Solution

a). The data has been linearized and a best fit line is used, as shown in Exhibit 2. Given the form of the best fit line for a zero-order reaction is $(C_A)_f = (C_A)_i - k_0 \cdot t$, the first-order rate constant is 7 mg/L·min.

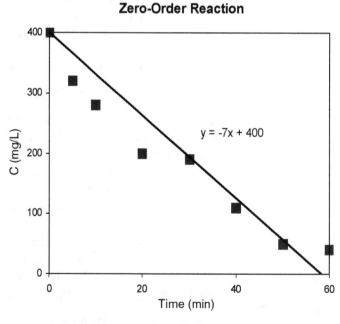

Exhibit 2 Best fit line for zero-order decay

b). As shown in Table 1.5, $\ln(C)$ vs. time should produce a nearly linear plot for a first-order equation. This plot is shown in Exhibit 3. From this plot, the reaction rate constant k_1 can be estimated to be 0.04 min^{-1}.

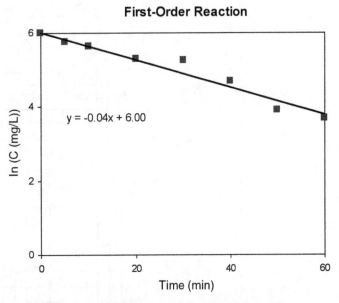

Exhibit 3 Linearization of data for first-order reaction

It is very important to note that the units differ depending on the order of the reaction. Also, in an exam situation, you might not have the luxury of plotting the data and fitting a trendline; however, a reaction rate constant of suitable accuracy can still be obtained by using a straightedge.

Example 8.9

Reaction kinetics

Given the following set of data, determine the initial concentration of species A and predict the concentration of species A after two hours.

Time [min]	15	30	45	60	90
C_A [mg/L]	256	164	102	67	29

Solution

First, the order of the reaction is needed to establish the appropriate rate expression. This can be done by plotting the linearized forms of the rate expressions. This can be immediately done for zero-order reactions (plot C vs. t), but transformation of the data is required for first($\ln[C]$ vs. t)-order and second($1/C$ vs. t)-order reactions. The transformed data is shown below.

	Zero Order	1st Order	2nd Order
t	C	$\ln(C)$	$1/C$
15	256	5.545	0.0039
30	164	5.100	0.0061
45	102	4.625	0.0098
60	67	4.205	0.0149
90	29	3.367	0.0345

Now plot the data on rectilinear graph paper and find the best linear fit.

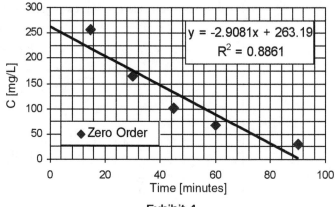

Exhibit 4

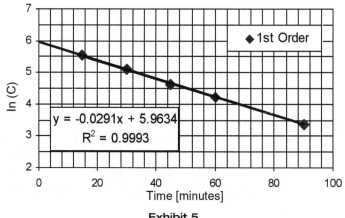

Exhibit 5

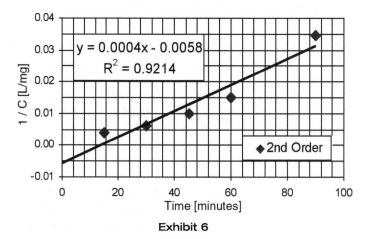

Exhibit 6

From these plots, the best linear fit is obviously the first-order expression with an R^2 of 0.9993. From the y-intercept (at $t = 0$ minutes), we can determine the initial concentration of species A as

$$\ln[(C_A)_i] = 5.9634 \quad \text{or} \quad (C_A)_i = \exp[5.9634] = 389 \frac{\text{mg}}{\text{L}}$$

The concentration at 120 minutes (2 hours) can now be estimated as

$$(C_A)_f = (C_A)_i \, e^{[-k_1 t]} = \left(389 \frac{\text{mg}}{\text{L}}\right) \exp\left[(-0.0291 \text{ min}^{-1})(120 \text{ min})\right] = 11.84 \approx 12 \frac{\text{mg}}{\text{L}}$$

While spreadsheet software makes this problem easier to solve, it cannot be used on the FE exam. Examinees, therefore, need to be able to recognize the most appropriate linear fit and determine the slope and intercept by taking the difference between two points that fall on the line that was fit to the data. A reasonable estimate would be that the y-intercept was at a value of 6, therefore $(C_A)_i$ would be determined as

$$\ln[(C_A)_i] = 6 \quad \text{or} \quad (C_A)_i = \exp[6] = 403 \frac{mg}{L}$$

Now, look for another point on the linear fit that crosses clear axis values, such as $\ln(C)$ equal to 4 at a time of 68 minutes. Now determine the rate constant as

$$\ln\left[\frac{(C_A)_f}{(C_A)_i}\right] = -k_1 t \quad \text{or} \quad \frac{\ln(C_A)_f - \ln(C_A)_i}{t_f - t_i} = -k_1 = -\frac{4-6}{68-0} = 0.0294 \text{ min}^{-1}$$

Plugging this into the rate expression as before yields

$$(C_A)_f = (C_A)_i e^{[-k_1 t]} = \left(403 \frac{mg}{L}\right) \exp\left[(-0.0294 \text{ min}^{-1})(120 \text{ min})\right] = 11.85 \approx 12 \frac{mg}{L}$$

EQUILIBRIUM

Equilibrium is the condition at which the steady state concentration of chemical reactant species and products does not change with time, as reactions are proceeding at equal rates in both directions. The condition of equilibrium is characterized by the equilibrium constant, K, which is given for a certain temperature. Both reactants and products must be at this temperature for the equilibrium constant to be valid. In general terms, if the reaction is $A + B = C + D$, the equilibrium constant is

$$K = \frac{[C][D]}{[A][B]}$$

A, B, C, and D appear in brackets as the values of the concentrations that, in the appropriate units, are inserted into the equation. Thus the concentration of any reactant or product is a function of the concentration of the other reactants or products, and the concentration of a single species can be affected by addition or removal of a different species by external means.

Example 8.10

Hydrogen (H_2) and carbon dioxide (CO_2) react at 100°C to form water (H_2O) and carbon monoxide (CO):

$$CO_2(g) + H_2(g) = CO(g) + H_2O(g)$$

Assume the reactants and products are at equilibrium and the number of moles of each gas in a 1.00-L reaction vessel is 0.050, 0.025, 0.040, and 0.055 for CO_2, H_2, CO, and H_2O, respectively. Calculate the equilibrium constant, K, for the reaction at this temperature.

Solution

$$K = \frac{[0.040][0.055]}{[0.050][0.025]} = 1.8$$

The units of each term are mol/liter; units cancel.

For reversible reactions, the calculation of an equilibrium constant (K_{eq}) from the molar concentrations of reactive species and their products in dilute solutions, like in natural waters, may be expressed as follows:

$$a\text{A} + b\text{B} \leftrightarrow c\text{C} + d\text{D}$$

$$K_{eq} = \frac{k_f}{k_b} = \frac{[C]^c[D]^d}{[A]^a[B]^b}$$

where k_f is the rate constant for the forward reaction, k_b is the rate constant for the backward reaction, and the bracketed quantities are molar concentrations of the target species. This equilibrium relationship is invaluable to environmental engineers, as will be further explored in the following section.

Example 8.11

Equilibrium

A solution has a Ca^{2+} concentration of 50 mg/L. Determine the concentration of $Ca(OH)_2$ in mg/L if the solution pH is 8.5. You may assume a K_{eq} for $Ca(OH)_2$ of 5×10^{-9}.

Solution

We start with the equilibrium expression for dissociation of calcium hydroxide as follows:

$$Ca(OH)_2 \leftrightarrow Ca^{2+} + 2\left[OH^-\right]$$

From this we may express the equilibrium constant as

$$K_{eq} = \frac{\left[Ca^{2+}\right]\left[OH^-\right]^2}{\left[Ca(OH)_2\right]} = 5 \times 10^{-9}$$

We can determine the hydroxide ion concentration from the pH data as follows:

$$pH + pOH = 14 \quad \Rightarrow \quad pOH = 14 - pH = 14 - 8.5 = 5.5 = -\log\left[OH^-\right]$$

$$\text{therefore} \quad \left[OH^-\right] = 10^{-5.5} = 3.16 \times 10^{-6} \frac{\text{mol}}{\text{L}}$$

We can convert the concentration units for Ca^{2+} from mg/L to mol/L as follows:

$$\left[Ca^{2+}\right] = 50 \frac{\text{mg}}{\text{L}} \times \frac{1 \text{ mmol}}{40.078 \text{ mg}} = 1.25 \frac{\text{mmol}}{\text{L}} = 1.25 \times 10^{-3} \frac{\text{mol}}{\text{L}}$$

Substituting these values into the equilibrium constant expression yields

$$5 \times 10^{-9} = \frac{(1.25 \times 10^{-3})(3.16 \times 10^{-6})^2}{X} \quad \Rightarrow \quad X = \left[Ca(OH)_2\right] = 2.5 \times 10^{-6} \frac{\text{mol}}{\text{L}}$$

Since our answer must have units of mg/L, we can convert mol/L as follows:

$$2.5 \times 10^{-6} \frac{\text{mol}}{\text{L}} = 2.5 \times 10^{-3} \frac{\text{mmol}}{\text{L}} \times 74.092 \frac{\text{mg}}{\text{mmol}} = 0.185 \frac{\text{mg}}{\text{L}}$$

Several solution equilibrium relationships based on fundamental chemical principles may be important to environmental engineers. One example is in the determination of species solubility and evaluating the potential for using precipitation reactions to remove unwanted compounds. This tool for removing undesired constituents in water is used extensively in water treatment as coagulants or in the removal of hardness, Fe/Mn, heavy metals, and phosphate. The solubility product (K_{sp}) of a compound allows for the estimation of the equilibrium concentration of the target species after using reaction chemistry to produce the new species. It may be expressed as the equilibrium constant defined previously; however, the solid reactive species is assumed to have a value of 1. This yields an expression as follows:

$$aA \leftrightarrow bB + cC$$

$$K_{sp} = \frac{[B]^b[C]^c}{[A]^a} = \frac{[B]^b[C]^c}{1} = [B]^b[C]^c$$

Values of K_{sp} for typical water treatment reactions are given in Table 8.7.

Table 8.7 Solubility products for select reactions common in environmental engineering

Equilibrium Equation	K_{sp} @ 25°C	Significance
$Al(OH)_3(s) \leftrightarrow Al^{3+} + 3OH^-$	1.26×10^{-33}	Coagulation
$CaCO_3(s) \leftrightarrow Ca^{2+} + CO_3^{2-}$	4.95×10^{-9}	Hardness removal
$Ca(OH)_2(s) \leftrightarrow Ca^{2+} + 2OH^-$	7.88×10^{-6}	Hardness removal
$CaSO_4(s) \leftrightarrow Ca^{2+} + SO_4^{2-}$	4.93×10^{-5}	Flue gas desulfurization
$Ca_3(PO_4)_2(s) \leftrightarrow 3Ca^{2+} + 2PO_4^{3-}$	2.02×10^{-33}	Phosphate removal
$Cu(OH)_2(s) \leftrightarrow Cu^{2+} + 2OH^-$	2.0×10^{-19}	Heavy metal removal
$Fe(OH)_3(s) \leftrightarrow Fe^{3+} + 3OH^-$	2.67×10^{-39}	Coagulation, iron removal
$Fe(OH)_2(s) \leftrightarrow Fe^{2+} + 2OH^-$	4.79×10^{-17}	Coagulation, iron removal
$Ni(OH)_2(s) \leftrightarrow Ni^{2+} + 2OH^-$	5.54×10^{-16}	Heavy metal removal

Another important use of equilibrium constants is in solving acid/base neutralization reactions. While strong acids and bases are assumed to completely ionize in solution, several important acids and bases in the environment are considered weak, as they only partially ionize in aqueous solution. Ionization constants for common weak acids (K_A) and weak bases (K_B) are provided in Table 8.8 and allow for the calculation of solution pH for known molar concentrations of the target species.

Table 8.8 Ionization constants for acids and bases common in environmental engineering

Acid or Base	Equilibrium Equation	K_A or K_B	Significance
Acetate	$CH_3COO^- + H_2O \leftrightarrow CH_3COOH + OH^-$	5.56×10^{-10}	Organic wastes
Acetic acid	$CH_3COOH \leftrightarrow H^+ + CH_3COO^-$	1.8×10^{-5}	Organic wastes
Ammonia	$NH_3 + H_2O \leftrightarrow NH_4^+ + OH^-$	1.8×10^{-5}	Nutrient
Ammonium	$NH_4^+ \leftrightarrow H^+ + NH_3$	5.56×10^{-10}	Nitrification

Acid or Base	Equilibrium Equation	K_A or K_B	Significance
Calcium hydroxide	$CaOH^+ \leftrightarrow Ca^{2+} + OH^-$	3.5×10^{-2}	Softening
Carbonic acid	$H_2CO_3 \leftrightarrow H^+ + HCO_3^-$	4.5×10^{-7}	Corrosion, coagulation
	$HCO_3^- \leftrightarrow H^+ + CO_3^{2-}$	4.7×10^{-11}	
Hypochlorous acid	$HOCl \leftrightarrow H^+ + OCl^-$	2.9×10^{-8}	Disinfection
Phosphoric acid	$H_3PO_4 \leftrightarrow H^+ + H_2PO_4^-$	7.6×10^{-3}	Nutrient, phosphate removal
	$H_2PO_4^- \leftrightarrow H^+ + HPO_4^{2-}$	6.3×10^{-8}	
	$HPO_4^{2-} \leftrightarrow H^+ + PO_4^{3-}$	4.8×10^{-13}	
Magnesium hydroxide	$MgOH^+ \leftrightarrow Mg^{2+} + OH^-$	2.6×10^{-3}	Softening

Example 8.12

Ionization constants and pH

The ionization constant (K_A) for acetic acid is 1.8×10^{-5}. Estimate the pH of a solution containing 0.6 g of acetic acid in 1.0 L of water.

Solution

First, the concentration is needed in molar units. Since the formula for acetic acid is CH_3COOH, it has a MW of 60 g/mol, and therefore 0.6 g = 10 mmol. In 1.0 L of water, it would have a concentration of 10 mM = 0.01 M, and we can write the acetate (Ac) balance as

$$0.01 = HAc + Ac^- \text{ or equivalently } HAc = 0.01 - Ac^-$$

Since the equilibrium expression for the dissociation is

$$HAc \leftrightarrow H^+ + Ac^-$$

we can write the ionization constant as

$$K_A = 1.8 \times 10^{-5} = \frac{[H^+][Ac^-]}{[HAc]}$$

In order to satisfy solution electroneutrality, we know that

$$[H^+] = [OH^-] + [Ac^-]$$

If we assume a pH < 6, we know that $[H^+] \gg [OH^-]$, and we can neglect $[OH^-]$, and may assume

$$[H^+] \approx [Ac^-]$$

Setting an unknown value x for $[H^+]$ and $[Ac^-]$, our equilibrium expression can be written as

$$1.8 \times 10^{-5} = \frac{x^2}{0.01 - x}$$

This expression may be solved for x as

$$1.8 \times 10^{-7} - 1.8 \times 10^{-5} x = x^2 \text{ or equivalently } x = 4.15 \times 10^{-4} = [H^+]$$

This is the molar hydrogen ion concentration and can be converted to pH using the relationship pH = –log[H⁺] as follows

$$pH = -\log[4.15 \times 10^{-4}] = 3.38$$

Since the pH is so low, our assumption that [H⁺] » [OH⁻] is valid.

OXIDATION AND REDUCTION

Oxidation refers to the loss of electrons by an atom, which is the case when the metallic elements react with oxygen. The element that gains the electrons has been reduced, reduction has taken place. Thus, in the reaction of magnesium (Mg) with oxygen (O_2), there is a transfer of two electrons from the Mg, which then exists as the positive ion, whereas oxygen becomes O^{2-}, the oxide ion:

$$2\,Mg + O_2 = 2\,MgO$$

The product, MgO, is an ionic lattice, and there is ionic bonding in the lattice between the Mg^{2+} ions and the O^{2-} ions.

Although the reaction with elemental oxygen is a common example of oxidation, there are other oxidation reactions in which oxygen is not involved. For example, the burning of hydrogen (H_2) in chlorine (Cl_2) is an example of oxidation of the H_2, as it loses electrons to the chlorine in the process:

$$H_2 + Cl_2 = 2\,HCl$$

Reduction occurs when an atom gains electrons. Oxidation and reduction can take place when an element changes in oxidation state, or the number of electrons in the outer shell, in a reaction. For example, in the reaction of sulfur dioxide (as a solution of sulfurous acid) with iodine in solution

$$SO_2 + 2\,I + H_2O = H_2SO_4 + 2\,HI$$

the sulfur atom, S, changes from an oxidation state of +4 to +6. Therefore it loses electrons in its outer shell and is oxidized. At the same time the iodine atom, I, has gained an electron and changed from a state of no charge to a reduced state of –1. The iodine therefore has gained an electron and is reduced. The **oxidation state** is also known as the **oxidation number**, and is a useful concept in balancing equations in which oxidation and reduction take place.

Example 8.13

A standard solution of sodium thiosulfate, $Na_2S_2O_3$, is made by dissolving 25 g of the hydrated salt $Na_2S_2O_3 \cdot 5\,H_2O$ in water, and diluting to 1.0 L of solution. The solution is standardized with iodine, I_2, according to the reaction

$$I_2 + 2\,Na_2S_2O_3 = 2\,NaI + Na_2S_4O_6$$

where the iodine, I, is reduced to iodide, I⁻, and the thiouslfate, $S_2O_3^{2-}$, is oxidized to tetrathionate, $S_4O_6^{2-}$. If 0.500 g of pure iodine is dissolved for the standardization and requires 39.40 ml of solution to reach the end point (disappearance of the iodine color), what is the normality of the thiosulfate solution in this reaction?

Solution

From the equation, two moles of thiosulfate react with one mole of iodine (I_2). The 0.500 g of pure iodine is

$$\frac{0.500 \text{ g}}{(2 \times 126.90 \text{ g/mol})} = 0.00197 \text{ mol } I_2$$

and 0.00197 mol I_2 reacts with (2 × 0.00197) = 0.00394 mol thiosulfate. The molarity of thiosulfate solution is

$$\frac{0.00394 \text{ mol}}{0.0394 \text{ L}} = 0.100 \text{ mol/L}$$

Since there is change of one transferred electron per thiosulfate molecule in the oxidation $2\, S_2O_3^{2-} - 2e^- = S_4O_6^{2-}$, the equivalent weight of thiosulfate is the same as the molecular weight and the solution is also 0.100 N. The original 25 g of sodium thiosulfate weighed out would be

$$M = \frac{g}{m.w} = \frac{25.0}{(2 \times 23.00) + (2 \times 32.07) + (3 \times 15.00) + (5 \times 18.02)}$$
$$= \frac{25.0}{248.24} = 0.101 \text{ M}$$

so the titration showed that the hydrated salt was slightly impure.

Redox Reactions

Redox reactions require that one species is oxidized as another is simultaneously reduced, such that oxidizing agents are reduced, and reducing agents are oxidized in the reaction. Generally, the simplest way to write a balance redox reaction is the use of half-reactions, which are usually written as reduction reactions as shown in Table 8.9. Oxidation reactions are written as the reverse of the reduction reaction found in the table, and complete reactions are formed by adding the two.

Table 8.9 Half-reactions (reductions) common in environmental engineering

Element	Half-Reaction
C	$\frac{1}{4} CO_2(g) + \frac{7}{8} H^+ + e^- = \frac{1}{8} CH_3COO^- + \frac{1}{4} H_2O$
Cl	$\frac{1}{2} Cl_2(aq) + e^- = Cl^-$
Fe	$Fe^{3+} + e^- = Fe^{2+}$
Fe	$\frac{1}{2} Fe^{2+} + e^- = \frac{1}{2} Fe(s)$
Mn	$\frac{1}{3} MnO_4^- + \frac{4}{3} H^+ + e^- = \frac{1}{3} MnO_2 + \frac{2}{3} H_2O$
Mn	$\frac{1}{2} MnO_2(s) + 2 H^+ + e^- = \frac{1}{2} Mn^{2+} + H_2O$
N	$\frac{1}{5} NO_3^- + \frac{6}{5} H^+ + e^- = \frac{1}{10} N_2(g) + \frac{3}{5} H_2O$
O	$\frac{1}{4} O_2(g) + H^+ + e^- = \frac{1}{2} H_2O$
S	$\frac{1}{8} SO_4^{2-} + \frac{5}{4} H^+ + e^- = \frac{1}{8} H_2S(aq) + \frac{1}{2} H_2O$

Oxidation and reduction reactions can be considered quantitatively from the viewpoint of an electrochemical cell. Such a cell has two electrodes and oxidation takes place at the negative pole (electrons are given up to the electrode by the species being oxidized) and reduction takes place at the positive pole (electrons are taken from the electrode by the species being reduced). The total cell potential is the sum of the two reactions:

$$E_{cell} = E_{ox} + E_{red}$$

The potential of the half-cell $H^+ + 2e^- = H_2$ is assigned the value 0.0 volts at standard conditions. If this reaction takes place at one of the electrodes, then the measured cell potential when another reaction takes place at the other electrode is the half-cell or single electrode potential for the reaction of interest. This is at standard conditions, which means unit **activity** or approximately unit concentration (i.e., one mole per liter solution concentration or one atmosphere gas pressure) and 25°C.

Tables are available for reduction potentials, which is the cell potential when the reduction of the species being considered takes place at the positive electrode and the other electrode is the hydrogen electrode reaction with 0 volts half-cell potential (by definition). When the reduction takes place spontaneously the resultant half-cell potential is positive. All the positive potential half-reactions take place spontaneously when the other electrode is the hydrogen electrode, and each reaction at a higher potential takes place more readily than one at a lower potential. Therefore, if a reduction reaction in the table is linked in a cell to an electrode with a lower reduction potential, the initially considered reaction should still take place (at standard conditions).

Below the $2H^+ + 2e^- = H_2$ reaction in the table the potential values are negative, indicating that the reduction of the species shown takes place less readily than that for the hydrogen electrode reaction. The values for reduction potential continue negative to the bottom of the table, where the most difficult species to reduce are listed. As in the top part of the table, each species is more easily reduced than the ones below it (more negative reduction potential) and presumably would take place when coupled in a cell to a reaction with a more negative reduction potential.

Although the half-cell potentials may be measured, in principle, as indicated above, this is not always experimentally feasible. Therefore some data on single electrode potentials is calculated from other thermodynamic data.

A partial list of species and reactions in order of their electrochemical single electrode reduction potential is shown in Table 8.10.

ELECTROCHEMISTRY

A chemical reaction that involves an electron transfer can, when properly connected in a cell, be a source of an electric potential and an electric current. However, a single chemical reaction will only function as a **half-cell**, so that two chemical reactions must be coupled to complete an electrochemical cell. In order to be electrically coupled, the chemical reaction at the electrode must have (1) one species that is conducting and connected to an external circuit, or (2) electrical contact between the reacting species and a conducting phase that can be connected to the external circuit. There must also be a continuity of the reacting solution between the electrodes, which can be provided by a **salt bridge**, porous barrier, or even an interface between two solutions of difference densities.

In order to have data that can be used to compare a number of combinations of chemical reactions in cells, it has become conventional to list the open circuit potential of reactions versus the potential of a hydrogen electrode. A table of this type was shown in the previous section on oxidation and reduction. The half-cell potential of the reaction

$$H_2 - 2e^- = 2H^+$$

is taken to be 0 volts, and half-cell potentials of all other combinations giving electrode reactions are measured against this standard. As the concentration of the species involved also affects the emf, the concentrations considered are, by convention, that of gas phases at one atmosphere and solution phases at a concentration of one mole per liter.

Table 8.10 Partial electrochemical series

Oxidizing Agents	Reaction	Voltage	Reducing Agents
(strong oxidizing)			(weak reducing)
F_2	$F_2 + 2e^- = 2\,F^-$	2.9	F^-
Au^+	$Au^+ + e^- = Au$	1.7	Au
MnO_4^-	$MnO_4^- + 8\,H^+ + 5e^- = Mn^{2+} + 4\,H_2O$	1.5	Mn^{2+}
N_2O_4	$N_2O_4 + 4\,H^+ + 4e^- = 2\,NO + H_2O$	1.0	NO
I_2	$I_2 + 2e^- = 2\,I^-$	0.5	I^-
Cu^{2+}	$Cu^{2+} + 2e^- = Cu$	0.3	Cu
H^+	$2\,H^+ + 2e^- = H_2$	0.0	H_2
Ni^{2+}	$Ni^{2+} + 2e^- = Ni$	−0.2	Ni
S	$S + 2e^- = S^{2-}$	−0.5	S^{2-}
Be^{2+}	$Be^{2+} + 2e^- = Be$	−1.7	Be
Cs^+	$Cs^+ + e^- = Cs$	−2.9	Cs
(weak oxidizing)			(strong reducing)

Example 8.14

$Zn - 2e^- = Zn^{2+}$ standard electrode potential = + 0.76 volts, and $Cl_2 + 2e^- = 2\,Cl^-$ standard electrode potential = +1.36 volts. What is the open circuit potential at standard conditions when the two half-cells are combined?

Solution

When the two half-cells are combined as an electrochemical cell, the open circuit emf will be 0.76 + 1.36 = 2.12 volts. This is for standard concentration conditions, which in this case would be 1 mole/liter Z^{2+} and 1 atm Cl_2.

The effect of the concentration of reactant species on the emf of the cell can be calculated from the Nernst equation, which, for the reaction $A + B = C + D$, is

$$E = E_0 - \frac{RT}{nF} \ln \frac{[C]\,[D]}{[A]\,[B]}$$

where

$$\frac{[C]\,[D]}{[A]\,[B]}$$

is the **reaction quotient**, sometimes designated by Q. The equation is a thermodynamic derivation and includes the gas constant, R; the Faraday constant, F; the absolute temperature, T; and the number of transferred elections, n. At 25°C, with $n = 1$, and converting to $\log_{10}$, the factor RT/nF in the equation is 0.059.

Example 8.15

For the Zn and Cl_2 reaction

$$Zn + Cl_2(g) = Zn^{2+} + 2\,Cl^-$$

$$E = E_0 - \frac{0.059}{2} \log \frac{[Zn^{2+}][Cl^-]^2}{[Zn][Cl_2]}$$

If the $ZnCl_2$ is 0.1 molar and the Cl_2 gas is 0.01 atmosphere, what is the electrochemical potential of the cell?

Solution

$$\begin{aligned}
E &= 2.12 - \frac{0.059}{2} \log \frac{(0.1)(0.2)^2}{(1)(0.01)} \\
&= 2.12 - 0.030 \log \frac{0.004}{0.01} \\
&= 2.12 - 0.030 \log 0.4 \\
&= 2.12 + 0.012 = 2.14 \text{ volts}
\end{aligned}$$

COMBUSTION

All materials that can burn, and therefore produce heat, are called fuels. Most of these fuels are combinations of carbon and hydrogen atoms and are thus called hydrocarbons. Hydrocarbon fuels react with air and produce heat, carbon dioxide, and water.

Consider a simple hydrocarbon fuel, namely methane, and burn it in oxygen.

$$CH_4 + O_2 \rightarrow CO_2 + H_2O$$

Now, we must set up a balance across the equation so that we have equal numbers of C, O, and H on each side of the arrow.

C:	1	=	1
H:	4	=	2(2) = 4
O:	2(2)	=	2 + 2(1) = 4

Therefore, the equation becomes a stoichiometric equation of methane burning in oxygen:

$$CH_4 + 2O_2 \rightarrow CO_2 + 2H_2O$$

What about the case where the fuel is burned in air? We must then look at the major constituents of air. Air consists of 21% oxygen and 79% nitrogen. Therefore, for each mole of oxygen entering the combustion process, there are 3.76 moles of nitrogen entering the process (.79/.21).

Now, taking our methane and burning it in a complete combustion process in air, the equation becomes:

$$CH_4 + 2O_2 + 2(3.76)N_2 \rightarrow CO_2 + 2H_2O + 7.52N_2$$

A term often used in combustion processes is the *air/fuel ratio*. This term indicates the mass of air in the combustion process divided by the mass of fuel in the process. Consider the previous methane equation where one mole of methane is burned in air. The air/fuel ratio may be calculated as:

$$AF = \frac{\text{mass air}}{\text{mass fuel}} = \frac{NM_{air}}{NM_{fuel}}$$

where

N = number of moles

M = molar mass

Therefore:

$$AF = \frac{2(1+3.76)(28.92)}{(12+4)} = 17.2$$

Example 8.16

Octane is burned in air, and the combustion is complete. Determine (i) the stoichiometric mixture equation, and (ii) the air/fuel ratio.

Solution

$$C_8H_{18} + \alpha(O_2 + 3.76N_2) \rightarrow \beta CO_2 = \gamma H_2O + \alpha(3.76)N_2$$

C: $8 = \beta = 8$

H: $18 = 2\gamma \quad \gamma = 9$

O: $2\alpha = 2(\beta) + \gamma = 2(8) + 9 = 25 \quad \alpha = 12.5$

(i) Therefore, the stoichiometric equation is

$$C_8H_{18} + 12.5(O_2 + 3.76N_2) = 8CO_2 + 9H_2O + 47N_2$$

(ii) The air/fuel ratio is

$$AF = \frac{NM_{air}}{NM_{fuel}} = \frac{(12.5)(4.76)(28.92)}{(12)(8)+18} = 15.09$$

Example 8.17

Ethylene C_2H_4 is burned in 300% theoretical air. Assume complete combustion and a total pressure of combustion products is 75 kPa. Determine (i) the stoichiometric equation and (ii) the air/fuel ratio.

Solution

$$C_2H_4 + \alpha(O_2 + 3.76N_2) \rightarrow \beta CO_2 + \gamma H_2O + \alpha 3.76N_2$$

$$C: 2 = \beta = 2$$
$$H: 4 = 2\gamma \quad \gamma = 2$$
$$O: 2\alpha = 2(\beta) + \gamma = 2(2) + 2 = 6 \quad \alpha = 3$$

(i) Therefore, the stoichiometric equation is

$$C_2H_4 + 3(O_2 + 3.76N_2) \rightarrow 2CO_2 + 2H_2O + 11.28N_2$$

(ii) For the air/fuel ratio with 300% theoretical air, first write the actual combustion equation.

$$C_2H_4 + 9(O_2 + 3.76N_2) \rightarrow 2CO_2 + 2H_2O + 3O_2 + 33.84N_2$$

$$AF = \frac{9(4.76)(28.92)}{2(12) + 4} = 44.2$$

PROBLEMS

8.1 First ionization energy refers to:
  a. removal of an electron from a gas atom
  b. energy to form the most probable ion
  c. trapping an ion in a lattice structure
  d. formation of a –1 anion

8.2 From the periodic table predict the molecular formula of silicon (Si) oxide (O).
  a. SiO
  b. Si_2O
  c. SiO_4
  d. SiO_2

8.3 The formula for potassium aluminum sulfate is (not including water of hydration):
  a. $KAlSO_4$
  b. $KAl(SO_4)_2$
  c. K_2AlSO_4
  d. $KAl(SO_4)_3$

8.4 To indicate a compound is pentahydrate, you would write which of the following at the end of the molecular formula?
  a. • 5 (H^+)
  b. • 5 (H^-)
  c. • 5 (OH^-)
  d. • 5 H_2O

8.5 The common name of the oxide of nitrogen with the formula N_2O is:
  a. nitrogen dioxide
  b. nitrous oxide
  c. nitric oxide
  d. dinitrogen oxide

8.6 The name of $(NH_4)_2Cr_2O_7$ is:
  a. diammonium chromate
  b. ammonium chromate
  c. ammonium(II) chromate
  d. ammonium dichromate

8.7 A sample of 1.38 moles of manganese (IV) oxide contains

  a. how many moles of manganese ions?
  a. 1.00
  b. 2.76
  c. 1.38
  d. 4.00

  b. how many moles of oxygen ions?
  a. 1.38
  b. 2.76
  c. 2.00
  d. 1.76

  c. how many grams of material? (a.w. Mn = 54.9, O = 16.0)
  a. 1.38
  b. 86.9
  c. 2.76
  d. 120

8.8 A student in a chemistry laboratory wants to weigh out 2.00 moles of calcium carbonate ($CaCO_3$) but picks up sodium chloride (NaCl) by mistake. How many moles of NaCl will the student weigh out? (m.w. $CaCO_3$ = 40.1 + 12.0 + (3 × 16.0) = 100.1; NaCl = 23.0 + 35.5 = 58.5)
  a. 3.42
  b. 1.87
  c. 1.16
  d. 2.00

8.9 A protein molecule is known to bind one molecule of oxygen (O_2) per molecule of protein. If 12.2 g of protein bind 9.8 mg of O_2, what is the molecular weight of the protein? (m.w. $O_2 = 2 \times 16.0 = 32.0$)
    a. 25,700   c. 38,900
    b. 79,600   d. 39,800

8.10 An impure sample of $FeSO_4 \cdot 7H_2O$ weighing 1.285 g is analyzed for Fe(II) content by titration with 0.03820 M $KMnO_4$ in acid solution. The endpoint is 21.83 ml. What is the weight percent of Fe(II) in the impure sample? (a.w. Fe = 55.85, K = 39.10, S = 32.07, Mn = 54.94, O = 16.0, H = 1.008)
    a. 18.12%   c. 9.060%
    b. 36.76%   d. 20.09%

8.11 A student wishes to check the concentration of a bottle of hydrogen peroxide, H_2O_2. To do so, the student carried out a redox titration of the H_2O_2 with $KMnO_4$, as shown in the reaction

$$5\ H_2O_2(aq) + 2\ KMnO_4(aq) + 3\ H_2SO_4(aq) \rightarrow$$
$$K_2SO_4(aq) + 2\ MnSO_4(aq) + 8\ H_2O(l) + 5\ O_2(g)$$

It required 39.7 ml of 0.0103 M $KMnO_4$(aq) to react with 1.546 g of the hydrogen peroxide solution. What was the percent of H_2O_2 by mass in the solution? (a.w. K = 39.10, H = 1.008, O = 16.00, Mn = 54.94, S = 32.07)
    a. 1.80%   c. 1.65%
    b. 2.70%   d. 2.24%

8.12 What weight in g of $SnCl_2$ is needed to react with 40 mL of a 0.10 normal I_2 solution as follows?

$$Sn^{2+} + I_2 \rightarrow Sn^{4+} + 2\ I^- \quad \text{(a.w. Sn = 118.70, I = 126.90, Cl = 35.45)}$$

    a. 0.38 g   c. 3.8 g
    b. 0.76 g   d. 0.19 g

8.13 A solution is prepared by dissolving 5.88 g $K_2Cr_2O_7$ in dilute acid and diluting with water to 1.000 L. Calculate the normality of the solution assuming that the half-reaction for the $Cr_2O_7^{2-}$ in solution is

$$14\ H^+ + Cr_2O_7^{2-} \rightarrow 2\ Cr^{3+} + 7\ H_2O$$

    a. 0.163 N   c. 0.120 N
    b. 0.060 N   d. 0.050 N

8.14 Given the following standard electrode potentials:

$$Al^{3+} + 3\ e^- \rightarrow Al;\ E^0 = -1.66\ V$$
$$Cd^{2+} + 2\ e^- \rightarrow Cd;\ E^0 = -0.40\ V$$

determine the standard cell potential for the reaction

$$Al + 3\ Cd^{2+} \rightarrow 2\ Al^{3+} + 3\ Cd$$

    a. $E^0 = 2.06$ V   c. $E^0 = 1.23$ V
    b. $E^0 = -1.26$ V   d. $E^0 = 1.26$ V

8.15 Calculate the standard cell potential, E^0, for the reactions below, using the correct half-reactions and the corresponding standard electrode potentials, E^0. State whether each reaction is spontaneous or nonspontaneous at standard conditions.

Table of standard reduction potentitals

Half-Reaction	E^0
$Fe^{2+} + 2\ e^- = Fe(s)$	−0.41 V
$Sn^{2+} + 2\ e^- = Sn(s)$	−0.14 V
$2\ H^+ + 2\ e^- = H_2(g)$	0.00 V
$Sn^{4+} + 2\ e^- = Sn^{2+}$	0.15 V
$Cu^{2+} + e^- = Cu^+$	0.16 V
$Fe^{3+} + 2\ e^- = Cu(s)$	0.34 V
$Fe^{3+} + e^- = Fe^{2+}$	0.77 V

a. $Fe(s) + 2\ H^+ \rightarrow Fe^{2+} + H_2(g)$
  a. −0.41 V (nonspontaneous)
  b. 0.20 V (spontaneous)
  c. 0.41 V (spontaneous)
  d. 0.20 V (nonspontaneous)

b. $Cu(s) + 2\ H^+ \rightarrow Cu^{2+} + H_2(g)$
  a. 0.34 V (nonspontaneous)
  b. 0.17 V (spontaneous)
  c. −0.34 V (spontaneous)
  d. −0.34 V (nonspontaneous)

c. $Sn^{2+} + 2\ Fe^{3+} \rightarrow Sn^{4+} + 2\ Fe^{2+}$
  a. 0.62 V (spontaneous)
  b. 0.92 V (spontaneous)
  c. 0.62 V (nonspontaneous)
  d. −0.62 V (spontaneous)

8.16 Given the cell Cu|Cu^{2+} || Fe^{3+}, Fe^{2+}| Pt and the standard reduction potentials

$$Cu^{2+} + 2\ e^- \rightarrow Cu(s);\ E^0 = 0.34\ V$$
$$Fe^{3+} + e^- \rightarrow Fe^{2+};\ E^0 = 0.77\ V$$

calculate the cell potential (emf) of the cell at 25°C when [Fe^{2+}] = 0.040 M, [Fe^{3+}] = 0.0020 M, and [Cu^{2+}] = 0.050 M.
  a. 0.43 V
  b. 0.35 V
  c. 1.07 V
  d. 0.39 V

8.17 Using a current of 3.75 A (amperes), how long (in seconds) would it take to electroplate 6.19 g metallic chromium from a solution in which the chromium was in the +3 oxidation state (Cr^{3+}). (a.w. Cr = 52.00, Faraday constant, F = 96,485 C • mol^{-1})
  a. 2.70×10^5 s
  b. 1.02×10^3 s
  c. 1.29×10^5 s
  d. 9.18×10^3 s

8.18 A solution of nickel sulfate (NiSO$_4$) was electrolyzed for 0.75 hr between inert electrodes. If 17.5 g of nickel metal was deposited, what was the average current? (a.w. Ni = 58.69, Faraday constant, F = 96,485 C • mol^{-1})
  a. 1.1×10^1
  b. 1.3×10^3
  c. 2.1×10^1
  d. 1.6×10^1

8.19 A spoon, with a surface of 45 cm², is suspended in a cell filled with a 0.10 M solution of gold (III) chloride, $AuCl_3$. A current of 0.52 A has been passed through the cell, until a coating of gold 0.10 mm thick, has plated on the spoon. How long did the current run? (density Au = 19.3 g/cm³, a.w. Au = 196.97, Faraday constant, F = 96,485)
    a. 1.3×10^4 sec (3.6 hr)   c. 4.1×10^4 sec (11 hr)
    b. 2.5×10^4 sec (6.9 hr)   d. 8.2×10^3 sec (2.3 hr)

8.20 Assuming 100% dissociation, calculate the molarity of the H^+ (H_3O^+) ions in a solution made by diluting 10.0 mL of 0.10 m HNO_3(aq) to 500.0 mL.
    a. 0.0010 M   c. 0.020 M
    b. 0.010 M   d. 0.0020 M

8.21 How many mL of water need to be added to 30.0 mL of a 12.0 molar (M) solution of HCl to give a 3.00 M solution?
    a. 120 mL   c. 90 mL
    b. 133 mL   d. 60 mL

8.22 A stock solution is prepared by dissolving 30.0 g of NaOH in water and diluting to a final volume of 500 mL. How many mL of this stock solution are necessary to prepare 1.00 L of 0.100 M NaOH? (a.w. Na = 22.99, O = 16.00, H = 1.008)
    a. 66.7 mL   c. 150 mL
    b. 37.5 mL   d. 167 mL

8.23 Calculate the volume of 0.250 N $Ca(OH)_2$ needed to completely neutralize 75.0 mL of 0.150 N H_3PO_4.
    a. 45.0 mL   c. 22.5 mL
    b. 125 mL   d. 41.7 mL

8.24 Metallic elements are found where in the periodic table?
    a. In the far left-hand and far right-hand groups
    b. In the middle of the table and Group VIIIA
    c. In the left-hand and middle groups
    d. Only in groups IA and IIA

8.25 Metals have:
    a. both high electrical and high thermal conductivity
    b. high electrical but low thermal conductivity
    c. low cohesive strength and high luster
    d. high luster and low ductility

8.26 Nonmetals are:
    a. malleable but not ductile
    b. very reactive with acids
    c. good conductors of electricity
    d. able to form halides, which react with water to give an oxyacid

8.27 The following metals are in a single group in the periodic table:
a. lithium, sodium, potassium, strontium, and cesium
b. iron, cobalt, nickel, platinum, and gold
c. boron, aluminum, gallium, indium, and thallium
d. beryllium, magnesium, calcium, barium, and radium

8.28 The halogens:
a. will not react with each other
b. are strong electron donors
c. form strong oxyacids of the formula HOX_3
d. form strong covalent bonds with Group IA metals

8.29 The dissolution of sulfur dioxide (SO_2) in water produces:
a. a weak solution of sulfuric acid, H_2SO_4
b. a weak solution of pyrosulfuric acid (disulfuric acid), $H_2S_2O_7$
c. a solution used as an analytical reagent to precipitate metal cations
d. a weak solution of sulfurous acid, H_2SO_3

8.30 Silicon (Si) is important in the semiconductor industry because:
a. of its outer electron configuration of two electrons in *s* orbitals and two electrons in *p* orbitals
b. the energy band gap between valence electrons and conductance electrons in the crystal is relatively small
c. its melting point is low enough that it can be melted and cast into chips
d. it is the most dense of the Group IVA elements and this allows a high electron density in devices

8.31 The reaction below is carried out in a 5.00-L reaction vessel at 600 K.

$$CO(g) + H_2O(g) = CO_2(g) + H_2(g)$$

At equilibrium it is found that 0.020 mol CO, 0.0215 mol H_2O, 0.070 mol CO_2, and 2.00 mol H_2 are present. Evaluate K_c for this reaction.
a. 65.1   c. 236
b. 0.00307   d. 326

8.32 Consider the following reaction at equilibrium:

$$N_2O_4(g) = 2\,NO_2(g)$$

If a 5.00-L reaction vessel, held at constant temperature, is initially filled with 10.0 mol pure $N_2O_4(g)$, and if 3.5 mol $NO_2(g)$ are found in the vessel once equilibrium has been established, what is the value of the equilibrium constant, K_c, for this reaction (at the temperature of the experiment)?
a. 0.297 M   c. 0.424 M
b. 1.48 M   d. 0.0594 M

8.33 At a given temperature, the equilibrium concentrations in a reactor were found to be $PCl_3 = 0.025$ M, $Cl_2 = 0.25$ M, and $PCl_5 = 0.125$ M. Calculate the equilibrium constant at the given temperature for the reaction $PCl_5 = PCl_3 + Cl_2$.
    a. 0.125 M   c. 0.50 M
    b. 2.0 M   d. 0.25 M

8.34 What is the molarity of a solution prepared by dissolving 15.0 g of $La(NO_3)_4$ in 800 mL of water? (a.w. La = 138.91, N = 14.01, O = 16.00)
    a. 0.093 M   c. 0.048 M
    b. 0.031 M   d. 0.039 M

8.35 If 15.0 g of 26% LiBr solution is diluted to a volume of 80.0 ml, what is the molarity of the resultant solution? (a.w. Li = 6.9, Br = 79.9)
    a. 0.71 M   c. 0.44 M
    b. 0.36 M   d. 0.56 M

8.36 The following initial rate data were obtained for the reaction: $2B \rightarrow C$.

[B] M	$r = \dfrac{d[B]}{dt}$ M • sec^{-1}
(1) 0.245	2.92×10^{-4}
(2) 0.490	4.13×10^{-4}

Find the exponential coefficient of the rate law $r = k[B]^n$ for this reaction.
    a. $n = -0.5$ c. $n = 1.0$
    b. $n = 0.5$ d. $n = 2.0$

8.37 Consider the following gas phase reaction:

$$CH_3CHO \rightarrow CH_4 + CO$$

Can the order of this reaction be determined from the above balanced equation? If so, determine the order and explain; if not, explain.
    a. Yes. The reaction is second order because each molecule of reactant must collide with a second molecule of reactant to decompose.
    b. Yes. The reaction is third order because each molecule of reactant collides with other reactant species as well as with each of two species of product molecules.
    c. No. The reaction order can't be determined from balanced equation and reaction coefficients alone; experimental data is required.
    d. Yes. It is a first order reaction because the only reactant is CH_3CHO, which can only react with itself.

8.38 Consider the following set of data:

Set	Rate, mol^3/L^3 • s	[A], mol/L	[B], mol/L
(1)	0.020	0.10	0.20
(2)	0.080	0.10	0.40
(3)	0.040	0.20	0.20
(4)	0.060	0.30	0.20

Write the rate equation (R = rate), the overall order of the reaction, and the specific rate constant, k, in proper units.
a. $R = k [A] [B]^2$, order = 3, $k = 5.0$ s^{-1}
b. $R = k [A]^2 [B]$, order = 3, $k = 10$ s^{-1}
c. $R = k [A]^2 [B]^2$, order = 4, $k = 50$ L $\cdot$ mol^{-1} $\cdot$ s^{-1}
d. $R = k [A] [B]^0$, order = 1, $k = 0.20$ mol^2 $\cdot$ L^{-2} $\cdot$ s^{-1}

8.39 An analysis of a 2.147-g sample of a hydrocarbon produced 7.260 g of carbon dioxide and 1.485 g of water. (a.w. C = 12.01, O = 16.00, H = 1.008)

Calculate the percentage composition of the hydrocarbon.
a. 97.27% C, 2.73% H
b. 92.27% C, 7.73% H
c. 93.73% C, 6.27% H
d. 91.27% C, 8.73% H

8.40 Cyclopropane is 85.7% carbon. Cyclohexane has:
a. somewhat more than 85.7% carbon
b. somewhat less than 85.7% carbon
c. exactly 85.7% carbon
d. exactly half the carbon percentage of cyclopropane

8.41 An organic compound has an empirical formula of C_3H_8O. This formula can represent:
a. three alcohols
b. one organic acid
c. two alcohols and one ether
d. two ethers and one alcohol

8.42 Name the *type* of compound shown in each of the following structural formulas.

a. CH$_3$—CH$_2$—OH

    a. aldehyde   c. organic acid
    b. alcohol   d. ketone

b.
$$\text{CH}_3-\overset{\overset{\displaystyle H}{|}}{\text{C}}=\text{O}$$

    a. alcohol   c. alkyne
    b. ester   d. aldehyde

c.
$$\text{CH}_3\overset{\overset{\displaystyle O}{\|}}{\text{C}}-\text{O}-\text{CH}_3$$

    a. ester   c. organic acid
    b. aldehyde   d. glycol

d. CH$_3$—CH$_2$—NH$_2$

    a. nitrile   c. amide
    b. alkyne   d. amine

e. CH$_3$—C≡N

    a. amine   c. amide
    b. nitrile   d. azide

f.
$$\text{CH}_3-\overset{\overset{\displaystyle O}{\|}}{\text{C}}-\text{CH}_3$$

    a. aldehyde   c. ketone
    b. alkane   d. ether

Use the following information for problems 8.43 and 8.44. Ethane is burned in 150% theoretical air. Assume complete combustion and that the process takes place in atmospheric air.

8.43 The air/fuel ratio in 150% theoretical air is most nearly:

    a. 32 kg air/kg fuel   c. 24 kg air/kg fuel
    b. 28 kg air/kg fuel   d. 16 kg air/kg fuel

8.44 The air/fuel ratio for stoichiometric conditions is:

    a. 10 kg air/kg fuel   c. 22 kg air/kg fuel
    b. 16 kg air/kg fuel   d. 24 kg air/kg fuel

SOLUTIONS

8.1 a. Choice (a) is the definition of ionization energy. Choice (b) is incorrect since the second and third ionization energy may be pertinent to forming the most probable ion. Choice (c) is incorrect since the ion formed does not have to be trapped in a lattice structure, and the energy is unrelated. Choice (d) is incorrect since a + 1 cation forms.

8.2 d. Oxygen (atomic no. 8) is in the second period and Group VIA of the periodic table. Silicon (atomic no. 14) is in the third period and Group IVA of the periodic table. The most probable electronic structure for oxygen in a compound is to pick up two electrons to fill its outermost shell, giving it the electron configuration of neon (atomic no. 10). Silicon, with four electrons in the outer shell, could either gain four electrons or lose four electrons to form a complete octet in the outer shell. However, since O wants to gain electrons, it is more likely that Si will lose them, making it also isoelectronic with neon. The appropriate number of Si and O atoms must then combine so that the number of electrons gained by the O atoms equals the number of electrons lost by the Si atoms. Therefore, there must be twice as many O atoms as Si atoms in the molecular formula, and the simplest molecular formula is SiO_2.

8.3 b. In this mixed salt (alum), the K has a valence of +1, the Al has a valence of +3, and (SO_4) is a group with a valence of –2. The formula is $KAl(SO_4)_2$. Alum also includes 12 H_2O of hydration in the crystal.

8.4 d. Penta means 5, **hydrate** means water. This is one of the few times a coefficient shows up inside a formula. Some prefer to write salt$(H_2O)_5$.

8.5 b. There are six oxides of nitrogen with oxidation numbers of the nitrogen ranging from +1 to +5. The formulas are N_2O, NO, N_2O_3, NO_2, and N_2O_5. Their common names are nitrous oxide, nitric oxide, dinitrogen trioxide, nitrogen dioxide, and dinitrogen pentaoxide (or pentoxide), respectively. N_2O_4 is the dimmer of NO_2 and N is +4 in both species. The systematic name of N_2O is dinitrogen oxide and of NO is nitrogen monoxide. Nitric oxide is the correct common name for N_2O. Although (d) is a correct name for N_2O, it is a systematic name, and the question calls for the common name.

8.6 d. Chromium forms two oxyanions with Cr in the +6 oxidation state. These are the chromate ion, $(CrO_4)^{2-}$, and the dichromate ion, $(Cr_2O_7)^{2-}$. Choices (a), (b), and (c) are chromates, which is incorrect. Choice (d) is systematically named but ignores the fact that the name of an ionic compound is built from the names of the ions present.

8.7

a. c. 1 mol MnO_2 = 1 mol Mn^{4+} + 2 mol O^{2-}; 1.38 mol MnO_2 × 1.0 mol Mn^{4+}/1.0 mol MnO_2 = 1.38 mol Mn^{4+}

b. b. 1.38 mol MnO_2 × 2.0 mol O^{2-}/1.0 mol MnO_2 = 2.76 mol O^{2-}

c. d. m.w. MnO_2 = 54.9 + (2 × 16.0) = 86.9; 1.38 mol MnO_2 × (86.9 g MnO_2/1 mol MnO_2) = 120 g MnO_2

8.8 a. Thinking he or she has $CaCO_3$, the student will weigh out 2.00 mol × (100.1 g $CaCO_3$/mol) = 200.2 g. But since it was NaCl that was weighed, the number of moles of NaCl is 200.2 g × (1 mol NaCl/58.5 g) = 3.42 NaCl.

8.9

$$9.8 \text{ mg } O_2 \times 10^{-3} \text{ g/1 mg} = 9.8 \times 10^{-3} \text{ g } O_2$$
$$1 \text{ molecule } O_2 / 1 \text{ molecule protein} = 1 \text{ mol } O_2 / 1 \text{ mol protein}$$
$$1 \text{ mol } O_2 = 32.0 \text{ g } O_2$$

$$\frac{12.2 \text{ g protein}}{9.8 \times 10^{-3} \text{ g } O_2} \times \frac{32.0 \text{ g } O_2}{1 \text{ mol } O_2} \times \frac{1 \text{ mol } O_2}{1 \text{ mol protein}} = \frac{39.8 \times 10^3 \text{ g protein}}{1 \text{ mol protein}}$$

m.w. protein = 39.8×10^3 = 39,800

or

$$\frac{9.8 \times 10^{-3} \text{ g } O_2}{32.0 \text{ g/1 mol } O_2} = 0.3062 \times 10^{-3} \text{ mol } O_2 = 0.3062 \times 10^{-3} \text{ mol protein}$$

$$g = \text{mol} \times \text{m.w.}$$

$$\text{m.w.} = \frac{g}{\text{mol}}$$

$$\frac{12.2 \text{ g protein}}{0.3062 \times 10^{-3} \text{ mol protein}} = 39.8 \times 10^3 = 39,800 \text{ m.w. protein}$$

8.10 a. First calculate moles of $KMnO_4$ to reach the endpoint:

$$21.83 \text{ mL } KMnO_4 \times \frac{1 \text{ L}}{10^3 \text{ mL}} \times \frac{0.03820 \text{ mol}}{1 \text{ L}} = 8.339 \times 10^{-4} \text{ mol } KMnO_4$$

We need the balanced equation to relate moles of $KMnO_4$ to moles of Fe(II):

$$5 \text{ Fe}^{2+} + MnO_4^- + 8 \text{ H}^+ \rightarrow 5 \text{ Fe}^{3+} + Mn^{2+} + 4 \text{ H}_2O$$

since MnO_4^- = moles of $KMnO_4$

$$8.399 \times 10^{-4} \text{ mol } Mn_4^- \times \frac{5 \text{ mol Fe}^{2+}}{1 \text{ mol } MnO_4^-} = 4.170 \times 10^{-3} \text{ mol Fe}^{2+}$$

$$4.170 \times 10^{-3} \text{ mol Fe}^{2+} \times \frac{5{,}585 \text{ g}}{\text{mol Fe}} = 0.239 \text{ g Fe}^{2+} \text{ in sample}$$

$$\frac{0.2329 \text{ g Fe}^{2+}}{1.285 \text{ g sample}} \times 100 = 18.12\% \text{ Fe}^{2+} \text{ in sample}$$

8.11 d. Percent H_2O_2 by mass is equal to

$$\frac{\text{g } H_2O_2}{\text{g sample}} \times 100$$

so we must find g H_2O_2. This information is supplied by the titration and the balanced equation

$$\text{moles KMnO}_4 = \frac{0.0103 \text{ ml}}{1 \text{ L}} \times \frac{0.0397 \text{ L}}{\text{titre}} = 4.09 \times 10^{-4} \text{ mol}$$

From the balanced equation the reacting ratio is

$$\frac{5 \text{ mol } H_2O_2}{2 \text{ mol KMnO}_4}$$

$$\frac{5 \text{ mol } H_2O_2}{2 \text{ mol KMnO}_4} \times \frac{4.09 \times 10^{-4} \text{ mol KMnO}_4}{\text{sample}} = 1.02 \times 10^{-3} \text{ mol } H_2O_2$$

$$\text{g } H_2O_2 = \text{mol} \times \text{m.w.} = 1.02 \times 10^{-3} \text{ mol} \times \frac{34.02 \text{ g}}{1 \text{ mol}} = 3.47 \times 10^{-2}$$

$$\text{percent } H_2O_2 = \frac{0.0347 \text{ g } H_2O_2}{1.546 \text{ g solution}} \times 100 = 2.24\% \ H_2O_2$$

8.12 a. First calculate the number of equivalents of I_2:

$$\text{Equivalents} = \text{volume (L)} \times \text{normality (N)}$$
$$= 0.040 \text{ L} \times 0.10 = 0.0040 \text{ equiv. } I_2$$

Since the values of normality and equivalents are taken with respect to the particular reaction being considered,

$$\text{equivalents SnCl}_2 = \text{equivalents } I_2 = 0.040$$

The half-reaction for Sn^{2+} is

$$Sn^{2+} \rightarrow Sn^{4+} + 2\ e^-$$

As the oxidation number of Sn changes from +2 to +4, a change of two, the equivalent weight of $SnCl_2$ is

$$\frac{\text{formula weight}}{2} = \frac{189.60}{2} = 94.80$$

$$\text{weight SnCl}_2 = \frac{94.80 \text{ g}}{\text{equivalent}} \times 0.0040 \text{ equiv.} = 0.38 \text{ g}$$

8.13 c. In the half-cell reaction for $Cr_2O_7^{2-}$, the oxidation state of Cr changes from +6 to +3. Since two Cr atoms are in the $Cr_2O_7^{2-}$ formula, the change for the species is six and the equivalent weight is the molecular weight divided by six. This can also be seen by balancing the half-cell charges by the addition of six e^- to the left side of the equation.

$$\text{equivalent weight } K_2Cr_2O_7 = \frac{294.20}{6} = 49.03$$

(Note: Since we are relating this figure to the weight of dissolved $K_2Cr_2O_7$, it is the molecular weight—or formula weight—of the K salt that is used in the calculation.)

$$\text{equivalents } K_2Cr_2O_7 = \frac{5.88 \text{ g}}{49.03}$$

$$\frac{\text{g equivalent}}{\text{g}} = 0.120 \text{ equivalent}$$

$$\text{normality} = \frac{\text{equivalents}}{\text{liter}} = \frac{0.120 \text{ equivalent}}{1.00 \text{ L}} = 0.120 \text{ N}$$

8.14 d. The cell reaction shows Al as oxidized (loss of electrons). Therefore, the equation for the standard electrode potential must be rewritten as an oxidation reaction, and the sign of E^0 will be positive:

$$Al \rightarrow Al^{3+} + 3 \ e^-; E^0 = 1.66 \text{ V}$$

The cell reaction for Cd is written correctly, as Cd^{2+} is reduced (gains electrons). However, the equations must be multiplied by the appropriate factors so that each has the same number of electrons. The factor is 2 for the Al equation and 3 for the Cd equation, to give

$$2 \ Al \rightarrow 2 \ Al^{3+} + 6 \ e^-; E^0 = 1.66 \text{ V}$$
$$3 \ Cd^{2+} + 6 \ e^- \rightarrow 3 \ Cd; E^0 = -0.40 \text{ V}$$

The equations are added to give:

$$2 \ Al + 3 \ Cd^{2+} = 2 \ Al^{3+} + 3 \ Cd \text{ (the electrons cancel)}; E^0 = 1.26 \text{ V}$$

Note that the values of E^0 are not multiplied by the coefficients used to balance the electron transfer in the equation.

8.15

a. c.
$$Fe(s) \rightarrow Fe^{2+} + 2\ e^-:$$
+0.41 V (oxidation half-reaction, reversed from reduction half-reaction table, E^0 sign reversed)

$$2\ H^+ + 2\ e^- \rightarrow H_2(g):$$
0.00 V (reduction half-reaction from table, same sign of E^0)

$$Fe(s) + 2\ H^+ = Fe^{2+} + H_2(g):$$
0.41 V (add half-reactions, electrons cancel, E^0 additive)
Reaction is spontaneous at standard conditions when E^0 positive.

b. d.
$$Cu(s) \rightarrow Cu^{2+} + 2\ e^-:$$
–0.34 V (oxidation half-reaction, reversed from reduction half-reaction table, E^0 sign reversed)

$$2\ H^+ + 2\ e^- \rightarrow H_2(g):$$
0.00 V (reduction half-reaction, from table, same sign of E^0)

$$Cu(s) + 2\ H^+ \rightarrow Cu^{2+} + H_2(g):$$
–0.34 V (add half-reactions, electrons cancel, E^0 additive)
Reaction is nonspontaneous at standard conditions when E^0 negative.

c. a.
$$Sn^{2+} \rightarrow Sn^{4+} + 2\ e^-:$$
–0.15 V (oxidation half-reaction, reversed from reduction half-reaction table, E^0 sign reversed)

$$Fe^{3+} + e^- \rightarrow Fe^{2+}:$$
0.77 V (reduction half-reaction from table, same sign of E^0)

$$2\ Fe^{3+} + 2\ e^- \rightarrow 2\ Fe^{2+}:$$
0.77 V (double equation so electrons will cancel; note E^0 does not double)

$$Sn^{2+} + 2\ Fe^{3+} \rightarrow Sn^{4+} + 2\ Fe^{2+}:$$
0.62 V (add half-reactions, electrons cancel, E^0 additive)
Reaction is spontaneous at standard conditions when E^0 positive.

8.16 d.
$$Cu(s) \rightarrow Cu^{2+} + 2\ e^-:$$
$E^0 = -0.34$ V (anode half-reaction; reverse equation above and change sign of E^0 to give oxidation potential)

$$2\ Fe^{3+} + 2\ e^- \rightarrow 2\ Fe^{2+}:$$
$E^0 = 0.77$ V (cathode half-reaction; double above equation so equal number electrons in each half-reaction; E^0, reduction potential, remains the same)

$$Cu(s) + 2\ Fe^{3+} \rightarrow Cu^{2+} + 2\ Fe^{2+}:$$
$E^0 = 0.43$ V (add half-reactions, electrons cancel, cell E^0 is $E^0_{ox} + E^0_{red}$)

Cell potential at concentrations given is calculated using the Nernst equation:

$$E = E^0 - \frac{0.0592}{n} \log \frac{[Cu^{2+}][Fe^{2+}]^2}{[Fe^{3+}]^2}$$

$$= E^0 - \frac{0.0592}{n} \log \frac{[0.050][0.040]^2}{[0.0020]^2}$$

$$= 0.43 - \frac{0.0592}{n} \log 20 = 0.43 - (0.0296)(1.301)$$

$$= 0.43 - 0.038 = 0.39 \text{ V}$$

8.17 d. The reaction for electroplating Cr from Cr^{3+} in solution is $Cr^{3+}(aq) + 3\ e^- \rightarrow Cr(s)$.

$$\text{factor} = \frac{1 \text{ mol Cr}}{3 \text{ mol } e^-}$$

$$6.19 \text{ g Cr} \times \frac{1 \text{ mol Cr}}{52.00 \text{ g Cr}} = 0.1190 \text{ mol Cr}$$

$$0.1190 \text{ mol Cr} \times \frac{3 \text{ mol } e^-}{1 \text{ mol Cr}} = 0.3570 \text{ mol } e^- \text{ used to electroplate } 6.19 \text{ g Cr}$$

$$0.357 \text{ mol } e^- \times \frac{96,465 \text{ C}}{1 \text{ mol } e^-} = 34,445 \text{ C}$$

$$\text{Coulombs} = \text{amperes} \times \text{seconds} \qquad \text{seconds} = \frac{C}{\text{amperes}} = \frac{34445 \text{ C}}{3.75 \text{ A}}$$

$$= 9185 \text{ s} = 9.18 \times 10^3 \text{ s}$$

8.18 c. Moles of Ni = $17.5 \text{ g} \times \frac{1 \text{ mol}}{58.69 \text{ g}} = 2.98 \times 10^{-1}$ mol

Reaction is $Ni^{2+}(aq) + 2\ e^- = Ni(s)$

Factor is $\frac{1 \text{ mol Ni}}{2 \text{ mol } e^-} = \frac{1 \text{ mol Ni}}{2 \text{ F}}$

$$\text{Coulombs} = 2.98 \times 10^{-1} \text{ mol Ni} \times \frac{2 \text{ F}}{1 \text{ mol Ni}} \times \frac{96,485}{1 \text{ F}} = 5.75 \times 10^4$$

C = amp × sec

Number of seconds = 0.75 hr × 3600 = 2.70×10^3 s

$$\text{amperes} = \frac{5.75 \times 10^4 \text{ C}}{2.70 \times 10^3 \text{ sec}} = 2.1 \times 10^1 \text{ A}$$

8.19 b. The electrode reaction is $Au^{3+}(aq) + 3\,e^- = Au(s)$

Factor is $\dfrac{1 \text{ mol Au}}{3 \text{ mol } e^-}$

The volume of Au deposited $= 45 \text{ cm}^2 \times 0.10 \text{ mm} \times \dfrac{1.00 \text{ cm}}{10 \text{ mm}} = 0.45 \text{ cm}^3$

The weight of Au deposited is $= 0.45 \text{ cm}^3 \times \dfrac{19.3 \text{ g}}{1 \text{ cm}^3} = 8.68 \text{ g Au}$

The coulombs passed $= 8.68 \text{ g Au} \times \dfrac{1 \text{ mol Au}}{196.97 \text{ g Au}} \times \dfrac{3 \text{ mol } e^-}{1 \text{ mol Au}} \times \dfrac{96{,}485 \text{ C}}{\text{mol } e^-}$

$= 1.28 \times 10^4 \text{ C}$

$C = \text{amp} \times \text{sec} \qquad \text{sec} = \dfrac{C}{\text{amp}}$

Time of run $= \dfrac{1.28 \times 10^4 \text{ C}}{0.52 \text{ amp}} = 2.46 \times 10^4 \text{ sec} = 2.5 \times 10^4 \text{ sec} \ (=6.9 \text{ hr})$

8.20 d. Molarity (M) $= \dfrac{\text{moles solute}}{\text{liters (L) solution}}$

Therefore, moles solute = M × L solution

moles solute = 0.10 × 0.010 L = 0.0010 mol HNO_3

A quantity of 0.0010 mol HNO_3 is diluted to 500 mL. The molarity (M) of the solution is given by

$$M = \dfrac{0.0010 \text{ moles } HNO_3}{0.500 \text{ L}} = 0.0020$$

Since $HNO_3(aq)$ is 100% dissociated into ions, $HNO_3 \rightarrow H^+ + NO_3^-$, the 0.0020 M solution will contain 0.0020 M $H^+(H_3O^+)$ and 0.0020 M (NO_3^-).

8.21 c. Since the number of moles of HCl remains the same, that is, moles HCl = moles HCl, we can use the relationship (volume$_1$) × (molarity$_1$) = (volume$_2$) × (molarity$_2$).

(30.0 ml) × (12.0 M) = (x mL) × (3.00 M)

x = 120 mL for volume of solution$_2$

Since the original volume (solution$_1$) is 30.0 mL, the volume of water to be added is 120 − 30 = 90 mL.

8.22 a. First calculate the number of moles of NaOH dissolved, and the molarity (M) of the stock solution.

$$\frac{30.0 \text{ g NaOH}}{40.0 \text{ g/mol}} = 0.750 \text{ mol NaOH}$$

$$M = \frac{\text{moles}}{\text{liter}} = \frac{0.750 \text{ mol}}{0.500 \text{ L}} = 1.50$$

In 1.00 L of 0.100 M NaOH, there are

$$\frac{0.100 \text{ mol}}{\text{L}} \times 1.00 \text{ L} = 0.100 \text{ mol NaOH}$$

Now calculate the volume of 1.50 M NaOH solution that contains 0.100 mol NaOH.

$$M = \frac{\text{moles}}{\text{L}}$$

$$L = \frac{0.100}{1.50} = 0.0667 \text{ L} = 66.7 \text{ mL}$$

$$(3.77 \times 10^{-4} \text{ mol NaOH}) \times \frac{1 \text{ mol HCl}}{1 \text{ mol NaOH}} \times \frac{36.46 \text{ g HCl}}{1 \text{ mol HCl}} \times \frac{100 \text{ mg HCl}}{1 \text{ g HCl}}$$
$$= 13.7 \text{ mg HCl}$$

8.23 a. By definition one equivalent of acid is equal to one equivalent of base in a reaction. The number of equivalents of Ca(OH)$_2$ at the neutralization point must be equal to the number of equivalents of H$_3$PO$_4$ at the neutralization point. In a two-step solution find the number of equivalents of H$_3$PO$_4$ first, from equivalents = volume in L × normality:

$$\text{equiv.} = \text{vol(L)} \times N = 0.0750 \times 0.150 = 0.01125$$

Since the same number of equivalents of Ca(OH)$_2$ were used in the reaction,

$$0.01125 = 0.250x$$
$$x = 0.0450 \text{ L Ca(OH)}_2 \text{ solution}$$

A one-step solution is

$$\text{equivalents Ca(OH)}_2 = \text{equivalents H}_3\text{PO}_4$$

Since normality $(N) = \dfrac{\text{equivalents}}{\text{volume solution (L)}}$

$$N_1 \times \text{vol(L)}_1 = N_2 \times \text{vol(L)}_2$$
$$(0.250) \text{ vol(L)}_1 = (0.150)(0.0750 \text{ L})$$

$$\text{vol(L)}_1 = 0.0450 \text{ L} = 45.0 \text{ mL Ca(OH)}_2$$

It is also convenient to multiply $N_1 \times \text{vol(L)}_1 = N_2 \times \text{vol(L)}_2$ by 1000, so that $N_1 \times \text{vol(mL)}_1 = N_2 \times \text{vol(mL)}_2$ and

$$\text{vol(mL)}_1 = \frac{(0.150)(75.0)}{0.250} = 45.0 \text{ mL Ca(OH)}_2$$

8.24 c. The metallic elements are found primarily in groups IA and IIA at the left of the table and groups IIIB–VIIIB and IB–IIB in the middle. They are also in the higher-period elements of groups IIIA–VIA. There are no metals in the far right groups and VIIIA, ruling out (a) and (b). There are metals in a number of groups besides IA and IIA, ruling out (d).

8.25 a. The mobile electron cloud of outer electrons gives metals both high electrical and high thermal conductivity. The mobile electrons and accompanying crystal structure also give metals high luster, cohesive strength, and high ductility. The chemical reactivity of metals varies from high (Li, Na) to low (Pt, Au).

8.26 d. The nonmetals are generally neither malleable nor ductile. They do not react with acids and are poor conductors of electricity. Often they have low melting points and are volatile. However, they do form covalent halides, which react with water to give oxyacids.

8.27 d. Strontium is a member of Group IIA, not IA. Gold is in Group IB, while the other metals in choice (b) are in Group VIIIB. All the elements in choice (c) are in IIIA, but boron is not considered a metal. The elements in choice (d) are all metals in Group IIA.

8.28 c. Halogens form a number of interhalogen compounds, such as ClF and ICl. They are noted as strong electron acceptors. They do form strong oxyacids of the formula HOX_3, as well as HOX_4. Halogens form ionic bonds in compounds with Group IA metals. The halogens bond with some nonmetals, such as sulfur and phosphorous, as well as other halogens.

8.29 d. Sulfuric acid forms from the reaction of SO_3 with water. Pyrosulfuric acid is formed from SO_3 and concentrated sulfuric acid. SO_2 solution is used as a reducing agent, but it is H_2S that is used in analytical determination of metal cations. SO_2 can be converted to SO_3 by oxidation in the presence of a catalyst. A weak and unstable solution of sulfurous acid is formed upon dissolution in water.

8.30 b. The relatively small energy band gap makes silicon an intrinsic semiconductor and leads to its industrial use.

8.31 d. The chemical equation is balanced so:

$$K_c = \frac{[CO_2(g)][H_2(g)]}{[CO_2(g)][H_2O(g)]}$$

$$\frac{\left[\frac{0.070 \text{ mol } CO_2}{5 \text{ L}}\right]\left[\frac{2.00 \text{ mol } H_2}{5 \text{ L}}\right]}{\left[\frac{0.020 \text{ mol } CO}{5 \text{ L}}\right]\left[\frac{0.0215 \text{ mol } H_2O}{5 \text{ L}}\right]} = 326$$

Note that the units cancel, leaving K_c dimensionless.

8.32 a. A good general approach to equilibrium problem is (1) write the balanced equilibrium reaction equation, (2) write the equilibrium constant expression in terms of the balanced equation, and (3) identify in a simple tabular form what is known about the concentrations of species initially and at equilibrium, and what changes occur. For the above problem, these procedures give

(1) (as given) $N_2O_4(g) = 2\,NO_2(g)$

(2) $K_c = \dfrac{[NO_2]^2}{[N_2O_4]}$

(3)

Species:	mol N_2O_4	mol NO_2
Initial amt.	10.0	0
Change	(−1.75)	(+3.50)
Equilibrium amt.	(8.25)	3.50

The values *not* in parentheses are given with the problem and are filled in first. The blanks are filled in with the values in parentheses from the calculations below.

In order to produce 3.50 mol NO_2, the equation shows us that half this amount of N_2O_4 has decomposed, or

$$3.50\text{ mol }NO_2 \times \dfrac{1\text{ mol }N_2O_4}{2\text{ mol }NO_2} = 1.75\text{ mol }N_2O_4$$

Therefore, at equilibrium, 10.0 − 1.75 = 8.25 mol N_2O_4 remain. The equation for the equilibrium expression is then

$$K_c = \dfrac{\left[\dfrac{3.50\text{ mol }NO_2}{5.00\text{ L}}\right]^2}{\left[\dfrac{8.25\text{ mol }N_2O_4}{5.00\text{ L}}\right]} = \dfrac{[0.70\text{ M}]^2}{[1.65\text{ M}]} = 0.297\text{ M}$$

Note that the presence of the square term dictates the equilibrium constant units (mol/L = M). Note also that the vessel size (5.00 L) must be included in the calculations.

8.33 c. The expression for the equilibrium constant is

$$K_c = \dfrac{[PCl_3][Cl_2]}{[PCl_5]} = \dfrac{[0.25\text{ M}][0.25\text{ M}]}{[0.125\text{ M}]} = \dfrac{[0.0625\text{ M}]^2}{[0.125\text{ M}]} = 0.50\text{ M}$$

8.34 c. First, calculate the molecular weigh of $La(NO_3)_4$.

$$1 \times (\text{a. w. La}) + 4 \times (\text{a.w. N}) + 12 \times (\text{a.w. O})$$
$$138.91 + (4 \times 14.01) + (12 \times 16.00) = 386.9$$

$$\text{Number of moles in 15.0 g La} = \dfrac{15.0\text{ g}}{386.9\text{ g/mol}} = 0.0388\text{ mol}$$

$$\text{Molarity} = M\dfrac{0.0388\text{ mol}}{0.800\text{ L}} = 0.048\text{ M}$$

8.35 d. The moles of solute before dilution equal the moles of solute after dilution.

$$\text{moles} = \frac{g}{\text{m.w.}} = \text{volume (L)} \times \text{molarity (M)}$$

$$\frac{0.26 \times 15.0 \text{ g}}{86.8 \text{ g/mol}} = 0.080 \text{ L} \times \text{M}$$

$$M = 0.56$$

8.36 b. Divide equation for data (1) by equation for data (2) to find a single value of n.

$$\frac{r_1}{r_2} = \frac{k[B]_1^n}{k[B]_2^n} \quad \text{or} \quad \frac{r_1}{r_2} = \frac{[B]_1^n}{[B]_2^n} \quad (k \text{ cancels})$$

Take the log of both sides of the equation:

$$\log \frac{r_1}{r_2} = \log \frac{[B]_1^n}{[B]_2^n} = n \log \frac{[B]_1}{[B]_2}$$

Solve for n:

$$n = \frac{\log \frac{r_1}{r_2}}{\log \frac{[B]_1}{[B]_2}} = \frac{\log \frac{2.92 \times 10^{-4} \text{ M} \cdot \text{s}^{-1}}{4.13 \times 10^{-4} \text{ M} \cdot \text{s}^{-1}}}{\log \frac{0.245 \text{ M}}{0.490 \text{ M}}} = \frac{-0.151}{-0.301} = 0.5 \quad r = k[B]^{0.5}$$

8.37 c. The reaction order can be obtained for a reaction by a balanced stoichiometric equation only by considering adequate experimental data. Such data would come from experiments where the concentration of reactants is varied one at a time and compared with the rate of disappearance of a reactant or appearance of a product.

There is possible confusion because the reaction written as a stoichiometric equation can derive from one or more steps called **elementary reactions**. The elementary reactions can be unimolecular, bimolecular, or termolecular. When an elementary reaction step is established, the reaction order of that particular elementary step follows from its equation.

8.38 a. From set (1) and set (2) of data, when the concentration of $[B]$ doubles, the rate quadruples. Therefore, the reaction is second order in $[B]$. From set (1) and set (3) of data, when the concentration of $[A]$ doubles, the rate doubles. Therefore, the reaction is first order in $[A]$.

$$R = k[A][B]^2$$

Overall order is $1 + 2 = 3$.

The specific rate constant, k, can be calculated from any set of data:

Set	Rate, mol³/L³ · s	[A], mol/L	[B], mol/L	k, s⁻¹
(1)	0.020	0.10	0.20	5.0
(2)	0.080	0.10	0.40	5.0
(3)	0.040	0.20	0.20	5.0
(4)	0.060	0.30	0.20	5.0

8.39 b. All the carbon in the carbon dioxide and all the hydrogen in the water come from the hydrocarbon.

$$\% \text{ C in } CO_2: \quad \frac{12.01 \text{ g C/mol}}{44.01 \text{ g } CO_2/\text{mol}} \times 100 = 27.29\% \text{ C}$$

$$\text{g C} = 7.260 \text{ g } CO_2 \times 0.2729 = 1.981 \text{ g C}$$

$$\% \text{ H in } H_2O: \quad \frac{2.016 \text{ g H/mol}}{18.016 \text{ g } H_2O/\text{mol}} \times 100 = 11.19\% \text{ H}$$

$$\text{g H} = 1.485 \text{ g } H_2O \times 0.1119 = 0.166 \text{ g H}$$

Calculate elemental percentages:

$$\% \text{ C} = \frac{1.981 \text{ g C}}{2.147 \text{ g compound}} \times 100 = 92.27\% \text{ C}$$

$$\% \text{ H} = \frac{0.166 \text{ g H}}{2.147 \text{ compound}} \times 100 = 7.73\% \text{ H}$$

Note: Since the compound is defined as a hydrocarbon, the percent of each element calculated independently should add to 100%.

8.40 c. Both compounds contain only carbon and hydrogen. Cyclopropane is a ring structure of three carbons singly bonded to each other and two hydrogens attached to each carbon. The formula is C_3H_6. The ratio of carbon to hydrogen is 1:2. Cyclohexane is also a ring structure with six carbons singly bonded to each other and two hydrogens attached to each carbon. The formula is C_6H_{12}. The ratio of carbon to hydrogen in cyclohexane is 1:2. Therefore, cyclohexane must have the same percentage of carbon as cyclopropane.

Check: 6 × 12.01 = 72.06 g C per moel cyclohexane; (84.16 m.w. cyclohexane)

12 × 1.008 = 12.096 g H per mole cyclohexane

$$\frac{72.1}{84.1} = 0.857 \times 100 = 85.7\% \text{ C in cyclohexane}$$

8.41 c. Only alcohols (R—OH) and ethers (R—O—R′) fit the general formula C_3H_8O. Three structures are possible:

the alcohols

$$\begin{array}{cc} CH_3CH_2CH_2 & \text{and} \quad CH_3CHCH_3 \\ | & | \\ OH & OH \end{array}$$

and the ether CH_3—O—CH_2CH_3

8.42

a. b. The compound is an alcohol, characterized by the OH group bonded to a carbon.

b. d. The compound is an aldehyde, characterized by the CHO group bonded to a carbon.

c. a. The compound is an ester, characterized by the COO in which the characteristic carbon is double bonded to an O atom and single bonded to another C, and an O atom. In the ester, the second O atom is also bonded to another C.

d. d. The compound is an amine, as the carbon chain ends in a carbon bond to the N of an NH_2 group.

e. b. The compound is a nitrile, characterized by the triple bond between the end C atom and an N atom.

f. c. The compound is a ketone, characterized by the C—O group in which the C is bonded to each of two other C atoms.

8.43 c. Ethane $C_2H_6 + \alpha(O_2 + 3.76N_2) = 2CO_2 + 3H_2O + 3.76\alpha N_2$

$$C: \quad 2 = 2$$
$$H: \quad 6 = 6$$
$$O: \quad 2\alpha = 4 + 3 \quad \alpha = 3.5$$

The equation at 150% theoretical air is:

$$C_2H_6 + (1.5)(3.5)(O_2 + 3.76N_2) = 2CO_2 + 3H_2O + .5(3.5)O_2 + 1.5(13.16)N_2$$

Air/fuel ratio at 150% air = $(5.25)(28.92)(4.76)/(2 \times 12 + 6)$
= 24.06 kg air/kg fuel

8.44 b. The air/fuel ratio at stoichiometric conditions:

$$C_2H_6 + 3.5(O_2 + 3.76N_2) = 2CO_2 + 3H_2O + 13.16N_2$$

Air/fuel ratio = $(3.5)(28.92)(4.76)/(2 \times 12 + 6) = 16$ kg air/kg fuel

CHAPTER 9

Material Science

OUTLINE

ATOMIC ORDER IN ENGINEERING MATERIALS 320
Atoms, Ions, and Electrons ■ Molecules ■ Crystallinity ■
Directions and Planes ■ Characteristics of Ordered Solids

ATOMIC DISORDER IN SOLIDS 329
Point Imperfections ■ Lineal Imperfections ■ Boundaries ■
Grain Boundaries ■ Solutions ■ Amorphous Solids

MICROSTRUCTURES OF SOLID MATERIALS 332
Atomic Movements in Materials ■ Structures of Single-Phase Solids ■
Phase Diagrams ■ Microstructures of Multiphase Solids

MATERIALS PROCESSING 339
Extraction and Compositional Adjustments ■ Shaping Processes ■
Deformation Processes ■ Annealing Processes ■ Time-Dependent
Processes ■ Surface Modification

MATERIALS IN SERVICE 345
Mechanical Failure ■ Thermal Failure ■ Radiation Damage ■
Chemical Alteration

COMPOSITES 353
Reinforced Materials ■ Mixture Rules ■ Directional Properties ■ Preparation

SELECTED SYMBOLS AND ABBREVIATIONS 357

REFERENCE 357

PROBLEMS 358

SOLUTIONS 369

All engineering products are made of materials. Thus, engineers become directly involved with materials, whether they be design engineers, production engineers, or applications engineers. Their familiarity with a wide spectrum of materials becomes particularly important as they advance through management and into administration, where they must oversee the activities of additional engineers on their technical staffs.

The way that an engineering product performs in service is a consequence of the combination of the components of the product. Thus, a cellular phone must have the diodes, resistors, capacitors, and other components that function together to meet its design requirements. Likewise, a competitively produced car must possess a carefully designed engine with its numerous parts, as well as safety features and operating characteristics that meet customer approval. Materials are pertinent to each and every design consideration.

Just as it is to be expected that the internal circuitry of a four-function hand calculator will differ from the internal circuitry of its multifunctional scientific counterpart, the internal structure of a steel gear differs from that of the sheet steel to be used in an automotive fender. Their roles, and therefore their properties, are designed to be different.

The variations in the internal structures of materials that lead to property differences include variations in atomic coordination and electronic energies, differences in internal geometries (microstructures), and the incorporation of larger structures, sometimes called macrostructures. Each of these is considered in the following sections, along with procedures for obtaining desired structures and properties.

ATOMIC ORDER IN ENGINEERING MATERIALS

Atoms, Ions, and Electrons

There is an order within atoms. Each atom has an integer number of protons. That number is called the **atomic number**. The natural elements possess, progressively, 1 to 92 of these protons, which carry a positive charge. A neutral atom has a number of electrons equal to the number of protons. Each electron is negative with a charge of 1.6×10^{-17} coulombs. Electrons are only allowed in given orbitals (called shells) that correspond to specific allowed energy levels.

With the exception of the principal isotope of hydrogen, each atom possesses neutrons. While these are charge-neutral, they add to the mass of each atom. The protons and neutrons reside in the nucleus of the atom. Figure 9.1 shows the Bohr model of a sodium atom. The lighter elements contain approximately equal numbers of protons and neutrons; however, in heavier elements the number of neutrons exceeds the number of protons. Furthermore, the number of neutrons per atom is not fixed. Thus, we encounter several **isotopes** for most atoms.

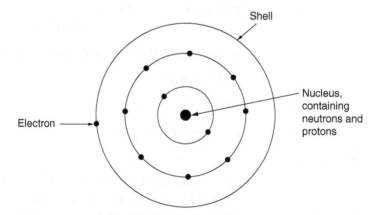

Figure 9.1 Bohr model of a sodium atom

Table 9.1 Data for selected atoms

Element	Protons Electrons Atomic No.	Neutrons (in natural isotopes)	Atomic Mass Unit	Grams per Avogadro's Number*
Hydrogen	1	0 or 1	1.008	1.008
Carbon	6	6 or 7	12.011	12.011
Oxygen	8	8, 9, or 10	15.995	15.995
Chlorine	17	18 or 20	35.453	35.453
Iron	26	28, 30, 31, or 32	55.847	55.847
Gold	79	118	196.97	196.97
Uranium	92	142, 143, or 146	238.03	238.03

*Avogadro's number = 6.022×10^{23}

Since the mass of an electron is appreciably less than 1% of that of protons and neutrons, the mass of an atom is directly related to the combined number of the latter two. By definition, an **atomic mass unit** (amu) is 1/12 of the mass of a carbon isotope that has six protons and six neutrons, C^{12}. The **atomic mass** of an element is equal to the number of these atomic mass units. (Selected values are listed in Table 9.1.) Thus while there are integer numbers of neutrons, protons, and electrons in each atom, the mass generally is not an integer, because more than one isotope is typically present.

A limited number of electrons may be accepted or released by an atom, thus introducing a charge on the atom (due the difference in the number of protons and electrons). A charged atom is an **ion**. Negative ions that have accepted extra electrons are called **anions**. Positive ions that have released electrons are called **cations**. Because they are charged, ions respond to electric fields. These fields may involve macroscopic dimensions (in electroplating baths); or they may involve interatomic distances (in molecules). Unlike charges attract; like charges repel.

Energy is required to remove an electron from a neutral atom. Figure 9.2 shows this schematically for a sodium atom. Conversely, fixed quantities of energy are released when electrons are captured by a positive ion. These energy levels (**states**) associated with an atom are fixed. Furthermore each state may accept only two electrons, and these must have opposite magnetic characteristics. Electronic, magnetic, and optical properties of materials must be interpreted accordingly.

Molecules

Atoms can join to one another; this is called **bonding**. Strong attractive forces can develop between atoms by three mechanisms; (1) coulombic attraction between oppositely charged ions, forming ionic bonds; (2) sharing of electrons to fill outer shells, creating covalent bonds; and (3) formation of ion cores surrounded by valence electrons that have been excited above the Fermi level and have become free electrons, forming metallic bonds. More detailed examples of these three primary types of bonds follow.

Ionic bonds form between metallic and nonmetallic atoms. The metallic atoms release their valence electrons to become cations (which are positively charged) and the nonmetallic atoms accept them to become anions. Ionic bonds are nondirectional. For example, a sodium ion that has lost an electron (Na^+) associates with as many negatively charged chlorine ions (Cl^-) as space will allow. And each

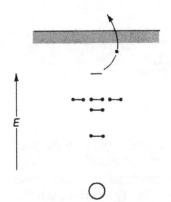

Figure 9.2 Ionization energy (schematic for sodium). Electrons reside at specific energy **states** (levels). Energy must be supplied to remove an electron from an atom, producing a positive ion. Two electrons (of opposite magnetic spins) may reside in each state.

Cl⁻ ion will become *coordinated* with as many Na⁺ ions as necessary to balance the charge. The resulting structure will continue to grow in three dimensions until all available ions are positioned. Energy is released with each added ion.

Covalent bonds form between atoms that share valence electrons in order to fill their outer shells. In the simplest case, two hydrogen atoms release energy as they combine to produce a hydrogen molecule:

$$2H \rightarrow H_2 \quad \text{or} \quad 2H \rightarrow H-H \tag{9.1}$$

The bond between the two involves a pair of shared electrons. This mechanism is common among many atomic pairs. In this case only one pair of atoms is involved. The covalent bonds of molecules are stereospecific; that is, they are between specific atoms and are therefore directional bonds.

In polymers a string or network of thousands or millions of atoms is bonded together. Examples include polyethylene, which has the structure shown in Figure 9.3(a), in which there is a backbone of carbon atoms that are covalently bonded; that is, they share pairs of valence electrons. Since each carbon atom has four valence electrons, it can form four covalent bonds, thus adding two hydrogens at the side of the chain in addition to the two bonds along the chain. Polyvinyl chloride, Figure 9.3(b), is related but has one of the four hydrogen atoms replaced by a chlorine atom.

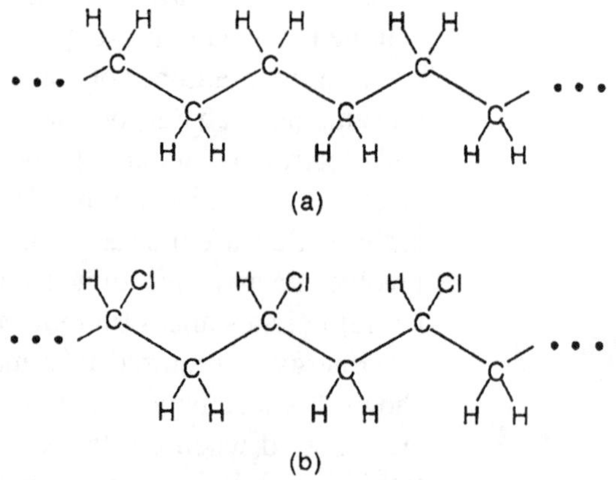

Figure 9.3 Covalent bonds: (a) polyethylene; (b) polyvinyl chloride

Metallic bonds form between the elements on the left side of the periodic table known as metals. Metallic bonds can form between atoms of the same element, to form a pure metal such as gold (Au), or between different metal atoms to form an alloy, such as brass, a mixture of copper (Cu) and zinc (Zn). The basis for the metallic bond is the formation of ion cores created when the metal's valence electrons are no longer associated with a specific atom. These electrons (called free electrons) move freely around surrounding metal ion cores, shielding them from each other. These freely moving electrons are often referred to as a *sea of electrons*. Metallic bonds are nondirectional.

The type and strength of bonding between atoms and molecules determines many properties of materials. As a general rule (and with other things being equal), the stronger the atomic bonding, the higher the melting temperature, hardness, and elastic modulus of materials. Ionically bonded materials are usually electrical and thermal insulators, while materials with metallic bonds have high electrical and thermal conductivities.

Crystallinity

The repetition of atomic coordinations in three dimensions produces a periodic structure that is called a **crystal**. The basic building block of a crystal is called a unit cell. Figure 9.4 illustrates the unit cell structure of iron. Each atom in this metal has eight nearest neighbors, which are symmetrically coordinated to give a **body-centered cubic** (BCC) crystal. About 30% of the metals have this structure. The atoms of aluminum and copper, among another 30% of the metals, become coordinated with 12 nearest neighbors with the result that **face-centered cubic** (FCC) crystal lattices are formed, as shown in Figure 9.5. A third group of metals form **hexagonal crystals** with **close packing** (HCP), as shown in Figure 9.6. As in the FCC crystals, each atom is coordinated with 12 nearest neighbors in HCP crystals.

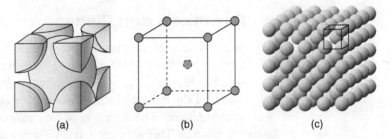

Figure 9.4 Body-centered cubic crystal structure: (a) a hard-sphere unit cell representation; (b) a reduced-sphere unit cell; (c) an aggregate of many atoms

It is possible to have a very high degree of perfection in crystals. For example, the repetition dimension (**lattice constant**) of the FCC lattice of pure copper is constant to the fifth significant figure (and to the sixth if the thermal expansion is factored in). This high degree of ordering provides a quantitative base for anticipating properties. Included are density calculations, certain thermal properties, and some of the effects of alloying.

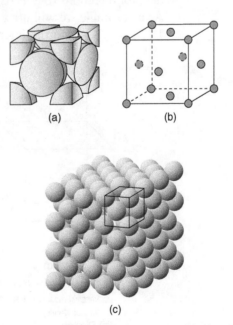

Figure 9.5 Face-centered cubic crystal structure: (a) a hard-sphere unit cell representation; (b) a reduced-sphere unit cell; (c) an aggregate of many atoms

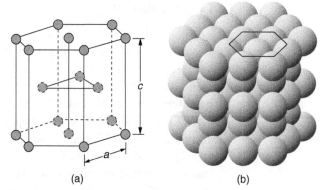

Figure 9.6 Hexagonal close-packed crystal structure: (a) a reduced-sphere unit cell (a and c represent the short and long edge lengths, respectively); (b) an aggregate of many atoms

Directions and Planes

Many properties are **anisotropic**; that is, they differ with direction and orientation. We can identify crystal directions by selecting any zero location (the origin) and determining the x, y, and z coordinates for any point along the direction ray. A corner of the unit cell is often used as the origin. The unit length is the edge length of the unit cell. The direction of easy magnetization in iron is parallel to one of the crystal axes. This is labeled the [100] direction, because that direction passes from the origin through a point that is one unit along the x-axis and zero units along the other two axes. Figure 9.7(a) shows a ray in the [120] direction and a ray in the $[1\bar{1}0]$ direction, where the overbar indicates a negative direction. Parallel directions carry the same label. We use square brackets, [ ], for closures for direction rays.

In cubic crystals, each of the four directions that are diagonal through the cube are identical (because the three axes are identical). We label these four directions as a *family* with pointed arrows for closures, <111>.

Figure 9.7(b) identifies the three shaded planes as (001), (210), and (111). Here the labeling procedure (known as indexing) is somewhat more complicated than for directions (but leads to simplified mathematics for complex calculations). As an example we will draw the (210) plane. We first invert the three indices, 1/**2**, 1/**1**, 1/**0**. These are the intercept dimensions of the plane across the three axes;

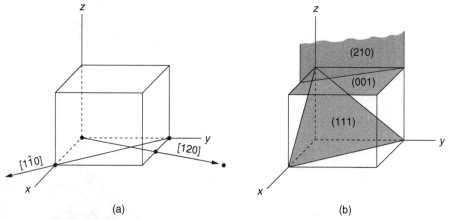

Figure 9.7 Crystal notation. (a) Directions, [120] and $[1\bar{1}0]$. The x, y, and z coordinates are used for crystal directions. Square brackets are used as closures. Negative coordinates are indicated with an overbar. (b) Planes, (001), (210), and (111). The reciprocals of the axial intercepts are used for crystal planes. Parentheses are used as closures. Any point may be selected as an arbitrary origin.

specifically, 0.5 on the x-axis, 1.0 on the y-axis, and infinity along the z-axis. An adjacent parallel plane with intercepts of 1, 2, and ∞ carries the same (210) index. We use parentheses, ( ), as closures for the indices of individual planes, and braces, { }, as closures for a family of comparable planes. To index an unknown plane, the procedure is reversed: (1) Choose an origin that the plane does not pass through. (2) Determine the intercepts of the plane on the x, y, and z axes. (3) Take reciprocals of these intercepts. (4) Clear fractions and enclose in parentheses ( ).

Characteristics of Ordered Solids

There are several useful properties of unit cells that can be determined through geometric relations and 3-D visualization. These include:

- Number of atoms per unit cell
- Number of nearest-neighbor atoms (coordination number)
- Lattice parameter (spacing of atoms)
- Distance of nearest approach of atoms
- Atomic packing factor (the volume of atoms per unit volume of the solid)
- Density

Some of these relationships are given in Table 9.2 and in the NCEES *Fundamentals of Engineering Supplied-Reference Handbook* and therefore do not need to be memorized, but it is useful to see how these are determined. The following examples will illustrate these relationships.

Table 9.2 Characteristics of selected crystal structures

Unit Cell	Number of Atoms per Unit Cell	Coordination Number	Lattice Parameter	Packing Factor
BCC	2	8	$a = 4R/\sqrt{3}$	0.68
FCC	4	12	$a = 2R/\sqrt{2}$	0.72
HCP		12		0.72

Figure 9.4 showed a unit cell of α iron with an atom at its center. Inasmuch as each of the eight corner atoms is shared by the eight adjacent unit cells, we can note that there are (1 + 8/8 = 2) atoms per unit cell. From Table 9.1, each iron atom has a mass of 55.85 g per 0.6022×10^{24} atoms. Since iron forms a body-centered cubic crystal, its unit cell has a mass of 1.855×10^{-26} g. X-ray diffraction techniques give a lattice constant value of 2.866×10^{-18} m. As a result, the mass per unit volume, that is, the **density**, may be calculated to be nearly 7.88 g/cm³. Careful density measurements give a value of slightly more than 7.87 g/cm³ at ambient temperatures. The close agreement for the simple property of density implies that the concept of the crystal structure is valid.

The relationship between the size of the atoms (atomic radius, r) and lattice parameter (a) for BCC crystals is demonstrated in Figure 9.8. Note that the atoms touch along the body diagonal. By inspection this gives a known length of the body diagonal of $4r$.

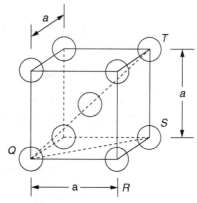

Figure 9.8 BCC unit cell

Using the triangle QRS

$$\|\overline{QS}\|^2 = a^2 + a^2 = 2a^2$$

and for triangle QST

$$\|\overline{QT}\|^2 = \|\overline{TS}\|^2 + \|\overline{QS}\|^2$$

But $\overline{QT} = 4r$, r being the atomic radius. Also, $\overline{TS} = a$. Therefore,

$$(4r)^2 = a^2 + 2a^2$$

or

$$a = \frac{4r}{\sqrt{3}}$$

The relationship between r and a for FCC crystals can be similarly determined by noting that the atoms touch along the face diagonal as shown in Figure 9.9.

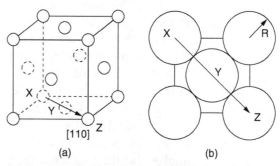

Figure 9.9 (a) Reduced-sphere FCC unit cell with the [110] direction indicated. (b) The bottom face-plane of the FCC unit cell in (a) on which is shown the atomic spacing in the [110] direction, through atoms labeled X, Y, and Z.

The plastic deformation of these solids is also related to crystal structure. To illustrate, in Figure 9.10 the {111} plane of an FCC structure is shown. The {111} planes of aluminum and other FCC metals are the most densely packed planes; each atom of those planes is surrounded by six other atoms in the same plane. There is no arrangement where more atoms could have been included. This is not true for other planes within the FCC crystal. Since there is a fixed number of atoms per unit volume, it is apparent that the interplanar spacings between parallel (111) planes through the centers of atoms must be greater than between planes of other orientations. It is thus not surprising that sliding (**slip**) occurs there at lower stresses than on other planes.

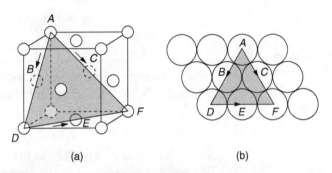

Figure 9.10 (a) A {111}<110> slip system shown within an FCC unit cell. (b) The (111) plane from (a) and three <110> slip directions (as indicated by arrows) within that plane comprise possible slip systems.

Not only is the required shear stress for slip low on the {111} planes, the <110> directions on those planes require less stress for slip than other directions. This is because the step distance between like crystal positions is the shortest, specifically, $2r$ where r is the atomic radius. We speak of the <110>{111} **slip system** for FCC metals. In a BCC metal, <111> {110} is the prominent slip system. Note especially that the {111} planes and the <110> directions are operative in FCC metals, whereas it is the {110} planes and the <111> directions in BCC metals. Hexagonal metals such as Mg, Zn, and Ti do not deform as readily as BCC and FCC metals because there are fewer combinations of directions and planes for slip by shear stresses.

Example 9.1

Carbon (12.011 amu) contains C^{12} and C^{13} isotopes with masses of 12.00000 amu and 13.00335 amu, respectively. What are the percentages of each?

Solution

$$12.011 = x(12.00000) + (1-x)(13.00335)$$
$$1.00335\,x = 0.99335$$
$$x = 98.9\%$$

Carbon is 98.9% C^{12} and 1.1% C^{13}.

Example 9.2

Aluminum has a face-centered cubic unit cell, that is, an atom at each corner of the unit cell and an atom at the center of each face (see Figure 9.5). The Al–Al distance ($= 2r$) is 0.2863 nm. Calculate the density of aluminum. (The mass of an aluminum atom is 26.98 amu.)

Solution

$$\text{Volume} = [2(0.2863 \times 10^{-9} \text{ m})/\sqrt{2}]^3 = 6.638 \times 10^{-29} \text{ m}^3$$

$$\text{Mass} = (8/8 + 6/2 \text{ atoms})(26.98 \text{ g}/6.022 \times 10^{23} \text{ atoms}) = 1.792 \times 10^{-22} \text{ g}$$

$$\text{Density} = (1.792 \times 10^{-22} \text{ g})/(6.638 \times 10^{-29} \text{ m}^3) = 2.700 \times 10^6 \text{ g/m}^3 = 2.700 \text{ g/cm}^3$$

$$\text{Actual density} = 2.699 \text{ g/cm}^3$$

Example 9.3

What is the repeat distance along a <211> direction of a copper crystal that is face-centered cubic and has a unit cell dimension (lattice constant) of 0.3615 nm?

Solution

Select the center of any atom as the origin. Make a sketch of a cubic unit cell with that origin arbitrarily set at the lower left rear corner, as shown in Exhibit 1. One of the <211> directions, with coefficients of 2, 1, and 1, exits the first unit cell through the center of its front face, where another atom is centered (with no other intervening atoms).

$$d^2 = a^2 + \left(\frac{a}{2}\right)^2 + \left(\frac{a}{2}\right)^2 = \frac{6}{4}(0.3615 \text{ nm})^2; \; d = 0.4427 \text{ nm}$$

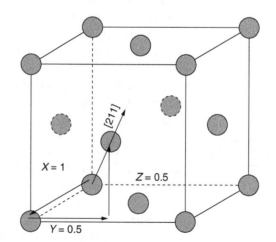

Exhibit 1 FCC unit cell

Example 9.4

Assuming spherical atoms, calculate the packing factor of a BCC metal.

Solution

The packing factor is the volume of atoms per unit volume of the solid. Based on Figure 9.4, there are (8/8 + 1) atoms/unit cell.

$$\text{Volume of 2 atoms in the BCC unit cell} = 2 \times (4\pi/3)r^3 = 8.38r^3$$

Since the cube diagonal is $4r$,

$$\text{Volume of unit cell} = a^3 = \left(\frac{4r}{\sqrt{3}}\right)^3 = 12.32r^3$$

$$\text{Packing factor} = \frac{8.38r^3}{12.32r^3} = 68\%$$

Example 9.5

How many atoms are there per mm² on one of the {110} planes of copper (FCC)?

Solution

From Example 9.3, $a_{Cu} = 0.3615$ nm. A {110} plane lies diagonally through the unit cell and is parallel to one of the axes. There are (4/4 + 2/2 atoms) in an area measuring a by $a\sqrt{2}$. The number of atoms is 2 atoms/[$(0.3615 \times 10^{-6}$ mm$)^2 \sqrt{2}$] = 10.8×10^{12}/mm2.

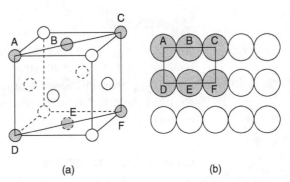

Exhibit 2 (a) Reduced-sphere FCC unit cell with (110) plane. (b) Atomic packing of an FCC (110) plane. Corresponding atom positions from (a) are indicated.

ATOMIC DISORDER IN SOLIDS

In the previous section, we paid attention to the orderly combinations that can exist in engineering materials. A variety of properties and behaviors are closely related to that ordering. Examples that were cited included density, slip systems, and molecular melting.

However, no solid has perfect order. There are always imperfections present, and these may be highly significant. A few missing potassium atoms in a compound such as KBr do not detectably affect the density; however, their absence introduces color. Likewise, the absence of a partial plane of atoms in a metal significantly modifies the shear stress required by a slip system for plastic deformation.

Also, a rubber is vulcanized by the joining (**crosslinking**) of adjacent molecules with only a minor compositional change (sulfur addition). As a final example, the thermal conductivity is doubled in diamond if the 1% of naturally present C^{13} has been removed.

Crystal defects can be characterized as point, line (one-dimensional), plane (two-dimensional), or volume (three-dimensional) defects.

Point Imperfections

Imperfections may be atomically local in nature. Missing atoms (**vacancies**), extra atoms (**interstitials**), **displaced** atoms, and impurity atoms are called **point imperfections** (Figure 9.11). Their existence facilitates the transport of atoms (**diffusion**), thus becoming important in materials processing. In service, the presence of point imperfections scatters internal waves and thus reduces energy transport. These include elastic waves for thermal conductivity, light waves for optical transparency, and electron waves for electrical conduction.

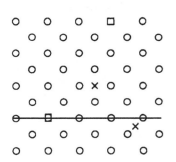

○ Atom
□ Vacancy
× Interstitialcy

Figure 9.11 Point imperfections. These defects can originate from imperfect crystallization, or through a relocation of energized atoms.

Lineal Imperfections

The principal defect of this type is a **dislocation**. It is most readily visualized (Figure 9.12) as a partial displacement along a slip system or an extra half-plane of atoms. Dislocations facilitate plastic deformation by slip; however, increased numbers of dislocations lead to dislocation tangles or *traffic snarls* and therefore to interference of slip. Thus, ductility decreases and strength increases. Dislocations may develop during initial crystal growth as well as from plastic deformation.

Boundaries

No liquid or solid is infinite; each has a **surface**. The resulting two-dimensional boundary has a different structure and bonding than that encountered in the underlying material. Since atoms are absent from one side of the boundary, the atoms at these surfaces possess additional energy, and subsurface distortions are introduced.

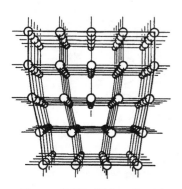

Figure 9.12 Dislocation, ⊥ (schematic). These lineal imperfections facilitate slip within crystals. Excessive numbers of defects, however, lead to their entanglement and a resistance to deformation. The resulting increase in strength is called **work hardening**.

Grain Boundaries

Boundaries also occur where two growing crystals meet. These are called grain boundaries. There are atoms on each side of the boundary; however, any misorientation between the two crystal grains leads to local inefficiencies in atomic packing. As a result some atom-to-atom distances are compressed; others are stretched. Both distortions increase the energy of the atoms along the grain boundaries. This **grain-boundary energy** introduces reactive sites for structural modification during processing and in service. The imperfect grain boundary is also an *avenue* for atomic diffusion within solid materials (Figure 9.13). Further, the mismatch at a grain boundary blocks slip that might otherwise occur, particularly at ambient temperatures where atom-by-atom mobility is limited.

Solutions

Both sugar and salt dissolve in water, each producing a solution (commonly called a syrup and a brine, respectively). There are many familiar solutions. Lower melting temperatures, increased conductivity, and altered viscosities commonly result.

Solid Solutions

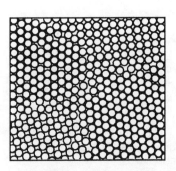

Figure 9.13 Grain boundaries (schematic). Most materials contain a multitude of grains, each of which is a separate crystal. The boundary between grains is a zone of mismatch. Atoms along grain boundaries possess added +energy because they are not as efficiently coordinated with their neighbors.

Impurities, both unwanted and intentional, may also dissolve into a solid. A crystal cannot be perfect when foreign atoms are present. Common brass (70Cu–30Zn) is a familiar example. Zinc atoms simply substitute for copper atoms to produce a **substitutional solid solution**. It has the face-centered cubic crystal structure of pure copper; however, with approximately one-third of the copper atoms replaced by zinc atoms, we can anticipate certain changes. First, the size of the unit cell is increased because zinc atoms are approximately 4% larger than copper atoms. (Fifteen percent mismatch is the practical limit for extensive substitutional solid solution.) Second, charge transport is greatly reduced because the electrons are scattered as they travel toward the positive electrode through locally varying electrical fields. Thus, the electrical and thermal **conductivities** are decreased. Also, atoms of a different size immobilize dislocation movements and in turn produce **solution hardening**.

There are some important situations in which small atoms can be positioned among larger atoms. The most widely encountered example is the solution of carbon in face-centered cubic iron at elevated temperatures. The result is an **interstitial solid solution**. Iron changes to body-centered cubic at ambient temperatures. The BCC iron cannot accommodate many carbon atoms in its interstices. This loss of interstitial solubility plays a major role in the various heat treatments of steel.

Amorphous Solids

Materials lose their crystallinity when they melt. The long-range order of the crystals is not maintained. As a liquid, the material is amorphous (literally, without form). Those materials that are closely packed—for example, metals—will expand on melting [Figure 9.14(a)]. Being thermally agitated, the atoms do not maintain close coordination with neighboring atoms (or molecules). A few materials, such as ice, which has stereospecific bonds, lose volume on melting and become more dense [Figure 9.14(b)].

In certain cases, crystallization may be avoided during cooling. An amorphous solid can result. Two common examples are window glass and the candy part of peanut brittle. In neither case is there time for the relatively complex crystalline structures to form. Very rapid cooling rates are required to avoid crystallization when the crystalline structures are less complex. For example, metals must be quenched a thousand degrees in milliseconds to avoid crystallization. The amorphous materials that result are considered to be **vitreous**, or glass-like.

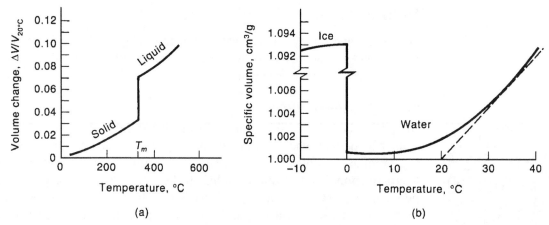

Figure 9.14 Volume changes during heating. (a) Lead (FCC); (b) H2O. Melting destroys the efficient packing of metallic atoms within solids, so most metals expand when melted. The crystalline structure of ice, silicon, and a number of related materials with stereospecific bonds have low, inefficient atomic packing within solids. Therefore, they lose volume when melted.

Example 9.6

Sterling silver contains 92.5% silver and 7.5% copper by weight. What percentage of the atoms on a (111) plane are silver? Copper?

Solution

The alloy is a random solid solution; therefore the percentage of atoms on the (111) plane or any other plane will be the same as throughout the alloy. Change weight percent to atom percent. The atomic masses are 107.87 and 63.54 amu, respectively.

$$\text{Basis: } 1000 \text{ amu} = 925 \text{ amu Ag} + 75 \text{ amu Cu}$$

$$\text{Ag: } 925 \text{ amu}/(107.87 \text{ amu/Ag}) = 8.58 \text{ Ag atoms}$$

$$\text{Cu: } 75 \text{ amu}/(63.54 \text{ amu/Cu}) = \underline{1.18} \text{ Cu atoms}$$

$$\text{Total: } = 9.76 \text{ atoms}$$

$$\text{Ag atoms} = 8.58/9.76 = 87.9\%; \text{ Cu atoms} = 12.1\%$$

MICROSTRUCTURES OF SOLID MATERIALS

The atomic coordination within solids is on the nanometer scale. The resulting structures involve either crystalline solids, or amorphous solids such as the glasses. As discussed, certain properties arise from these atom-to-atom relationships. Other properties arise from longer-range structures, generally with micrometer to millimeter dimensions, called **microstructures**.

Atomic Movements in Materials

Our initial examination of crystals implied that an atom becomes permanently coordinated with adjacent atoms and remains fixed in position. This is not entirely true. In the first place, there is thermal vibration of the atoms within the crystal. Thus, while the lattice constant and the mean interatomic distances are fixed to several significant figures, the instantaneous interatomic distances vary. The amplitude of vibration increases with temperature. At the melting temperature, the crystal is literally *shaken apart*. As the melting point is approached, a measurable fraction of atoms jump out of their crystalline positions. They may return, or they may move to other sites, producing the vacancies and interstitials discussed in the previous section.

Diffusion

Within a single-component material, such as pure copper, there is equal probability that like numbers of copper atoms will jump in each of the coordinate directions. Thus, there is no net change.

Imagine, however, one location, x_1, in nickel containing 2000 atoms of copper for every mm^3 of nickel, whereas 1 mm to the right at x_2 there are 1000 atoms of copper for every mm^3. Although all copper atoms have the same probability for jumping in either direction, there is a net movement of copper atoms to the right simply because there are unequal numbers of copper atoms in the two locations. There is a copper **concentration gradient**, $\Delta C/\Delta x$. In this case

$$\Delta C/\Delta x = (C_2 - C_1 \text{ Cu/mm}^3)/(x^2 - x^1 \text{ mm}) = -(1000 \text{ Cu/mm}^3)/\text{mm} \quad (9.2)$$

The rate of diffusion, called the **flux**, J, is proportional to the concentration gradient

$$J = -D\frac{dC}{dx} \quad (9.3)$$

where D is the *diffusivity*, also called the **diffusion coefficient**. (Its units are m^{2}s^{-1} since the units for flux and concentration gradient are m^{-2}s^{-1} and m^{-4}, respectively.)

The diffusion coefficient can be calculated by

$$D = D_o e^{-Q/(RT)}$$

where D_o (with units of m^{2}s^{-1}) is the proportionality constant, Q (with units of J(mole)$^{-1}$) is the activation energy, R is the gas constant, and T is absolute temperature.

Among the various factors that affect the diffusivity are (1) the size of the diffusing atom, (2) the crystal structure of the matrix, (3) bond strength, and (4) temperature. Comparisons are made in Table 9.3. The diffusivity of the C in Fe$_{FCC}$ is higher than for the Fe in Fe$_{FCC}$ because the diffusing carbon atom is smaller than the iron atom. The diffusivity of the C in Fe$_{BCC}$ is higher than for the C in Fe$_{FCC}$ because the latter contains more iron atoms per m^3. The FCC packing factor is higher than the BCC form. The diffusivity for Fe in Fe$_{FCC}$ is lower than for Cu in Cu$_{FCC}$ because the iron atoms are more strongly bonded than the copper atoms. (The melting temperatures provide evidence of this.) In each case, higher diffusion coefficients accompany higher temperatures.

The process engineer obtains many of the properties required of a material by heat-treating procedures. Diffusion plays the predominant role in achieving the required microstructures.

Table 9.3 Selected diffusion coefficients

Diffusing Atom	Host Structure	Diffusion Coefficient, D, m^2/s	
		500°C (930°F)	1000°C (1830°F)
Fe	FCC Fe	2×10^{-23}	2×10^{-16}
C	FCC Fe	5×10^{-15}	3×10^{-11}
C	BCC Fe	1×10^{-12}	2×10^{-9}
Cu	FCC Cu	1×10^{-18}	2×10^{-13}

Structures of Single-Phase Solids

Many materials possess only one structure. Examples include copper wire, transparent polystyrene cups, and Al_2O_3 substrates for electronic circuits. The wire contains only face-centered-cubic crystals. The polystyrene is an amorphous solid with minimal crystallinity. The substrates have numerous crystals, all with the same crystalline structure. We speak of single phases because none of these materials contains a second structure.

Grains

Each of the individual crystals in a copper wire is called a **grain**. Recall from the previous section that adjacent crystals may be misoriented with respect to each other, and that there is a boundary between them. This is shown schematically in Figure 9.15.

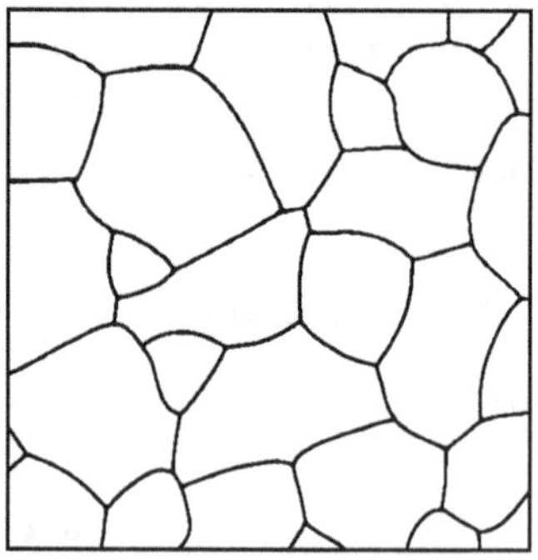

The **grain size** is an important structural parameter, because the grain boundary area varies inversely with the grain size. Diffusion is faster along grain boundaries because there is less-perfect packing of the atoms and, consequently, a more open structure. At ambient temperatures, grain boundaries interfere with plastic deformation, thus increasing the strength. At elevated temperatures the grain boundaries contribute to creep and therefore are to be minimized. Grain boundaries also serve as locations that initiate structural changes within a solid.

Grain growth may occur in a single-phase material. The driving force is the fact that the atoms at the grain boundary possess extra energy. Grain growth reduces this excess energy by minimizing the boundary area. Higher temperatures increase the rate of grain growth because the atoms migrate faster. However, since the growth rate is inversely related to grain size, we see a decrease in the rate of growth with time.

The texture of a single-phase solid can also depend on **grain shape** and **orientation**. Even a noncrystalline material may possess a structure. For example, the molecular chains within a nylon fiber have been aligned during processing to provide greater tensile strength.

Figure 9.15 Grains (schematic). Each grain is a separate crystal. There is a surface of mismatch between grains because adjacent grains have unlike orientations. The **grain boundary area** is inversely related to grain size.

Phase Diagrams

Many materials possess more than one phase. A simple and obvious example is a cup containing both ice and water. While ice and water have the same composition, the structures of the two phases are different. There is a **phase boundary** between the two phases. A less obvious, but equally important material is the steel used as a bridge beam. The steel in the beam contains two phases: nearly pure iron (body-centered cubic), and an iron carbide, Fe_3C.

In these two examples, we have a mixture of two phases. Solutions are phases with more than one component. In the previous section, we encountered brass, a solid solution with copper and zinc as its components.

Although the steel beam just cited contains a mixture of two phases at ambient temperatures, it contained only one phase when it was red-hot during the shaping process. At that temperature, the iron was face-centered-cubic and was thereby able to dissolve all of the carbon into its interstices. No Fe_3C remained. The phases within a material can be displayed on a **phase diagram** as a function of temperature and composition.

Phase diagrams are useful to the engineer because they indicate the temperature and composition requirements for attaining the required internal structures and accompanying service properties. The phase diagram shows us (1) *what* phases to expect, (2) the *composition* of each phase that is present, and (3) the *quantity* of each phase within a mixture of phases.

What Phases?

Sterling silver contains 92.5Ag–7.5Cu (weight percent). Using the Ag–Cu phase diagram of Figure 9.16, we observe that this composition is liquid above 910°C; below 740°C, it contains a mixture of the α and β solid structures. The former is a solid solution of silver plus a limited amount of copper; the latter is a solid solution of copper plus a limited amount of silver. Between 740 and 810°C, only one phase is present, α. In that temperature range, all of the copper can be dissolved in the α solid solution. From 810 to 910°C, the alloy changes from no liquid to all liquid.

Phase Compositions?

Pure silver has a face-centered-cubic structure. Copper forms the same crystalline structure. Not surprisingly, copper atoms can be substituted for silver. But the solid solution in α is limited, because the silver atom is 13% larger in diameter than is copper. However, the solubility increases with temperature to 8.8 weight-percent copper at 780°C. This is shown by the boundary of the silver-rich shaded area of Figure 9.16. Within that shaded area, we have a single phase called α.

Conversely, the FCC structure of copper can dissolve silver atoms in solid solution. At room temperature the solubility is very small. Again, as the temperature is raised the solubility limit increases to 8 weight-percent silver (92 wt. % Cu) at 780°C. This copper-rich phase is called β.

Silver and copper are mutually soluble in a liquid solution. Above 1100°C there is no limit to the solubility. As the temperature is reduced below the melting point of copper (1084°C), the copper solubility limit decreases from 100% to 28%. Excess copper produces the β phase. Likewise, below the melting point of silver (962°C), the silver solubility limit decreases from 100% to 72%. Excess silver produces the α phase. The two solubility curves cross at approximately 72Ag–28Cu and 780°C. We call this low-melting liquid **eutectic**.

How Much of Each Phase in a Mixture?

At 800°C, a 72Ag–28Cu alloy is entirely liquid. At the same temperature, but with added copper, the solubility limit is reached at 32% Cu. Beyond that limit, still at 800°C, any additional copper precipitates as β. Halfway across the two-phase field of (L + β), there will be equal quantities of the two phases. Additional copper in the alloy increases the amount of β. Within this two-phase region the liquid remains saturated with 32% Cu, and the solid β is saturated with 8% Ag (thus, 92% Cu).

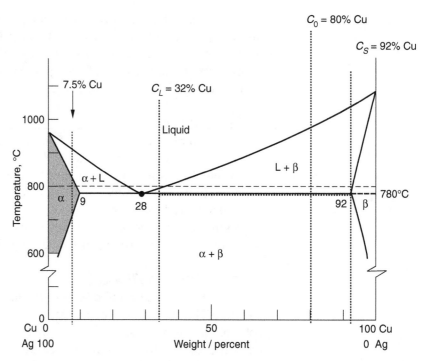

Figure 9.16 Silver-copper phase diagram. Sterling silver contains 7.5% copper; therefore, it has a single phase, α, in the 740–810°C range.

The *composition* of each phase is dictated by the solubility limits. (In a one-phase field, the solubility limits are not factors; the alloy composition and the phase composition are identical.) The *amount* of each phase in a two-phase field is determined by interpolation between the solubility limits using what is called the lever law. For example, at 800°C, an alloy consisting of 80% Cu and 20% Ag would consist of a liquid phase containing 32% Cu and a solid phase of 92% Cu. The weight fraction of liquid phase, W_L, and solid phase, W_s, is determined by $W_L + W_S = 1$.

$$W_L = \frac{C_S - C_O}{C_S - C_L} \quad \text{and} \quad W_S = \frac{C_O - C_L}{C_S - C_L}$$

Reaction Rates in Solids

A phase diagram is normally an equilibrium diagram; that is, all reactions have been completed. The time required to reach equilibrium generally increases with decreasing temperature. Thus, a material may not always possess the expected phases with predicted amounts or compositions. Even so, the phase diagram is valuable. For example, sterling silver (92.5Ag–7.5Cu) contains only one phase when equilibrated at 775°C (Figure 9.16). Slow cooling to room temperature precipitates β as a minor second phase, as would be expected when plenty of time is available. Rapid cooling, however, traps the copper atoms within the α solid solution. This situation is used to advantage, because the solid solution is stronger than the (α + β) combination. Also, a single-phase alloy corrodes less readily. This explains why the *impure* sterling silver is commonly preferred over pure silver.

The selection of compositions and processing treatments is generally based on a knowledge of equilibrium diagrams plus a knowledge of how equilibrium is circumvented.

Microstructures of Multiphase Solids

The microstructure of a single-phase, crystalline solid includes the *size*, *shape*, and *orientation* of the grains. Variations in these properties are also found in multiphase solids. In addition, the microstructure of a multiphase solid may also vary in the *amount* of each phase and the *distribution* of the phases. In an equilibrated microstructure, the amounts of the phases may be predicted directly from the phase diagrams using the lever law. From Figure 9.17, a 1080 steel (primarily iron, with 0.80 percent carbon) will have twice as much Fe_3C (W_{Fe3C}) at room temperature as a 1040 steel (with 0.40 percent carbon). The Fe_3C is a hard phase. Therefore, with all other factors equal, we expect a 1080 steel to be harder than a 1040 steel, and it is.

$$W_{Fe_3C} = \frac{C_0 - C_\alpha}{C_{Fe_3C} - C_\alpha}$$

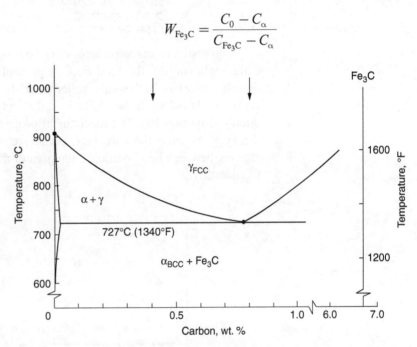

Figure 9.17 Portion of the Fe–Fe_3C phase diagram. Most steels are heat treated by initially forming austenite, γ, which is Fe_{FCC}. It changes to a mixture of ferrite, α, and carbide at lower temperatures.

The distribution of phases within microstructures is more difficult to quantify, so only descriptive examples will be cited. Similar to sterling silver, aluminum will dissolve several percent of copper in solid solution at 550°C. If it is cooled slowly, the copper precipitates as a minor, hard, brittle compound ($CuAl_2$) along grain boundaries of the aluminum. The alloy is weak and brittle and has little practical use. If the same alloy is cooled rapidly from 550°C, trapping the copper atoms within the aluminum grains as shown in Figure 9.18(a), the quenched solid solution is stronger and more ductile and has commercial uses. If the quenched alloy is reheated to 100°C, the $CuAl_2$ precipitates, as expected from the phase diagram shown in Figure 9.18(b) and (c). In this case, however, the precipitate is very finely dispersed within the grains of aluminum. The alloy retains its toughness, because the brittle $CuAl_2$ does not form a network for fracture paths. In addition, the strength is increased because the submicroscopic hard particles interfere with the deformation of the aluminum along slip planes. Alloys of this type are used in airplane construction because they are light, strong, and tough. Observe that all of the examples in this paragraph are for the same alloy. The properties have been varied through microstructural control.

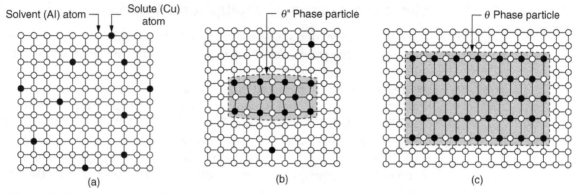

Figure 9.18 Schematic depiction of several stages in the formation of the equilibrium precipitate (q) phase. (a) A supersaturated a solid solution; (b) a transition, θ'', precipitate phase; (c) the equilibrium θ' phase, within the a-matrix phase. Actual phase particle sizes are much larger than shown here.

Two distinct microstructures may be produced in a majority of steels. In one, called **spheroidite**, the hard Fe_3C is present as rounded particles in a matrix of ductile ferrite (α). In the other, called **pearlite**, Fe_3C and ferrite form fine alternating layers, or lamellae. Spheroidite is softer but tougher; pearlite is harder and less ductile (Figure 9.19). The mechanical properties are controlled by the spacing of the Fe_3C, because the hard Fe_3C phase stops dislocation motion. The closer the spacing between Fe_3C particles, the greater the strength or hardness, but the lower the ductility.

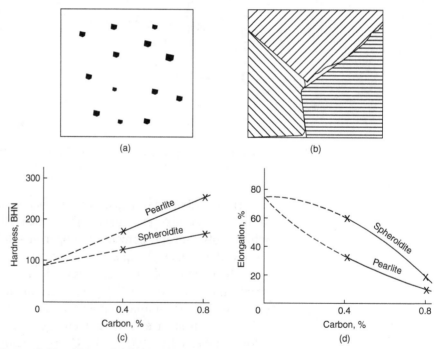

Figure 9.19 Phase distributions (schematic). The two samples are of the same steel, both containing ferrite (white) and carbide (black) but heat treated differently. Spheroidite is shown in (a), and pearlite in (b). The hardness of the two structures is shown in (c), and ductility is shown in (d).

Example 9.7

At 780°C, an Ag–Cu liquid of eutectic composition (Figure 9.16) solidifies to solid α and solid β. (i) What are the compositions of the two solid phases just below the eutectic temperature? (ii) How much of each of these two phases exists per 100 grams of alloy?

Solution

(i) From Figure 9.16, the α phase is 91Ag–9Cu, and the β phase is 8Ag–92Cu.

(ii) Interpolation along the 780°C tie-line between the solubility limits yields

$$\beta: \text{(using Cu) } [(28 - 9)/(92 - 9)](100 \text{ g}) = 23 \text{ g}$$

$$\alpha: \text{(using Cu) } [(92 - 28)/(92 - 9)](100 \text{ g}) = 77 \text{ g}$$

or

$$\alpha: \text{(using Ag) } [(72 - 8)/(92 - 9)](100 \text{ g}) = 77 \text{ g}$$

Example 9.8

The phases and microstructures of an SAE 1040 steel may be related to the Fe-Fe_3C phase diagram (Figure 9.17). (i) Assume equilibrium. What are the phases and their weight percents at 728°C? At 726°C? (ii) Pearlite has the eutectoid composition of 0.77% C. What percent of the steel will *not* be contained in the pearlite?

Solution

(i) At 728°:

$$\alpha\ (0.02\% \text{ C}): (0.77 - 0.40)/(0.77 - 0.02) = 49\%\ \alpha$$

$$\gamma\ (0.77\% \text{ C}): (0.40 - 0.02)/(0.77 - 0.02) = 51\%\ \gamma$$

$$Fe_3C \text{ contains } 12/[12 + 3(55.85)] = 6.7\% \text{ C}$$

At 726°:

$$\alpha\ (0.02\% \text{ C}): (6.7 - 0.40)/(6.7 - 0.02) = 94\%\ \alpha$$

$$Fe_3C: (0.40 - 0.02)/(6.7 - 0.02) = 6\%\ Fe_3C$$

(ii) All of the pearlite, $\alpha + Fe_3C$, comes from the γ. This FCC phase changes to pearlite at 727°C. The α that was present at 728°C remains unchanged. Therefore, the answer is $P = 51\%$; and unchanged $\alpha = 49\%$.

MATERIALS PROCESSING

For many engineers, materials are first encountered in terms of handbook data or stockroom inventories. However, materials always have a prior history. They must be obtained from natural sources, then subjected to compositional modifications and complex shaping processes. Finally, specific microstructures are developed to achieve the properties that are necessary for extended service.

Extraction and Compositional Adjustments

Few materials are used in their natural form. Wood is cut and reshaped into lumber, chipboard, or plywood; most metals must be extracted from their ores; rubber latex is useless unless it is vulcanized.

Extraction from Ores

Most ores are oxides or sulfides that require chemical reduction. Commonly, oxide ores are chemically reduced with carbon- or CO-containing gas. For example,

$$Fe_2O_3 + 3\ CO \rightarrow 2\ Fe + 3\ CO_2 \quad (9.4)$$

Elevated temperatures are used to speed up the reactions and more completely reduce the metal. If the metal is melted, it is more readily separated from the accompanying gangue materials. The reduced product is a carbon-saturated metallic iron. As such, it has only limited applications. Normally, further processing is required.

Refining

Dissolved impurities must be removed. Even if the above ore were of the highest quality iron oxide, it would be necessary to refine it, because the metallic product is saturated with carbon. In practice, small but undesirable quantities of silicon and other species are also reduced and dissolved in the iron. They are removed by closely controlled **reoxidation** at chemically appropriate temperatures (followed by **deoxidation**). The product is a **steel**. Alloying additions are made as specified to create different types of steel.

Chemicals from which a variety of plastics are produced are refined from petroleum. The principal step of petroleum refinement involves selective distillation of liquid petroleum. Lightweight fractions are removed first. Controlled temperatures and pressures distill the molecular fractions that serve as precursors for polymers. Residual fractions are directed to other products.

Polymerization makes macromolecules out of the smaller molecules that are the product of distillation. **Addition polymerization** involving a C=C→C—C reaction is encountered in the polymerization of ethylene, $H_2C=CH_2$; vinyl chloride, $H_2C=CHCl$; and styrene, $H_2C=CH(C_6H_5)$.

$$C=C \rightarrow C-C- \quad (9.5)$$

Shaping Processes

The earliest cultural ages of human activities produced artifacts that had been shaped from stone, bronze, or iron. In modern technology, we speak of casting, deformation, cutting, and joining in addition to more specialized shaping procedures.

Casting

The concept of casting is straightforward. A liquid is solidified within a mold of the required shape. For metal casting, attention must be given to volume changes; in most cases there is shrinkage. In order to avoid porosity, provision must be made for feeding molten metal from a **riser**. **Segregation** may occur at the solidifying front because of compositional differences in the $(\alpha + L)$ range. (See the earlier discussion of phase diagrams and the discussion of annealing processes in a following subsection.)

A number of ceramic products are made by **slip casting**. The slip is a slurry of fine powders suspended in a fluid, usually water. The mold is typically of gypsum plaster with a porosity that absorbs water from the adjacent suspension. When the shell forming inside the mold is sufficiently thick, the remaining slip is drained. Subsequent processing steps are **drying** and **firing**. The latter high-temperature step bonds the powder into a coherent product.

The casting process is also used in forming polymeric products. Here the solidification is accomplished by polymerization. There is a chemical reaction between the small precursor molecules of the liquid to produce macromolecules and a resulting solid.

Deformation Processes

Deformation processes include forging, rolling, extrusion, and drawing, plus a number of variants. In each case, a force is applied, and a dimensional change results. **Forging** involves shape change by impact. **Rolling** may be used for sheet products as well as for products with constant cross sections, for example, structural beams. **Extrusion** is accomplished through open or closed dies. The former requires that the product be of uniform cross section, such as plastic pipe or siderails for aluminum ladders. Closed dies are molds into which the material is forced. These forming processes can be done at ambient temperature (cold working) or elevated temperatures ($T > 0.4T_m$ in Kelvin; hot working). T_m is the melting temperature. The forming temperature has a strong effect on mechanical properties. Products formed at ambient temperature have high dislocation densities and hence greater strength and hardness but lower ductility than hot-worked products.

In general, ceramics do not lend themselves to the above deformation processes because they lack ductility. Major exceptions are the glasses, which deform not by crystalline slip but by viscous flow.

Cutting

Chiseling and sawing are cutting processes that predate history. Current technology includes **machining** in which a cutting tool and the product move with respect to each other. Depth and rates of cut are adjustable to meet requirements. **Grinding** is a variant of machining that is used for surface removal.

Joining

The process of **welding** produces a joint along which the abutting material has been melted and filled with matching metal. **Soldering** and **brazing** processes use fillers that have a lower melting point than the adjoining materials, which remain solid. There are glass solders as well as metallic ones. Adhesives have long been used for joining wood and plastic components, and many have now been developed to join metals and ceramics.

Annealing Processes

Annealing processes involve reheating a material sufficiently that internal adjustments may be made between atoms or between molecules. The temperature of annealing varies with (1) the material, (2) the amount of time available, and (3) the structural changes that are desired.

Homogenization

The dynamics of processing will produce segregation. For example, when an 80Cu–20Ni alloy starts to solidify at 1200°C, the first solid contains 30% Ni. When solidification is completed, the final liquid has only 12% Ni. Uniformity can be obtained if the alloy is reheated to a temperature at which the atoms can relocate by diffusion. There is a time-temperature relationship (log t vs. $1/T$). In this case an increase in temperature from 500 to 550°C reduces the necessary annealing time by a factor of eight.

Recrystallization

Networks of dislocations are introduced when most metals are plastically deformed at ambient temperatures (cold working). The result is a work hardening and loss of ductility. Whereas the resulting increase in strength is often desired, the property changes resulting from dislocations make further deformation processing more difficult. Annealing will remove the dislocations and restore the initial workable characteristics by forming new, strain-free crystals.

A one-hour heat treatment is a common shop practice because it allows for temperature equalization as well as scheduling requirements. For that time frame, it is necessary to heat a metal to approximately 40 percent of its melting temperature (on the absolute scale). Thus, copper that melts at 1085°C (1358 K) may be expected to recrystallize in the hour at 270°C (545 K). The recrystallization time is shorter at higher temperatures and longer at lower temperatures.

Grain Growth

Extended annealing, beyond that required for recrystallization, produces grain growth and therefore coarser grains. Normally, this is to be avoided. However, grain growth has merit in certain applications because grain boundaries hinder magnetic domain boundary movements, reducing creep. So, coarse grains are preferred in the sheet steel used in transformers, for example. At high temperatures, grain boundaries permit creep to occur under applied stresses, producing changes in dimensions.

Residual Stresses

Expansions and contractions occur within materials during heat-treating operations. These are isotropic for many materials; or they may vary with crystal orientation. Also, differential expansions exist between the two or more phases in a multiphase material. The latter, plus the presence of thermal gradients, introduce internal stresses, which can lead to delayed fracture if not removed.

It is generally desirable to eliminate these residual stresses by an annealing process called **stress relief**. The required temperature is less than that for recrystallization because atomic diffusion is generally not necessary; rather, adjustments are made through the local movement of dislocations. Stress relief is performed on metals before the final machining or grinding operation. Annealing is always performed on glass products, because any residual surface tension easily activates cracks in this nonductile material.

Induced Stresses

In apparent contradiction to the last statement, residual compressive stresses may be prescribed for certain glasses, since glass like most nonductile materials is strong in **compression** but weak in **tension**. As an example, a familiar dinnerware product is made from a *sandwich* glass sheet in which the *bread* layers have a

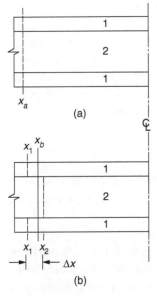

Figure 9.20 Induced stresses (sandwich glass). (a) The previously bonded composite containing glasses 1 and 2 is annealed. There are no internal stresses at x_a. (b) If the three layers were separate, the outer layers, which have a lower thermal coefficient, would contract less during cooling than the center glass (x_1 vs. x_2). Since the layers are bonded together, the restricted contraction of the center (x_a to x_b rather than x_a to x_2) induces compression within the surface layers (x_1 to x_b).

lower thermal expansion than does the *meat* layer. The processing involves heating the dinnerware to relieve all stresses (annealing). As the dinnerware pieces are cooled, the center layer tries to contract more than the surface layers, placing the surfaces under compression (Figure 9.20). Any tension encountered in service must overcome the residual compression before a crack can propagate.

Time-Dependent Processes

We have seen in a previous section that sterling silver is solution treated to dissolve all of the copper within the silver-rich α phase. The single phase is preserved by rapidly cooling the alloy. This avoids the precipitation of the copper-rich β phase, as required for equilibrium. The cooling rate need not be drastic for sterling silver because it takes a minute or more to nucleate and grow the precipitate, β.

Even more time is available for the production of a **silicate glass**—a supercooled liquid. The necessary bond breaking and rearrangements are very slow. The available processing times can approach an hour or more. However, in order to produce a **metallic glass**, the cooling rate must approach 1000°C per *millisecond*. Otherwise, individual metal atoms rapidly order themselves into one of the crystalline patterns described earlier.

Martensitic Reactions

Most steel processing treatments initially heat the steel to provide a single-phase microstructure of γ or **austenite** (Fe_{FCC}). This is face-centered-cubic and dissolves all of the carbon that is present. Normal cooling produces **pearlite**, a lamellar microstructure containing layers of α or ferrite (Fe_{BCC}) and **carbide** (Fe_3C), as shown previously on the Fe–Fe_3C phase diagram (Figure 9.17).

$$Fe_{FCC} \xrightarrow{\text{cooling}} Fe_{BCC} + Fe_3C \tag{9.6}$$

If **quenched**, a different structure forms:

$$Fe_{FCC} \xrightarrow{\text{quenching}} \text{Martensite} \tag{9.7}$$

Martensite is a transition phase. It offers an interesting possibility for many applications because it is much harder than pearlite. Unfortunately, martensite is also very brittle, and its usefulness in steel is severely limited. Martensite will exist almost indefinitely at ambient temperatures, but reheating the steel provides an opportunity for the completion of the reaction of Equation (9.6).

$$\text{Martensite} \xrightarrow{\text{tempering}} Fe_3C \tag{9.8}$$

The reheating process is called **tempering**. The product, which has a microstructure of finely dispersed carbide particles in a ferrite matrix, is both hard and tough. It is widely used and is called **tempered martensite**.

Hardenability

Tempered martensite is the preferred microstructure of many high-strength steels. Processing requires a sufficiently rapid quench to obtain the intermediate martensite, Equation (9.7), followed by tempering, Equation (9.8). This means severe quenching for products of Fe–C steels that are larger than needles or razor blades. Even then, martensite forms only at the quenched surface. The subsurface metal transforms directly to the ferrite and carbide, Equation (9.6).

The reaction rate of Equation (9.6) can be decreased by the presence of various alloying elements in the steel. Thus, gears and similar products commonly contain fractional percentages of nickel, chromium, molybdenum, or other metals. Quenching severity can be reduced, and larger components can be hardened throughout. These alloying elements delay the formation of carbide because, not only must small carbon atoms relocate, it is also necessary for the larger metal atoms to choose between residence in the ferrite or in the carbide.

Surface Modification

Products may be treated so that their surfaces are modified and therefore possess different properties than the original material. Chrome plating is a familiar example.

Carburizing

Strength and hardness of a steel increase with carbon content. Concurrently, ductility and toughness decrease. With these variables, the engineer must consider trade-offs when specifying steels for mechanical applications. An alternate possibility is to alter the surface zone. A common example is the choice of a low-carbon, tough steel that has had carbon diffused into the subsurface (<1 mm) to produce hard carbide particles. Wear resistance is developed without decreasing bulk toughness.

Nitriding

Results similar to carburizing are possible for a steel containing small amounts of aluminum. Nitrogen can be diffused through the surface to form a subscale containing particles of aluminum nitride. Since AlN has structure and properties that are related to diamond, wear resistance is increased for the steel.

Shot Peening

Superficial deformation occurs when a ductile material is impacted by sand or by hardened shot. A process employing shot peening places the surface zone in local compression and therefore lowers the probability for fracture initiation during tensile loading.

Example 9.9

Assume a single spherical shrinkage cavity forms inside a 2-kg lead casting as it is solidified at 327°C. What is the initial diameter of the cavity after solidification? (The greatest density of molten lead is 10.6 g/cm^3.)

Solution

Refer to Figure 9.14(a). Based on a unit volume at 20°C, lead shrinks from 1.07 to 1.035 during solidification.

The volume of molten lead is (2000 g)/(10.6 g/cm^3) = 188.7 cm^3.

Volume of solid lead at 327°C: (188.7 cm^3)(1.035/1.07) = 182.5 cm^3

Shrinkage: 188.7 cm^3 − 182.5 cm^3 = $\pi d^3/6$; d = 2.28 cm

Example 9.10

How much energy is involved in polymerizing one gram of ethylene (C_2H_4) into polyethylene, Equation (9.5). The double carbon bond possesses 162 kcal/mole, and the single bond has 88 kcal/mole.

Solution

As shown in Equation (9.5), one double carbon bond changes to two single bonds/mer:

$$(1 \text{ g})/(24 + 4 \text{ g/mole}) = 0.0357 \text{ moles}$$

The energy required to break 0.0357 moles of double bonds is

$$(1)(0.0357)(162 \text{ kcal}) = 5.79 \text{ kcal}$$

The energy released in joining twice as many single bonds is

$$(2)(0.0357)(-88 \text{ kcal}) = -6.29 \text{ kcal}$$

The net energy change is $-6.29 + 5.79$ kcal = 500 cal released/g.

Example 9.11

There are 36 equiaxed grains per mm^2 observed at a magnification of 100 in a selected area of copper. The copper is heated to double the average grain diameter. (i) How many grains exist per mm^2? (ii) What will be the percentage (increase, decrease) in grain boundary area?

Solution

(i) Doubling a lineal dimension decreases the number of grains by a factor of four, so there are 9 grains per mm^2.

(ii) Surface area is a function of the lineal dimension squared:

$$a_1/a_2 \propto d_1^2/d_2^2$$

Thus, $a_1/a_2 = 0.25$, or a 75% decrease in grain boundary area.

MATERIALS IN SERVICE

Products of engineering are made to be used. Conditions that are encountered in service most often vary tremendously from those present in the stockroom: static and dynamic loads, elevated and subambient temperatures, solar and nuclear radiation exposure, and many other reactions with the surrounding environment. All of these situations can lead to deterioration and even to failure. The design engineer should be able to anticipate the conditions of failure.

Mechanical Failure

Under ambient conditions, excessive loads can lead to bending or to cracking, the principal modes of mechanical failure. The former depends upon geometry and the stress level. **Stress** is defined as load divided by cross-sectional area, $\sigma = F/A$. Cracking (and succeeding fracture) includes those two considerations, plus the loading rate.

Yield Strength

When solids are stressed, strain occurs. Strain is the change in length divided by the original length, $\varepsilon = \Delta L/L_0$. The ratio of stress to strain is called the **elastic modulus**, $E = \sigma/\varepsilon$, and is initially constant in most solids as shown in Figure 9.21. The interatomic spacings are altered as the load is increased. Initial slip starts at a threshold level called the **yield strength**, σ_y, shown in Figure 9.22. Higher stresses will produce a permanent distortion, which will be called failure if the product was designed to maintain its initial shape. The toughness of a material is related to the energy or work required for fracture, a product of both strength and ductility. In a tensile test, the energy is the area under the stress-strain curve. Ductile fracture involves significant plastic deformation and hence absorbs much more energy than brittle fracture, which has little or no plastic deformation. A complete stress-strain curve is shown in Figure 9.23.

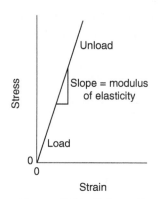

Figure 9.21 Schematic stress-strain diagram showing linear elastic deformation for loading and unloading cycles

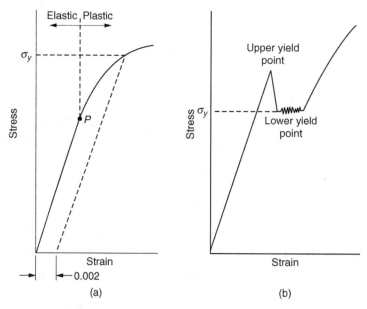

Figure 9.22 (a) Typical stress-strain behavior for a metal, showing elastic and plastic deformations, the proportional limit P, and the yield strength σ_y, as determined using the 0.002 strain offset method; (b) representative stress-strain behavior found for some steels, demonstrating the yield point phenomenon

Fracture

Breakage always starts at a location of **stress concentration**. This may be at a *flaw* of microscopic size, such as an abrasion scratch produced while cleaning eyeglasses, or it may be of larger dimensions, such as a hatchway on a ship.

With a crack, the **stress intensity factor**, K_I, is a function of the applied stress, σ, and of the square root of the crack length a:

$$K_I = y\sigma\sqrt{\pi a} \qquad (9.9)$$

The proportionality constant, y, relates to cross-sectional dimensions and is generally near unity.

Fracture toughness is the **critical** stress intensity factor, K_{Ic}, to propagate fracture and is a property of the material. This corresponds to the yield strength, σ_y, being the critical stress to initiate slip. However, stronger materials generally have lower fracture toughness and *vice versa*.

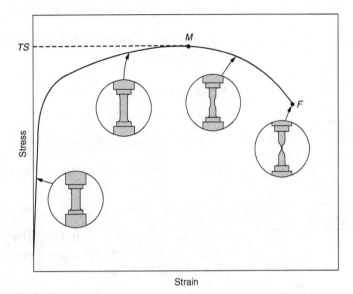

Figure 9.23 Typical engineering stress-strain behavior to fracture, point F. The tensile strength TS is indicated at point M. The circular insets represent the geometry of the deformed specimen at various points along the curve.

To illustrate the relationship between strength and toughness, consider a steel that has a yield strength, σ_y, of 1200 MPa, and a critical stress intensity factor, K_{Ic}, of 90 MPa • m$^{1/2}$.

In the presence of a 2-mm crack, a stress of 1135 MPa would be required to propagate a fracture, according to Equation (9.9). This is below the yield stress. If the value of K_{Ic} had been 100 MPa • m$^{1/2}$, fracture would not occur; rather, the metal would deform at 1200 MPa. A 2.2-mm crack would be required to initiate fracture without yielding.

Fatigue

Cyclic loading reduces permissible design stresses, as illustrated in Figure 9.24. Minute structural changes that introduce cracks occur during each stress cycle. The crack extends as the cycles accumulate, leading to eventual fracture. When this delayed behavior was first observed, it was assumed that the material got tired; hence the term *fatigue*. Steels and certain other materials possess an *endurance limit*, below which unlimited cycling can be tolerated.

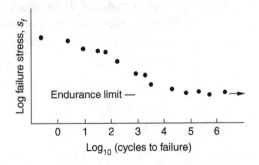

Figure 9.24 Cyclic fatigue. The stress for failure, sf, decreases as the number of loading cycles is increased. Most steels exhibit an endurance limit, a stress level for unlimited cycling. (A static tensile test is only one-fourth of a cycle.)

Example 9.12

(i) What is the maximum static force that can be supported without permanent deformation by a 2-mm-diameter wire that has a yield strength of 1225 MPa?

(ii) The elastic deformation at this threshold stress is 0.015 m/m. What is its elastic modulus?

Solution

(i) $\sigma_y = F/A$

$F = \pi(2 \text{ mm}/2)^2(1225 \text{ MPa}) = 3800 \text{ N}$

(ii) $E = \sigma/\varepsilon = 1225 \text{ MPa}/0.015 \text{ m/m} = 82{,}000 \text{ MPa}$

Example 9.13

The value of K_{Ic} for steel is 186 MPa·m$^{1/2}$. What is the maximum tolerable crack length, a, when the steel carries a nominal stress of 800 MPa? (Assume 1.1 as the proportionality constant.)

Solution

$$186 \text{ MPa·m}^{1/2} = 800 \text{ MPa } (1.1)(\pi a)^{1/2}$$

$$a = 0.014 \text{ m} = 14 \text{ mm}$$

Thermal Failure

Melting is the most obvious change in a material at elevated temperatures. Overheating, short of melting, can also introduce microstructural changes. For example, the tempered martensite of a tool steel is processed so that it has a very fine dispersion of carbide particles in a tough ferrite matrix. It is both hard and tough and serves well for machining purposes. However, an excessive cutting speed raises the temperature of the cutting edge, causing the carbide particles to grow and softening the steel; if heating continues, failure eventually occurs by melting at the cutting edge. *High-speed* tools incorporate alloy additions, such as vanadium and chromium, that form carbides that are more stable than iron carbide. Thus, they can tolerate the higher temperatures that accompany faster cutting speeds.

Creep

As the name implies, **creep** describes a slow (<0.001%/hr) dimensional change within a material. It becomes important in long-term service of months or years. Slow viscous flow is commonly encountered in plastic materials. Refractories (temperature-resistant ceramics) will slowly slump when small amounts of liquid accumulate.

In metals, creep occurs when atoms become sufficiently mobile to migrate from compressive regions of the microstructure into tensile regions. Grain boundary areas are heavily involved. For this reason, coarser grained metals are advantageous for high-temperature applications. Three stages of creep are identified in Figure 9.25. Following the initial elastic strain, Stage 1 of creep is fairly rapid as stress variations are equalized. In Stage 2, the creep rate, dL/dt, is essentially constant. Design considerations are focused on this stage. Stage 3, which accelerates when the cross-sectional area starts to be reduced, leads to eventual rupture.

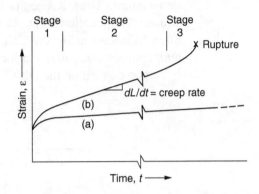

Figure 9.25 Creep. (*a*) Low stresses and/or low temperatures; (*b*) high stresses, high temperatures. The initial strain is elastic, followed by rapid strain adjustments (Stage 1). Design calculations are commonly based on the steady-state strain rate (Stage 2). Strain accelerates (Stage 3) when the area reduction becomes significant. Rupture occurs at *x*.

Spalling

Spalling is thermal cracking. It is the result of stress caused by differential volume changes during processing or service. As discussed earlier, stresses can be introduced into a material (1) by thermal gradients, (2) by anisotropic volume changes, or (3) by differences in expansion coefficients in multiphase materials. Cyclic heating and cooling lead to **thermal fatigue** when the differential stresses produce localized cracking.

The spalling resistance index (SRI) of a material is increased by higher thermal conductivities, k, and greater strengths, S; it is reduced with greater values of the thermal expansion coefficients, α, and higher elastic moduli. In functional form,

$$\text{SRI} = f(kS/\alpha E) \tag{9.10}$$

Low-Temperature Embrittlement

Many materials display an abrupt drop in ductility and toughness as the temperature is lowered. In glass and other amorphous materials, this change is at the temperature below which atoms or molecules cannot relocate in response to the applied stresses. This is the **glass-transition temperature**, T_g. Metals are crystalline and do not have a glass transition. However, steels and a number of other metals have a **ductility-transition temperature** below which fracture is nonductile. The impact energy required for fracture can drop by an order of magnitude at this transition temperature. Thus it becomes a very significant consideration in design for structural applications. The **ductile-to-brittle transition temperature** (DBTT) is often measured using the Charpy Impact test, with samples soaked at different temperatures immediately prior to testing. Representative curves and the influence of carbon content on the DBTT in steel are shown in Figure 9.26.

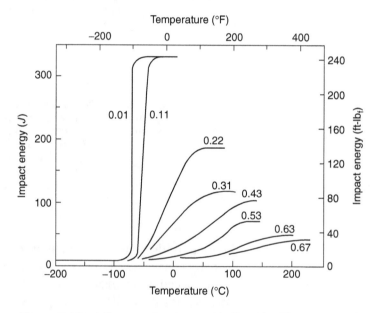

Figure 9.26 Influence of carbon content on the Charpy V-notch energy-versus-temperature behavior for steel

Radiation Damage

Unlike heat, which energizes all of the atoms and molecules within a structure, radiation introduces energy at *pinpoints* called **thermal spikes**. Individual bonds are broken, specific atoms are dislodged, molecules are ruptured, and electrons are energized. Each of these actions disorders the structure and alters the properties of the material. As expected, slip and deformation are resisted. Therefore, while strength and hardness increase, ductility and toughness decrease. Electrical and thermal conductivities drop within metals, because there is more scattering of the electrons as they move along the voltage or thermal gradient. These property changes, among others, are considered to be damaging changes, especially when they contradict carefully considered design requirements.

Damage Recovery

Partial correction of radiation damage is possible by annealing. Reheating a material allows internal readjustments, since atoms that have been displaced are able to relocate into a more ordered structure. It is similar to recrystallization after work hardening, where dislocations involving lineal imperfections composed of many atoms must be removed, except that radiation damage involves *pinpoint* imperfections, which means that it can be removed at somewhat lower temperatures than those required for recrystallization.

Chemical Alteration
Oxidation
Materials can be damaged by reacting chemically with their environments. All metals except gold oxidize in ambient air. Admittedly, some oxidize very slowly at ordinary temperatures. Others, such as aluminum, form a protective oxide surface that inhibits further oxidation. However, all metals—including gold—will oxidize significantly at elevated temperatures or in chemical environments that consume the protective oxidation.

Several actions are required for oxide scale to accumulate on a metal surface. Using iron as an example, the iron atom must be ionized to Fe^{2+} before it or the electrons move through the scale to produce oxygen ions, O^{2-}, at the outer surface. There, FeO or Fe_3O_4 accumulates. As the scale thickens, the oxidation rate decreases. However, exceptions exist. For example, the volume of MgO is less than the volume of the original magnesium metal. Therefore, the scale cracks and admits oxygen directly to the underlying metal. Also, an Al_2O_3 scale is insulating so the ionization steps are precluded.

Moisture
Moisture can produce chemical **hydration**. As examples, MgO reacts with water to produce $Mg(OH)_2$, and Fe_2O_3 can be hydrated to form $Fe(OH)_3$. Water can be absorbed into materials. Small H_2O molecules are able to diffuse among certain large polymeric molecules. Consequently, polymers such as the aramids, which we normally consider to be very strong, are weakened—a fact that the design engineer must consider in specifications.

Corrosion
Metallic corrosion is familiar to every reader who owns a car, since rust—$Fe(OH)_3$—is the most obvious product of corrosion. Oxidation produces positive ions and electrons:

$$M \rightarrow M^{n+} + ne^- \tag{9.11}$$

The reaction stops unless the electrons are removed. Oxygen accompanied by water (Figure 9.27) is a common consumer of the electrons:

$$O_2 + 2\,H_2O + 4e^- \rightarrow 4\,(OH)^- \tag{9.12}$$

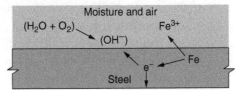

Figure 9.27 Rust formation. Electrons are removed from iron atoms and react with water and oxygen to produce $Fe(OH)_3$—rust.

Alternatively, if ions of a metal with a low oxidation potential are present, they can be reduced, consuming electrons from the preceding corrosion reaction, Equation (9.11). Copper is cited as a common example:

$$Cu^{2+} + 2e^- \rightarrow Cu^0 \qquad (9.13)$$

Electroplating uses this reaction advantageously to deposit metals from a solution by the addition of electrons. The relative reactivity of metals with respect to standard electrodes is represented by the electromotive force (emf) series given in Table 9.4. However, in real environments such as sea water the galvanic series, as shown in Table 9.5, is more commonly used to determine the likelihood of corrosion.

Table 9.4 The standard emf series

	Electrode Reaction	Standard Electrode Potential V^0(V)
↑ Increasingly inert (cathodic)	$Au^{3+} + 3e^- \rightarrow Au$	+1.420
	$O_2 + 4H^- + 4e^- \rightarrow 2H_2O$	+1.229
	$Pl^{2+} + 2e^- \rightarrow Pl$	+1.200
	$Ag^+ + e^- \rightarrow Ag$	+0.800
	$Fe^{3+} + e^- \rightarrow Fe^{2+}$	+0.771
	$O_2 + 2H_2O + 4e^- \rightarrow 4(OH^-)$	+0.401
	$Cu^{2+} + 2e^- \rightarrow Cu$	+0.340
	$2H^+ + 2e^- \rightarrow H_2$	0.000
	$Pb^{2+} + 2e^- \rightarrow Pb$	–0.126
	$Sn^{2+} + 2e^- \rightarrow Sn$	–0.136
	$Ni^{2+} + 2e^- \rightarrow Ni$	–0.250
	$Co^{2+} + 2e^- \rightarrow Co$	–0.277
	$Cd^{2+} + 2e^- \rightarrow Cd$	–0.403
	$Fe^{2+} + 2e^- \rightarrow Fe$	–0.440
Increasingly inert (cathodic) ↓	$Cr^{3+} + 3e^- \rightarrow Cr$	–0.744
	$Zn^{2+} + 2e^- \rightarrow Zn$	–0.763
	$Al^{2+} + 3e^- \rightarrow Al$	–1.662
	$Mg^{2+} + 2e^- \rightarrow Mg$	–2.363
	$Na^+ + e^- \rightarrow Na$	–2.714
	$K^+ + e^- \rightarrow K$	–2.924

Corrosion Control

Corrosion is minimized by a variety of means. Some involve the avoidance of one or more of the above reactions. Feed water for steam-power boilers is deaerated; surfaces are painted to limit the access of water and air; junctions between unlike metals are electrically insulated. Other control procedures induce a reverse reaction: sacrificial anodes such as magnesium are attached to the side of a ship; iron sheet is galvanized with zinc. Each corrodes preferentially to the underlying steel and forces the iron to assume the role of Equation (9.13). Corrosion may also be restricted by an impressed voltage. Natural gas lines utilize this procedure by connecting a negative dc voltage to the pipe.

COMPOSITES

Composites are not new. Straw in brick and steel reinforcing rods in concrete have been used for a long time. But there is current interest in the development of new composites by designing materials appropriately. It is possible to benefit from the positive features of each of the contributors in order to optimize the properties of the composite.

The internal structures of composites may be viewed as enlarged poly-component micro-structures, which were previously discussed. Attention is given to size, shape, amounts, and distribution of the contributing materials. However, there is commonly a significant difference in processing composites. Typically, the internal structure of a composite is a function of mechanical processing steps—mixing, emplacement, surface deposition, and so on—rather than thermal processing. These processing differences suggest a different approach in examining property-structure relationships.

Table 9.5 The galvanic series

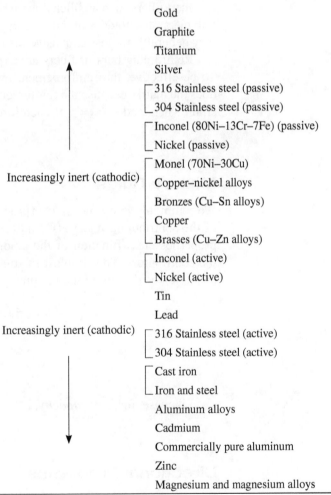

Reinforced Materials

In familiar composites, such as reinforced concrete, the steel carries the tensile load. Also, there must be bonding between the reinforcement and the matrix. In a *glass* fishing rod, the glass fibers are all oriented longitudinally within the polymer matrix. If it is fractured, the break exhibits a splintered appearance with a noticeable amount of fiber pull-out.

Steel reinforcing bars (rebar) are used to reinforce concrete to meet strength and cost factors. Positioning is dictated by stress calculations. Rebar surfaces are commonly merloned (ridged circumferentially) for better anchorage.

Glass is widely used in fiber-reinforced plastics (FRP). The positioning of the fibers varies with the product. Continuous fibers are used in such structures as fuel storage tanks or rocket casings. However, chopped fibers are required when FRP are processed within molds. A matrix-to-glass bond must be achieved through chemical coatings at the interfaces.

Silicon carbide (SiC) and alumina (Al_2O_3) fibers are used increasingly in high-temperature composites with either metallic or ceramic matrices.

Inert fillers such as silica flour and wood flour serve multiple purposes in many polymers. They add strength, rigidity, and hardness to the product. In addition, they generally are less expensive than the matrix polymer.

Reinforcing bars or fibers are expected to carry the bulk of the tensile load. To be effective, the reinforcement must have a higher elastic modulus, E_r, than does the matrix because the reinforcement and the matrix undergo the same tensile strain when loaded ($\varepsilon_r = \varepsilon_m$). Therefore, $(E/\sigma)_r = (E/\sigma)_m$; and

$$\sigma_r / \sigma_m = E_r / E_m \tag{9.14}$$

Mixture Rules

We commonly study and analyze properties of composites in terms of the properties of the contributing materials. Mixture rules can then be formulated in which the properties are a function of the amounts and geometric distributions of each of the contributors. The simplest mixture rules are based on volume fraction, f. For example, the density of the mixture, ρ_m, is the volume-weighted average:

$$\rho_m = f_1\rho_1 + f_2\rho_2 + \ldots \tag{9.15a}$$

or simply

$$\rho_m = \sum f_i \rho_i \tag{9.15b}$$

Likewise, for *heat capacity*, c:

$$c_m = \sum f_i c_i \tag{9.16}$$

Directional Properties

Many composites are anisotropic. Laminates of two or more layers are bidirectional, as is a matted FRP that uses chopped fibers. The previously cited *glass* fishing rod has a structure with uniaxial anisotropy.

When considering conductivity, elastic moduli, strength, or other properties that are directional, attention must be given to the anisotropy of the composite. Consider the electrical resistance across a laminate that contains alternate layers of two materials with different resistances, R_1 and R_2. In a direction perpendicular to the laminate, the plies are in series. Thus the relationship is

$$R_\perp = L_1 R_1 + L_2 R_2 + \cdots \tag{9.17}$$

where L_i is the thickness of each ply. For unit dimensions, volume fractions, f_i, and the resistivities, ρ_i, are applicable:

$$\rho_\perp = f_1 \rho_1 + f_2 \rho_2 + \cdots \tag{9.18}$$

In contrast, for directions parallel to the laminate, resistivities follow the physics analog for parallel circuits:

$$\frac{1}{\rho_\parallel} = \frac{f_1}{\rho_1} + \frac{f_2}{\rho_2} + \cdots \tag{9.19}$$

The mixture rules for conductivities, either thermal, k, or electrical, σ, are inverted to those for resistivities:

$$\frac{1}{k_\perp} = \frac{f_1}{k_1} + \frac{f_2}{k_2} + \cdots \tag{9.20}$$

and

$$k_\parallel = f_1 k_1 + f_2 k_2 + \cdots \tag{9.21}$$

More elaborate mixture rules must be used when the structure of the composite is geometrically more complex, such as the use of particulate fillers or chopped fibers. In general, however, the property of the composite falls between the calculated values using the parallel and series versions of the preceding equations.

Preparation

Composites receive their name from the fact that two or more distinct starting materials are combined into a unified material. The application of a protective coating to metal or wood is one form of composite processing. The process is not as simple as it first appears because priming treatments are commonly required after the surface has been initially cleared of contaminants and moisture; otherwise, peeling and other forms of deterioration may develop. Electroplating, Equation (9.13), commonly requires several intermediate surface preparations so that the final plated layer meets life requirements.

Wood paneling combines wood and polymeric materials into large sheets. **Plywood** not only has dimensional merit, it transforms the longitudinal anisotropy into a more desirable two-dimensional material. Related products include chipboard and similar composites. Composite panels include those with veneer surfaces where appearance and technical properties are valued.

The concept of uniformly mixing particulate **fillers** into a composite is simple. The resulting product is isotropic.

Several considerations are required for the use of **fibers** in a composite. Must the fiber be continuous? What are the directions of loading? What is the shear strength between the fiber and the matrix? Chopped fibers provide more reinforcement than do particles, and at the same time permit molding operations. The **aspect ratio** (L/D) must be relatively low for die molding. Higher ratios, and therefore more reinforcement, are used in sheet molding. (Sheet molded products are used where strength is not critical in the third dimension.) **Continuous fibers** maximize the mechanical properties of FRP composites. Their uses, however, are generally limited to products that permit parallel layments.

Example 9.14

A rod contains 40 volume percent longitudinally aligned glass fibers within a plastic matrix. The glass has an elastic modulus of 70,000 MPa; the plastic, 3500 MPa. What fraction of a 700-N tensile load is carried by the glass?

Solution

Based on Equation (9.14):

$$(F_{gl}/0.4A)/(F_p/0.6A) = (70{,}000 \text{ MPa}/3500 \text{ MPa}) = 20$$

$$F_{gl} = 20(0.4/0.6)\, F_p = 13.3\, F_p$$

$$F_{gl}/(F_{gl} + F_p) = (13.3\, F_p)/(13.3\, F_p + F_p) = 93\%$$

Example 9.15

An electric highline cable contains one cold-drawn steel wire and six annealed aluminum wires, all with a 2-mm diameter. (The steel provides the required strength; the aluminum, the conductivity.) Using the following data, calculate (i) the resistivity and (ii) the elastic modulus of the composite wire.

$$\text{Steel: } \rho = 17 \times 10^{-6}\ \Omega \cdot \text{cm},\ E = 205{,}000 \text{ MPa}$$

$$\text{Aluminum: } \rho = 3 \times 10^{-6}\ \Omega \cdot \text{cm},\ E = 70{,}000 \text{ MPa}$$

Solution

(i) From Equation (9.19):

$$1/\rho_\| = (1/7)/(17 \times 10^{-6}\ \Omega \cdot \text{cm}) + (6/7)/(3 \times 10^{-6}\ \Omega \cdot \text{cm})$$
$$= 0.294 \times 10^6\ \Omega^{-1} \cdot \text{cm}^{-1}$$

$$\rho_\| = 3.4 \times 10^{-6}\ \Omega \cdot \text{cm}$$

(ii) We must write a mixture rule for the elastic modulus of a composite in parallel. Let A be the area of one wire.

Since $\varepsilon_c = \varepsilon_{Al} = \varepsilon_{St}$, $[(F/A)/E]_C = [(F/A)/E]_{Al} = [(F/A)/E]_{St}$

Also, $F_C = F_{Al} + F_{St} = F_C(f_{Al} A_C/A_C)(E_{St}/E_C) + F_C(f_{St} A_C/A_C)(E_{St}/E_C)$

Canceling,

$$E_C = f_{Al} E_{Al} + f_{St} E_{St} = (6/7)(70{,}000 \text{ MPa}) + (1/7)(205{,}000 \text{ MPa}) = 89{,}000 \text{ MPa}$$

The apparent modulus will be lower because there will also be cable extension by the straightening of the cable wire.

SELECTED SYMBOLS AND ABBREVIATIONS

Symbol or Abbreviation	Definition
A	area
a	unit cell length
a	crack length
amu	atomic mass unit
D	diffusion coefficient
D_o	proportionality constant
E	elastic modulus
ε	engineering strain
F	force
f	volume fraction
K_I	stress intensity factor
P	density
Q	activation energy
R	gas constant
r	atomic radius
σ	stress
σ_y	yield strength
SRI	Spalling Resistance Index
Tg	glass-transition temperature
Tm	melting temperature
V^0	standard electrode potential
W_L	weight fraction of liquid phase
W_S	weight fraction of solid phase
y	proportionality constant

REFERENCE

Van Vlack, L. *Elements of Materials Science and Engineering*, 6th ed. Addison-Wesley, 1989.

PROBLEMS

9.1 For a neutral atom:
  a. the atomic mass equals the mass of the neutrons plus the mass of the protons
  b. the atomic number equals the atomic mass
  c. the number of protons equals the atomic number
  d. the number of electrons equals the number of neutrons

9.2 All isotopes of a given element have:
  a. the same number of protons
  b. the same number of neutrons
  c. equal numbers of protons and neutrons
  d. the same number of atomic mass units

9.3 Which of the following statements is *FALSE*?
  a. An anion has more electrons than protons.
  b. Energy is released when water is solidified to ice.
  c. Energy is required to remove an electron from a neutral atom.
  d. Energy is released when a H_2 molecule is separated into two hydrogen atoms.

9.4 Select the correct statement.
  a. Crystals possess long-range order.
  b. Within a crystal, like ions attract and unlike ions repel.
  c. A body-centered cubic metallic crystal (for example, iron) has nine atoms per unit cell.
  d. A face-centered cubic metallic crystal (for example, copper) has 14 atoms per unit cell.

9.5 In a cubic crystal, a is the edge of a unit cell. The shortest repeat distance in the [111] direction of a body-centered cubic crystal is:
  a. $a\sqrt{2}$
  c. $a\sqrt{3}/2$
  b. $2a$
  d. $a\sqrt{3}/4$

9.6 All but which of the following data are required to calculate the density of aluminum in g/m^3?
  a. Avogadro's number, which is 6.0×10^{23}
  b. atomic number of Al, which is 13
  c. crystal structure of Al, which is face-centered cubic
  d. atomic mass of Al, which is 27 amu

9.7 The atomic packing factor of gold, an FCC metal, is:
  a. $(4\pi r^3/3)/(4r/\sqrt{3})^3$
  b. $4(4\pi r^3/3)/(4r/\sqrt{2})^3$
  c. $4(4\pi r^3/3)/(r/\sqrt{2})^3$
  d. $4(2r/\sqrt{2})^3/(4\pi r^3/3)$

9.8 Ethylene is C_2H_4. To meet bonding requirements, how many bonds are present?
   a. 6 single
   b. 4 single and 2 double
   c. 1 double and 4 single
   d. 12 single

9.9 Gold is FCC and has a density of 19.3 g/cm³. Its atomic mass is 197 amu. Its atomic radius, r, may be calculated using which of the following?
   a. $19.3 \text{ g/cm}^3 = (197)(6.02 \times 10^{23})/[(4r/)\sqrt{2}^3]$
   b. $19.3 \text{ g/cm}^3 = 2(197/6.02 \times 10^{23})/[(4r/)\sqrt{2}^3]$
   c. $19.3 \text{ g/cm}^3 = 4(197/6.02 \times 10^{23})/[(4r/)\sqrt{2}^3]$
   d. $19.3 \text{ g/cm}^3 = 6(197)(6.02 \times 10^{23})/[(4r/)\sqrt{2}^3]$

9.10 Each of the following groups of plastics is thermoplastic *EXCEPT*:
   a. polyvinyl chloride (PVC) and a polyvinyl acetate
   b. phenolics, melamine, and epoxy
   c. polyethylene, polypropylene, and polystyrene
   d. acrylic (Lucite) and polyamide (nylon)

9.11 Styrene resembles vinyl chloride, C_2H_3Cl, except that the chlorine is replaced by a benzene ring. The mass of each mer is:
   a. $8(12) + 9(1)$ amu
   b. $26 + 78$ amu
   c. $27 + 6(12) + 6(1)$ amu
   d. $2(12) + 3(1) + 77$ amu

9.12 The $<1\bar{1}0>$ family of directions in a cubic crystal include all but which of the following? (An overbar is a negative coefficient.)
   a. [110]
   b. [0$\bar{1}$1]
   c. [101]
   d. [1$\bar{1}$1]

9.13 The {112} family of planes in a cubic crystal includes all but which of the following directions?
   a. (212)
   b. (211)
   c. (1$\bar{1}$2)
   d. (121)

9.14 Crystal imperfections include all but which of the following?
   a. Dislocations
   b. Displaced atoms
   c. Interstitials
   d. Dispersions

9.15 A dislocation may be described as a:
   a. displaced atom
   b. shift in the lattice constant
   c. slip plane
   d. lineal imperfection

9.16 A grain within a microstructure is:
  a. a particle the size of a grain of sand
  b. the nucleus of solidification
  c. a particle the size of a grain of rice
  d. an individual crystal

9.17 Which of the following does NOT apply to a typical brass?
  a. An alloy of copper and zinc
  b. A single-phase alloy
  c. An interstitial alloy of copper and zinc
  d. A substitutional solid solution

9.18 Sterling silver, as normally sold:
  a. is pure silver
  b. is a supersaturated solid solution of 7.5% copper in silver
  c. is 24-carat silver
  d. has higher conductivity than pure silver

9.19 Atomic diffusion in solids matches all but which of the following generalities?
  a. Diffusion is faster in FCC metals than in BCC metals.
  b. Smaller atoms diffuse faster than do larger atoms.
  c. Diffusion is faster at elevated temperatures.
  d. Diffusion flux is proportional to the concentration gradient.

9.20 The proportionality constant for a particular gas diffusing through copper at 1000°C is 0.022 cm^2/s and the activation energy is 97 kJ/mole. Find the diffusion coefficient.
  a. 1.6×10^{-6} cm^2/s
  b. 1.9×10^{-6} cm^2/s
  c. 2.3×10^{-6} cm^2/s
  d. 4.4×10^{-6} cm^2/s

9.21 Grain growth involves all but which of the following?
  a. Reduced growth rates with increased time
  b. An increase in grain boundary area per unit volume
  c. Atom movements across grain boundaries
  d. A decrease in the number of grains per unit volume

9.22 Imperfections within metallic crystal structures may be any but which of the following?
  a. Lattice vacancies and extra interstitial atoms
  b. Displacements of atoms to interstitial site (Frenkel defects)
  c. Lineal defects or slippage dislocations caused by shear
  d. Ion pairs missing in ionic crystals (Shottky imperfections)

9.23 All but which of the following statements about solid solutions are correct?
  a. In metallic solid solutions, larger solute atoms occupy the interstitial space among solvent atoms in the lattice sites.
  b. Solid solutions may result from the substitution of one atomic species for another, provided radii and electronic structures are compatible.
  c. Defect structures exist in solid solutions of ionic compounds when there are differences in the oxidation state of the solute and solvent ions, because vacancies are required to maintain an overall charge balance.
  d. Order-to-disorder transitions that occur at increased temperatures in solid solutions result from thermal agitation that dislodges atoms from their preferred neighbors.

9.24 In ferrous oxide, $Fe_{1-x}O$, 2% of the cation sites are vacant. What is the Fe^{3+}/Fe^{2+} ratio?
  a. 2/98
  b. 0.04/0.94
  c. 0.04/0.96
  d. 0.06/0.94

9.25 A solid solution of MgO and FeO contains 25 atomic percent Mg^{2+} and 25 atomic percent Fe^{2+}. What is the weight fraction of MgO? (Mg: 24; Fe: 56; and O: 16 amu)
  a. 40/(40 + 72)
  b. 24/(24 + 56), or 4/80
  c. 25/(25 + 25)
  d. (25 + 25)/(50 + 50)

9.26 The boundary between two metal grains provides all but which of the following?
  a. An impediment to dislocation movements
  b. A basis for an increase in the elastic modulus
  c. A site for the nucleation of a new phase
  d. Interference to slip

9.27 If 5% copper is added to silver:
  a. the hardness is decreased
  b. the strength is decreased
  c. the thermal conductivity is decreased
  d. the electrical resistivity is decreased

9.28 All but which of the following statements about diffusion and grain growth are correct?
  a. Atoms can diffuse both within grains and across grain (crystal) boundaries.
  b. The activation energy for diffusion through solids is inversely proportional to the atomic packing factor of the lattice.
  c. Grain growth results from local diffusion and minimizes total grain boundary area. Large grains grow at the expense of small ones, and grain boundaries move toward their centers of curvature.
  d. Net diffusion requires an activation energy and is irreversible. Its rate increases exponentially with temperature. It follows the diffusion equation, in which flux equals the product of diffusivity and the concentration gradient.

9.29 Refer to the accompanying Mg-Zn phase diagram, Exhibit 9.29. Select an alloy of composition C (71Mg–29Zn) and raise it to 575°C so that only liquid is present. Change the composition to 60Mg–40Zn by adding zinc. When this new liquid is cooled, what will be the first solid to separate?
a. A solid intermetallic compound
b. A mixture of solid intermetallic compound and solid eutectic C (71Mg–29Zn)
c. A solid eutectic C (71Mg–29Zn)
d. A solid solution containing less than 1% intermetallic compound dissolved in Mg

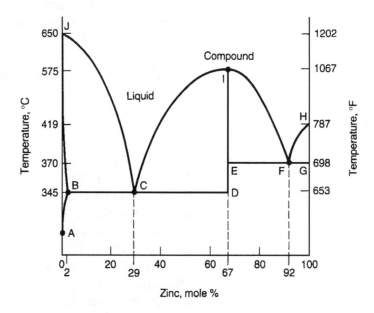

Exhibit 9.29 Magnesium-zinc phase diagram

9.30 Refer to the Mg-Zn phase diagram of Exhibit 9.29. Which of the following compounds is present?
a. Mg_3Zn_2 c. $MgZn$
b. Mg_2Zn_3 d. $MgZn_2$

9.31 Refer to Exhibit 9.31, a schematic sketch of the Fe-Fe₃C phase diagram. All but which of the following statements are *TRUE*?
a. A eutectoid reaction occurs at location C, 727°C (1340°F).
b. The eutectic composition is 99.2 weight percent Fe and 0.8 weight percent C.
c. A peritectic reaction occurs at K, 1500°C (2732°F).
d. A eutectic reaction occurs at G, 1130°C (2202°F).

9.32 Refer to Exhibit 9.31, the Fe-Fe₃C phase diagram. Pearlite contains ferrite (α) and carbide (Fe_3C). The weight fraction of carbide in pearlite is:
a. 0.8%   c. CD/BD
b. BC/CD   d. BC/BD

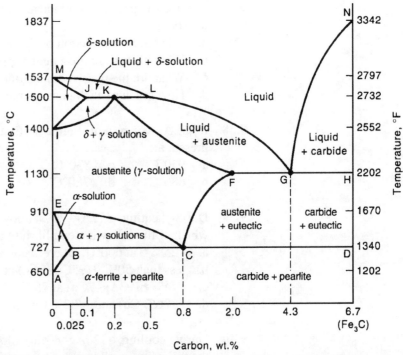

Exhibit 9.31 Iron-iron carbide phase diagram (schematic)

9.33 Consider the Ag-Cu phase diagram (Figure 9.16). Silver-copper alloys can contain *approximately* half liquid and half solid at all but which of the following situations?
   a. 40Ag–60Cu at 781°C
   b. 81Ag–19Cu at 781°C
   c. 20Ag–80Cu at 910°C
   d. 50Ag–50Cu at 779°C

9.34 Refer to the Ag-Cu phase diagram of Figure 9.16. Which of the following statements is *FALSE*?
   a. The solubility limit of copper in the liquid at 900°C is approximately 62%.
   b. The solubility limit of silver in b at 700°C is approximately 5%.
   c. The solubility limit of silver in b at 900°C is approximately 7%.
   d. The solubility limit of copper in b at 800°C is approximately 92%.

9.35 Other factors being equal, diffusion flux is facilitated by all but which of the following?
   a. Smaller grain sizes
   b. Smaller solute (diffusing) atoms
   c. Lower concentration gradients
   d. Lower-melting solvent (host structure)

9.36 Consider copper. All but which of the following statements are applicable for grain growth?
   a. Atoms jump the boundary from large grains to small grains.
   b. Grain size varies inversely with boundary area.
   c. Grain growth occurs because the boundary atoms possess higher energy than interior atoms.
   d. Grain growth occurs because larger grains have less boundary area.

9.37 A phase diagram can provide answers for all but which of the following questions?
  a. What are the direction of the planes at a given temperature?
  b. What phases are present at a given temperature?
  c. What are the phase compositions at a given temperature?
  d. How much of each phase is present at a given temperature?

9.38 In the Ag-Cu system (Figure 9.16), three equilibrated phases may be present at:
  a. 930°C   c. 830°C
  b. 880°C   d. 780°C

9.39 During heating, a 72Ag–28Cu alloy (Figure 9.16) may have any but which of the following equilibrium relationships?
  a. ~25% β at 600°C, where β contains ~2% Ag
  b. less than 50% β (95Cu–5Ag) at 700°C
  c. ~74% α(93Ag–7Cu) at 750°C
  d. α(26% Cu) and β(32% Cu) at 800°C

9.40 During cooling, a (20Ag–80Cu) alloy (Figure 9.16) has all but which of the following equilibrium situations?
  a. The first solid forms at 980°C
  b. The first solid contains 5% Ag
  c. The second solid appears at 780°C
  d. The last liquid contains 32% Cu

9.41 Add copper to 100 g of silver at 781°C (Figure 9.16). Assume equilibrium. Which statement is *TRUE*?
  a. Liquid first appears with the addition of exactly 9 g of copper.
  b. The last α disappears when approximately 39 g of copper has been added.
  c. The solubility limit of copper in solid silver (α) is 28% copper.
  d. Solid β first appears with the addition of 92 g of copper.

9.42 All but which of the following statements about strain hardening is correct?
  a. Strain hardening is produced by cold working.
  b. Strain hardening is relieved during annealing above the recrystallization temperature.
  c. With more strain hardening, more time-temperature exposure is required for relief.
  d. Strain hardening is relieved during recrystallization. Recrystallization produces less strained and more ordered structures.

9.43 Which process is used for the high-temperature shaping of many materials?
  a. Reduction   c. Polymerization
  b. Recrystallization d. Extrusion

9.44 All but which of the following processes strengthens metals?
   a. Precipitation processes that produce submicroscopic particles during a low-temperature heat treatment
   b. Increasing the carbon content of low-carbon steels
   c. Annealing above the recrystallization temperature
   d. Mechanical deformation below the recrystallization temperature (cold working)

9.45 All but which of the following statements about the austentite-martensite-bainite transformations are correct?
   a. Pearlite is a stable lamellar mixture consisting of BCC ferrite (α) plus carbide (Fe_3C). It forms through eutectoid decomposition during slow cooling of austenite. Most alloying elements in steel retard this transformation.
   b. Martensite has a body-centered structure of iron that is tetragonal and is supersaturated with carbon. It forms by shear during the rapid quenching of austenite (FCC iron).
   c. Tempering of martensite is accomplished by reheating martensite to precipitate fine particles of carbide within a ferrite matrix, thus producing a tough, strong structure.
   d. Bainite and tempered martensite have distinctly different microstructures.

9.46 Steel can be strengthened by all but which of the following practices?
   a. Annealing
   b. Quenching and tempering
   c. Age or precipitation hardening
   d. Work hardening

9.47 Residual stresses can produce any but which of the following?
   a. Warpage
   b. Distortion in machined metal parts
   c. Cracking of glass
   d. Reduced melting temperatures

9.48 The reaction ($\gamma \rightarrow \alpha + Fe_3C$) is most rapid at:
   a. the eutectoid temperature
   b. 10°C above the eutectoid temperature
   c. the eutectic temperature
   d. 100°C below the eutectoid temperature

9.49 Grain growth, which reduces boundary area, may be expected to:
   a. decrease the thermal conductivity of ceramics
   b. increase the hardness of a solid
   c. decrease the creep rate of a metal
   d. increase the recrystallization rate

9.50 Rapid cooling can produce which one of the following in a material such as sterling silver?
   a. Homogenization
   b. Phase separation
   c. Grain boundary contraction
   d. Supersaturation

9.51 Martensite, which may be obtained in steel, is a:
   a. supersaturated solid solution of carbon in iron
   b. supercooled iron carbide, Fe_3C
   c. undercooled FCC structure of austenite
   d. superconductor with zero resistivity at low temperature

9.52 Alloying elements produce all but which one of the following effects in steels?
   a. They alter the number of atoms in a unit cell of austenite.
   b. They increase the depth of hardening in quenched steel.
   c. They increase the hardness of ferrite in pearlite.
   d. They retard the decomposition of austenite.

9.53 Hardenability may be defined as:
   a. resistance to indentation
   b. the hardness attained for a specified cooling rate
   c. another measure of strength
   d. rate of increased hardness

9.54 The linear portion of the stress-strain diagram of steel is known as the:
   a. irreversible range   c. modulus of elasticity
   b. scant modulus   d. elastic range

9.55 The ultimate (tensile) strength of a material is calculated from:
   a. the applied force divided by the true area at fracture
   b. the applied force times the true area at fracture
   c. the tensile force at the initiation of slip
   d. the applied force and the original area

9.56 All but which one of the following statements about slip are correct?
   a. Slip occurs most readily along crystal planes that are least densely populated.
   b. Slip, or shear along crystal planes, results in an irreversible plastic deformation or permanent set.
   c. Ease of slippage is directly related to the number of low-energy slip planes within the lattice structure.
   d. Slip is impeded by solution hardening, with odd-sized solute atoms serving as anchor points around which slippage does not occur.

9.57 When a metal is cold worked more severely, all but which one of the following generally occur?
 a. The recrystallization temperature decreases.
 b. The tensile strength increases.
 c. Grains become equiaxed.
 d. Slip and/or twinning occur.

9.58 All but which of the following statements about the rusting of iron are correct?
 a. Contact with water or oxygen is required for rusting to occur.
 b. Halides aggravate rusting, a process that involves electrochemical oxidation-reduction reactions.
 c. Contact with a more electropositive metal restricts rusting.
 d. Corrosion occurs in oxygen-rich areas.

9.59 All but which of the following statements about mechanical and thermal failure is *TRUE*?
 a. Creep is time-dependent, plastic deformation that accelerates at increased temperatures. Stress rupture is the failure following creep.
 b. Ductile fracture is characterized by significant amounts of energy absorption and plastic deformation (evidenced by elongation and reduction in cross-sectional area).
 c. Fatigue failure from cyclic stresses is frequency-dependent.
 d. Brittle fracture occurs with little plastic deformation and relatively low energy absorption.

9.60 The stress intensity factor is calculated from:
 a. yield stress and crack depth
 b. applied stress and crack depth
 c. tensile stress and strain rate
 d. crack depth and strain rate

9.61 Service failure from applied loads can occur in all but which of the following cases?
 a. Cyclic loading, tension to compression
 b. Glide normal to the slip plane
 c. Cyclic loading, low tension to higher tension
 d. Stage 2 creep

9.62 Brittle failure becomes more common when:
 a. the endurance limit is increased
 b. the glass-transition temperature is decreased
 c. the critical stress intensity factor is increased
 d. the ductility-transition temperature is increased

9.63 Where applicable, all but which of the following procedures may reduce corrosion?
 a. Avoidance of bimetallic contacts
 b. Sacrificial anodes
 c. Aeration of feed water
 d. Impressed voltages

9.64 A fiber-reinforced rod contains 50 volume percent glass fibers ($E = 70$ GPa, $\sigma_y = 700$ MPa) and 50 volume percent plastic ($E = 7$ GPa). The glass carries what part of a 5000-N tensile load?

a. $[(7000 \text{ MPa})(0.5)]/[70{,}000 \text{ MPa})(0.5)] = 0.1$; $F_{gl} = (0.1)(5000 \text{ N}) = 500$ N
b. $[(70{,}000)(0.5) + 7000(0.5)] = 5000/x$; $x = 0.0002$
c. $[(F_{gl}/0.5A)/(F_p / 0.5A)] = (70/7) = 10$; $F_{gl}/(F_p + F_{gl}) = 10F_p/(10F_p + F_p) = 0.91$
d. $(70)/[70 + 2(7)] = 0.83$

SOLUTIONS

9.1 **c.** Each step through the periodic table introduces an additional proton and electron to a neutral atom.

9.2 **a.** The number of protons are fixed for an individual element. If the number of protons (and electrons) were varied, the chemical properties would be affected.

9.3 **d.** To separate H_2 into hydrogen atoms, the H-to-H bond would have to be broken, thus requiring energy.

9.4 **a.** Unlike ions attract. FCC metals possess four atoms per unit cell; BCC metals have two.

9.5 **c.** The [111] direction passes diagonally through the unit cell. The distance is $a\sqrt{3}$, which equals two repeat distances.

9.6 **b.** The mass is determined from 27 amu per 6.0×10^{23} atoms. Each cell of four atoms has a volume of $(4r/\sqrt{2})^3$.

9.7 **b.** Assuming spherical atoms, there are four atoms of radius r per unit cell. The cube edge is $4r/\sqrt{2}$.

9.8 **c.** There is a double bond between the two carbons. Each hydrogen is held with a single bond.

9.9 **c.** Density is mass/volume. The mass per FCC unit cell is 4 Au × (197 g/6.02×10^{23} Au). The volume per FCC unit cell of a metal is (face diagonal/$\sqrt{2}$)³ or $(4r/\sqrt{2})^3$.

9.10 **b.** Thermoplastic materials are polymerized but soften for molding at elevated temperatures. The polymeric molecules are linear. Thus they include the ethylene-type compounds that are bifunctional (two reaction sites per mer).

Thermosetting materials develop three-dimensional structures that become rigid during processing. For example, phenol is trifunctional and thus forms a network structure. Reheating does not soften them.

9.11 **d.** Benzene is C_6H_6; however, in styrene, one hydrogen is absent at the connection to the C_2H_3–base.

9.12 **d.** Since a cubic crystal has interchangeable x-, y-, and z-axes, the <110> family includes all directions with permutations of 1, 1, and 0 (either + or –). (This is not necessarily true for noncubic crystals.)

9.13 **a.** Since a cubic crystal has interchangeable x-, y-, and z-axes, the {112} family includes all planes with index permutations of 1, 1, and 2 (either + or –). (This is not necessarily true for noncubic crystals.)

9.14 d. (b) and (c) involve individual atoms (point imperfections). (a) is a lineal imperfection. Boundaries result from a two-dimensional mismatch of crystal structures.

9.15 d. There are two types of dislocations: (1) an edge dislocation may be described as an edge of a missing half-plane of atoms; (2) a screw dislocation is the core of a helix.

9.16 d. Unless special efforts are made to grow single crystals, many crystals are nucleated and grow until they encounter neighboring crystals. Each grain is individually oriented.

9.17 c. Zinc is sufficiently near copper in size and electrical behavior to proxy for copper in the crystal structure. It is too big for the interstices.

9.18 b. The 7.5% copper replaces silver atoms. If it is cooled rapidly, the copper is retained in solid solution. The copper atoms interfere with electron movements within the silver.

9.19 a. FCC metals have a higher packing factor than do BCC metals; therefore, with other factors equal, diffusion is reduced.

9.20 c.

$$D = D_o e^{-Q/(RT)} = (0.022 \text{ cm}^2/\text{s}) e^{\frac{-97000 \text{ J/mol}}{8.314 \text{ J/(mol-K)}(1000+273 \text{ K})}} = 2.3 \times 10^{-6} \text{ cm}^2/\text{s}$$

9.21 b. As the grains grow, their volume increases by the third power. Their surface area increases by the square.

9.22 d. Metallic crystals are not ionic and do not have discrete ions.

9.23 a. The interstitial sites are smaller than the atoms in metals.

9.24 b. To balance the charge, each missing Fe^{2+} ion must be compensated by two Fe^{3+} ions. Therefore, out of 100 cation sites, two are vacant, four are Fe^{3+}, and thus 94 are Fe^{2+}.

9.25 a. Using a computational basis of four atoms, $(1 \text{ Mg}^{2+} + 1 \text{ O}^{2-}) + (1 \text{ Fe}^{2+} + 1 \text{ O}^{2-}) = (24 + 16) + (56 + 16) = (40 + 72)$.

9.26 b. The elastic strains between atoms along the boundary follow the same relationships as the strains among atoms within the grains.

9.27 c. Solid solution increases strength (solution hardening). It also decreases conductivity (and increases resistivity). Sterling silver is 92.5Ag–7.5Cu.

9.28 **b.** When atoms are moved from one site to another, bonds are broken and reconstituted. During transition, an activation energy is required to distort the lattice. Small solute atoms, low-melting-point solvents, and lower atomic packing factors in a lattice all require a lower activation energy. Hence activation energy for diffusion is *directly* proportional to the packing factor.

9.29 **a.** On cooling, curve CI is encountered at approximately 420°C (790°F). That curve is the solubility limit of Zn in that liquid. Zinc in excess of the solubility limit separates as the intermetallic compound, $MgZn_2$, which is plotted as the vertical line EI.

9.30 **d.** A ratio of 67 Zn atoms to 33 Mg atoms is 2-to-1; therefore, $MgZn_2$.

9.31 **b.** The eutectic composition is that of a low-melting liquid saturated with two solids. The 0.8 weight percent composition is a solid, not a liquid, at 727°C.

9.32 **d.** The (ferrite + carbide) area extends across the lower part of the phase diagram from nil carbon to 6.7 carbon. At the left side there is no Fe_3C; at the right side there is only carbide (Fe_3C contains 6.7% carbon). The amount of carbide between the two extremes may be determined by linear interpolation.

9.33 **d.** The eutectic temperature is 780°C (1445°F). That is the lowest temperature at which liquid can exist of equilibrium. At 779°C there are approximately equal amounts of α and β, the two solid solutions.

9.34 **d.** The curves of a phase diagram are solubility limits for the phases within the single-phase regions. Since copper is the solvent for the β structure, β has no upper limit of copper solubility (other than 100%).

9.35 **c.** The diffusion flux is proportional to the concentration gradient. (The other choices cited reduce the activation energy needed for diffusion.)

9.36 **a.** Boundary atoms possess higher energy. Therefore, the boundary is reduced and the grain size is increased at temperatures where the atoms can move. The net movement of the atoms is to the larger grains (with less boundary area).

9.37 **a.** Phase diagrams cannot be used to predict crystalline properties.

9.38 **d.** Above the eutectic temperature, (α + L) can be present concurrently, as can (L + β), but not (α + β). Below the eutectic temperature, (α + β) may be present, but no liquid. As the eutectic temperature is passed during cooling or heating, all three phases may coexist.

9.39 **d.** At 800°C, a 72Ag–28Cu alloy is fully liquid, which therefore has 72–28 composition.

9.40 d. During equilibrium cooling, the final liquid for this alloy does not disappear until the eutectic temperature is reached at 780°C. At that temperature, the liquid composition is 72Ag–28Cu.

9.41 b. All α disappears on the right side of the (α + L) field, where the composition is 72Ag–28Cu, or 100g Ag to 39 g Cu.

9.42 c. As the temperature is increased, the atoms gain additional energy and can relocate, eliminating the strain energy that accompanies dislocations. *Less time* is required at higher temperature. *Less time* is also required for a highly cold-worked material because there is additional stored energy present.

9.43 d. Reduction and polymerization involve chemical reactions. Tempering and recrystallization involve reheating but no shape change.

9.44 c. Strength and hardness are increased at the expense of ductility and toughness (opposite of brittleness). The increase is facilitated by microstructures that interfere with dislocation movements. These include a high density of dislocations from plastic deformation, and the presence of many fine, hard particles. Annealing removes dislocations and permits the agglomeration of particles into fewer large particles.

9.45 d. The production of tempered martensite is indicated in (c) above. Bainite is formed by isothermally decomposing austenite directly to a microstructure of fine carbide particles within a ferrite matrix. Although the processing differs, the resulting microstructure and properties are nearly identical.

9.46 a. Annealing removes the hardness that was introduced by cold work. Quenched and tempered steels are harder with higher carbon contents, because more hard carbide particles are present. Alloying elements perform several hardening functions: They solution-harden the ferrite matrix; they slow down grain growth; and they delay the formation of pearlite, thus permitting more martensite with slower cooling rates (in turn, more tempered martensite may be realized farther below the quenched surface).

9.47 d. Stresses will relax below the melting temperature. Tensile stresses facilitate the cracking of brittle materials; compression limits cracking.

9.48 d. The reaction occurs only below the eutectoid temperature.

9.49 c. While grain boundaries interfere with slip at low temperatures, they facilitate creep at elevated temperatures.

9.50 d. The processing step of rapid cooling, such as quenching, retains the structures that existed at higher temperatures, even though a solubility limit is exceeded.

9.51 a. Martensite is a transition phase between austenite and ferrite, which retains carbon interstitially. Given an opportunity, the carbon separates as Fe_3C.

9.52 a. Alloying elements can dissolve substitutionally in austenite, which remains FCC.

9.53 b. Hardenability may be described as the ability, or the ease, by which martensitic hardness is obtained at various cooling rates (as located on an end-quenched, or Jominy, test).

9.54 d. The ratio of stress-to-strain is defined as the elastic modulus.

9.55 d. $\sigma_u = F/A_0$

9.56 a. Slip occurs most readily in directions that have the shortest steps, and along planes that are farthest apart. The latter are automatically the planes that are most densely populated.

9.57 c. Cold working—such as rolling, forging, drawing, or extrusion—deforms the material at temperatures below the recrystallization temperature. Strain hardening occurs, increasing both the yield and ultimate strength. Internal strains and minute cracks are introduced as slip or twinning occurs. Ductility, elongation, and the recrystallization temperature are decreased. A preferred grain orientation is introduced in the direction of elongation, and the grains are flattened.

9.58 d. Since oxygen is required for rust formation, oxygen-depleted areas become the anode and are corroded. This may lead to pitting, particularly if rust or other corrosion products are accumulated locally to prevent the access of oxygen. Iron and other metals may be protected from corrosion by the presence of a more electropositive metal such as magnesium or zinc. This is the reason for coating steel with zinc to produce galvanized sheet.

9.59 c. Although fatigue strength is not sensitive to temperature or loading rates, it is very sensitive to surface imperfections from which cracks originate and propagate.

9.60 b. The stress intensity factor is proportional to the applied stress and the square root of the crack depth.

9.61 b. Glide occurs parallel to the slip plane.

9.62 d. Brittleness exists below T_{DT}.

9.63 c. Corrosion commonly occurs in the combined presence of oxygen and water. Protection may be obtained by making a cathode out of the critical part or by avoiding air.

9.64 c. With equal strains, $(\sigma_{gl}/\sigma_p) = E_{gl}/E_p = 10$. Likewise with equal areas, $F_{gl} = 10 F_p$, and $F_{gl}/(F_{gl} + F_p) = 10/(10 + 1)$.

CHAPTER 10

Thermodynamics and Phase Equilibrium

OUTLINE

THERMODYNAMIC SYSTEMS 376
Terminology ■ Thermodynamic Properties ■ Energy Definitions

FIRST LAW OF THERMODYNAMICS 377
Closed System ■ Open System

SECOND LAW OF THERMODYNAMICS 379
Carnot Cycle ■ Entropy

PROPERTIES OF PURE SUBSTANCES 382
Use of Steam Tables ■ Use of R-134a and NH_3 Tables and P-h Diagrams

IDEAL GASES 389
The Ideal Gas Law ■ Ideal Gas Mixtures and Partial Pressure ■ Enthalpy and Internal Energy Changes

PHASE EQUILIBRIA 395
Vapor-Liquid Equilibria ■ Binary and Multicomponent Systems ■ The Phase Rule and Duhem's Theorem ■ T-xy and x-y Diagrams for Binary Systems ■ Immiscible Liquids ■ Ideal Mixtures (Raoult's Law) ■ Relative Volatility α_{ij} ■ Henry's Law

PROCESSES 402

CYCLES 404
Rankine Cycle (Steam) ■ Vapor Compression Cycle (Refrigeration) ■ Psychrometrics ■ Otto Cycle (Gasoline Engine)

HEAT TRANSFER 415
Conduction ■ Convection ■ Radiation

SELECTED SYMBOLS AND ABBREVIATIONS 419

PROBLEMS 420

SOLUTIONS 429

Thermodynamics deals with the transformation of energy from one form to another in macro systems. Experience has shown that all energy transformations occur under certain restrictions, which are known as laws of thermodynamics. These laws together with definitions of the properties of materials allow a chemical engineer to study a wide variety of problems connected with physical and chemical processes. The most important problems are the estimation of heat and work requirements of a process, determination of chemical reaction equilibrium, and equilibrium for the transfer of chemical species between phases.

THERMODYNAMIC SYSTEMS

Terminology

We begin this review of thermodynamics with some basic definitions.

System—a portion of the universe (e.g., a substance or a group of substances) set apart for study.

Closed system—a system with constant mass. There is no exchange of matter between the system and the surroundings. A closed system is thermally isolated when its enclosing walls allow no flow of heat into or out of the system. It is mechanically isolated if it is enclosed by rigid walls. It is completely isolated if neither matter nor energy in any form can be added to or removed from the system.

Open system—a system with variable mass (mass is transferred across the boundaries of the system).

Reversible process—one that begins from an equilibrium state of a system and proceeds under conditions of balanced forces in such a manner that its direction can be reversed by applying an infinitesimal driving force in the opposite direction, restoring both the system and its surroundings to their initial equilibrium state. In a reversible process, there is no degradation of energy taking place and availability of the energy of the combined system and its surroundings is the same before or after the process as the system and its surroundings are restored to their original equilibrium state.

As an example of a reversible process, one may consider the vaporization of a liquid (enclosed in a cylinder with a frictionless piston and in contact with an isothermal reservoir) under its own vapor pressure. At any time, an infinitely small increase in pressure on the piston will cause condensation, whereas an infinitely small decrease in pressure will result in vaporization of the liquid. This is, however, an ideal situation, as the concepts of a frictionless piston and an isothermal reservoir are assumptions only and are far removed from actual situations.

Actual processes are invariably irreversible and therefore there is in reality a decrease in available energy. A few examples of irreversible processes are (a) flow of heat from one body to the other under the influence of a finite temperature difference, (b) free expansion of a gas, (c) mixing of hot and cold fluids, and (d) a spontaneous chemical reaction.

In a later section, reversibility will again be reviewed when the second law of thermodynamics and the concept of entropy are postulated.

Thermodynamic Properties

Any measurable characteristic of a system in terms of fundamental dimensions is called its property, for example, temperature, pressure, mass, area, volume, surface tension, and so on.

Intensive properties, such as pressure and temperature, are independent of mass.

Extensive properties depend on the mass of the system. Volume, internal energy, and enthalpy are examples of extensive properties. However, when these extensive properties are based on a unit such as a unit mass or mol, they become intensive properties because their values do not depend upon the total material actually present in the system. Specific volume or density, internal energy per unit mass or mol, and enthalpy per unit basis are also examples of intensive properties.

State properties depend upon the thermodynamic state of the system and are independent of the path the system takes to reach that state. Some state properties are: pressure P, temperature T, specific volume V, internal energy U, enthalpy H, entropy S, Gibbs free energy G, Helmholtz free energy A, and heat capacities at constant pressure C_P and at constant volume C_V.

Energy Definitions

Definitions of some energy terms are needed before the energy balance for a system is developed. The energy terms are

Potential energy, PE—external energy possessed by a system due to its position in a gravitational field

Kinetic energy, KE—all energy of a system associated with the macroscopic motion of its mass $= M(u^2/2g_c)$

Internal energy, U—total energy possessed by a system due to its component molecules, their relative positions, and their movement; for a closed system (constant mass), it is a state property

Heat energy, Q—energy transferred across the boundaries of a system because of the temperature difference

Work, W—work is defined as W = force × distance; it is a form of energy transferred across the boundaries of a system and is not system state property

Enthalpy, H—defined as $H = U + PV$

Both U and H are determined relative to a selected reference state. The reference state of zero enthalpy is chosen as 1 atm and 0 °C. For water, it is taken as 0 °C and its own vapor pressure at 0 °C.

FIRST LAW OF THERMODYNAMICS

The first law of thermodynamics is a bookkeeping system to keep track of the energy quantities defined above. As for notation, variables with capital letters generally refer to extensive properties, while lowercase letters generally refer to intensive properties.

Closed System

For a typical closed thermodynamic system (see Figure 10.1), the kinetic and potential energy are not important, and since there is no flow, the bookkeeping is simple and reduces to

$$u_1 + q = u_2 + w$$

Open System

For an open system (Figure 10.2), the bookkeeping system yields

$$u_1 + p_1 v_1 + \frac{V_1^2}{2g_c} + \frac{Z_1 g}{g_c} + q = u_2 + p_2 v_2 + \frac{V_2^2}{2g_c} + \frac{Z_2 g}{g_c} + w$$

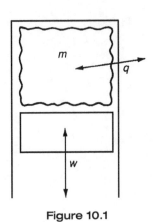

Figure 10.1

In many common open-system processes, the kinetic energy ($V^2/2g_c$) and the potential energy (Zg/g_c) are not important. For convenience, u and pv are combined into h (enthalpy), and the equation then reduces to

$$h_1 + q = h_2 + w$$

The thermodynamic sign convention for heat and work is that q_{in} is positive and w_{out} is positive. This is the normal flow of heat and work in an engine or power plant.

Note also that a "change in" a property or state refers to the second point value minus the first point value.

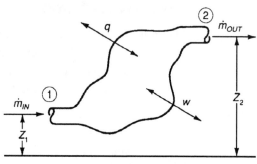

Figure 10.2

Care must be taken to keep the units consistent. For example, the normal units for internal energy, heat, and work are kJ/kg. In the SI system, $V^2/2g_c$ and Zg/g_c must be divided by 1000. Remember that g_c in the SI system has a value of 1 but is 32.2 in the U.S. system.

Example 10.1

A piston-cylinder contains 5 kg of air. During a compression process 100 kJ of heat is removed while 250 kJ of work is done on the air. Find the change in internal energy of the air.

Solution

The system is a closed system since the mass is fixed. So

$$u_1 + q = u_2 + w$$

Here, $q = -100$ kJ, since heat leaving a system is negative, and $w = -250$ kJ, since work done *on* a system is negative. Thus

$$u_2 - u_1 = -100 - (-250) = 150 \text{ kJ}, \quad \frac{150 \text{ kJ}}{5 \text{ kg}} = 30 \text{ kJ/kg}$$

Example 10.2

Heated air enters a turbine at a flow rate of 5 kg/s. The entering and leaving conditions are shown in the following table. The heat loss from the turbine is 50 kW. Find the power produced.

	Inlet	Exit
Pressure, kPa	1000	100
Temperature, K	800	500
Velocity, m/s	100	200
Specific internal energy, kJ/kg	137	85
Specific volume, m³/kg	0.23	1.44
Elevation, m	3	10

Solution

The system has mass flowing across the boundaries, so it is an open system.

$$u_1 + p_1 v_1 + \frac{Z_1 g}{g_c} + \frac{V_1^2}{2g_c} + q = u_2 + p_2 v_2 + \frac{Z_2 g}{g_c} + \frac{V_2^2}{2g_c} + w$$

$$w = (u_1 - u_2) + (p_1 v_1 - p_2 v_2) + \frac{(Z_1 - Z_2)g}{g_c} + \frac{V_1^2 - V_2^2}{2g_c} + q$$

$$= (137 - 85) + (1000)(0.23) - 100(1.44)$$

$$+ \frac{(3-10)}{1000} \frac{9.81}{1.0} + \frac{100^2 - 200^2}{2(1)(1000)} + \left(-\frac{50}{5}\right)$$

$$= 52 + 86 - 0.069 - 15 - 10 = 112.9 \ \frac{kJ}{kg}$$

$$\dot{W} = w(\dot{m}) = 112.9 \times 5 = 564.5 \text{ kW}$$

Example 10.3

Air at 27°C is heated to 927°C. Find the change in enthalpy and internal energy, treating air as a perfect gas (c_p and c_v constant).

Solution

For air at room temperature, $c_p = 1.00$ and $c_v = 0.718$ kJ/kg, so

$$h_2 - h_1 = (927 - 27) \times 1.00 = 900.0 \text{ kJ/kg}$$
$$u_2 - u_1 = (927 - 27) \times 0.718 = 646.2 \text{ kJ/kg}$$

SECOND LAW OF THERMODYNAMICS

The concept of reversibility of a process is the basis on which the second law of thermodynamics is postulated. There are various statements of this law. One statement is that *all the spontaneous processes are invariably irreversible to some extent and are accompanied by a degradation of energy.*

Heat cannot be converted into work quantitatively and cannot be transferred from a lower temperature to a higher temperature without the aid of an external agency. This principle is covered by another statement of the second law of thermodynamics: *No practical engine can convert heat into work quantitatively nor it is impossible for a self-acting machine unaided by an external agency to transfer heat from a lower temperature to a higher one.*

Carnot Cycle

Useful statements of the second law are as follows:

1. Whenever energy is transferred, some energy is reduced to a lower level.
2. No heat cycle is possible without the rejection of some heat.
3. A Carnot cycle converts the maximum amount of heat into work; it has the highest thermal efficiency.
4. All Carnot cycles operating between two temperature reservoirs have the same efficiency.
5. A Carnot machine's efficiency, or coefficient of performance (COP), is a function of only the two reservoir temperatures.

A Carnot cycle consists of the following four processes:

4–1 Reversible adiabatic compression
1–2 Reversible adiabatic constant temperature heat addition
2–3 Reversible adiabatic expansion
3–4 Reversible constant temperature heat rejection

The normal property diagrams are shown in Figures 10.3 and 10.4. If the processes proceed in a clockwise direction, the Carnot engine operates as a power-producing engine; if in a counterclockwise direction, the engine is a refrigerator or a heat pump.

The efficiencies, or COPs, are

$$\eta = \frac{W_{out}}{Q_{in}} = 1 - \frac{T_C}{T_H}$$

$$COP_{REFR} = \frac{Q_{in}}{W_{in}} = \frac{T_C}{T_H - T_C}$$

$$COP_{HEATPUMP} = \frac{Q_{out}}{W_{in}} = \frac{T_H}{T_H - T_C}$$

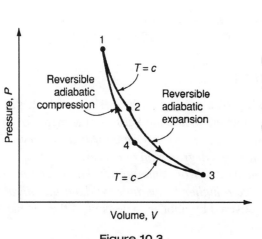

Figure 10.3

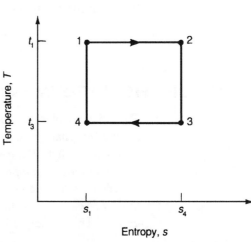

Figure 10.4

Example 10.4

A Carnot machine operates between a hot reservoir at 200°C and a cold reservoir at 20°C. (i) When operated as an engine, it receives 1000 kJ/kg; find the work output. (ii) Find the COP when the machine is operated as a refrigerator and as a heat pump.

Solution

(i)
$$\eta = \frac{T_H - T_C}{T_H} = \frac{200 - 20}{473} = 0.381$$
$$W = \eta (Q_{in}) = 0.381(1000) = 381 \text{ kJ/kg}$$

(ii)
$$\text{COP}_{REFR} = \frac{T_C}{T_H - T_C} = \frac{293}{200 - 20} = 1.63$$
$$\text{COP}_{HEATPUMP} = \frac{T_H}{T_H - T_C} = \text{COP}_{REFR} + 1.0 = 2.63$$

Entropy

Entropy is a state property used in the evaluation of the second law of thermodynamics that accounts for heat flow but not work crossing the system boundaries. The following statements are useful in solving problems:

1. Natural processes (which typically involve friction) result in an increase in entropy.
2. Entroy will always *increase* when heat is added.
3. Entropy will remain *constant* when processes are reversible and adiabatic.
4. Entropy can *decrease* only when heat is removed.

For reversible processes,
$$ds = \frac{dq}{T}$$
$$T\,ds = du + P\,dv = dh - v\,dP$$

For an ideal gas,
$$s_2 - s_1 = c_v \ln \frac{T_2}{T_1} + R \ln \frac{v_2}{v_1}$$
$$= c_p \ln \frac{T_2}{T_1} - R \ln \frac{p_2}{p_1}$$

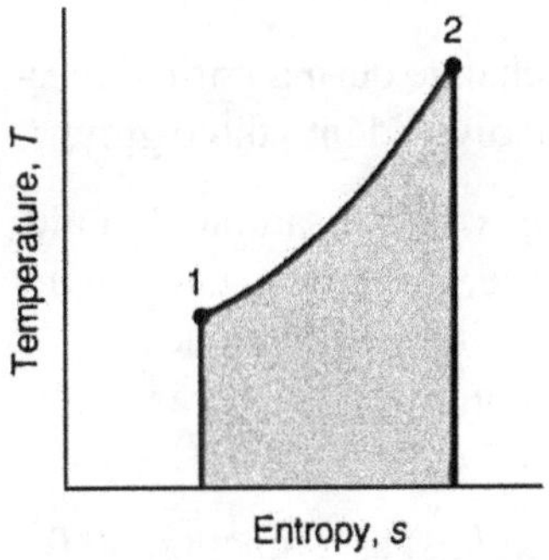

Just as work for a closed system,
$$W = \int_1^2 P\,dv$$

can be shown as an area on a *P-V* diagram (Figure 10.5), so can heat,
$$Q = \int_1^2 T\,ds$$

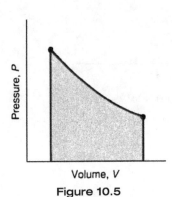

Figure 10.5

Figure 10.6

be shown as an area on a *T-s* diagram (Figure 10.6), and the area *enclosed* by the process lines on a *T-s* diagram shows the *net* heat flow in a cycle. This, of course, is equal to the *net* work.

An **isentropic process** is defined as one that is reversible and adiabatic. Of course, on a property diagram showing entropy (*s*), the process would appear as a straight line. For several important thermodynamic devices, the isentropic process is a standard of comparison and is used in the calculation of the component efficiency (turbine, compressor, pump, nozzle).

Since all natural processes produce an increase in entropy, the ideal (isentropic) and the actual processes can be compared, as shown in Figures 10.7 and 10.8.

$$\eta_{\text{turbine nozzle}} = \frac{h_1 - h_3}{h_1 - h_2}$$

$$\eta_{\text{compr., pump}} = \frac{h_2 - h_1}{h_3 - h_1}$$

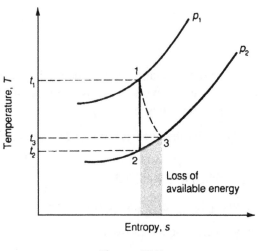

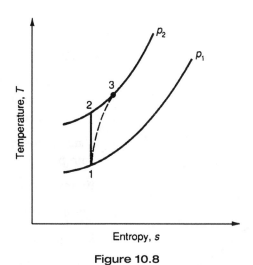

Figure 10.7

Figure 10.8

PROPERTIES OF PURE SUBSTANCES

Use of Steam Tables

Property tables for substances that go through a phase change during normal thermodynamic processes, such as H_2O, R-12, and NH_3, are divided into three groups:

- Saturated Tables. These show the properties (*v*, *u*, *h*, *s*) of the saturated liquid R-134a at *P* = 2 and saturated vapor. For convenience, there are usually two sets: one using *T* as the entering argument and one using *P*. The highest temperature/pressure entry is usually the critical point. The quality *x* is needed to define properties in the mixture region.

- Superheated Tables. These show the properties (*v*, *u*, *h*, *s*) as a function of *T* and *P* in the superheated area to the right of the saturated vapor curve and are usually grouped by pressure. *Any two* properties (*v*, *u*, *h*, *s*, *T*, *P*) may be used to define the state. The saturated state is usually noted. At moderate pressures and temperatures well away from saturation, perfect gas relationships may be used as an approximation.

- Subcooled or Compressed Liquid Tables. These tables show properties (v, u, h, and s) as a function of T and P in the area to the left of the saturated liquid curve and are usually grouped by pressure. As with the Superheated Tables, any two properties may be used to define the state, and the saturated state is usually noted. For points in the region that are below the tabulated pressures, the properties are approximated as those of the saturated liquid at the same temperature (v, u, h, and s are weak functions of pressure).

The procedure for finding the state-point properties for given data where the condition of the substances is not defined is best done in a structured manner:

1. Always look at the saturation tables first to determine whether the state point is liquid, vapor, or "wet" (Figure 10.9).

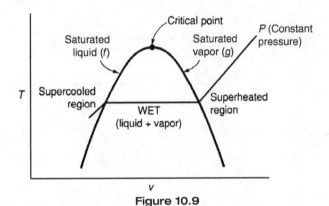

Figure 10.9

2. When one is given T or P and h, v, or u:

   a. If h, v, or u is between f and g, the point is in the wet region (inside the dome). Calculate the quality x using saturation properties and given properties. For example,

   $$x = \frac{u - u_f}{u_{fg}}$$

   b. If h, v, or u is greater than the saturated-vapor value, the point is in the superheated region. Locate the properties in the superheated tables.

   c. If h, v, or u is less than the saturated-liquid value, the point is in the subcooled liquid region. Locate the properties in the liquid tables and/or calculate from the saturated-liquid properties.

3. Given T or P and x:

   Go directly to the saturated tables, since an intermediate value of x implies it is in the "wet" region.

4. Given T and P:

   Compare the saturation temperature with the given temperature. The saturation temperature is read at the given pressure.

   If T is greater than T_{sat}, the point is superheated vapor.
   If T is less than T_{sat}, the point is subcooled liquid.

If x is given, then v, u, h, and s can be calculated using values from the steam tables inserted into the following equations:

$$v = xv_g + (1-x)v_f \quad v = v_f + xv_{fg}$$
$$u = xu_g + (1-x)u_f \quad u = u_f + xu_{fg}$$
$$h = xh_g + (1-x)h_f \quad h = h_f + xh_{fg}$$
$$s = xs_g + (1-x)s_f \quad s = s_f + xs_{fg}$$

Example 10.5

Fill in the following table for steam (water) using the extracted tabulations.

	T, °C	P, kPa	x, %	h, kJ/kg	u, kJ/kg	v, m³/kg
a).	200	—	—	852.45	—	—
b).	—	143.3	—	—	1000	—
c).	300	800	—	—	—	—
d).	200	5000	—	—	—	—
e).	—	400	—	—	—	0.85
f).	300	—	80	—	—	—

Superheated Water Tables

T Temp. °C	v m³/kg	u kJ/kg	h kJ/kg	s kJ/(kg·K)	v m³/kg	u kJ/kg	h kJ/kg	s kJ/(kg·K)
	\multicolumn{4}{c}{$P = 0.01$ MPa (45.81°C)}	\multicolumn{4}{c}{$P = 0.05$ MPa (81.33°C)}
Sat.	14.674	2437.9	2584.7	8.1502	3.240	2483.9	2645.9	7.5939
50	14.869	2443.9	2592.6	8.1749				
100	17.196	2515.5	2687.5	8.4479	3.418	2511.6	2682.5	7.6947
150	19.512	2587.9	2783.0	8.6882	3.889	2585.6	2780.1	7.9401
200	**21.825**	**2661.3**	**2879.5**	**8.9038**	**4.356**	**2659.9**	**2877.7**	**8.1580**
250	24.136	2736.0	2977.3	9.1002	4.820	2735.0	2976.0	8.3556
300	26.445	2812.1	3076.5	9.2813	5.284	2811.3	3075.5	8.5373
400	31.063	2968.9	3279.6	9.6077	6.209	2968.5	3278.9	8.8642
500	35.679	3132.3	3489.1	9.8978	7.134	3132.0	3488.7	9.1546
600	**40.295**	**3302.5**	**3705.4**	**10.1608**	**8.057**	**3302.2**	**3705.1**	**9.4178**
700	44.911	3479.6	3928.7	10.4028	8.981	3479.4	3928.5	9.6599
800	49.526	3663.8	4159.0	10.6281	9.904	3663.6	4158.9	9.8852
900	54.141	3855.0	4396.4	10.8396	10.828	3854.9	4396.3	10.0967
1000	58.757	4053.0	4640.6	11.0393	11.751	4052.9	4640.5	10.2964
1100	**63.372**	**4257.5**	**4891.2**	**11.2287**	**12.674**	**4257.4**	**4891.1**	**10.4859**
1200	67.987	4467.9	5147.8	11.4091	13.597	4467.8	5147.7	10.6662
1300	72.602	4683.7	5409.7	11.5811	14.521	4683.6	5409.6	10.8382
	\multicolumn{4}{c}{$P = 0.10$ MPa (99.63°C)}	\multicolumn{4}{c}{$P = 0.20$ MPa (120.23°C)}
Sat.	1.6940	2506.1	2675.5	7.3594	0.8857	2529.5	2706.7	7.1272
100	1.6958	2506.7	2676.2	7.3614				
150	1.9364	2582.8	2776.4	7.6134	0.9596	2576.9	2768.8	7.2795
200	2.172	2658.1	2875.3	7.8343	1.0803	2654.4	2870.5	7.5066
250	**2.406**	**2733.7**	**2974.3**	**8.0333**	**1.1988**	**2731.2**	**2971.0**	**7.7086**
300	2.639	2810.4	3074.3	8.2158	1.3162	2808.6	3071.8	7.8926
400	3.103	2967.9	3278.2	8.5435	1.5493	2966.7	3276.6	8.2218
500	3.565	3131.6	3488.1	8.8342	1.7814	3130.8	3487.1	8.5133
600	4.028	3301.9	3704.4	9.0976	2.013	3301.4	3704.0	8.7770
700	**4.490**	**3479.2**	**3928.2**	**9.3398**	**2.244**	**3478.8**	**3927.6**	**9.0194**
800	4.952	3663.5	4158.6	9.5652	2.475	3663.1	4158.2	9.2449
900	5.414	3854.8	4396.1	9.7767	2.705	3854.5	4395.8	9.4566
1000	5.875	4052.8	4640.3	9.9764	2.937	4052.5	4640.0	9.6563
1100	6.337	4257.3	4891.0	10.1659	3.168	4257.0	4890.7	9.8458
1200	**6.799**	**4467.7**	**5147.6**	**10.3463**	**3.399**	**4467.5**	**5147.5**	**10.0262**
1300	7.260	4683.5	5409.5	10.5183	3.630	4683.2	5409.3	10.1982
	\multicolumn{4}{c}{$P = 0.40$ MPa (143.63°C)}	\multicolumn{4}{c}{$P = 0.60$ MPa (158.85°C)}
Sat.	0.4625	2553.6	2738.6	6.8959	0.3157	2567.4	2756.8	6.7600
150	0.4708	2564.5	2752.8	6.9299				
200	0.5342	2646.8	2860.5	7.1706	0.3520	2638.9	2850.1	6.9665
250	0.5951	2726.1	2964.2	7.3789	0.3938	2720.9	2957.2	7.1816
300	**0.6548**	**2804.8**	**3066.8**	**7.5662**	**0.4344**	**2801.0**	**3061.6**	**7.3724**
350					0.4742	2881.2	3165.7	7.5464
400	0.7726	2964.4	3273.4	7.8985	0.5137	2962.1	3270.3	7.7079
500	0.8893	3129.2	3484.9	8.1913	0.5920	3127.6	3482.8	8.0021
600	1.0055	3300.2	3702.4	8.4558	0.6697	3299.1	3700.9	8.2674
700	**1.1215**	**3477.9**	**3926.5**	**8.6987**	**0.7472**	**3477.0**	**3925.3**	**8.5107**
800	1.2372	3662.4	4157.3	8.9244	0.8245	3661.8	4156.5	8.7367
900	1.3529	3853.9	4395.1	9.1362	0.9017	3853.4	4394.4	8.9486
1000	1.4685	4052.0	4639.4	9.3360	0.9788	4051.5	4638.8	9.1485
1100	1.5840	4256.5	4890.2	9.5256	1.0559	4256.1	4889.6	9.3381
1200	**1.6996**	**4467.0**	**5146.8**	**9.7060**	**1.1330**	**4466.5**	**5146.3**	**9.5185**
1300	1.8151	4682.8	5408.8	9.8780	1.2101	4682.3	5408.3	9.6906
	\multicolumn{4}{c}{$P = 0.80$ MPa (170.43°C)}	\multicolumn{4}{c}{$P = 1.00$ MPa (179.91°C)}
Sat.	0.2404	2576.8	2769.1	6.6628	0.194 44	2583.6	2778.1	6.5865
200	0.2608	2630.6	2839.3	6.8158	0.2060	2621.9	2827.9	6.6940
250	0.2931	2715.5	2950.0	7.0384	0.2327	2709.9	2942.6	6.9247
300	0.3241	2797.2	3056.5	7.2328	0.2579	2793.2	3051.2	7.1229
350	**0.3544**	**2878.2**	**3161.7**	**7.4089**	**0.2825**	**2875.2**	**3157.7**	**7.3011**
400	0.3843	2959.7	3267.1	7.5716	0.3066	2957.3	3263.9	7.4651
500	0.4433	3126.0	3480.6	7.8673	0.3541	3124.4	3478.5	7.7622
600	0.5018	3297.9	3699.4	8.1333	0.4011	3296.8	3697.9	8.0290
700	0.5601	3476.2	3924.2	8.3770	0.4478	3475.3	3923.1	8.2731
800	**0.6181**	**3661.1**	**4155.6**	**8.6033**	**0.4943**	**3660.4**	**4154.7**	**8.4996**
900	0.6761	3852.8	4393.7	8.8153	0.5407	3852.2	4392.9	8.7118
1000	0.7340	4051.0	4638.2	9.0153	0.5871	4050.5	4637.6	8.9119
1100	0.7919	4255.6	4889.1	9.2050	0.6335	4255.1	4888.6	9.1017
1200	0.8497	4466.1	5145.9	9.3855	0.6798	4465.6	5145.4	9.2822
1300	**0.9076**	**4681.8**	**5407.9**	**9.5575**	**0.7261**	**4681.3**	**5407.4**	**9.4543**

Saturated Water - Temperature Table											
		Specific Volume m³/kg		Internal Energy kJ/kg			Enthalpy kJ/kg			Entropy kJ/(kg·K)		
Temp. °C T	Sat. Press. kPa P_{sat}	Sat. liquid v_f	Sat. vapor v_g	Sat. liquid u_f	Evap. u_{fg}	Sat. vapor u_g	Sat. liquid h_f	Evap. h_{fg}	Sat. vapor h_g	Sat. liquid s_f	Evap. s_{fg}	Sat. vapor s_g
0.01	0.6113	0.001 000	206.14	0.00	2375.3	2375.3	0.01	2501.3	2501.4	0.0000	9.1562	9.1562
5	0.8721	0.001 000	147.12	20.97	2361.3	2382.3	20.98	2489.6	2510.6	0.0761	8.9496	9.0257
10	1.2276	0.001 000	106.38	42.00	2347.2	2389.2	42.01	2477.7	2519.8	0.1510	8.7498	8.9008
15	1.7051	0.001 001	77.93	62.99	2333.1	2396.1	62.99	2465.9	2528.9	0.2245	8.5569	8.7814
20	2.339	0.001 002	57.79	83.95	2319.0	2402.9	83.96	2454.1	2538.1	0.2966	8.3706	8.6672
25	3.169	0.001 003	43.36	104.88	2304.9	2409.8	104.89	2442.3	2547.2	0.3674	8.1905	8.5580
30	4.246	0.001 004	32.89	125.78	2290.8	2416.6	125.79	2430.5	2556.3	0.4369	8.0164	8.4533
35	5.628	0.001 006	25.22	146.67	2276.7	2423.4	146.68	2418.6	2565.3	0.5053	7.8478	8.3531
40	7.384	0.001 008	19.52	167.56	2262.6	2430.1	167.57	2406.7	2574.3	0.5725	7.6845	8.2570
45	9.593	0.001 010	15.26	188.44	2248.4	2436.8	188.45	2394.8	2583.2	0.6387	7.5261	8.1648
50	12.349	0.001 012	12.03	209.32	2234.2	2443.5	209.33	2382.7	2592.1	0.7038	7.3725	8.0763
55	15.758	0.001 015	9.568	230.21	2219.9	2450.1	230.23	2370.7	2600.9	0.7679	7.2234	7.9913
60	19.940	0.001 017	7.671	251.11	2205.5	2456.6	251.13	2358.5	2609.6	0.8312	7.0784	7.9096
65	25.03	0.001 020	6.197	272.02	2191.1	2463.1	272.06	2346.2	2618.3	0.8935	6.9375	7.8310
70	31.19	0.001 023	5.042	292.95	2176.6	2569.6	292.98	2333.8	2626.8	0.9549	6.8004	7.7553
75	38.58	0.001 026	4.131	313.90	2162.0	2475.9	313.93	2321.4	2635.3	1.0155	6.6669	7.6824
80	47.39	0.001 029	3.407	334.86	2147.4	2482.5	334.91	2308.8	2643.7	1.0753	6.5369	7.6122
85	57.83	0.001 033	2.828	355.84	2132.6	2488.4	355.90	2296.0	2651.9	1.1343	6.4102	7.5445
90	70.14	0.001 036	2.361	376.85	2117.7	2494.5	376.92	2283.2	2660.1	1.1925	6.2866	7.4791
95	84.55	0.001 040	1.982	397.88	2102.7	2500.6	397.96	2270.2	2668.1	1.2500	6.1659	7.4159
	MPa											
100	0.101 35	0.001 044	1.6729	418.94	2087.6	2506.5	419.04	2257.0	2676.1	1.3069	6.0480	7.3549
105	0.120 82	0.001 048	1.4194	440.02	2072.3	2512.4	440.15	2243.7	2683.8	1.3630	5.9328	7.2958
110	0.143 27	0.001 052	1.2102	461.14	2057.0	2518.1	461.30	2230.2	2691.5	1.4185	5.8202	7.2387
115	0.169 06	0.001 056	1.0366	482.30	2041.4	2523.7	482.48	2216.5	2699.0	1.4734	5.7100	7.1833
120	0.198 53	0.001 060	0.8919	503.50	2025.8	2529.3	503.71	2202.6	2706.3	1.5276	5.6020	7.1296
125	0.2321	0.001 065	0.7706	524.74	2009.9	2534.6	524.99	2188.5	2713.5	1.5813	5.4962	7.0775
130	0.2701	0.001 070	0.6685	546.02	1993.9	2539.9	546.31	2174.2	2720.5	1.6344	5.3925	7.0269
135	0.3130	0.001 075	0.5822	567.35	1977.7	2545.0	567.69	2159.6	2727.3	1.6870	5.2907	6.9777
140	0.3613	0.001 080	0.5089	588.74	1961.3	2550.0	589.13	2144.7	2733.9	1.7391	5.1908	6.9299
145	0.4154	0.001 085	0.4463	610.18	1944.7	2554.9	610.63	2129.6	2740.3	1.7907	5.0926	6.8833
150	0.4758	0.001 091	0.3928	631.68	1927.9	2559.5	632.20	2114.3	2746.5	1.8418	4.9960	6.8379
155	0.5431	0.001 096	0.3468	653.24	1910.8	2564.1	653.84	2098.6	2752.4	1.8925	4.9010	6.7935
160	0.6178	0.001 102	0.3071	674.87	1893.5	2568.4	675.55	2082.6	2758.1	1.9427	4.8075	6.7502
165	0.7005	0.001 108	0.2727	696.56	1876.0	2572.5	697.34	2066.2	2763.5	1.9925	4.7153	6.7078
170	0.7917	0.001 114	0.2428	718.33	1858.1	2576.5	719.21	2049.5	2768.7	2.0419	4.6244	6.6663
175	0.8920	0.001 121	0.2168	740.17	1840.0	2580.2	741.17	2032.4	2773.6	2.0909	4.5347	6.6256
180	1.0021	0.001 127	0.194 05	762.09	1821.6	2583.7	763.22	2015.0	2778.2	2.1396	4.4461	6.5857
185	1.1227	0.001 134	0.174 09	784.10	1802.9	2587.0	785.37	1997.1	2782.4	2.1879	4.3586	6.5465
190	1.2544	0.001 141	0.156 54	806.19	1783.8	2590.0	807.62	1978.8	2786.4	2.2359	4.2720	6.5079
195	1.3978	0.001 149	0.141 05	828.37	1764.4	2592.8	829.98	1960.0	2790.0	2.2835	4.1863	6.4698
200	1.5538	0.001 157	0.127 36	850.65	1744.7	2595.3	852.45	1940.7	2793.2	2.3309	4.1014	6.4323
205	1.7230	0.001 164	0.115 21	873.04	1724.5	2597.5	875.04	1921.0	2796.0	2.3780	4.0172	6.3952
210	1.9062	0.001 173	0.104 41	895.53	1703.9	2599.5	897.76	1900.7	2798.5	2.4248	3.9337	6.3585
215	2.104	0.001 181	0.094 79	918.14	1682.9	2601.1	920.62	1879.9	2800.5	2.4714	3.8507	6.3221
220	2.318	0.001 190	0.086 19	940.87	1661.5	2602.4	943.62	1858.5	2802.1	2.5178	3.7683	6.2861
225	2.548	0.001 199	0.078 49	963.73	1639.6	2603.3	966.78	1836.5	2803.3	2.5639	3.6863	6.2503
230	2.795	0.001 209	0.071 58	986.74	1617.2	2603.9	990.12	1813.8	2804.0	2.6099	3.6047	6.2146
235	3.060	0.001 219	0.065 37	1009.89	1594.2	2604.1	1013.62	1790.5	2804.2	2.6558	3.5233	6.1791
240	3.344	0.001 229	0.059 76	1033.21	1570.8	2604.0	1037.32	1766.5	2803.8	2.7015	3.4422	6.1437
245	3.648	0.001 240	0.054 71	1056.71	1546.7	2603.4	1061.23	1741.7	2803.0	2.7472	3.3612	6.1083
250	3.973	0.001 251	0.050 13	1080.39	1522.0	2602.4	1085.36	1716.2	2801.5	2.7927	3.2802	6.0730
255	4.319	0.001 263	0.045 98	1104.28	1596.7	2600.9	1109.73	1689.8	2799.5	2.8383	3.1992	6.0375
260	4.688	0.001 276	0.042 21	1128.39	1470.6	2599.0	1134.37	1662.5	2796.9	2.8838	3.1181	6.0019
265	5.081	0.001 289	0.038 77	1152.74	1443.9	2596.6	1159.28	1634.4	2793.6	2.9294	3.0368	5.9662
270	5.499	0.001 302	0.035 64	1177.36	1416.3	2593.7	1184.51	1605.2	2789.7	2.9751	2.9551	5.9301
275	5.942	0.001 317	0.032 79	1202.25	1387.9	2590.2	1210.07	1574.9	2785.0	3.0208	2.8730	5.8938
280	6.412	0.001 332	0.030 17	1227.46	1358.7	2586.1	1235.99	1543.6	2779.6	3.0668	2.7903	5.8571
285	6.909	0.001 348	0.027 77	1253.00	1328.4	2581.4	1262.31	1511.0	2773.3	3.1130	2.7070	5.8199
290	7.436	0.001 366	0.025 57	1278.92	1297.1	2576.0	1289.07	1477.1	2766.2	3.1594	2.6227	5.7821
295	7.993	0.001 384	0.023 54	1305.2	1264.7	2569.9	1316.3	1441.8	2758.1	3.2062	2.5375	5.7437
300	8.581	0.001 404	0.021 67	1332.0	1231.0	2563.0	1344.0	1404.9	2749.0	3.2534	2.4511	5.7045
305	9.202	0.001 425	0.019 948	1359.3	1195.9	2555.2	1372.4	1366.4	2738.7	3.3010	2.3633	5.6643
310	9.856	0.001 447	0.018 350	1387.1	1159.4	2546.4	1401.3	1326.0	2727.3	3.3493	2.2737	5.6230
315	10.547	0.001 472	0.016 867	1415.5	1121.1	2536.6	1431.0	1283.5	2714.5	3.3982	2.1821	5.5804
320	11.274	0.001 499	0.015 488	1444.6	1080.9	2525.5	1461.5	1238.6	2700.1	3.4480	2.0882	5.5362
330	12.845	0.001 561	0.012 996	1505.3	993.7	2498.9	1525.3	1140.6	2665.9	3.5507	1.8909	5.4417
340	14.586	0.001 638	0.010 797	1570.3	894.3	2464.6	1594.2	1027.9	2622.0	3.6594	1.6763	5.3357
350	16.513	0.001 740	0.008 813	1641.9	776.6	2418.4	1670.6	893.4	2563.9	3.7777	1.4335	5.2112
360	18.651	0.001 893	0.006 945	1725.2	626.3	2351.5	1760.5	720.3	2481.0	3.9147	1.1379	5.0526
370	21.03	0.002 213	0.004 925	1844.0	384.5	2228.5	1890.5	441.6	2332.1	4.1106	0.6865	4.7971
374.14	22.09	0.003 155	0.003 155	2029.6	0	2029.6	2099.3	0	2099.3	4.4298	0	4.4298

Solution

Fill in the following table for steam (water) using the extracted tabulations.

	T, °C	P, kPa	x, %	h, kJ/kg	u, kJ/kg	v, m³/kg
a).	200	1553.8	0	**852.45**	850.65	0.0012
b).	110	**143.3**	26.2	1046	**1000**	0.318
c).	300	800	—	3056.5	2797.2	0.3241
d).	200	5000	—	852	851	0.0012
e).	466	400	—	3413	3075	**0.85**
f).	300	8581	80	2467.9	2316.8	0.0176

a). This state falls directly on the saturated liquid line.

b). The given value for internal energy falls between u_f and u_g, so the state is inside the vapor dome. Calculate x and use x to find h and v.

c). The temperature of 300°C is higher than $T_{sat} = 170.4°C$, so the state is superheated.

d). The temperature of 200°C is less than $T_{sat} = 264°C$, so the state is subcooled. Find the properties in the saturation tables at 200°C (good approximation).

e). The specific volume is greater than $v_g = 0.6058$ at the saturation pressure of 300 kPa, so it is superheated (interpolation required).

f). A quality value (x) is specified, so the state is obviously in the vapor dome. Use x to find h, u, and v.

Use of R-134a and NH₃ Tables and P-h Diagrams

Typically, the four working fluids used in thermodynamic texts and on the FE/EIT exam are H_2O, R-134a, NH_3, and air. The refrigerant R-22 has characteristics similar to R-134a, so it is not customarily tabled. The refrigerants R-134a and NH_3 exhibit the same phase-diagram characteristics as water, so they have the same type of tabled data, that is, subcooled liquid, saturated mixture, and superheated vapor. The guidelines given for finding one's way around the steam tables apply equally well to R-134a and NH_3 tables.

By tradition and for convenience, refrigerant properties are shown in a P-h diagram; this is especially useful when analyzing a vapor compression refrigeration cycle, since it operates at two basic pressure levels. A skeleton P-h chart is shown in Figure 10.10. A larger P-h diagram on the next page will be used in the next example and in the section on the vapor compression refrigeration cycle.

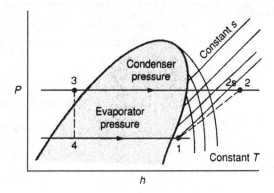

Figure 10.10

P-h DIAGRAM FOR REFRIGERANT HFC-134a
(metric units)
(Reproduced by permission of the DuPont Company)

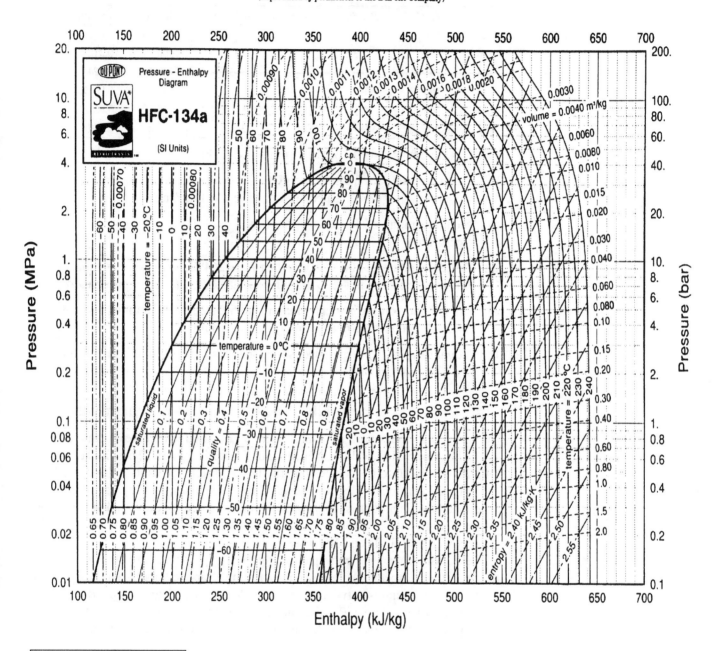

Example 10.6

Saturated liquid R-134a at $P = 2$ bars is heated to 100°C at constant pressure. Find the original and final state-point properties.

Solution

At $P = 2$ bars (.2 MPa) on the saturated liquid line,

$$v_f = .00075 \text{ m}^3/\text{kg}$$
$$h_f = 186 \text{ kJ/kg}$$
$$T_f = -10°C$$
$$s_f = 0.95 \text{ kJ/(kg} \cdot \text{K)}$$

At $P = 2$ bars and 100°C, the R-134a is superheated.

$$v = .148 \text{ m}^3/\text{kg}$$
$$h = 491 \text{ kJ/kg}$$
$$s = 2.0484 \text{ kJ/(kg} \cdot \text{K)}$$

IDEAL GASES

In this section we review the behavior of ideal and real gases and their relationships. We will also review *PVT* behavior of real gases in terms of equations of state, and the processes of vaporization and condensation, and then proceed to review the methods of solving material balance problems involving these processes.

The Ideal Gas Law

A gas is termed an ideal gas when the space or volume occupied by it is very large so that the intramolecular forces are negligible and the volume of the molecules of the gas is very small compared to the volume occupied by the gas. Ideal gas is a conceptual gas that meets the following conditions:

1. The gas obeys the following equation of state:

$$PV = nRT \qquad (10.1)$$

where

$P =$ absolute pressure of the gas
$V =$ total volume occupied by the gas
$n =$ number of moles of the gas
$R =$ ideal gas constant in proper units
$T =$ absolute temperature of the gas

2. The internal energy is independent of both the pressure and volume and is a function of the temperature alone.

For practical purposes, actual gases obey the ideal gas law only at low pressures and high temperatures within reasonable accuracy. It has been found to apply to all gases within certain pressure and temperature ranges, the accuracy increasing as the pressure decreases and temperature increases.

Sometimes the ideal gas law is written in terms of specific volume per mole of gas as

$$P\hat{V} = RT \qquad (10.1a)$$

where $\hat{V} =$ specific volume of the gas per mole. Equation 10.1 can be applied to a pure component or a mixture of two or more components.

Equation 10.1 represents the *equation of state* for an ideal gas. Two other relations for ideal gases are expressed as follows:

Boyle's law: $PV =$ Constant at constant temperature
Charles's law: $P/T =$ Constant at constant volume

As mentioned previously, some conditions of temperature and pressure are used as standard conditions. Table 10.1 lists the standard conditions for an ideal gas in different systems of units.

Table 10.1 Standard conditions for the ideal gas

System of Units	T	p	$\hat{V}$
SI	273.15 K	101.35 kPa	22.415 m^3/kg mol
Scientific	0.0°C	760 mm Hg	22.415 liters/g mol
American engineering	32.0°F	14.696 psia	359.05 ft^3/lb mol
Natural gas industry	60°F	1 atm	379.48 ft^3/lb mol

The values of ideal gas law constant R for different systems of units are listed in Table 10.2.

Table 10.2 Values of ideal gas law constant $R = \dfrac{PV}{nT}$

Numerical Value of R	Corresponding Unit of R
1.987	$\dfrac{\text{cal}}{\text{g mol} \cdot \text{K}}$
1.987	$\dfrac{\text{Btu}}{\text{lb mol} \cdot °\text{R}}$
82.06	$\dfrac{\text{cm}^3 \cdot \text{atm}}{\text{g mol} \cdot \text{K}}$
0.08206	$\dfrac{\text{liter} \cdot \text{atm}}{\text{g mol} \cdot \text{K}}$
10.731	$\dfrac{\text{psia} \cdot \text{ft}^3}{\text{lb mol} \cdot °\text{R}}$
0.7302	$\dfrac{\text{atm} \cdot \text{ft}^3}{\text{lb mol} \cdot °\text{R}}$
21.848	$\dfrac{\text{in Hg} \cdot \text{ft}^3}{\text{lb mol} \cdot °\text{R}}$
1545.3	$\dfrac{\text{ft} \cdot \text{lb}_f}{\text{lb mol} \cdot °\text{R}}$
7.8045×10^{-4}	$\dfrac{\text{hp} \cdot \text{h}}{\text{lb mol} \cdot °\text{R}}$
5.8198×10^{-4}	$\dfrac{\text{kWh}}{\text{lb mol} \cdot °\text{R}}$
8.314	$\dfrac{\text{J}}{\text{mol} \cdot \text{K}} = \dfrac{\text{N} \cdot \text{m}}{\text{mol} \cdot \text{K}}$

Numerical Value of R	Corresponding Unit of R
8.314	$\dfrac{Pa \cdot m^3}{mol \cdot K} = \dfrac{kPa \cdot m^3}{kmol \cdot K}$
8314	$\dfrac{Pa \cdot m^3}{kmol \cdot K}$
8314	$\dfrac{J}{kmol \cdot K}$

P = pressure, V = volume, T = temperature, n = mol

Example 10.7

Use of Standard Condition

Calculate the volume of 50 kg of NO_2 gas in cubic meters at standard conditions.

Solution

Basis: 50 kg of NO_2

$$\text{Volume of 50 kg } NO_2 = \frac{50 \text{ kg } NO_2}{} \left| \frac{1 \text{ kg mol } NO_2}{46 \text{ kg } NO_2} \right| \frac{22.42 \text{ m}^3 \text{ } NO_2}{1 \text{ kg mol } NO_2}$$

$$= 24.37 \text{ m}^3 \text{ } NO_2 \text{ at standard conditions}$$

Example 10.8

Calculate, from the ideal gas law, the value of universal gas law constant R if the units are to be pressure in psia, $\hat{V}$ in ft³, and T in °R.

Solution

The standard conditions are

$$P = 14.696 \text{ psia}$$
$$T = 492°R$$
$$\hat{V} = 359.05 \text{ ft}^3/\text{lb mol}$$

For 1 mol of gas,

$$R = \frac{P\hat{V}}{T} = \frac{14.696 \text{ psia}}{492°R} \left| \frac{359.05 \text{ ft}^3}{1 \text{ lb mol}} \right. = 10.73 \frac{\text{ft}^3 \cdot \text{psia}}{\text{lb mol} \cdot °R}$$

In many processes a gas goes from one initial state to other. In this case final and initial states can be related by the equation

$$\frac{P_2 V_2}{P_1 V_1} = \frac{n_2 R T_2}{n_1 R T_1} \quad \text{or} \quad \left(\frac{P_2}{P_1}\right)\left(\frac{V_2}{V_1}\right) = \left(\frac{n_2}{n_1}\right)\left(\frac{T_2}{T_1}\right) \quad (10.2)$$

and if $n_2 = n_1$,

$$\left(\frac{P_2}{P_1}\right)\left(\frac{V_2}{V_1}\right) = \frac{T_2}{T_1} \quad (10.2a)$$

In the preceding relations, the subscripts 2 and 1 refer to final and initial states. In all these equations, both the temperature and pressure are in absolute units and you must use the same units for both states of the gas. In Equations 10.2 and 10.2a, the ideal gas law, R is eliminated in taking the ratio of final to initial state.

It should be particularly noted that in the preceding equations the temperature is in K or °R, and P is the absolute (not the gauge) pressure in appropriate units. The values of gas constant R will depend on the units chosen for P, V, and T. They can be calculated as in Example 10.8 or simply read from the table of R values in handbooks or textbooks. Some examples will illustrate the use of the ideal gas law.

Example 10.9

Calculate a). the volume occupied by 20 kg of O_2 at 25°C and 750 mm Hg pressure. b). What is its density in SI units at these conditions? c). What is its specific gravity compared to air at 68°F and 1 atm?

Solution

a). At the conditions of the problem, both O_2 and air can be considered ideal. We will take mol wts of oxygen and air as 32 and 29, respectively.

Volume of O_2: From Table 10.1 of standard conditions, 1 kg mol occupies 22.415 m³/kg mol at standard conditions.

$$\text{Mols of } O_2 = 20/32 = 0.625 \text{ kg mol}$$
$$T = 273.15 + 25 = 298.15 \text{ K} \quad P = 750 \text{ mm Hg}$$

$$V = \frac{0.625 \text{ kg mol}}{} \left| \frac{22.415 \text{ m}^3}{1 \text{ kg mol}} \right| \frac{760 \text{ mm Hg}}{750 \text{ mm Hg}} \left| \frac{298.15 \text{ K}}{273.15 \text{ K}} \right| = 15.5 \text{ m}^3$$

b). Density of $O_2 = \dfrac{20 \text{ kg}}{15.5 \text{ m}^3} = 1.29 \text{ kg/m}^3$

c). Basis: kg mol of air

Temperature of air = 68°F = (68 − 32)/1.8 = 20°C = 273.15 + 20 = 293.15 K

$$V_{air} = \frac{1 \text{ kg mol air}}{} \left| \frac{22.415 \text{ m}^3}{1 \text{ kg mol air}} \right| \frac{298.15 \text{ K}}{273.15 \text{ K}} \left| \frac{1 \text{ atm}}{1 \text{ atm}} \right| = 24.056 \text{ m}^3$$

1 kg mol air = 29 kg
density of air (reference substance) = 29/24.056 = 1.206 kg/m³

Therefore specific gravity of O_2 relative to that of air $= \dfrac{1.29 \text{ kg/m}^3 O_2}{1.206 \text{ kg/m}^3 \text{ air}} = 1.07.$

Ideal Gas Mixtures and Partial Pressure

Frequently, you may have to deal with mixtures of gases instead of a single gas. For such calculations, the partial pressure p_i of a component i in a gaseous mixture is defined as the pressure that the component i would exert if it occupied the volume of the mixture alone at the same temperature.

Thus by definition,
$$p_i V_m = n_i R T_m \qquad (10.3)$$

where

V_m = volume of gaseous mixture at temperature T_m
T_m = temperature of the gaseous mixture
n_i = number of moles of component i

The ideal gas law gives for the total mixture
$$PV_m = n_m R T_m \qquad (10.3a)$$

where

P = total pressure, absolute

The subscript m refers to the total mixture. By dividing Equation 10.3 by Equation 10.3a, one gets
$$\frac{p_i V_m}{PV_m} = \frac{n_i R T_m}{n_m R T_m}$$

which results in the relation
$$\frac{p_i}{P} = \frac{n_i}{n_m} \qquad (10.3b)$$

Since $\dfrac{n_i}{n_m} = y_i,$
$$p_A = y_i P \qquad (10.3c)$$

where y_i = mol fraction of component i in the mixture, and p_i is partial pressure of component i.

Dalton's law states that the total pressure exerted by a gaseous mixture at a given temperature is the sum of the component partial pressures. Thus
$$P = p_A + p_B + p_C + \ldots \qquad (10.3d)$$
$p_A = y_i P$ and similar relations for other components

Amagat's law states that the total volume of an ideal gaseous mixture equals the sum of the pure component volumes at the same temperature and pressure. Thus
$$V = V_A + V_B + V_C + \ldots \qquad (10.4)$$
$V_A = y_A V$ and similar equations for other pure components

Using the ideal gas law,
$$PV_A = n_A R T_m, \qquad PV_B = n_B R T_m, \qquad PV_C = n_C R T_m \qquad (10.4a)$$

By addition,
$$P(V_A + V_B + V_C + \ldots) = R T_m (n_A + n_B + n_C + \ldots) = n R T_m = PV_m$$

Therefore,
$$\frac{V_A}{V_M} = \frac{\dfrac{n_A R T_m}{P}}{\dfrac{n R T_m}{P}} = \frac{n_A}{n} = y_i \qquad \therefore V_A = y_i V_m$$

Example 10.10

A natural gas has the following composition by volume: methane 94%, ethane 3.1%, and nitrogen 2.9%. The gas is at 80°F and 50 psia. Calculate, assuming ideal gas behavior:

a). Partial pressure of nitrogen

b). Pure component volume of nitrogen per 100 ft³ of gas mixture

c). Density of gas in lb_m/ft^3

Solution

Basis: 1 lb mol of gas

a). Volume % = mol % for ideal gases in a mixture

$$\text{partial pressure of } N_2 = (0.029) \times 50 \text{ psia} = 1.45 \text{ psia}$$

b). Pure component volume of nitrogen $= y_i V_m$ where V_m is volume of mixture
$$= 0.029 \times 100 = 2.9 \text{ ft}^3$$

c). Volume of mixture by ideal gas law

$$V_m = \frac{359.05 \text{ ft}^3}{1 \text{ lb mol}} \left| \frac{(460+80)°R}{(460+32)°R} \right| \frac{14.7 \text{ psia}}{50 \text{ psia}} = 115.9 \text{ ft}^3/\text{lb mol}$$

Mol wt. of gas $= 0.94 \times 16 + 0.031 \times 30 + 0.029 \times 28 = 16.782$

$$\therefore \text{density} \rho = \frac{16.782 \text{ lb}}{\text{lb mol}} \left| \frac{1}{115.9 \text{ ft}^3/\text{lb mol}} \right. = 0.1448 \frac{\text{lb}}{\text{ft}^3}$$

Example 10.11

How many kg mol of O_2 will occupy 1000 m³ at 350 K and 110 kPa?

Solution

$P = 110$ kPa $V = 1000$ m³ $T = 350$ K $R = 8.314$ kPa · m³/kg mol · K

$V = 1000$ m³ at 350 K and 110 kPa

$$PV = nRT$$

$$\text{kg mol of oxygen} = n = \frac{PV}{RT} = \frac{110 \times 1000}{8.314 \times 350} = 37.8 \text{ kg mol}$$

We can also solve this problem by calculating volume of the gas at standard conditions and knowing that at standard conditions, 1 kg mol occupies 22.41 m³.
Since n remains unchanged,

$$\frac{P_1 V_1}{T_1} = \frac{P_2 V_2}{T_2}$$

where subscript 1 refers to standard conditions.

$$V_1 = \frac{P_2 V_2 T_1}{T_2 P_1} = \frac{110 \times 1000 \times 273.15}{350 \times 101.3} = 847.46 \text{ m}^3$$

$$\text{Number of kg mol of } O_2 = \frac{847.46 \text{ m}^3}{22.41 \text{ m}^3/1 \text{ kg mol}} = 37.8 \text{ kg mol}$$

Notice that in this calculation you don't have to look for the value of R, the gas law constant, as it is eliminated in division.

Enthalpy and Internal Energy Changes

The other condition that is usually considered to be part of the definition of ideal gas is that the specific heats or heat capacities (c_p and c_v) are constant. This allows a number of working relationships to be developed:

$$h_2 - h_1 = c_p (T_2 - T_1) \qquad c_p = \frac{k\text{R}}{k-1}$$

$$u_2 - u_1 = c_v (T_2 - T_1) \qquad c_v = \frac{\text{R}}{k-1}$$

$$c_p = c_v + \text{R} \quad \text{(always true)}$$

$$k = \frac{c_p}{c_v} = \text{constant}$$

PHASE EQUILIBRIA

So far we have reviewed mostly properties of pure substances or mixtures of constant compositions. Mass transfer operations such as distillation, absorption, and extraction involve contact of different phases but also changes in composition. When two or more phases come in contact, their compositions change because of mass transfer between them. The degree of change and the rate of transfer depend on how far the system is away from its equilibrium state. Thus phase equilibria are important in quantitative treatment of mass transfer operations. In practice, environmental engineers have to deal with vapor-liquid, liquid-liquid, solid-liquid, and vapor-solid systems, although vapor-liquid systems are encountered more often than the others.

Vapor-Liquid Equilibria

Equilibrium is a state or condition of a system in which there is an absence of change as well as the absence of any tendency to change. Separation of components requires that the composition of vapor is different from the composition of the liquid when the two are in equilibrium. The basic data needed for design calculations are the phase equilibria between the liquid and vapor phases. Such data may be obtained from the literature, by experimental determination, or by using theoretical methods. In this section we review vapor-liquid equilibria and the calculation of phase compositions at equilibrium. We will also review Raoult's and Henry's laws that allow the calculation of temperature, pressure, and phase compositions in vapor-liquid equilibrium.

Binary and Multicomponent Systems

A binary system consists of two components. A multicomponent system consists of more than two components. Both multicomponent and binary systems can be either ideal or nonideal. Ideal behavior of systems is found at low pressures and normal distilling temperatures. Nonideal behavior is exhibited by systems consisting of substances of dissimilar nature or where temperature and pressure conditions are severe.

The Phase Rule and Duhem's Theorem

According to phase rule for nonreacting systems, the number of variables that may be independently fixed for a system at equilibrium is the difference between the total number of variables that describe the intensive state of the system and the number of independent equations that can be written relating the variables. If the system contains N chemical components at given T and P, the phase rule variables are $N - 1$ mole fractions and T and P. The degrees of freedom F are given by (Smith et al.)

$$F = 2 - \pi + N \tag{10.5}$$

Duhem's theorem pertains to a closed system at equilibrium for which both extensive and intensive states of the system are fixed. In this case, there are $2 + (N - 1)\pi$ phase rule intensive variables and π extensive variables expressing the masses or moles of the phases. Therefore total number of variables is $(\pi-1)N+N=\pi N$. We can write N additional independent material balance equations (one each for each component), giving the total number of independent equations $= (\pi -1) N + N = \pi N$. The difference between the number of variables and the number of independent equations is then $2 + N \pi - \pi N = 2$. Thus for a closed system of constant component masses, the equilibrium state is completely determined when two independent variables are fixed. The phase rule, however, fixes the number of independent intensive variables. Therefore, depending on the degrees of freedom F for a system, the number of extensive variables that can be chosen will also be determined.

For a binary system at equilibrium comprising two phases (vapor and liquid), $N = 2$ and $\pi = 2$. Therefore, $F = 2$. Thus, fixing two variables determines binary system equilibrium. A ternary system with two phases in equilibrium has $F = 3$ and will be defined by three variables. For an n component system, and two phases in equilibrium, there are n degrees of freedom.

T-xy and *x-y* Diagrams for Binary Systems

For binary systems, the differences in compositions of the liquid and vapor phases are graphically represented by either boiling point (temperature-composition, *T–xy*) or equilibrium (composition, *y* vs. *x* only) diagrams at constant pressure. In the former, boiling point temperature is plotted against liquid and vapor compositions, whereas in the latter the equilibrium vapor phase compositions of more volatile components are plotted against the corresponding liquid phase compositions as in Figure 10.11a and b.

Systems that obey Raoult's law are called ideal systems. Those systems that do not obey Raoult's law are nonideal systems. Boiling point and equilibrium diagrams such as shown in Figure 10.11a and b are typical of ideal binary systems.

Boiling point diagrams and equilibrium diagrams for nonideal systems exhibit different behavior such as shown in Figure 10.12a and b. For example, an acetone-chloroform system is a maximum boiling point system, whereas the ethyl acetate-ethanol system shows a minimum boiling point. At the azeotropic point, vapor and liquid composition are the same ($x_{Az} = y_{Az}$). For maximum boiling point systems, this happens at T_{max} and for minimum boiling point systems at T_{min}.

Such diagrams are typical of binary systems that obey Raoult's law.

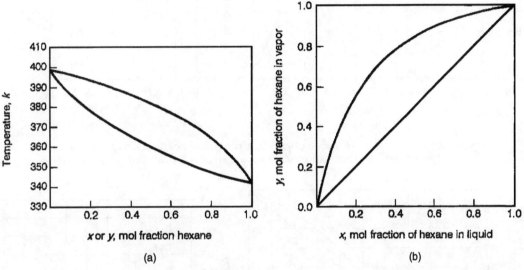

Figure 10.11 (a) *T-xy* diagram or boiling point diagram, (b) equilibrium or *x-y* diagram

For a liquid and vapor mixture at equilibrium, the ratio of the moles of the liquid phase to moles of the vapor phase is given by the inverse lever rule:

$$\frac{L}{V} = \frac{y_{AV} - x_{AF}}{z_{AF} - x_{AL}}$$

where
- L = mols of liquid phase
- V = mols of vapor phase in equilibrium with the liquid
- y_{AV} = mol fraction of component A in vapor phase
- z_{AF} = mol fraction of component A in the feed before vaporization
- x_{AL} = mol fraction of component A in the liquid phase at equilibrium

The boiling point and equilibrium diagrams of Figure 10.11a and b are typical of ideal binary systems that follow Raoult's law. Nonideal systems that do not follow Raoult's law show a different behavior, such as a maximum or a minimum boiling point.

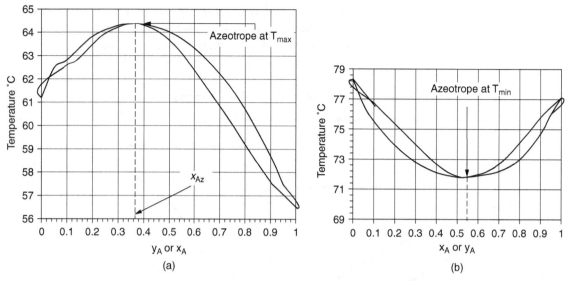

Figure 10.12 (a) System acetone chloroform, maximum boiling point azeotrope, and (b) system ethyl acetate-ethanol, minimum boiling point azeotrope

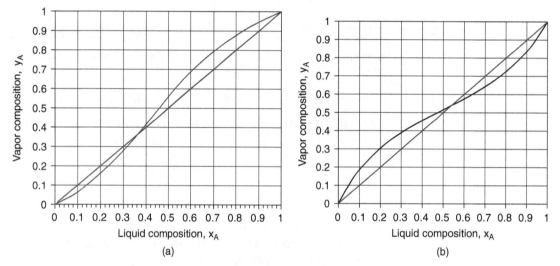

Figure 10.13 (a) System acetone-chloroform x-y diagram and (b) system ethyl acetate-ethanol x-y diagram

On the *x-y* diagram, the equilibrium curves cross the 45° line at the maximum or minimum boiling point (Figure 10.13a and b).

Azeotropic mixtures are very common. They cannot be completely separated by ordinary distillation methods since at the azotropic composition $y_A = x_A$. One of the most important is the ethanol-water azeotrope, which occurs at 78.4°C and 1 atm. Azeotropism disappears in this system at pressures below 70 mm Hg.

Some substances exhibit very large positive deviations from ideality in that they are not completely soluble in the liquid state and form a pair of partially immiscible components. Isobutanol-water is such a system. The boiling point and equilibrium diagrams for this system are shown in Figure 10.14a and b.

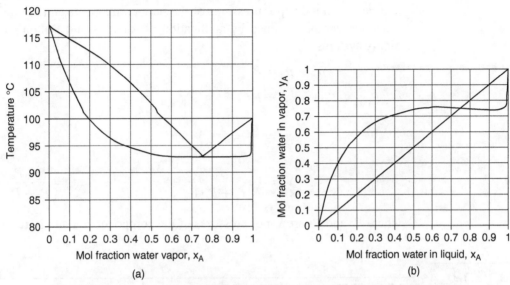

Figure 10.14 (a) System water-l-butanol and (b) heterogeneous azeotrope

Immiscible Liquids

If the liquids are completely insoluble in each other, each exerts its own vapor pressure at the prevailing temperature. When the sum of the two vapor pressures equals the total pressure, the mixture boils. The vapor composition is then easily calculated by application of the ideal gas law:

$$P_A + P_B = P \quad \text{and} \quad y = \frac{P_A}{P} \tag{10.6}$$

Ideal Mixtures (Raoult's Law)

Ideal gas, ideal liquid, and ideal gas and liquid mixtures are the basis of many quantitative equilibrium calculations. Raoult's law relates the partial pressure of a component in the vapor phase of a gaseous mixture to its concentration (mol fraction) in the liquid phase and its vapor pressure at the temperature of the mixture in contact with the liquid at equilibrium. Mathematically it can be stated as follows:

$$p_i = x_i P_i \tag{10.7}$$

where
p_i = partial pressure of the component i in the vapor phase
x_i = mol fraction of component i in the liquid phase
P_i = vapor pressure of pure component i at the temperature of the liquid-vapor mixture

By combining Raoult's law with Dalton's law of partial pressures, we obtain an expression relating mixtures of ideal vapors and liquids in equilibrium:

$$P = \sum_i^n p_i = y_i \quad P = \sum_i^n x_i P_i \tag{10.8}$$

For a single component, $\quad y_i P = x_i P_i \tag{10.8a}$

where P = total pressure.

The vapor pressure of a pure component is a unique property of the pure component and is a function of temperature. It increases with an increase in temperature. A component that has a higher vapor pressure at a given temperature than another component is called more volatile.

For systems that obey Raoult's law it is possible to calculate the boiling point diagram from the vapor pressures of the pure components and the following relations that can be obtained by application of Raoult's and Dalton's laws to ideal binary systems:

$$p_A + p_B = P \tag{10.9}$$

$$P_A x_A + P_B (1 - x_A) = P$$

$$y_A = \frac{p_A}{P} = \frac{P_A x_A}{P}$$

$$y_A = \frac{P_A x_A}{P_A x_A + P_B(1-x_A)} = \frac{P_A x_A}{P} \qquad x_A = \frac{P - P_B}{P_A - P_B} \tag{10.10}$$

where P_A and P_B are vapor pressures of the components A and B, respectively.

Example 10.12

Use of Raoult's Law

The vapor pressure data for the system hexane-octane are given in the following table. The system can be assumed to be ideal and obeying Raoult's law. Plot the T-xy and x-y diagrams for hexane-octane system at 101.3 kPa pressure.

	Vapor Pressure	
$T°C$	n-Hexane kPa	n-Octane kPa
67.8	101.3	16.1
79.4	136.7	23.1
93.3	197.3	39.2
107.2	284.0	57.9
125.7	456.0	101.3

Solution

First liquid compositions are calculated using Equation 10.9. The relation

$$x_A = \frac{P - P_B}{P_A - P_B}$$

is used for this purpose. Then vapor composition is calculated by

$$y_A = \frac{p_A}{P} = \frac{P_A x_A}{P}$$

Calculations done on a spreadsheet are given in the following table. The boiling point and equilibrium diagrams are plotted in Exhibit 1a and b, respectively.

$T°C$	$P-P_A$	P_A-P_B	$x_A = \dfrac{P-P_B}{P_A-P_B}$	$y_A = \dfrac{P_A x_A}{P}$
68.7	85.2	85.2	1	1
79.4	78.2	113.6	0.68838	0.928940
93.3	64.2	160.2	0.40075	0.780531
107.2	43.4	226.1	0.19195	0.538143
125.7	0	354.7	0	0

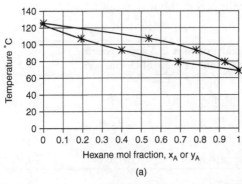
(a)

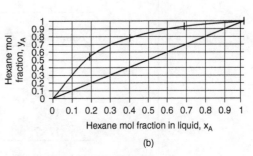
(b)

Exhibit 1 T-x, x-y, or equilibrium diagrams for Example 10.12

Relative Volatility α_{ij}

For a vapor phase in equilibrium with its liquid phase, the relative volatility of a component i with respect to component j is given by

$$\alpha_{ij} = \frac{y_i/x_i}{y_j/x_j} \tag{10.11}$$

where y is the mole fraction of a component in the vapor phase and x is the mole fraction of the same component in the liquid phase. The relative volatility is the measure of separability of the two components by distillation. The larger the value of α_{ij} above unity, the greater the degree of separability.

For a binary system, $y_j = 1 - y_i$ and $x_j = 1 - x_i$; therefore,

$$\alpha = \left(\frac{y_i}{1-y_i}\right)\left(\frac{1-x_i}{x_i}\right) \tag{10.12}$$

In general,

$$\alpha = \left(\frac{y}{1-y}\right)\left(\frac{1-x}{x}\right) \tag{10.13}$$

from which

$$y = \frac{\alpha x}{1+(\alpha-1)x} \qquad x = \frac{y}{\alpha + y(1-\alpha)} \tag{10.13a}$$

For ideal mixtures, α is the ratio of the vapor pressures or

$$\alpha_{ij} = \frac{P_i}{P_j} \tag{10.14}$$

and for a binary system Equation 10.10 applies.

Example 10.13

Calculation of Relative Volatility

From the data of Example 10.12, calculate the relative volatilities of *n*-hexane–*n*-octane system at various temperatures.

Solution

For *n*-hexane–*n*-octane system, which follows Raoult's law, relative volatility

$$\alpha = \frac{P_A}{P_B}$$

The calculations of α are presented in the following table:

Temperature °C	Vapor Pressure of Hexane kPa	Vapor Pressure of Octane kPa	$\alpha = P_A/P_B$
68.7	101.3	16.1	6.3
79.4	136.7	23.1	5.92
93.3	197.3	39.2	5.32
107.2	284.0	57.9	4.91
125.7	456.0	101.3	4.50

Henry's Law

If the pressure is low enough so that the vapor phase can be assumed to be an ideal gas, Henry's law can be applied. For a component present in liquid phase at very low concentrations, Henry's law states that the partial pressure of the component in the vapor phase is directly proportional to its mol fraction in the liquid phase.

Thus,
$$p_A = y_i P = H_i x_i \quad (10.15)$$

where H = Henry's law constant characteristic of the system. Values of H_i are to be obtained experimentally. Henry's law applies to the vapor pressure of a solute in a dilute solution (Raoult's law applies to the vapor pressure of the solvent).

PROCESSES

Thermodynamic processes usually involve a "working fluid" such as a pure substance (like water) or a gas (like air), so tables of properties or ideal gas relationships are used. For the process path to be known, the process must be reversible. If the process involves friction or turbulence and is irreversible, then only the first law of thermodynamics applies.

So, generally, the processes are considered to be reversible (no friction or turbulence) and are one of the following:

- Constant pressure (Isobaric)
- Constant volume (Isometric)
- Constant temperature (Isothermal)
- No heat flow (adiabatic) (Isentropic)

Table 10.3 shows the applicable relationships for the various processes.

Table 10.3 First and second thermodynamics law formulas for reversible processes of an ideal gas*

Process	Closed System (nonflow)	Open System (steady flow)
General $(Pv = RT)$ $\dfrac{p_1 v_1}{T_1} = \dfrac{p_2 v_2}{T_2}$	$q = c_p (T_2 - T_1) + w$ $w = \int_1^2 P\, dv$ $s_2 - s_1 = c_v \ln\dfrac{T_2}{T_1} - R \ln\dfrac{v_2}{v_1}$	$q = c_p (T_2 - T_1) + w$ $w = \int_1^2 P\, dv$ $s_2 - s_1 = c_v \ln\dfrac{T_2}{T_1} - R \ln\dfrac{v_2}{v_1}$
	($s_2 - s_1$ for closed or open systems)	
Polytropic $Pv^n =$ constant $\dfrac{p_2}{p_1} = \left(\dfrac{T_2}{T_1}\right)^{n/(n-1)} = \left(\dfrac{v_1}{v_2}\right)^n$ $\dfrac{T_2}{T_1} = \left(\dfrac{p_2}{p_1}\right)^{(n-1)/n} = \left(\dfrac{v_1}{v_2}\right)^{n-1}$ $\dfrac{v_2}{v_1} = \left(\dfrac{p_2}{p_1}\right)^{1/n} = \left(\dfrac{T_1}{T_2}\right)^{1/n}$	$q = \dfrac{k-n}{1-n} c_v (T_2 - T_1)$ $w = \dfrac{k-1}{1-n} c_v (T_2 - T_1)$ $s_2 - s_1 = c_p \ln\dfrac{T_2}{T_1} - R \ln\dfrac{p_2}{p_1}$ $s_2 - s_1 = \phi_2 - \phi_1 - R \ln\dfrac{p_2}{p_1} = \phi_2 - \phi_1 + R \ln\dfrac{v_2}{v_1}$	$q = \dfrac{k-n}{1-n} c_v (T_2 - T_1)$ $w = n\dfrac{k-1}{1-n} c_v (T_2 - T_1) - \Delta KE - \Delta PE$ $s_2 - s_1 = c_v \ln\dfrac{T_2}{T_1} - R \ln\dfrac{v_2}{v_1}$
Constant with volume (isometric) $v_2 = v_1,\quad n = \infty$ $\dfrac{p_2}{T_2} = \dfrac{p_1}{T_1}$	$w = 0$ $s_2 - s_1 = c_v \ln(T_2/T_1)$	$q = c_v (T_2 - T_1) \quad q = cv(T_2 - T_1)$ $w = -v(p_2 - p_1) - \Delta KE - \Delta PE$ $s_2 - s_1 = c_v \ln(T_2/T_1)$
Constant pressure (isobaric) $p_2 = p_1,\quad n = 0$ $\dfrac{v_2}{T_2} = \dfrac{v_1}{T_1}$	$w = p(v_2 - v_1)$ $w = R(T_2 - T_1)$ $s_2 - s_1 = c_p \ln(T_2/T_1)$	$q = c_p(T_2 - T_1) \quad q = c_p(T_2 - T_1)$ $w = -\Delta KE - \Delta PE$ $s_2 - s_1 = c_p \ln(T_2/T_1)$
Constant temperature (isothermal) $T_2 = T_1,\quad n = 1$ $p_2 v_2 = p_1 v_1$	$q = w = T(s_2 - s_1)$ $q = w = RT\, \ln\dfrac{v_2}{v_1}$ or $\dfrac{p_1}{p_2}$ $s_2 - s_1 = R \ln\dfrac{v_2}{v_1}$ or $\dfrac{p_1}{p_2}$	$q = T(s_2 - s_1) = RT \ln\dfrac{v_2}{v_1}$ or $\dfrac{p_1}{p_2}$ $w = RT \ln\dfrac{v_2}{v_1} - \Delta KE - \Delta PE$ or $w = RT \ln\dfrac{p_1}{p_2} - \Delta KE - \Delta PE$ $s_2 - s_1 = R \ln\dfrac{v_2}{v_1} = R \ln\dfrac{p_1}{p_2}$
Adiabatic (isentropic) $n = k$ $s_2 = s_1$	$q = 0$ $w = c_v (T_1 - T_2)$ $w = \dfrac{p_1 v_1 - p_2 v_2}{k-1}$ $w = R(T_1 - T_2)/(k-1)$ $s_2 - s_1 = 0$	$q = 0$ $w = c_p (T_1 - T_2) - \Delta KE - \Delta PE$ $w = \dfrac{k(p_1 v_1 - p_2 v_2)}{k-1} - \Delta KE - \Delta PE$ $w = kR(T_1 - T_2)/(k-1) - \Delta KE - \Delta PE$ $s_2 - s_1 = 0$

*Per-unit mass basis and constant (average) specific heats (c_v, c_p) assumed. $R = c_p - c_v$, $k = c_p/c_v$, $c_p = kR/(k-1)$, $C_v = R/(k-1)$.
$\Delta u = u_2 - u_1 = c_v (T_2 - T_1)$, $\Delta h = h_2 - h_1 = c_p (T_2 - T_1)$
ΔKE (S.I.) $= \dfrac{v_2^2 - v_1^2}{2000 \times g_c}$, ΔPE (S.I.) $= \dfrac{g(z_2 - z_1)}{1000 \times g_c}$
ΔKE and ΔPE may be negligible for many open systems.

CYCLES

Rankine Cycle (Steam)

An ideal Rankine cycle with superheated steam flowing into the turbine is shown in Figure 10.15. The four open-system components are analyzed as follows:

1. Boiler

$$q_{in} = h_2 - h_1 \left(\frac{kJ}{kg}\right)$$
$$\dot{Q}_{in} = \dot{m}_{stm}(h_2 - h_1) \text{ (kW)}$$

2. Turbine

$$w_T = h_3 - h_2 \text{ (kJ/kg)}$$
$$\dot{W}_T = \dot{m}_{stm}(h_3 - h_2) \text{ (kW)}$$

3. Condenser

$$q_{out} = h_3 - h_4 = c_p \Delta T \text{ (kJ/kg)}$$
$$\dot{Q}_{out} = \dot{m}_{stm}(h_3 - h_4) = \dot{m}_{cw} c_p \Delta T_{cw} \text{ (kW)}$$

4. Pump

$$w_p = h_1 - h_4 = v_4(p_1 - p_4) \text{ (kJ/kg)}$$
$$\dot{W}_p = \dot{m}_{stm}(h_1 - h_4) \text{ (kW)}$$

State points are found in property (steam) tables. State 2 is usually given by pressure and temperature and can be either saturated or (normally) superheated. State 2 can be found in the steam tables. State 3 is found by using an isentropic process, so that entropy (State 2) and pressure (condensing) are known. State 4 is saturated liquid found in the tables at the condensing pressure.

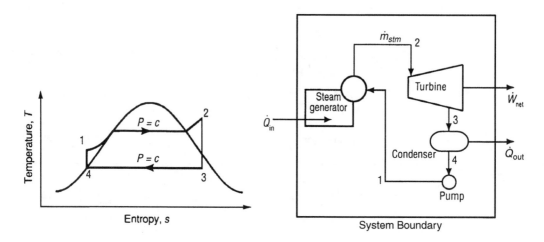

Figure 10.15

Also, State 1—using an isentropic process—can be found by the tables but is more easily found by

$$h_1 = h_4 + v_4(p_1 - p_4)$$

The thermal efficiency is

$$\eta = \frac{W_T - W_P}{Q_{in}} = \frac{h_2 - h_3 - (h_1 - h_4)}{h_2 - h_1}$$

Adding $h_1 - h_4$ (the pump work) to both the numerator and denominator results in

$$\eta_{approx.} = \frac{h_2 - h_3}{h_2 - h_4}$$

Example 10.14

A Rankine cycle using steam has turbine inlet conditions of $P = .4$ MPa, $T = 300°C$, and a condenser temperature of $70°C$. The turbine efficiency is 90%, and the pump efficiency is 80%.

For both the ideal cycle and the cycle considering the component efficiencies, find (a) the thermal efficiency (η), (b) the turbine discharge quality (x), and (c) the steam flow rate ($\dot{m}$) for 1 MW of net power.

Properties of water (SI units): superheated-vapor table

Temp °C	v	u	h	s
	0.4 MPa ($T_{sat} = 143.6 °C$)			
300	.6548	2805	3067	7.566

v, m³/kg; u, kJ/kg; h, kJ/kg; s, kJ/(kg · K)

Properties of saturated water (SI units): pressure table

		Specific volume		Internal energy		Enthalpy			Entropy	
Press. Bars kPa	Temp. °C T	Sat. Liquid v_f	Sat. Vapor v_g	Sat. Liquid u_f	Sat. Vapor u_g	Sat. Liquid h_f	Evap. h_{fg}	Sat. Vapor h_g	Sat. Liquid s_f	Sat. Vapor s_g
31.2	70	.00102	5.04	293	2570	293	2334	2627	9549	7.7553

v, m³/kg; u and h, kJ/kg; s, kJ/(kg · K)

Solution

Starting with State 2 (turbine inlet), the properties h and s can be found with the steam tables.

$$h_2 = 3067 \frac{kJ}{kg} \qquad s_2 = 7.566 \frac{kJ}{kg \cdot K}$$

The ideal turbine discharge (State 3) is found at $T_3 = 70°C$ and $s_3 = 7.566$:

$$x = \frac{s - s_f}{s_{fg}} = \frac{7.566 - .9549}{7.7553 - .9549} = 0.972$$

$$h_3 = h_f + x h_{fg} = 293 + .972 \times 2334 = 2562 \text{ kJ/kg}$$

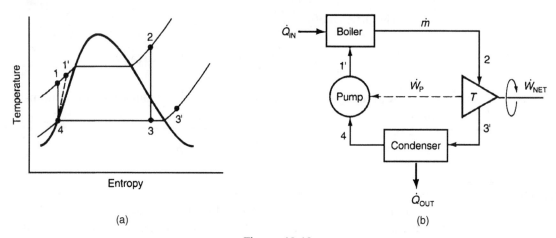

Figure 10.16

The saturated liquid state leaving the condenser (State 4) is read from the tables:

$$h_4 = 293 \text{ kJ/kg}$$
$$s_4 = .9549 \frac{\text{kJ}}{\text{kg} \cdot \text{K}}$$
$$v_4 = 0.00102 \text{ m}^3/\text{kg}$$

The compressed (subcooled) liquid leaving the pump (State 1) can be found in the tables if values are available for the condition. The usual approximation is to calculate

$$h_1 = h_4 + v_4(p_4 - p_1)$$
$$= 293 + .00102(400 - 31.2)$$
$$= 293 + .4 = 293.4 \frac{\text{kJ}}{\text{kg}}$$

For the ideal cycle,

a). $\eta = \dfrac{W_{net}}{Q_{in}} = \dfrac{W_T - |W_p|}{Q_{in}} = \dfrac{(3067 - 2562) - .4}{3067 - 293.4}$

$= \dfrac{505 - .4}{2773.6} = 0.182$

b). $x = 0.972$

c). $\dot{m} = \dfrac{1000 \text{ kW}}{W_{net}} = \dfrac{1000}{505} = 2 \text{ kg/s}$

For the cycle considering the component efficiencies (Figure 10.16),

$$\eta_{turb.} = \frac{h_2 - h_{3'}}{h_2 - h_3}$$

$$h_{3'} = 3067 - .9(3067 - 2562) = 2613 \text{ kJ/kg, and } P = 31.2 \text{ kPa}$$

$$h_{1'} = h_4 + \frac{h_1 - h_4}{\eta_p} = 293 + \frac{.4}{.8} = 293.5 \frac{\text{kJ}}{\text{kg}}$$

$$x = \frac{h_{3'} - h_f}{h_{fg}} = \frac{2613 - 293}{2334} = 0.99$$

$$\eta = \frac{W_T - W_P}{Q_{in}} = \frac{h_2 - h_{3'} - h_{1'} - h_4}{h_2 - h_{1'}}$$

$$= \frac{(3067 - 2613) - \frac{.4}{.8}}{3067 - \left(293 + \frac{.4}{.8}\right)} = \frac{453.5}{2773.5} = .164$$

$$\dot{m} = \frac{1000}{453.5} = 2.2 \frac{\text{kg}}{\text{s}}$$

The pump work makes little numerical difference and can usually be ignored in the calculation of thermal efficiency and steam flow rate.

Vapor Compression Cycle (Refrigeration)

An ideal vapor compression cycle is shown in Figure 10.17. The four open-system components are analyzed below:

1. Compressor

$$W_{in} = h_2 - h_1$$

$$\dot{W}_{in} = \dot{m}(h_2 - h_1)$$

2. Condenser

$$Q_{out} = h_2 - h_3$$
$$\dot{Q}_{out} = \dot{m}(h_2 - h_3)$$

3. Expansion valve

$$h_3 = h_4$$

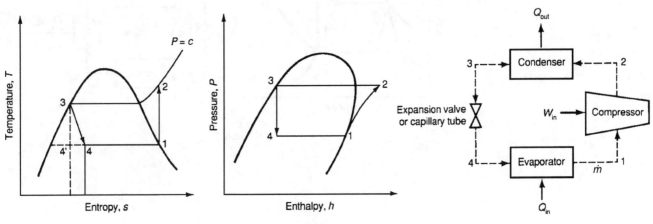

Figure 10.17

4. Evaporator

$$Q_{in} = h_1 - h_4$$
$$\dot{Q}_{in} = \dot{m}(h_1 - h_4)$$

The coefficient of performance (COP), if the cycle is used as a refrigerator, is

$$\text{COP}_{REFR} = \frac{Q_{in}}{W_{in}} = \frac{h_1 - h_4}{h_2 - h_1}$$

If the cycle is used as a heat pump,

$$\text{COP}_{HEATPUMP} = \frac{Q_{out}}{W_{in}} = \frac{h_2 - h_3}{h_2 - h_1} = \text{COP}_{REFR} + 1.0$$

The state points are found in property tables and/or property diagrams (P-h). State 1 is usually given as saturated vapor at a given pressure or temperature. State 2 is found by assuming an isentropic process so that the entropy and pressure (or corresponding condensing temperature) are known. This is best done on a P-h diagram. State 3 is saturated liquid at the given condensing pressure. State 4 is found at the same enthalpy as State 3 and at the evaporating pressure.

Example 10.15

An ideal vapor compression refrigeration cycle using R-134a operates between 100 kPa and 1000 kPa. Find the COP and the mass flow rate required for 100 kW of cooling.

Solution

Refer to the P-h diagram on page 475 and a schematic of the components (Exhibit 2). State 1 (saturated vapor at 100 kPa), from P-h diagram or from property tables:

$$h_1 = 383 \text{ kJ/kg}$$
$$s_1 = 1.746 \text{ kJ/(kg} \bullet \text{K)}$$
$$T_1 = -26.31°C$$

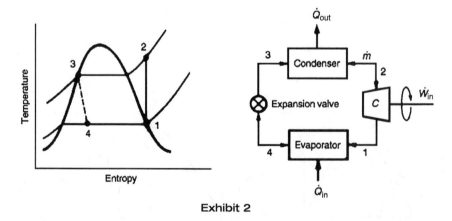

Exhibit 2

State 2 ($P_2 = 1000$ kPa)

$$s_1 = s_2 = 1.746 \text{ kJ/(kg} \cdot \text{K)}$$
$$h_2 = 431 \text{ kJ/kg}$$
$$T_2 = 49.8°\text{C}$$

State 3 (saturated liquid)

$$h_3 = 255.6 \text{ kJ/kg}$$
$$T_3 = 39.3°\text{C}$$

State 4 (liquid + vapor)

$$h_3 = h_4 = 225.6 \text{ kJ/kg}$$
$$\text{COP}_R = \frac{\dot{q}_{evap}}{w_c} = \frac{h_1 - h_4}{h_2 - h_1} = \frac{383 - 255.6}{431 - 383} = \frac{127.4}{48} = 2.65$$
$$\dot{q}_{evap} = \dot{m}(h_1 - h_4), \quad \dot{m} = \frac{100}{127.4} = 0.478 \frac{\text{kg}}{\text{s}}$$

Psychrometrics

Our atmosphere is a mixture of noncombustible gases, namely air and water vapor. The state of this mixture at a particular pressure may be specified by two intensive properties. We may determine the conditions of the air-water mixture by using equations or by the use of the psychrometric chart. Note, however, that to use the psychrometric chart, the pressure of the air-water mixture must be 1 atm, or 14.7 lbf/in² in the English system, or 101.35 kPa in the SI system.

The psychrometric chart provides easily readable values of the properties of moist air at various conditions. Refer to the NCEES *Fundamentals of Engineering Supplied-Reference Handbook* for a complete psychrometric chart. Figure 10.18 shows a schematic diagram of the psychromatic chart.

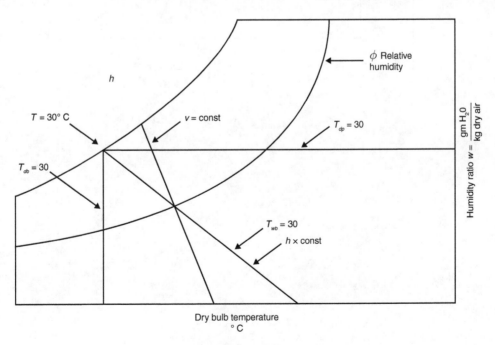

Figure 10.18

Example 10.16

A room at atmospheric pressure contains an air-water mixture at 25°C and relative humidity of 50%. Using the psychrometric chart, determine the humidity ratio, wet bulb temperature, dew point temperature, enthalpy, and specific volume.

Solution

Exhibit 3 is what the solution looks like on a psychrometric chart.

The solution to the problem is:

Humidity ratio (w)	=	10 grams water/kg dry air
Wet bulb temperature	=	17°C
Dew point temperature	=	14°C
Enthalpy (h)	=	51 kJ/kg
Specific volume (v)	=	0.8575 kg/m³

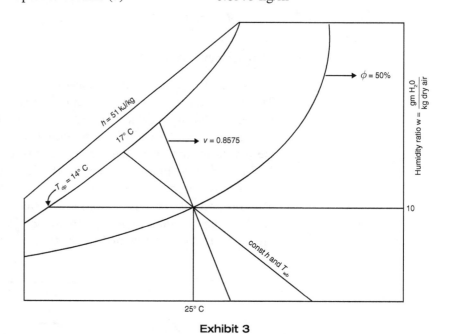

Exhibit 3

Example 10.17

Outside air at 10°C and relative humidity of 20% enters a heating duct. The air is heated to 25°C. Steam is then injected into the air stream, and the air leaves the unit at 30°C and 60% relative humidity.

Using the psychrometric chart, determine the amount of heat and water added, relative humidity, and wet bulb temperature at State 2.

Exhibit 4 shows the schematic diagram of the system.

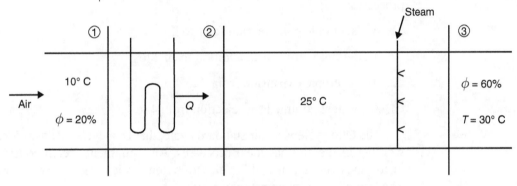

Exhibit 4

Solution

Using the psychrometric chart in Exhibit 5, obtain the values.

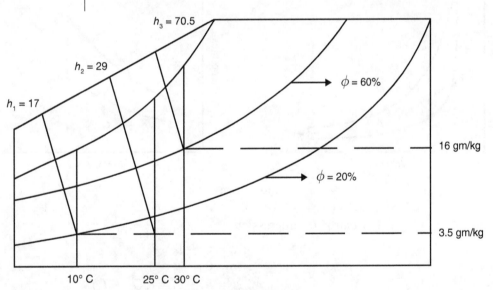

Exhibit 5

The heat added is

$$h_1 + ({}_1q_2) = h_2$$
$$({}_1q_2) = 29 - 17 = 12 \text{ kJ/kg}$$

The water added in the complete process:

$$(\omega_3 - \omega_2) = 16 \text{ gm water/kg dry air} - 3.5 \text{ gm water/kg dry air} = 12.5 \text{ gm/kg}$$

Note that an additional amount of energy is added to the air in terms of steam. Determine, then, the output of enthalpy and the wet bulb temperature of the exiting air.

$$h_3 = 70.5 \text{ kJ/kg} \qquad T_{wb3} = 23.4°C$$

Otto Cycle (Gasoline Engine)

An ideal Otto cycle is shown in Figure 10.19. It consists of the following four processes:

1. An isentropic compression for 1 to 2
2. A constant volume heat addition from 2 to 3
3. An isentropic expansion from 3 to 4
4. A constant volume heat rejection from 4 to 1

The Otto cycle is an air standard cycle, that is, a cycle that uses ideal air as the working media and has ideal processes. An equipment sketch consists of only a piston and cylinder, since it is a closed-system cycle using a fixed quantity of mass. The four closed-system processes reduce to

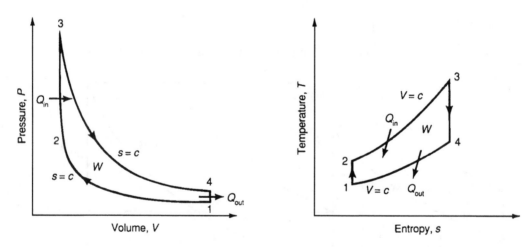

Figure 10.19

1. Isentropic compression

$$u_1 + q = u_2 - W$$
$$q = 0$$
$$W_{comp} = u_2 - u_1 = c_v(T_2 - T_1)$$

2. Heat addition

$$u_2 + q_{in} = u_3 + w$$
$$w = 0$$
$$q_{in} = u_3 - u_2 = c_v(T_3 - T_2)$$

3. Isentropic expansion

$$u_3 + q = u_4 + w$$
$$q = 0$$
$$W_{exp} = u_3 - u_4 = c_v(T_3 - T_4)$$

4. Heat rejection
$$u_4 + q_{out} = u_1 + w$$
$$w = 0$$
$$q_{out} = u_1 - u_4$$

The thermal efficiency is

$$\eta = \frac{W_{net}}{Q_{in}} = \frac{W_{exp} - W_{comp}}{Q_{in}} = \frac{u_3 - u_4 - (u_2 - u_1)}{u_3 - u_2} = \frac{T_3 - T_4 - (T_2 - T_1)}{T_3 - T_2} = 1 - \frac{1}{r_c^{k-1}}$$

Note that r_c is the compression ratio, a ratio of the *volume* at the bottom of the piston stroke (bottom dead center) to the *volume* of the top of the stroke (top dead center). This is also equal to v_1/v_2.

The state points are found by ideal gas laws or air tables: State 1 is usually given by T and P. State 2 is found by using an isentropic process; for an ideal gas,

$$\frac{T_2}{T_1} = \left(\frac{v_1}{v_2}\right)^{k-1} = r_c^{k-1}$$

If air tables are used:

$$\frac{v_{r_2}}{v_{r_1}} = \frac{v_2}{v_1} = \frac{1}{r_c}$$

State 3 is usually found by knowing the heat addition:

$$Q_{in} = u_3 - u_2 = c_v(T_3 - T_2)$$
$$T_3 = \frac{q}{c_v} + T_2$$

State 4 is found by an isentropic process, for an ideal gas,

$$\frac{T_3}{T_4} = \left(\frac{V_4}{V_3}\right)^{k-1} = r_c^{k-1}$$
$$T_4 = \frac{T_3}{r_c^{k-1}}$$

Example 10.18

An engine operates on an air standard Otto cycle with a temperature and pressure of 27°C and 100 kPa at the beginning of compression. The compression ratio is 8.0, and the heat added is 1840 kJ/kg. Find the state-point properties and the thermal efficiency. The properties for ideal air are c_p = 1.008 kJ/(kg · K), c_v = 0.718 kJ/(kg · K), R = 0.287 kJ/(kg · K), and k = 1.4.

Solution

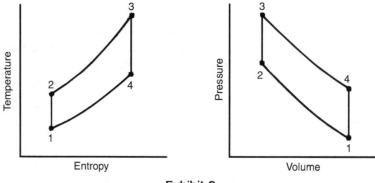

Exhibit 6

Refer to the *T-s* diagram and *P-V* diagram in Exhibit 6.

State 1

$$P_1 = 100 \text{ kPa}$$
$$T_1 = 300 \text{ K}$$
$$v_1 = \frac{RT_1}{P_1} = 0.287 \frac{300}{100} = 0.861 \text{ m}^3/\text{kg}$$

State 2

$$v_2 = \frac{v_1}{8} = 0.1076 \text{ m}^3/\text{kg}$$
$$T_2 = T_1 \left(\frac{v_1}{v_2}\right)^{k-1} = 300\,(8.0)^{0.4} = 689 \text{ K}$$
$$P_2 = P_1 \left(\frac{v_1}{v_2}\right)^{k} = 100\,(8.0)^{1.4} = 1838 \text{ kPa}$$

State 3

$$v_3 = v_2 = 0.1076 \text{ m}^3/\text{kg}$$
$$u_3 = u_2 + q_{in}$$
$$q_{in} = u_3 - u_2 = c_v(T_3 - T_2)$$
$$T_3 = \frac{1840}{0.718} + 689 = 3255 \text{ K}$$
$$P_3 = P_2 \left(\frac{T_3}{T_2}\right) = 1838 \frac{3255}{689} = 8683 \text{ kPa}$$

State 4

$$v_4 = v_1 = 0.861 \text{ m}^3/\text{kg}$$

$$T_4 = \frac{T_3}{r_c^{k-1}} = \frac{3255}{(8.0)^{0.4}} = 1417 \text{ K}$$

$$P_4 = \frac{P_3}{r_c^{k}} = \frac{8676}{(8.0)^{1.4}} = 472 \text{ kPa}$$

$$\eta_{TH} = \frac{W_{net}}{Q_{in}} = \frac{W_{3-4} - W_{1-2}}{Q_{in}} = \frac{c_v(T_3 - T_4) - c_v(T_2 - T_1)}{1840}$$

$$= \frac{0.718(3255 - 1416) - 0.718(689 - 300)}{1840}$$

$$= \frac{1318 - 279}{1840} = \frac{1039}{1840} = 0.565$$

Check:

$$\eta_{TH} = \frac{Q_{in} - Q_{out}}{Q_{in}} = \frac{1840 - c_v(T_4 - T_1)}{1840} = \frac{1840 - 800.9}{1840} = 0.565$$

$$\eta_{TH} = 1 - \frac{1}{r_c^{k-1}} = 1 - \frac{1}{(8.0)^{0.4}} = 0.565$$

HEAT TRANSFER

The three modes of heat transfer are conduction, convection, and radiation. The heat transfer "laws" are based on both empirical observations and theory but are consistent with the first and second laws of thermodynamics. That is, energy is conserved and heat flows from hot to cold.

Conduction

Conduction occurs in all phases of materials (Figure 10.20). The equation for one-dimensional, planar, steady-state conduction heat transfer is

$$\dot{Q} = kA \frac{T_H - T_C}{x} \quad \text{(watts)}$$

The **conductivity**, k, is a property of the material and is evaluated at the average temperature of the material. The **heat flow rate**, q, is sometimes expressed as a heat flux $\dot{Q}/A$.

For multiple layers of different materials (Figure 10.21), as in composite structures, it is usually best to use an electrical analogy:

$$\dot{Q}_1 = \frac{T_H - T_{x_1}}{R_1} \qquad R_1 = \frac{x_1}{A_1 k_1}$$

$$\dot{Q} = \dot{Q}_1 = \dot{Q}_2 = \dot{Q}_3$$

$$\dot{Q}_2 = \frac{T_{x_1} - T_{x_2}}{R_2} \qquad R_2 = \frac{x_2}{A_2 k_2}$$

$$\dot{Q}_3 = \frac{T_{x_2} - T_C}{R_3} \qquad R_3 = \frac{x_3}{A_3 k_3}$$

$$\dot{Q}_T = \frac{T_H - T_C}{R_1 + R_2 + R_3}$$

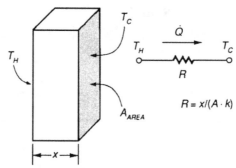

Figure 10.20

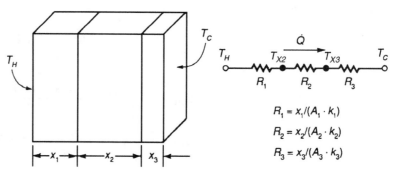

Figure 10.21

Example 10.19

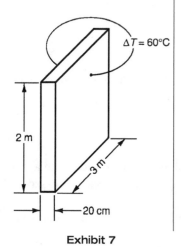

Exhibit 7

A plane wall is 2 m high by 3 m wide and is 20 cm thick. It is made of material that has a thermal conductivity of 0.5 W/(m • K). A temperature difference of 60°C is imposed on the two large faces. Find the heat flow, the heat flux, and the conductive resistance.

Solution

Refer to Exhibit 7.

$$\dot{Q} = \frac{kA(T_H - T_C)}{x} = \frac{0.5 \times 3 \times 2 \times 60}{0.20} = 900 \text{ W}$$

$$\frac{\dot{Q}}{A} = \frac{900}{3 \times 2} = 150 \frac{\text{W}}{\text{m}^2}$$

$$R = \frac{x}{kA} = \frac{0.2}{0.5 \times 3 \times 2} = 0.0667 \frac{\text{K}}{\text{W}}$$

Convection

Convection occurs at the boundary of a solid and a fluid (liquid or gas) when there is a temperature difference. The mechanism is complex and can be evaluated analytically only for a few simple cases; most situations are evaluated empirically. The equation for convective heat transfer is

$$q = hA(T_{\text{surface}} - T_{\text{fluid}})$$

The evaluation of h, the heat transfer coefficient, normally involves use of data correlated in the form of dimensionless parameters, for example, Nussult number, Reynolds number, Prandtl number.

The conduction and convection mechanisms can be combined as shown in Figure 10.22. So the temperature of the surface, T_s, is dependent on the relative magnitude of the two resistances.

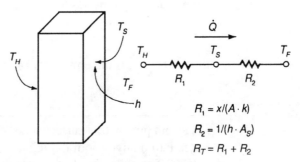

Figure 10.22

Example 10.20

Water at an average temperature of 20°C flows through a 5-cm-diameter pipe that is 2 m long. The pipe wall is heated by steam and is held at 100°C. The convective heat transfer coefficient is 2.2×10^4 W/(m² · K). Find the heat flow, the heat flux, and the convective resistance.

Solution

Refer to Exhibit 8.

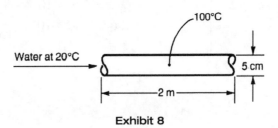

Exhibit 8

$$\dot{Q} = hA(T_H - T_C) = 2.2 \times 10^4 (\pi)(0.05)(2)(100 - 20) = 5.53 \times 10^5 \text{ W}$$

$$\frac{\dot{Q}}{A} = \frac{5.53 \times 10^5}{\pi(0.05)(2)} = 1.76 \frac{\text{MW}}{\text{m}^2}$$

$$R = \frac{1}{hA} = \frac{1}{2.2 \times 10^4 (\pi)(0.05)(2)} = 1.45 \times 10^{-4} \frac{\text{K}}{\text{W}}$$

Radiation

Radiation heat transfer occurs between two surfaces via electromagnetic waves and *does not* require an intervening medium to permit the energy flow. In fact, it travels best through a vacuum as radiant energy does from the sun. The equation for radiation energy exchange between two surfaces is

$$q = \sigma A_1 F_e F_s \left(T_1^4 - T_2^4\right)$$

where the Stefan-Boltzmann constant is $\sigma = 5.67 \times 10^{-8}$ W/(m² · K⁴), F_e is a factor that is a function of the emissivity of the two surfaces with a value from 0 to 1.0, and F_s is a modulus that is a function of the relative geometries of the two surfaces with a value from 0 to 1.0.

Note that the heat flow is not proportional to the linear temperature difference but is a function of the temperature of the surfaces to the fourth power.

The simplest, and by far the most common, case of radiation energy exchange occurs in the case of a small surface radiating to large surroundings. In this case, the equation simplifies to

$$q = \sigma A_1 \varepsilon_1 \left(T_1^4 - T_2^4\right)$$

where ε_1 is the emissivity of the radiating surface.

Example 10.21

A steam pipe with a surface area of 5 m² and a surface temperature of 600°C radiates into a large room (which acts as a black body), the surfaces of which are at 25°C. The pipe gray-body surface emissivity is 0.6. Find the heat flow and heat flux from the surface to the room.

Solution

Refer to Exhibit 9.

$$\dot{Q}_{1-2} = \sigma A F_e F_s \left(T_1^4 - T_2^4\right)$$

For a gray body radiating to a black-body enclosure,

$$F_e F_s = \varepsilon_1$$
$$\dot{Q}_{1-2} = 5.67 \times 10^{-8} \times 5 \times 0.6 \times (873^4 - 298^4) = 9.75 \times 10^4 \text{ W}$$
$$\frac{\dot{Q}}{A} = \frac{9.75 \times 10^4}{5} = 1.95 \times 10^4 \ \frac{\text{W}}{\text{m}^2}$$

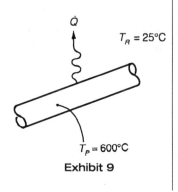

Exhibit 9

SELECTED SYMBOLS AND ABBREVIATIONS

Symbol or Abbreviation	Description
c_p	specific heat at constant pressure
c_v	specific heat at constant volume
H	enthalpy
h	specific enthalpy
h	heat transfer coefficient
k	thermal conductivity
m	mass
P	absolute pressure
p	partial pressure
P_r	relative pressure
Q	heat taken in or given off
q	heat per mass
r_c	compression ratio
R	gas constant
$\overline{R}$	universal gas constant
S	entropy
s	specific entropy
T	absolute temperature (in Kelvin)
T_H	high or hot temperature
T_L, T_C	low temperature, cold temperature
U	internal energy
u	specific internal energy
V	total volume
v	specific volume
V_r	relative volume
W	work
w	specific work
Z	elevation
Z	compressibility factor

PROBLEMS

10.1 Given the following data for a fluid, what is its state at 40°C and 3 kPa?

Saturated property table

T, °C	P, kPa	v_f, m³/kg	v_g, m³/kg	h_f, kJ/kg	h_g, kJ/kg	s_f, kJ/(kg • K)	s_g, kJ/(kg • K)
40	7.38	.001008	19.52	167.57	2574.3	.5725	8.257
80	47.39	.001029	3.407	334.9	2643.7	1.1343	7.5445
120	198.5	.001060	.8919	503.71	2706.3	1.5276	7.1296

  a. Saturated liquid
  b. Superheated vapor
  c. Compressed liquid
  d. Saturated vapor

10.2 Using the data table in Problem 10.1, what is the fluid's entropy in kJ/(kg • K) at 120°C and 80% quality?
  a. 1.53
  b. 6.009
  c. 7.13
  d. 28.8

10.3 Using the refrigerant data table in Problem 10.1, what is its latent heat (heat of vaporization) in kJ/kg at 80°C?
  a. 198.5
  b. 2706
  c. 1306
  d. 2308.8

10.4 A nonflow (closed) system contains 1 kg of an ideal gas ($c_p = 1.0$, $c_v = .713$). The gas temperature is increased by 10°C while 5 kJ of work are done by the gas. What is the heat transfer in kJ?
  a. −3.3
  b. −2.6
  c. +12.1
  d. +7.4

10.5 Shaft work of −15 kJ/kg and heat transfer of −10 kJ/kg change the enthalpy of a system by:
  a. −25 kJ/kg
  b. −15 kJ/kg
  c. −10 kJ/kg
  d. +5 kJ/kg

10.6 A quantity of 55 cubic meters of water passes through a heat exchanger and absorbs 2,800,000 kJ. The exit temperature is 95°C. The entrance water temperature in °C is nearest to:
  a. 49
  b. 56
  c. 68
  d. 83

10.7 A fluid at 690 kPa has a specific volume of .25 m³/kg and enters an apparatus with a velocity of 150 m/s. Heat radiation losses in the apparatus are equal to 25 kJ/kg of fluid supplied. The fluid leaves the apparatus at 135 kPa with a specific volume of .9 m³/kg and a velocity of 300 m/s. In the apparatus, the shaft work done by the fluid is equal to 900 kJ/kg. Does the internal energy of the fluid increase or decrease, and how much is the change?
  a. 858 kJ/kg (increase)
  b. 858 kJ/kg (decrease)
  c. 908 kJ/kg (increase)
  d. 908 kJ/kg (decrease)

10.8 Exhaust steam from a turbine exhausts into a surface condenser at a mass flow rate of 4000 kJ/hr, 9.59 kPa, and 92% quality. Cooling water enters the condenser at 15°C and leaves at the steam inlet temperature.

Properties of saturated water (US units): Temperature table

Temp. °C T	Press. kPa P	Specific volume		Internal energy		Enthalpy			Entropy	
		Sat. Liquid v_f	Sat. Vapor v_g	Sat. Liquid u_f	Sat. Vapor u_g	Sat. Liquid h_f	Evap. h_{fg}	Sat. Vapor h_g	Sat. Liquid s_f	Sat. Vapor s_g
15	1.705	.001	77.9	62.99	2396	62.99	2466	2529	.2245	8.781

v, m³/kg; u and h, kJ/kg; s, kJ/(kg • K)

The cooling water mass flow rate in kg/hr is closest to:

a. 157,200   c. 95,000
b. 70,200   d. 88,000

10.9 The mass flow rate of a Freon refrigerant through a heat exchanger is 5 kg/min. The enthalpy of entry Freon is 238 kJ/kg, and the enthalpy of exit Freon is 60.6 kJ/kg. Water coolant is allowed to rise 6°C. The water flow rate in kg/min is:

a. 24   c. 83
b. 35   d. 112

10.10 The maximum thermal efficiency that can be obtained in an ideal reversible heat engine operating between 833°C and 170°C is closest to:

a. 100%   c. 78%
b. 60%   d. 40%

10.11 A 2.2-kW refrigerator or heat pump operates between −17°C and 38°C. The maximum theoretical heat that can be transferred from the cold reservoir is nearest to:

a. 7.6 kW   c. 15.6 kW
b. 4.7 kW   d. 10.2 kW

10.12 In any nonquasistatic thermodynamic process, the overall entropy of an isolated system will:

a. increase and then decrease   c. stay the same
b. decrease and then increase   d. increase only

10.13 For spontaneously occurring natural processes in an isolated system, which expression best expresses ds?

a. $ds = \dfrac{dq}{T}$ c. $ds > 0$
b. $ds = 0$ d. $ds < 0$

10.14 Which of the following statements about entropy is *FALSE*?
a. The entropy of a mixture is greater than that of its components under the same conditions.
b. An irreversible process increases the entropy of the universe.
c. The entropy of a crystal at 0°C is zero.
d. The net entropy change in any closed cycle is zero.

10.15 A high-velocity flow of gas at 250 m/s possesses kinetic energy nearest to which of the following?
a. 3.13 kJ/kg
b. 313 kJ/kg
c. 31,300 kJ/kg
d. 31.3 kJ/kg

10.16 $(u + Pv)$ is a quantity called:
a. flow energy
b. shaft work
c. entropy
d. enthalpy

10.17 In flow process, neglecting *KE* and *PE* changes, $-\int v\, dP$ represents which item below?
a. Heat transfer
b. Shaft work
c. Closed system work
d. Flow energy

10.18 Power may be expressed in units of:
a. joules
b. watts
c. kJ
d. newtons

10.19 The temperature-entropy diagram in Exhibit 10.19 represents a:
a. Rankine cycle with superheated vapor
b. Carnot cycle
c. diesel cycle
d. refrigeration cycle

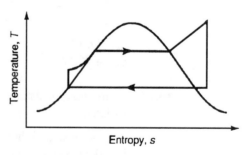

Exhibit 10.19

10.20 Entropy is the measure of:
a. the change in enthalpy of a system
b. the internal energy of a gas
c. the heat capacity of a substance
d. randomness or disorder

10.21 A Carnot heat engine cycle is represented on the *T-s* and *P-V* diagrams in Exhibit 10.21. Which of the several areas bounded by numbers or letters represents the amount of heat rejected by the fluid during one cycle?

a. Area 1–2–6–5
b. Area B–C–H–G
c. Area 3–4–5–6
d. Area D–A–E–F

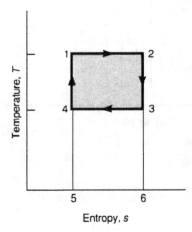
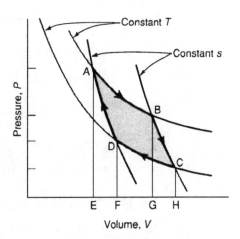

Exhibit 10.21

10.22 A Carnot engine operating between 70°C and 2000°C is modified solely by raising the high temperature by 150°C and raising the low temperature by 100°C. Which of the following statements is *FALSE*?

a. The thermodynamic efficiency is increased.
b. More work is done during the isothermal expansion.
c. More work is done during the isentropic compression.
d. More work is done during the reversible adiabatic expansion.

10.23 In the ideal heat pump system represented in Exhibit 10.23, the expansion valve 4–1 performs the process that is located on the *T-s* diagram between points:

a. A and B
b. B and C
c. C and D
d. E and A

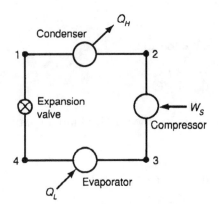
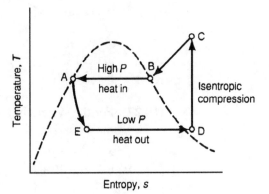

Exhibit 10.23

Use the psychrometric chart in solving problems 10.24 through 10.25.

10.24 Air is flowing through a duct and is heated using an electric coil (see Exhibit 10.24). The entering air is 20°C and $\phi = 40\%$. The exiting air is 35°C. The relative humidity of the exiting air is most nearly:

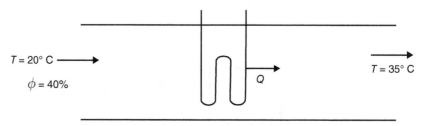

Exhibit 10.24

a. 60%   c. 25%
b. 40%   d. 17%

10.25 The humidity ratio of the exiting air in Problem 10.24 is most nearly:

a. 5.8   c. 4.5
b. 20   d. 16

10.26 An industrial air-conditioning system (see Exhibit 10.26) consists of 5 kg/min of outside air at 10°C and $\phi = 80\%$ mixing with 5 kg/min of inside air at T = 30°C and $\phi = 40\%$. The temperature at Position 3 is most nearly:

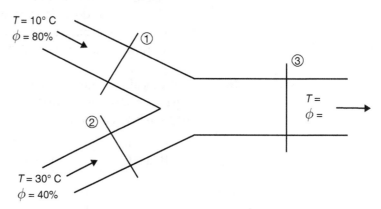

Exhibit 10.26

a. 30°C   c. 0°C
b. 25°C   d. 15°C

10.27 For the system in Problem 10.26, the relative humidity at Position 3 is most nearly:

a. $\phi = 58\%$ c. $\phi = 45\%$
b. $\phi = 40\%$ d. $\phi = 64\%$

10.28 Data in the following table describe two states of a working fluid that exist at two locations in a piece of hardware:

	P, kPa	v, m³/kg	T, °C	h, kJ/kg	s, kJ/(kg·K)
State 1	25	0.011	20	19.2	0.0424
State 2	125	0.823	180	203.7	0.3649

Which of the following statements about the path from State 1 to 2 is *FALSE*?

a. The path results in an expansion.
b. The path determines the amount of work done.
c. The path is indeterminate from these data.
d. The path is reversible and adiabatic.

10.29 Name the process that has no heat transfer.

a. Isentropic
b. Isothermal
c. Quasistatic
d. Reversible

10.30 In a closed system with a moving boundary, which of the following represents work done during an isothermal process?

a. $W = P(V_2 - V_1)$

b. $W = 0$

c. $W = P_1 V_1 \ln\left(\dfrac{P_1}{P_2}\right) = P_1 V_1 \ln\left(\dfrac{V_2}{V_1}\right) = mRT \ln\left(\dfrac{P_1}{P_2}\right)$

d. $W = \dfrac{P_2 V_2 - P_1 V_1}{1-k} = \dfrac{mR(T_2 - T_1)}{1-k}$

10.31 The work of a polytropic ($n = 1.21$) compression of air ($c_p/c_v = 1.40$) in a system with moving boundary from $P_1 = 15$ kPa, $V_1 = 1.0$ m³ to $P_2 = 150$ kPa, $V_2 = 0.15$ m³ is:

a. −35.7 kJ
b. −324 kJ
c. 1080 kJ
d. 5150 kJ

10.32 The isentropic compression of 1 m³ of air, $c_p/c_v = 1.40$, from 20 kPa to a pressure of 100 kPa gives a final volume of:

a. 0.16 m³
b. 0.20 m³
c. 0.32 m³
d. 0.40 m³

10.33 An ideal gas at a pressure of 500 kPa and a temperature of 75°C is contained in a cylinder with a volume of 700 m³. Some of the gas is released so that the pressure in the cylinder drops to 250 kPa. The expansion of the gas is isentropic. The specific heat ratio is 1.40, and the gas constant is .287 kJ/(kg•K). The mass of the gas (in kg) remaining in the cylinder is nearest to:

a. 900
b. 1300
c. 1500
d. 2140

10.34 The theoretical power required for the isothermal compression of 800 m³/min of air from 100 to 900 kPa is closest to:

a. 70
b. 90
c. 130
d. 290

10.35 Which of the following statements is *FALSE* concerning the deviations of real gases from ideal gas behavior?

a. Molecular attraction interactions are compensated for in the ideal gas law.
b. Deviations from ideal gas behavior become large near the saturation curve.
c. Deviations from ideal gas behavior become significant at pressures above the critical point.
d. Molecular volume becomes significant as specific volume is decreased.

10.36 There are 3 kg of air in a rigid container at 250 kPa and 50°C. The gas constant for air is .287 kJ/(kg • K). The volume of the container, in m³, is nearest to:

a. 2.2
b. 1.1
c. 2.8
d. 3.1

10.37 A mixture at 100 kPa and 20°C that is 30% by weight CO_2 (m.w. = 44) and 70% by weight N_2 (m.w. = 28) has a partial pressure of CO_2 in kPa that is nearest to:

a. 21.4
b. 31.5
c. 68.3
d. 78.6

10.38 Dry air has an average molecular weight of 28.79, consisting of 21 mole-percent O_2, 78 mole-percent N_2, and 1 mole-percent Ar (and traces of CO_2). The weight-percent of O_2 is nearest to:

a. 21.0
b. 22.4
c. 23.2
d. 24.6

10.39 The temperature difference between the two sides of a solid rectangular slab of area A and thickness L, as shown in Exhibit 10.39, is ΔT. The heat transferred through the slab by conduction in time, t, is proportional to:

a. $AL\Delta Tt$
b. $AL\dfrac{\Delta T}{t}$
c. $AL\dfrac{t}{\Delta T}$
d. $\dfrac{A\Delta Tt}{L}$

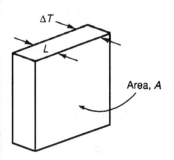

Exhibit 10.39

10.40 The composite wall in Exhibit 10.40 has an outer temperature $T_1 = 20°C$ and an inner temperature $T_4 = 70°C$. The temperature T_3, in °C, is nearest to:

a. 38   c. 58
b. 46   d. 69

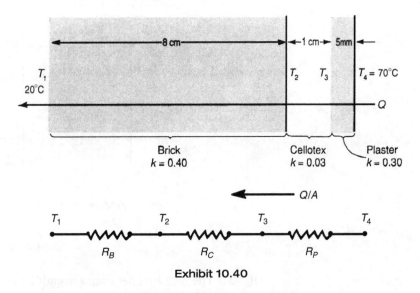

Exhibit 10.40

10.41 In Exhibit 10.41, the inner wall is at 30°C, and the outer wall is exposed to ambient wind and surroundings at 10°C. The film coefficient, h, for convective heat transfer in a 7-m/s wind is about 20 W/m² • °C. Ignoring any radiation losses, an overall coefficient (in the same units) for the conduction and convection losses is most nearly:

a. 1.4   c. 12.5
b. 2.6   d. 7.1

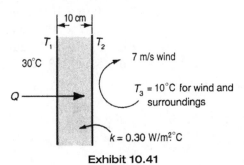

Exhibit 10.41

10.42 Heat is transferred by conduction from left to right through the composite wall shown in Exhibit 10.42. Assume the three materials are in good thermal contact and that no significant thermal resistance exists at any of the interfaces. The overall coefficient U in W/(m² • °C) is most nearly:

a. 0.04
b. 0.20
c. 0.35
d. 0.91

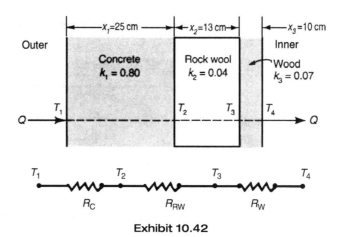

Exhibit 10.42

10.43 The heat loss per hour through 1 m² of furnace wall that is 40 cm thick is 520 W. The inside wall temperature is 1000°C, and its average thermal conductivity is 0.61 W/(m² • °C).

The outside surface temperature of the wall is nearest to:

a. 100°C
b. 300°C
c. 700°C
d. 1000°C

10.44 Which of the following is the usual expression for the power/unit-area Stefan-Boltzmann constant for black-body radiation?

a. 1.36×10^{-12} cal/(s • cm² • K⁴)
b. 5.67×10^{-5} ergs/(s • cm² • K⁴)
c. 5.67×10^{-8} watts/(m² • K⁴)
d. 5.67×10^{-8} coulombs/(s • m² • K⁴)

SOLUTIONS

10.1 b. At 40°C, equilibrium between liquid and gas exists at 7.38 kPa. Below 7.38 kPa superheated vapor exists, and above 7.38 kPa only pressurized liquid exists.

10.2 b. At 120°C, $s_f = 1.5276$ and $s_{fg} = 7.1296 - 1.5276 = 5.602$. Here s_f is saturated liquid at 0% quality and s_g is saturated vapor of 100% quality. Thus s at 80% quality $= s_f + (0.80) s_{fg} = 1.5276 + .8 \times 5.602 = 6.009$ kJ/(kg • K).

10.3 d. Here, $h_{fg} = h_g - h_f = 2643.7 - 334.9 = 2308.8$ kJ/kg.

10.4 c. The thermodynamic sign convention is + for heat in and + for work out of a system. Apply the first law for a closed system and an ideal gas working fluid:

$$\Delta U = mc_v \, \Delta T = q - w$$
$$.713(10) = q - (+5), \qquad 7.13 = q - 5, \qquad q = 12.13$$

10.5 d. The first law applied to a flow system is
$h = q - w_s = -10 - (-15) = +5$.

10.6 d. For liquid water, $c_p = 4.18$ kJ/(kg • °C):

$$Q = mc_p \Delta T = mc_P(T_2 - T_1)$$
$$2{,}800{,}000 = (55 \, \text{m}^3) \left(\frac{1000 \, \text{kg}}{\text{m}^3} \right) (4.18)(95 - T_1)$$
$$12.2 = 95 - T_1 \qquad T_1 = 82.8°C$$

10.7 d. The basis of the calculation will be 1 kg. Use the thermodynamic sign convention that heat in and work out are positive. The first-law energy balance for the flow system is $h_2 + KE_2 - h_1 - KE_1 = Q - W_s$. Since the working fluid is unspecified and the internal energy change is desired, use the definition $h = u + Pv$. Then

$$u_2 + P_2 v_2 + KE_2 - u_1 - P_1 v_1 - KE_1 = Q - W_s \quad \text{or}$$

$$u_2 - u_1 = Q - W_s + P_1 v_1 + KE_1 - P_2 v_2 - KE_2$$

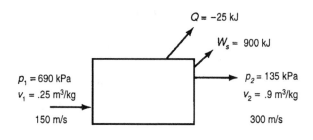

Exhibit 10.7

Now calculate numerical values for all terms except $u_2 - u_1$:

$$P_2 v_2 = 135 \times .9 = 121.5 \text{ kJ/kg} \qquad P_1 v_1 = 690(.25) = 172.5 \text{ kJ/kg}$$

$$KE_2 = \frac{V^2}{2gJ} = \frac{300^2}{2000} = 45 \text{ kJ/kg}$$

$$KE_1 = \frac{V^2}{2gJ} = \frac{(150)^2}{2000} = 11.3 \text{ kJ/kg}$$

$$W_s = +900 \text{ kJ/kg}$$

Therefore,

$$u_2 - u_1 = -25 - 900 + 172.5 + 11.3 - 121.5 - 45 = -907.7 \text{ kJ/kg}$$

10.8 Saturated steam table data at 9.59 kPA are

T, °C	h_f, kJ/kg	h_{fg}, kJ/kg	h_g, kJ/kg
45	188.45	2394.8	2583.2

The enthalpy of steam at 92% quality = $h_1 = h_f + 0.92 h_{fg}$ = 188.45 + 92 × 2394.8 = 2391.7. The enthalpy of liquid water at 45°C = h_2 = 188.45 kJ/kg. The enthalpy of liquid water at 15°C = h_3 = 62.99 kJ/kg above reference of 0°C.

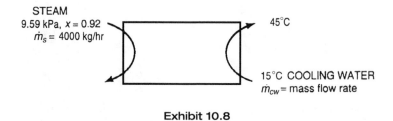

Exhibit 10.8

In the absence of data, assume that the steam condensate leaves at 45°C; if a heat balance is written over a 1-hour period, then the heat from steam = heat to cooling water, or

$$\dot{m}_s(h_1 - h_2) = \dot{m}_{cw}(h_2 - h_3)$$
$$4000(2391.7 - 188.45) = \dot{m}_{cw}(188.45 - 62.99)$$
$$\dot{m}_{cw} = 70,245 \text{ kg/hr}$$

10.9 b. Over a 1-minute period, the heat gain by water equals heat loss by Freon:

$$\dot{m}_{cw} c_p \Delta T = \dot{m}_F (h_1 - h_2)$$
$$\dot{m}_{cw} \times 4.2 \times 6 = 5(238 - 60.6)$$
$$\dot{m}_{cw} = \frac{887}{25.2} = 35.2 \text{ kg/min}$$

10.10 b. Maximum efficiency is achieved with a Carnot engine.

$$T_L = 170 + 273 = 443 \text{ K} \qquad T_H = 833 + 273 = 1106 \text{ K}$$

$$\eta_{TH} = \frac{w}{q_H} = \frac{q_H - q_L}{q_H}$$
$$= 1 - \frac{Q_L}{Q_H} = 1 - \frac{T_L}{T_H}$$
$$= 1 - \frac{443}{1106} = 1 - 0.40 = 0.60 = 60\%$$

10.11 d. The coefficient of performance of a Carnot refrigerator or heat pump is

$$\text{COP} = \frac{T_L}{T_H - T_L} = \frac{256 \text{ K}}{311 \text{ K} - 256 \text{ K}} = 4.65 = \frac{q_L}{w} = 4.65$$

$$\text{COP} = \frac{q_L}{q_H - q_L} = \frac{q_L}{w} = \frac{q_L}{2.2}; \qquad q_L = 4.65(2.2) = 10.2 \text{ kW}$$

10.12 d. Quasistatic means infinitely slow, lossless, hypothetical, by differential increments. The overall entropy will increase for an isolated system or for the system plus surroundings.

10.13 c. $ds > 0$ is the correct answer because all naturally occurring spontaneous processes are irreversible and result in an entropy increase.

(a) $ds = \frac{dq_{\text{rev}}}{T}$ only. The reversible requirement is necessary to generate the exact height vs. rectangular area equivalence on the Carnot cycle T-s diagram.

(b) Only a reversible adiabatic process is isentropic by definition.

(d) An energy input from the surroundings is required to reduce the entropy.

10.14 c. All are true *except* (c). The entropy of a perfect crystal at absolute zero (0 K or 0°R) is zero. This is the third law of thermodynamics. There is presumably no randomness at this temperature in a crystal without flaws, impurities, or dislocations.

10.15 d. Per 1 kg of flowing fluid,

$$KE = \frac{V^2}{2g_c} \quad \text{where } V \text{ is in m/s, and } g_c = 1.0$$

Use 1000 to convert J to kJ:

$$KE = \frac{250^2}{2(1000)} = 31.3 \text{ kJ/kg}$$

10.16 d. Flow energy is Pv. Shaft work, W_s, is $-\int v\,dP$. Entropy is s. Internal energy is u. Enthalpy h is defined as $u + Pv$, the sum of internal energy plus flow energy.

10.17 b. Shaft work is work or mechanical energy crossing the fixed boundary (control volume) of a flow (open) system. Shaft work W_s is defined, in the absence of PE and KE changes, by $dh = T\,ds + v\,dP$, where $T\,ds = dq_{rev}$ and $-v\,dP$ is dW_s. In integrated form, $\Delta h = \int T\,ds + \int v\,dP = q_{rev} - W_s$, where W_s is represented by $-\int v\,dP$. Closed system work W is defined by $du = T\,ds - P\,dv$, or $\Delta u = \int T\,ds - \int P\,dv = q_{rev} - W$. Thus, closed system work is $+\int P\,dv$. Flow energy is the Pv term, and enthalpy change is ΔH.

10.18 b. Power is energy per unit time. The usual power units are watts.

10.19 a.

10.20 d.

10.21 c. The table below gives the significance of each area of the diagrams:

Process	T-s Diagram: Area Representing Heat	P-V Diagram: Area Representing Work
Isothermal expansion, 1–2 and A–B	1–2–6–5 = heat in from high-temp. reservoir	A–B–G–E = work done by fluid
Isentropic expansion, 2–3 and B–C	2–3–6 = 0 heat transfer	B–C–H–G = work done by fluid
Isothermal compression, 3–4 and C–D	3–4–5–6 = heat out to low-temp. reservoir	C–D–F–H = work done on fluid
Isentropic compression, 4–1 and D–A	4–1–5 = 0 heat transfer	D–A–E–F = work done on fluid
Net result of process	1–2–3–4 = net heat converted to work	A–B–C–D = net work done by process

10.22 **a.** The Carnot cycle efficiency is originally

$$\eta = \frac{T_H - T_L}{T_H} = \frac{2273 - 343}{2273} = 0.849$$

After the change,

$$\eta = \frac{2423 - 443}{2423} = .817 \quad \text{(efficiency is reduced)}$$

On the *T-s* and *P-V* diagrams in Exhibit 10.22, the original cycle is shown as ABCD, and the modified cycle is shown as A′B′C′D′.

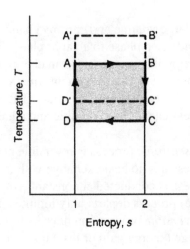
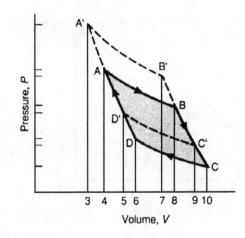

Exhibit 10.22

Compare the work done during the isothermal expansion (A to B vs. A′ to B′):
 Original: area A–B–8–4
 Modified: area A′–B′–7–3 is larger
Compare the work done during the isentropic compression (D to A vs. D′ to A′):
 Original: area D–A–4–6
 Modified: area D′–A′–3–5 is larger
Compare the work during the reversible (isentropic) expansion (B to C vs. B′ to C′):
 Original: area B–C–10–8
 Modified: area B′–C′–9–7 is larger
Compare the work during the isothermal compression (C to D vs. C′ to D′):
 Original: area C–D–6–10
 Modified: area C′–D′–5–9 is larger
Statements (b), (c), and (d) are correct.

10.23 **d.** The vapor compression-reversed Rankine cycle is conducted counterclockwise on both the schematic and the *T-s* diagram. Numbers on the schematic and letters on the *T-s* diagram are related: 1 = A, 2 = B, 3 = D, and 4 = E. Process C–B–A occurs in the condenser between 2 and 1. The expansion process A–E occurs between 1 and 4.

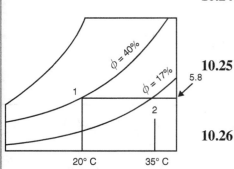

Exhibit 10.24a

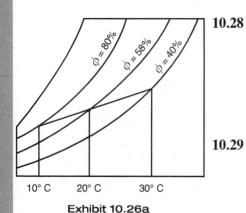

Exhibit 10.26a

10.24 d. On the psychrometric chart, locate point 1, which is $T = 20°C$ and $\phi = 40\%$ (see Exhibit 10.24a). Draw a horizontal line to point 2, which is at $35°C$. Now, read ϕ from the chart to find $\phi = 17\%$.

10.25 a. From the psychrometric chart for the preceding solution, extend the line from point 1 to point 2 over to the humidity ratio to find that humidity ratio = 5.8 gm water/kg dry air.

10.26 c. On the psychrometric chart, locate the points at which $T = 10°C$, $\phi = 80\%$ and $T = 30°C$, $\phi = 40\%$. Draw a line between the two points and bisect it to find $T = 20°C$ (see Exhibit 10.26a).

10.27 a. From Exhibit 10.26a, $\phi = 58\%$.

10.28 d. The large volume and entropy changes indicate a change from a condensed phase to a vapor phase. Temperature, pressure, and enthalpy increases require an energy input. The path from 1 to 2 is indeterminate because no information on intermediate states is given. Work is always path dependent. The entropy increase means the process cannot be reversible and adiabatic (isentropic).

10.29 a. An *isentropic* process is reversible and adiabatic. An *adiabatic* process has no heat exchange with its surroundings. An *isothermal* process is conducted at constant temperature. A *quasistatic* (almost static) process departs only infinitesimally from an equilibrium state. A *reversible* process can have its initial state restored without any change (energy gain or loss) taking place in the surroundings.

10.30 c. For a closed system (piston-cylinder type, nonrepetitious) the work done is $W = \int P \, dV$. The equations in the problem are valid for ideal gases in the following processes, respectively:

a. constant pressure c. isothermal process
b. constant volume   d. isentropic process

10.31 a. The work of a closed system (moving boundary) polytropic process for an ideal gas is

$$W = \frac{P_2 V_2 - P_1 V_1}{1-n} = \frac{[150(0.15) - 15(1.0)]}{1-1.21} = -35.7 \text{ kJ}$$

which is work done on the gas.

10.32 c. An isentropic process for an ideal gas follows the path

$$PV^k = P_1 V_1^k = P_2 V_2^k = \text{constant} \qquad \text{where } k = c_p/c_v$$
$$20(1)^{1.4} = 100(V_2)^{1.4}; \qquad V_2^{1.4} = 0.20; \qquad \text{hence, } V_2 = 0.317 \text{ m}^3$$

10.33 d. Given:

$$k = c_p/c_v = 1.40 \qquad R = .287 \frac{kJ}{kg \bullet K}$$

$$P_1 = 500 \text{ kPa} \qquad P_2 = 250 \text{ kPa}$$
$$V_1 = 700 \text{ m}^3 \qquad V_2 = 700 \text{ m}^3$$
$$T_1 = 75°C + 273 = 348 \text{ K} \qquad T_2 = ?$$
$$w_2 = ?$$

Basis: The ideal gas law may be written $PV = mRT$, and the basic equation for reversible adiabatic (isentropic) expansion is

$$\frac{T_2}{T_1} = \left(\frac{P_2}{P_1}\right)^{(k-1)/k}$$

The gas remaining in the tank cools as it expands; the new temperature is

$$T_2 = T_1 \left(\frac{P_2}{P_1}\right)^{(k-1)/k} = 348\left(\frac{250}{500}\right)^{(1.4-1)/1.4} = 348\left(\frac{1}{2}\right)^{0.2857} = 285 \text{ K}$$

Now apply the gas law at State 2, $P_2 V_2 = m_2 R T_2$:

$$250 \times 700 = m_2(.287)(285)$$

$$m_2 = \frac{(250)(700)}{(.287)(285)} = 2139 \text{ kg}$$

10.34 d. Since a volume flow rate is specified, the process is a flow process. The work of isothermal compression of an ideal gas is numerically the same in a steady flow process as in a closed system:

$$PV = \text{constant} = P_1 V_1 = P_2 V_2 = mRT$$

In a closed system,

$$W = \int_{V_1}^{V_2} P\, dV = P_1 V_1 \ln \frac{V_2}{V_1} = P_1 V_1 \ln \frac{P_1}{P_2} = mRT \ln \frac{V_2}{V_1} = mRT \ln \frac{P_1}{P_2}$$

In a flow system,

$$W_s = \int_{P_1}^{P_2} V\, dP = -P_1 V_1 \ln \frac{P_2}{P_1} = P_1 V_1 \ln \frac{P_1}{P_2} = P_1 V_1 \ln \frac{V_2}{V_1} = mRT \ln \frac{V_2}{V_1} = mRT \ln \frac{P_1}{P_2}$$

Over a 1-minute interval,

$$W_s = P_1 V_1 \ln \frac{P_1}{P_2} = (100)(800)\left(\frac{100}{900}\right) = -17{,}580 \text{ kJ/min}$$

$$W_s\left(\frac{-17{,}580}{60}\right) = -293 \text{ kW}$$

10.35 a. All statements except (a) are true. The ideal gas law does not consider the volume of the molecules or any interaction other than elastic collisions.

10.36 b. The ideal gas law is $PV = mRT$. Here $P = 250$ kPa and $T_1 = 50°C + 273 = 323$ K. Hence,

$$250V = 3 \times .287 \times 323$$

$$V = \frac{(3)(.287)(323)}{250} = 1.11 \text{ m}^3$$

10.37 a. The calculation is based on 1 kg of mixed gases. (1) Calculate the weight of each component and the number of moles of each that is present. (2) Compute the mole fraction of each, and apportion the total pressure in proportion to the mole fraction. The computations are in the following table:

Component	Weight, kg	Number of kg-mol	Mole Fraction	Partial Pressure, kPa
CO_2	0.30	$\frac{0.30}{44} = 0.00682$	$\frac{0.00682}{0.03182} = 0.214$	21.4
N_2	0.70	$\frac{0.70}{28} = 0.0250$	$\frac{0.0250}{0.03182} = 0.786$	78.6
Total	1.00	0.03182	1.000	100

Since the mole fraction of a gas is the same as the volume fraction, the composition of the mixture is 21.4% vol. CO_2 and 78.6% vol. N_2. From the table, the correct partial pressure of CO_2 is 21.4 kPa.

10.38 c. The calculation will be based on 1 kg-mol of dry air and arranged in the following table:

Component	m.w.	Mole Fraction	Weight, kg	Weight, %
O_2	32.0	0.21	6.72	23.2
N_2	28.0	0.78	21.80	75.4
Ar	40.0	0.01	0.40	1.4
Totals		1.00	28.92	100.0

10.39 d. The heat transfer rate through the slab by conduction is governed by the equation

$$Q = kA\Delta T/L$$

In time t the amount of heat transfer is proportional to

$$A\frac{\Delta T}{L}t$$

The symbol k is the coefficient of thermal conductivity of the material; hence, the heat transfer in a given material is proportional to the other variables.

10.40 d. At steady state the same Q flows across each material, and the temperatures descend in direct proportion to the thermal resistances (reciprocal of conductivity).

$$\text{Resistance of brick} = \frac{x}{k} = \frac{.08\,\text{m}}{.4\,\frac{w}{m \cdot °C}} = .2\,\frac{\text{m}^2}{\text{W} \cdot °C}$$

$$\text{Resistance of Cellotex} = \frac{x}{k} = \frac{.01}{.03} = .333\,\frac{\text{m}^2}{\text{W} \cdot °C}$$

$$\text{Resistance of plaster} = \frac{x}{k} = \frac{.005}{.3} = .017\,\frac{\text{m}^2}{\text{W} \cdot °C}$$

Total resistance $= .2 + .333 + .017 = .55\,\text{m}^2/(\text{W} \cdot °C)$

$$Q/A = \frac{\Delta T_{\text{total}}}{\text{total resistance}} = \left(\frac{\Delta T}{x/k}\right)_{\text{layer}}$$

Hence,

$$Q/A = \frac{50}{.55} = \frac{T_4 - T_3}{.017} = \frac{T_3 - T_2}{.333} = \frac{T_2 - T_1}{.2} = 90.9\ \text{W/m}^2$$

$T_4 - T_3 = 1.5°C,$ since $T_4 = 70°C$, $T_3 = 68.5°C$
$T_3 - T_2 = 30.3°C,$ since $T_3 = 68.5°C$, $T_2 = 38.2°C$
$T_2 - T_1 = 18.2°C,$ since $T_2 = 38.2°C$, $T_1 = 20°C$
                        (in agreement with given data)

10.41 b. Since conduction and convection are based on ΔT, absolute temperatures are not required. For steady state, the heat conducted through a wall must equal the heat lost by convection:

$$Q = \frac{kA(T_1 - T_2)}{x} = hA(T_2 - T_3) \tag{10.41}$$

In a similar way, Q can be expressed by an overall coefficient

$$Q = UA(T_1 - T_3) \tag{10.42}$$

Here, U is calculated in a manner analogous to that used for thermal conductivities in series:

$$U = \frac{1}{\frac{1}{h_1} + \frac{x_1}{k_1}} \tag{10.43}$$

In this case,

$$U = \frac{1}{\frac{1}{20} + \frac{.1}{.3}} = \frac{1}{.05 + .333} = 2.61\ \text{W}/(\text{m}^2 \cdot °C)$$

10.42 b. The overall coefficient U, the thermal conductivity k/x, and the film coefficient h are the reciprocals of their thermal resistances. Thermal resistances in series are handled analogously to series electrical resistances; hence

$$U = \left(\sum_i \frac{x_i}{k_i}\right)^{-1} = \frac{1}{R_T} = \frac{1}{R_1 + R_2 + R_3}$$

The overall coefficient U is then used in the simplified conduction equation $Q = UADT$.

In this problem

$$U = \frac{1}{\frac{.25}{.80} + \frac{.13}{.04} + \frac{.10}{.07}} = \frac{1}{.313 + 3.25 + 1.43} = 0.20 \text{ W/(m}^2 \cdot °C)$$

10.43 c. The heat conduction equation is

$$Q = k\frac{A}{L}(T_1 - T_2)$$

where $T_1 = 1000°C$, $T_2 =$ outside temperature, $k = 0.61$ W/(m$^2 \cdot °C$), $Q/A = 520$ W/m^2, and $L = .4$ m.

Solving for T_2, one has

$$T_2 = -\frac{Q}{A}\frac{L}{k} + T_1 = -520\frac{.4}{.61} + 1000 = 659°C$$

10.44 c. All are numerically correct conversions of the constant in terms of power per unit area. The units of watts/(m$^2 \cdot$ K^4), however, are normally used in heat transfer.

CHAPTER 11

Fluid Mechanics

OUTLINE

FLUID PROPERTIES 440
Density ■ Specific Gravity ■ Specific Weight ■
Viscosity ■ Pressure ■ Surface Tension

FLUID STATICS 443
Hydrostatic Pressure ■ Manometers ■ Forces on Flat
Submerged Surfaces ■ Buoyancy

CONSERVATION LAWS 449
Mass ■ Energy ■ Friction and Minor Losses ■ Hazen-Williams Equation ■
Series and Parallel Flow ■ Flow in Noncircular Conduits

VELOCITY AND FLOW MEASURING DEVICES 458
Pitot Tubes ■ Flow Meters ■ Flow from a Tank

PUMP SELECTION 461
Types of Pumps ■ Pump Characteristics ■
NPSH ■ Scaling Laws ■ Specific Speed

TURBINES 469

OPEN CHANNEL FLOW 470
Energy Considerations ■ Manning's Equation
for Uniform Flow ■ Weir Equations

INCOMPRESSIBLE FLOW OF GASES 477

FANS AND BLOWERS 478
Fan Laws ■ Duct-Fan Characteristics ■ Fans in Series ■ Fans in Parallel

AIR AND GAS COMPRESSORS 483
Work of Compressor without Clearance ■ Isothermal Compression
Horsepower ■ Polytropic and Isentropic Compression Horsepower ■
Gas Compressor with Clearance ■ Volumetric Efficiency ■ Capacity ■
Effect of Clearance on Compressor Performance ■ Efficiency

PROBLEMS 492

SOLUTIONS 500

Fluid mechanics is the study of fluids at rest or in motion. The topic is generally divided into two categories: *liquids* and *gases*. Liquids are considered to be incompressible, and gases are compressible. The treatment of incompressible fluids and compressible fluids each has its own group of equations. However, there are times when a gas may be treated as incompressible (or at least uncompressed). For example, the flow of air through a heating duct is one such case. This chapter will concentrate on incompressible fluids.

FLUID PROPERTIES

Thermodynamic properties are important in incompressible fluid mechanics. Those of particular importance are density, specific gravity, specific weight, viscosity, and pressure. Temperature is also important but is primarily used in finding other properties such as density and viscosity in tables or graphs.

Density

The **density**, ρ, is the mass per unit volume and is the reciprocal of the specific volume, a property used in thermodynamics:

$$\rho = \frac{m}{V} = \frac{1}{v}\frac{\text{kg}}{\text{m}^3}$$

Specific Gravity

The **specific gravity**, SG, is defined by the following equation:

$$SG = \frac{\rho\left(\frac{\text{kg}}{\text{m}^3}\right)}{1000\ \frac{\text{kg}}{\text{m}^3}}$$

where 1000 kg/m3 is the density of water at 4°C.

In many cases the specific gravity of a liquid is known or found from tables and must be converted to density using this equation.

Specific Weight

The **specific weight**, γ, of a fluid is its weight per unit volume and is related to the density as follows:

$$\gamma = \frac{W}{V} = \rho\left(\frac{g}{g_c}\right)\frac{\text{N}}{\text{m}^3}$$

where g = local acceleration of gravity, $\frac{\text{m}}{\text{s}^2}$, and g_c = gravitational constant:

$$g_c = \frac{\text{kg} \cdot \text{m}}{\text{N} \cdot \text{s}^2}$$

The density of water at 4°C is 1000 kg/m³. Its specific weight at sea level (g = 9.81 m/s²) is calculated as follows:

$$\gamma = \rho\frac{g}{g_c} = 1000\,\frac{\text{kg}}{\text{m}^3} = \frac{9.81\,\frac{\text{m}}{\text{s}^2}}{\frac{\text{kg} \cdot \text{m}}{\text{N} \cdot \text{s}^2}} = 9810\,\frac{\text{N}}{\text{m}^3}$$

The density, specific gravity, and specific weight of a liquid are generally considered to be constant, with little variation, over a wide temperature range.

Viscosity

The **viscosity** of a fluid is a measure of its resistance to flow; the higher the viscosity the more resistance to flow. Water has a relatively low viscosity, and heavy fuel oils have a high viscosity. The **dynamic (absolute) viscosity**, μ, of a fluid is defined as the ratio of shearing stress to the rate of shearing strain. In equation form:

$$\mu = \frac{\tau}{\frac{dV}{dy}} \quad \frac{N \bullet s}{m^2} \left(\frac{kg}{m \bullet s} \right)$$

where τ = shearing stress (force per unit area), N/m², and dV/dy = rate of shearing strain, $1/s$.

Fluids may be classified as Newtonian or non-Newtonian. Newtonian fluids are those in which dV/dy in the above equation can be considered to be constant for a given temperature. Thus the shearing stress, τ (horizontal force divided by the surface area), of a plate on a thin layer, δ, of a Newtonian fluid, as shown in Figure 11.1, may be found from

$$\tau = \mu \frac{dV}{dy} = \frac{\mu V}{\delta}$$

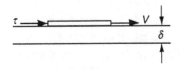

Figure 11.1

where V = velocity, m/s, and δ = thickness, m.

Most common fluids (liquids), such as water, oil, gasoline, and alcohol, are classified as Newtonian fluids.

The **kinematic viscosity** is defined by

$$v = \frac{\mu}{\rho} \quad \frac{m^2}{s}$$

Both the dynamic and kinematic viscosities are highly dependent on temperature. The viscosity of most liquids decreases significantly (orders of magnitude) with increases in temperature, while the viscosity of gases increases mildly with increases in temperature. The viscosity of any gas is less than the viscosity of any liquid. Viscosities are generally found in tables and graphs.

The definition of viscosity assists in the development of the engineering definition of a fluid as follows:

> A fluid is a substance that will deform readily and continuously when subjected to a shear force, no matter how small the force.

Pressure

Pressure, p, is the force per unit area of a fluid on its surroundings or vice versa. Pressure may be specified using two different datums. Absolute pressure, P_{abs}, is measured from absolute zero or a complete vacuum (void). At absolute zero there are no molecules and a negative absolute pressure does not exist. Absolute pressures are needed for ideal gas relations and in compressible fluid mechanics. Gage pressure, p_{gage}, on the other hand, uses local atmospheric pressure as its datum. Gage pressures may be positive (above atmospheric pressure) or negative (below atmospheric pressure). Negative gage pressure is also called vacuum. A complete vacuum occurs at a negative gage pressure that is equivalent to the atmospheric pressure or at absolute zero.

The relationship between absolute pressure and gage pressure is as follows:

$$P_{abs} = p_{gage} + p_{atm} \frac{N}{m^2} \text{ (Pa)}$$

Actually, the pressure is usually expressed in kN/m² or kPa but should be converted to these units for use in most equations.

Example 11.1

A pressure gage measures 50 kPa vacuum in a system. What is the absolute pressure if the atmospheric pressure is 101 kPa?

Solution

Change vacuum to a negative gage pressure:

$$p_{abs} = -50 \text{ kPa} + 101 \text{ kPa} = 51 \text{ kPa} \cdot 1000 = 51{,}000 \text{ Pa}$$

Most pressure-measuring devices measure gage pressure. For incompressible fluid dynamics, gage pressure may be used in most equations. This capability simplifies equations significantly when one or more pressures in the system are atmospheric or $p_{gage} = 0$.

Surface Tension

Surface tension is another property of liquids. It is the force that holds a water droplet or mercury globule together, since the cohesive forces of the liquid are more than the adhesive forces of the surrounding air. The surface tension (or surface tension coefficient), σ, of liquids in air is available in tables and can be used to calculate the internal pressure, p, in a droplet from

$$p = \frac{4\sigma}{d}$$

where σ = surface tension of the liquid, kN/m, and d = droplet diameter, m. Values of surface tension for various liquids are found in tables as a function of the surrounding medium (air, etc.) and the temperature.

Surface tension is also the property that causes a liquid to rise (or fall) in a capillary tube. The amount of rise (or fall) depends on the liquid and the capillary tube material. When *adhesive* forces dominate, the liquid will rise—as with water. When cohesive forces dominate, it will fall—as with mercury. The capillary rise, h, can be calculated from the following equation:

$$h = \frac{4\sigma \cos \beta}{\gamma d}$$

where β = angle made by the liquid with the tube wall, and d = diameter of capillary tube, as shown in Figure 11.2.

The angle, β, varies with different liquid/tube material combinations and is found in tables. β is within the range 0 to 180°. When $\beta > 90°$, h will be negative.

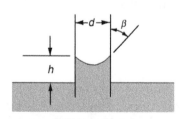

Figure 11.2

FLUID STATICS

Hydrostatic Pressure

For a static liquid, the pressure increases with depth (decreases with height) according to the following relationship

$$p_2 - p_1 = -\gamma(Z_2 - Z_1) = \gamma h$$

where h = depth from Point 1 to Point 2.

If p_1 is at the surface of a liquid that is open to the atmosphere, then the gage pressure at Point 2 is found from

$$p_2 = p = \gamma h$$

Example 11.2

Calculate the gage pressure at a depth of 100 meters in seawater, for which $\gamma = 10.1$ kN/m³.

Solution

$$p = \gamma h = \left(\frac{10.1 \text{ kN}}{\text{m}^3}\right)(100 \text{ m}) = 1010 \frac{\text{kN}}{\text{m}^2} = 1010 \text{ kPa}$$

Manometers

A manometer is a device used to measure moderate pressure differences using the pressure-height relationship. The simplest manometer is the U-tube shown in Figure 11.3. The pressure difference between System 1 and System 2 is found from

$$p_1 - p_2 = \gamma_m h_m + \gamma_2 h_2 - \gamma_1 h_1$$

where γ_m, γ_1, and γ_2 = specific weight of manometer fluid, fluid in System 1, and fluid in System 2, respectively, and h_m, h_1, h_2 = depths as shown. If the fluids in both systems are gases and the manometer fluid is any liquid, then $\gamma_m \gg \gamma_1$ or γ_2 and the equation simplifies to

$$p_1 - p_2 = \gamma_m h_m = \gamma h$$

If System 2 were the atmosphere ($p_2 = 0_{\text{gage}}$) then

$$p_1 = \gamma h = \text{gage pressure in System 1}$$

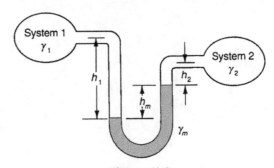

Figure 11.3

If System 1 were the atmosphere ($p_1 = 0$) then

$$p_2 = -\gamma h = \text{gage pressure in System 2}$$

The gage pressure in System 2 would be negative or a vacuum. Manometers are commonly used to measure system pressures between −101.3 kPa and +101.3 kPa. In many cases, particularly where the gage pressure is negative, the pressure may be given in millimeters of a fluid, and the equation above used to convert it to standard units.

Example 11.3

A system gage pressure is given as 500 millimeters of mercury vacuum (mm Hg vac). What is the gage pressure in kPa? The specific gravity of mercury is 13.6.

Solution

The pressure is $p = p_2 = -\gamma h$ since vacuum is a negative gage pressure:

$$\gamma_m = (13.6)\left(9.81 \frac{\text{kN}}{\text{m}^3}\right) = 133.4 \frac{\text{kN}}{\text{m}^3}$$

$$p = -\gamma h = -133.4 \frac{\text{kN}}{\text{m}^3} \times 0.5\,\text{m} = -66.7\,\text{kPa}$$

The conversion factor from millimeters of mercury to N/m² (pascals) is 133.4.

A barometer is a special type of mercury manometer. In this case one leg of the U-tube is very wide. If we can adjust the scale on the narrow leg so that zero is at the level of the large leg, then the narrow leg will read the atmospheric pressure impinging on the wide leg corrected by the vapor pressure of the mercury. A barometer is shown in Figure 11.4.

There are several other types of manometers. A compound manometer consists of more than one U-tube in series between one system and another. The equation for $p_1 - p_2$ may be developed by starting at System 2 and adding γh's going downward and subtracting γh's going upward until System 1 is reached as follows:

$$p_2 + \sum \gamma h \text{ (downward)} - \sum \gamma h \text{ (upward)} = p_1$$

An inclined manometer is used to measure small pressure differentials. The measurement along the manometer must be multiplied by the sine of the angle of incline. An inclined manometer is generally "single leg," similar to the barometer previously described, and is shown in Figure 11.5. The pressure difference is found from $p_1 - p_2 = \gamma_m L \sin \alpha$, where L = length along manometer leg and α = angle of inclination.

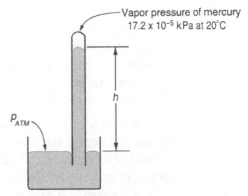

Figure 11.4

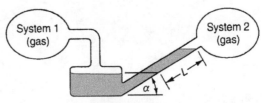

Figure 11.5

Forces on Flat Submerged Surfaces

A flat surface of arbitrary shape below a liquid surface is shown in Figure 11.6. Thew resultant force, F, on one side of the flat surface acts perpendicular to the surface. Its magnitude and location may be calculated from the following equations:

$$F = (p_0 + \gamma h_c)A$$

and

$$h_p = h_c + \frac{I_c \sin^2 \alpha}{\left(\dfrac{p_0}{\gamma} + h_c\right)A}$$

Figure 11.6

where
    F = resultant force on the flat surface, N
    p_0 = gage pressure on the surface, Pa
    γ = specific weight of the fluid, N/m³
    h_c = vertical distance from fluid surface to the centroid of the flat surface area, m

A = area of flat surface, m²
h_p = vertical distance from fluid surface to the center of pressure of the flat surface (where the equivalent, concentrated force acts), m
I_c = moment of inertia of the flat surface about a horizontal axis through its centroid, m⁴
α = angle that the inclined flat surface makes with the horizontal surface

The values of h_c, the location of the centroid axis from the base $\bar{y}$, and the moment of inertia about the centroid I_c for common geometric shapes, such as rectangles, triangles, and circles, may be determined from existing tables. Typical values are presented in the *FE Supplied Reference Handbook* in the Statics section.

For the common case when p_0 is atmospheric pressure ($p_0 = 0$), the equations simplify to

$$F = \gamma h_c A$$

and

$$h_p = h_c + \frac{I_c \sin^2 \alpha}{h_c A}$$

From the above equations it is apparent that the center of pressure is always below the centroid except when the surface is horizontal ($\alpha = 0$). In that case the center of pressure is at the centroid. The deeper the flat surface is located below the fluid surface, the closer the center of pressure is to the centroid.

The pressure profile on the flat surface is generally trapezoidal (triangular, if the flat surface pierces the surface of a fluid exposed to atmospheric pressure). The slope of the pressure profile is equivalent to the specific weight of the fluid. Examples are shown in the *FE Supplied Reference Handbook* in the Statics section.

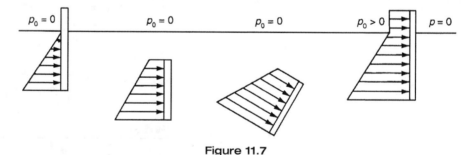

Figure 11.7

Example 11.4

A vertical side of a saltwater tank contains a round viewing window 60 cm in diameter with its center 5 meters below the liquid surface. If the specific weight of the saltwater is 10 kN/m3, find the force of the water on the window and where it acts.

Solution

Assume atmospheric pressure on the liquid surface, $p_0 = 0$.

$$d = 60 \text{ cm} = 0.6 \text{ m}, \quad A = \frac{\pi(0.6)^2}{4} = 0.283 \text{ m}^2$$

$$F = \gamma h_c A = 10 \frac{\text{kN}}{\text{m}^3} \times 5 \text{ m} \times 0.283 \text{ m}^2 = 14.14 \text{ kN}$$

$$I_c = \frac{\pi d^4}{64} = \frac{\pi (0.6 \text{ m})^4}{64} = 0.00636 \text{ m}^4$$

$$\alpha = 90°, \quad \sin \alpha = 1$$

$$h_p = h_c + \frac{I_c \sin^2 \alpha}{h_c A} = 5 \text{ m} + \frac{0.00636 \text{ m}^4 (1)^2}{5 \text{ m} \bullet 0.283 \text{ m}^2} = 5.0045 \text{ m}$$

Example 11.5

In many cases problems involving fluid forces on flat surfaces are combined with a statics problem. The fluid force is just another force to be added into the statics equation.

In the previous example, suppose the circular window were hinged at the top with some sort of clamp at the bottom (Exhibit 1). What force, P, would be required of the clamp to keep the window closed?

Solution

From Example 11.4 calculations, the force of the water is 14.14 kN located 5.0045 m below the fluid surface. The hinge is located 5 m − 0.6 m/2 = 4.7 m below the water surface. Thus the force location is 5.0045 m − 4.7 m = 0.3045 m below the hinge.

Summing moments about hinge,

$$\sum M_H = 0.6 \text{ m} \bullet P - 0.3045 \text{ m} \bullet 14.14 \text{ kN} = 0$$

$$P = \frac{0.3045 \text{ m} \bullet 14.14 \text{ kN}}{0.6 \text{ m}} = 7.18 \text{ kN}$$

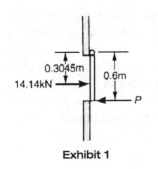

Exhibit 1

Buoyancy

In addition to the force of gravity, or weight, all objects submerged in a fluid are acted on by a buoyant force, F_B. The buoyant force acts upward and is equal to the weight of the fluid displaced by the object. This is known as Archimedes' principle. The upward buoyant force also acts through the center of gravity (or centroid) of the displaced volume, known as the center of buoyancy, B. Thus

$$F_B = \gamma_f V_D$$

where F_B = buoyant force, N; γ_f = specific weight of the fluid, N/m³; and V_D = volume displaced by the object, m³.

For a freely floating object (no external forces) the weight of the object (acting downward) is equal to the buoyant force on the object (acting upward) or

$$W = F_B = \gamma_f V_D$$

This equation is useful in determining what part of an object will float below the surface of a liquid. For objects partially submerged in a liquid and a gas, the

Example 11.6

A wooden cube that is 15 centimeters on each side with a specific weight of 6300 N/m³ is floating in fresh water ($\gamma = 9{,}810$ N/m³) (Exhibit 2). What is the depth of the cube below the surface?

Solution

$$W = F_B = \gamma_f V_D$$

There are actually two buoyant forces on the cube, that of the water on the volume below the surface and that of the air on the volume above the surface. Neglecting the buoyant force of the air and rearranging the equation:

$$V_D = \frac{W}{\gamma_f} = \frac{\gamma_C V_C}{\gamma_f} = \frac{(6300 \text{ N/m}^3)(0.15 \text{ m})^3}{9810 \text{ N/m}^3} = 0.00217 \text{ m}^3$$

$$V_D = (0.15)^2 \cdot d = 0.00217 \text{ m}^3$$

$$d = \frac{0.00217 \text{ m}^3}{.0225 \text{ m}^2} = .0964 \text{ m} = 9.64 \text{ cm}$$

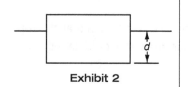

Exhibit 2

Neutral buoyancy exists when the buoyant force equals the weight when an object is completely submerged. The object will remain at whatever location it is placed below the fluid surface.

In the case of an object floating at the interface of two liquids, the total buoyant force is equal to the sum of the buoyant forces on the object created by each fluid on that part that is immersed. When external forces also act on a submerged or partially submerged object, they must be included in the force balance on the object. The force balance equation then becomes

$$W + \sum F_{\text{ext}}(\text{down}) = F_B + \sum F_{\text{ext}}(\text{up})$$

If weight is added to an object internally, or possibly on top of a partially submerged object, it will only affect the weight of the object. But if the weight is added externally, beneath the surface of the fluid, its buoyant force as well as its weight must be considered.

Example 11.7

If, in Example 11.6, a concrete weight (anchor) is added to the bottom of the cube externally, what anchor volume, V_A, would be required to make the cube float neutrally (below the surface). The specific weight of the concrete, γ_c, is 24 kN/m³.

Solution

Let the subscript C denote the properties of the cube and subscript A denote those of the anchor. Summing forces vertically,

$$W_C + W_A = F_{BC} + F_{BA}$$

$$\gamma_C V_C + \gamma_A V_A = \gamma_f V_D + \gamma_f V_A$$

Solving for V_A,

$$V_A = \frac{\gamma_f V_D - \gamma_C V_C}{\gamma_A - \gamma_f}$$

But for neutral buoyancy, the displaced volume, V_D, is equal to the total volume of the cube, V_C, and

$$V_A = \frac{(\gamma_f - \gamma_C) \cdot V_C}{\gamma_A - \gamma_f} = \frac{(9810 - 6300)\frac{N}{m^3} \cdot (0.15\ m)^3}{(24{,}000 - 9810)\frac{N}{m^3}} = 8.34 \times 10^{-4}\ m^3$$

CONSERVATION LAWS

All steady state, incompressible, one-dimensional (1-D) flow systems must satisfy the conservation laws for mass and energy. Before looking at the laws in more detail, it will be useful to review some basic descriptions of fluid flow.

The vast majority of all fluid flow problems can be described as *steady*, where fluid or system properties do not change with time, and *uniform*, where fluid or system properties do not change with location. A further simplification that applies in most systems is the assumption of *one-dimensional* flow, where the velocity vector is zero in all dimensions of the coordinate system except one. Most often, the coordinate system is either rectilinear (that is, x, y, z), and 1-D flow is generally expressed in the x-direction, or cylindrical (that is, r, z, θ), where 1-D flow is usually expressed in the z-dimension and termed *axial* flow.

Mass

The most basic of the conservation laws is applied to the *conservation of mass* and is generally referred to as the *Equation of Continuity* (*EOC*). Because mass cannot be created or destroyed in natural environments, the EOC simply states that the mass flow rate of any material into and out of a system at steady state must be equal. In its most general form, the EOC can be expressed as

$$\dot{m}_{in} = \dot{m}_{out}$$

where $\dot{m}$ is the system mass flow rate (kg/s). Often, this is written as

$$\rho_1 Q_1 = \rho_2 Q_2$$

or

$$\rho_1 V_1 A_1 = \rho_2 V_2 A_2$$

where

ρ = the fluid density (kg/m³)

Q = the volumetric flow rate (m³/s)

V = the average fluid velocity through the conduit (m/s)

A = the conduit cross-sectional area (perpendicular to the flow path) (m²)

The subscripts 1 and 2 are used to evaluate mass flow at any two locations in the system, usually selected where flow crosses the system boundary. Multiple inlets or outlets may be summed to complete a balance on the entire system, or several subsystems can be defined and evaluated with only one inlet and one outlet. For an incompressible fluid, as in the case of water flow problems, density is assumed constant, and the EOC simplifies to

$$Q_1 = Q_2$$

or

$$V_1 A_1 = V_2 A_2$$

Energy

In addition to the EOC, all steady state, incompressible, 1-D flow systems must also satisfy the *energy equation* (sometimes called the *field equation*), which is an expression that ensures conservation of mechanical energy between two specified points within a system. This equation is most often expressed in units of length (also called *head*) and can be represented as

$$z_1 + \frac{V_1^2}{2g} + \frac{P_1}{\gamma} + h_P = z_2 + \frac{V_2^2}{2g} + \frac{P_2}{\gamma} + h_L + h_T$$

where

z = the elevation (m)

V = the average velocity (m/s)

g = the acceleration due to gravity (equal to 9.81 m/s²)

P = the pressure (Pa, or equivalently N/m²)

γ = the specific weight of the fluid (N/m³)

h_P = the pump head (m)

h_L = the loss head (m)

h_T = the turbine head (m)

Again, subscripts 1 and 2 are used to evaluate energy at any two locations in the system, often selected where flow crosses the system boundary, in which case subscript 1 usually specifies energy entering the system and subscript 2 usually specifies energy exiting the system. The pump and turbine head terms are often used to determine power required (for pumps) or delivered (for turbines) from a system, when the remaining terms are known or can be estimated.

Many flow systems do not contain a pump or turbine, and often these terms are dropped from the equation. If the flow is also assumed to be *inviscid* (that is, the fluid viscosity is negligible, or, equivalently, there are no energy losses due to friction), the loss head is also zero, and the energy equation reduces to the familiar *Bernoulli equation*, which describes ideal flow as

$$z_1 + \frac{V_1^2}{2g} + \frac{P_1}{\gamma} = z_2 + \frac{V_2^2}{2g} + \frac{P_2}{\gamma}$$

where

z is called the *elevation* (or *potential*) *head*

$V^2/(2g)$ is called the *velocity head*

P/γ is called the *pressure* (or *static*) *head*

The sum of the elevation and pressure heads is called the *hydraulic grade line* (*HGL*) and can be measured with a *piezometer* tube. The sum of the HGL and the velocity head, the total energy of the system, is called the *energy grade line* (*EGL*) and can be measured with a *pitot* tube, often called a *stagnation* tube.

Friction and Minor Losses

Energy loss in a system is generally due to viscous shear at the wall (called *friction loss* and given the symbol h_f), or due to fittings or other variations in the pipe network (called *minor losses* and given the symbol h_m). These can be shown to be a function of fluid properties (specifically, viscosity and density), flow properties (velocity), and system geometry (pipe diameter and length, and pipe material or surface roughness). Often the two losses are combined and expressed as the loss head as

$$h_L = h_f + h_m$$

The most common equation used to calculate friction loss is the *Darcy-Weisbach equation*, which can be expressed as

$$h_f = f \frac{L}{D} \frac{V^2}{2g}$$

where

f = the *friction factor* (unitless)

L = the pipe length (m)

D = the pipe diameter (m)

V and g are as defined previously

The friction factor is dependent upon a dimensionless quantity called the *Reynolds number* (*Re*), a ratio of inertial forces to viscous forces in fluid flow, and the ratio of the pipe surface *roughness* (ε) to pipe diameter (D), called the *relative roughness*. The Reynolds number and the relative roughness can be expressed as

$$\text{Re} = \frac{\rho D V}{\mu} = \frac{DV}{\nu}$$

$$\text{relative roughness} = \frac{\varepsilon}{D}$$

where μ is the fluid dynamic (or absolute) viscosity (N · s/m²), ν is the kinematic viscosity (m²/s), and all other variables are as defined previously. While equations exist to determine f for each of the flow regimes (laminar, transitional, and turbulent), it is far more common to use the Moody diagram to estimate the value of the friction factor, as seen in the *FE Supplied Reference Handbook*. Values for pipe roughness are usually supplied by manufacturers, specified in problem statements, or may be found in tables accompanying the Moody diagram for a variety of materials, as seen in the *FE Supplied Reference Handbook*.

The estimation of minor losses is determined experimentally and is generally expressed as

$$h_m = K_L \frac{V^2}{2g}$$

where K_L is called the loss coefficient (unitless) and is generally tabulated for a variety of pipe fittings, as presented in Table 11.1.

Table 11.1 Loss coefficients (K_L) for select pipe fittings

Component	K_L
Entrance, sharp	0.5
Entrance, rounded	0.2
Exit	1.0
Sudden contraction	0.0–0.5
Sudden expansion	0.0–1.0
90° elbow, flanged	0.3
90° elbow, threaded	1.5
Long radius 90°, threaded	0.7
45° bend	0.2–0.4
180° return bend, flanged	0.2
180° return bend, threaded	1.5
Threaded coupling	0.1
Through tee, flanged	0.2
Through tee, threaded	0.9
Globe valve, open	10
Gate valve, open	0.15
Gate valve, ¾ open	0.25
Gate valve, ½ open	2.1
Gate valve, ¼ open	17
Ball valve, open	0.05
Ball valve, ⅔ open	5.5
Ball valve, ⅓ open	210

Finally, total losses for a system are determined by summing all of the friction and minor losses for the flow line under evaluation. For the case where pipe diameter is constant throughout the system, the system head loss can be expressed as

$$h_L = \left(f \frac{\Sigma L}{D} + \Sigma K_L \right) \frac{V^2}{2g}$$

Example 11.8

Energy equation

Determine the maximum flow rate of 60°F water (in gallons per minute) through 250 feet of a ¾-inch diameter garden hose if the elevation at location 2 is 16 feet higher than at location 1, and the pressure at the inlet is 85 psi. You may assume a friction factor of 0.02.

Solution

We start with the energy equation as follows:

$$h_P + Z_1 + \frac{V_1^2}{2g} + \frac{P_1}{\gamma} = Z_2 + \frac{V_2^2}{2g} + \frac{P_2}{\gamma} + h_f + h_T$$

This may be simplified by recognizing that there are no pumps or turbines, the flow exits to the atmosphere, the velocity through the system is constant, and the elevation change is given. This allows us to write

$$h_P = h_T = P_2 = (V_2 - V_1) = 0 \quad \text{and} \quad Z_2 - Z_1 = 16 \text{ ft} \quad \Rightarrow \quad \frac{P_1}{\gamma} = 16 - h_f$$

Using the data given in the problem statement and correcting for units yields

$$\frac{\left(85 \frac{\text{lb}}{\text{in}^2}\right)\left(144 \frac{\text{in}^2}{\text{ft}^2}\right)}{\left(62.4 \frac{\text{lb}}{\text{ft}^3}\right)} = 16 + h_f \quad \Rightarrow \quad h_f = 180.2 \text{ ft}$$

Now that the head loss is known, we may calculate velocity using the Darcy-Weisbach equation as follows:

$$h_f = f \frac{L}{D} \frac{V^2}{2g} = \frac{(0.02)(250 \text{ ft}) V^2}{\left(\frac{0.75}{12} \text{ ft}\right)(2)\left(32.2 \frac{\text{ft}}{\text{s}^2}\right)} = 180.2 \text{ ft} \quad \Rightarrow \quad V = 12.0 \frac{\text{ft}}{\text{s}}$$

Finally, volumetric flow rate can be calculated from velocity and hose diameter as follows:

$$Q = VA = \left(12.0 \frac{\text{ft}}{\text{s}}\right)\left(\frac{\pi}{4}\right)\left(\frac{0.75}{12} \text{ ft}\right)^2 = 0.037 \frac{\text{ft}^3}{\text{s}} \times \frac{7.48 \text{ gal}}{\text{ft}^3} \times \frac{60 \text{ s}}{\text{min}} = 16.5 \text{ gpm}$$

Let's check the assumption of $f = 0.02$. First we will need to calculate Re for water at 60°F as

$$\text{Re} = \frac{DV}{\nu} = \frac{(0.75 \text{ in})\left(\frac{1 \text{ ft}}{12 \text{ in}}\right)\left(12 \frac{\text{ft}}{\text{s}}\right)}{1.21 \times 10^{-5} \frac{\text{ft}^2}{\text{s}}} = 62{,}000$$

Assuming the garden hose is considered smooth (plastic), the Moody diagram gives a friction factor at Re = 62,000 for smooth pipe of 0.02. So, the assumption in the problem statement was valid. If the problem did not specify a friction factor, an initial guess for f could have been made, and that guess validated, as was demonstrated here. If the assumption was not correct, the newly calculated value of f would be used to find a new velocity, and then validated as before. The process would be completed iteratively until the verified value closely matched the assumed value.

Hazen-Williams Equation

Another method to determine head loss in a system, commonly used in water supply design, is the *Hazen-Williams equation*, which can be expressed as

$$V = 0.849 \, C R_H^{0.63} S^{0.54} \quad \text{(SI units)}$$

$$V = 1.318 \, C R_H^{0.63} S^{0.54} \quad \text{(English units)}$$

where

V = the fluid velocity (m/s)

C = the Hazen-Williams roughness coefficient

R_H = the hydraulic radius (m)

S = the slope of the EGL (m/m)

The Hazen-Williams coefficient is a function of pipe material (like the value for ε discussed previously) and is often tabulated as seen in Table 11.2.

Table 11.2 Hazen-Williams coefficient values for a variety of pipe materials

Material	C_{new}	C_{old}
Brick	na	100
Cast iron	130–140	80–120
Concrete	130	120
Plastic	150	130
Riveted steel	110–130	100–110
Vitrified clay	110	110
Welded steel	120–140	110–120
Wood stave	120	110

The *hydraulic radius* (R_H) is defined as area divided by wetted perimeter, and expressed as

$$R_H = \frac{A}{P_w}$$

where A is the cross-sectional area of flow (m²) and P_w is the wetted perimeter (m), which is the portion of the conduit perimeter that is in contact with the fluid. Note that a pipe with a circular cross section that is flowing full has $R_H = D/4 = r/2$.

The slope of the EGL can be expressed as

$$S = \frac{h_L}{L}$$

where h_L is the energy loss term described previously (m), and L is the total length of conduit in the system being evaluated (m). Due to the widespread use in water distribution networks where water flows full through circular pipes, it is often desired to determine the volumetric flow rate for circular cross sections, in which case the Hazen-Williams equation can be expressed as

$$Q = 0.278 C D^{2.63} S^{0.54} \quad \text{(SI units)}$$

$$Q = 0.432 C D^{2.63} S^{0.54} \quad \text{(English units)}$$

where Q is the volumetric flow rate (m³/s), and D is the pipe diameter (m). It should be noted that the use of the Hazen-Williams equation is limited to water near room temperature under turbulent flow. Use of the Hazen-Williams equation is greatly enhanced through the use of the Hazen-Williams nomograph, as presented in Figure 11.8.

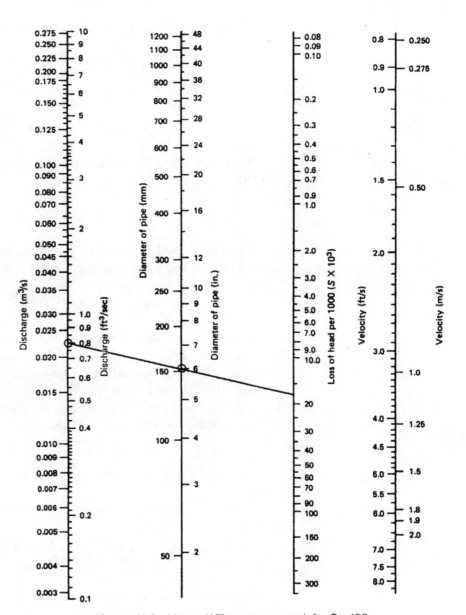

Figure 11.8 Hazen-Williams nomograph for C = 100

Source: Reprinted by permission of Waveland Press, Inc., from Ram Gupta, *Hydrology and Hydraulic Systems*, 2nd Edition. Long Grove, IL: Waveland Press, Inc., 2001. All rights reserved.

Example 11.9

Hazen-Williams equation

Water at 1 cfs is conveyed through 8-inch cast iron ($C = 120$) pipe over a distance of two miles. Determine the pressure drop (in units of psi) over this distance.

Solution

Because a value for the Hazen-Williams coefficient (C) is given, start with the Hazen-Williams equation for volumetric flow:

$$Q = 0.432 \, C D^{2.63} S^{0.54}$$

With Q, C, and D given, rearrange to solve for the slope of the EGL as

$$S^{0.54} = \frac{Q}{0.432\, CD^{2.63}} = \frac{1}{(0.432)(120)\left(\frac{8}{12}\right)^{2.63}} = 0.056$$

$$S = (0.056)^{1/0.54} = (0.056)^{1.852} = 0.0048\, \frac{\text{ft}}{\text{ft}}$$

Since slope is feet of head loss per foot of pipe length, head loss can be calculated as

$$S = \frac{h_L}{L} \Rightarrow h_L = SL = 0.0048\, \frac{\text{ft}}{\text{ft}} \times 2\, \text{mi} \times \frac{5280\, \text{ft}}{\text{mi}} = 50.7\, \text{ft}$$

The units of feet here refer to feet of water column, or ft H_2O. Since the problem statement asked for units of psi, convert ft H_2O to psi as follows:

$$H = 50.7\, \text{ft}\, H_2O \times \frac{14.7\, \text{psi}}{33.9\, \text{ft}\, H_2O} = 22.0\, \text{psi}$$

The alternative method is to use the Hazen-Williams nomograph. Start by drawing a straight line through the given flow and diameter on the nomograph and extend this line to the loss column. This yields a value for S of 6.7 ft per 1000 ft. However, this value is for $C = 100$. To convert this value to an equivalent value for this problem where $C = 120$, we need to evaluate the ratios of the slopes to Hazen-Williams coefficient with constant Q and D as

$$\frac{Q_1}{Q_2} = \frac{0.432\, C_1\, D_1^{2.63}\, S_1^{0.54}}{0.432\, C_2\, D_2^{2.63}\, S_2^{0.54}} \quad \text{or} \quad \left(\frac{S_2}{S_1}\right)^{0.54} = \frac{C_1}{C_2} \Rightarrow \frac{S_2}{S_1} = \left(\frac{C_1}{C_2}\right)^{1.85}$$

This may be rearranged to solve for S_2 using the information calculated previously as

$$S_2 = S_1 \left(\frac{C_1}{C_2}\right)^{1.85} = 6.7 \left(\frac{100}{120}\right)^{1.85} = 4.8\, \frac{\text{ft}}{1000\, \text{ft}}$$

This is exactly the same answer as calculated above, giving a pressure drop of 22 psi.

Series and Parallel Flow

For the case where several pipes of different diameters or material properties are present in the system, connected in a *series configuration*, the equation of continuity dictates that the volumetric flow rates in each segment are equal, which can be expressed as

$$Q_1 = Q_2 = \cdots = Q_n$$

where n is the number of different pipes in the series, and the energy equation requires the total head loss for the system to be calculated as

$$(h_L)_{\text{TOTAL}} = (h_L)_1 + (h_L)_2 + \cdots + (h_L)_n$$

When several pipes are in a system connected in a *parallel configuration*, which can be defined as two or more pipes that diverge at some point in the system and converge at another point, the equation of continuity dictates that the volumetric flow rates can be expressed as

$$Q_{\text{TOTAL}} = Q_1 + Q_2 + \cdots + Q_n$$

and the energy equation requires the head loss for each parallel branch in the system to be equal, which can be expressed as

$$(h_L)_1 = (h_L)_2 = \cdots = (h_L)_n$$

Example 11.10

A simple flow network

Water flows through the pipes in the simple network shown in Exhibit 3 below with $Q_{AB} = 1$ cfs. Determine Q_{BCD} given the following pipe data: $D_{BC} = D_{ED} = 10$ in; $D_{CD} = 6$ in; $D_{BE} = 8$ in; $L_{BC} = 1000$ ft; $L_{CD} = 400$ ft; $L_{BE} = 800$ ft; $L_{ED} = 600$ ft; $f = 0.018$.

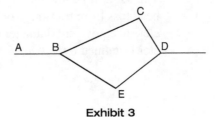

Exhibit 3

Solution

Based on the rules of parallel pipe networks, head loss in each branch must be equal, or, equivalently,

$$(h_L)_{BCD} = (h_L)_{BED}$$

Further, each parallel branch is comprised of two pipes in series; therefore, we can write

$$(h_L)_{BC} + (h_L)_{CD} = (h_L)_{BE} + (h_L)_{ED}$$

Substituting the expression for h_L above and adding the appropriate subscripts, we may write

$$h_L = \frac{f L V^2}{2 g D} = \frac{f L Q^2}{2 g D A^2} = \frac{8 f L Q^2}{g \pi^2 D^5}$$

$$\frac{8 f L_{BC} Q_{BC}^2}{g \pi^2 D_{BC}^5} + \frac{8 f L_{CD} Q_{CD}^2}{g \pi^2 D_{CD}^5} = \frac{8 f L_{BE} Q_{BE}^2}{g \pi^2 D_{BE}^5} + \frac{8 f L_{ED} Q_{ED}^2}{g \pi^2 D_{ED}^5}$$

This may be simplified to

$$\frac{L_{BC} Q_{BC}^2}{D_{BC}^5} + \frac{L_{CD} Q_{CD}^2}{D_{CD}^5} = \frac{L_{BE} Q_{BE}^2}{D_{BE}^5} + \frac{L_{ED} Q_{ED}^2}{D_{ED}^5}$$

We also know that flow through pipes in series must be equal; therefore,

let $Q_{BC} = Q_{CD} = Q_1$ and $Q_{BE} = Q_{ED} = Q_2$

Substituting the values given we may write

$$\frac{(1000)Q_1^2}{\left(\frac{10}{12}\right)^5} + \frac{(400)Q_1^2}{\left(\frac{6}{12}\right)^5} = \frac{(800)Q_2^2}{\left(\frac{8}{12}\right)^5} + \frac{(600)Q_2^2}{\left(\frac{10}{12}\right)^5}$$

Simplifying, we can express Q_2 as a function of Q_1:

$$15{,}288\, Q_1^2 = 7{,}568\, Q_2^2 \quad \Rightarrow \quad Q_2 = 1.4213\, Q_1$$

Since total system flow is 1 cfs, we may write

$$Q_1 + Q_2 = 1\text{ cfs} \quad \Rightarrow \quad Q_1 + 1.4213\, Q_1 = 2.4213\, Q_2 = 1\text{ cfs} \quad \Rightarrow \quad Q_1 = 0.413\text{ cfs}$$

Therefore the flow through the upper branch is approximately 0.4 cfs, or 40% of the total flow.

Flow in Noncircular Conduits

The same fundamental equations for Reynolds number, relative roughness, and head loss from friction may be used for noncircular conduits. In place of the diameter, an equivalent diameter (or characteristic length) is used. The equivalent diameter is defined by

$$d_e = 4 R_H = 4 \frac{A}{WP}$$

where d_e = equivalent diameter, R_H = hydraulic radius = $\dfrac{A}{WP}$, A = cross-sectional area, and WP = wetted perimeter.

Example 11.11

Calculate the equivalent diameter of a rectangular conduit 0.6 meters wide and 0.3 meters high.

Solution

$$A = 0.6\text{ m} \bullet 0.3\text{ m} = 0.18\text{ m}^2$$
$$WP = 2(0.6\text{ m} + 0.3\text{ m}) = 1.8\text{ m}$$
$$d_e = 4 \frac{A}{WP} = 4 \bullet \frac{0.18\text{ m}^2}{1.8\text{ m}} = 0.4\text{ m}$$

VELOCITY AND FLOW MEASURING DEVICES

Pitot Tubes

For liquids flowing at relatively low pressures the mean static pressure may be measured using a piezometric tube indicated in Figure 11.9 by h_1. The stagnation pressure (i.e., the pressure at which the velocity is zero) is indicated by h_2.

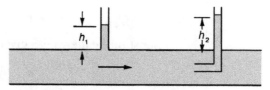

Figure 11.9

The relationship of each measurement to pressure, specific weight, and velocity is shown in the following two equations:

$$h_1 = \frac{p}{\gamma}, \quad h_2 = \frac{p_s}{\gamma} = \frac{p}{\gamma} + \frac{V^2}{2g}$$

Combining these relations, the velocity in the duct is

$$V = \sqrt{2g\left(\frac{p_s - p}{\gamma}\right)} = \sqrt{2g(h_2 - h_1)}$$

where V = velocity, γ = specific weight, p = static pressure, and p_s = stagnation pressure.

The combination of the two tubes as a single device is known as a pitot tube. If a manometer is connected between the static and stagnation pressure taps, velocities at moderate pressures may be calculated using the following equation:

$$V = \sqrt{2gh_m\left(\frac{\gamma_m}{\gamma} - 1\right)}$$

where h_m = height indicated by the manometer, and γ_m = specific weight of manometer fluid.

Example 11.12

A pitot tube is used to measure the mean velocity in a pipe where water is flowing. A manometer containing mercury is connected to the pitot tube and indicates a height of 150 mm. The specific weights of the water and mercury are 9810 N/m³ and 133,400 N/m³, respectively. Calculate the velocity of the water.

Solution

$$V = \sqrt{2gh_m\left(\frac{\gamma_m}{\gamma} - 1\right)}$$

$$V = \sqrt{2 \bullet 9.81\frac{\text{m}}{\text{s}^2} \bullet 0.15\text{m} \bullet \left(\frac{133{,}400\frac{\text{N}}{\text{m}^3}}{9810\frac{\text{N}}{\text{m}^3}} - 1\right)} = 6.09\frac{\text{m}}{\text{s}}$$

The pitot tube equation may also be used for compressible fluids with Mach numbers less than or equal to 0.3.

Flow Meters

There are three commonly used meters that measure flow rate in fluid systems: venturi meters, flow nozzles, and orifice meters. All three operate on the same basic principle, their equations being developed by combining the Bernoulli and continuity equations. The three meters are shown in Figure 11.10.

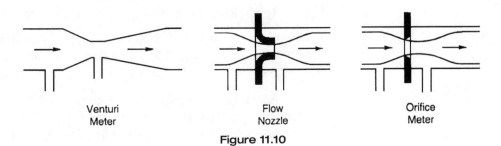

Venturi Meter   Flow Nozzle   Orifice Meter

Figure 11.10

The equation for flow rate is given by

$$Q = \frac{c_v c_c}{\sqrt{1 - c_c^2 \left(\frac{A_2}{A_1}\right)^2}} \cdot A_2 \cdot \sqrt{2g\left(\frac{p_1 - p_2}{\gamma} + Z_1 - Z_2\right)}$$

or if a manometer is used between the pressure taps:

$$Q = \frac{c_v c_c}{\sqrt{1 - c_c^2 \left(\frac{A_2}{A_1}\right)^2}} \cdot A_2 \cdot \sqrt{2gh_m\left(\frac{\gamma_m}{\gamma} - 1\right)}$$

where
Q = flow rate
$p_1 - p_2$ = pressure difference between a point before the entrance to the meter, and the point of narrowest flow cross section in the meter
$Z_1 - Z_2$ = height difference between a point before the entrance to the meter and the point of narrowest flow cross section in the meter
A_1 = area of entrance
A_2 = area of narrowest flow cross section, except in the orifice, where it is the orifice area
h_m = height indicated by manometer
γ_m = specific weight of manometer fluid
γ = specific weight of fluid
c_v = coefficient of velocity
c_c = coefficient of contraction

For the orifice meter (and sometimes the flow nozzle) the coefficient terms in the equation are combined as follows:

$$c = \frac{c_v c_c}{\sqrt{1 - c_c^2 \left(\frac{A_2}{A_1}\right)^2}}$$

and the flow rate equation is then written as

$$Q = cA_2 \sqrt{2g\left(\frac{p_1 - p_2}{\gamma} + Z_1 - Z_2\right)} = cA_2 \sqrt{2gh_m\left(\frac{\gamma_m}{\gamma} - 1\right)}$$

where c = orifice (or flow nozzle) coefficient.
The following values of c_c, c_v, and c are used for the various flow meters:

Venturi: $c_c = 1$, $0.95 < c_v < 0.99$; c_v (nominal) = 0.984
Flow nozzle: $c_c = 1$, $0.95 < c_v < 0.99$; c_v (nominal) = 0.98
Orifice: $c_c = 0.62$, $c_v = 0.98$; c (nominal) = 0.61

Actual values of c for orifice meters (and flow nozzles) vary with the diameter ratio, $d_0:d_1$, and the Reynolds number and are found on existing graphs. Curves for the values of c_v for venturi meters and flow nozzles are also available.

Flow from a Tank

The flow from a tank through various types of exit configurations can be calculated by using the energy equation and experimentally determined configuration coefficients. Consider a tank as shown in Figure 11.11. For frictionless flow, the flowrate can be calculated using the energy and continuity equations, which reduce to

$$Q = AV = A\sqrt{2gh}$$

Considering friction, the flow rate may be calculated from

$$Q = cA\sqrt{2gh}, \quad c = c_v c_c$$

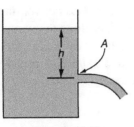

Figure 11.11

where Q = flow rate from the tank, h = height of water level above the exit, and c = coefficient of discharge for the exit.

If the friction in the exit is neglected, then $c = 1$.

For a sharp-edged orifice, $c_v = 0.98$, $c_c = 0.62$, and $c = 0.61$.

For a rounded exit, $c_v = 0.98$, $c_c = 1.00$, and $c = 0.98$.

For a short tube exiting from the tank, $c_v = 0.80$, $c_c = 1.00$, and $c = 0.80$.

For a re-entrant pipe, $c_v = 0.98$, $c_c = 0.52$, and $c = 0.51$.

For the special case when the flow from the tank discharges beneath the surface of the same fluid outside the tank, the flow rate from the tank is given by

$$Q = cA\sqrt{2gh(h_1 - h_2)}$$

where h_1 = height of fluid above exit in tank, and h_2 = height of fluid above exit outside tank.

PUMP SELECTION

Pumps are machines that add energy to liquid systems (as compared to compressors, which add energy to gases), and may be classified as static-type or dynamic-type. Static-type pumps are often called *positive displacement* (or piston-style) pumps and produce flow through the static forces involved with changing the volume of the pump chamber. This type of pump is relatively uncommon in environmental applications, so it will not be discussed in detail in this review.

Types of Pumps

Dynamic-type pumps generally use a constant volume chamber, and flow is generated through the energy added by a set of blades (vanes, impellers) that are attached to a rotating shaft, which is turned by a motor. The most common dynamic device is the *centrifugal pump*, comprised of an *impeller* attached to a rotating shaft and a fixed *housing* (*casing*) enclosing the impeller.

Centrifugal pumps are further classified based upon the predominant direction of fluid flow within the housing, usually radial-flow, axial-flow, or mixed-flow. In *radial-flow pumps*, fluid inlet into the pump occurs at the center (*eye*) of the impeller, and the curved blades accelerate the liquid radially toward the end of the blade, where the fluid kinetic energy is converted to pressure head as the fluid impacts against the housing. This energy conversion is capable of developing large pressure increases, but flow rates are limited by the eye diameter. For a *single-stage pump*, fluid is then directed through an increasing volume channel toward the discharge opening. For systems where there is a demand for large pressure head, a *multistage pump* may be used where the discharge from one impeller is directed to the eye of a second impeller, and additional head is developed. *Axial-flow pumps* deliver fluid energy without substantial change in the fluid flow path, which is along the pump's primary axis. In this way, high flow rates can be accommodated; however, the pressure rise is limited. *Mixed-flow pumps* employ both radial and axial flow regimes to deliver reasonable flow rates at moderate pressure increases.

Pump Characteristics

Pump performance is usually based upon head delivered, pump efficiency, and brake horsepower, which are determined as a function of volumetric flow rate. The power gained by the fluid can be expressed as

$$\dot{P} = Q\gamma h_p$$

where $\dot{P}$ is the power gained by the fluid (N · m/s), γ is the fluid specific weight (N/m³), and Q and h_p are as defined previously. When calculated in English units (lb · ft/s) and divided by 550, the power has units of horsepower and is often referred to as the *water horsepower*. The overall pump efficiency (η) is a ratio of the power gained by the fluid to the power delivered to the pump by the rotating shaft and can be expressed as

$$\eta = \frac{\text{power gained by fluid}}{\text{shaft power delivered to pump}} = \frac{\dot{P}}{\dot{W}_S} = \frac{Q\gamma h_p / 550}{bhp}$$

where *bhp* is the pump *brake horsepower*, often supplied by pump manufacturers.

Example 11.13

Pump efficiency

Determine the pump efficiency for the flow through the garden hose in Example 11.8 if the brake horsepower of the pump is 1.25 hp.

Solution

Pump efficiency can be defined as follows:

$$\eta = \frac{\text{power gained by fluid}}{\text{shaft power delivered to pump}} = \frac{\dot{P}}{\dot{W}_S}$$

The shaft power delivered to the pump is also known as brake horsepower (bhp). The power gained by the fluid can be determined from the following expression:

$$P = Q\gamma h_p$$

where h_p is the pump head, which is equal to the total dynamic head (TDH). In this case, TDH may be calculated as either the sum of the friction head plus the elevation head, or it may be calculated from the pressure loss from the water inlet to the outlet. Since we know the friction head (180.2 ft) and the elevation head (16 ft), we may solve for the fluid power:

$$P = Q\gamma H = \left(0.037\,\frac{\text{ft}^3}{\text{s}}\right)\left(62.4\,\frac{\text{lb}}{\text{ft}^3}\right)(196.2\,\text{ft}) = 453\,\frac{\text{ft}\cdot\text{lb}}{\text{s}}$$

This can be converted to units of horsepower as

$$P = 453 \frac{\text{ft} \cdot \text{lb}}{\text{s}} \times \frac{1 \frac{\text{ft} \cdot \text{lb}}{\text{s}}}{550 \text{ hp}} = 0.824 \text{ hp}$$

The pump efficiency can now be calculated as

$$\eta = \frac{0.824 \; hp}{1.25 \; hp} = 65.9\,\%$$

Performance characteristics for a given pump geometry and operating speed are usually presented graphically as *performance (characteristic) curves*, as presented in Figure 11.12.

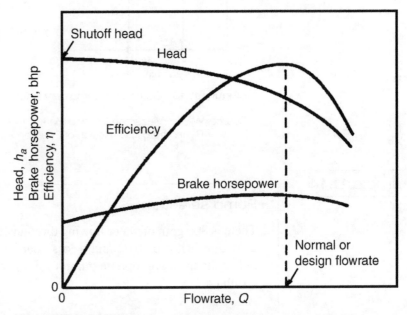

Figure 11.12 Typical centrifugal pump performance (characteristic) curve

Source: Munson, Young, and Okiishi, *Fundamentals of Fluid Mechanics*, 5th ed., © 2006 John Wiley & Sons, Inc. Reprinted by permission.

Variable speed drives may allow one pump to operate over a large range of flows without significant efficiency loss, offering an economical means to deliver varied demands. However, it is often necessary to use more than one pump in a system to provide the necessary head or flow rate, and multiple pumps offer redundancy for required or emergency maintenance. Analogous to the pipe flow analysis, it can be shown that multiple pumps in a *series* configuration must have equivalent flow rates, while the total head delivered is calculated as the sum of the individual heads delivered by each pump. Further, pumps in a *parallel* configuration must have equivalent head delivered, but the total volumetric flow rate is calculated as the sum of the individual flow rates delivered by each pump. The effect of pump configuration can be seen by examining the composite pump curves in Figure 11.13.

464 Chapter 11 Fluid Mechanics

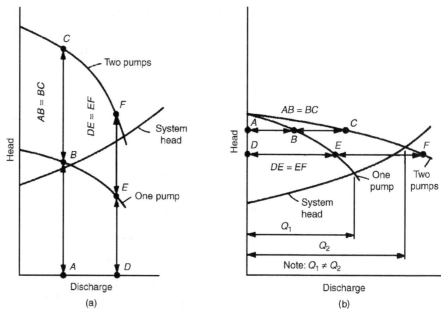

Figure 11.13 Composite pump curves for (a) series, and (b) parallel configurations

Source: Warren Viessman, Jr., and Mark J. Hammer, *Pollution Control*, 7th ed., © 2005. Reprinted by permission of Pearson Education, Inc., Upper Saddle River, N.J.

Example 11.14

Pump selection

There is 800 gpm of water flowing through 1000 feet of 6-inch-diameter pipe with a friction factor of 0.02 and a loss coefficient for all minor losses equal to 10. Select the most appropriate pump from the choices offered on the pump curve in Exhibit 4.

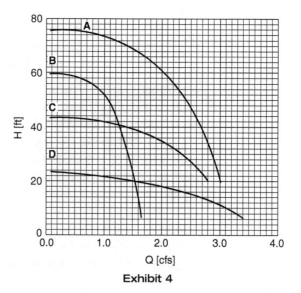

Exhibit 4

Solution

First, we need volumetric flow rate in cfs:

$$Q = \frac{800}{(7.48)(60)} = 1.78 \text{ cfs}$$

From the energy equation, if we assume there are no turbines, the elevation change is zero, and the velocity is constant in the pipe, then the head losses must be equal to the pressure drop, which is equal to the energy required by the pump to maintain flow. The head loss in the system, and therefore the head delivered by pump, may be calculated as

$$h_L = \frac{V^2}{2g}\left(f\frac{L}{D} + \sum K\right)$$

Problem statement gives f, L, D, and ΣK, but we need to calculate V:

$$V = \frac{Q}{A} = \frac{1.78}{\frac{\pi}{4}(0.5)^2} = 9.07\,\frac{\text{ft}}{\text{s}}$$

Now we can solve for head loss as

$$h_L = \frac{(9.07)^2}{(2)(32.2)}\left[(0.02)\left(\frac{1000}{0.5}\right) + 10\right] = 63.9\,\text{ft}$$

Now we can go to the pump curves and find a pump that can supply a flow rate of 1.8 cfs at a head of 64 ft. The correct choice is pump A.

NPSH

The inlet side of a pump is characterized by low pressure developed as liquid is drawn into the pump, which has the potential to cause *cavitation*. Cavitation is a phenomenon where the system pressure is less than or equal to the vapor pressure of the liquid, causing the liquid to form small bubbles of vapor, potentially damaging the structural components of the pump and substantially reducing pump efficiency. The minimum pressure head at the inlet to the pump is described by the required *net positive suction head* (*NPSH*) and can be expressed by the energy equation as

$$NPSH = \frac{P_i}{\gamma} + \frac{V_i^2}{2g} - \frac{P_v}{\gamma}$$

where P_i is the inlet pressure to the pump, V_i is the velocity at the inlet to the pump, and P_v is the vapor pressure of the fluid being pumped. Note that all pressures in this expression are required to be absolute instead of gage pressures. Applying the energy equation to the NPSH equation between the reservoir surface and the pump inlet yields

$$NPSH = \frac{P_{atm}}{\gamma} - \frac{P_v}{\gamma} - h_L - z_i$$

where P_{atm} is the prevailing atmospheric pressure, z_i is the elevation of the pump inlet, and all other variables are as defined previously.

Sometimes these two equations are denoted as *available NPSH*, or *NPSHA*. These equations are most often used to locate the pump inlet when all other parameters are known quantities. Values for NPSHA are to be compared to pump manufacturers' data that list the *required NPSH* (*NPSHR*) for a specific pump. For horizontal shaft pumps, NPSH is measured from the shaft for small pumps and from the top of the pump for large pumps. Care should be taken in locating any pump to ensure the NPSHR is exceeded at all times to ensure the absence of cavitation and therefore maximize the longevity of the pump.

Example 11.15

NPSH

Determine the maximum elevation of the pump inlet for a pump that has an NPSHR specified by the manufacturer of 13.5 ft H_2O. You may assume head losses on the suction side of the pump are 8 ft H_2O, and the vapor pressure for water at the operating temperature of 50°F is 0.18 psi. Atmospheric pressure is 100 kPa.

Solution

With the data given in the problem statement, we only need to input the values, corrected for consistent units, into the NPSH equation, which incorporates the energy balance as follows:

$$\text{NPSH} = \frac{P_{atm}}{\gamma} - \frac{P_v}{\gamma} - h_L - z_i$$

$$\text{NPSH} = \frac{(100\text{ kPa})\left(\frac{14.696\frac{\text{lb}}{\text{in}^2}}{101.325\text{ kPa}}\right)\left(144\frac{\text{in}^2}{\text{ft}^2}\right)}{\left(62.4\frac{\text{lb}}{\text{ft}^3}\right)} - \frac{\left(0.18\frac{\text{lb}}{\text{in}^2}\right)\left(144\frac{\text{in}^2}{\text{ft}^2}\right)}{\left(62.4\frac{\text{lb}}{\text{ft}^3}\right)} - 8\text{ ft} - z_i$$

$$= 13.5\text{ ft }H_2O$$

Solving for z_i yields a maximum elevation of 11.55 ft.

Scaling Laws

Primary pump characteristics depend upon a variety of geometric, fluid, and pump operating variables. Using dimensional analysis, a series of dimensionless groups can be developed to assist in the prediction of large pump performance based on laboratory observations of smaller, geometrically similar pumps. Although purely theoretical, these relationships are often referred to as pump *similitude* and can be represented by a series of *scaling laws,* which may be expressed as

$$\left(\frac{Q}{ND^3}\right)_1 = \left(\frac{Q}{ND^3}\right)_2 \quad \text{Flow coefficient}$$

$$\left(\frac{\dot{m}}{\rho ND^3}\right)_1 = \left(\frac{\dot{m}}{\rho ND^3}\right)_2 \quad \text{Mass flow coefficient}$$

$$\left(\frac{H}{N^2 D^2}\right)_1 = \left(\frac{H}{N^2 D^2}\right)_2 \quad \text{Head rise coefficient}$$

$$\left(\frac{P}{\gamma N^2 D^2}\right)_1 = \left(\frac{P}{\gamma N^2 D^2}\right)_2 \quad \text{Pressure rise coefficient}$$

$$\left(\frac{\dot{W}}{\rho N^3 D^5}\right)_1 = \left(\frac{\dot{W}}{\rho N^3 D^5}\right)_2 \quad \text{Power coefficient}$$

where

Q = volumetric flow rate
N = the pump rotational speed
D = the impeller diameter
$\dot{m}$ = mass flow rate
ρ = fluid density
H = the pump head (represented previously in this review as h_p)
P = the pressure rise through the pump
$\dot{W}$ = the shaft power

It is convenient to note that units must be consistent on both sides of the equality but need not be consistent within a single grouping. Also, as seen in the scaling laws, the flow coefficient is related to the mass flow coefficient by density, and the head rise coefficient is related to the pressure rise coefficient by the specific weight. However, it is often the case that the density or specific weight changes in the system and is usually dropped from the equality.

Example 11.16

Scaling laws

An 8-inch-diameter centrifugal pump provides 2200 gpm of water at 42 ft of head, with an efficiency of 72% when rotating at 2400 rpm. Determine the shaft work, in hp, of a geometrically similar pump with a 10-inch impeller that provides the same volumetric flow rate.

Solution

In order to find the shaft work, we need to use the power coefficient equation, with the assumption that the fluid density is the same for both systems:

$$\left(\frac{\dot{W}}{\rho N^3 D^5}\right)_1 = \left(\frac{\dot{W}}{\rho N^3 D^5}\right)_2 \quad \Rightarrow \quad \dot{W}_2 = \dot{W}_1 \left(\frac{N_2}{N_1}\right)^3 \left(\frac{D_2}{D_1}\right)^5$$

First, we must determine the shaft work for pump 1 as follows:

$$Q = \frac{2200}{(7.48)(60)} = 4.9 \text{ cfs}$$

$$\dot{W}_1 = \frac{P_1}{\eta} = \frac{Q\gamma H}{\eta \, 550} = \frac{(4.9)(62.4)(42)}{(0.72)(550)} = 32.44 \text{ hp}$$

Now, using the flow coefficient scaling relationship:

$$\left(\frac{Q}{ND^3}\right)_1 = \left(\frac{Q}{ND^3}\right)_2$$

Setting the volumetric flow rate the same between the two pumps, we write

$$Q_1 = Q_2 \quad \text{or, equivalently,} \quad N_1 D_1^3 = N_2 D_2^3$$

Using the data given for D_1, D_2, and N_1, we can solve for N_2 as follows:

$$N_2 = N_1 \left(\frac{D_1}{D_2}\right)^3 = (2400)\left(\frac{8}{10}\right)^3 = 1230 \text{ rpm}$$

Finally, substituting values for $\dot{W}_1$, N_1, N_2, D_1, and D_2 into the power coefficient equation:

$$\dot{W}_2 = (32.44)\left(\frac{1230}{2400}\right)^3 \left(\frac{10}{8}\right)^5 = 13.33 \text{ hp}$$

Specific Speed

One dimensionless group derived from the scaling laws commonly used in pump evaluations is the ratio of the flow coefficient to the head rise coefficient, which is called the *specific speed* (N_s) and is expressed as

$$N_S = \frac{NQ^{1/2}}{(gH)^{3/4}}$$

While the above expression is dimensionally homogeneous, it is more common in the United States to express specific speed in *US customary units* as follows:

$$N_S = \frac{NQ^{1/2}}{H^{3/4}}$$

where

N = rotational speed in revolutions per minute (rpm)

Q = volumetric flow rate in gallons per minute (gpm)

H = head in feet (ft)

Calculation of the specific speed in US customary units assists in the selection of the type of pump best suited for a particular system. Radial-flow pumps provide high head at low flow rates and are best suited to specific speeds less than 3000 or 4000, depending on the reference text you are reading. High capacity, low-head pumps are described by specific speeds greater than 8000 to 10,000 (again, depending on reference source), and generally require the use of axial-flow pumps. Mixed-flow pumps are usually selected for specific speeds in the range between the radial-flow and axial-flow regimes.

Example 11.17

Specific speed

A pump with an 8-inch impeller delivers 1 cfs of water with a delivered head of 55 ft when rotating at 2800 revolutions per minute. Is this pump most likely a radial-flow, axial-flow, or mixed-flow pump?

Solution

The equation for specific speed in US customary units as given above is

$$N_S = \frac{NQ^{1/2}}{H^{3/4}}$$

However, Q is given in units of cfs, not gpm as required. Converting to the correct units yields

$$N_S = \frac{NQ^{1/2}}{H^{3/4}} = \frac{(2800 \text{ rpm})\left[\left(1\frac{\text{ft}^3}{\text{s}}\right)\left(7.48\frac{\text{gal}}{\text{ft}^3}\right)\left(60\frac{\text{s}}{\text{min}}\right)\right]^{1/2}}{55^{3/4}} = 2937$$

Since this is less than 3000, it is safe to assume that this system would require a radial-flow pump. It might have been safe to assume a radial-flow pump without calculating the specific speed, given the relatively high head delivered (55 ft) at a relatively low flow rate (1 cfs is approximately 450 gpm). However, the calculation confirms the intuitive guess.

TURBINES

Turbine specific speed can be used to classify the major types of hydropower turbines:

N_s	Type	ϕ
1–10	Impulse (Pelton)	~0.47
15–110	Francis	0.6–0.9
100–250	Propeller (Brightwood)	1.4–2.0

The quantity ϕ is the **peripheral speed factor**, which is the ratio of a typical runner speed to a typical fluid speed for the turbine. The specific speed allows one to identify clearly the different turbine types.

Impulse turbines are low-discharge, high-head devices in which one or more high-speed jets of water at atmospheric pressure act on carefully shaped vanes or "buckets" on the periphery of the rotating turbine wheel. Although its specific speed range is already narrow, it probably should be still narrower because efficiency drops rapidly at both ends of the range; for $N_s \leq 2$ the impulse wheel is relatively large and cumbersome with large electrical losses, and at $N_s > 8$ the fluid jet is not handled well by the buckets. The equations for force on a bucket F and for power P are

$$F = \rho Q v_r (1 - \cos\beta) \qquad P = Fu \tag{11.1}$$

in which the fluid speed relative to the runner is $v_r = V - u$, the speed of the fluid jet is V, the runner speed itself is $u = \omega r$, the angular velocity of the wheel is ω, the midbucket radius from the axis of rotation (the "pitch circle") is r, and β is the angle through which the fluid is turned on impact with the bucket. Normally β is 165° or a bit more.

Both Francis and propeller turbines are reaction turbines. The turbine runner is fully enclosed and acted upon by water under pressure.

Francis turbines are for moderate-head, moderate-discharge applications. At the lower values of N_s the flow through the rotating turbine runner is in the radial direction, whereas at the higher end of the N_s range the flow direction is mixed, being partly radial and partly axial.

For propeller turbines, the predominant flow direction through the runner is axial in order to handle a high discharge at low head. For all reaction turbines, the flow leaves the delivery pipe, called a penstock, and enters the turbine through a scroll case that wraps around the central turbine unit, gradually decreasing its cross section to force the flow toward the rotating impeller through fixed guide vanes and adjustable wicket gates.

The power, torque, and discharge for a reaction turbine are all related to the interaction of the flow with the turbine runner blades. Figure 11.14 depicts a pair of velocity diagrams for flow at the inlet, section 1, and the outlet, section 2, of a typical turbine blade. The runner angular velocity is ω. The absolute velocity, V, of the fluid can be viewed in two ways. It is composed of radial and tangential components V_r and V_t, respectively, but it is also composed of the blade velocity $u = \omega r$ and the fluid velocity w relative to the moving blade. Two angles help to define the velocity diagram: The angle at which the fluid enters the runner region is α, measured from the tangent to the circle surrounding the runner—also $\tan \alpha = V_r/V_t$; the second angle, β, is measured form u to w. The power and torque are

$$P = T\omega \qquad T = \rho Q [V_{t1} r_1 - V_{t2} r_2] \qquad (11.2)$$

The discharge is the product of the radial velocity component and the area through which this velocity flows, or

$$Q = V_r (2\pi r b) \qquad (11.3)$$

in which b is the thickness of the section. Useful velocity relations from Figure 11.14 are $V_r = w \sin\beta$ and $V_t = u + w \cos\beta = u + V_r \cot\beta$.

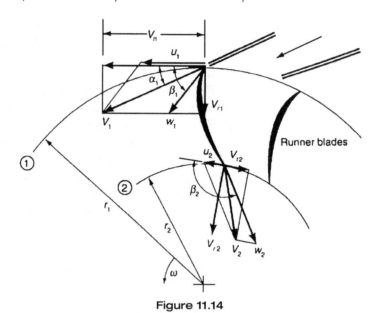

Figure 11.14

OPEN CHANNEL FLOW

Open channel flow is characterized by the presence of a *free surface*, or an interface between two fluids, where the fluid below is usually water and the fluid above is usually air. The fluid above is open to the atmosphere, and, as such, the driving force for flow is from gravity (the weight of the fluid) rather than pressure. The most common examples of open channel flow are rivers and streams; however, the environmental engineer has used gravity-induced flow for the benefit of mankind for centuries, as demonstrated by the Roman aqueduct and evidence of the use of irrigation channels in primitive societies.

Open channel flow is called *uniform* if the depth does not vary with distance. This is the case when the energy gained by the system due to the elevation change of the stream is directly offset by the friction losses due to friction at the channel bottom. *Varied flow* describes a flow condition where the depth of flow is not constant along a length of channel. Varied flow may be further categorized as *rapidly varying flow* when the change in depth is of the same order of magnitude as the change in distance along the channel, or *gradually varying flow* when there is a change in depth, but it is small or occurs over long distances.

Energy Considerations

If we apply the energy equation to open channel flow, we can assume that there are no pumps or turbines and set h_p and h_T equal to zero. Further, if we use the bottom of the channel as our basis for elevation, the slope of the channel (S_0) can be expressed as

$$S_0 = \frac{\text{rise}}{\text{run}} = \frac{\Delta z}{\Delta x} = \frac{\Delta z}{L}$$

where L is the length of the channel. Further, we know the pressure at the bottom of the channel is due to hydrostatic pressure and can be expressed as the depth of the water, usually given the symbol y to distinguish it from the elevation term z. We may now write the energy equation as

$$y_1 - y_2 = \left(\frac{V_2^2 - V_1^2}{2g}\right) + \left(S_f - S_0\right)L$$

where S_f is often called the *friction slope* and is equal to h_L/L. If we define the *specific energy* (E) of a fluid as the sum of the hydrostatic pressure head and velocity head, we may express E as

$$E = y + \frac{V^2}{2g} = y + \frac{Q^2}{2gA^2}$$

The energy equation may now be rewritten as

$$E_1 = E_2 + \left(S_f - S_0\right)L$$

If we further assume that most rivers can be approximated by a rectangular cross section of width b, we may express the flow rate per unit width (q) as

$$q = \frac{Q}{b} = \frac{Vyb}{b} = Vy$$

Note that the quantity q is constant for a given stream width, even if y varies. We may now express the specific energy for a stream of rectangular cross section and constant width as

$$E = y + \frac{q^2}{2gy^2}$$

Example 11.18

Specific energy

A stream that is 50 ft wide possesses the same specific energy when the depth of flow is 4 ft deep as it does when the depth of flow is 1 ft deep. Determine the volumetric flow rate (Q) of the stream in cfs.

Solution

First, we know that the specific energies are equal, so we can write

$$E_1 = E_2 \quad \Rightarrow \quad y_1 + \frac{q^2}{2gy_1^2} = y_2 + \frac{q^2}{2gy_2^2}$$

Substituting in the values given in the problem statement:

$$4 + \frac{q^2}{(2)(32.2)(4)^2} = 1 + \frac{q^2}{(2)(32.3)(1)^2}$$

Solving for the flow rate per unit width, we get

$$3 = 0.01456 q^2 \quad \Rightarrow \quad q = 14.36 \, \frac{\text{ft}^2}{\text{s}}$$

Finally, solve for the volumetric flow rate as follows:

$$q = \frac{Q}{b} \quad \therefore \quad Q = qb = (14.36)(50) = 718 \text{ cfs}$$

Solution of this expression for E with respect to y is a cubic function in y, requiring three roots. For values of specific energy above a threshold (E_{min}), there is one negative root, which has no significant meaning and may be ignored in our analysis. To determine E_{min}, the derivative of the previous equation with respect to y is set equal to zero, and the critical value of y (y_c), defined as the *depth corresponding to the minimum energy state*, can be found as

$$y_c = \left(\frac{q^2}{g} \right)^{1/3}$$

Substitution of y_c into the specific energy expression yields

$$E_{min} = \frac{3}{2} y_c$$

Values of E that are above this minimum have two positive roots and correspond to the two possible flow depths that yield that specific energy value. One root is greater than y_c, indicating a stream that is deep and therefore moving at a relatively lower velocity in conformance with the EOC. This flow regime is termed *subcritical*, and the depth is given the symbol y_{sub}. The other root is smaller than y_c, indicating a stream that is shallow and therefore moving at a relatively greater velocity in conformance with the EOC. This flow regime is termed *supercritical*, and the depth is given the symbol y_{sup}. The *specific energy diagram* is a graphical representation of the specific energy as a function of depth, as shown in Figure 11.15, and is extremely useful in evaluating the potential flow conditions of a particular stream.

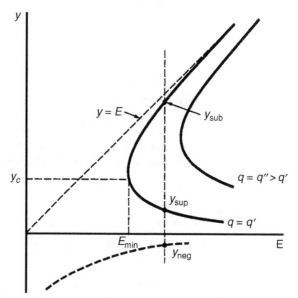

Figure 11.15 Specific energy diagram

Source: Munson, Young, and Okiishi, *Fundamentals of Fluid Mechanics*, 5th ed., © 2006, John Wiley & Sons, Inc. Reprinted by permission.

A useful dimensionless parameter to determine flow criticality is the *Froude number* (*Fr*), which is a ratio of inertial forces to gravitational forces, and may be expressed as

$$Fr = \frac{V}{(gy)^{1/2}} = \frac{q}{g^{1/2} y^{3/2}}$$

where

V = the stream velocity at a flow per unit width q

g = the gravity constant

y = the depth of flow that corresponds to that velocity

When $Fr < 1$ the flow is subcritical, when $Fr > 1$ the flow is supercritical, and when $Fr = 1$ the flow is critical. Transition from supercritical to subcritical flow regimes is a common phenomenon and is known as a *hydraulic jump*, while transition from subcritical to supercritical flow is often characterized by flow under a sluice gate. In either case, the transition must occur along the specific energy curve and, therefore, must pass through the critical flow (as seen in the specific energy diagram). As such, both changes would be classified as rapidly varying flow.

Example 11.19

Critical flow

Determine the critical flow rate of a stream 20 ft wide if the critical velocity is 9.6 fps.

Solution

Since we are given critical velocity, we know that $Fr = 1$ at this point, allowing us to calculate the critical depth as follows:

$$Fr = \frac{V}{(g\, y_c)^{1/2}} = 1 = \frac{9.6\, \frac{ft}{s}}{\left[\left(32.3\, \frac{ft}{s^2}\right) y_c\right]^{1/2}} \Rightarrow y_c = 2.86\, ft$$

Given a critical depth, we may calculate volumetric flow per unit stream width as follows:

$$y_c = \left(\frac{q^2}{g}\right)^{1/3} = 2.86\, ft = \left(\frac{q^2}{32.2\, \frac{ft}{s^2}}\right)^{1/3} \Rightarrow q = 27.45\, \frac{ft^2}{s}$$

Finally, this allows us to calculate the stream's critical flow rate as:

$$Q = q\, b = \left(27.45\, \frac{ft^2}{s}\right)(20\, ft) = 549\, \frac{ft^3}{s}$$

Manning's Equation for Uniform Flow

Most open channel flow encountered in environmental engineering is uniform flow in sewers and channels within the wastewater treatment plant. The most common relationship used to express stream velocity during uniform flow is *Manning's equation*, which can be expressed as

$$V = \frac{1.00}{n} R_H^{2/3} S^{1/2} \quad \text{(SI units)}$$

$$V = \frac{1.49}{n} R_H^{2/3} S^{1/2} \quad \text{(English units)}$$

where n is Manning's roughness coefficient, S is the slope of the EGL (m/m), which is often expressed as the slope of the channel bottom, and V and R_H are as defined previously. The volumetric flow rate is calculated by multiplying the above velocity expressions by the cross-sectional area of flow. Manning's coefficient is a function of pipe material and is often tabulated (like the value for C in the previous discussion) as presented in Table 11.3.

Table 11.3 Values of Manning's coefficient, n, for select materials

Material	n
Plastic (PVC)	0.009–0.011
Vitrified clay	0.010–0.017
Steel pipe	0.012–0.015
Concrete	0.012–0.016
Brick sewers	0.012–0.017
Earth, clean	0.018–0.022
Corrugated metal	0.022–0.028
Earth, with grass	0.025–0.035

To aid in the design of sewer systems that utilize circular pipes, Manning's equation may also be represented graphically as a nomograph, as shown in Figure 11.16. This version of the nomograph requires the use of a center pivot line to differentiate $Q/V/D$ values from n/S values, as shown on the figure. Therefore, it is necessary to have data that includes two of the three $Q/V/D$ quantities, plus either n or S to complete the evaluation. Usually, Q is calculated from flow data, and n is given based on material of construction. Values for V are then assumed from maximum and minimum limits, and a range of acceptable slopes and diameters is developed.

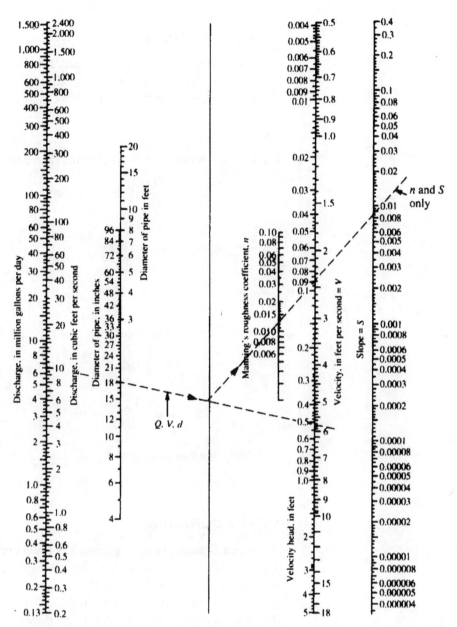

Figure 11.16 Manning's nomograph for all values of *n* for circular pipes flowing full

Source: Warren Viessman, Jr., and Mark J. Hammer, *Pollution Control*, 7th ed., ©2005. Reprinted by permission of Pearson Education, Inc., Upper Saddle River, N.J.

Often, the nomograph is simplified by representing a single value of *n* (usually 0.013), and allowing for slope and diameter to be calculated for given flow rates and velocity ranges using the ratio method demonstrated in Example 11.9 for the Hazen-Williams nomograph.

It should be noted that values from the nomograph represent the pipe flowing at capacity and are often designated by a subscript *f* to designate full flow rate (Q_f) or full velocity (V_f). Properly designed sewer systems will always flow at less than capacity, but, due to the circular cross section, calculating values for hydraulic elements may be difficult. This process is simplified through the use of the *partial flow nomograph*, or plot of hydraulic elements for circular cross sections. This nomograph allows the easy calculation of all hydraulic elements as a function of the full-flow value and a known depth of flow, and is provided in the *FE Supplied Reference Handbook*. Additional coverage on sewer hydraulics can be found in Chapter 12.

Example 11.20

Manning's equation

A 24-inch sanitary sewer line currently flows at 6.7 fps with a volumetric flow rate that is 60 percent of the value for a full pipe. If Q_{min} is 10 percent of the current flow, determine the velocity at Q_{min}. You may assume n varies with depth.

Solution

First, we know that $Q/Q_f = 60$ percent. Therefore, using the partial flow nomograph:

$$\frac{Q}{Q_f} = 0.6 \quad \text{therefore,} \quad \text{depth} = 63\% \quad \text{and} \quad \frac{V}{V_f} = 91\%$$

Now we may calculate V_f as

$$V_f = \frac{6.7 \frac{\text{ft}}{\text{s}^2}}{0.91} = 7.36 \frac{\text{ft}}{\text{s}}$$

Now we can find depth of flow and V/V_f at the new flow rate:

$$\frac{Q_{min}}{Q_f} = 10\% \frac{Q}{Q_f} = (0.1)(0.6) = 0.06 \quad \text{therefore} \quad \text{depth} = 21\%$$

and $\quad \dfrac{V_{min}}{V_f} = 49\%$

Finally, we can determine V_{min} as

$$V_{min} = 49\% \, V_f = (0.49)\left(7.36 \tfrac{\text{ft}}{\text{s}}\right) = 3.6 \text{ fps}$$

Weir Equations

Flow measurement in open channels is often conducted through the use of weirs. *Sharp-crested weirs* are vertical, flat plates placed perpendicular to the flow path, often with notches cut into the plate to allow for a greater range of flow measurements with improved accuracy. *Broad-crested weirs* are obstructions, placed on the channel bottom, of specific relative dimensions in order to create critical flow over the weir.

Although more complex equations have been developed to estimate volumetric flow rate using weirs of many orientations, it is safe to combine weir parameters and simplify flow estimations as

$$Q = CbH^{3/2} \quad \text{Rectangular and trapezoidal}$$

$$Q = CH^{5/2} \quad \text{V-notch}$$

where

Q = the volumetric flow rate (m³/s)

H = the depth of water over the weir, measured at an upstream location (m)

b = the width of the base in the opening of the rectangular/trapezoidal weir (m) (that is, the width at the bottom of the rectangle or trapezoid)

C = the flow constant, which is dependent upon the type of weir and system of units

The flow constant is generally given the following values

$C = 1.84$ Rectangular, SI units

$C = 3.33$ Rectangular, English units

$C = 1.86$ Trapezoidal, SI units

$C = 3.367$ Trapezoidal, English units

$C = 1.40$ 90° V-notch, SI units

$C = 2.54$ 90° V-notch, English units

Example 11.21

Weir equation

You desire to measure flow rates ranging from 0.01 to 0.45 cfs using a 90° V-notch weir. Determine the maximum depth of flow above the bottom of the notch.

Solution

Start with the weir equation, and if we assume that maximum depth will occur at maximum flow rate, we may write

$$Q = CH^{5/2} = 0.45 \text{ cfs}$$

Find the weir coefficient for a 90° V-notch weir in English units, and rearrange to solve for H as

$$C_{90} = 2.54 \quad \Rightarrow \quad H^{5/2} = \frac{0.45}{2.54} \quad \Rightarrow \quad H = \left(\frac{0.45}{2.54}\right)^{2/5} = 0.50 \text{ ft} = 6.0 \text{ in}$$

INCOMPRESSIBLE FLOW OF GASES

The relationships thus far developed are primarily for the flow of incompressible fluids (liquids). Many are also applicable to the flow of compressible fluids (gases), for example, the continuity equation, equation for the Reynolds number, and so forth. The energy equation for incompressible flow also may be used under the following conditions:

1. The change in pressure in the pipe length is less than 10 percent of the inlet pressure. The density and specific weight at inlet conditions (pressure and temperature) should be used.

2. The change in pressure in the pipe length is between 10 and 40 percent of the inlet pressure. The density and specific weight at the average of the inlet and outlet conditions should be used. In some problems the outlet (or inlet) pressure is sought. In this case, the inlet (or outlet) conditions are used to find initial values of density and specific weight, and the approximate outlet (or inlet) pressure is then calculated. An iterative process ensues.

It may be necessary to utilize the perfect gas law from thermodynamics to calculate various properties, particularly the density of the gas given the pressure and temperature. The perfect gas law may be written in the following form

$$\rho = \frac{P}{RT}$$

where $R = \overline{R}/MW$ = gas constant
MW = gas molecular weight
$\overline{R}$ = universal gas constant

It should also be noted that the speed of sound, c, in a perfect gas is given by

$$c = \sqrt{kRT}$$

where k = ratio of specific heats, c_p/c_v
c_p = specific heat at constant pressure
c_v = specific heat at constant volume

It is apparent from the equation above that the speed of sound (acoustic velocity) in a gas depends only on its temperature.

The mach number Ma is the ratio of the actual fluid velocity to the speed of sound:

$$Ma = V/c$$

The accuracy of utilizing incompressible fluid flow equations for the flow of gases decreases with increasing velocities and their use is not recommended for mach numbers greater than 0.2.

FANS AND BLOWERS

A fan/blower is a device that moves gasses or vapors from one location to another. Since fans are usually low-velocity devices, the gas, which usually is air, can be considered to be incompressible for the majority of engineering calculations. The general characteristics of fan operation are the following:

a). Volumetric fan output varies directly with the fan speed of rotation for a given fan.

b). The pressure, or head, of the fan varies directly with the square of the speed of the fan.

c). The power required to run a given fan varies directly with the cube of the speed.

d). For a given installation and constant fan speed, the pressure output and the operating power required will be proportional to the density of the gas.

e). For a constant mass flow rate of gas, the fan speed, the volumetric output, and the pressure vary inversely with the density of the gas. In addition, the power required varies inversely with the square of the density of the gas.

f). At a constant pressure, the speed, volumetric output, and power vary inversely with the square root of the density of the gas.

We begin our review of fans and blowers with two solved examples that employ two of the basic equations of interest.

Example 11.22

Find the motor size needed to provide the forced-draft service to a boiler that burns coal at the rate of 10 tons per hr. The air requirements are 59,000 cfm, air is being provided under 6 in. water gauge (WG) by the fan, which has a mechanical efficiency of 60%. Assume the fan to deliver at a total pressure of 6 in. WG.

Solution

The horsepower is determined by the basic formula given as

$$\text{hp} = \frac{\text{cfm} \times \text{pressure, psf}}{(33{,}000 \text{ ft-lb/hp-min})(\text{efficiency})}. \tag{11.4}$$

We must convert the 6 in. of water to represent a pressure in pounds per square foot.

$$\frac{6}{12} \times 62.4 = 31.2 \text{ psf}$$

Finally,

$$\text{hp} = \frac{59{,}000 \times 31.2}{33{,}000 \times 0.60} = 93 \text{ required; use 100-hp motor.}$$

Example 11.23

A blower with the inlet open to the atmosphere delivers 3000 cfm of air at a pressure of 2 in. WG through a duct 11 in. in diameter, the manometer being attached to the discharge duct at the blower. Air temperature is 70°F, and the barometer pressure is 30.2 in. Hg. Calculate the air horsepower.

Solution

$$\dot{W} = \frac{\dot{m}v\Delta P}{\eta_f} \quad \text{Air hp} = \dot{m}v\Delta P \tag{11.5}$$

$$\text{bhp} = \frac{(\dot{m} \text{ lbm/s})(v \text{ ft}^3/\text{lbm})(\Delta P \text{ lbf/ft}^2)}{550 \frac{\text{ft-lbf}}{\text{hp-s}} \eta_f \text{ (decimal eff.)}}$$

$$\frac{\text{Correct standard air density}}{\text{to actual air density}} = \left(0.075 \text{ lbm/ft}^3\right)\left(\frac{30.2 \text{ actual pr}}{29.92 \text{ stand pr}}\right)$$

$$\rho = 0.0757 \quad \text{so sp. vol } v = 13.21 \text{ ft}^3/\text{lb}$$

Find total head knowing that it is the sum of static head and velocity head.

$$\text{Static head} = \left(\frac{2 \text{ in. H}_2\text{O}}{12 \text{ in./ft}}\right)\left(\frac{62.4 \text{ lb/ft}^3 \text{ H}_2\text{O}}{0.0757 \text{ lb/ft}^3 \text{ air}}\right) = 137.4 \text{ ft air}$$

$$\text{Velocity} = \frac{Q}{A} = \frac{3000 \text{ ft}^3/\text{min}}{60 \text{ s/min}\left(\frac{\pi}{4}\right)\left(\frac{11}{12}\right)^2 \text{ ft}^2} = 75.76 \text{ ft/s}$$

$$\text{Velocity head} = \frac{(75.76)^2 \text{ ft}^2/\text{s}^2}{2(32.2) \text{ ft/s}^2} = 89.1 \text{ ft air}$$

$$\text{so air hp} = \frac{\left(\frac{3000 \text{ ft}^3/\text{min}}{60 \text{ s/min}}\right)(137.4 + 89.1)\text{ft}\,(0.0757 \text{ ft}^3/\text{lb})}{(550 \text{ ft-lbf/hp-s})} = 1.56 \text{ hp}$$

Fan Laws

In order to determine the effect of changes in the conditions of fan operation, certain fan laws are used and apply to all types of fans. Their application is necessarily restricted not only to fans of the same shape but also to the same point of rating on the performance curve. In the majority of cases, system resistance is so small that there is no need to correct horsepower for difference in pressure. However, if there is a great difference or change in temperature through a system, the fan exhausting hot gases may require more power than a blower furnishing the same weight of air to the system. It would be helpful to review the laws of affinity on pumps.

The following constitute the several fan laws:

- Air or gas capacity varies directly as the fan speed.
- Pressure (static, velocity, and total) varies as the square of the fan speed.
- Power demand varies as the cube of the fan speed.

The above apply to a fan having a constant wheel diameter. When air or gas density varies, the following apply:

- At constant speed and capacity the pressure and power vary directly as the air or gas density.
- At constant pressure the speed, capacity, and power vary inversely as the square root of density.
- For a constant weight of air or gas, the speed, capacity, and pressure vary inversely as the density. Also, the horsepower varies inversely as the square of the density.

Example 11.24

A certain fan delivers 12,000 m³/m at a static pressure of 1 cm. WG when operating at a speed of 400 rpm, and requires an input of 4 kW. If 15,000 m³/m are desired in the same installation, what will be the new speed, new static pressure, and new power needs?

Solution

$$\text{New speed} = 400 \times \frac{15,000}{12,000} = 500 \text{ rpm}$$

$$\text{New static pressure} = 1 \times \left(\frac{500}{400}\right)^2 = 1.56 \text{ cm}$$

$$\text{New power} = 4 \times \left(\frac{500}{400}\right)^3 = 7.81 \text{ kW}$$

Example 11.25

A certain fan delivers 12,000 cfm at 70°F and normal barometric pressure at a static pressure of 1 in. WG when operating at 400 rpm, and requires 4 hp. If the air temperature is increased to 200°F (density 0.06018 lb per cu ft) and the speed of the fan remains the same, what will be the new static pressure and power?

Solution

$$\text{New static pressure} = 1 \times \frac{0.06081}{0.075} = 0.80 \text{ in.}$$

$$\text{New power} = 4 \times \frac{0.06081}{0.075} = 3.2 \text{ hp}$$

Example 11.26

If the speed of the fan in Example 11.25 is increased so as to produce a static pressure of 1 in. WG at 200°F, what will be the new speed, new capacity, and new power needs?

Solution

$$\text{New speed} = 400 \times \sqrt{\frac{0.075}{0.06018}} = 446 \text{ rpm}$$

$$\text{New capacity} = 12,000 \times \sqrt{\frac{0.075}{0.06018}} = 13,392 \text{ cfm (at 200°F)}$$

$$\text{New power} = 4 \times \sqrt{\frac{0.075}{0.06018}} = 4.46 \text{ hp}$$

Example 11.27

If the speed of the fan of the previous examples is increased so as to deliver the same weight of air at 200°F as at 70°F, what will be the new speed, new capacity, new static pressure, and new power?

Solution

$$\text{New speed} = 400 \times \frac{0.075}{0.06018} = 498 \text{ rpm}$$

$$\text{New capacity} = 12,000 \times \frac{0.075}{0.06018} = 14,945 \text{ cfm}$$

$$\text{New static pressure} = 1 \times \frac{0.075}{0.06018} = 1.25 \text{ in.}$$

$$\text{New power} = 4 \times \left(\frac{0.075}{0.06018}\right)^2 = 6.20 \text{ hp}$$

Duct-Fan Characteristics

As in pumping, all system resistance curves pass through the origin. The curves will intersect all fan performance curves at some point. However, where the intersect takes place, operation must be stable and the efficiency must be high.

A static pressure curve can easily be drawn through a given point based on the fact that the pressure required to overcome system resistance to flow varies for all practical purposes as the square of the flow rate. When a certain fan is connected to a given system, the system characteristic may be used to learn what will happen. Suppose we want to know what such an arrangement will look like when handling 12,500 cfm against a static pressure of 0.95 in. WG (see Fig. 11.17). We see that but one condition satisfies both fan and system. This is point A. A higher-speed fan having a definite wheel diameter when attached to the same system would show higher cfm capacity, higher static pressure, and greater horsepower requirements. In order to realize the needed capacity at the static pressure resulting from this flow, the duct system must be dampered or the fan speed reduced. Normally, in practice the fan is selected to produce the correct flow and develop the desired static pressure at the selection point where both curves cross. Because dampering is a waste of power when carried on in the duct system, this mode of operation should be avoided as much as possible. For best total results, radial dampering at fan inlet is suggested.

Fans in Series

When low-pressure fans are in series, the weight of flow is substantially that of one fan but the total pressure is the sum of the total pressures of the fans in series. The installation of two identical fans in series obviously does not double the quantity of flow through a given system. Double the air quantity would require some four times the pressure and eight times the horsepower. Placing two identical fans in series a little more than doubles the horsepower and increases the capacity and static pressure as indicated in Figure 11.18a. This placing of two identical low-pressure fans in series is often termed staging.

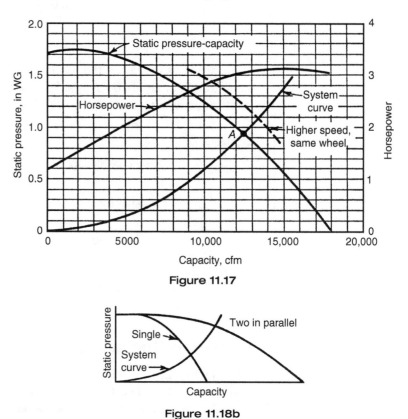

Figure 11.17

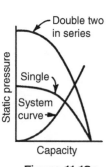

Figure 11.18a

Figure 11.18b

Fans in Parallel

Previously we passed over paralleling of fans. We shall now show how two identical fans can be hooked up characteristically. The characteristic of the combined set is such that the capacity is the sum of the separate capacities for a given static or total pressure (see Fig. 11.18b). It is well to note that the placing of fans in parallel has been a more common practice than placing them in series.

AIR AND GAS COMPRESSORS

Standard types of stationary reciprocating compressors include (a) vertical and V-type single-acting units in sizes up to 100-hp single-stage, and two-stage, either air- or water-cooled; (b) vertical or V-type semi-radial and right-angle type double-acting machines, available in sizes of 60 hp and above, single-stage and multistage, and water-cooled only; (c) single-frame horizontal or vertical double-acting compressors, available in sizes up to 125 hp, single-stage and multistage, and water-cooled only; (d) duplex, horizontal, or vertical double-acting machines in sizes of 75 hp and over, single-stage and multistage, and water-cooled only.

Choice between single- and two-stage compression depends on many varying factors such as size of cylinders, speed, ratio of compression, discharge temperature limitation, cost of power, continuity of service, method of cooling, permanence of installation, etc. In general, the dividing line between single- and two-stage air compression for double-acting compressors may be drawn as follows:

For pressures below 60 psig, single-stage

For pressures above 100 psig, two-stage

For pressures between 60 and 100 psig and capacities below 300 cfm use single stage, for greater requirements use two stage.

Cost of power is an important factor in selecting the type of compressor. Because of the relatively long life of compressor equipment, higher efficiency with consequent lower power cost often justifies a higher initial investment.

Treatment of compressor problems will be confined to the reciprocating type. The student is referred to standard texts for the many other types.

Example 11.28

Air-compressor capacity is the quantity of air compressed and delivered per unit of time. It is usually expressed in cfm at intake pressure and temperature. Assuming that the intake temperature is at 19°C and 100 kPa, how much compressed air at 1500 kPa and 25°C is delivered by a compressor with a rated capacity of 200 m³/min?

Solution

Assuming that the perfect-gas law applies, we can proceed as follows:

$$\frac{P_1 V_1}{T_1} = \frac{P_2 V_2}{T_2} \quad \text{so} \quad \frac{(100 \text{ kPa})(200 \text{ m}^3/\text{min})}{(19 + 273) \text{ K}} = \frac{(1500 \text{ kPa}) V_2}{(25 + 273) \text{ K}}$$

Solving for V, we find it to be equal to 13.6 m³/min.

Work of Compressor without Clearance

Refer to Figure 11.19. The work done on the air (or gas) is the area enclosed within the diagram and is expressed by the following equation for polytropic processes on ideal gases:

$$W = \frac{nwR(T_2 - T_1)}{1 - n} \text{ ft-lb.} \tag{11.6}$$

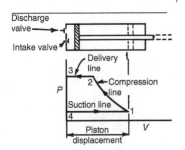

Figure 11.19

A more convenient form is

$$W = \frac{nP_1V_1}{1-n}\left[\left(\frac{P_2}{P_1}\right)^{(n-1)/n} - 1\right] \text{ft-lb} \qquad (11.7)$$

where V_1 is volume drawn into cylinder measured in cubic feet and w is the weight of this charge of gas, which passes through the compressor. For isentropic compression

$$W = \frac{kP_1V_1}{1-k}\left[\left(\frac{P_2}{P_1}\right)^{(k-1)/k} - 1\right] \text{ft-lb.} \qquad (11.8)$$

For isothermal compression

$$W = wRT_1 \ln\frac{P_1}{P_2} \text{ ft-lb.} \qquad (11.9)$$

Example 11.29

Calculate the power required to compress 5000 cu ft of "free" air per hour, initially at 14.5 psia and 70°F, to 100 psia, assuming nonclearance compressor (Fig. 11.19) and single-stage compression, a). isothermal compression and b). isentropic compression.

Solution

a). We need to use the isothermal work equation so weight of air needs to be calculated using the ideal gas law:

$$PV = WRT$$

$$\dot{W} = \frac{P\dot{V}}{RT} = \frac{(14.5 \text{ psia})(144 \text{ in.}^2/\text{ft}^2)(5000 \text{ ft}^3/\text{hr})}{53.3 \frac{\text{ft-lb}}{\text{lb °R}}(70+460)°\text{R}}$$

$$\dot{W} = 369.6 \text{ lb/hr}$$

Using isothermal work equation:

$$\dot{W} = \dot{w}RT_1 \ln\frac{P_2}{P_1} = (369.6 \text{ lb/hr})\left(53.3\frac{\text{ft-lb}}{\text{lb°R}}\right)(530°\text{R})\ln\frac{100 \text{ psia}}{14.5 \text{ psia}}$$

$$\dot{W} = 20.2 \times 10^6 \text{ ft-lb/hr} = \frac{20.2 \times 10^6 \text{ ft-lb/hr}}{(3600 \text{ s/hr})(530 \text{ ft-lb/hps})}$$

$$\dot{W} = 10.2 \text{ hp}$$

b). Now we use the equation for isentropic work.

$$\dot{W} = -\dot{w}\left(\frac{KRT_1}{K-T_1}\right)\left[\left(\frac{P_2}{P_1}\right)^{\frac{k-1}{k}} - 1\right]$$

$$\dot{W} = -369.6 \text{ lb/hr}\left[\frac{1.4\left(53.3 \frac{\text{ft-lb}}{\text{lb}°\text{R}}\right)(530°\text{R})}{1.4-1}\right]\left[\left(\frac{100 \text{ psia}}{14.7 \text{ psia}}\right)^{\frac{1.4-1}{1.4}} - 1\right]$$

$$\dot{W} = 26 \times 10^6 \frac{\text{ft-lb}}{\text{hr}} \quad \text{or} \quad \frac{26 \times 10^6 \text{ ft-lb/hr}}{(3600 \text{ s/hr})(550)\frac{\text{ft-lb}}{\text{hp-s}}} = 13.1 \text{ hp}$$

Isothermal Compression Horsepower

Although isothermal compression is never attained in practice, the work and horsepower required by the process are quite often used as ideal conditions with which the actual performance of a compressor may be compared.

As we have indicated previously, the net work represented by an ideal PV diagram is equal to the area enclosed in the diagram. By integration, it can be shown that the area is equal to the work expressed in foot-pounds.

$$W = wRT_1 \ln\frac{P_2}{P_1} = P_1V_1 \ln\frac{P_2}{P_1} \qquad (11.10)$$

where
W = Work of isothermal compression, ft-lb
P_1 = Intake pressure, lb per sq ft abs
V_1 = Volume of gas compressed measured at intake conditions, cu ft.

As we know, gas volumes are ordinarily measured in standard cubic feet so that Equation 11.10 can be made into a more useful form by substituting pressure and volume at standard conditions 14.7 psia and cubic feet measured at 60°F. Then Equation 11.10 becomes

$$W = (144)\text{in.}^2/\text{ft}^2(14.7)\text{lb/in.}^2(V_{ST}) \ln\frac{P_2}{P_1}. \qquad (11.11)$$

Combining all constants,

$$W = (2116.8 \text{ lb/ft}^2)V_s \text{ ft}^3/\text{lb} \ln\frac{P_2}{P_1}. \qquad (11.12)$$

This equation can be used to calculate the work required to compress any volume of gas isothermally or the horsepower according to the following:

$$\text{hp} = \frac{2116.8}{33000} V_s \ln\frac{P_2}{P_1} \qquad (11.13)$$

$$\text{hp} = 0.064 V_s \ln\frac{P_2}{P_1} \qquad (11.14)$$

Example 11.30

What is the isothermal horsepower required for compressing 1000 scf of air per minute from an intake pressure of 14.0 psia to a discharge pressure of 60 psia?

Solution

You may either use Equation 11.10 or Equation 11.14. Equation 11.14 will be used here because it is simpler.

$$\text{hp} = 0.064 V_s \ln\frac{P_2}{P_1} = 0.064(1000) \ln\frac{60}{14.7}$$

hp = 93.1 power required to compress the gas

Example 11.31

A compressor operating isothermally compresses 2000 m³/min of air from 100 kPa to 1600 kPa. Entering temperature is 17°C. Determine hp required.

Solution

Use the isothermal compressor equation:

$$\text{hp} = wRT_1 \ln\frac{P_2}{P_1} = P_1 V_1 \ln\frac{P_2}{P_1} \tag{11.15}$$

$$\text{hp} = (100 \text{ kPa})(2000 \text{ m}^3/\text{min}) \ln\frac{100}{1600}$$

hp = –5545 kNm/min = –5545 kJ/min

$$\text{hp} = \left(\frac{-5545 \text{ kJ/min}}{60 \text{ s/min}}\right) 1.341 \text{ hp/kW} = 123.9 \text{ hp}$$

Polytropic and Isentropic Compression Horsepower

With a given value of *n*, the actual compressor cylinder horsepower required for the compression of a given quantity of gas in a given length of time can be calculated. This indicated horsepower of the compressor should not be confused with the brake hp of compression. The brake horsepower of compression is somewhat larger than the indicated horsepower and represents the brake horsepower that an engine must deliver to a compressor not only to compress the gas but also to overcome the friction in the moving parts of the compressor.

Because of the similarity of the calculations involved, the determination of polytropic and isentropic hp will be discussed jointly.

The relation for work is given by

$$W = \frac{n}{n-1} P_1 V_1 \left[\left(\frac{P_2}{P_1}\right)^{\frac{n-1}{n}} - 1\right]. \tag{11.16}$$

For isentropic work substitute *k* for *n*.

If V_1 is the volume handled per minute, the preceding equation will give foot-pounds of work done per minute.

Because a common standard pressure for gas measurement is 14.7 psia, Equation 11.16 can be changed to a more convenient form by substituting this standard pressure in the equation as follows:

$$W = \frac{n}{n-1} \times 144 \times 14.7 V_s \left[\left(\frac{P_2}{P_1}\right)^{\frac{n-1}{n}} - 1 \right] \quad (11.17)$$

$$\text{or} \quad W = 2117 V_s \frac{n}{n-1} \left[\left(\frac{P_2}{P_1}\right)^{\frac{n-1}{n}} - 1 \right] \quad (11.18)$$

where V_s is the volume of gas handled measured at 14.7 psia. When V_s is in terms of cubic feet per minute, the horsepower required for compression is

$$\text{hp} = \frac{2117}{33,000} V_s \frac{n}{n-1} \left[\left(\frac{P_2}{P_1}\right)^{\frac{n-1}{n}} - 1 \right]. \quad (11.19)$$

Gas Compressor with Clearance

Refer to Figure 11.20. The events shown there are the same as for those in the case of no clearance, except that because the piston does not force all the gas from the cylinder at pressure P_2, the remaining gas must reexpand to the intake pressure, process 3-4, before intake starts again.

Because the value of n on the expansion curve has little effect on the results, it is taken as being the *same for both compression and expansion*, although actually the values are different. Without clearance, the volume of air drawn into the cylinder is the same as the piston displacement. The work of compression required is given by

$$W = \frac{n}{n-1} w' R T_1 \left[\left(\frac{P_2}{P_1}\right)^{\frac{n-1}{n}} - 1 \right] \text{ ft-lb} \quad (11.20)$$

where w' is the weight of actual volume drawn into cylinder along 4-1. In an actual compressor with clearance, the piston displacement must be greater than volume drawn in, for a given capacity; this means a larger machine than without clearance, costing more and having greater mechanical friction.

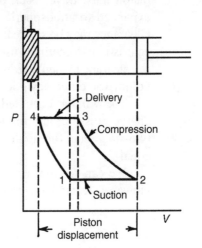

Figure 11.20

Volumetric Efficiency

Volumetric efficiency ranges from 65 to 85% and is determined from the following:

$$E_v = \frac{\text{Actual volume drawn into compressor, cu ft}}{\text{Piston displacement, cu ft}} \quad (11.21)$$

$$\text{or } E_v = 1 + c - \left[c \left(\frac{P_2}{P_1} \right)^{1/n} \right] \quad (11.22)$$

where c is clearance percent as a decimal.

Capacity

Let us assume the gas is air. The capacity is actual volume of "free" air delivered. Altitude and ambient air temperature affect capacity. The higher above sea level and the greater the ambient temperature, the lower the delivered capacity. In these cases, a booster compressor is required to bring the main compressor up to capacity. In an actual compressor installation the air must be filtered and drawn from a cool environment to realize best performance.

Example 11.32

An air compressor with 6% clearance handles 50 lb per min for air between the pressures of 14.7 and 64.7 psia with n equal to 1.33. What is the weight in the cylinder per minute?

Solution

The volumetric efficiency is

$$E_v = 1 + 0.06 - \left[0.06 \left(\frac{64.7}{14.7} \right)^{1/1.33} \right] = 0.877, \text{ or } 87.7\%.$$

Piston displacement weight = 50/0.877 = 57.1 lb. The weight corresponding to total volume including clearance volume is 57.1 × 1.06 = 60.5 lb.

Effect of Clearance on Compressor Performance

In a single-stage compressor, clearance reduces volumetric efficiency. The percent of capacity reduction is greater than the percent of cylinder clearance because the piston must travel back part of its return stroke before clearance-space air has expanded to atmospheric pressure, permitting free air to flow into the cylinder. Clearance may be so great that no air is discharged from the compressor. This characteristic is sometimes used to control the output of the compressor by increasing the clearance when a reduced output is desired. Observe from Equation 11.22 that volumetric efficiency goes down as the pressure ratio goes up.

The volume occupied by expanded clearance air is in proportion to its discharge pressure; and the loss in compressor capacity due to clearance is less for two-stage than for single-stage compression, volume and terminal pressure being equal.

We can see that it is desirable to have the clearance as small as practicable. However, it has been shown that because there is no significant variation in the actual horsepower required for small variations of the clearance, there is no need to increase the cost of manufacturing just to reduce the clearance by a small amount. Clearances vary from about 1% in some very large compressors to 8% or more in other compressors, with clearances of 4% to 8% being common. Neither clearance nor volumetric efficiency is a reliable indicator of the quality of a compressor. The user is most concerned about the power consumed for a given capacity. An increase in clearance requires a larger compressor to deliver the same amount of gas, hence requires more power.

Example 11.33

Air is compressed in an adiabatic compressor at a rate of 500 ft³/s. The air enters at 14.7 psia and 60°F and exits at 380°F. Isentropic efficiency = 90%.

Determine:

a). Exit air pressure if the air has a variable special heat

b). Power input to compressor

Solution

a). Obtain values from air tables.

$$520°R - h_1 = 124.27 \quad Pr_1 = 1.2147$$
$$840°R - h_{2g} = 201.56 \quad Pr_{2g} = 6.573$$

$$n_c = \frac{h_{2g} - h_1}{h_{2a} - h_1}$$

$$0.90 = \frac{201.56 - 124.27}{h_{2a} - 124.27}$$

$$h_{2a} = 210.14 \quad \text{from air tables } Pr_{2a} = 2.614$$

so knowing: $\dfrac{P_1}{P_2} = \dfrac{Pr_1}{Pr_{2a}} = \dfrac{1.2147}{7.614} = \dfrac{14.7}{P_2}$

$$P_2 = 92 \text{ psia}$$

b). $\quad \text{hp} = \dot{m}(h_{2a} - h_1) \quad \dot{m} = \dfrac{P_1 V_1}{RT_1} = \dfrac{(14.7 \text{ psia})(144 \text{ in.}^2/\text{ft}^2)(500 \text{ ft}^3/\text{s})}{53.3 \frac{\text{ft-lbf}}{\text{lbm°R}}(520°R)}$

$\text{hp} = 38.2 \text{ lb/s } (210.14 - 124.27) \text{ Btu/lb}$

$\text{hp} = \dfrac{3280 \text{ Btu/s}}{0.70696 \text{ Btu/hp-s}} \quad \dot{m} = 38.2 \text{ lb/s}$

$\text{hp} = 4640$

Efficiency

Compression efficiency is found from the air horsepowers we have been calculating and given by previous equations.

$$\text{Compression efficiency} = \frac{\text{Air hp}}{\text{ihp}} \tag{11.23}$$

For mechanical efficiency:

$$E_m = \frac{\text{ihp}}{\text{bhp}} \qquad (11.24)$$

For compressor efficiency:

$$E_c = \text{Compression eff.} \times \text{Mech eff.} = \frac{\text{Air hp}}{\text{bhp}} \qquad (11.25)$$

Air horsepower may be required for an isothermal, isentropic, or polytropic compressor. There are times when the efficiency given in an examination problem may be called adiabatic efficiency. If such a problem involves a reciprocating compressor and mechanical efficiency is mentioned, the adiabatic efficiency is the compression efficiency referred to isentropic compression. If the problem involves a centrifugal compression, or other rotary compressor, as in a gas-turbine application and where mechanical friction is small and mechanical efficiency is high and indicated horsepower is not mentioned, then adiabatic efficiency is compressor efficiency referred to isentropic compression.

The indicated horsepower for a compressor of the reciprocating type may be determined from the relation very familiar to all,

$$\text{ihp} = \frac{PLAN}{33{,}000} \qquad (11.26)$$

where
  P = mean effective pressure, psi
  L = length of stroke, ft
  A = piston area, sq in.
  N = number of working strokes per minute (equal to two times the rpm for a double-acting single-cylinder (compressor).

Example 11.34

Air is compressed from an inlet state of 1 kPa and 18°C to an exit condition of 10 kPa. Find work of compression per unit mass.

a). Isothermal process

b). Isentropic process

c). Polytropic process, $n = 1.2$

Solution

a). Isothermal compression

$$\text{work} = P_1 V_1 \ln \frac{P_2}{P_1} = R_1 T_1 \ln \frac{P_2}{P_1}$$

$$\text{work} = \left(\frac{8.314 \text{ kJ/kmol K}}{29 \text{ kg/mol}}\right)(18+273) \text{ K } \ln\left(\frac{10}{1}\right)$$

$$\text{work} = 192 \text{ kJ/kg}$$

b). Isentropic process

$$\text{work} = \frac{KRT_1}{K-1}\left[\frac{P_2}{P_1}^{\frac{k-1}{k}} - 1\right] = \frac{(1.4)(0.287 \text{ kJ/kg K})(291 \text{ K})}{1.4-1}\left[\frac{10}{1}^{\frac{1.4-1}{1.4}} - 1\right]$$

work = 272 kJ/kg

c). Polytropic process, $n = 1.2$

$$\text{work} = \frac{nRT_1}{n-1}\left[\frac{P_2}{P_1}^{\frac{n-1}{n}} - 1\right] = \frac{(1.2)(0.287 \text{ kJ/kg K})(291 \text{ K})}{1.2-1}\left[\frac{10}{1}^{\frac{1.2-1}{1.2}} - 1\right]$$

work = 234 kJ/kg

Note: as heat transfer increases, work decreases.

PROBLEMS

11.1 Kinematic viscosity may be expressed in units of:
  a. m^2/s
  b. s^2/m
  c. kg • s/m
  d. kg/s

11.2 The absolute viscosity of a fluid varies with pressure and temperature and is defined as a function of:
  a. density and angular deformation rate
  b. density, shear stress, and angular deformation rate
  c. density and shear stress
  d. shear stress and angular deformation rate

11.3 An open chamber rests on the ocean floor in 50 m of sea water (SG = 1.03). The air pressure in kilopascals that must be maintained inside to exclude water is nearest to:
  a. 318
  b. 431
  c. 505
  d. 661

11.4 What is the static gage pressure in pascals in the air chamber of the container in Exhibit 11.4? The specific weight of the water is 9810 N/m³.
  a. −14,700 Pa
  b. −4500 Pa
  c. 0
  d. +4500 kPa

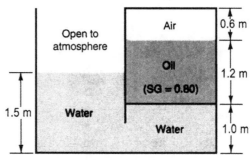

Exhibit 11.4

11.5 The pressure in kilopascals at a depth of 100 meters in fresh water is nearest to:
  a. 268 kPa
  b. 650 kPa
  c. 981 kPa
  d. 1620 kPa

11.6 What head, in meters of air, at ambient conditions of 100 kPa and 20 °C, is equivalent to 15 kPa?
  a. 49
  b. 131
  c. 257
  d. 1282

11.7 With a normal barometric pressure at sea level, the atmospheric pressure at an elevation of 1200 meters is nearest to:
  a. 87.3 kPa
  b. 83 kPa
  c. 115.3 kPa
  d. 101.3 kPa

11.8 The funnel in Exhibit 11.8 is full of water. The volume of the upper part is 0.165m³ and of the lower part is 0.057m³. The force tending to push the plug out is:
a. 1.00 kN
c. 1.63 kN
b. 1.47 kN
d. 2.00 kN

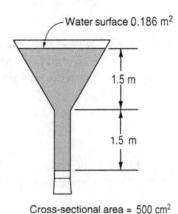

Exhibit 11.8

11.9 An open-topped cylindrical water tank has a horizontal circular base 3 meters in diameter. When it is filled to a height of 2.5 meters, the force in Newtons exerted on its base is nearest to:
a. 17,340
c. 100,000
b. 34,680
d. 170,000

11.10 A cubical tank with 1.5 meter sides is filled with water (see Exhibit 11.10). The force, in kilonewtons, developed on one of the vertical sides is nearest to:
a. 4.1
c. 16.5
b. 8.3
d. 33.0

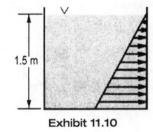

Exhibit 11.10

11.11 A conical reducing section (see Exhibit 11.11) connects an existing 10-centimeter-diameter pipeline with a new 5-centimeter-diameter line. At 700 kPa under no-flow conditions, what tensile force in kilonewtons is exerted on the reducing section?
a. 5.50
c. 1.37
b. 2.07
d. 4.13

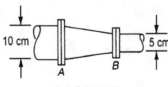

Exhibit 11.11

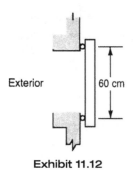

Exhibit 11.12

11.12 A circular access (see Exhibit 11.12) port 60 cm in diameter seals an environmental test chamber that is pressurized to 100 kPa above the external pressure. What force in newtons does the port exert upon its retaining structure?
 a. 7100
 b. 9500
 c. 14,100
 d. 28,300

11.13 A gas bubble rising from the ocean floor is 2.5 centimeters in diameter at a depth of 15 meters. Given that the specific gravity of seawater is 1.03, the buoyant force in newtons being exerted on the bubble at this instant is nearest to:
 a. 0.0413
 b. 0.0826
 c. 0.164
 d. 0.328

11.14 The ice in an iceberg has a specific gravity of 0.922. When floating in seawater (SG = 1.03), the percentage of its exposed volume is nearest to:
 a. 5.6
 b. 7.4
 c. 8.9
 d. 10.5

11.15 A cylinder of cork is floating upright in a container partially filled with water. A vacuum is applied to the container that partially removes the air within the vessel. The cork will:
 a. rise somewhat in the water
 b. sink somewhat in the water
 c. remain stationary
 d. turn over on its side

11.16 A floating cylinder 8 cm in diameter and weighing 9.32 newtons is placed in a cylindrical container that is 20 cm in diameter and partially full of water. The increase in the depth of water when the float is placed in it is:
 a. 10 cm
 b. 5 cm
 c. 3 cm
 d. 2 cm

11.17 A block of wood floats in water (see Exhibit 11.17) with 15 centimeters projecting above the water surface. If the same block were placed in alcohol of specific gravity 0.82, the block would project 10 centimeters above the surface of the alcohol. The specific gravity of the wood block is:
 a. 0.67
 b. 3.00
 c. 0.55
 d. 0.60

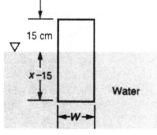

Exhibit 11.17

11.18 The average velocity in a full pipe of incompressible fluid at Section 1 in Exhibit 11.18 is 3 m/s. After passing through a conical section that reduces the stream's cross-sectional area at Section 2 to one-fourth of its previous value, the velocity at Section 2, in m/s, is:
a. 1.0
b. 1.5
c. 3
d. 12

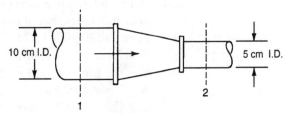

Exhibit 11.18

11.19 Refer to Exhibit 11.18. If the static pressure at Section 1 is 700 kPa and the 10-cm-diameter pipe is full of water undergoing steady flow at an average velocity of 10 m/s at A, the mass flow rate in kg/s at Section 2 is nearest to:
a. 10.0
b. 19.5
c. 78.5
d. 98.6

11.20 Air flows in a long length of 2.5-cm-diameter pipe. At one end the pressure is 200 kPa, the temperature is 150°C, and the velocity is 10 m/s. At the other end, the pressure has been reduced by friction and heat loss to 130 kPa. The mass flow rate in kg/s at any section along the pipe is nearest to:
a. 0.008
b. 0.042
c. 0.126
d. 0.5

11.21 Water flows through a long 1.0 cm I.D. hose at 10 liters per minute. The water velocity in m/s is nearest to:
a. 1
b. 2.12
c. 4.24
d. 21.2

11.22 Gasoline ($\rho = 800$ kg/m³) enters and leaves a pump system with the energy in N • m/N of fluid that is shown in the following table:

	Entering	Leaving
Potential energy, Z meters above datum	1.5	4.5
Kinetic energy, $V^2/(2g_c)$	1.5	3.0
Flow energy, p/γ	9.0	45
Total energy	12.0	52.5

The pressure increase in kPa between the entering and leaving streams is nearest to:
a. 283
b. 566
c. 722
d. 803

11.23 Use the data of Problem 11.22. If the volume flow rate of the gasoline (800 kg/m³) is 55 liters per minute, the theoretical pumping power, in kW, is nearest to:
a. 0.3
b. 0.5
c. 3.2
d. 300

11.24 Water flowing in a pipe enters a horizontal venturi meter whose throat area at B is 1/4 that of the original and final cross sections at A and C, as shown in Exhibit 11.24. Continuity and energy conservation demand that which one of the following be *TRUE*?
a. The pressure at B is increased.
b. The velocity at B is decreased.
c. The potential energy at C is decreased.
d. The flow energy at B is decreased.

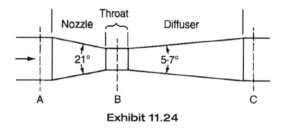

Exhibit 11.24

11.25 Given the energy data in N•m/N shown below existing at two sections across a pipe transporting water in steady flow, what frictional head loss in feet has occurred?

	Section A	Section B
Potential energy	20	40
Kinetic energy	15	15
Flow energy	100	75
Total	135	130

a. 0
b. 5
c. 130
d. 265

11.26 The power in kilowatts required in the absence of losses to pump water at 400 liters per minute from a large reservoir to another large reservoir 120 meters higher is nearest to:
a. 5.85
b. 7.85
c. 15.70
d. 30.00

11.27 The theoretical velocity generated by a 10-meter hydraulic head is:
a. 3 m/s
b. 10 m/s
c. 14 m/s
d. 16.4 m/s

11.28 What is the static head corresponding to a fluid velocity of 10 m/sec?
a. 5.1 m
b. 10.2 m
c. 16.4 m
d. 50 m

11.29 The elevation to which water will rise in a piezometer tube is termed the:
 a. stagnation pressure
 b. energy grade line
 c. hydraulic grade line
 d. friction head

11.30 For the configuration in Exhibit 11.30, compute the velocity of the water in the 300-meter branch of the 15-cm-diameter pipe. Assume the friction factors in the two pipes are the same and that the incidental losses are equal in the two branches. The velocity in m/s is:
 a. 10.0
 b. 4.2
 c. 1.8
 d. 3.7

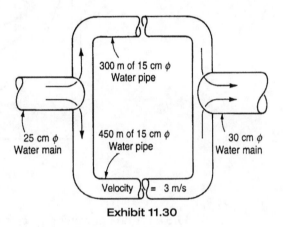

Exhibit 11.30

11.31 Which of the following statements most nearly approximates conditions in turbulent flow?
 a. Fluid particles move along smooth, straight paths.
 b. Energy loss varies linearly with velocity.
 c. Energy loss varies as the square of the velocity.
 d. Newton's law of viscosity governs the flow.

11.32 For turbulent flow of a fluid in a pipe, all of the following are true *EXCEPT*:
 a. the average velocity will be nearly the same as at the pipe center.
 b. the energy lost to turbulence and friction varies with kinetic energy.
 c. pipe roughness affects the friction factor.
 d. the Reynolds number will be less than 2300.

11.33 If the fluid flows in parallel, adjacent layers and the paths of individual particles do not cross, the flow is said to be:
 a. laminar   c. critical
 b. turbulent   d. dynamic

11.34 Which of the following constitutes a group of parameters with the dimensions of power?
 a. ρAV c. $\dfrac{DV\rho}{\mu}$
 b. pAV d. $\dfrac{\rho V^2}{P}$

11.35 At or below the critical velocity in small pipes or at very low velocities, the loss of head from friction:
a. varies linearly with the velocity
b. can be ignored
c. is infinitely large
d. varies as the velocity squared

11.36 The Moody diagram in Exhibit 11.36 is a log-log plot of friction factor vs. Reynolds number. Which of the lines A–D represents the friction factor to use for turbulent flow in a smooth pipe of low roughness ratio (ε/D)?
a. A   c. C
b. B   d. D

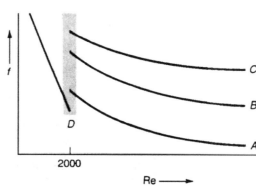

Exhibit 11.36

11.37 A 60-cm water pipe carries a flow of 0.1 m³/s. At Point A the elevation is 50 meters and the pressure is 200 kPa. At Point B, 1200 meters downstream from A, the elevation is 40 meters and the pressure is 230 kPa. The head loss, in feet, between A and B is:
a. 6.94   c. 20.88
b. 15.08   d. 100.2

11.38 Entrance losses between tank and pipe, or losses through elbows, fittings, and valves are generally expressed as functions of:
a. kinetic energy
b. pipe diameter
c. friction factor
d. volume flow rate

11.39 A 5-cm-diameter orifice discharges fluid from a tank with a head of 5 meters. The discharge rate, Q, is measured at 0.015 m³/s. The actual velocity at the *vena contracta* (v.c.) is 9 m/s. The coefficient of discharge, C_D, is nearest to:
a. 0.62   c. 0.99
b. 0.77   d. 0.86

11.40 At normal atmospheric pressure, the maximum height in meters to which a nonvolatile fluid of specific gravity 0.80 may be siphoned is nearest to:
a. 4.0   c. 10.3
b. 6.4   d. 12.9

11.41 The water flow rate in a 15-centimeter-diameter pipe is measured with a differential pressure gage connected between a static pressure tap in the pipe wall and a pitot tube located at the pipe centerline. Which volume flow rate Q in cubic meters per second results in a differential pressure of 7 kPa?
  a. 0.005
  b. 0.066
  c. 0.50
  d. 1.00

11.42 The hydraulic formula $CA\sqrt{2gh}$ is used to find the:
  a. discharge through an orifice
  b. velocity of flow in a closed conduit
  c. length of pipe in a closed network
  d. friction factor of a pipe

11.43 The hydraulic radius of an open-channel section is defined as:
  a. the wetted perimeter divided by the cross-sectional area
  b. the cross-sectional area divided by the total perimeter
  c. the cross-sectional area divided by the wetted perimeter
  d. one-fourth the radius of a circle with the same area

11.44 To calculate a Reynolds number for flow in open channels and in cross-sections, one must utilize hydraulic radius, R, and modify the usual expression for circular cross sections, which is

$$\text{Re} = \frac{DV\rho}{\mu} = \frac{VD}{\nu}$$

where D = diameter, V = velocity, ρ = density, μ = absolute viscosity, and ν = kinematic viscosity.

Which of the following modified expressions for Re is applicable to flow in open or noncircular cross sections?

  a. $\dfrac{RD}{\nu}$
  b. $\dfrac{RV\rho}{\mu}$
  c. $\dfrac{2RD}{\nu}$
  d. $\dfrac{4RV}{\nu}$

SOLUTIONS

11.1 a. Kinematic viscosity $v = \dfrac{\text{Absolute viscosity}}{\text{Density}} = \dfrac{\mu}{\rho}$

Units of absolute viscosity: N·s/m² or kg/m-s

Units of density: kg/m³

The dimensions of kinematic viscosity would be

$$v = \dfrac{\mu \ (\text{kg/m}\cdot\text{s})}{\rho \ (\text{kg/m}^3)} = \dfrac{\text{m}^2}{\text{s}}$$

11.2 d. The absolute viscosity is proportional to the shear stress (τ) divided by the angular deformation rate. Density is not involved in the definition. The rate of angular deformation $\cong \dfrac{dV}{dy}$.

$$\text{Thus, } \tau = \mu \dfrac{dV}{dy}.$$

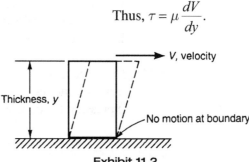

Exhibit 11.2

11.3 c. The internal pressure must equal the local external pressure. Externally, the pressure is

$$p = \gamma h = (\text{SG})(\gamma_{\text{water}})(h)$$

$$p = (1.03)\left(9.81 \dfrac{\text{kN}}{\text{m}^3}\right)(50 \text{ m}) = 505 \text{ kPa (gage)}$$

11.4 b. Since the situation is static, gage pressure at the base is 1.5 m of water. In the air chamber it is 1.5 m of water, less 1 m of water less 1.2 m of oil.

$$p = \gamma h$$

$$p = 1.5 \ (9810) - 1 \ (9810) - 1.2 \ (0.8) \ (9810) = -4513 \text{ Pa}$$

11.5 c. Pressure $= \gamma h = 9810 \ (100) = 981{,}000 \text{ Pa} = 981 \text{ kPa}$

11.6 d. The density of air can be calculated from the ideal gas law using

$$R = 0.286 \frac{kN \bullet m}{kg \bullet K}$$

$$\rho = \frac{p}{RT} = \frac{100 \text{ kN/m}^3}{0.286 \frac{kN \bullet m}{kg \bullet K}(20+273)K} = 1.19 \frac{kg}{m^3}$$

The specific weight of air $\gamma = 1.19 \frac{kg}{m^3} \bullet \left(\frac{9.81 \frac{m}{s^2}}{1 \frac{kg \bullet m}{N \bullet s^2}} \right) = 11.7 \frac{N}{m^3}$

$$p = \gamma h \text{ or } h = \frac{p}{\gamma} \text{ and } h = \frac{15{,}000 \frac{N}{m^2}}{11.7 \text{ N/m}^3} = 1282 \text{ m}$$

11.7 a. Assuming atmospheric pressure at sea level at 101.3 kPa and a constant specific weight of air at 11.7 N/m³ (as previously calculated):

$$p = p_{SL} - \gamma h = 101.3 - \frac{11.7(1200)}{1000} = 87.26 \text{ kPa}$$

11.8 b.

$$\text{Force} = PA = \gamma h A = 9.81 \frac{kN}{m^3} \times 3 \text{ m} \times 500 \text{ cm}^2 \times \left(\frac{1 \text{ m}}{100 \text{ cm}}\right)^2 = 1.47 \text{ kN}$$

11.9 d. The pressure at the tank base $= p = \gamma_w h = (9810)(2.5) = 24{,}325 \text{ N/m}^3$

Area of tank base, $A = \frac{\pi}{4}(3)^2 = 7.07 \text{ m}^2$

Force on tank base $= pA = 24{,}325 \ (7.07) = 171{,}978 \text{ N}$

11.10 c. The average pressure exerted on one side is the pressure that exists at the centroid of the side times the area of the side.

$$F = \gamma h_c A$$

where h_c = the depth in meters from the fluid-free surface to the centroid of the area, and A = area. Since the sides are square, $h_c = 1.5/2 = 0.75$ m.

$$F = 9.81(0.75)(1.5 \times 1.5) = 16.55 \text{ kN}$$

11.11 d. The static force at $A = \left(700 \frac{kN}{m^2}\right)\left[\frac{\pi}{4}(0.1 \text{ m})^2\right] = 5.50 \text{ kN}$ tension on the bolts at A.

The static force at $B = \left(700 \frac{kN}{m^2}\right)\left[\frac{\pi}{4}(0.05 \text{ m})^2\right] = 1.37 \text{ kN}$ tension on the bolts at B.

The end restraint by the pipes opposes a net force of 5.50 − 1.37 = 4.13 kN to the right on the reducing section.

11.12 d. Area of port $= \frac{\pi}{4}D^2 = \frac{\pi}{4}(0.6)^2 = 0.283 \, \text{m}^2$

Pressure $= 100 \, \text{kPa} = 100{,}000 \, \frac{\text{N}}{\text{m}^2}$

$F = pA = 100{,}000 \, (0.283) = 28{,}300 \, \text{N}$

11.13 b. The volume of the bubble equals the volume of the displaced seawater, which equals

$$V_D = \frac{4}{3}\pi r^3 = \frac{4}{3}\pi(0.0125)^3 = 8.18 \times 10^{-6} \, \text{m}^3$$

Since the specific weight of seawater is

$$(SG)(\gamma_W) = 1.03\left(9810 \, \frac{\text{N}}{\text{m}^2}\right) = 10{,}104 \, \frac{\text{N}}{\text{m}^2}$$

The buoyant force, B, is

$$B = \gamma V_D (10{,}104)(8.18 \times 10^{-6}) = 0.0826 \, \text{N}$$

11.14 d. A buoyant force is equal to the weight of fluid displaced. At equilibrium, or floating, the weight downward is equal to the buoyant force.

Let V_1 = total volume of the iceberg in m³. Its weight is $V_1(9810)(0.922) = 9045(V_1)$ N.

Let V_2 = immersed volume of the iceberg, which equals the volume of seawater displaced. The weight of seawater displaced is then $V_2(9810)(1.03) = 10{,}104(V_2)$ N.

Hence $\dfrac{V_2}{V_1} = \dfrac{9045}{10{,}104} = 0.895$ is the volume fraction of the iceberg immersed, and the volume fraction exposed is $1 - 0.895 = 0.105 = 10.5\%$.

11.15 b. Archimedes' principle applies equally well to gases. Thus a body located in any fluid, whether liquid or gaseous, is buoyed up by a force equal to the weight of the fluid displaced. A balloon filled with a gas lighter than air readily demonstrates the buoyant force.

Thus the weight of the cork is equal to the weight of water displaced plus the weight of air displaced. When the air within the vessel is removed, the cork is no longer provided a buoyant force equal to the weight of air displaced. For equilibrium, the cork will sink somewhat in the water.

11.16 c. $V_D = \dfrac{W}{\gamma} = \dfrac{9.32 \, \text{N}}{9810 \, \frac{\text{N}}{\text{m}^2}} = 0.00095 \, \text{m}^3 = 950 \, \text{cm}^3$

The change in total volume, ΔV, beneath the water surface equals the area of the cylindrical container, A, times the change in water level, dh, or $dV = A \, dh$. The depth of the water will increase

$$dh = \frac{dV}{A} = \frac{950 \, \text{cm}^3}{\frac{\pi}{4}(20)^2} = 3.02 \, \text{cm}$$

11.17 d. Let x = height of wood block, W = width of wood block, L = length of wood block, and γ = specific weight of water, N/m³. The weight of the block is equal to the weight of the liquid displaced.

Weight of the block in water = $(x - 15)WL\gamma(1.0)$

Weight of the block in alcohol = $(x - 10)WL\gamma(0.82)$

Since the weight of the block is constant,

$$(x-15)WL\gamma = 0.82(x-10)WL\gamma$$
$$x - 15 = 0.82x - 8.2$$
$$x = \frac{6.8}{0.18} = 37.8 \text{ cm}$$

The specific gravity of the wood block is, by definition,

$$\frac{\text{Volume of water displaced}}{\text{Total volume}} = \frac{(x-15)WL}{xWL} = \frac{37.8-15}{37.8} = 0.603$$

11.18 d. Continuity requires that

$$Q = A_1 V_1 = A_2 V_2 = A_1(3) = \frac{A_1}{4} V_2, \qquad V_2 = 12 \text{ m/s}$$

11.19 c. Continuity requires that the mass flow rate be the same at all sections in steady flow. Calculate $\dot{m}$ at Section 1, where the velocity is given, using $\dot{m} = \rho A V$. This will also be the mass flow rate at Section 2.

Cross-sectional area at Section 1:

$$\frac{\pi}{4}(0.10)^2 = .00785 \text{ m}^2$$

$$\dot{m} = \rho A V \left(1000 \frac{\text{kg}}{\text{m}^3}\right)(.00785 \text{ m}^2)(10 \text{ m/s}) = 78.5 \text{ kg/s}$$

11.20 a. The mass flow rate $\dot{m} = \rho Q = \rho A V$. The density of air at 200 kPa and 150°C (423 K) is obtained from the ideal gas law:

$$\frac{P}{\rho} = RT$$

Use $R = 286 \dfrac{\text{N} \bullet \text{m}}{\text{kg} \bullet \text{K}}$ for air.

$$\rho = \frac{P}{RT} = \frac{200{,}000 \frac{\text{N}}{\text{m}^2}}{286 \frac{\text{N} \bullet \text{m}}{\text{kg} \bullet \text{K}} \bullet 423 \text{ K}} = 1.65 \frac{\text{kg}}{\text{m}^3}$$

The cross-sectional area = $A = \dfrac{\pi}{4} D^2 = 0.785(.025 \text{ m})^2$
= 490×10^{-6} m², and

$$\dot{m} = \rho A v = \left(1.65 \frac{\text{kg}}{\text{m}^3}\right)(490 \times 10^{-6} \text{ m}^2)\left(10 \frac{\text{m}}{\text{s}}\right) = 0.00809 \frac{\text{kg}}{\text{s}}$$

11.21 b.

$$Q = 10\frac{L}{min} \times \frac{m^3}{1000\,L} \times \frac{min}{60\,s} = 167 \times 10^{-6}\,\frac{m^3}{s}$$

The cross-sectional area $A = \frac{\pi}{4}D^2 = 0.785(0.01)^2 = 78.5 \times 10^{-6}\,m^2$

$$V = \frac{Q}{A} = \frac{167 \times 10^{-6}}{78.5 \times 10^{-6}} = 2.13\,m/s$$

11.22 a. The pressure (flow) energy change is $45 - 9 = 36$ N•m/N, or

$$\gamma = \frac{g}{g_c}\rho = \frac{9.81}{1.0}(800) = 7848\,N/m^3$$

$$\frac{\Delta p}{\gamma} = 36, \quad \Delta p = 36\gamma = 36(7848) = 282{,}500\,Pa = 282.5\,kPa$$

11.23 a. The volume flow rate is

$$55\frac{L}{min} \bullet \frac{m^3}{1000\,L} \bullet \frac{min}{60\,s} = 917 \times 10^{-6}\,m^3/s$$

Ignoring the head loss, h_L, from friction, the required energy input is $52.5 - 12 = 40.5\,\frac{N\bullet m}{N}$.

$$\text{Power} = \gamma Q h_A \left(7848\,\frac{N}{m^3}\right)(917 \times 10^{-6}\,m^3/s)\left(40.5\,\frac{N-m}{N}\right)$$
$$= 291\,W = 0.291\,kW$$

11.24 d. In a venturi throat, the increased velocity required by continuity results in a *KE* (velocity) increase that occurs at the expense of pressure (flow) energy. Since the system is horizontal, no change in potential energy has occurred. At *B* the pressure (flow energy) decreases and *KE* increases. For a well-designed venturi, the conditions existent at *A* are essentially restored at *C*.

11.25 b. Apply an energy balance of the fluid flowing: Total energy in = Total energy out + Energy losses − Energy inputs. Thus, $135 = 130 + h_L - 0$. The head loss $h_L = 5$ N•m/m, or 5 meters.

11.26 b. Ignoring frictional losses, pump inefficiency, and noting that any changes in *KE* or pressure are essentially 0, pumping power is equal to the increase in potential energy between the reservoirs.

The potential energy increase per lb_m is Z or h = 120 meters or $\frac{N\bullet m}{N}$

The volume flow rate, Q, is

$$400\,\frac{L}{min} \bullet \frac{m^3}{1000\,L} \bullet \frac{min}{60\,s} = 6.67 \times 10^{-3}\,m^3/s$$

The power required is

$$P = \gamma Q h_a = 9.81\,\frac{kN}{m^3} \bullet 6.67 \times 10^{-3}/s \bullet 120\,m = 7.85\,kW$$

11.27 c.

$$h = \frac{V^2}{2g} \quad \text{or} \quad V = (2gh)^{1/2} = (2 \times 9.81 \times 10)^{1/2} = 14 \text{ m/s}$$

11.28 a. The head is

$$h = \frac{V^2}{2g} = \frac{10^2}{2(9.81)} = 5.10 \text{ m}$$

11.29 c. A **piezometer tube** indicates static pressure and is equivalent to a static pressure gage.

Stagnation pressure is an increased pressure developed at the entrance to a pitot tube when the velocity locally becomes zero.

The **hydraulic grade line** is a flow energy or pressure head in meters, which can be plotted vertically above the pipe centerline along the pipe.

The **energy grade line** is the total mechanical energy (flow energy or pressure head, plus kinetic energy or dynamic head, plus potential energy or height above datum) in meters, which may be plotted vertically above the datum along the pipe.

The **friction head** is the head loss h_f in meters caused by fluid friction.

The **critical depth** above the channel floor in open channels is the depth for minimum potential and kinetic energy for the given discharge. Tranquil flow (low *KE* and high *PE*) exists when the actual flow is above critical depth, and rapid flow (high *KE* and low *PE*) exists when the actual flow is below critical depth.

11.30 d. There is a drop in the energy line from the 25-cm main to the 30-cm main. This head loss must be equal in both 15-cm lines, or

$$h_{f300} = h_{f450}$$

The Darcy equation is

$$h_f = f \frac{L}{d} \frac{V^2}{2g}$$

where h_f = head loss in meters, f = friction factor, L = length of pipeline in meters, d = diameter of pipe in meters, and g = 9.81 m/s². Thus, in this situation,

$$f \frac{300}{0.15} \frac{V_{300}^2}{2(9.81)} = f \frac{450}{0.15} \frac{V_{450}^2}{2(9.81)}$$

which reduces to

$$300 V_{300}^2 = 450(3)^2$$

$$V_{300} = \sqrt{\frac{4050}{300}} = 3.67 \text{ m/s}$$

11.31 c. Laminar (streamline, viscous) flow is compared with turbulent flow in the following table:

	Laminar Flow	**Turbulent Flow**
Motion of fluid particles	Parallel to stream velocity. Paths of particles do not cross.	Particle paths cross and move in all directions.
Energy loss, h_f	$h_f = f\left(\dfrac{L}{D}\right)\left(\dfrac{V^2}{2g}\right)$ f is independent of surface roughness and decreases with Re. $f = \dfrac{64}{\text{Re}}$	$h_f = f\left(\dfrac{L}{D}\right)\left(\dfrac{V^2}{2g}\right)$ f varies with surface roughness, decreases with Re to a constant value. See Moody diagram.
Velocity distribution in pipe	Average is $1/2$ of maximum at centerline. parabolic distribution. Zero at wall.	Essentially same throughout, except for thin boundary layer at wall. Follows 1/7 power law.
Reynolds number $\text{Re} = DV\rho/\mu$	Less than 2300	Greater than 2300

Newton's law of viscosity defines μ on the basis of shear stress and the rate of fluid angular deformation. The Reynolds number contains μ as a contributing parameter. Very viscous liquids usually move in laminar flow.

On the basis of the above data, select (c). Do not confuse the energy loss, h_f, with the friction factor, f.

11.32 d. In turbulent flow the Reynolds number is *greater* than 2300.

11.33 a. Turbulent flow is highly agitated flow with individual particles crossing paths and colliding; critical flow is a point at which some property of the fluid—or some parameter related to it—changes; dynamic flow is redundant; uniform flow is of constant rate.

11.34 b. Choice (a) is mass flow rate, $\dot{m}$, in kg/s.

Choice (b) has these dimensions:

$$pAV = \left(\frac{\text{N}}{\text{m}^2}\right)(\text{m}^2)\left(\frac{\text{m}}{\text{s}}\right) = \frac{\text{N}\bullet\text{m}}{\text{s}} = \text{W}$$

Choice (c) is the Reynolds number, Re; it is the dimensionless ratio of inertial force to viscous force.

Choice (d) is the Euler number, Eu; it is the dimensionless ratio of inertial force to pressure force.

11.35 a. Below the critical velocity (Re < 2300) flow is laminar, and $f = 64/\text{Re}$. Substitution of this term into the Darcy equation for friction loss in pipes, $h_f = f \dfrac{L}{D} \dfrac{V^2}{2g}$, yields the Hagen-Poiseuille equation,

$$h_f = \frac{32 \mu L V}{\gamma d^2}.$$

11.36 a. Line D applies to all roughness ratios in laminar flow (Re < 2300) because the boundary layer at the wall makes the friction factor independent of roughness ratio:

$$f = \frac{64}{\text{Re}}$$

In turbulent flow (Re > 2300), increasing roughness is represented by A for a smooth pipe to C for a very rough pipe; moreover, only the thinnest boundary layer exists in a turbulent flow, so the friction factor is very dependent on surface roughness.

11.37 a. Use an energy balance to determine h_f. Upon substituting the given data, the resulting equation is

$$Z_A + \frac{V_A^2}{2g} + \frac{P_A}{\gamma} = Z_B + \frac{V_B^2}{2g} + \frac{P_B}{\gamma} + h_f$$

Since the pipe diameter is unchanged, continuity requires that V be the same at both points. Thus the kinetic energy terms can be deleted from both sides of the equation.

$$Z_A + \frac{P_A}{\gamma} = Z_B + \frac{P_B}{\gamma} + h_f$$

$$h_f = \frac{200 - 230}{9.81} + 50 - 40 = 6.94 \text{ m}$$

11.38 a. Typical head losses for the above items are expressed as an empirical average constant, K or C times the kinetic energy, $V^2/2g$:

$$h_f = K \frac{V^2}{2g}$$

11.39 b. The discharge coefficient is $C_D = C_c C_v$, where C_c = coefficient of contraction = (area of v.c.)/(area of orifice) and C_v = coefficient of velocity = (actual velocity at v.c.)/(theoretical velocity at v.c.) ignoring losses.

The theoretical velocity at the v.c. is

$$V = \sqrt{2gh} = \sqrt{2(9.81)(5)} = 9.9 \text{ m/s}$$

$$C_v = \frac{9.0}{9.9} = 0.909$$

The area of the v.c. is

$$A = \frac{Q}{V} = \frac{.015}{9} = 0.00167 \, m^2$$

The area of the orifice $= \frac{\pi}{4}D^2 = \frac{\pi}{4}(.05)^2 = 0.00196 \, m^2$. Thus,

$$C_c = \frac{0.00167}{0.00196} = 0.852 \quad \text{and} \quad C_D = C_c C_v = (0.852)(0.909) = 0.774$$

11.40 d. The maximum height to which a fluid may be siphoned is determined when the pressure of the fluid column plus its vapor pressure equals the external pressure. The minimum pressure at the highest point is 0 kPa plus vapor pressure.

Ignoring the vapor pressure (small),

$$p = \gamma h \quad \text{or} \quad h = \frac{P}{\gamma}$$

$$h = \frac{-101.3 \, kN/m^2}{(0.8)9.81 \, kN/m^3} = 12.91 \, m \text{ in depth (or height)}$$

11.41 b. A pitot tube generates a stagnation pressure as fluid kinetic energy is converted to pressure head. Hence

$$V = \sqrt{2g\left(\frac{\Delta p}{\gamma}\right)}$$

$$V = \sqrt{2(9.81)\frac{m}{s^2}\left(\frac{7000 \, N/m^3}{9810 \, N/m^3}\right)} = 3.74 \, m/s$$

$$Q = AV = \frac{\pi}{4}D^2 V = \frac{\pi}{4}(0.15)^2(3.74) = 0.0661 \, m^2/s$$

11.42 a. For a static head orifice discharging freely into the atmosphere

$$Q = CA\sqrt{2gh}$$

11.43 c. Hydraulic radius, R, is defined as cross-sectional area, divided by wetted perimeter.

11.44 d. Choices (a) and (c) are not dimensionless, as required for a Reynolds number. Since the hydraulic radius R = (cross sectional area)/(wetted perimeter), for a circular cross section,

$$R = \frac{\frac{\pi D^2}{4}}{\pi D} = \frac{D}{4} \quad \text{or} \quad D = 4R$$

Therefore,

$$Re = \frac{4RV\rho}{\mu} = \frac{4RV}{\nu}$$

CHAPTER 12

Water Resources Engineering

OUTLINE

WATER RESOURCES PLANNING 509

HYDROLOGIC ELEMENTS 509
Precipitation ■ Evapotranspiration ■ Infiltration

WATERSHED HYDROGRAPHS 512
Unit Hydrograph ■ Change of Unitgraph Duration

PEAK DISCHARGE ESTIMATION 518

HYDROLOGIC ROUTING 521
Reservoir Routing ■ River Routing

WATER DEMAND 525
Population Forecasting ■ Storage Structures and Reservoirs

WATER DISTRIBUTION NETWORKS 529

WASTEWATER FLOWS 531

SEWER SYSTEM DESIGN 532

PROBLEMS 536

SOLUTIONS 539

WATER RESOURCES PLANNING

Available water resources vary by geographic region; however, the need to effectively manage those available resources is the same for each municipality. The ability of an engineer to properly design water and wastewater treatment facilities, as well as distribution and collection networks, is wholly dependent on accurate predictions of the quantities of water demanded and the expected return. Further, planning for the long-term is not an exact science, and appropriate assumptions must be made in order to provide for the future, while remaining cost-effective in the present.

HYDROLOGIC ELEMENTS

Hydrology is in general a multidisciplinary subject that is the study of water movement and distribution on earth. This movement is a closed loop called the hydrologic cycle in which water is first evaporated primarily from the oceans, then transported as vapor by the atmosphere, and, under proper circumstances, precipitated to the earth's surface as rain or snow. The surface water may return to the atmosphere again as evaporation, it may infiltrate into the soil and reach the groundwater or be taken up by plants and transpired back into the air, or it may flow over the land surface and find its way into streams, rivers, or lakes, eventually flowing back into the oceans to complete the cycle.

Precipitation

The most common form of precipitation is liquid rain; when the amounts of other forms, such as snow, must be quantified, they are often melted first, and the amount is reported in terms of its liquid equivalent. The most common precipitation gage in the United States is the Weather Service 8-in.-diameter cylindrical gage, which directs captured rain into a measuring tube that is one tenth the cross-sectional area of the collector, and depths are then measured to 0.01 in. within it. Three types of recording gages are also in common use; they are the tipping-bucket gage, the weighing-type gage, and the float recording gage.

The average precipitation $\bar{p}$ over a region can be obtained from point data in one of three ways, all of which fit the formula

$$\bar{p} = \frac{1}{A_T}\Sigma A_i P_i \qquad A_T = \Sigma A_i \qquad (12.1)$$

in which p_i is a point precipitation value, A_i is a weighting factor, and A_T is the sum of the weighting factors:

1. The simple arithmetic mean is appropriate when the individual values are all similar. In this case set each $A_i = 1$, and then A_T is just the number of points.

2. The widely used Thiessen average is a weighted average that in effect assumes that the value p_i best represents the true precipitation at all locations that are closer to gage i than to any other gage. Each A_i is the area surrounding gage i, and A_T is the total gaged area. The boundary of each A_i is formed by lines that are the perpendicular bisectors of lines drawn between the gages themselves.

3. The isohyetal method is the only method that allows a knowledge of basin topography to enter the calculation. One begins the computation by drawing contour lines of equal precipitation (isohyets) throughout the region. Then in Equation (12.1) A_i is the area between adjacent isohyets, p_i is the average precipitation between these adjacent isohyets, and A_T is the total gaged area.

Evapotranspiration

The quantification of evaporation or transpiration amounts (or of evapotranspiration, ET, the sum of the two) can become important to engineers who conduct water supply studies. There are several computational approaches and one primary experimental method of estimating evaporation; each approach has its problems and leads to imprecision in the result:

1. The water budget or mass conservation method attempts to account for all flows of water to and from the water body under study, including inflow, outflow, direct precipitation, and even seepage to the groundwater.

2. The energy budget is like the water budget, except the energy flows rather than mass flows are the basic accounting medium.

3. Direct empirical meteorological correlations are used to avoid the uncertainties of the first two methods, but attempts to avoid excessive complexity here usually lead to incomplete, and thus inaccurate, results.

4. A combination of the above methods has been relatively successful. For example, the Penman equation, when used with a set of charts, has become popular when all the required data can be obtained or estimated.

5. The National Weather Service Class A pan is 4 feet in diameter, 10 inches deep, made of unpainted galvanized iron, and used to measure evaporation by direct experiment. Multiplication of this result by a pan coefficient, typically about 0.7, then gives the evaporation from the adjacent larger water body. Difficult correlation studies are needed to ensure that the coefficient is appropriate to a particular application.

Infiltration

Infiltration of water into the soil is important in some studies. Horton's infiltration equation is widely used for this purpose, which is

$$f = f_c + (f_o - f_c)e^{-kt} \tag{12.2}$$

Here f is infiltration rate (in./hr); f_o and f_c are the initial and final infiltration rates, respectively; t is time (hr); and k (1/hr) is an empirically determined constant. Another common way of characterizing infiltration is via the ϕ index method. In this method one plots the overall precipitation rate versus time; a horizontal line called the ϕ index is drawn on the plot, such that the volume of rainfall excess above this line is equal to the actual volume of observed runoff. Thus the index indicates the average infiltration rate for the storm event.

Example 12.1

Exhibit 1 is a histogram that describes hourly rainfall for a 5-hour storm.

a). Previous experience has determined that the Horton infiltration parameters for the soil in this region are $f_o = 0.4$ in./hr, $f_c = 0.2$ in./hr, and $k = 0.5$/hr. Determine the volume of rainfall that infiltrates during the 5-hr period.

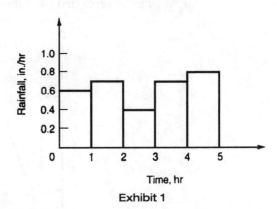

Exhibit 1

b). If the net runoff from this storm is known to be 1.8 in., compute the ϕ index for this event.

Solution

a). Using the Horton equation, the infiltrated volume V_I is

$$V_I = \int_0^5 f dt = 0.2 \int_0^5 \left[1 + e^{-0.5t}\right] dt$$

$$V_I = 0.2 \left[t - \frac{1}{0.5} e^{-0.5t}\right]_0^5 = 0.2 \left[5 - \frac{e^{-2.5}}{0.5} + \frac{1}{0.5}\right]$$

$$V_I = 1.37 \text{ in.}$$

b). The net runoff is the area in Exhibit 1 that lies above the ϕ index. Assuming $\phi < 0.4$ in, then one can write

$$1.8 = \sum_i [p_i - \phi] = (0.6 - \phi) + (0.7 - \phi) + (0.4 - \phi) + (0.7 - \phi) + (0.8 - \phi)$$

$$\phi = 0.28 \text{ in.}$$

WATERSHED HYDROGRAPHS

A watershed or drainage basin is the region drained by a stream or river. When a precipitation event (a storm) occurs over the watershed, it causes several processes within the basin. First is the initial moistening of the land surface and the vegetation, followed by the local filling of small surface indentations (depression storage) and the buildup of some depth of water on the land surface (initial detention storage) before the flow of water over the land begins; at the same time infiltration begins. For the larger storms some, possibly even most, of the precipitation enters a stream and flows out of the basin. The discharge past this outflow point is a time-variant process. A plot of the outflow versus time is called a hydrograph; Figure 12.1 is a definition sketch of a hydrograph.

There are two components to any perennial streamflow, a relatively short-term component, which is the storm-induced surfaced water outflow, and a longer-term, slowly varying component called base flow, which is the contribution from the groundwater to the flow. In Figure 12.1 the storm hydrograph is caused by an effective storm precipitation of D hr, causing first the increasing discharge on the rising limb of the hydrograph from A to the crest C, and then the recessional limb from C to B when the storm-related discharge ceases. The base flow, below line AB in the figure, can be separated from the storm flow in any of several ways:

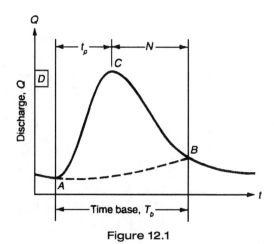

Figure 12.1

1. From the low point A before the storm outflow begins, simply draw a horizontal line.

2. From point A extend the upstream line to a point directly below the crest C. From that point draw a straight line to B, which is located a distance N (days) after point C. The value N is empirically found from

$$N = A^{0.2} \quad\quad A \text{ in square miles}$$
$$ = 0.8A^{0.2} \quad A \text{ in square kilometers} \quad\quad (12.3)$$

The area of the drainage basin is A. Sometimes N is adjusted to the nearest full day.

3. Sometimes an attempt is made to mimic near B the character of the flow near A by patching the slope of the recession curve from A into the base-flow separation near B. This method leaves unanswered the choice of the remainder of the separation curve under the rising limb and crest. It must be drawn arbitrarily.

When the base flow has been removed from the original storm hydrograph, the remainder is direct storm runoff.

Unit Hydrograph

A unit hydrograph (unitgraph) has a volume of 1 in. (or 1 cm) of direct runoff over the drainage basin as a result of a storm of D hours' effective duration. Effective duration is the time interval when excess rainfall exists and direct runoff occurs. Any direct storm runoff has a volume

$$V = \int Q dt \quad\quad (12.4)$$

which may also be written as $V = xA$, in which x is in inches (cm) and A is the basin area. Determination of the volume V, which is the area ABC in Figure 12.1, can be computed efficiently and accurately by use of the trapezoidal rule. Assume the time base T_b is divided into m intervals $\Delta t = T_b/m$ and the direct runoff ordinates Q_i, $i = 1$ to $m + 1$, are known with $Q_1 = Q_{m+1} = 0$. By the trapezoidal rule,

$$V = \int Q dt = (Q_1 + Q_2)\frac{\Delta t}{2} + (Q_2 + Q_3)\frac{\Delta t}{2} + \ldots + (Q_m + Q_{m+1})\frac{\Delta t}{2}$$

or

$$V = \Delta t \sum_{i=2}^{m} Q_i \quad\quad (12.5)$$

Normally x will not be 1.0 in. (cm). The unit hydrograph is simply obtained by dividing each of the ordinates Q_i of the direct storm runoff plot by x. The unitgraph has a variety of applications.

The suitability of the unitgraph for these uses, however, depends on the appropriateness of several assumptions, including these:

- Rainfall excesses of one duration D will always produce hydrographs with the same time base, independent of the intensity of the excess.

- The time distribution of the runoff does not change from storm to storm, so long as D is unchanged; thus an increase in runoff volume by $P\%$ increases each hydrograph ordinate Q_i by $P\%$. Moreover, the distribution is not affected by prior precipitation.

The development of a unit hydrograph that produces reliable results in applications will be enhanced if one follows some experience rules:

- Basin sizes should be between 1000 acres and 1000 square miles.
- The direct storm runoff should preferably be within a factor of 2 of 1.0 inch, and the storm structure should be relatively simple.
- The unitgraph should be derived from several storms of the same duration. In other words, compute several unit hydrographs, and then average them.

If one does not have sufficient storm data to derive a unitgraph, then theoretical or empirical methods may be used to develop a "synthetic" unitgraph based on information such as peak flow values and basin characteristics. Numerous such methods have been proposed. Two of the more commonly used synthetic methods are Snyder's method, originally developed for Appalachian watersheds, and the SCS method, developed by the Soil Conservation Service. They must be applied with care for best results; space does not permit an explanation here of these methods in the detail that is needed, so the reader may consult the chapter references for the complete methods.

Change of Unitgraph Duration

Each unit hydrograph is associated with an effective storm duration D. If one wants a unitgraph for some other effective storm duration without developing it directly from storm data, this can be done. (If the new storm duration differs from the existing one by no more than 25%, then normal practice is to use the existing one without alteration.) Two methods are used:

1. *Lagging.* This method can be used to construct a new unitgraph for a storm of effective duration nD, given the unitgraph for the storm having effective duration D, where n is an integer only. Simply add together n of the original unit hydrographs, starting each successive unitgraph D hours after the beginning of the preceding one. This step produces a hydrograph associated with an effective duration of nD hours and having a runoff volume of n inches over the basin. Now divide all the hydrograph ordinates Q_i by n to obtain the new unitgraph. The method is easily set up in a table.

2. *S-curve.* This method is much more general and can be used to construct a unitgraph for either a shorter or longer effective storm duration than the original. Say the desired new effective storm duration is D_{new}. First one constructs the S-curve (it is a summation curve, that is, a sum of unitgraphs, and it also takes the general shape of an S) by successively lagging by D hours and summing (adding together) the ordinates of a total of T_b/D original unitgraphs. Next draw a second S-curve, lagged D_{new} hours after the first S-curve. The differences in ordinates of these two S-curves, each multiplied by the ratio D/D_{new}, will be the ordinates of the new unitgraph for the storm of effective duration D_{new}.

Example 12.2

Stream runoff from a 1500-acre watershed is plotted in Exhibit 2 for a storm having an effective duration D of 2 hours.

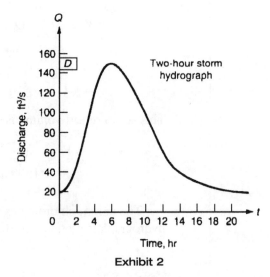

Exhibit 2

a). Compute the ordinates of, and plot, the 2-hour unit hydrograph.

b). Use the information for the 2-hour unit hydrograph to construct a 3-hour unit hydrograph.

c). Construct the composite storm hydrograph caused by 1.5 inches of excess precipitation falling in the first 2 hours, followed immediately by 0.7 inch of excess precipitation in the next 2 hours. Assume a base flow of 10 ft³/s.

Solution

a). The computations are presented in Exhibit 3. First the amount of the base flow must be identified and separated from the overall runoff. Since little information is available in this problem and also because the runoff duration is relatively short, it is assumed that the base flow is a constant 20 ft³/s.

Exhibit 3

Time, hr (1)	Stream Flow, ft³/s (2)	Storm Flow Q_i, ft³/s (3)	Unitgraph Ord. U_i, ft³/s (4)
0	20	0	0
2	60	40	58
4	113	93	135
6	150	130	188
8	127	107	155
10	96	76	110
12	65	45	65
14	43	23	33
16	27	7	10
18	20	0	0

The data in column 2 come directly from the hydrograph, Exhibit 2. The storm flow Q_i, column 3, is the stream flow minus the base flow. Selecting a time interval $\Delta t = 2$ hr for use in Equation (12.5), the storm runoff volume is

$$V = \Delta t \sum_{i=2}^{m} Q_i = (2 \text{ hr})(521 \text{ ft}^3/\text{s}) = 1042 \frac{\text{ft}^3}{\text{s}}\text{-hr}$$

$$V = \left[1042 \frac{\text{ft}^3}{\text{s}}\text{-hr}\right]\left[60^2 \frac{\text{s}}{\text{hr}}\right] = 3.75 \times 10^6 \text{ ft}^3$$

This storm runoff volume is equivalent to a depth x of water over the basin of

$$x = \frac{V}{A} = \frac{(3.75 \times 10^6 \text{ ft}^3)(12 \text{ in./ft})}{(1500 \text{ acres})(43{,}560 \text{ ft}^2/\text{acre})} = 0.69 \text{ in.}$$

The unitgraph ordinates $U_i = Q_i/x$ are tabulated in column 4, and the unit hydrograph is plotted in Exhibit 4.

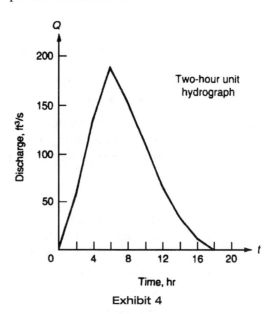

Exhibit 4

b). One constructs the S-curve by repeatedly lagging the 2-hour unitgraph, whose ordinates are listed in column 4 in Exhibit 3, and adding together all the values that are associated with each time instant. The individual ordinates S_i of the S-curve are

$$S_i = \sum_{n=1}^{i} U_n$$

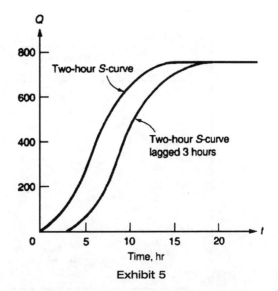

Exhibit 5

The 2-hr S-curve is plotted in Exhibit 5. Also shown is this same S-curve lagged 3 hours; the differences in ordinates of these two S-curves are then multiplied by the ratio $D/D_{new} = 2/3$ to scale the volume of the new hydrograph properly to end with the 3-hr unitgraph plotted in Exhibit 6.

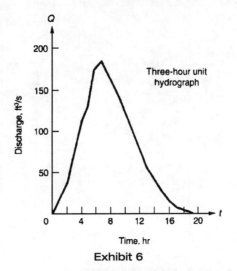

Exhibit 6

Scrutiny of this computational sequence shows that the peak discharge in the new unitgraph is slightly smaller than the peak of the 2-hr unitgraph, as one would expect.

c). Computations are tabulated in Exhibit 7. Time is measured from the start of the storm. The 2-hr unitgraph is multiplied by 1.5 for the first portion of the runoff, followed by a second unitgraph scaled by 0.7. Finally, the base flow is added.

Exhibit 7

Time, hr	Unitgraph Ord., U_i, ft³/s	1.5 × U_i, ft³/s	0.7 × U_i, lag 2 hr, ft³/s	Sum, with BF, ft³/s
0	0	0	—	10
2	58	87	0	97
4	135	203	41	254
6	188	282	95	387
8	155	233	132	375
10	110	165	109	284
12	65	98	77	185
14	33	50	46	106
16	10	15	23	48
18	0	0	7	17
20			0	10

The composite storm hydrograph is plotted in Exhibit 8.

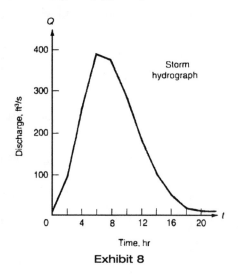

Exhibit 8

PEAK DISCHARGE ESTIMATION

Hydrographs convey a multitude of information. The volume of runoff over a time period is useful in water supply, flood control, and reservoir and detention basin studies. In other studies it is the peak discharge rate that is important—for example, in selecting pipe or culvert sizes or channel dimensions—and the other additional information is not needed.

The rational method is the most widely used method for the estimation of peak discharge Q_p (ft³/s) from runoff over small surface areas. In using it one assumes that a spatially and temporally uniform rainfall occurs for a time period that allows the entire catchment area to contribute simultaneously to the outflow. Clearly the satisfaction of these limitations becomes more difficult as the basin size increases, so this equation is normally limited to basins that are below 1 square mile (640 acres) in size. The equation is

$$Q_p = CiA \tag{12.6}$$

in which C is a nondimensional runoff coefficient that indicates the fraction of the incident rain that runs off the surface, i is the appropriate storm intensity (in./hr), and A is the watershed area (acres). Some add a dimensional conversion factor to this equation, but since 1 ft^3/s = 1.008 acre-in./hr, the conversion factor is usually ignored, as the other factors in the equation are not known with such accuracy. Table 12.1, adapted from Reference 1, gives reasonable ranges for C for various surfaces, as well as some guidance in selecting a value in the range.

Table 12.1 Runoff coefficients, C

Description of Area	Runoff Coefficients
Business	
Downtown	0.70 to 0.95
Neighborhood	0.50 to 0.70
Residential	
Single-family	0.30 to 0.50
Multiunits, detached	0.40 to 0.60
Multiunits, attached	0.60 to 0.75
Residential (suburban)	0.25 to 0.40
Apartment	0.50 to 0.70
Industrial	
Light	0.50 to 0.80
Heavy	0.60 to 0.90
Parks, cemeteries	0.10 to 0.25
Playgrounds	0.20 to 0.35
Railroad yard	0.20 to 0.35
Unimproved	0.10 to 0.30

It often is desirable to develop a composite runoff coefficient based on the percentage of different types of surface in the drainage area. This procedure often is applied to typical "sample" blocks as a guide to selection of reasonable values of the coefficient for an entire area. Coefficients with respect to surface type currently in use are:

Character of Surface	Runoff Coefficients
Pavement	
Asphaltic and concrete	0.70 to 0.95
Brick	0.70 to 0.85
Roofs	0.75 to 0.95
Lawns, sandy soil	
Flat, 2%	0.05 to 0.10
Average, 2 to 7%	0.10 to 0.15
Steep, 7%	0.15 to 0.20
Lawns, heavy soil	
Flat, 2%	0.13 to 0.17
Average, 2 to 7%	0.18 to 0.22
Steep, 7%	0.25 to 0.35

The coefficients in these two tabulations are applicable for storms of five- to ten-year frequencies. Less frequent, higher-intensity storms require the use of higher coefficients because infiltration and other losses have a proportionally smaller effect on runoff. The coefficients are based on the assumption that the design storm does not occur when the ground surface is frozen.

Source: *Design and Construction of Sanitary and Storm Sewers*, Manual No. 37, 1986; reproduced by permission of ASCE.

The intensity factor must also be chosen carefully. It is normally defined as the intensity of rainfall of a chosen frequency that lasts for a duration equal to the time of concentration t_c for the basin. Sometimes the frequency will be dictated by policy (one-year, five-year, or ten-year). After the frequency has been chosen, one usually consults an intensity-duration-frequency (IDF) plot to obtain i once the time of concentration has been picked. Conceptually this time is the time required for flow from the most remote point in the basin to reach the outlet, but in some cases it is simply estimated to be in the 5- to 15-minute range. Picking a shorter time usually leads to a higher-intensity i and a larger Q_p; in one sense this is conservative, but it may also be wasteful by causing one to design for an excessively large flow. The IDF plot, if developed properly, reports information that is the result of long-term statistical averages of many individual storms, not just the result of a compilation of relatively few data.

When the basin surface is not homogeneous, one should either subdivide the basin into smaller regions that are (nearly) uniform or compute a weighted average value for C, the weights being the areas.

Several other approaches to the estimation of peak discharge exist, the SCS methods being among the most prominent. If one wants to apply these methods properly, however, a lengthy description of the method and the supporting data and charts are required. One should consult the references at the end of this chapter for an adequate description of the procedures.

Example 12.3

A storm drain is to be extended to serve two developing areas in a suburb. Exhibit 9 presents the intensity-duration-frequency plot for this region as well as a schematic diagram of the developments. Area A consists of 40 acres of mostly single-family residential units, with some multiple-family units; the time of concentration is 15 min. Area B drains to inlet 2 and contains several small businesses. The transit time for storm water to move from inlet 1 to inlet 2 is $T = 5$ min. Assuming a five-year return period, estimate the peak discharges expected at the two inlets.

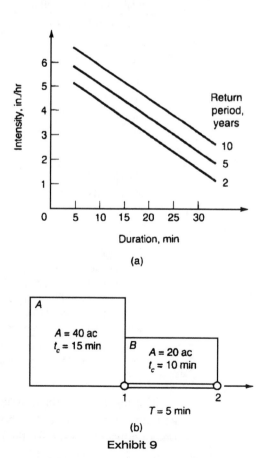

Exhibit 9

Solution

For area A one assumes a 15-min duration and finds $i = 4.50$ in./hr for a 5-yr return period from Exhibit 9(a). Referring to Table 12.1, it appears that $C = 0.45$ is reasonable for this residential area. For point 1 the peak discharge should be about

$$Q_p = CiA = (0.45)(4.50)(40) = 81 \text{ ft}^3/\text{s}$$

This peak is expected to appear at the second inlet location at $15 + 5 = 20$ min after the storm begins.

If area B is considered separately, then a 10-min duration leads, via Exhibit 9(a), to $i = 5.17$ in./hr, the runoff coefficient may be nearly $C = 0.70$, and

$$Q_p = (0.70)(5.17)(20) = 72 \text{ ft}^3/\text{s}$$

at inlet 2 from area B. However, the two computed peak discharges do not both arrive at point 2 at the same instant. The peak flow from B arrives 10 min before the flow from A arrives.

To compensate for the fact that the two peak discharges do not coincide in time, the usual approximate procedure is to use an area-weighted coefficient C_w and a time of concentration that applies to the combination of the areas. Here the time of concentration is 20 min. Thus

$$C_w = \Sigma C_i A_i / \Sigma A_i = [0.45(40) + 0.70(20)]/[40 + 20] = 0.53$$

For the 5-yr return period and a 20-min duration Exhibit 9(a) gives $i = 3.83$ in./hr and

$$Q_p = (0.53)(3.83)(60) = 122 \text{ ft}^3/\text{s}$$

which is lower by some 30 ft³/s than the sum of the individual peak flows.

HYDROLOGIC ROUTING

Routing methods track water masses as a function of time as they course through streams, rivers, and reservoirs. Hydrologic routing is based on conservation of mass, supplemented by a relation between storage and discharge; it is an incomplete, approximate computation since it ignores momentum considerations, but it is often used because it can produce sufficiently accurate results with far less computational effort than is required in hydraulic routing, which does include the momentum equation. In this section the hydrologic routing of flows through reservoirs and rivers will be reviewed.

When the inflow hydrograph to either a reservoir or river reach is compared with the subsequent outflow hydrograph at the other end, two characteristic features are normally present: (1) the peak discharge of the inflow is attenuated—that is, reduced—in the outflow, and (2) the peak outflow occurs later than—that is, lags—the peak inflow. The difference between inflow I and outflow Q at any instant is equal to the rate of change of the storage S of water in the region between the inflow and outflow stations, or

$$I - Q = \frac{dS}{dt} \tag{12.7}$$

Usually this equation is integrated between two time instants t_n and t_{n+1} and the trapezoidal rule is applied over the interval $\Delta t = t_{n+1} - t_n$ to obtain

$$(I_{n+1} + I_n)\frac{\Delta t}{2} - (Q_{n+1} + Q_n)\frac{\Delta t}{2} = S_{n+1} - S_n \tag{12.8}$$

The typical routing problem begins with an inflow hydrograph given (a set of values I_n, $n = 1, N$). The value of the initial outflow must also be known. The remaining two unknowns in the equation are Q_{n+1} and S_{n+1}. Once the relation between storage and outflow is specified, the new outflow can be computed, and the computation can progress to the next time increment. This storage relation differs, however, depending on the application.

Reservoir Routing

Reservoir outflow either is controlled by gages and/or valves or is not controlled, owing to their absence. In uncontrolled reservoirs the storage relation is of the form $S = f(Q)$ when the reservoir water surface has no slope, as in short or deep reservoirs, or $S = f(Q, I)$ when the surface does slope, as in shallow reservoirs. For controlled reservoirs the storage representation may again be of either type, with the added problem that a separate storage relation must be determined for each combination of gate/valve settings. When $S = f(Q, I)$ the routing method is similar to river routing.

The storage indication, or Puls, method of hydrologic routing is commonly applied to reservoirs. When storage is assumed to be a function only of outflow, the method uses the following steps:

- Equation (12.8) is rearranged to give

$$I_n + I_{n+1} + \left(\frac{2}{\Delta t} S_n + Q_n\right) = \left(\frac{2}{\Delta t} S_{n+1} + Q_{n+1}\right) \quad (12.9)$$

- From whatever data are given, a table or graph of $(2S/\Delta t + Q)$ versus Q is prepared; it is called a storage indication curve.

- The storage indication curve and inflow data are used in applying Equation (12.9) sequentially over time increments until the outflow has been computed as a function of time.

The Puls method is applied in Example 12.4.

Example 12.4

Some elevation-discharge and elevation-area data for a small reservoir with an ungated spillway are given below. An inflow sequence to the reservoir for part of a flood is given in a second table.

Elev., ft	0	1	2	3	4	5	6
Area, acres	1000	1020	1040	1050	1060	1080	1100
Outflow, ft³/s	0	525	1490	2730	4200	5880	7660

Date	Hour	Inflow, ft³/s
4/23	12 PM	1500
4/24	12 AM	1600
	12 PM	3100
4/25	12 AM	9600

Determine by routing the outflow discharge and reservoir water surface elevation at 12 AM on 25 April. Arrange the computations in a tabular form. Use a 12-hour routing period, and assume that the reservoir water level just reaches the spillway crest (elevation 0.0) at 12 PM on 23 April.

Solution

Since the outflow Q is given directly as a function of elevation, the first task is to determine the reservoir storage S as a function of elevation also. The given areas are the surface areas of the reservoir water surface; integrating these areas over the incremental elevation changes produces the incremental changes in storage. This computation will be tabulated along with the compilation of data points for the storage indication curve. Elevation values will also be used as the index n in the equations. The equations used in computing the table entries are

$$\bar{A} = \frac{1}{2}(A_n + A_{n+1}) \quad \Delta S = \bar{A}\Delta h \quad S_{AF} = \sum \Delta S$$

$$\frac{2}{\Delta t} S + Q = \frac{2 S_{AF}(43,560)}{12(60^2)} + Q = 2.02 S_{AF} + Q, \text{ft}^3/\text{s}$$

Elev., n, ft	Area A, acres	Avg. Area, $\bar{A}$, acres	S_{AF}, acre-ft	$\frac{2}{\Delta t} S + Q$, ft^3/s
0	1000		0	0
		1010		
1	1020		1010	2560
		1030		
2	1040		2040	5600
		1045		
3	1050		3085	8950
		1055		
4	1060		4140	12,550
		1070		
5	1080		5210	16,400
		1090		
6	1100		6300	20,400

The resulting storage indication curve is plotted in Exhibit 10.

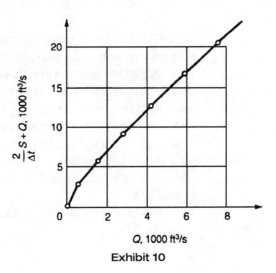

Exhibit 10

Now Equation (12.9) can be applied sequentially in the next table, with all flows in ft³/s:

n (a)	Time (b)	I (c)	$\dfrac{2}{\Delta t}S - Q$ (d)	$\dfrac{2}{\Delta t}S + Q$ (e)	Q (f)
1	4/23 12 PM	1500	0	—	0
2	4/24 12 AM	1600	1700	3100	700
3	4/24 12 PM	3100	2800	6400	1800
4	4/25 12 AM	9600		15,500	5500

All inflows were given data. Also $Q_1 = 0$ was given. Thus the value $(2S/\Delta t - Q)_1$ can be computed to be zero. Now all terms on the left side of Equation (12.9) are known for $n = 1$, and this equation gives $2S/\Delta t + Q = 3100$ for $n + 1 = 2$ in column (e). Entering the storage indication curve, Exhibit 10, with this value gives $Q = 700$ ft³/s [$n = 2$, column (f)]. Since $(2S/\Delta t + Q) - 2Q = 2S/\Delta t - Q$, column (d) with $n = 2$ is $3100 - 2(700) = 1700$. Applying Equation (12.9) with $n = 2$ then yields $1600 + 3100 + 1700 = 6400$ in column (e) for $n = 3$. And these operations are cyclically repeated until the solution is completed. Thus the outflow from the reservoir at 12 AM, 25 April, is $Q = 5500$ ft³/s. Using this discharge and the outflow-discharge data, the water surface elevation E at that time is, using interpolation,

$$E = 4.00 + \left(\frac{5500 - 4200}{5880 - 4200}\right) \times 1.00 = 4.77 \text{ ft}$$

above the spillway crest.

River Routing

All forms of hydrologic river routing begin with the assumption of some relation between storage in the river section and the inflow and outflow at the ends of the section. The most common of these methods is the Muskingum method, which assumes that this relation is a weighted linear relation between storage, inflow, and outflow taking the form

$$S = K[xI + (1 - x)Q] \quad (12.10)$$

in which K is a proportionality factor with units of time, and x is the weighting factor giving the relative importance of the inflow and outflow contributions to storage. For example, for a simple reservoir one expects $S = f(Q)$ only so $x = 0$ could be chosen; if inflow and outflow are of equal importance, then $x = 0.5$ should be selected. For most streams x is between 0.2 and 0.3. The parameters K and x can be determined for a specific routing application if suitable data are available so that $[xI + (1 - x)Q]$ can be plotted versus storage S for several values of x between 0 and 0.5. The value of x that most nearly collapses the plotted data onto a single fitted straight line is used in the routing application, and $1/K$ is the slope of that fitted line.

The final form of the Muskingum routing equation is

$$Q_{n+1} = C_0 I_{n+1} + C_1 I_n + C_2 Q_n \quad (12.11)$$

in which

$$C_0 = (\Delta t/2 - Kx)/D \quad (12.12)$$

$$C_1 = (\Delta t/2 + Kx)/D \quad (12.13)$$

$$C_2 = (K - Kx - \Delta t/2)/D \quad (12.14)$$

and

$$D = K - Kx + \Delta t/2 \quad (12.15)$$

Observe that one must always have $C_0 + C_1 + C_2 = 1$. These equations can be derived by using Equation (12.10) to express S_n and S_{n+1}, inserting the results in Equation (12.9), and rearranging the terms.

WATER DEMAND

Estimating water demand is generally a process of reviewing historical water use trends for the geographic region of interest. Most states and municipal water plants publish water use data on a periodic basis, which is a function of climate, economics, type of users, as well as conservation and management practices. In the United States, the greatest quantities of water are used for thermoelectric power generation and irrigation. However, for this review, we will focus primarily on the domestic potable water supply. The equations used for population forecasting in this section are found in the environmental engineering section of the *FE Supplied-Reference Handbook*.

Often, water use data is obtained from national, regional, or local averages, or is based upon house size or lot size. In the absence of more specific data, the national average per capita water use is generally considered to be 180 gpcd (gallons per capita day, or gallons per person per day). Estimating demand and planning treatment and distribution systems becomes a simple task of multiplying average per capita use by the population served. This quantity is usually called the *average annual* demand.

Other demands may be more appropriate for the design of water distribution networks and in the design of water treatment plant process units. Water use varies throughout the day, as the commonality of human schedules increase demand during meal times and in the early evening, while demand is low during sleeping hours. It is useful to define the *peak hourly* flow to correspond with the highest demand for one hour in an average year. Human schedules are also dependent on day of the week, or season of the year. These fluctuations are described by the *maximum daily* flow, which is assumed to be the highest daily flow in an average year and is used to design capacities for water treatment plant process units.

To estimate these fluctuations in demand, flows may be calculated as a percentage of average annual flow. This is often done through the use of nomographs or scaling equations, such as

$$Q_{\text{max daily}} = 1.8 \, Q_{\text{avg annual}}$$

$$Q_{\text{peak hourly}} = 1.5 \, Q_{\text{max daily}}$$

Specific events that have high demands must also be accounted for, primarily demands required to fight fires, called *fire flow*. Two common equations used to estimate fire flow are

$$Q = 1020 P^{1/2}(1 - 0.01 P^{1/2})$$

$$Q = 18 C A^{1/2}$$

where Q is the fire-flow requirement in gallons per minute (gpm), P is the population in thousands (i.e., population/1000), C is a structure coefficient (1.5 for timber frame, 1 for ordinary, 0.8 for noncombustible, and 0.6 for fire resistant), and A is the total floor area excluding the basement (ft^2). The *combined draft* is defined as the maximum daily flow plus the fire flow. The design of distribution networks are usually based on the greater of the peak hourly or the combined draft. While per capita water use trends change slightly with time, it is more likely that population growth will have a more profound effect on total water demand.

Population Forecasting

Population forecasting is not an exact science. While the census provides population data for every municipality, past growth is not a perfect indicator of future growth. One method of population forecasting is the *graphical comparison* method, where cities similar in geography, industry, and culture, but larger in population are compared to the target city. It is assumed that the target city will grow in a similar manner to the comparison cities when those cities had similar populations. An example of the graphical comparison is presented in Figure 12.2.

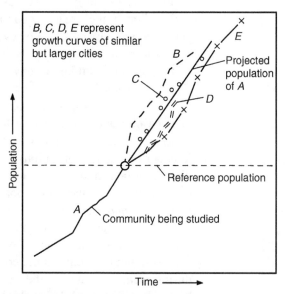

Figure 12.2 Graphical comparison for population growth estimation (Viessman, Jr., Warren, and Hammer, Mark J., *Water Supply and Pollution Control*, 7th Edition, © 2005. Reprinted by permission of Pearson Education, Inc., Upper Saddle River, NJ.)

Another population forecasting technique is based upon the assumption that most growth patterns are natural and, therefore, follow a similar trend. Since graphical representation of the data exhibits an S-shaped curve, as seen in Figure 12.3, mathematical models can be assigned to the three portions of the curve. The shape of the curve assumes that a given geographic area can support a limited number of people, known as the *saturation population*, and that this value must be determined through comparison with similar, but larger cities.

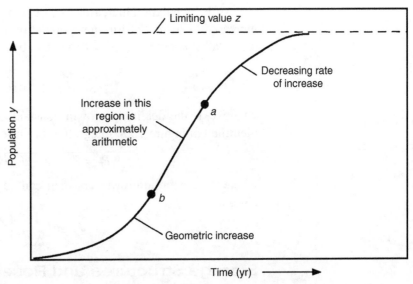

Figure 12.3 Graphical representation of population growth (Viessman, Jr., Warren, and Hammer, Mark J., *Water Supply and Pollution Control*, 7th Edition, © 2005. Reprinted by permission of Pearson Education, Inc., Upper Saddle River, NJ.)

At low population levels (less than 20 percent of the saturation population), relatively unlimited resources encourage *exponential growth*, also called *geometric* or *constant percent growth*. Mathematically, this may be expressed as

$$\frac{dP}{dt} = k_g t$$

where P is the population, k_g is the geometric growth constant, and t is time. This expression is integrated to determine the population at time $T(P_T)$ as a function of initial population (P_0) as

$$P_T = P_0 \exp[k_g \Delta t]$$

where Δt is the time interval. The geometric growth constant is determined from data of similar cities using the following expression:

$$k_g = \frac{\ln(P_2/P_1)}{\Delta t} = \frac{\ln P_2 - \ln P_1}{t_2 - t_1}$$

where P_2 and P_1 are populations of similar cities evaluated over time period Δt.

At moderate population levels (between 20 and 80 percent of the saturation population), the S-shaped curve is assumed to be *linear*, and the *arithmetic (algebraic) growth* model can be expressed as

$$\frac{dP}{dt} = k_a$$

where k_a is the arithmetic growth constant. Integration of this expression leads to

$$P_T = P_0 + k_a \Delta t$$

where the algebraic growth constant can be determined from data of similar cities as

$$k_a = \frac{\Delta P}{\Delta t} = \frac{P_2 - P_1}{t_2 - t_1}$$

where ΔP is the change in population of the similar cities over a period of Δt.

As the population increases to a value near the saturation population level, the increase slows, as the limiting population is approached asymptotically. This is often called *decreasing growth*, and can be expressed as

$$\frac{dP}{dt} = k_d(Z - P)$$

where k_d is the declining growth constant and Z is the saturation population as identified in Figure 12.3. Integration of this expression leads to

$$P_T = Z - (Z - P_0)\exp[-k_d \Delta t]$$

where the algebraic growth constant can be determined from the data of similar cities as

$$k_d = -\frac{1}{\Delta t}\ln\left(\frac{Z - P_2}{Z - P_1}\right)$$

Storage Structures and Reservoirs

When identifying water resources for an area, it is possible that the calculated demand is less than the available resources at all times, and withdrawals from the resource are taken as needed. It is also possible that the identified resources will never meet the demand; in which case, additional resources must be sought to supplement the identified sources. It is sometimes the case that total water available over a period of time, usually evaluated annually, is sufficient to meet the demand; however, the rates of supply are seasonal and may not meet the demand in the dry months.

In the final case above, it may be possible to construct structures that are designed to retain water during periods of high flow, which are later available for use during low flow periods. The capacity of that structure is called *storage*, and the structure is called a *reservoir*. Further, because water treatment plants are designed to meet maximum daily demand, additional storage may be required in the pipe network to meet short-term high demands, such as fire demands. In such circumstances, enclosed reservoirs, such as clear wells and water towers, are designed and constructed to meet this storage requirement.

Required storage capacity is determined through the evaluation of cumulative inflows and outflows of a reservoir over the course of several years if the data is available. This is usually accomplished through graphical techniques or through the use of computer spreadsheets. It should be noted that a few years of data rarely represent best- or worst-case scenarios, and often the amount of storage available in a good design will exceed the estimate provided by a limited data set. Wherever possible, extended historical flow data should be sought and used for reservoir design.

To use the *graphical technique*, a plot is made of cumulative inflow and cumulative outflow on the same set of axes. Assuming draft as a constant rate, the cumulative outflow curve will be a straight line. To determine required storage, a line, parallel to the draft curve, is drawn from any peak in inlet flow to a later point in time where it intersects the inlet flow curve, as seen in Figure 12.4. The maximum vertical distance between the cumulative inflow curve and any of the parallel lines is the required storage.

Two spreadsheet methods both calculate cumulative inflows and outflows and the disparity between the two for each date. In the first method, the difference is calculated as inflow minus outflow, which are summed over the time period to determine the cumulative difference. The required storage is determined by finding the maximum value from the cumulative difference column and subtracting the subsequent minimum value. This is called the *difference technique*. In the second method, the deficiency is calculated as outflow minus inflow. In the cumulative deficiency column, any negative values are set equal to zero, and the summation is continued. The required storage is taken directly from the maximum value in the cumulative deficiency column. This is called the *deficiency technique*.

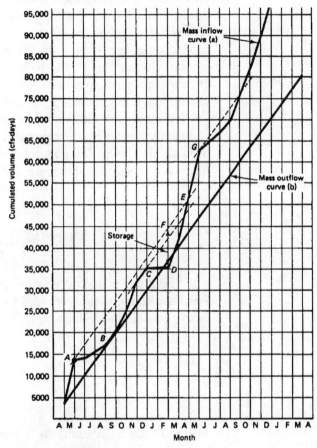

Figure 12.4 Graphical technique for estimation of storage requirement (Reprinted by permission of Waveland Press, Inc. from Ram Gupta, *Hydrology and Hydraulic Systems*, 2nd Edition (Long Grove, IL: Waveland Press, Inc., 2001). All rights reserved.)

WATER DISTRIBUTION NETWORKS

Water distribution networks are the piping systems that provide water service from a municipal water treatment plant. Most water distribution networks pump at a constant rate through *primary lines,* or *mains,* to storage tanks strategically placed throughout the service area. Mains are spaced approximately every kilometer, or 0.6 miles, through the distribution area, often in large looping networks. *Secondary lines,* or *sub-mains,* are slightly smaller in diameter and usually run through the center of each loop created by the mains. *Distribution mains* are smaller than the sub-mains and generally run down each street in the service area, looping when possible, but often ending at a terminus of service, also called a *branched* distribution main. Lines that run to each customer are called *service connections* and originate at a distribution main.

Often in the design of a water distribution network, demand is first calculated as discussed earlier in this review. The mains, sub-mains, and distribution mains are represented on a distribution grid, and demands from the service connections are divided between the nodes that are created at the intersection of each line in the distribution grid. In this way, the entire system demand is distributed throughout the service area. The design capacity for the system should be based on the higher of the (1) peak hourly flow, or (2) combined draft.

The evaluation of a distribution network is necessary to determine if the lines are capable of meeting the highest demands while maintaining a minimum pressure throughout the system, which is usually set in the 40- to 50-psi range. Maximum pressure throughout the system should not exceed 100 psi in order to maintain connection integrity and to minimize loss of water through imperfect fittings, and pressure fluctuations should be kept at a minimum. An overall system evaluation at current demands may be able to identify specific lines that have high head loss values, suggesting the possible need for upgrade to a larger diameter line.

The most common evaluation technique is the single-path adjustment method, commonly known by the originator's name as the *Hardy Cross method*. In this method, the Hazen-Williams equation is rearranged to express the head loss as a function of volumetric flow rate, as follows:

$$h_f = \frac{10.7L}{C^{1.85} D^{4.87}} Q^{1.85} \quad \text{(SI units)}$$

$$h_f = \frac{4.73L}{C^{1.85} D^{4.87}} Q^{1.85} \quad \text{(English units)}$$

with SI units of m and m³/s, and English units of ft and ft³/s. Similarly, the Darcy-Weisbach equation is rearranged to express the head loss as a function of volumetric flow rate, as follows:

$$h_f = \frac{fL}{12.1D^5} Q^2 \quad \text{(SI units)}$$

$$h_f = \frac{fL}{39.7D^5} Q^2 \quad \text{(English units)}$$

where SI and English units are as stated previously. Each of these expressions can be expressed in the general form

$$h_f = KQ^n$$

where n has a value of 1.85 for the Hazen-Williams equation, and a value of 2 for the Darcy-Weisbach equation. Further, we know that the sum of all head losses around any closed loop must be equal to zero; therefore, we may write

$$\Sigma h_f = 0$$

In order to obtain the correct volumetric flow rate in each pipe in the loop, the EOC requires that any correction in flow rate must be applied to all pipes in that loop. If δ is the volumetric flow-rate correction applied to an initial, assumed value Q_i for each pipe in the loop, then we may write

$$\Sigma K(Q_i + \delta)^n = 0$$

Finally, we may now calculate a value for δ for each loop in a larger network as

$$\delta = -\frac{\Sigma h_f}{n \Sigma \left| \frac{h_f}{Q_i} \right|}$$

This correction must be applied to every pipe in the loop, and pipes that are common to more than one loop must have the correction from both loops applied.

WASTEWATER FLOWS

Wastewater flows are usually estimated using water use data and an assumed *return rate*, which is defined as *the percentage of water use directed toward the sanitary sewer*. Return rates are dependent on the types of human activities utilizing water and the relative proportion of domestic to industrial uses, but they are often higher in urban areas (~90%–95%) than in suburban (~70%–85%) or rural (~60%–80%) areas. The most common units used in water and wastewater system design are gallons per day (gpd), or for large systems million gallons per day (Mgal/day), often represented as mgd.

Wastewater flow estimation for new or replacement systems should use projected demand based on population forecasting. As in water demand, consideration must be made for the variation in flow that occurs throughout the day due to human schedules. This is often accomplished by estimating the maximum hourly flow (Q_{max}) and minimum hourly flow (Q_{min}) using scaling factors derived from equations or graphical data, as provided in Figure 12.5.

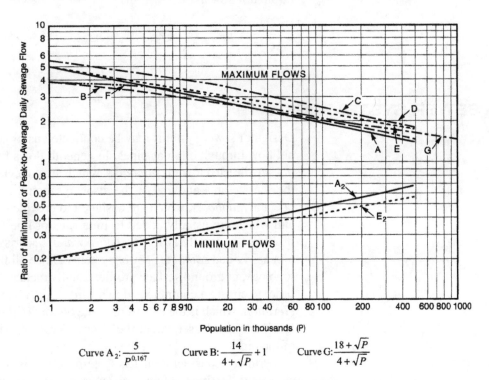

Curve A_2: $\dfrac{5}{P^{0.167}}$ Curve B: $\dfrac{14}{4+\sqrt{P}}+1$ Curve G: $\dfrac{18+\sqrt{P}}{4+\sqrt{P}}$

Figure 12.5 Sewer flow ratio curves

Source: *Design and Construction of Sanitary and Storm Sewers*, 1970, American Society of Civil Engineers. Reprinted by permission.

In addition to direct flows, consideration must be made for extraneous waters that end up in the sanitary sewer line. This contribution is often called *infiltration and inflow (I&I)* and is due to groundwater entering through breeches in the pipe or stormwater runoff directed (often illegally) to the sanitary sewer from sump pumps and roof gutter downspouts. The amount of I&I is a function of infiltration waters present, age and diameter of the sewer line, and amount of inflow, for which data is rarely available. In the absence of any other data, the general practice is to assign a value of 30,000 gpd/mi (gallons per day per mile of sewer line), which is multiplied by the total length of sewer line in the service area.

Example 12.5

Minimum/Maximum sewage flows

A city with an average sanitary sewage flow of 11.6 mgd has a maximum flow of 23.2 mgd. Determine the minimum anticipated flow. You may assume that the variations are described by the A and A_2 sewage flow ratio curves.

Solution

We start by finding the ratio of maximum to average flow as follows:

$$\frac{Q_{max}}{Q_{ave}} = \frac{23.2}{11.6} = 2$$

From curve A on the sewage flow ratio curves (Figure 12.5), a maximum-to-average ratio of 2 occurs at a population of 80,000. Using curve A_2 at this population gives a minimum-to-average ratio of 0.47. This allows us to estimate the minimum flow as:

$$\frac{Q_{min}}{Q_{ave}} = 0.47 \quad \Rightarrow \quad Q_{min} = (0.47)Q_{ave} = (0.47)\left(11.6 \cdot 10^6 \text{ mgd}\right) = 5.45 \cdot 10^6 \text{ mgd}$$

SEWER SYSTEM DESIGN

Design of sewer systems may be conducted using the nomographs for open channel flow through circular pipes. In general, flows from smaller lines, called *laterals*, are directed toward *submains*, which are collected in *mains* that run throughout the service area and end at the discharge point. All intersections for sewer lines occur at *manholes*, which are designed to provide service access to the sewer lines and which are usually added to long runs of line for maintenance purposes. The discharge is usually a river for *stormwater sewer* systems or the WWTP for *sanitary sewer* and *combined sewer* systems. Since all flow is gravity induced, the area topology, or ground surface profile, usually dictates layout and direction of flow. In order to ensure a continuous downward slope, it is necessary to evaluate the *invert elevation*, which is the elevation of the bottom of the pipe usually referenced to sea level, and to a lesser extent the *crown elevation*, which is the elevation of the top of the pipe.

Primary consideration for system design is to determine pipe diameter and slope at the estimated flows, given constraints for minimum and maximum velocity. Minimum velocities (V_{min}) are generally established at 2 ft/s and are necessary to ensure self-cleansing (that is, conveying all solids with the liquid flow). Maximum velocities (V_{max}) vary depending on pipe material but range from 10 ft/s to 20 ft/s to maintain pipe integrity by minimizing scour and pipe erosion. Another consideration is minimum slope, which varies depending on pipe diameter and anticipated depth of flow. In general, pipes 8 to 24 inches in diameter require minimum slopes of 0.4–0.1%, while larger pipes may have minimum slopes as small as 0.05% (0.5 ft per 1000 feet of pipe).

Since Q_{max} and Q_{min} can vary by as much as an order of magnitude, standard practice is to design for Q_{max} plus *I&I* near V_{max}, then check the velocity at Q_{min} plus *I&I* to ensure compliance with V_{min}. In order to provide sufficient capacity for extreme flow events and for the potential of population growth in the service area, standard practice for sanitary sewers is to design for Q_{max} plus *I&I* at 50% depth of flow for lateral lines and at 75% depth of flow for main lines. In general, this level of overdesign will still provide a sewer system that is capable of meeting V_{min} requirements.

The other design constraint is *depth of cover*, which is the amount of soil over the sewer line and is generally a design driving force due to the cost of trench excavation. Minimum depths are established to prevent freezing, while maximum depths are usually a function of the ability to excavate the trench, usually 25–35 feet. In flat areas, it is desired to keep the slope as small as possible to avoid *lift stations*, which are pump stations in a manhole designed to add energy to the water by pumping from the maximum depth of cover to the minimum depth of cover, where gravity flow is allowed to continue. Although highly discouraged in common practice, in high slope areas a *drop manhole* may be used to dissipate energy. This is required where pipe slopes required to maintain a minimum depth of cover would be excessive and would cause extremely high velocities.

The design of a large system generally begins with a layout determined from topographic maps of the intended service area. Lateral lines are then designed with the upstream (inlet) side at the minimum ground cover. Downstream (outlet) invert elevation is determined by multiplying section length by installed slope, and depth of cover is determined by accounting for any change in surface elevation. Once all laterals for a given submain are designed, the submain is designed based on the lateral with the lowest invert elevation. Once the depth of the submain is determined, laterals are redesigned with the slope required by their outlet connection to the submain. This process is repeated for the main/submain connections. Sewer system design in this manner is labor intensive and cumbersome. Thankfully, new computer software has greatly enhanced the design process, and entire cities can be evaluated with relative ease.

Example 12.6

Sewer pipe design

Determine the diameter and slope of a sewer line flowing at 50% depth at maximum daily flow for a population of 40,000 with a per capita flow of 140 gpcd. You may assume a velocity at 50% depth of 8.2 fps, I&I is negligible, n is 0.013 and varies with depth, and that curves A and A_2 on the sewage flow ratio curves apply.

Solution

We start by calculating the average flow rate as follows:

$$Q_{ave} = (40,000 \text{ people})\left(140 \frac{\text{gal}}{\text{person} \cdot \text{day}}\right) = 5.6 \frac{\text{Mgal}}{\text{day}} \times \frac{1.547 \text{ ft}^3/\text{s}}{\text{Mgal/day}} = 8.7 \frac{\text{ft}^3}{\text{s}}$$

Using curve A on the sewage flow ratio curves for a population of 40,000, $Q_{max}/Q_{ave} = 2.3$; therefore, Q_{max} may now be determined as

$$Q_{max} = (2.3)(8.7) = 20.0 \text{ cfs}$$

Now we can determine the full pipe flow rate by using the partial flow nomograph, or plot of hydraulic elements for circular cross sections. From the nomograph at 50% depth of flow with variable n we see the following:

$$\frac{Q}{Q_f} = 0.4 \text{ @ 50\% depth of flow}$$

Therefore, the full pipe flow rate is
$$Q_f = \frac{Q_{max}}{0.4} = \frac{20.0}{0.4} = 50.0 \frac{\text{ft}^3}{\text{s}}$$

Also, from the partial flow nomograph at a depth of 50%, we can determine the full pipe velocity as follows:
$$\frac{V}{V_f} = 0.8 \Rightarrow V_f = \frac{V}{0.8} = \frac{8.2 \text{ fps}}{0.8} = 10.25 \text{ fps}$$

Finally, with full pipe flow rate and velocity, we can size the pipe as follows:
$$Q = VA = V\frac{\pi}{4}D^2 \Rightarrow D = \left[\frac{4Q}{\pi V}\right]^{1/2} = \left[\frac{(4)\left(50.0 \frac{\text{ft}^3}{\text{s}}\right)}{(\pi)\left(10.25 \frac{\text{ft}}{\text{s}}\right)}\right]^{1/2}$$
$$= 2.49 \text{ ft} \times \frac{12 \text{ in}}{\text{ft}} = 29.9 \text{ in}$$

Note that the final diameter will need to be increased to the next common pipe size, in this case, 30 inches. Now, with a known value for n and V for a full pipe, we can write the Manning's equation as
$$Q = \frac{1.486}{n} A R_H^{2/3} S^{1/2} \Rightarrow \frac{Q}{A} = \frac{1.486}{0.013} R_H^{2/3} S^{1/2} = 10.25 \frac{\text{ft}}{\text{s}}$$

The hydraulic radius for a circular pipe flowing full is $D/4$, or equivalently,
$$R_H = \frac{D}{4} = \frac{30 \text{ in}}{4} = 7.5 \text{ in} \times \frac{1 \text{ ft}}{12 \text{ in}} = 0.625 \text{ ft}$$

Substituting this value into the expression above yields
$$S^{1/2} = \frac{\left(10.25 \frac{\text{ft}}{\text{s}}\right)(0.013)}{(1.486)(0.625 \text{ ft})^{2/3}} = 0.123 \Rightarrow S = 0.015 = 1.5\%$$

Example 12.7

Sewer pipe depth of cover

Pipe A is 18-in diameter and enters manhole 1 with a crown elevation of 672 ft. Pipe B is 12-in diameter and enters manhole 1 with a crown elevation of 674 ft. The exiting pipe (Pipe C) has a diameter of 24 in, a slope of 0.2%, and is 1000 ft from manhole 2. Assuming the exiting pipe crown must be at or below the inlet invert to keep from restricting flow, determine the minimum depth of cover for the exit pipe at manhole 2 if the ground elevation there is 680 ft.

Solution

Assuming the outlet pipe crown is at the elevation of the lowest inlet pipe invert, the exiting pipe crown elevation can be determined as
$$\text{elevation} = 672 \text{ ft} - \frac{18}{12} \text{ ft} = 670.5 \text{ ft}$$

Next, we need to determine the elevation change of that pipe over its length as follows:

$$S = \frac{\Delta y}{\Delta x} \quad \Rightarrow \quad \Delta y = S\,\Delta x = \left(0.002\,\frac{\text{ft}}{\text{ft}}\right)(1000\text{ ft}) = 2\text{ ft}$$

This elevation change is subtracted from the initial elevation of the pipe to find the crown elevation at manhole as follows:

$$\text{elevation} = 670.5 - 2 = 668.5\text{ ft}$$

Finally, cover at this location is determined by difference in crown and ground elevation as

$$\text{cover} = 680 - 668.5 = 11.5\text{ ft}$$

PROBLEMS

12.1 A small Midwestern city has a population of 37,638, with an average water use of 165 gpcd. Recent growth trends indicate exponential growth at a rate of 1.5 percent per year. If the current water treatment plant has a capacity of 7.5 mgd, what percent reduction in water use is necessary 20 years from now?
   a. 7.5 percent
   b. 10.5 percent
   c. 13.5 percent
   d. 16.5 percent

12.2 Using the cumulative demand curve provided in Exhibit 12.2, determine the storage requirement assuming a 24-hr pumping period.
   a. 107,000 ft^3
   b. 72,000 ft^3
   c. 51,000 ft^3
   d. 35,000 ft^3

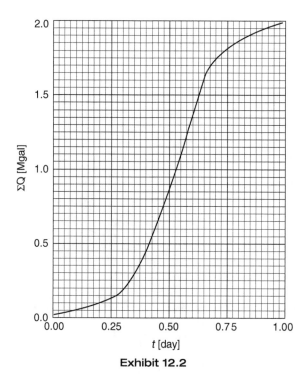

Exhibit 12.2

12.3 Estimate the maximum daily wastewater flow based on the current population and water use for the city described in Problem 12.1 if the return rate is 80 percent.
   a. 1.99 mgd
   b. 4.97 mgd
   c. 13.2 mgd
   d. 24.8 mgd

12.4 A recent storm surge from a hurricane in the Gulf of Mexico dropped 2.25 inches of rain in southeastern Ohio in 1.5 hours. Using the intensity-duration-frequency curve provided in Exhibit 12.4, determine the historic frequency of this storm.

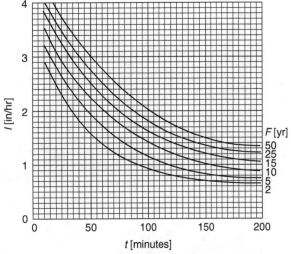

Exhibit 12.4

a. 50 yrs
b. 25 yrs
c. 15 yrs
d. 10 yrs

12.5 The runoff coefficient for a particular 50-acre industrial area is determined to be 0.8. A storm produces 0.5 inches of rain per hour. A 3-ft-diameter storm sewer transports the water from the area. Estimate the velocity of the water, in fps, if the sewer pipe flows full.

a. 2.1 fps
b. 2.8 fps
c. 3.5 fps
d. 4.2 fps

12.6 Flow enters the distribution network shown in Exhibit 12.6 at point A (1000 gpm) and point C (500 gpm). Demand occurs at point D (500 gpm) and point F (1000 gpm). Initial flows were determined by setting $Q_{AB} = 600$ gpm and $Q_{CB} = 300$ gpm. Determine the corrected flow rate in pipe BE after one iteration of the Hardy-Cross method if $\Delta Q_{ABED} = -42.3$ gpm and $\Delta Q_{BCFE} = 223$ gpm.

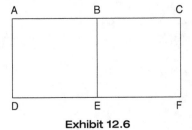

Exhibit 12.6

a. 635 gpm
b. 677 gpm
c. 782 gpm
d. 858 gpm

12.7 A flow of 1.5 cfs enters manhole 1, which has a ground elevation of 768 ft and an invert elevation of 760 ft above sea level. Manhole 2 is connected by 500 ft of 15-inch sewer line ($n = 0.013$, varies with depth) and has a ground elevation of 762 ft. If the depth of flow is 32 percent, find the depth of cover (to the pipe crown) at manhole 2.
    a.   16.5 ft
    b.   14.5 ft
    c.   12.5 ft
    d.   10.5 ft

SOLUTIONS

12.1 b. Start with the exponential population growth expression to get an estimate of the population 20 years from now if the growth rate is 1.5 percent per year (i.e., k_g 0.015):

$$P_T = P_0 e^{kt} = (37{,}638)\exp[(0.015)(20)] = 50{,}806$$

Determine the water use rate for that population assuming the plant is at its capacity:

$$\text{use} = \frac{Q}{P_T} = \frac{7.5 \times 10^6 \text{ gpd}}{50{,}806} = 147.6 \text{ gpcd}$$

Now we can determine the required reduction in water use as follows:

$$\text{reduction} = \frac{165 - 147.6}{165} = 10.5\%$$

12.2 a. Start by drawing the Q_{ave} cumulative curve as shown in Exhibit 12.2a, which is the 45° line from (0,0) to (1,2). Then draw a line tangent to the cumulative demand curve given, parallel to Q_{ave} at the maximum and minimum points on the curve.

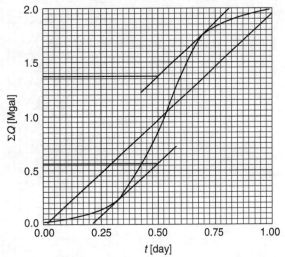

Exhibit 12.2a

Now, find Δy by drawing horizontal lines where the tangent lines cross any value of t. Note the value shown in the solution is $t = 0.5$.

$$\Delta y = \text{storage} = 1.375 - 0.575 = 0.8 \text{ Mgal}$$

$$\text{storage} = \frac{800{,}00 \text{ gal}}{7.48 \frac{\text{gal}}{\text{ft}^3}} = 106{,}952 \text{ ft}^3$$

12.3 c. First, determine the average daily flow as

$$Q = (\text{population})(\text{use rate})(\text{return rate}) = (37{,}638)(165)(0.08)$$
$$= 4{,}968{,}216 \text{ gpd}$$

Using the "Sewage Flow Ratio Curves" in the civil engineering section of the *FE Supplied-Reference Handbook*, you need to select a curve. Without other information available, it is usually best to select an average value. At a population of 38,000, curve A has a value of 2.3 and curve C has a value of 3.0. Using the average value of 2.65, we can calculate the maximum daily flow as

$$Q_{\text{max daily}} = (2.65)(4.97 \text{ mgd}) = 13.17 \text{ mgd}$$

12.4 d. First, determine the storm intensity from the data given in the problem statement as follows:

$$I = \frac{2.25 \text{ in}}{1.5 \text{ hr}} = 1.5 \frac{\text{in}}{\text{hr}}$$

Next, find the intersection of I of 1.5 in/hr and D of 90 minutes (Exhibit 12.4a).

$$F = 10 \text{ yr}$$

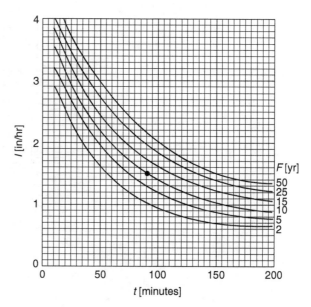

Exhibit 12.4a

12.5 b. First, determine the peak discharge from the site using the rational equation:

$$Q = CIA = (0.8)(0.5)(50) = 20 \text{ cfs}$$

Now calculate the velocity in the pipe using the following relationship:

$$Q = VA \Rightarrow V = \frac{Q}{A} = \frac{Q}{\frac{\pi}{4}D^2} = \frac{20 \text{ cfs}}{\frac{\pi}{4}(3 \text{ ft})^2} = 2.83 \text{ fps}$$

12.6 **a.** First, we need to estimate the flow in branch BE based on the original flow directions and the data given in the problem statement (Exhibit 12.6a). Since it was given that

$$Q_{AB} = 600 \text{ gpm} \quad \text{and} \quad Q_{CB} = 300 \text{ gpm}$$

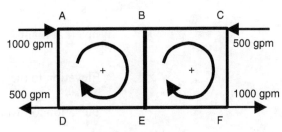

Exhibit 12.6a

and since flow from A to B is positive and flow from C to B is positive, we can assume flow from B to E is the sum of the two, or

$$Q_{BE} = 900 \text{ gpm}$$

Now, since the loop ABED correction is a negative value and since the flow direction for Q_{BE} in that loop is positive, we must add the negative value to Q_{BE}. Further, the loop BCFE correction is a positive value, and the flow direction for Q_{BE} in that loop is negative; therefore, we must subtract the positive value from Q_{BE}. The corrected value for Q_{BE} can therefore be expressed as

$$(Q_{BE})_{new} = 900 + (-42.3) - (223) = 634.7$$

12.7 **d.** We start with Manning's equation in English units, solving for slope:

$$Q = \frac{1.486}{n} AR^{2/3} S^{1/2} \quad \Rightarrow \quad S^{1/2} = \frac{Qn}{(1.486)AR^{2/3}}$$

$$\Rightarrow \quad S = \left[\frac{Qn}{(1.486)AR^{2/3}}\right]^2$$

Now, from the partial flow nomograph at a depth of 32 percent we can find each element:

$$\frac{A}{A_f} = 26\% \quad \Rightarrow \quad A = 26\% \, A_f = 0.26 \frac{\pi}{4} D^2 = (0.26)\frac{\pi}{4}\left(\frac{15}{12} \text{ ft}\right)^2 = 0.32 \text{ ft}^2$$

$$\frac{R}{R_f} = 73\% \quad \Rightarrow \quad R = 73\% \, R_f = 0.73\frac{D}{4} = (0.73)\left(\frac{15/12}{4} \text{ ft}\right) = 0.23 \text{ ft}$$

$$\frac{n}{n_f} = 126\% \quad \Rightarrow \quad n = (1.26)(0.013) = 0.0164$$

Substituting these values, along with the value for Q given in the problem statement, into the expression for S above yields

$$S = \left[\frac{(1.5)(0.0164)}{(1.486)(0.32)(0.23)^{2/3}}\right]^2 = 0.019 \frac{\text{ft}}{\text{ft}}$$

Now, elevation change can be calculated from the slope and pipe length as

$$\Delta y = SL = (0.019 \text{ ft/ft})(500 \text{ ft}) = 9.5 \text{ ft}$$

The elevation of the invert at manhole 2 can be calculated as follows:

$$\text{invert at manhole 2} = \text{invert at manhole 1} - \text{elevation change}$$
$$= 760 - 9.5 = 750.5 \text{ ft}$$

Elevation to the crown can be calculated as

$$\text{elevation to invert} + \text{pipe diameter} = 750.5 + \frac{15}{12} = 751.75 \text{ ft}$$

Finally, the depth of cover is determined by the difference between ground elevation and crown elevation as

$$\text{cover} = 762 - 751.75 = 10.25 \text{ ft}$$

CHAPTER 13

Soils and Groundwater

OUTLINE

SOIL CLASSIFICATION 543
Particle Size ■ Specific Gravity of Soil Solids, G_s ■ Weight-Volume Relationships ■ Relative Density ■ Consistency of Clayey Soils

GROUNDWATER 549
Aquifers ■ Permeability ■ Flow Nets ■ Effective Stress

CONSOLIDATION 556

GROUNDWATER FLOW 557
Well Hydraulics ■ Steady Flow (Thiem Solution) ■ Unsteady Flow (Theis and Cooper-Jacob Solutions)

GROUNDWATER CONTAMINATION 566
NAPLs in the Environment ■ Subsurface Contaminant Modeling ■ Introduction to Soil and Groundwater Remediation

PROBLEMS 574

SOLUTIONS 576

SOIL CLASSIFICATION

Primarily due to the world's reliance on groundwater resources, and our dependence on vegetative matter as a primary food source and a food source for the animals we eat, the soil contained in the near surface of the Earth's crust plays a critical role in environmental engineering. Waste management decisions must take into account the interactions between soil and water in natural systems, as well as the impact that deposition of airborne pollutants, either directly or through precipitation (e.g., rain, snow, etc.) events, has on soil quality. In the case of land disposal of waste materials, care must be taken to ensure that the stored material does not pose a threat to the subsurface environment, and therefore to the health of those who rely upon that soil for food or water.

The purpose of this section is to review the basic terminology and properties of soils to aid in the understanding of the role soils play in environmental engineering. Topics in geotechnical engineering can be found in the civil engineering section of the *FE Supplied-Reference Handbook*.

Soils are masses of particles of various sizes and compositions, packed in such a way as to provide void spaces that may be filled with gas or liquid. The particles are derived primarily from the weathering of rock, with the addition of organic compounds that are the product of the decomposition of plant or animal matter. The chemical composition of soils varies by location and soil type, but has the following major elemental constituents (listed in decreasing concentration): O, Si, Al, Fe, C, Ca, K, Na, Mg. Soil chemistry impacts the quality of the water that flows through it, and is often described by *soil pH* and *cation exchange capacity* (CEC), which is a measure of the soil affinity for, and retention of, various cations. Additionally, the potential exists for nonnative chemical species (contaminants) to enter the soil column through human (anthropogenic) activities including, agricultural use of fertilizers and pesticides, municipal services activities, and industrial emissions, discharges, and accidental spills.

Particle Size

Based on the size of the particles present, soils can be described as gravel, sand, silt, or clay. Following are two grain size classification systems generally used by geotechnical engineers:

- System of the American Association of State Highway and Transportation Officials (AASHTO)
  Gravel: 75 mm to 2 mm
  Sand: 2 mm to 0.075 mm
  Silt and clay: Less than 0.075 mm

- Unified System
  Gravel: 76.2 mm to 4.75 mm
  Sand: 4.75 mm to 0.075 mm
  Silt and clay: Less than 0.075 mm

The particle size distribution in a given soil is determined in the laboratory by sieve analysis and hydrometer analysis.

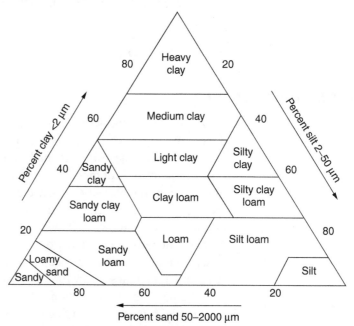

Figure 13.1 Triangular representation of soil texture classifications (Kiely, Gerard, *Environmental Engineering*, © 1996, McGraw-Hill. Reproduced with permission of the McGraw-Hill Companies.)

A typical particle size distribution curve is shown in Figure 13.2. For classification purposes in coarse-grained soils, the following two parameters can be obtained from a particle size distribution curve:

- Uniformity coefficient:

$$C_c = \frac{D_{60}}{D_{10}} \tag{13.1}$$

- Coefficient of gradation:

$$C_z = \frac{D_{30}^2}{D_{10} \times D_{60}} \tag{13.2}$$

The coefficient of gradation is also sometimes referred to as the coefficient of curvature. The definitions of D_{10}, D_{30}, and D_{60} are shown in Figure 13.2.

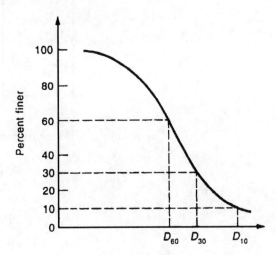

Figure 13.2

Specific Gravity of Soil Solids, G_s

The specific gravity of soil solids is defined as

$$G_s = \frac{\text{unit weight of soil solids only}}{\text{unit weight of water}} \tag{13.3}$$

The general range of G_s for various soils is given in Table 13.1.

Table 13.1 General range of G_s for various soils

Soil Type	Range of G_s
Sand	2.63–2.67
Silt	2.65–2.7
Clay and silty clay	2.67–2.8
Organic soil	Less than 2

Weight-Volume Relationships

Soils are three-phase systems containing soil solids, water, and air (Figure 13.3). Referring to Figure 13.3,

$$W = W_s + W_w \qquad (13.4)$$

$$V = V_s + V_v = V_s + V_w + V_a \qquad (13.5)$$

where W = total weight of the soil specimen, W_s = weight of the solids, W_w = weight of water, V = total volume of the soil, V_s = volume of soil solids, V_v = volume of voids, V_w = volume of water, and V_a = volume of air.

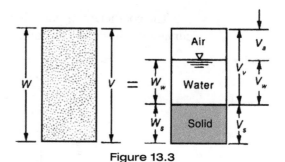

Figure 13.3

The *volume relationships* can then be given as follows:

$$\text{Void ratio} = e = \frac{V_v}{V_s} \qquad (13.6)$$

$$\text{Porosity} = n = \frac{V_v}{V} \qquad (13.7)$$

$$\text{Degree of saturation} = S = \frac{V_w}{V_v} \qquad (13.8)$$

Similarly, the *weight relationships* are

$$\text{Moisture content} = w = \frac{W_w}{W_s} \qquad (13.9)$$

$$\text{Moist unit weight} = \gamma = \frac{W}{V} \qquad (13.10)$$

$$\text{Dry unit weight} = \gamma_d = \frac{W_s}{V} \qquad (13.11)$$

Consider a soil sample with a unit volume of soil solids (Figure 13.4) to derive the following relationships:

1. $e = \dfrac{n}{1-n}$

2. $n = \dfrac{e}{1+e}$

3. $\gamma = \dfrac{G_s \gamma_w + w G_s \gamma_w}{1+e} = \dfrac{G_s \gamma_w (1+w)}{1+e}$

4. $\gamma_d = \dfrac{G_s \gamma_w}{1+e}$

5. $S = \dfrac{w G_s}{e}$

For *saturated soils*, $V_a = 0$ and $V_v = V_w$. Hence,

1. $S = 100\%$
2. $\gamma = \gamma_{sat} = \dfrac{\gamma_w(G_s + e)}{1 + e} = \dfrac{\gamma_w(G_s + wG_s)}{1 + wG_s}$
3. $\gamma_d = \dfrac{G_s \gamma_w}{1 + e}$

In the preceding relationships, γ_w = unit weight of water = 62.4 lb/ft³ (or 9.81 KN/m³).

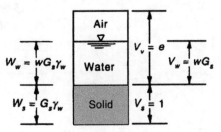

Figure 13.4

Example 13.1

A soil has a volume of 0.3 ft³ and weighs 36 lb. Given $G_s = 2.67$ and moisture content $(w) = 18\%$, determine a). moist unit weight (γ), b). dry unit weight (γ_d), c). void ratio (e), d). porosity (n), and e). degree of saturation (S).

Solution

a). $\gamma = \dfrac{W}{V} = \dfrac{36}{0.3} = 120 \text{ lb/ft}^3$

b). $\gamma_d = \dfrac{W_s}{V} = \dfrac{W}{(1+w)V} = \dfrac{36}{\left(1 + \dfrac{18}{100}\right)0.3} = 101.7 \text{ lb/ft}^3$

c). $\gamma_d = \dfrac{G_s \gamma_w}{1 + e}$

$e = \dfrac{G_s \gamma_w}{\gamma_d} - 1 = \dfrac{(2.67)(62.4)}{101.7} - 1 = 0.64$

d). $n = \dfrac{e}{1 + e} = \dfrac{0.64}{1 + 0.64} = 0.39$

e). $S = \dfrac{wG_s}{e} = \dfrac{(0.18)(2.67)}{0.64} = 0.75 \,(75\%)$

Relative Density

In granular soils, the degree of compaction is generally expressed by a nondimensional parameter called *relative density*, D_r, or

$$D_r = \dfrac{e_{max} - e}{e_{max} - e_{min}} \qquad (13.12)$$

where
- e = actual void ratio in the field
- e_{max} = void ratio in the loosest state
- e_{min} = void ratio in the densest state

In many practical cases, a granular soil is qualitatively described by its relative density in the following manner.

D_r (%)	Description
0–15	Very loose
15–50	Loose
50–70	Medium
70–85	Dense
85–100	Very dense

Example 13.2

A sand has the following maximum and minimum dry unit weights:

$$\gamma_{d(max)} = 17.29 \text{ kN/m}^3$$
$$\gamma_{d(min)} = 15.41 \text{ kN/m}^3$$

Given $G_s = 2.66$ and dry unit weight in the field = 16.51 kN/m³, determine the relative density in the field.

Solution

$$\gamma_{d(max)} = \frac{G_s \gamma_w}{1 + e_{min}}$$

$$e_{min} = \frac{G_s \gamma_w}{\gamma_{d(max)}} - 1 = \frac{(2.66)(9.81)}{17.29} - 1 = 0.51$$

Similarly,

$$\gamma_{d(min)} = \frac{G_s \gamma_w}{1 + e_{max}}$$

$$e_{min} = \frac{G_s \gamma_w}{\gamma_{d(min)}} - 1 = \frac{(2.66)(9.81)}{15.41} - 1 = 0.69$$

Also,

$$e = \frac{G_s \gamma_w}{\gamma_{d(field)}} - 1 = \frac{(2.66)(9.81)}{16.51} - 1 = 0.58$$

$$D_r = \frac{e_{max} - e}{e_{max} - e_{min}} = \frac{0.69 - 0.58}{0.69 - 0.51} = 0.61 = 61\%$$

Consistency of Clayey Soils

When a cohesive soil is mixed with an excessive amount of water, it will be in a somewhat liquid state and flow like a viscous liquid. However, when this viscous liquid is gradually dried, with the loss of moisture it will pass into a plastic state. With further reduction of moisture, the soil will pass into a semisolid and then into a solid state. This is shown in Figure 13.5. The moisture content, in percent, at which the cohesive soil will pass from a liquid state to a plastic state is called the *liquid limit*. Similarly, the moisture contents at which the soil changes from a plastic state to a semisolid state and from a semisolid state to a solid state are referred to as the *plastic limit* and the *shrinkage limit*, respectively. These limits are referred to as the *Atterberg limits*.

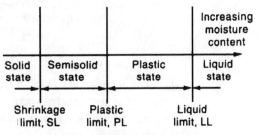

Figure 13.5

The liquid limit (LL) is the moisture content, in percent, at which the groove in a Casagrande liquid limit device closes for a distance of 0.5 in. after 25 blows. The plastic limit (PL) is the moisture content, in percent, at which the soil, when rolled into a thread of ⅛-in. diameter, crumbles. The *plasticity index* (PI) is defined as

$$PI = LL - PL \qquad (13.13)$$

The shrinkage limit (SL) is the moisture content, in percent, at or below which the volume of the soil mass no longer changes from drying.

GROUNDWATER

Groundwater refers to water that is stored beneath the Earth's surface, as contrasted to *surface water*, which is stored above the Earth's surface in ponds, lakes, rivers, and oceans. Groundwater is stored in geologic formations called *aquifers*. An aquifer is defined as a geologic formation that is able to hold large amounts of water *and* is able to conduct that water.

Aquifers

Aquifers may be either confined or unconfined. A *confined aquifer* is one that is bounded on top by a confining layer, sometimes called an *aquitard*. Thus, a confined aquifer is under pressure; the pressure at the top of the confined aquifer is due to the head of water above the top of the aquifer. An unconfined aquifer is one in which the upper level of the aquifer is at atmospheric pressure. The surface of an unconfined aquifer is called the *water table*. Figure 13.6 demonstrates these two types of aquifers.

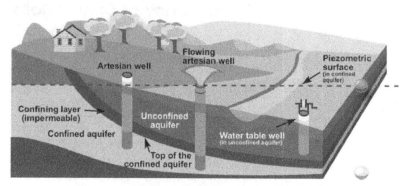

Figure 13.6 Confined and unconfined aquifers

Source: Environment Canada's Freshwater Web site (*www.ec.gc.ca/water*). Reproduced with the permission of the Minister of Public Works and Government Services, 2007.

Figure 13.6 also shows the *piezometric surface*. The piezometric surface is the imaginary level to which the water in an aquifer would rise. For an unconfined aquifer, the piezometric surface is the water surface. For a confined aquifer, the piezometric surface is some distance above the bottom of the confining layer, and this distance is equal to the pressure head in the confined aquifer.

The pressure head in a confined aquifer can be large enough such that the piezometric surface is above the ground surface. If a pathway exists from the confined aquifer to the ground surface, the groundwater will flow from the aquifer to above the ground surface. Such a phenomenon is known as an *artesian well* (Figure 13.7).

Of increasing importance to environmental engineers is the *recharge area,* also shown in Figure 13.7. This is the area in which water seeps into the ground to supply the confined aquifer. The recharge area is significant in that it provides an opportunity for potential contamination of the aquifer. Also, increasing the imperviousness of the recharge area via land development can negatively impact the ability of the aquifer to be recharged.

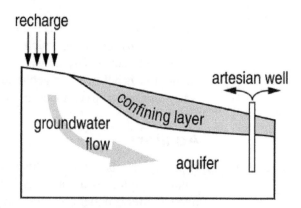

Figure 13.7 Artesian well

Source: "Water Quantity Issues at Chickasaw National Recreation Area," *Park Science* 19, no. 2, December 1999. Reprinted by permission of the National Park Service.

As previously defined, aquifers are characterized by their ability to transport water and to store water. The ability to store and transport water is characterized by the following properties:

- *Porosity, n*, is defined as $n = \dfrac{V_v}{V_t}$

  where V_v = the volume of voids

  V_t = the total volume

- *Void ratio, e* is defined as $e = \dfrac{V_v}{V_s} = \dfrac{n}{1-n}$

  where V_s = the volume of the solids (that is, soil particles)

- *Specific yield* is the percentage of an aquifer's water that will drain due to gravity.

- The *storage coefficient* (or *storativity*) is the volume of water that an aquifer gains or loses in response to a unit change in head. The values of storativity range from 10^{-5} to 10^{-3} for confined aquifers, and between 10^{-2} and 0.35 for unconfined aquifers. Storativity is illustrated for confined and unconfined aquifers in Figure 13.8.

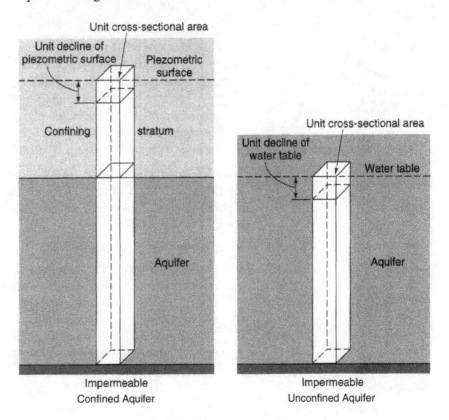

Figure 13.8 Storativity

Source: Todd, *Groundwater Hydrology*, 2nd ed., © 1980, John Wiley & Sons, Inc. Reprinted by permission.

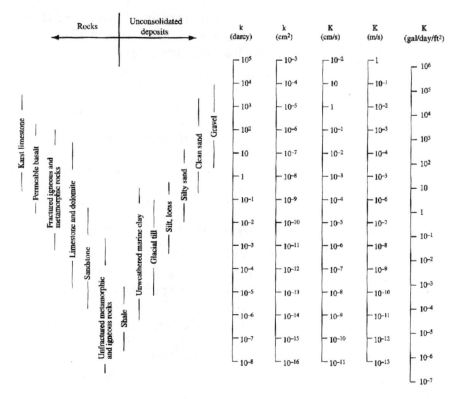

Figure 13.9 Values of hydraulic conductivity and permeability

Source: Freeze and Cherry, *Groundwater*, © 1979. Reprinted by permission of Pearson Education, Inc., Upper Saddle River, N.J.

- The *hydraulic gradient* is the slope of the piezometric surface. The rate at which groundwater travels is directly related to this gradient.

- *Hydraulic conductivity* is the ability of a geologic formation to transport water. It is the rate of water flow through a cross section of an aquifer in response to a unit hydraulic gradient. Values for hydraulic conductivity K are provided in Figure 13.9.

- *Permeability* is a function of the geologic formation and not a function of the fluid properties. Permeability is related to the square of the grain size diameter. Values for permeability k are provided in Figure 13.9.

Permeability

The rate of flow of water through a soil of gross cross-sectional area A can be given by the following relationships (Figure 13.10);

$$v = ki \qquad (13.14)$$

where
  v = discharge velocity
  k = coefficient of permeability
  i = hydraulic gradient = h/L (see Figure 13.9)

$$q = vA = kiA \qquad (13.15)$$

where
- q = flow through soil in unit time
- A = area of cross section of the soil at a right angle to the direction of flow

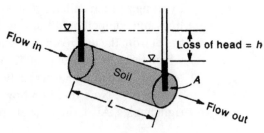

Figure 13.10

Equation (13.14) is known as *Darcy's law*. For granular soils, the coefficient of permeability can be estimated as

$$k \propto e^2 \qquad (13.16)$$

and

$$k \propto \frac{e^3}{1+e} \qquad (13.17)$$

where e is the void ratio. The range of the coefficient of permeability in various types of soil is given in Table 13.2.

Table 13.2 Range of k in various soils

Soil Type	Range of k (cm/sec)
Coarse sand	1–10^{-2}
Fine sand	10^{-2}–10^{-3}
Silt	10^{-3}–10^{-5}
Clay	$<10^{-6}$

Example 13.3

A sandy soil has a coefficient of permeability of 0.006 cm/sec at a void ratio of 0.5. Estimate k at a void ratio of 0.7.

Solution

From Equation (13.16),

$$\frac{k_1}{k_2} = \frac{e_1^2}{e_2^2}$$

so

$$\frac{0.006}{k_2} = \frac{(0.5)^2}{(0.7)^2}$$

$$k_2 = \frac{(0.006)(0.7)^2}{(0.5)^2} = 0.0118 \text{ cm/s}$$

Flow Nets

In many cases, flow of water through soil varies in direction and in magnitude over the cross section. In those cases, calculation of rate of flow of water can be made by using a graph called a *flow net*. A flow net is a combination of a number of flow lines and equipotential lines. A flow line is a line along which a water particle will travel from the upstream to the downstream side. An equipotential line is one along which the potential head at all points is the same. Figure 13.11 shows an example of a flow net in which water flows from the upstream to the downstream around a sheet pile. Note that in a flow net the flow lines and equipotential lines cross at right angles. Also, the flow elements constructed are approximately square.

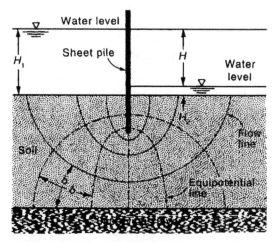

Figure 13.11

Referring to Figure 13.11, the flow in unit time (q) per unit length normal to the cross section shown is

$$q = k \frac{N_f}{N_d} H \qquad (13.18)$$

where
N_f = number of flow channels
N_d = number of drops
H = head difference between the upstream and downstream side

(*Note:* In Figure 13.11, $N_f = 4$ and $N_d = 6$.)

Example 13.4

Refer to the flow net shown in Figure 13.11. Given $k = 0.001$ ft/min, $H_1 = 30$ ft, and $H_2 = 5$ ft, determine the seepage loss per day per foot under the sheet pile structure.

Solution

$$q = k \frac{N_f}{N_d} H = (0.001 \times 60 \times 24 \text{ ft/day}) \left(\frac{4}{6}\right)(30-5) = 24 \text{ ft}^3/\text{day/ft}$$

Effective Stress

The total stress, σ, at a point in a soil mass is the sum of two components: (1) pore water pressure, u, and (2) effective stress, σ'. Thus

$$\sigma = \sigma' + u \qquad (13.19)$$

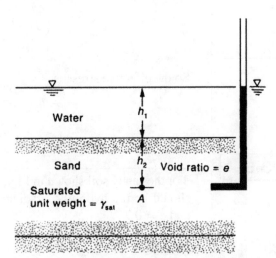

Figure 13.12

The effective stress is the sum of the vertical components of the forces developed at the points of contact of the solid particles per unit cross section of the soil mass. Referring to Figure 13.12, at point A

$$\sigma = h_1 \gamma_w + h_2 \gamma_{sat}$$
$$u = \gamma_w (h_1 + h_2)$$

so

$$\sigma' = \sigma - u = (h_1 \gamma_w + h_2 \gamma_{sat}) - \gamma_w (h_1 + h_2) = h_2 (\gamma_{sat} - \gamma_w) = h_2 \gamma'$$

In the preceding relationships, γ_w = unit weight of water, γ_{sat} = saturated unit weight of soil, and $\gamma' = \gamma_{sat} - \gamma_w$ = effective unit weight of soil. From the section on weight-volume relationships,

$$\gamma_{sat} = \frac{\gamma_w (G_s + e)}{1 + e}$$

so

$$\gamma' = \gamma_{sat} - \gamma_w = \frac{\gamma_w (G_s + e)}{1 + e} - \gamma_w = \frac{\gamma_w (G_s - 1)}{1 + e} \qquad (13.20)$$

For a quicksand condition, for example, when the effective stress $\sigma' = 0$, the hydraulic gradient is given as

$$i = i_{cr} = \frac{\gamma'}{\gamma_w} = \frac{G_s - 1}{1 + e} \qquad (13.21)$$

Example 13.5

Refer to Figure 13.12. For the soil: void ratio $e = 0.5$, $G_s = 2.67$, $h_1 = 1.5$ m, $h_2 = 3.05$ m, determine the effective stress at A.

Solution

$$\gamma' = \frac{\gamma_w(G_s - 1)}{1 + e} = \frac{9.81(2.67 - 1)}{1 + 0.5} = 10.92 \text{ kN/m}^3$$

So the effective stress is

$$\sigma' = h_2\gamma' = (3.05)(10.92) = 33.31 \text{ kN/m}^2$$

Example 13.6

For the sandy soil shown in Figure 13.12, if there is an upward flow of water, what should be the hydraulic gradient for the quicksand condition? Given: $G_s = 2.65$ and $e = 0.7$.

Solution

For the quicksand condition

$$i_{cr} = \frac{\gamma'}{\gamma_w} = \frac{G_s - 1}{1 + e} = \frac{2.65 - 1}{1 + 0.7} = 0.97$$

CONSOLIDATION

Consolidation settlement is the result of volume change in saturated clayey soils due to the expulsion of water occupied in the void spaces. In soft clays the major portion of the settlement of a foundation may be due to consolidation. Based on the theory of consolidation, a soil may be divided into two major categories: (a) normally consolidated and (b) overconsolidated. For *normally consolidated* clay, the *present effective overburden pressure* is the maximum to which the soil has been subjected in the recent geologic past.

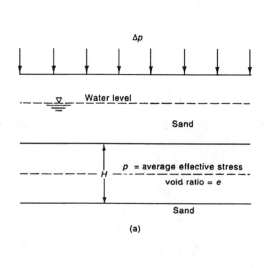
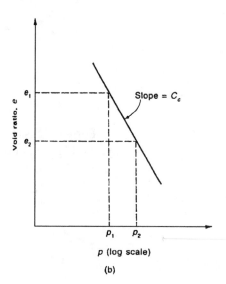

Figure 13.13

Figure 13.13(a) shows a normally consolidated clay deposit of thickness H. The consolidation settlement (ΔH) due to surcharge Δp can be determined as

$$\Delta H = \frac{C_c H}{1+e_i} \log\left(\frac{p_i + \Delta p}{p_i}\right) \tag{13.22}$$

where e_i = initial void ratio, p_i = initial *average* effective pressure, and C_c = compression index.

The compression index can be determined from the laboratory as (see Figure 13.13(b))

$$C_c = \frac{e_1 - e_2}{\log\left(\frac{p_2}{p_1}\right)} \tag{13.23}$$

GROUNDWATER FLOW

As described previously, the simplest equation for groundwater flow is the one-dimensional Darcy's law:

$$Q = -K \cdot A \frac{dh}{dx} \tag{13.24}$$

where

K = hydraulic conductivity [L/T]

A = cross-sectional area of aquifer [L^2]

dh/dx = hydraulic gradient [L/L]

The negative sign in Darcy's law arises from the fact that the slope of the piezometric surface is negative in the direction of flow.

The variables found in Darcy's law are illustrated in Figure 13.14. In this diagram, the hydraulic gradient could be written as $\frac{dh}{dx} = \frac{h_1 - h_2}{L}$. The cross-sectional area is the product of b, the thickness of the aquifer, and the distance of the aquifer into the paper. Figure 13.14 shows Darcy's law for a confined aquifer, but the analysis is the same for an unconfined aquifer. For the unconfined aquifer, the hydraulic gradient is the slope of the actual water surface.

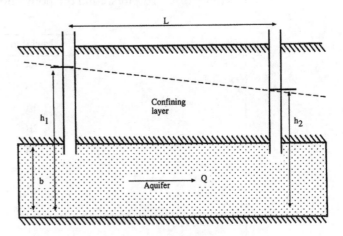

Figure 13.14 Darcy's law

Source: ENV 302, Environmental Hydrogeology, Northern Arizona University.

The *Darcy velocity* is defined as:

$$v_{darcy} = \frac{Q}{A} = -K \cdot \frac{dh}{dx} \qquad (13.25)$$

where Q is the flow rate calculated by the Darcy equation. The Darcy velocity is much lower than the true velocity, as the cross-sectional area through which the groundwater flows is much smaller than A; that is, the groundwater is flowing through the pores between the soil particles. Of more use to environmental engineers is the *pore velocity*, as this better characterizes the flow of water within the pores.

$$v_{pore} = \frac{v_{darcy}}{n} \qquad (13.26)$$

Example 13.7

Two observation wells are drilled 2000 feet apart as shown in Exhibit 1.

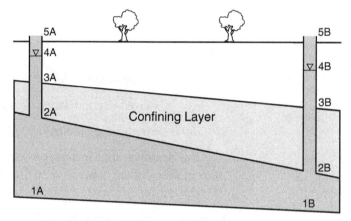

Not to scale

Exhibit 1

Soil boring data were obtained when drilling the wells, and some elevations are provided in Exhibit 2. The soil has a porosity of 22% and a hydraulic conductivity of $2 \cdot 10^{-3}$ m/s. Find the pore velocity.

Location	Elevation
1A	881
1B	871
2A	970
2B	846
3A	1031
3B	1010
4A	1069
4B	1042
5A	1123
5B	1135

Exhibit 2 Soil boring data

Solution

A large amount of extraneous information is provided from the soil boring results. The only data relevant to this example problem are the elevations of the piezometric surface at the upstream and downstream wells (1069 ft and 1042 ft, respectively). Given this information, the Darcy velocity can be calculated as:

$$v_{darcy} = -K \cdot \frac{dh}{dx} = -2 \cdot 10^{-3} \frac{\text{ft}}{\text{sec}} \cdot \frac{-(1069 \text{ ft} - 1042 \text{ ft})}{2000 \text{ ft}} = 2.7 \cdot 10^{-5} \text{ ft/s}$$

The pore velocity is

$$v_{pore} = \frac{v_{darcy}}{n} = \frac{2.7 \cdot 10^{-5} \text{ ft/s}}{0.22} = 1.23 \cdot 10^{-4} \text{ ft/s}$$

Thus, the groundwater in this instance travels approximately 10.6 feet each day.

Well Hydraulics

The basic equation describing local, steady groundwater movement is Darcy's law, which can be written

$$V = -K_i \qquad (13.27)$$

In this equation, V is the average velocity of a discharge Q that occurs through a soil cross-sectional area A. Darcy's law indicates that V is the product of the local hydraulic conductivity K, which depends on the local soil or rock properties, and the local gradient i of the piezometric head $H = p/\gamma + z$, or $i = dH/dL$. This may also be interpreted as a difference in fluid energy between points, because the kinetic energy associated with groundwater flow is negligible. To obtain the actual fluid velocity, called the seepage velocity, in the subsurface saturated zone, divide the average velocity by the local porosity. A variety of units are used in describing groundwater parameters, so you should take care to use consistent units in all computations.

Steady Flow (Thiem Solution)

Equations for steady flow from a well in either an unconfined or a confined aquifer can be derived from Darcy's law. These simple equations only have meaning and are accurate when several simplifying assumptions are valid, including the following: (1) the aquifer, which is a geologic formation that contains enough saturated permeable material to yield significant quantities of water, must be large in extent and have uniform hydraulic properties (for example, K is constant); (2) the pumping must occur at a constant rate for an extremely long time so that startup transients no longer exist; (3) the well fully penetrates the aquifer; (4) the well depth is much larger than the drawdown near the well; and (5) the estimate of the gradient i is a good one.

An aquifer is called unconfined if the upper edge of the saturated zone (ignoring capillary effects) is at atmospheric pressure; this edge is called the water table. Figure 13.15 is a schematic cross section of a well in an unconfined, horizontal aquifer. A cylindrical coordinate system (x, y) is placed at the base of the well; the drawdown of the undisturbed water table is s; the gradient of the piezometric head is $i = dy/dx$ at the water table and is assumed to apply to the entire water column below it. The radius of the well is $x = r_w$. Applying Darcy's law gives

$$V = \frac{Q}{Q} = \frac{Q}{2\pi xy} = -K \frac{dy}{dx} \qquad (13.28)$$

Rearranging this expression and integrating it between points (r_1, h_1) and (r_2, h_2) along the water table yields

$$Q = \frac{\pi K \left[h_2^2 - h_1^2 \right]}{\ln(r_2/r_1)} \qquad (13.29)$$

as the expression for the steady pumping rate, or discharge, for this case.

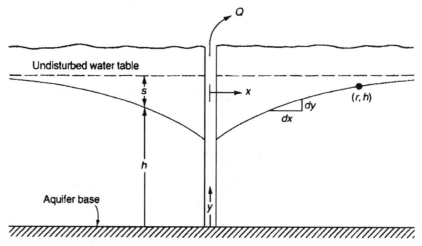

Figure 13.15 Cross section of a well in a confined aquifer

The case for steady pumping from a confined, horizontal aquifer of thickness m is similar to the first case, as shown in Figure 13.16. However, the gradient i is determined from the local slope of the piezometric head curve (shown dashed), which is no longer the same as the edge of the saturated zone. Equation 13.28 still applies to this case if the area through which flow occurs is corrected to $A = 2\pi xb$. Now, the integration between points (r_1, h_1) and (r_2, h_2) on the piezometric surface results in

$$Q = \frac{2\pi Kb(h_2 - h_1)}{\ln(r_2/r_1)} \qquad (13.30)$$

for the discharge. Sometimes, the transmissivity $T = Kb$ is introduced into this equation.

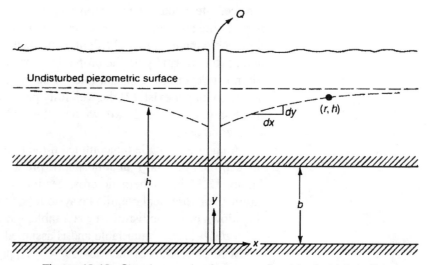

Figure 13.16 Steady pumping from a confined, horizontal aquifer

Unsteady Flow (Theis and Cooper-Jacob Solutions)

The first significant solution for unsteady flow to a well was originally developed by Theis for a confined aquifer. It expresses the drawdown s as

$$s = \frac{Q}{4\pi T} W(u) \qquad (13.31)$$

in which

$$u = \frac{r^2 S}{4Tt} \qquad (13.32)$$

and

$$W(u) = \int_u^\infty \frac{e^{-u} du}{u} = -0.5772 - \ln(u) + u - \frac{u^2}{2 \times 2!} + \cdots \qquad (13.33)$$

is called the well function of u. Table 13.3 presents tabulated values for this function. The discharge Q is constant over the pumping period, and r is the radius at which s is computed (to find the drawdown at the well, use $r = r_w$) at a time t after pumping began. The solution depends on a knowledge of two aquifer properties: the transmissivity T and the storage constant S. The storage constant is the amount of water removed from a unit volume of the aquifer when the piezometric head is lowered one unit. Two methods for the determination of these aquifer properties will be described next.

Table 13.3 Values of the function $W(u)$ for various values of u

u	$W(u)$	u	$W(u)$	u	$W(u)$	u	$W(u)$
1×10^{-10}	22.45	7×10^{-8}	15.90	4×10^{-5}	9.55	1×10^{-2}	4.04
2	21.76	8	15.76	5	9.33	2	3.35
3	21.35	9	15.65	6	9.14	3	2.96
4	21.06	1×10^{-7}	15.54	7	8.99	4	2.68
5	20.84	2	14.85	8	8.86	5	2.47
6	20.66	3	14.44	9	8.74	6	2.30
7	20.50	4	14.15	1×10^{-4}	8.63	7	2.15
8	20.37	5	13.93	2	7.94	8	2.03
9	20.25	6	13.75	3	7.53	9	1.92
1×10^{-9}	20.15	7	13.60	4	7.25	1×10^{-1}	1.823
2	19.45	8	13.46	5	7.02	2	1.223
3	19.05	9	13.34	6	6.84	3	0.906
4	18.76	1×10^{-6}	13.24	7	6.69	4	0.702
5	18.54	2	12.55	8	6.55	5	0.560
6	18.35	3	12.14	9	6.44	6	0.454
7	18.20	4	11.85	1×10^{-3}	6.33	7	0.374
8	18.07	5	11.63	2	5.64	8	0.311
9	17.95	6	11.45	3	5.23	9	0.260
1×10^{-8}	17.84	7	11.29	4	4.95	1×10^{0}	0.219
2	17.15	8	11.16	5	4.73	2	0.049

u	W(u)	u	W(u)	u	W(u)	u	W(u)
3	16.74	9	11.04	6	4.54	3	0.013
4	16.46	1×10^{-5}	10.94	7	4.39	4	0.004
5	16.23	2	10.24	8	4.26	5	0.001
6	16.05	3	9.84	9	4.14		

Source: Bedient and Huber, *Hydrology and Floodplain Analysis*, 3d ed., © 2002. Reprinted by permission of Pearson Education, Inc., Upper Saddle River, N.J.

The first method of determining T and S is by using the original Theis equations. An examination of these equations shows that a plot of $W(u)$ vs. u, called a type curve, will have the same shape as a plot of s vs. r^2/t on log-log graph paper. The two curves are plotted, and one graph is laid over the other so the curves lie on one another. Then, a so-called match point, which is a set of data for u, $W(u)$, s, and r^2/t, is taken from the plots, inserted in Equations 13.31 and 13.32, and the resulting relations are solved for S and T. Since the match point is used to establish a connection between the data plot and the other plots, the match point need not be on the curve itself, although most practitioners do choose the match point atop the superimposed curves.

The second method, called the Cooper-Jacob method, is appropriate when u is small (for example, $u \leq 0.01$ is a common rule). In this method, s is plotted against pumping time on semi-logarithmic paper; the curve eventually becomes linear. A fitted straight line is then extended to the point $s = 0$, where the value $t = t_0$ is noted. You then solve for the aquifer properties from

$$T = \frac{2.3Q}{4\pi(s_2 - s_1)} \log 10 \left(\frac{t_2}{t_1} \right) \quad (13.34)$$

and

$$S = \frac{2.25 T t_0}{r^2} \quad (13.35)$$

Use of these equations is simplified if points 1 and 2 are chosen so that $t_2/t_1 = 10$; of course, $s_2 - s_1$ is the difference in drawdown over this same time interval. The result in Equation 13.35 must be nondimensional.

Several approaches are possible for unsteady, unconfined well flow, but the simplest is to use the Theis method, Equation 13.31, with modified definitions of T and S. Now $T = Kb$ is based on the saturated thickness when pumping commences, and S is the specific yield, the volume of water released when the water table drops one unit. This approach is accurate when the drawdown is small in comparison with the saturated thickness of the aquifer.

Example 13.8

A well has been pumped at a steady rate for a very long time. The well has a 12-inch diameter and fully penetrates an unconfined aquifer that is 150 feet thick. Two small observation wells are 70 and 150 feet from the well, and the corresponding observed drawdowns are 24 and 20 feet. If the estimated hydraulic conductivity is 10 ft/day (sandstone), what is the discharge?

Solution

The saturated aquifer thicknesses at the observation wells are $h_1 = 150 - 24 = 126$ ft, $h_2 = 150 - 20 = 130$ ft, and the use of Equation 13.29 leads directly to

$$Q = \frac{\pi K \left[h_2^2 - h_1^2 \right]}{\ln(r_2 / r_1)} = \frac{\pi(10)\left[(130)^2 - (126)^2\right]}{\ln(150/70)} = 42,200 \text{ ft}^3/\text{day}$$

This is equivalent to 0.49 ft³/s or 220 gal/min.

Example 13.9

Calculating transmissivity and specific yield I

Data on time t since pumping began versus drawdown s were collected from an observation well located 400 ft from a well that fully penetrated a confined aquifer that is 80 ft thick and is pumped at 200 gal/min. The data are presented in Exhibit 3.

t, min	s, ft	t, min	s, ft
35	2.82	103	4.43
41	3.12	131	4.60
48	3.25	148	5.00
60	3.60	205	5.35
80	3.98	267	5.80

Exhibit 3

Determine the aquifer properties T and S.

Solution

The data in Exhibit 3 have been used with $r = 400$ ft to compute s vs. r^2/t, which have been plotted in Exhibit 4 on a sheet of log-log paper and the plot placed on top of a log-log plot of $W(u)$ vs. u (the type curve). The two plots have been moved around until the closest fit between the curves was found, taking care that the coordinate axes are parallel. If the match point is chosen as shown on the figure, then $s = 5$ ft,
$r^2/t = 10^2$ ft³/min, $u = 0.0175$, and $W(u) = 350$. Rearranging Equation 13.31,

$$T = Q \frac{W(u)}{4\pi s} = \frac{200 \text{ gal/min} \cdot (3.50)}{7.48 \text{ gal/ft}^3 \cdot 4\pi(5 \text{ ft})} = 1.49 \text{ ft}^2/\text{min}$$

Then from Equation 13.32

$$S = \frac{4Tu}{r^2/t} = \frac{4(1.49)(0.0175)}{10^3} = 1.04 \times 10^{-4}$$

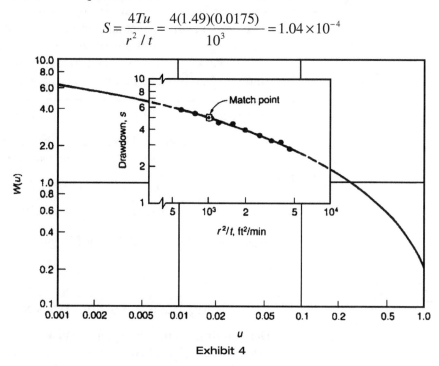

Exhibit 4

Example 13.10

Calculating transmissivity and specific yield II

Last April pumping began at an 8-inch-diameter well at a steady rate of 300 gal/min, while observations were made at a well 100 ft away. Values of elapsed time and drawdown were taken for 16 hours and plotted (see Exhibit 5). Use the plotted data to determine values for the transmissivity and storage constant of this aquifer. In addition, estimate the drawdown in the observation well after four months of steady pumping at 300 gal/min.

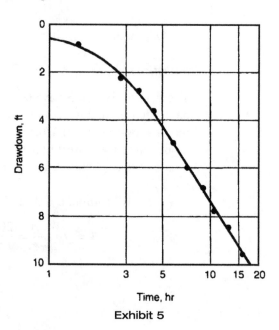

Exhibit 5

Solution

Exhibit 5 is a semi-logarithmic plot of data in the form needed to apply the Cooper-Jacob (or modified Theis, as it is also called) method. If you extend the straight-line portion of the plot to $s = 0$, as shown in Exhibit 6, you can read from the plot the value $t_0 = 2$ hrs. You also need a pair of data points (s_1, t_1) and (s_2, t_2) for use in Equation 13.34. For example, at $t_1 = 4.0$ hr, $s_1 = 3.3$ ft, and at $t_2 = 10.0$ hr, $s_2 = 7.4$ ft. Then

$$T = \frac{2.3Q}{4\pi(s_2 - s_1)} \log_{10}\left(\frac{t_2}{t_1}\right) = \frac{2.3(300)}{4\pi(7.4 - 3.3)} \log_{10}\left(\frac{10.0}{4.0}\right) = 5.33 \text{ gal/min/ft}$$

$$T = \frac{5.33 \frac{\text{gal/min}}{\text{ft}}}{7.48 \frac{\text{gal}}{\text{ft}^3}} \left(60 \frac{\text{min}}{\text{hr}}\right)\left(24 \frac{\text{hr}}{\text{day}}\right) = 1026 \text{ ft}^2/\text{day}$$

The storage constant can then be found as

$$S = \frac{2.25 T t_0}{r^2} = \frac{2.25(1026 \text{ ft}^2/\text{day})(\frac{2}{24} \text{ day})}{(100 \text{ ft})^2} = 0.0192$$

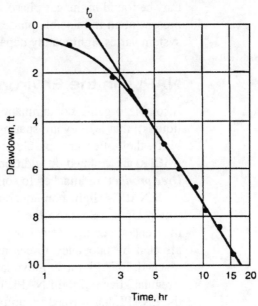

Exhibit 6

Equation 13.34 can also be used to find the drawdown after four months. If s_2 is the drawdown after four months, then t_2 is four months, or approximately 122 days. If you pick the other point to be $s_1 = 0$ and $t_1 = t_0 = 2.0$ hr, then

$$s_2 = \frac{2.3Q}{4\pi T} \log_{10}\left(\frac{t_2}{t_1}\right) = \frac{2.3(300)}{4\pi(5.33)} \log_{10}\left(\frac{122(24)}{2}\right) = 32.6 \text{ ft}$$

is the predicted drawdown.

GROUNDWATER CONTAMINATION

Contaminants may be transported in air, water, and soil. Exposure to water-borne contaminants will occur if people drink contaminated groundwater or surface water or accidentally ingest it while swimming, or if it comes into contact with their skin by any means. Humans will be exposed to hazardous substances in soil, sediment, or dust if they accidentally ingest it, inhale it (in the form of dust), or by direct dermal contact. Children are highly susceptible to exposure through soil pathways.

Transport via groundwater occurs according to the principles of groundwater flow described above. However, several additional important concepts related to contaminant flow in groundwater require further discussion.

Groundwater resources are relatively easy to contaminate as compared to surface water, and given the large quantity of groundwater extracted for drinking water, groundwater contamination is one of the most pressing issues facing environmental engineers. The problem is further intensified by recognizing that very low concentrations of these contaminants can have significant health risks. Moreover, removal or destruction of these contaminants is a complicated and costly undertaking, further complicated by the fact that modeling contaminant movement in the groundwater is not nearly as straightforward as modeling the movement *of* the groundwater.

Contaminants can take one or more forms in the subsurface environment. They may be found in the gas phase, adsorbed to soil particles, dissolved in solution, or present in an immiscible phase. Transport of the contaminant in the groundwater system varies significantly depending on which form the contaminants take.

NAPLs in the Environment

Insoluble organic contaminants may be present as NAPLs (non-aqueous phase liquids); that is, they are sparingly soluble in water. Although they have low solubility, they often are soluble enough such that the maximum contaminant levels (MCLs) are violated. In addition to being in solution, NAPLs may be found in bulk (*free product*) or attached to soil particles as *residual* NAPL.

LNAPLs (light non-aqueous phase liquids) are NAPLs that are lighter than water. Upon release to the environment, LNAPLs will migrate downward until they encounter a physical barrier (for example, low permeability strata) or are affected by buoyancy forces near the water table. Once the capillary fringe is reached, LNAPLs may move laterally as a continuous, free-phase layer. Note the irregular shape of the LNAPL in Figure 13.17 and the fact that a vapor plume of the LNAPL has formed. Examples of LNAPLs include gasoline and various types of oils. A common source of LNAPLs is a leaking underground storage tank.

Dense NAPLs (DNAPLs) have a specific gravity greater than 1 and will tend to sink to the bottom of surface waters and groundwater aquifers. The movement of DNAPLs in the subsurface is extremely complicated. DNAPLs will form a vapor phase in the unsaturated zone, will form pools based on soil heterogeneities in the saturated zone, will dissolve (albeit sparingly) and form an aqueous plume, and will eventually form a pool at the "bottom" of the aquifer but will continue to move from there into fractures in the confining layer and move en masse in the direction of groundwater flow. Clearly, modeling the movement of this contaminant in the subsurface and removing it for treatment is a tremendously complicated undertaking. Many chlorinated solvents, such as those used in dry cleaning operations, are DNAPLs, as are creosote, coal tar, and PCB oils. Sources include accidental spills or improper disposal practices (for example, unlined evaporation ponds or lagoons) in industries such as metal degreasing, pharmaceutical production, and pesticide formulation.

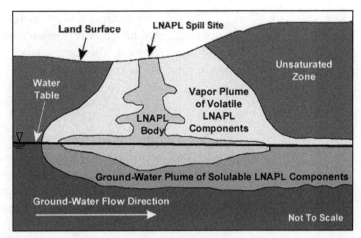

Figure 13.17 LNAPL in the subsurface
Source: U.S. Geological Survey, *http://toxics.usgs.gov/definitions/lnapls.html*.

Subsurface Contaminant Modeling

Subsurface contaminant modeling can vary in complexity, often based upon moisture classification of the soil bed. The *saturated zone* is the least complex, as groundwater flow is measurable, and transport models can estimate physical, chemical, and biological influences, while the *capillary fringe* and the *unsaturated* (vadose) *zones* are more complex due to variable flow. Another level of complexity is added for flow in more than one dimension (direction), flows that are not uniform, or for flows that are not steady. This review will focus on the simplest of groundwater flow models, considering the uniform, steady flow in one dimension through saturated media.

The model used to describe the change in concentration of a specific compound in subsurface flow is referred to as the *advection-dispersion model*, and is expressed for one dimensional flow as

$$\frac{\partial C}{\partial t} = -u_x \frac{\partial C}{\partial x} + D_x \frac{\partial^2 C}{\partial x^2} \pm \sum_{m=1}^{n} r_m$$

where C is the species concentration, t is time, u_x is the velocity in the x-dimension, x is the distance along the flow path, D_x is the dispersion coefficient, and r_m is any chemical, physical, or biological mechanism that adds or removes contaminant mass to the system. The velocity term as expressed above is the actual velocity through the porous medium (length/time), and may be calculated from the *Darcy (superficial) velocity* (v_x) by dividing by the bed porosity (i.e., $u_x = v_x/\eta$ where η is calculated as void volume/total volume).

The dispersion term is a result of the tortuous flow path of the water through the soil bed, and is a measure of the amount of mixing occurring due to flow. The extent of dispersion is dependent on soil grain size distribution and is usually quantified by using a *conservative tracer* test in the field or in laboratory soil columns or by model fitting. The conservative tracer is a non-reactive species that accurately describes the movement of the water through the bed, and is quantified in the effluent and plotted as a function of time (or equivalently, volume of water pumped through the system). Profiles of the solution to the advection-dispersion model equation for a known flow velocity and various dispersion numbers are compared to the contaminant effluent profile, and a value of D_x is selected that best represents the data.

The reaction term can account for any number of biological and/or abiotic transformations, however it is often convenient to express the degradation term as a first-order reaction. This can be incorporated into the advection-dispersion equation as a loss term as follows:

$$\frac{\partial C}{\partial t} = -u_x \frac{\partial C}{\partial x} + D_x \frac{\partial^2 C}{\partial x^2} - kC$$

where k is the first-order degradation rate constant. It should be noted that advanced models can incorporate more complex biological kinetic terms, such as the Monod kinetics discussed in Chapter 7, but they are often coupled with models for bacterial cell transport and are beyond the scope of this review.

Another common phenomenon that occurs in contaminant transport in the subsurface is *adsorption* of the contaminant species to the soil. As some molecules of the contaminant are removed from the water phase and become associated with the soil phase, this has the effect of slowing down the migration of the contaminant plume. While this may be an advantage when considering the reduced area of impact upon initial migration, the adsorbed species are available for release at a future time and cause persistent groundwater contamination problems. Further, groundwater remediation efforts are often hampered by species which have a stronger preference for the soil phase.

Species preference for either the soil or water phase is quantified by the *partition coefficient*, which may be determined in the laboratory by adding a small amount of contaminant to a sealed bottle containing a soil slurry. After allowing time for equilibration, the concentration of contaminant is measured in the aqueous phase, and by difference, determined for the soil phase. The species partition coefficient (K_{sw}) may be calculated as

$$K_{sw} = \frac{X}{C}$$

where X is the concentration of contaminant in the soil phase (mg/kg) and C is the concentration of contaminant in the aqueous phase (mg/kg). In many cases, the contaminant adsorbed to the soil is associated with the organic component of the soil, and the *organic carbon* partition coefficient (K_{oc}) can be defined as

$$K_{oc} = \frac{C_{soil}}{C_{water}}$$

where C_{soil} is the concentration of contaminant in the organic portion of the soil (mg species per kg organic carbon) and C_{water} is the concentration of contaminant in the aqueous phase (mg/kg). If the fraction of the soil that is organic matter is known (f_{oc}), K_{sw} may be estimated from K_{oc} as

$$K_{sw} = K_{oc} f_{oc}$$

When partitioning data cannot be obtained directly from field soil samples, another method of estimating the distribution behavior is to use *octanol-water* partition coefficient (K_{ow}). A known mass of contaminant is added to a sealed bottle containing known volumes of octanol and water and allowed to equilibrate. Each phase is analyzed for the contaminant and K_{ow} is calculated as

$$K_{ow} = \frac{C_{oct}}{C_w}$$

where C_{oct} and C_w is the contaminant concentration in the octanol and water phase, respectively (mg/L). Since K_{ow} data is available in reference texts for many organic compounds, and octanol is an organic compound, it is convenient to estimate a value for K_{oc} from K_{ow}. Several models have been proposed and are generally specific to the class of compounds that best typifies the contaminant of interest. Examples that typify these relationships include

$$\log K_{oc} = 1.00 \log K_{ow} - 0.21 \quad \text{aromatics, polynuclear aromatics}$$

$$\log K_{oc} = 0.72 \log K_{ow} + 0.5 \quad \text{halogenated aliphatics/aromatics}$$

Finally, the adsorption of contaminant can be included in the advection-dispersion model through the *retardation factor* (R), which is a measure of how much a species is hindered relative to the water flow. This is a function of soil properties and partitioning behavior, and can be expressed as

$$R = \frac{\text{water velocity}}{\text{contaminant velocity}} = 1 + \frac{\rho}{\eta} K_{sw}$$

where ρ is the bulk density of the bed, η is the bed porosity, and K_{sw} is as defined previously. Assuming biodegradation of the contaminant occurs only in the aqueous phase, the retardation factor is included in each of the terms in the advection-dispersion equation as follows:

$$\frac{\partial C}{\partial t} = -\frac{u_x}{R}\frac{\partial C}{\partial x} + \frac{D_x}{R}\frac{\partial^2 C}{\partial x^2} - \frac{k}{R}C$$

Solution to the model equation requires two boundary conditions and one initial condition to be established, and several analytical and numerical techniques may be used. One common solution makes use of the *complementary error function* (erfc), which may be defined as

$$\text{erfc}(y) = 1 - \frac{2}{\sqrt{\pi}} \sum_{n=0}^{\infty} \frac{(-1)^n y^{2n+1}}{n!(2n+1)}$$

With boundary conditions $C = C_0$ at $x = 0$ and $\partial C/\partial x = 0$ at $x = \times$, the advection-dispersion equation may be expressed as follows:

$$C = \frac{C_0}{2}\exp\left[\frac{(u_x-v)x}{2D_x}\right]\text{erfc}\left[\frac{Rx-vt}{2\sqrt{D_x Rt}}\right] + \frac{C_0}{2}\exp\left[\frac{(u_x+v)x}{2D_x}\right]\text{erfc}\left[\frac{Rx+vt}{2\sqrt{D_x Rt}}\right]$$

where v is defined as

$$v = u_x\left(1 + \frac{4kD_x}{u_x^2}\right)^{1/2}$$

The expression for C above is usually solved for various time steps (t) at a specific (or several) downgradient location (x), and a time-concentration profile is used to predict contaminant migration. Fortunately, values for erfc are tabulated in several reference texts or may be called as a function in many computer programs.

Example 13.11

Retardation of contaminants I

This example problem is based on Exhibit 7. For this example, a study was conducted by injecting three contaminants into a well and measuring their concentrations 5 meters down gradient. The average groundwater velocity on the site was 30 m/yr. Results are provided in Exhibit 7. Estimate a value for the retardation factor for carbon tetrachloride and tetrachloroethene.

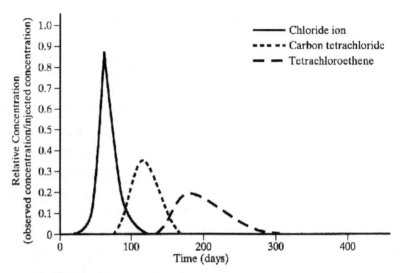

Exhibit 7 Concentration of contaminants in down-gradient well
Source: James R. Mihelcic, *Fundamentals of Environmental Engineering*, © 2001, John Wiley & Sons, Inc. Reprinted by permission.

Solution

The average velocity of each of the plumes is approximated by this relationship:

$$v = \frac{\text{distance}}{\text{time}} = \frac{5\text{ m}}{\text{time to peak}}$$

From the graphs, we estimate a time to peak for the carbon tetrachloride and tetrachloroethene to equal 116 days and 182 days, respectively. Therefore, the average velocities of these contaminants are 16 m/yr and 10 m/yr, respectively. Given the average groundwater velocity of 30 m/yr, the retardation factor for the carbon tetrachloride is 1.9 and 3 for the tetrachloroethene.

Exhibit 7 also demonstrates the use of a nonreactive tracer. The chloride peak occurs at approximately 70 days. This corresponds to a tracer velocity of 26 ft/s, demonstrating that the tracer moves nearly as fast as the groundwater.

This example problem further demonstrates that the relative concentration of contaminants from a source with multiple contaminants will vary with time and location.

Example 13.12

Retardation of contaminants II

Determine the retardation factor for the movement of benzene ($\log K_{oc} = 2.01$) and pyrene ($\log K_{oc} = 4.69$) through this soil matrix:

bulk density = 150 lb/ft³

porosity = 0.35

$f_{oc} = 0.5\%$

Solution

The partition coefficients for both contaminants may be calculated first:

$$K = K_{oc} \cdot f_{oc}$$
$$K_{benzene} = 10^{2.01} \cdot 0.005 = 0.51$$
$$K_{pyren} = 10^{4.69} \cdot 0.005 = 250$$

The retardation coefficients can be calculated now as:

$$R_{benzene} = 1 + \frac{150 \text{ lb/ft}^3}{0.35} 0.51 \frac{\text{cm}^3}{\text{g}} \frac{1000 \text{ g}}{2.205 \text{ lb}} \frac{1 \text{ ft}^3}{2.8 \cdot 10^4 \text{cm}^3} = 4.5$$

$$R_{pyrene} = 1 + \frac{150 \text{ lb/ft}^3}{0.35} 250 \frac{\text{cm}^3}{\text{g}} \frac{1000 \text{ g}}{2.205 \text{ lb}} \frac{1 \text{ ft}^3}{2.8 \cdot 10^4 \text{ cm}^3} = 1736$$

Consequently, the pyrene would take nearly 400 times longer to travel the same distance as the benzene.

Introduction to Soil and Groundwater Remediation

Contaminated soils and groundwater can be remediated using a wide array of technologies. One way to classify the treatment methods is *ex situ* or *in situ*. For *in situ* (or in place) treatment, the contaminant is not moved from the subsurface, while *ex situ* treatment involves the excavation of soil (or pumping of contaminated groundwater from the aquifer) for treatment.

Advantages of *in situ* treatment include the following:

- No costs for excavation or groundwater extraction
- Ability to treat soils under buildings and other structures without affecting the structure
- Avoidance of risks and costs associated with transportation
- Decreasing likelihood of spreading contaminants off-site

Advantages of *ex situ* treatment include:

- Shorter time periods for remediation
- Greater uniformity of treatment due to homogenization of solid phase (for example, soil or sludge) or the ability to monitor and continuously mix the groundwater

The *ex situ* treatment of groundwater is often termed "pump and treat." Pump-and-treat technologies are used at approximately three-quarters of all Superfund sites where groundwater is contaminated.

Pump-and-treat systems are used for two main purposes: treatment of the groundwater and/or hydraulic containment of contamination. Figure 13.18 illustrates a pump-and-treat system for remediation of a site contaminated by DNAPLs from a dry cleaning facility. Treated groundwater may be injected directly back into the aquifer as shown or may be discharged to surface water.

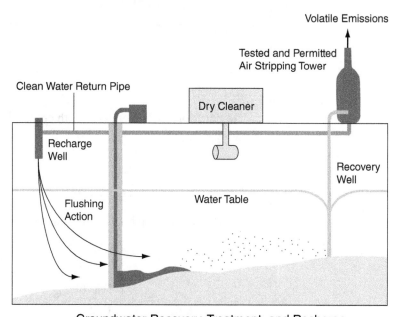

Figure 13.18 Pump-and-treat system example

Source: U.S. EPA, *www.epa.gov/OUST/graphics/cadnapl.htm*.

Two phenomena that complicate pump-and-treat technology are *tailing* and *rebound*. Tailing refers to the progressively slower rate of decline in dissolved contaminant concentration as pump-and-treat remediation continues. As a result of tailing, the volume of groundwater to be pumped, and therefore the cleanup time, can be on the order of ten times greater than if tailing did not occur. Rebound is a relatively rapid increase in concentration following cessation of pumping. These two concepts are illustrated in Figure 13.19.

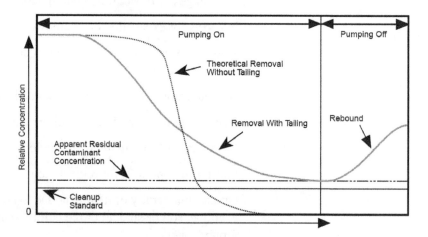

Figure 13.19 Tailing and rebound

Source: U.S. EPA, *Methods for Monitoring Pump-and-Treat Performance*, EPA/600/R-94/123, 1994.

The process for selecting a remediation strategy as part of a feasibility study begins with identifying the objectives of the remediation and screening technologies based on their technical feasibility, effectiveness, and cost. These two steps are typically followed by treatability studies to assess the effectiveness of the alternatives and to provide information for eventual optimization of the process. Following these initial steps, a remediation process is selected using the following nine criteria suggested by the U.S. EPA:

1. Overall protection of human health and the environment
2. Compliance with appropriate regulations
3. Long-term effectiveness and permanence
4. Reduction of toxicity, mobility, or volume
5. Short-term effectiveness
6. Implementability
7. Cost
8. State acceptance
9. Community acceptance

Treatment technologies for contaminated soils and groundwater, along with an array of hazardous waste treatment options, will be covered in detail in Chapter 16.

PROBLEMS

13.1 A moist soil specimen has a volume of 0.15 m³ and weighs 2.83 kN. The water content is 12%, and the specific gravity of soil solids is 2.69. Determine
  a. Moist unit weight, γ
  b. Dry unit weight, γ_d
  c. Void ratio, e
  d. Degree of saturation, S

13.2 For a soil deposit in the field, the dry unit weight is 14.9 kN/m³. From the laboratory, the following were determined: $G = 2.66$, $e_{max} = 0.89$, $e_{min} = 0.48$. Find the relative density in the field.

13.3 For a sandy soil, the maximum and minimum void ratios are 0.85 and 0.48, respectively. In the field, the relative density of compaction of the soil is 29.3 percent. Given $G = 2.65$, determine the moist unit weight of the soil at $w = 10\%$.

13.4 Refer to the flow net shown in Figure 1.4. Given $k = 0.03$ cm/min, $H_1 = 10$ m, and $H_2 = 1.8$ m, determine the seepage loss per day per meter under the sheet pile construction.

13.5 The results of a sieve analysis of a granular soil are as follows:

U.S. Sieve No.	Sieve Opening (mm)	Percent Retained on Each Sieve
4	4.75	0
10	2.00	20
40	0.425	20
60	0.25	30
100	0.15	20
200	0.075	5

Determine the uniformity coefficient and coefficient of gradation of the soil.

13.6 For a normally consolidated clay of 3.0 m thickness, the following are given:

Average effective pressure = 98 kN/m²
Initial void ratio = 1.1
Average increase of pressure in the clay layer = 42 kN/m²
Compression index = 0.27

Estimate the consolidation settlement.

13.7 An oedometer test in a normally consolidated clay gave the following results.

Average Effective Pressure (kN/m²)	Void Ratio
100	0.9
200	0.82

Calculate the compression index.

13.8 A soil core is 5 cm in diameter, 10 cm long, and weighs 365 g. If the porosity is 34 percent, estimate the degree of saturation given a specific gravity of the solids of 2.65.
  a. 30 percent
  b. 40 percent
  c. 50 percent
  d. 60 percent

13.9 Estimate the soil phase concentration of toluene in a contaminated aquifer that has an aqueous toluene concentration of 40 mg/L if the octanol-water partition coefficient for toluene is 490 and the soil organic fraction is 3.5 percent. You may assume that the relationship between K_{oc} and K_{ow} may be expressed as $K_{oc} = 1.00 \log K_{ow} - 0.21$.
  a. 302 mg/kg
  b. 423 mg/kg
  c. 548 mg/kg
  d. 686 mg/kg

13.10 Water is pumped at a rate of 250 gpm from an unconfined aquifer that is 75 ft deep. Wells located 100 and 150 ft from the pumping well experience a drawdown of 9 and 7 ft, respectively. Determine the hydraulic conductivity of the aquifer in ft/day.
  a. 46.3 ft/day
  b. 23.2 ft/day
  c. 20.7 ft/day
  d. 11.7 ft/day

13.11 Rework Problem 13.10 if the aquifer is confined with an aquifer thickness of 75 ft.
  a. 86.7 ft/day
  b. 46.3 ft/day
  c. 23.2 ft/day
  d. 20.7 ft/day

SOLUTIONS

13.1

a. $\gamma = \dfrac{W}{V} = \dfrac{2.83}{0.15} = 18.87 \text{ kN/m}^3$

b. $\gamma_d = \dfrac{\gamma}{1+w} = \dfrac{1887}{1+\left(\frac{12}{100}\right)} = 16.85 \text{ kN/m}^3$

c. $\gamma_d = \dfrac{G\gamma_w}{1+e}; \quad e = \dfrac{G\gamma_w}{\gamma_d} - 1 = \dfrac{(2.69)(9.81)}{16.85} - 1 = 0.566$

d. $S = \dfrac{wG}{e} \times 100 = \dfrac{(0.12)(2.69)}{0.566} \times 100 = 57.03\%$

13.2 In the field

$\gamma_d = \dfrac{G\gamma_w}{1+e}; \quad e = \dfrac{G\gamma_w}{\gamma_d} - 1 = \dfrac{(2.66)(9.81)}{14.9} - 1 = 0.75$

$D_d = \dfrac{e_{max} - e}{e_{max} - e_{min}} = \dfrac{0.85 - 0.75}{0.89 - 0.48} = 34\%$

13.3

$D_d = 0.293 = \dfrac{e_{max} - e}{e_{max} - e_{min}} = \dfrac{0.85 - e}{0.85 - 0.45}; \quad e = 0.733$

$\gamma = \dfrac{G\gamma_w(1+w)}{1+e} = \dfrac{(2.65)(9.81)(1+0.1)}{1+0.733} = 16.5 \text{ kN/m}^3$

13.4

$Q = k\dfrac{N_f}{N_d}H = \dfrac{(0.03 \times 60 \times 24 \text{cm/day})}{100}\left(\dfrac{4}{6}\right)(10-1.8) = 2.36 \text{ m}^3/\text{day/m}$

13.5

Sieve Opening (mm)	Cumulative Percent Passing
4.75	100
2.00	80
0.425	60
0.25	30
0.15	10
0.075	5

So $D_{60} = 0.425$ mm; $D_{30} = 0.25$ mm; $D_{10} = 0.15$ mm

$$c_u = \frac{D_{60}}{D_{10}} = \frac{0.425}{0.15} = 2.83$$

$$c_c = \frac{D_{30}^2}{D_{60} \times D_{10}} = 0.98$$

13.6

$$\Delta H = \frac{C_c H}{1+e_i} \log\left(\frac{p_i + \Delta p}{p_i}\right) = \frac{(0.27)(3)}{1+1.1} \log\left(\frac{98+42}{98}\right) = 0.0597 \text{ m} = 59.7 \text{ mm}$$

13.7

$$C_c = \frac{e_1 - e_2}{\log\left(\frac{p_2}{p_1}\right)} = \frac{0.9 - 0.82}{\log\left(\frac{200}{100}\right)} = 0.266$$

13.8 a. First, the volume of the cylinder may be calculated as

$$V = \pi r^2 l \quad \Rightarrow \quad V = \pi (0.025 \text{ m})^2 (0.1 \text{ m}) \quad \Rightarrow \quad V = 1.96 \times 10^{-4} \text{ m}^3$$

Next the weight of dry solids can be determined by multiplying the volume of solids by the specific weight of the solids as

$$\text{wt. dry solids} = (1 - 0.34)(1.96 \times 10^{-4} \text{ m}^3)(2.65)(1000 \text{ kg/m}^3)$$
$$= 0.343 \text{ kg} = 343 \text{ g}$$

Now, the weight of water may be determined by difference as

$$\text{wt. H}_2\text{O} = 365 \text{ g} - 343 \text{ g} = 22 \text{ g}$$

Next, the volume of water for the soil core can be calculated, as

$$\text{Volume H}_2\text{O} = \frac{22 \text{ g}}{\left(1 \frac{\text{g}}{\text{cm}^3}\right)} = 22 \text{ cm}^3 = 2.2 \times 10^{-5} \text{ m}^3$$

Finally, the degree of saturation (S) can be calculated as $S = V_w/V_v$, or

$$S = \frac{2.2 \times 10^{-5} \text{ m}^3}{(0.34)(1.96 \times 10^{-4} \text{ m}^3)} = 0.33 = 33\%$$

13.9 b. The organic carbon partition coefficient (K_{oc}) can be estimated from the octanol-water partition coefficient (K_{ow}) as follows:

$$\log K_{oc} = 1.0 \log K_{ow} - 0.21 = \log(490) - 0.21 = 2.48 \quad \Rightarrow \quad K_{oc} = 302$$

Now, the soil phase contaminant concentration (X) is related to the aqueous phase contaminant concentration (C) by K_{sw} as

$$K_{sw} = \frac{X}{C} \quad \text{or} \quad X = K_{sw} C$$

Further, the soil-water partition coefficient (K_{sw}) can be estimated using the K_{oc} and fraction of organic carbon in the soil (f_{oc}) as follows:

$$X = K_{oc} f_{oc} C = (302)(0.035)(40) = 422.8 \text{ mg/kg}$$

13.10 b. Starting with Darcy's law for an unconfined aquifer:

$$K = \frac{Q \ln(r_2/r_1)}{\pi \left(h_2^2 - h_1^2\right)}$$

we see that we need Q in units of cfs, which may be calculated as

$$Q = \frac{250 \text{ gpm}}{\left(7.48 \frac{\text{gal}}{\text{ft}^3}\right)\left(60 \frac{\text{s}}{\text{min}}\right)} = 0.557 \text{ cfs}$$

The distance to the observation wells was given in the problem statement as

$$r_1 = 100 \text{ ft} \quad \text{and} \quad r_2 = 150 \text{ ft}$$

Next, drawdown is defined as the aquifer thickness minus the depth of the water at the observation well. Therefore, depth of water (h) is found by difference as follows:

$$h_1 = 75 - 9 = 66 \text{ ft} \quad \text{and} \quad h_2 = 75 - 7 = 68 \text{ ft}$$

Finally, substituting all values into Darcy's law and correcting for the units given in the answer choices, we find

$$K = \frac{(0.557 \text{ cfs}) \ln(150/100)}{\pi(68^2 - 66^2 \text{ ft}^2)} = 0.000268 \frac{\text{ft}}{\text{s}} \times \frac{3600 \text{ s}}{\text{hr}} \times \frac{24 \text{ hr}}{\text{day}} = 23.18 \frac{\text{ft}}{\text{day}}$$

13.11 d. Starting with Darcy's law for a confined aquifer:

$$K = \frac{Q \ln(r_2/r_1)}{2\pi m(h_2 - h_1)}$$

Using the values for Q, r_1, r_2, h_1, and h_2 in Solution 13.10 along with the aquifer thickness given in the problem statement, we can calculate K as

$$K = \frac{(0.557 \text{ cfs}) \ln(150/100)}{(2)\pi(75 \text{ ft})(68 - 66 \text{ ft})} = 0.00024 \text{ fps} \times \frac{3600 \text{ s}}{\text{hr}} \times \frac{24 \text{ hr}}{\text{day}} = 20.7 \frac{\text{ft}}{\text{day}}$$

CHAPTER 14

Water and Wastewater

OUTLINE

WATER QUALITY INDICATORS 582
pH ■ Acidity, Alkalinity, and Hardness ■ Ion Balance ■ Solids, Turbidity, and Color ■ Dissolved Oxygen, BOD, and COD ■ Additional WQIs

HYDRAULIC CHARACTERISTICS OF REACTORS 594
Residence Time ■ Ideal Reactors—PFR vs. CSTR ■ Non-Ideal Reactors ■ Summary Tables for Reactors

WATER TREATMENT 601
Oxidation, Mixing, Coagulation, and Flocculation ■ Softening ■ Sedimentation ■ Gravity Filtration ■ Disinfection

ADVANCED WATER TREATMENT 612
Absorption ■ Membrane Filtration ■ Ion Exchange

WATER TREATMENT RESIDUALS MANAGEMENT 617
Sludge Properties ■ Sludge Quantity ■ Thickening ■ Coagulant Recovery ■ Conditioning ■ Dewatering ■ Drying ■ Disposal and Reuse

WASTEWATER TREATMENT 627
Wastewater Characteristics ■ Preliminary Treatment ■ Primary Treatment ■ Secondary Treatment ■ Secondary Clarification ■ Tertiary Treatment ■ Disinfection

WASTEWATER SLUDGE TREATMENT 649
Sludge Generation ■ Preconditioning ■ Conditioning through Anaerobic Digestion ■ Conditioning through Aerobic Digestion ■ Postconditioning

PROBLEMS 658

SOLUTIONS 663

If one assesses the value of a substance by the length of time they can survive without it, then water is second only to oxygen as the most valuable substance on Earth. All civilizations arose in locations where there was a ready source, or one able to be developed, of the fresh water that was necessary for human consumption, as well as farming and animal domestication. While the quality of the water resource varied by location, there is evidence that early civilizations adopted some treatment strategies to purify their water. Water purification today is based on the type of contaminants that are present in a particular source, with water quality standards setting the levels of purity required for human consumption.

Although water-borne diseases are relatively rare in the United States, they are responsible for millions of deaths worldwide. Some of the most common water-borne diseases are shown in Table 14.1.

Table 14.1 Common water-borne diseases

Disease	Effects	Scope
Arsenic poisoning	Cancers of skin, bladder, kidneys, and lungs	Currently no reliable worldwide estimate; up to 77 million in Bangladesh affected
Cholera	Severe diarrhea and vomiting leading to extreme dehydration and death	5000 deaths in year 2000
Campylobacteriosis	Diarrhea (often including the presence of mucus and blood), abdominal pain, malaise, fever, nausea, and vomiting	Approximately 5%–14% of all diarrhea worldwide is thought to be caused by Campylobacter
Malaria	Fever, chills, headache, muscle aches, tiredness, nausea and vomiting, diarrhea, anemia, and jaundice (yellow coloring of the skin and eyes). Convulsions, coma, severe anemia, and kidney failure can also occur.	300–500 million cases of malaria, with over 1 million deaths each year
Methaemoglobinemia	Reduced ability of blood to transport oxygen; highest risk for bottle-fed infants	Relatively rare
Typhoid	Symptoms can be mild or severe and include sustained fever as high as 39°C–40°C, malaise, anorexia, headache, constipation or diarrhea, rose-colored spots on the chest area, and enlarged spleen and liver.	17 million cases worldwide

Source: World Health Organization, *www.who.int/water_sanitation_health/diseases/diseasefact/en/index.html*.

WATER QUALITY INDICATORS

The U.S. EPA establishes quality standards for several water quality indicators (WQI) for both drinking water and surface waters. Maximum contaminant levels (MCLs) are enforceable limits, established as a result of comprehensive risk assessments (covered in Chapter 6), set to provide an adequate measure of safety to protect human health. Primary drinking water standards are MCLs that have been established for dozens of specific organic and inorganic chemicals, as well as biological contamination, disinfection by-products, radionuclides, and turbidity. As new information is forthcoming, the primary standards are continuously updated, both with respect to specific chemicals, as well as their MCLs, and should be researched prior to any remediation activity. Secondary contaminant standards are presented in Table 14.2 and are WQIs that are based more on aesthetic quality, rather than health hazard, and are not enforceable. However, they are generally accepted as appropriate targets and are used as an indicator as to general watershed health.

Table 14.2 Secondary contaminant standards for aesthetics of drinking water

Contaminant	Standard	Contaminant	Standard
Al	0.2 mg/L	Mn	0.05 mg/L
Cl$^-$	250 mg/L	odor	3 threshold #s
color	15 color units	pH	6.5–8.5
Cu	1.0 mg/L	Ag	0.1 mg/L
corrosivity	none	SO$_4^{2-}$	250 mg/L
F$^-$	2 mg/L	TDS	500 mg/L
foam	0.5 mg/L	Zn	5 mg/L
Fe	0.3 mg/L		

pH

Hydrogen ion activity (pH), often the most reported water quality indicator (WQI), is generally used to describe an aqueous environment as alkaline (basic) or acidic. The determination of pH is defined as

$$pH = -\log_{10}[H^+]$$

where the quantity [H$^+$] is the molar hydrogen ion concentration. An increase in [H$^+$] decreases pH (more acidic), while a decrease in [H$^+$] increases pH (more alkaline).

Although rarely used in practice, the hydroxide [OH$^-$] activity can be defined similarly as

$$pOH = -\log_{10}[OH^-]$$

where the quantity [OH$^-$] is the molar hydroxide ion concentration. Further, by definition

$$[H^+][OH^-] = 10^{-14}$$

or, equivalently

$$pH + pOH = 14$$

Solution neutrality dictates that hydrogen ion and hydroxide activity (and therefore concentration) must be equal, and this occurs at

$$[H^+] = [OH^-] = 10^{-7}$$

or equivalently

$$pH = pOH = 7$$

When hydrogen ion and hydroxide are present in solution at concentrations that yield a concentration product above 10^{-14}, it is most likely that they will combine to form water (H$_2$O) through the following neutralization reaction

$$H^+ + OH^- \rightarrow H_2O$$

Acidity, Alkalinity, and Hardness

Although the most widely known of the WQIs, pH only describes hydrogen acidity (or equivalently, hydroxide alkalinity) and neglects the probability that other chemical species may be present, providing neutralization (buffer) capacity to the system. These other chemical species possess the ability to neutralize added acid or alkali, with minimal change in system pH. Alkalinity can be expressed as *ability of a solution to neutralize [H⁺]*, while acidity can be expressed as *ability of a solution to neutralize [OH⁻]*.

Natural acidity generally arises from the release of CO_2 during biological activity that is not released to the atmosphere or is not neutralized by natural elements. However, some natural environments (such as the drainage from abandoned mine sites) may contain substantial amounts of mineral acidity, dissolved species that react with oxygen in aerobic environments or are transformed in biological processes. Compounds such as pyrite (FeS_2) may oxidize to produce sulfuric acid, which in turn increases the hydrogen acidity (i.e., lowers the pH). For example, most acid mine drainage treatment systems are designed to address this acidity through alkaline dosing and removal of the metal precipitates (iron salts) formed during the reaction.

Natural waters generally have a pH between 6 and 8.5, and, therefore, the primary source of alkalinity is bicarbonate (HCO_3^-), while higher pH waters may have a significant carbonate (CO_3^{2-}) and/or hydroxide (OH^-) contribution to alkalinity, as seen in Figure 14.1. Because many compounds may contribute to alkalinity, it is necessary to employ a standard unit for expressing this quantity. This is usually done by converting all species to equivalents of $CaCO_3$ and expressing the concentration of the sum of all species as mg/L $CaCO_3$. This is accomplished mathematically by dividing the species concentration in mg/L by the species EW and multiplying by the EW of $CaCO_3$, 50 mg/meq (see Example 14.2).

Hardness can be expressed as *the sum of all multivalent cations*, although the primary source of hardness is Ca^{2+} and Mg^{2+}. Hardness is a result of water contact with minerals in subsurface environments and is an issue due to the potential for scaling in high temperature process equipment for industrial customers and, to a lesser degree, soap consumption for residential and commercial customers. As was the case with alkalinity, many compounds contribute to hardness, and it is convenient to express hardness as mg/L $CaCO_3$. Waters containing up to 300 mg/L of hardness as $CaCO_3$ are usually considered "hard" and require treatment, although many customers object to "moderately hard" water (up to 150 mg/L as $CaCO_3$). Removal of hardness from municipal water supplies is called softening, and the process is detailed in this chapter.

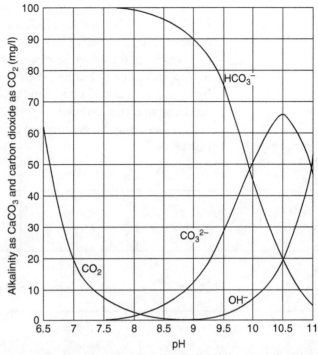

Figure 14.1 Alkalinity species dependence as a function of pH

Source: Warren Viessman, Jr., and Mark J. Hammer, *Pollution Control*, 7th ed., © 2005. Reprinted by permission of Pearson Education, Inc., Upper Saddle River, N.J.

Example 14.1

Hardness

Determine the hardness, in units of mg/L $CaCO_3$, of a solution with a Ca^{2+} concentration of 75 mg/L and a Mg^{2+} concentration of 20 mg/L.

Solution

Hardness is defined as the sum of the multivalent cations. In this case, we need to convert the Ca^{2+} and Mg^{2+} concentrations into meq/L, add the two together, and then convert those units back to mg/L of $CaCO_3$. This may be completed as follows:

$$\left[Ca^{2+}\right] = 75\,\frac{mg}{L} \times \frac{1\,meq}{20.0\,mg} = 3.75\,\frac{meq}{L} \text{ and } \left[Mg^{2+}\right] = 20\,\frac{mg}{L} \times \frac{1\,meq}{12.2\,mg} = 1.64\,\frac{meq}{L}$$

$$\text{hardness} = 3.75 + 1.64 = 5.39\,\frac{meq}{L} \times \frac{50.0\,mg\,CaCO_3}{1\,meq} = 270\,\frac{mg}{L} \text{ as } CaCO_3$$

Ion Balance

Water quality data is most often reported to the user as mg/L in a tabular format. In order to satisfy electroneutrality requirements for a solution, the sum of all cations must equal the sum of all anions when expressed in equivalence units, most often *milliequivalents of solute per liter solution* (meq/L or mN). Often, having a visual representation of possible ionic species combinations is useful. These ion balances (sometimes referred to as chemical bar graphs) are useful in the evaluation of softening requirements as seen in Figure 14.2. Cations are listed in the order of Ca^{2+}, Mg^{2+}, Na^+, and K^+, while anions are shown below the cations and are listed in the order of OH^- (if present), CO_3^{2-} (if present), HCO_3^-, SO_4^{2-}, Cl^-. Additional divalent or trivalent cations may be represented before Na^+ if they contribute significantly to sample hardness. Further, if CO_2 is present, it is often included in the bar graph by adding it as both a cation and anion to the left of the zero line. Again, solution electroneutrality requires both bars to be the same length, and hypothetical speciation can easily be determined by inspection as shown in the bottom portion of Figure 14.2.

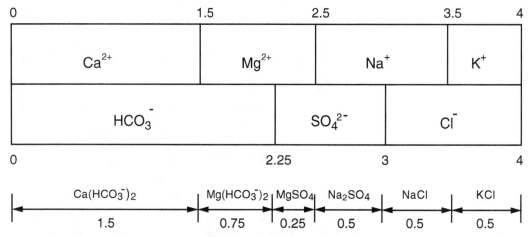

Figure 14.2 Example of ion bar graph; all units are meq/L

Example 14.2

Ion balance

Given the following concentrations of ions, determine the probable concentration of Na_2SO_4.

Ca^{2+} = 150 mg/L as $CaCO_3$; Mg^{2+} = 100 mg/L as $CaCO_3$; Na^+ = 46 mg/L; K^+ = 39 mg/L; HCO_3^- = 225 mg/L as $CaCO_3$; SO_4^{2-} = 72 mg/L; Cl^- = 71 mg/L

Solution

To generate a chemical bar graph, first determine the normality of each chemical species as

$$Ca^{2+} = \frac{150 \text{ mg/L as CaCO}_3}{50 \frac{\text{mg CaCO}_3}{\text{meq}}} = 3 \text{ meq/L}$$

$$Mg^{2+} = \frac{100 \text{ mg/L as CaCO}_3}{50 \frac{\text{mg CaCO}_3}{\text{meq}}} = 2 \text{ meq/L}$$

$$Na^+ = \frac{46 \text{ mg/L}}{23 \frac{\text{mg Na}^+}{\text{meq}}} = 2 \text{ meq/L}$$

$$K^+ = \frac{39 \text{ mg/L}}{39 \frac{\text{mg } K^+}{\text{meq}}} = 1 \text{ meq/L}$$

$$HCO_3^- = \frac{225 \text{ mg/L as CaCO}_3}{50 \frac{\text{mg CaCO}_3}{\text{meq}}} = 4.5 \text{ meq/L}$$

$$SO_4^{2-} = \frac{72 \text{ mg/L}}{48 \frac{\text{mg SO}_4^{2-}}{\text{meq}}} = 1.5 \text{ meq/L}$$

$$Cl^- = \frac{71 \text{ mg/L}}{35.5 \frac{\text{mg Cl}^-}{\text{meq}}} = 2 \text{ meq/L}$$

Next, draw the ion graph in units of meq/L, as shown in Exhibit 1.

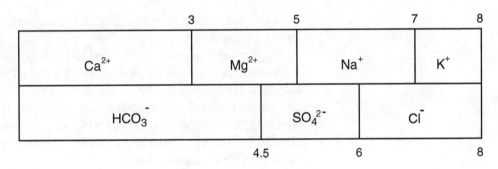

Exhibit 1

From the bar graph, the overlap for Na^+ and SO_4^{2-} occurs from 6 to 5 meq/L; therefore

$$Na_2SO_4 = 6 - 5 \text{ meq/L} = 1 \text{ meq/L}$$

Finally, this can be converted into mass concentration units by multiplying by the EW as

$$(1 \text{ meq/L}) \times (71 \text{ mg/meq}) = 71 \text{ mg/L } [Na_2SO_4]$$

Solids, Turbidity, and Color

As a WQI, total solids (TS) can be expressed as *residue on evaporation* or *residue upon drying at 105°C*. TS values are determined for an aqueous sample by placing a known volume of solution into a drying vessel and drying in a 105°C oven. The value for TS is calculated as

$$TS = \frac{\text{weight of residue}}{\text{volume of sample dried}}$$

and expressed in units of mg/L. Total solids are comprised of many compounds, and it is often necessary to know the forms of the solid to select appropriate treatment processes. Total suspended solids (TSS) are generally particles that are not in solution but rather are discrete particles contained in a sample. TSS are often subclassified into settleable or nonsettleable categories, based on the potential to use gravity as a means of solids removal. TSS are determined by filtering a sample through a 0.45 μm filter, which was previously dried at 105°C and weighed. The filter with the solids is then dried at 105°C and reweighed. The value for TSS is calculated as

$$TSS = \frac{\text{weight of residue on filter}}{\text{volume of sample filtered}}$$

and expressed in units of mg/L. Total dissolved solids (TDS) are the sum of all dissolved species and is determined in the same manner as TS, only using the filtrate from the TSS sample. The value for TDS is calculated as

$$TDS = \frac{\text{weight of residue from filtrate}}{\text{volume of sample dried}}$$

and expressed in units of mg/L. It should be noted that often, TSS or TDS can be determined by difference calculation using one of the two values and a known TDS concentration as follows:

$$TS = TSS + TDS$$

However, due to the simplicity of the method, validation by measurement is strongly suggested.

Two other important WQIs used to identify solids in water samples are total volatile solids (TVS) and total fixed solids (TFS). Fixed solids can be expressed as *residue upon igniting at 550°C*, and the volatile solids are often referred to as loss on ignition (LOI). TFS are determined by placing the TS vessel (after drying and weighing) into an ashing oven at 550°C for two hours. After ignition, the vessel is allowed to cool in a desiccator to room temperature prior to being weighed. The value for TFS is calculated as

$$TFS = \frac{\text{weight of residue after ignition}}{\text{original volume of TS sample}}$$

and expressed in units of mg/L. TVS is also expressed in units of mg/L and is determined by difference calculation using the following relationship:

$$TS = TFS + TVS$$

In the case of wastewater treatment, it is convenient to refer to the portion of suspended solids that are primarily biological organisms. This is determined by igniting the suspended solids, after determining TSS using a glass-fiber filter, and reporting the total volatile suspended solids (TVSS), which would be calculated in the same manner as TVS above.

Example 14.3

TDS

Determine the concentration of total dissolved solids from the data given below.

a). Volume of water sample = 100 mL

b). Empty dish = 56.345 g

c). Dish + dried sample = 56.612 g

d). Filter paper = 1.629 g

e). Filter paper + dried solids = 1.653 g

f). Dish + dried sample after filtering = 56.589 g

Solution

TDS is defined as

$$\text{TDS} = \frac{\text{weight of dry residue after filtering}}{\text{volume dried}}$$

From the data given, TDS is calculated as

$$\text{TDS} = \frac{56.589 - 56.345 \text{ g}}{100 \text{ mL}} = \frac{0.244 \text{ g}}{0.1 \text{ L}} = \frac{2.44 \text{ g}}{\text{L}} = \frac{2400 \text{ mg}}{\text{L}}$$

Example 14.4

Fixed solids

Seventy percent of the suspended solids in a wastewater sample are volatile. Determine the fixed solids of the sample if total solids are 1400 mg/L and total dissolved solids are 200 mg/L. You may assume that none of the dissolved solids are volatile.

Solution

We first need to determine the concentration of total suspended solids as follows:

$$\text{TS} = \text{TSS} + \text{TDS} \Rightarrow \text{TSS} = \text{TS} - \text{TDS} = 1400 - 200 = 1200 \frac{\text{mg}}{\text{L}}$$

Now we can find the fraction of TSS that are volatile as follows:

$$\text{TVSS} = (0.7)\left(1200 \frac{\text{mg}}{\text{L}}\right) = 840 \frac{\text{mg}}{\text{L}}$$

Further, if we assume that none of the dissolved solids is volatile, we may write:

$$\text{TVS} = \text{TVSS} + \text{TVDS} = \text{TVSS} + 0 = 840 \frac{\text{mg}}{\text{L}}$$

Finally, fixed solids may be determined as follows:

$$\text{TS} = \text{TVS} + \text{TFS} \Rightarrow \text{TFS} = \text{TS} - \text{TVS} = 1400 - 840 = 560 \frac{\text{mg}}{\text{L}}$$

One specific WQI that arises from nonsettleable solids is termed turbidity and generally describes waters that have suspended solids that scatter light as light is passed through the water. The standard unit of turbidity is the NTU, which is measured using a set of standard samples and a procedure called nephelometry (a measure of light scattering at 90° from the incident light). These solids may be inorganic (such as clays), organic, or biological in nature and are generally comprised of colloidal suspensions. These colloids are aesthetically unpleasing and may interfere with the disinfection process in water treatment plants. Turbidity is removed through coagulation, clarification (sedimentation), and filtration processes as described later in this chapter.

A WQI related to turbidity is color. Color is also caused by colloidal particles, most of which are organic in origin (for example, tannins or humic matter), although some dissolved inorganic chemicals may also contribute to color (such as Fe). Care should also be employed not to confuse true color with apparent color, the latter of which may arise from suspension of settleable solids in waters with high turbulence. As colloidal particles are the cause of color, treatment is the same as was the case for turbidity.

Dissolved Oxygen, BOD, and COD

An important WQI in wastewater treatment or the assessment of the quality of habitat for biological organisms is the dissolved oxygen (DO) concentration. Oxygen solubility in water is limited and is highly dependent on temperature, with maximum DO concentrations for a given temperature listed in saturation tables, as shown in Table 14.3. As with most gas-liquid systems, oxygen solubility is inversely proportional to temperature, with a high DO concentration of approximately 14.5 mg/L at 0°C, reducing to approximately 7 mg/L at 35°C for aqueous systems in contact with air at atmosphere pressure. In some processes, oxygen solubility can be increased by increasing total system pressure or by increasing the oxygen concentration in the gas phase, although both methods are generally expensive to implement. However, oxygen solubility in water decreases with increasing solute concentrations, such as the case of salt water or wastewaters, and must be accounted for.

Table 14.3 DO saturation values for clean water exposed to air at 760 mmHg

Temp. [°C]	DO [mg/L]	Temp. [°C]	DO [mg/L]
0	14.6	16	10.0
2	13.8	18	9.5
4	13.1	20	9.2
6	12.5	22	8.8
8	11.9	24	8.5
10	11.3	26	8.2
12	10.8	28	7.9
14	10.4	30	7.6

Due to the activity of biological agents, natural waters often do not possess the maximum oxygen concentration anticipated for the system temperature and DO concentrations are reported as percent of saturation. For these cases, DO concentrations would be calculated by multiplying the saturation value from the tables by the percent saturation. Further, the discharge of organic material from municipal or industrial wastewaters can have a strong impact on DO levels in the receiving stream, because aerobic biological activity consumes DO as organisms degrade the organic constituents. Modeling of the impact of wastewater discharges on receiving streams is covered under DO Sag Modeling in Chapter 7.

The biological transformation of organic constituents forms the basis for secondary treatment of municipal and industrial wastewaters. One analytical technique employed to determine the strength of a wastewater sample is biochemical oxygen demand (BOD). While BOD actually measures the amount of DO consumed in the biological utilization of substrate, it is related to the amount of biodegradable organic matter (OM) present. In short, the method places 300 mL of sample (including dilution water when necessary) into a sealed vessel and records the DO levels at time zero and at the end of five days of incubation at 20°C in a dark chamber. BOD_5 indicates a five-day incubation period, and the value measured in units of mg/L is calculated as

$$BOD_5 = \left(DO_i - DO_f\right) \times \left(DF\right)$$

where subscripts i and f are initial and final DO concentrations, and DF is the dilution factor, which is calculated as

$$DF = \frac{\text{volume of test bottle in mL}}{\text{volume of wastewater added in mL}}$$

Typical volumes for BOD test bottles are 300 mL. The dilution of a sample may be necessary to keep DO_f above 2 mg/L, which will ensure that any reduction in biological activity is due to substrate (OM) depletion and not depletion of the oxygen. Further, some samples require seeding, which is a process that introduces an active microbial population for testing water samples where adequate bacteria are not readily present. Presentation of the additional equations necessary to calculate the BOD in seeded bottles is beyond the scope of this review.

The five-day incubation period is chosen as a compromise between sufficient time to discern a significant change in DO and ending the test before organisms utilize some of the oxygen for nitrification, as seen in Figure 14.3. If a first-order reaction rate is assumed and DO concentrations from the BOD bottle are available for each day of the incubation period, a plot of BOD versus time can be regressed to determine the ultimate BOD (often represented as L) and reaction rate constant from the following equation:

$$BOD_t = L(1 - e^{-kt})$$

where k is the reaction rate in days^{-1}, and t is the time in days. The value of L is important in DO-sag modeling in streams as demonstrated in Chapter 7.

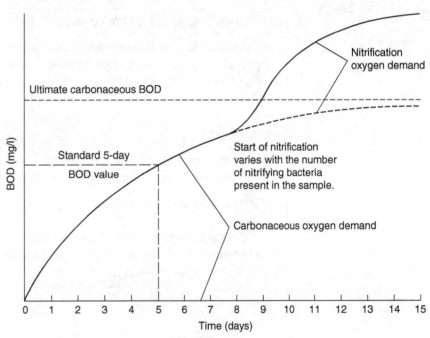

Figure 14.3 Example of BOD curve

Source: Warren Viessman, Jr., and Mark J. Hammer, *Pollution Control*, 7th ed., © 2005. Reprinted by permission of Pearson Education, Inc., Upper Saddle River, N.J.

Example 14.5

Ultimate BOD

Determine the ultimate BOD concentration of a wastewater sample that has a BOD_5 of 185 mg/L and a biological reaction rate of 0.37 day^{-1}.

Solution

Using the ultimate BOD equation and substituting a value of five days for t, we can write

$$BOD_5 = L(1 - e^{-k5}) = 185 \text{ mg/L}$$

Since $k = 0.37$ day^{-1}, then the product $kt = (0.37 \text{ day}^{-1})(5 \text{ days}) = 1.85$, or

$$L(1 - e^{-1.85}) = 185 \text{ mg/L}$$

Rearranging, this may be expressed as

$$\frac{185}{L} = 1 - e^{-1.85} \Rightarrow 1 - \frac{185}{L} = e^{-1.85} \Rightarrow 1 - \frac{185}{L} = 0.157$$

Solving for L gives

$$0.843 = \frac{185}{L} \quad \text{or equivalently} \quad L = 219.5 \text{ mg/L}$$

Example 14.6

Estimating BOD sample volume

You anticipate the five-day BOD concentration of a wastewater sample will be approximately 220 mg/L. Determine the maximum amount of sample that may be placed into a 300 mL BOD bottle if the initial dissolved oxygen concentration is 9.2 mg/L.

Solution

Using the expression for BOD and initial DO value given, and the fact that the final DO value must be at least 2 mg/L in order for the test to be valid, we may write

$$BOD = (DO_i - DO_f) \times DF = (9.2 - 2) \times DF = 7.2 \, DF = 220 \, \frac{mg}{L} \Rightarrow DF = 30.56$$

Using this value for the dilution factor and the equation given for DF, we may determine the volume of sample placed into the bottle as follows:

$$DF = \frac{300 \text{ mL}}{X} = 30.56 \Rightarrow X = \frac{300 \text{ mL}}{30.56} = 9.8 \text{ mL}$$

Example 14.7

BOD final DO

A wastewater with an ultimate BOD of 180 mg/L has a BOD_5 of 130 mg/L. Estimate the dissolved oxygen (DO) concentration in the test bottle after ten days if the initial DO was 8.9 mg/L and 10 mL of sample was placed in a 300 mL bottle.

Solution

From the data given in the problem statement, and the equation for BOD determination from ultimate BOD, the biological constant can be determined as

$$\text{BOD} = L\left(1 - e^{-kt}\right) \text{ for BOD}_5 \quad 130\,\frac{\text{mg}}{\text{L}} = \left(180\,\frac{\text{mg}}{\text{L}}\right)\left(1 - e^{-5k}\right) \text{ or } k = 0.256 \text{ day}^{-1}$$

Now the value for BOD_{10} can be determined as

$$\text{BOD}_{10} = \left(180\,\frac{\text{mg}}{\text{L}}\right)\left\{1 - \exp\left[-(10 \text{ days})\left(0.256 \text{ day}^{-1}\right)\right]\right\} = 166.1\,\frac{\text{mg}}{\text{L}}$$

Plugging this into the BOD equation based on DO, DO_f can be estimated as

$$\text{BOD} = \left(DO_i - DO_f\right)\left(\frac{V_{bottle}}{V_{sample}}\right) \quad \text{or} \quad 166.1\,\frac{\text{mg}}{\text{L}} = \left(8.9 - DO_f\right)\left(\frac{300 \text{ mL}}{10 \text{ mL}}\right)$$

$$\frac{166.1\,\frac{\text{mg}}{\text{L}}}{30} = 8.9 - DO_f \quad \text{or} \quad DO_f = 8.9 - 5.54 = 3.36\,\frac{\text{mg}}{\text{L}}$$

Often, in municipal and industrial wastewater discharges, five days is too long to wait for analytical results to be finalized, as flows generally are continuous and high contaminant levels could be discharged while waiting for test results. A rapid test related to BOD is chemical oxygen demand (COD), where a strong chemical oxidant (potassium dichromate, $K_2Cr_2O_7$) replaces the bacteria and DO in the oxidation of OM. After two hours in a heated mantle, samples are analyzed in a spectrometer for Cr^{3+}, which is generated from the reduction of the original Cr^{6+} as oxygen is consumed in the reaction with the OM. While the time required to obtain results is significantly reduced, the strength of the oxidant may overestimate the biodegradable fraction of the OM. This is due to the complete oxidation of all OM, including the fraction that is not biodegradable, and possibly some inorganic compounds as well.

This fact also makes the COD test especially appropriate for industrial wastewaters, where toxic compounds would otherwise inhibit biological activity and reduce or eliminate the usefulness of BOD results. However, for most domestic wastewater operations, relationships between BOD and COD can be developed from historical samples, and a weighting factor applied to the COD value can usually provide a reasonable estimate for BOD. Typical values for BOD_5 and COD are 200 mg/L and 300 mg/L, respectively, for domestic wastewaters.

Additional WQIs

Additional WQIs that may impact the quality of aquatic habitat include nutrients (generally various forms of N and P), chlorine residual (present due to disinfection processes), heavy metals (originating from geologic formations or industrial discharges), and trace organics (from runoff, landfill leachate, industrial discharges, or sanitary sewers). Each varies in its impact on aquatic systems and treatment processes necessary to reduce these constituents to levels that meet ambient water quality standards. Two additional WQIs that may arise from water contact with organic matter and biological activity are taste and odor. While it is difficult to identify a specific cause, it is generally accepted that municipal water supplies must be treated for objectionable taste and odor. This is usually accomplished through oxidation methods or carbon adsorption.

HYDRAULIC CHARACTERISTICS OF REACTORS

Most water and wastewater treatment processes are completed in tanks that serve as reactors, which possess varying fundamental hydraulic characteristics. Selection of the appropriate reactor system, and the operating parameters that drive it, is often the difference between successful and unsuccessful removal of the target contaminant. Reaction vessels may be operated in *batch mode,* where material is charged to a reactor, allowed to react, then subsequently discharged, or in *continuous* mode, where reactions occur during constant flow through the system. It is the continuous process units that are used, almost exclusively, in water and wastewater systems. The *FE Supplied-Reference Handbook* covers reactor properties in both the chemical engineering section and the environmental engineering section.

Residence Time

Continuous reactors are often sized based upon the *residence* (also called *retention* or *detention*) *time,* which can be defined as the time a molecule of water is present in the reactor vessel. Since there are an infinite number of potential pathways from the inlet to the outlet through a specific reactor, it is convenient to define the *theoretical mean residence* (detention) *time* (t_R), which is defined as the average residence time of all molecules, and can be expressed as

$$t_R = \frac{V}{Q}$$

where V is the volume of the reactor and Q is the volumetric flow rate through the reactor.

Ideal Reactors—PFR vs. CSTR

Ideal reactors describe theoretical behavior, and can rarely be achieved in practice. However, well operated reactors can come close, and the ideal assumptions are often used to describe reactor behavior. Two types of ideal reactors will be discussed in this review, the *continuous stirred tank reactor* (CSTR), often called the *completely mixed reactor,* and the *plug-flow reactor* (PFR).

The PFR assumes that, while reactions may take place that transform species and therefore change the concentration of any species, there is no mixing of molecules inside the reactor in the axial (longitudinal) dimension. Often this type of reactor is called a *tubular* reactor, and is best envisioned as a long cylinder or pipe (i.e., a tube). Any material in the inlet stream (*influent*) maintains its relative position throughout the reactor, and is discharged in the outlet stream (*effluent*) in the same sequence as it entered, and therefore, all molecules have a residence time equal to t_R.

Usually two cases are considered when completing a mass balance for a particular species in any reactor, and hydraulic analysis is simplified by the assumption of a conservative (non-reactive) species, often called a *tracer* chemical. The first case is called a *step input,* and assumes that a tracer, not originally present in the reactor (i.e., $C = 0$ for $t < t_0$), is added to the reactor at some time (t_0), continuously and at a constant concentration (C_0). If t_0 is set equal to zero, the effluent concentration of the tracer will be zero for all $t < t_R$, and will be equal to C_0 for all $t \geq t_R$. The second case is called a *pulse input,* and assumes that a defined quantity of the tracer, not originally present in the reactor (i.e., $C = 0$ for $t < t_0$), is added all at once to the reactor at some time (t_0). If t_0 is set equal to zero, the effluent concentration will be zero for all times, except t_R, which is when all of the material added will discharge at the same time. Effluent profiles for a PFR with step or pulse input, and a flow schematic for the PFR can be visualized in Figure 14.4.

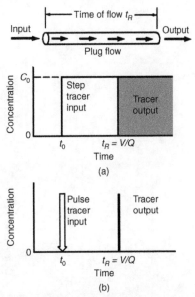

Figure 14.4 Schematic and effluent profiles for a non-reactive species in a PFR (Viessman, Jr., Warren, and Hammer, Mark J., *Pollution Control*, 7th Edition, © 2005. Reprinted by permission of Pearson Education, Inc., Upper Saddle River, NJ.)

The CSTR assumes that any material in the influent is instantaneously distributed evenly throughout the reactor, that every location inside the reactor has the same concentration distribution for all species, and that the concentration of any species in the effluent is the same as the concentration of that species inside the reactor. The *completely mixed* assumption is best envisioned as a set of high-speed impellers inside a tank, like the common household blender. The mass balance for this type of reactor states that the rate of accumulation of a species inside the reactor is equal to the mass entering the system minus the mass exiting the system. This can be expressed mathematically as

$$V \frac{dC}{dt} = QC_0 - QC = Q(C_0 - C)$$

All variables are as defined previously. For a step input, assuming that the value of t_0 is set equal to zero, this is integrated as follows:

$$C = C_0 \left[1 - \exp\left(\frac{-t}{t_R}\right)\right]$$

Alternatively, for a pulse input, integration with new boundary conditions yields

$$C = C_0 \exp\left(\frac{-t}{t_R}\right)$$

The effluent profiles, along with a flow schematic can be visualized in Figure 14.5. Remember, these profiles describe the reactor hydraulic characteristics, using a non-reactive species, and the equations shown assume t_0 is set equal to zero.

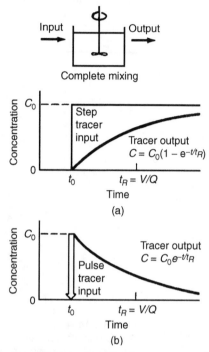

Figure 14.5 Schematic and effluent profiles for a non-reactive species in a CSTR (Viessman, Jr., Warren, and Hammer, Mark J., *Pollution Control*, 7th Edition, © 2005. Reprinted by permission of Pearson Education, Inc., Upper Saddle River, NJ.)

Non-Ideal Reactors

Non-ideal reactors are more realistic. However, as they possess hydraulic characteristics of both ideal reactor types, they are therefore more difficult to model. One approach is to assume primarily PFR behavior with a small amount of mixing, causing the concentration profile to distribute (disperse) around the centroid of the expected elution time, as shown in Figure 14.6. The *dispersion number* is a dimensionless measure of the amount of mixing present, and ranges from zero (PFR) to infinity (CSTR).

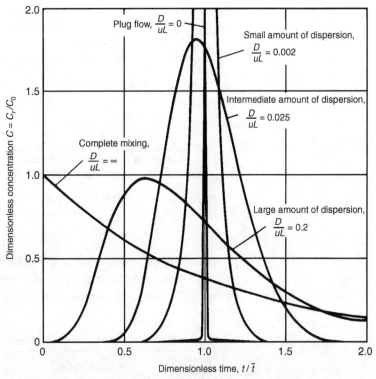

Figure 14.6 Effects of dispersion number (D/uL) in non-ideal PFR (Viessman, Jr., Warren, and Hammer, Mark J., *Pollution Control*, 7th Edition, © 2005. Reprinted by permission of Pearson Education, Inc., Upper Saddle River, NJ.)

Another method used in the laboratory to model PFR mixing behavior, is to represent the PFR as a series of CSTRs connected in series where the effluent of each reactor serves as the influent to the subsequent reactor. The concentration in the final reactor (C_n), and therefore the final effluent concentration can be described mathematically as

$$C_n = C_0 \frac{\left(\frac{t}{t_R}\right)^{n-1}}{(n-1)!} \exp\left[\frac{-t}{t_R}\right]$$

where n is the number of reactors in the series, C_0 is the influent concentration of the species into reactor 1, t is the time elapsed since the addition of the species into reactor 1, and t_R is the theoretical mean residence time of all n reactors (i.e., $n \cdot V/Q$).

Summary Tables for Reactors

Table 14.4 provides a series of equations, derived from the generic mass balance equation, that can be used for non–steady state applications in batch reactors and steady state reactions in PFRs and CSTRs.

Table 14.4 Comparison of performance* for decay reactions

Reaction Order	Ideal Batch	PFR	CSTR
0	$C_t = C_o - k_0 t$	$C_{eff} = C_o - k_0 \theta$	$C_{eff} = C_o - k_0 \theta$
1	$C_t = C_o[\exp(-k_1 \cdot t)]$	$C_{eff} = C_o[\exp(-k_1 \theta)]$	$C_{eff} = C_o/(1 + k_1 \cdot \theta)$
2	$C_t = \dfrac{C_o}{1 + k_2 \cdot t \cdot C_o}$	$C_{eff} = \dfrac{C_o}{1 + k_2 \cdot \theta \cdot C_o}$	$C_{eff} = \dfrac{(4k_2 \cdot \theta \cdot C_o + 1)^{0.5} - 1}{2k_2 \cdot \theta}$

*C_t = Concentration at time t for batch reactor (non–steady state).
C_{eff} = Concentration in effluent of PFR or CSTR (steady state).

Source: Adapted from M. L. Davis and D. A. Cornwell, *Introduction to Environmental Engineering*, 4th ed. (McGraw-Hill, 2006).

Alternatively, rather than solving for the effluent concentration, the mass balance equation could be manipulated to solve for detention time. A series of useful equations for the most common cases is shown in Table 14.5.

Table 14.5 Comparison of steady state mean retention times* for decay reactions

Reaction Order	Ideal Batch	PFR	CSTR
0	$\theta = (C_o - C_t)/k_0$	$\theta = (C_o - C_{eff})/k_0$	$\theta = (C_o - C_{eff})/k_0$
1	$\theta = \dfrac{\ln(C_o/C_t)}{k_1}$	$\theta = \dfrac{\ln(C_o/C_{eff})}{k_1}$	$\theta = \dfrac{(C_o/C_{eff}) - 1}{k_1}$
2	$\theta = \dfrac{(C_o/C_t) - 1}{k_2 \cdot C_o}$	$\theta = \dfrac{(C_o/C_{eff}) - 1}{k_2 \cdot C_o}$	$\theta = \dfrac{(C_o/C_{eff}) - 1}{k_2 \cdot C_{eff}}$

*C_t = Concentration at time t for batch reactor (non–steady state).
C_{eff} = Concentration in effluent of PFR or CSTR (steady state).

Source: Adapted from M. L. Davis and D. A. Cornwell, *Introduction to Environmental Engineering*, 4th ed. (McGraw-Hill, 2006).

Example 14.8

Decay in a batch reactor

A graduate student conducts a research project using a lab-scale, well-mixed, cylindrical tank (diameter = 8 cm, height = 18 cm). The contaminant (cyanide) is degraded using UV light according to first-order kinetics ($k = 3.2$ day^{-1}). If the concentration of cyanide in the tank is 30 mg/L at the beginning of the test, at what time will 99.9% removal be obtained?

Solution

Exhibit 2 shows a schematic of the problem.

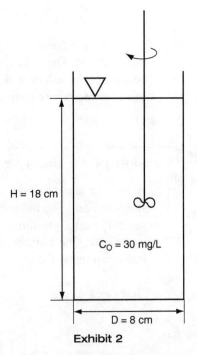

Exhibit 2

The solution begins by stating the governing equation for a batch reactor:

$$\left(\frac{dm}{dt}\right)_{cv} = \pm \dot{m}_{rxn}$$

Knowing that this is a first-order *decay* reaction, the right-hand side can be simplified to

$$\left(\frac{dm}{dt}\right)_{cv} = -V \cdot k_1 \cdot C$$

The rate of change of mass in the control volume can be represented as $V\frac{dC}{dt}$, which simplifies the equation to

$$V\frac{dC}{dt} = -V \cdot k_1 \cdot C$$

The volume terms cancel out from both sides and the equation can be solved by separation of variables:

$$\frac{dC}{C} = -k_1 \cdot dt$$

Integrating both sides yields $C_t = C_o \cdot \exp(-k_1 \cdot t)$, where

C_t = concentration of cyanide in the batch reactor at any time t

C_o = concentration of cyanide in the batch reactor at time 0

k_1 = first-order decay coefficient

Alternatively, this equation could be solved for t:

$$t = \ln\left(\frac{C_t}{C_o}\right) \cdot (-k_1^{-1})$$

Given that 99.9% reduction is desired, the quantity C/C_o is equal to 1/1000, and the time required is calculated as

$$t = \ln\left(\frac{1}{1000}\right) \cdot \left(\frac{-1}{3.2 \text{ day}^{-1}}\right) = 2.2 \text{ day}$$

Note that neither the volume nor the starting concentration is needed to solve this problem. The volume could be calculated and the starting concentration could be used in the solution, but these steps would have required additional time, which most people do not have when taking the FE exam.

Example 14.9

First-order decay in a CSTR

A 10-acre lake receives flow from an industrial facility. The lake is drained by a single stream. The industrial plant is legally discharging a 0.5 MGD effluent with a mercury concentration of 1.5 µg/L. The industrial facility is the only influent flow to the lake. The average depth of the lake is 7 feet, and the mercury decays with a rate constant of 1 day^{-1}. Find the steady state concentration in the lake.

Solution

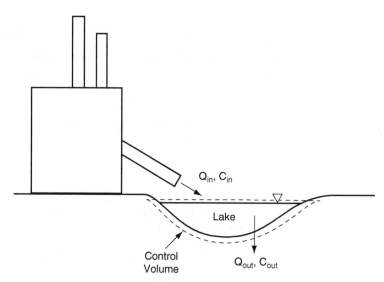

Exhibit 3 Schematic for Example 14.9

This problem is solved by beginning with the generic mass balance equation:

$$\left(\frac{dm}{dt}\right)_{cv} = \dot{m}_{in} - \dot{m}_{out} \pm \dot{m}_{rxn}$$

Given that the system is at steady state and that the Hg decays according to first-order kinetics, the generic mass balance equation can be simplified to

$$0 = \dot{m}_{in} - \dot{m}_{out} - V \cdot k_1 \cdot C_{out}$$

As in Example 14.8, the first order reaction rate law from Table 8.4 combined with the mass balance approach in Chapter 7 where mass flux is equal to flow rate times concentration (or equivalently, volume multiplied by the time rate of change of concentration) is used to make the appropriate substitution for the $\dot{m}_{rxn}$ term.

Note, that by the definition of a CSTR, the concentration anywhere in the lake is the same. Thus, the concentration of Hg in the lake is the same as the concentration of Hg exiting the lake (C_{out}).

The mass flux into the control volume is equal to $Q_{in} \cdot C_{in}$, while the mass flux exiting the control volume is equal to $Q_{out} C_{out}$. Thus, the equation simplifies to this:

$$0 = Q_{in} C_{in} - Q_{out} C_{out} - V \cdot k_1 \cdot C_{out}$$

Recognizing that $Q_{out} = Q_{in} = Q$, and rearranging this equation yields

$$C_{out} = \frac{-Q \cdot C_{in}}{-Q - V \cdot k_1} = C_{in} \frac{1}{1 + \frac{V}{Q} k_1} = C_{in} \frac{1}{1 + \theta \cdot k_1}$$

Values can now be substituted into this equation, being careful of units.

$$V = \text{area} \cdot \text{depth} = 10 \text{ acre} \cdot 7 \text{ ft} \frac{43{,}560 \text{ ft}^2}{\text{acre}}$$

Or, the volume equals $3.05 \cdot 10^6 \text{ ft}^3$.

The hydraulic residence time, θ, equals V/Q, or

$$\frac{3.05 \cdot 10^6 \text{ ft}^3}{0.5 \cdot 10^6 \frac{\text{gal}}{\text{day}}} \cdot \frac{7.48 \text{ gal}}{\text{ft}^3} = 45.6 \text{ day}.$$

Finally, the effluent concentration is found by substitution:

$$C_{out} = \left(1.5 \cdot 10^{-6} \frac{\text{g}}{\text{L}}\right) \frac{1}{\left(1 + 45.6 \text{ day} \cdot 1 \text{ day}^{-1}\right)} = 3.2 \cdot 10^{-8} \frac{\text{g}}{\text{L}} = 3.2 \cdot 10^{-2} \frac{\mu\text{g}}{\text{L}}$$

Discussion

1. The flow into the lake from the industrial discharge is the only flow given. Thus, its flow rate must equal the flow rate in the stream that empties the lake (neglecting evaporation, seepage, etc.).

2. The derived equation for this example *only* applies to a steady state application to a lake with a single input and a single outflow. Any other set of circumstances (for example non-steady state, multiple inputs, a PFR rather than a CMFR) requires a different equation to be derived from the governing mass balance equation.

WATER TREATMENT

Water treatment processes are those steps taken to transform a water resource into *potable* (drinkable) water. Not all of the processes covered in the sections to follow will be necessary in every water treatment plant (WTP). Those selected would depend upon the specific source of water to be treated. As discussed in detail in the following sections of this review, most water treatment utilizes chemical additions and mixing principles to transform soluble drinking water contaminants into an insoluble form followed by steps to remove the solids and subsequently, biological contamination is dealt with through disinfection.

Oxidation, Mixing, Coagulation, and Flocculation

Water that is developed from sources where oxygen is scarce (groundwater and deep reservoir water) may possess high levels of undesirable soluble species, particularly Fe^{2+} and Mn^{2+}. The exact sequence of removal is dependent on other treatment processes used (e.g., softening removes Fe^{2+} and Mn^{2+}), however targeted removal is accomplished through *oxidation*. The simplest oxidation process is *spray aeration,* where contact with air is sufficient to oxidize the iron to the insoluble Fe^{3+} species. Oxidation of Mn^{2+} to Mn^{4+} requires a pH over 9, and addition of an alkali is necessary if aeration is to be successful. Oxidation may also be accomplished using other oxidizing agents, such as *chlorine* and *potassium permanganate*. The permanganate reaction is faster and not as sensitive to pH, and treatment dosages may be estimated as follows:

$$[KMnO_4]_T = 0.94[Fe^{2+}] + 1.92[Mn^{2+}]$$

where quantities in brackets are concentrations in mg/L, and the subscript T denotes the theoretical concentration.

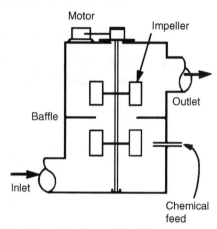

Figure 14.7 Example of a rapid mixing device used in WTP.

When chemicals need to be added to the water for treatment purposes, proper blending and distribution require consideration in designing the mixing process. Design of a *rapid mix* vessel (as shown in Figure 14.7) generally focuses on detention times of 10–60 seconds, use of square paddle impellers and a baffled chamber, and chemical feed that corresponds with the impeller location. Another device is the *static,* or *in-line,* mixer, which is placed inside process pipe and causes high turbulence while transporting water to the next process unit. Other methods of mixing may include *jet introduction* into a larger process unit, or introduction of a chemical just upstream of a process pump and allowing the pump impeller to perform the agitation.

Power requirements for rapid mixers assume complete turbulent flow has been established, and can be determined using the following expression:

$$P = K_T \rho n^3 D^5$$

where P has units of watts (W), ρ is the density of water (999 kg/m³ @ 25°C), n is the rotational speed of the impeller (rps, i.e., revolutions per second), D is the diameter of the impeller (m), and K_T is the impeller constant, which has values obtained from the environmental engineering section of the *FE Supplied-Reference Handbook*. Power requirements per unit volume of water can be used to assure adequate mixing, and should be in the range of 50–200 W/m³.

Coagulation is the process of adding chemicals to surface waters that aid in the building of particle aggregates, whereby the larger aggregates are more easily removed in subsequent treatment processes. The mechanisms employed to enhance particle aggregation are through the reduction in the electrostatic repulsion forces present in colloidal and particle suspensions, or through bridging of polymer coagulants. The most common coagulant is *alum,* or aluminum sulfate, which has the chemical formulation $Al_2(SO_4)_3 \cdot 14.3\ H_2O$. Dosages are in the 5–50 mg/L range, are usually determined in the laboratory by completion of a *jar test*. Often, *polymers,* which are high molecular weight soluble organic compounds, are also added in concentrations of 0.1–0.5 mg/L to increase the size and the durability of the aggregates. Polymers may be cationic, anionic, or nonionic, and jar tests are also used to determine which combinations of coagulants produce the most stable aggregates for a given particle suspension.

Example 14.10

Estimating alum requirements

Estimate the quantity of alum a 1 MGD treatment plant needs to purchase annually if it wishes to operate in the sweep floc range, assuming enough alkalinity is present to keep the pH at 8. Express your answer in units of

a). Tons

b). Gallons of 5% alum solution

Solution

An alum concentration of 35 mg/L should create sweep floc conditions at a pH of 8. The ability of this concentration to produce a sweep floc for the actual water supply at hand should be verified with a jar test.

a). the daily mass of alum required is

$$35 \text{ mg/L} \cdot 1 \text{ MGD} \cdot 8.34 = 292 \text{ lb/day}$$

Thus, the annual requirement is 53 tons. Notice the conversion factor 8.34 is used here. This is a great conversion factor to remember for use with wastewater treatment process unit design where flows are given in million gallons per day (mgd) and concentrations are given in mg/L. The factor 8.34 converts the product of (mgd) × (mg/L) into the units of lb/day.

b). Assuming that the 5% alum solution has the same density as water, its concentration may be expressed as 0.42 lb/gal, as follows:

$$\frac{5 \text{ g}}{100 \text{ g}} \cdot \frac{2.205 \cdot 10^{-3} \text{ lb}}{\text{gal}} \cdot \frac{1000 \text{ g}}{\text{L}} \cdot \frac{3.785 \text{ L}}{\text{gal}} = 0.42 \text{ lb/gal}$$

The daily requirement is

$$292 \text{ lb/day} \cdot (1 \text{ gal}/0.42 \text{ lb}) = 695 \text{ gal/day}$$

The annual requirement is then 254,000 gallons.

After chemical coagulants are added to modify particle aggregation potential, there is a need to encourage sufficient mixing to cause particles to collide, but not so vigorous as to break apart the aggregated particle masses. This process is called *flocculation,* and the aggregates are called *floc.* Flocculation occurs through the use of slowly rotating paddle or turbine mixers, usually at rotational speeds of 1–10 rpm, that keep the floc in suspension to allow more particles to combine, thus increasing floc size and removal percentage. Typical flocculation equipment and arrangement are presented in Figure 14.8.

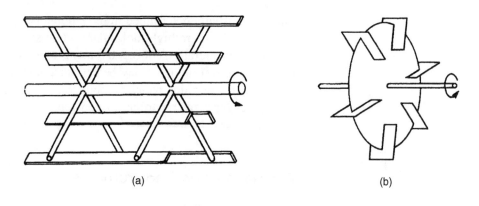

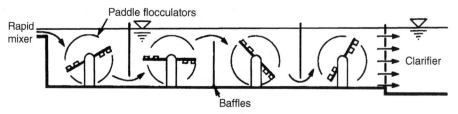

Figure 14.8 Flocculation equipment: (a) paddle flocculator, (b) flat-blade turbine, (bottom) typical flow through tank arrangement (Viessman, Jr., Warren, and Hammer, Mark J., *Pollution Control*, 7th Edition, © 2005. Reprinted by permission of Pearson Education, Inc., Upper Saddle River, NJ.)

The design of flocculation units typically examines the power dissipated (P) in the water, which has been related to the velocity gradient (G) as follows:

$$G = \left(\frac{P}{\mu V}\right)^{1/2}$$

where G is measured in units of s^{-1} (usually fps/ft), P is measured in units of ft·lb/s, μ is the dynamic viscosity of water (lb·s/ft^2), and V is tank volume (ft^3). Optimal values for G range from 30–60 s^{-1}, and the Gt product, where t is the amount of time in the mixing zone(s), should range from 10^4 to 10^5.

The power transferred to the water by a single board of a paddle flocculator can be determined as follows:

$$P_{\text{BOARD}} = \frac{C_D A_P \rho_f v_P^3}{2}$$

where P and ρ are as defined previously, C_D is the drag coefficient (usually 1.8), A_P is the paddle area (paddle length times paddle width), and v_P is the paddle velocity relative to the water, which can be expressed as

$$v_P = K_S 2\pi r n$$

where K_S is the slip coefficient (between 0.5 and 0.75, usually 0.7), r is the distance from the shaft to the center of the paddle board, and n is rotational speed (rps) as before. Each board on each arm of the paddle flocculator needs to be accounted for in the overall energy computation. Using the equations provided in the environmental engineering section of the *FE Supplied-Reference Handbook*, this is accomplished by determining the power for one board at a distance r from the shaft and multiplying that number by the total number of boards at that radius. This calculation is then repeated for boards at the remaining radii, and the total power is determined by summing the power from each board.

In practice, the total power can be determined from a single equation, although this is not provided in the *FE Supplied-Reference Handbook*. For a paddle flocculator that has N symmetrical paddle arms, each with two boards of area A located at distances of r_1 and r_2 from the shaft, and rotating at a speed of n rps, the total power dissipated (i.e., transferred to the water) can be calculated as

$$P = \frac{NC_D A_P \rho_f}{2}(K_S 2\pi n)^3 \left(r_1^3 + r_2^3\right)$$

Power dissipated by flocculators with additional boards on each arm may be calculated using the above expression and including an additional r^3 term at the appropriate distance. Design criteria for flocculator tanks generally suggest minimum t_R of 30 minutes with horizontal (flow-through) velocities between 0.5 and 1.5 ft/min, and paddle speeds (equal to v_P/K_S) that range from 0.5 to 3.0 ft/s.

Example 14.11

Preliminary design of a flocculation system

A flocculation system is required for a 5000 m³/day water treatment utility. The flocculation system is to consist of two treatment trains in parallel, with each train consisting of two compartments in series. Each compartment is desired to have a detention time of 10 minutes. A preliminary selection of vertical shaft flocculators has been made, and information describing the flocculators is as follows:

Area of paddles = 5 m²

Relative speed = 0.25 m/s

$C_d = 1.9$

a). Determine the size of each compartment

b). Estimate the power input to the water

Solution

a). Determine the size of each compartment using the definition of detention time and remembering that the flow is split between the two flocculation trains.

$$\text{Volume} = \theta \cdot Q$$
$$= 10 \text{ min} \cdot 2500 \text{ m}^3/\text{day} \cdot (1 \text{ day}/1440 \text{ min})$$
$$= 17.4 \text{ m}^3$$

b). First, compute the velocity gradient, G as:

$$G = \sqrt{\frac{C_d \cdot A \cdot v^3}{2\nu V}}$$

where

$C_d = 1.9$

$A = 5 \text{ m}^2$

$v = 0.25 \text{ m/s}$

$\nu = 10^{-6} \text{ m}^2/\text{s}$

$V = 17.4 \text{ m}^3$

This produces a velocity gradient of 65 sec⁻¹.

Obtain the power by rewriting the generic velocity gradient equation as

$$P = G^2 \cdot V \cdot \mu$$

$$P = (65 \text{ sec}^{-1})^2 \cdot (17.4 \text{ m}^3) \cdot (0.001 \text{ Pa} \cdot \text{s})$$

$$= 74 \text{ W}$$

Softening

The removal of excess hardness from water is completed by a process called *softening*, in which hydrated (or slaked) lime, $Ca(OH)_2$, and soda ash, Na_2CO_3, are used to precipitate the multivalent cations from solution. The process works by converting Ca^{2+} and Mg^{2+} species into insoluble forms, namely $CaCO_3$ and $Mg(OH)_2$. Reactions that are common in lime soda softening are presented in the environmental engineering section of the *FE Supplied-Reference Handbook*. Determination of the required lime and soda additions require water composition, which may be represented on an ion bar graph to greatly simplify solving the softening reactions. Using the stoichiometry from the reactions listed, the amount of lime and soda, in meq/L, are determined. Multiplying the meq/L of lime or soda by the respective equivalent weight will yield the required concentrations in mg/L. Total mass required for daily dosing is determined by multiplying the concentration in mg/L by the total volume of water treated daily.

Traditional lime soda softening cannot remove all hardness, due in part to the slight solubility of the $CaCO_3$ and $Mg(OH)_2$ species. While the *straight lime process* can be used to soften water with high Ca^{2+} and HCO_3^- hardness, removal of Mg^{2+} hardness requires *excess lime* treatment. In this process, an additional 1.25 meq/L of lime is added to the water to increase the pH to 11, the required value to precipitate $Mg(OH)_2$. This process is able to reduce calcium hardness to 30 mg/L as $CaCO_3$ and magnesium hardness to 10 mg/L as $CaCO_3$. The excess lime must be removed to prevent scaling and is usually done so via *recarbonation*, as described by the last process reaction listed in the *FE Supplied-Reference Handbook*.

Sedimentation

Gravity induced settling is called *sedimentation* or *clarification*, and is the process in which solid particles are allowed to fall out of suspension in relatively quiescent water. Tanks are designed to greatly reduce flow velocities, such that particles are not being agitated by movement of fluid and are allowed to collect at the bottom of the tank. Sedimentation in water treatment generally occurs for high turbidity surface waters as a first treatment step prior to chemical addition, or following coagulation and flocculation.

The rate at which particles move in the vertical direction is described by their *terminal settling velocity* (V_t), which can be expressed as

$$V_t = \left[\frac{4g(\rho_p - \rho_f)d}{3C_D \rho_f} \right]^{1/2}$$

where g is the gravity constant, ρ_p and ρ_f are the densities of the particle and the fluid, respectively, d is the diameter of the particle, and C_D is the drag coefficient.

C_D is dependent on Reynolds number (Re) and can be determined as

$$C_D = \frac{24}{\text{Re}} \qquad \text{Re} \leq 1$$

$$C_D = \frac{24}{\text{Re}} + \frac{3}{\sqrt{\text{Re}}} + 0.34 \qquad 1 \leq \text{Re} \leq 10^4$$

$$C_D = 0.4 \qquad \text{Re} > 10^4$$

where the Reynolds number is defined as

$$\text{Re} = \frac{\rho d V_t}{\mu}$$

Since V_t and Re are dependent on each other, an iterative solution is required where a velocity is assumed to determine Re, that value is used to determine C_D, which is used to calculate V_t. The Reynolds number is corrected and the process repeated until the error in V_t is small. To reduce the complexity of the solution method, it is often assumed that *Stoke's law* is applicable, which is used to describe external flow around spherical bodies at low values for Re. Low Re values are possible for very small particles (on the order of micrometers in diameter) and low settling velocities (on the order of centimeters per second) and simplifies the expression for V_t by setting $C_D = 24/\text{Re}$. In these cases, V_t can be expressed as

$$V_t = \frac{g(\rho_p - \rho_f)d^2}{18\mu}$$

where g, ρ_p, ρ_f, d, and μ are as defined previously.

The flow of water in the clarifier can be characterized by the fluid velocity in the vertical direction, called the *surface settling rate*, and is expressed as

$$V_0 = \frac{Q}{A}$$

where V_0 has units ft/s or m/s, Q is the volumetric flow rate into the clarifier (cfs or m³/s), and A is the surface area of the clarifier (ft² or m²). Any particles possessing $V_t > V_0$ are collected in the clarifier, while particles with $V_t < V_0$ are collected only if they enter the clarifier at a height less than the water depth in the basin.

For design purposes, the surface settling rate is more commonly expressed in the common English units for Q, gallons per day (gpd), and is called the *overflow rate*, with units of gpd/ft². Hydraulic residence time (t_R), as described previously, is also used for clarifier design. Another design parameter is the *side-water depth*, which usually describes the depth of water not including the sludge blanket that forms on the bottom of the tank. Because the three parameters above are not independent, care must be taken when using them in clarifier design to keep from overspecifying the design. An additional design parameter is the *weir loading* (wl), sometimes called the *weir overflow rate* (WOR), which is a measure of the rate of effluent flow over the discharge weir, and can be expressed as

$$wl = \frac{Q}{L_w}$$

where wl has common units of gpd/ft, L_w is the length of the outlet weir (ft), and Q is as above.

Clarifier shape is usually rectangular or circular, with the size of the tanks determined through the use of empirical design criteria, which are based on operating conditions that have been found to be most effective in practice. For circular tanks, influent is in the center of the tank, and water flows radially toward the weir, located at the tank perimeter, with weir length equal to the tank perimeter. Rectangular tanks generally have a $L{:}W$ ratio of 3:1 to 5:1, and weirs are set inside the tank at one end in a box or finger arrangement.

Design criteria are often listed in tables, and are provided for several water (and wastewater) treatment applications in the environmental engineering section of the *FE Supplied-Reference Handbook*. In general, V_0 for water treatment can range from 300 to 2,500 gpd/ft^2, (approximately equal to 12–100 m^3 per m^2 per day, or equivalently 12–100 m/d) and *horizontal* (or *flow-through*) *velocities* (V_h), defined as the volumetric flow rate divided by the cross-sectional area of the tank perpendicular to the velocity vector (i.e., width × depth), should be kept below 0.5 ft/min in all cases. Design criteria for detention times vary from 1 to 8 hours, side-water depths should be maintained greater than 10 ft, and weir loadings are generally less than 20,000 gpd/ft.

Gravity Filtration

Removal of non-settleable solids in water treatment is generally accomplished through *gravity filtration*, usually using a *granular media*. Filtration mechanisms vary, but characteristics of appropriate filter media are consistent: (1) fine enough to strain particles, (2) coarse enough to resist clogging, (3) deep enough to allow for large quantities of floc collection and long run times between cleaning cycles, and (4) shallow enough to minimize pressure drop and to allow for efficient backwash cleaning. Media characteristics are described with respect to bed depth and the diameter of the media particles, which may vary over a wide range of values. The standard notation to describe particle size distribution is the same as discussed in Chapter 13 where d_x was defined as the diameter that has X percent of the material mass below the stated size. The *effective size* may be defined as d_{10}, while the *uniformity coefficient* was previously defined as d_{60}/d_{10}.

Often, a *sand bed* is the media of choice, and can be effective as long as the uniformity coefficient is low. High uniformity coefficients can lead to size segregation during the backwash process, where coarse particles settle faster and leave a layer of fine sand on the top of the bed. This often leads to rapid clogging of the filter and the need for frequent cleaning. In order to avoid rapid surface plugging, it is common to use a *dual media* system where the sand bed is overlain with an *anthracite cap* layer. The specific gravity of the anthracite (~1.5) is lower than the specific gravity of the sand (2.65), which allows a sand media with a smaller d_{10} settle faster than the anthracite with a slightly larger d_{10}. This system allows for straining of larger floc in the large-diameter anthracite cap without media fouling, while smaller particles can be removed in the lower, small-diameter sand layers.

Design of a typical dual media rapid sand filter is based on hydraulic loading and depends on the characteristics of the particulate that is being removed. In water treatment, loading rates range from 2 to 10 gpm/ft^2 of bed area with a typical value of 5 gpm/ft^2, and maximum filter areas of around 1,000 ft^2 (30 ft × 30 ft) for each unit. Water from the coagulation and flocculation units is distributed over several rapid sand filters to a depth of 3–5 feet. The anthracite cap is usually 18 inches deep with a d_{10} of approximately 1 mm and a uniformity coefficient <1.7. The sand layer has the same uniformity coefficient, but has a d_{10} closer to 0.5 mm and a depth of 12 inches. The sand is supported above the underdrain system by 1.5–2.5 feet of sand and gravel, which is stratified in several layers of increasing grain size with increasing filter depth.

Pressure drop through a packed bed is due to the friction losses associated with the media, and affects the rate of flow that is possible through a filter bed for a given, fixed head (i.e., depth of water over the bed). Generally, clean media has a head loss on the order of 2–3 feet, while head loss of 8–10 feet may signal the need for backwashing. *Head loss* (h_f) through a clean filter containing a single diameter of filter media can be described by the Rose equation as follows:

$$h_f = \frac{1.067 V_S^2 L C_D}{g \eta^4 d}$$

where V_S is the approach velocity (volumetric flow rate divided by bed area, Q/A_S, where A_S is the area of the empty bed at plan view), L is the filter depth, C_D is the drag coefficient as defined in Chapter 12, g is the gravity constant, η is the bed porosity (void volume/total bed volume), and d is the mean diameter of the filter media grain. The mean diameter may be determined from either the arithmetic average ($[S_1 + S_2]/2$) or geometric average ($[S_1 \times S_2]^{1/2}$), where S_1 is the size of the sieve opening on which the particle was retained and S_2 is the next larger sieve opening.

Another expression for head loss through a clean filter containing a single diameter of filter media can be described by the Carmen-Kozeny equation as follows:

$$h_f = \frac{fL(1-\eta)V_S^2}{\eta^3 g d}$$

where L, η, V_S, g, and d are as defined previously, and f is the friction factor, which can be expressed as

$$f = \frac{150(1-\eta)}{\text{Re}} + 1.75$$

For the more realistic circumstance where filter media are not of a single type or size, the head loss equations above can be modified to account for the change in the friction loss due to different media types and grain sizes as follows:

$$h_f = \frac{1.067 V_S^2 L}{g \eta^4 d} \sum \frac{C_{Dij} x_{ij}}{d_{ij}} \quad \text{Rose Equation}$$

$$h_f = \frac{L(1-\eta)V_S^2}{\eta^3 g d} \sum \frac{f_{ij} x_{ij}}{d_{ij}} \quad \text{Carmen-Kozeny Equation}$$

where subscript i refers to the media type, subscript j refers to the size fraction, and x is the mass fraction of media of the specified type and size (m_{ij}/m_{tot}). Note that the Σx_{ij} must be equal to 1.

Particulate matter that becomes trapped within the filter bed decreases the pore volume available for flow, which increases frictional resistance (i.e., head loss). This particulate matter must be removed and the filter bed re-established in order for the filtration unit to be effective on a continuous basis. This is accomplished through a process called *backwashing,* where the flow of water to the filter is stopped and water and/or air is pumped through the bed in the upwards direction (i.e., opposite of the flow direction during filter operations) on a regular, periodic basis. This cleansing is made more effective by the *fluidization* of the filter media, where the flow velocity in the upwards direction is sufficient to lift and separate the media grains, and the mixing action of the particles scour trapped particulate matter.

The minimum velocity required to obtain fluidization has been calculated using various approaches in the literature. If fluid velocities during backwashing are too large, it is possible that the filter material may be flushed with the removed floc. One means to determine effective fluidization is to examine the bed *expansion*, which is a measure of the bed depth during backwashing relative to the bed depth during operation. An estimate of the depth of the fluidized bed (L_{fb}) during backwashing may be obtained from the following expression:

$$L_{fb} = \frac{L_0(1-\eta_0)}{1-\eta_{fb}} = \frac{L_0(1-\eta_0)}{1-\left(\frac{V_B}{V_t}\right)^{0.22}} \quad \text{Single type, diameter material}$$

$$L_{fb} = L_0(1-\eta_0) \sum \frac{x_{ij}}{1-\left(\frac{V_B}{V_{t,ij}}\right)^{0.22}} \quad \text{Multiple types, diameters of material}$$

where L_0 is the original bed depth of the fluidized fraction, η_0 is the initial bed porosity, η_{fb} is the porosity of the fluidized bed, V_B is the backwash velocity (defined as Q/A_S, where Q is the backwash flow rate and A_S is as defined above), V_t is the terminal settling velocity (calculated as described in the Sedimentation section above), and x_{ij} is as defined previously.

Expansion of the bed during backwashing is approximately 50 percent (i.e., $L_{fb} = 1.5 L_0$). Volumetric flow rates around 15 gpm/ft² of filter area maintained for 5–10 minutes have been shown to provide effective cleaning. Maximum backwashing flow rates are set to maintain fluidizing velocities that will not wash the filter media from the bed. After the backwashing flow is stopped, the filter media will settle by specific gravity and size, with the larger sand particles settling first, followed by the smaller sand particles, with the lighter anthracite particles on the top. Water that runs through the filter in the first 3–5 minutes after backwashing is often discarded because it may still contain particulates suspended during the wash.

Disinfection

Due to the potential presence of biological agents that are harmful to human health, a primary role of the water treatment plant is the *disinfection* of the water prior to release into the distribution network. By far, the most often used disinfectant is *chlorine* gas. The goal of disinfection is to inactivate any biological agent that could pose a threat to human health, while minimizing the *residual chlorine* levels that may cause a displeasing taste for consumers (usually assumed to be < 0.5 mg/L). This is balanced by the need to assure complete inactivation of specific persistent pathogens, and the desire to have a small (0.1–0.2 mg/L) residual chlorine level that will maintain disinfection throughout the distribution network.

Chlorine is generally shipped to the water treatment plant in compressed cylinders containing the chemical in its liquid state. Water flow through a mixing valve draws chlorine vapors from the headspace of the cylinder and dissolves the chlorine in the water at a fixed concentration. In the dissolved state, chlorine combines with water to form hypochlorous acid by the following reaction:

$$Cl_2 + H_2O \leftrightarrow HOCL + H^+ + Cl^-$$

Although the preceding equation is written as an equilibrium reaction, at the pH values anticipated in the water treatment plant, the reaction is nearly complete. While hypochlorous acid is a strong disinfectant, it will ionize in solution to a weaker disinfectant, hypochlorite ion, by the following equilibrium relationship:

$$HOCL \leftrightarrow H^+ + OCl^-$$

The equilibrium of this reaction is highly pH-dependent in the 6–9 range, with more than 95 percent of the material in the acid form at pH 6, more than 95 percent of the material in the ion form at pH 9, and a nearly linear distribution in between. *Free available residual chlorine* is the residual chlorine existing in the water as hypochlorous acid and the hypochlorite ion. This quantity is regulated due to the potential to generate chlorinated compounds that are toxic to fish.

The desired amount of treatment is often expressed as the *log removal* or *log inactivation,* and varies for each microbial species or class of species. One log removal would leave 10 percent of the original organisms, or equivalently, inactivate 90 percent of the present microbes. Two log removal would leave 1 percent of the original population, inactivating 99 percent, while three log removal and four log removal would inactivate 99.9 percent and 99.99 percent of the original organisms present, respectively. Although many species are of concern, chlorine doses are usually based on the species most resistant to disinfection.

In the case of surface waters, that species would be *Giardia lamblia* cysts, which requires 3 log removal to meet regulatory constraints. However, conventional water treatment (coagulation-flocculation-sedimentation-filtration) has been shown to offer 2–2.5 log removal of *Giardia lamblia*, leaving only a 0.5–1 log removal requirement from the disinfection process. Biological species present in ground-water sources depend on the depth to the aquifer and the aquifer's connectivity with surface waters, generally focusing on virus removal. Removal of viruses is generally regulated at 4 log removal, although conventional treatment has been shown to offer 2 log removal, leaving a 2 log removal requirement for the disinfection process.

Table 14.6 $C \cdot t$ values [(mg/L)Σmin] for 1 log removal of *Giardia lamblia* cysts at various water temperatures and pH values

pH	0.5°C	5°C	10°C	15°C
6	49	35	26	19
7	70	50	37	28
8	101	72	54	36
9	146	146	78	59

Source: Viessman and Hammer, Water Supply and *Pollution Control*, 7th Ed., adapted from Guidance Manual for Compliance with the Filtration and Disinfection Requirements for Public Water Systems Using Surface Water Sources (U.S. EPA, 1991).

The ability to achieve a specified removal of a microbial species is a function of both the *chlorine concentration* as well as the amount of *contact time* the organism must experience at that concentration. The combination of these two factors is often referred to as the $C \cdot t$ *product,* which carries units of (mg/L)·min, and is often tabulated for specific microorganisms at specified temperatures and pH values. $C \cdot t$ values for 1 log removal of *Giardia lamblia* are given in Table 14.6, and $C \cdot t$ values for 0.5 log removal are assumed to be one-half of the table values. As seen in the table, values are pH and temperature dependent and are assumed to produce a chlorine residual of 1.0 mg/L. While similar tables exist for viruses, $C \cdot t$ values less than 10 (mg/L)·min are usually sufficient for 4 log removal of viruses, while $C \cdot t$ values less than 5 (mg/L)·min are sufficient for 2 log removal. These values are much lower than those for *Giardia lamblia*, and therefore the *Giardia lamblia* disinfection concentration generally controls chlorine dosing.

Design of a chlorine contact basin is based on the contact time necessary to achieve the $C \cdot t$ value at a specified residual chlorine concentration, usually assumed to be 0.5–1.0 mg/L. Often a safety factor of 2–3 is applied to the calculated contact time to assure all water has sufficient contact. This is also enhanced through the use of a 40:1 or 50:1 $L:W$ ratio in the contact basin (plug flow tank), with square channel cross sections and placement of multiple tank baffles to encourage mixing and reduce short circuiting.

ADVANCED WATER TREATMENT

The water treatment methods discussed in the previous subsections are often considered *conventional treatment,* as they have been practiced effectively for many decades and work well for water resources with few undesired chemicals. As water resources become more scarce and water reuse from nontraditional sources increases in popularity, new treatment technologies are necessary to address contaminants that are difficult to remove using conventional technologies. The removal of unwanted chemicals in the aqueous phase can be accomplished through exchange with a solid phase in a process called *adsorption,* where the compound of interest attaches to a solid with an extremely high specific surface area. The fundamental principles of adsorption are identical for liquid-solid and gas-solid systems. Another process that takes advantage of mass transfer principles is absorption, while modern water treatment utilizes membrane filtration technology to obtain water quality that approaches pure water.

Absorption

One example of using preferential mass transfer for removing contaminants from fluids is called *absorption,* a phenomenon through which contaminant mass is transferred from the liquid phase to the gas phase (*air stripping*), or from the gas phase to the liquid phase (*gas scrubbing,* which will be discussed in Chapter 15). Both scenarios involve a column filled with a high surface area, high void volume packing material, often made of a plastic or ceramic compound, which is configured to provide uniform flow of both the liquid and gas phases. Mass transfer occurs at the interface between the liquid and gas and often a chemical is added to the liquid to further enhance the removal of the target species. For example, lime is a common additive that increases pH and shifts the equilibrium between NH_4^+ and NH_3 to allow ammonia to be removed by air stripping, or is available to react with the gaseous SO_2 present in coal-fired power plant flue gasses to form $CaSO_3(s)$.

The mass balance for an absorption system relies on the fact that environmental contaminants at low concentrations have a linear partitioning relationship between a liquid phase and a gas phase when the two phases are in equilibrium. This is known as *Henry's law,* and may be expressed mathematically for systems operating at atmospheric pressure as

$$y_{\text{out}} = H'C_{\text{in}} = \frac{H}{RT}C_{\text{in}}$$

where y_{out} is the molar concentration of contaminant in the gas exiting the column (given the symbol A_{out} in the *FE Supplied-Reference Handbook*), H' is the unitless form of the Henry's law constant, C_{in} is the molar concentration of contaminant in the liquid phase entering the column, H is the dimensioned form of the Henry's law constant (atm•m³/mole), R is the gas constant (8.026×10^{-5} atm•m³/mol•K), and T is the temperature (K). In general, the magnitude of H' is a good indicator of the direction of contaminant transfer, where values > 100 indicate species that have a strong preference for the air and values < 0.01 indicate species that have a strong preference for the aqueous phase.

For removing a pollutant species from water (air stripping), if we assume all of the contaminant entering in the liquid phase is removed by the gas, the steady-state mass balance is written as

$$Q_w C_{in} = Q_a y_{out} = Q_a H' C_{in}$$

where Q_w and Q_a is the volumetric flow rate of the water and air, respectively (m³/s), and C_{in}, y_{out}, and H' are as defined previously. Canceling C_{in} from both sides of the above expression, we see that the ratio Q_w/Q_a is equal to the unitless Henry's law constant (H'), which is often referred to as the *liquid-to-gas ratio* (L/G). Thus, for H' values > 100 (air preference species) we see that Q_a is less than 1 percent of Q_w and contaminant species are easily removed, whereas H' values < 0.01 (water preference species) require a extremely high volumetric flow rate of air relative to the liquid flow rate making contaminant species removal substantially more difficult.

The design of a stripping column often focuses on the height of the column necessary to reduce contaminant concentrations in the water to acceptable levels. While several models have been developed for this purpose, one of the simplest may be expressed as

$$Z = \text{HTU} \times \text{NTU}$$

where Z is the height of the stripping tower packing material (m), HTU is the *height of a transfer unit* (m), and NTU is the *number of transfer units* (unitless). The height of a transfer unit for an air stripping tower can be determined as follows:

$$\text{HTU} = \frac{Q_w}{K_L a A} = \frac{L}{M_W K_L a}$$

where Q_w is defined previously, $K_L a$ is the overall mass transfer coefficient (s⁻¹), A is the column cross sectional area (m²), L is the molar liquid loading rate (kmol/s per m² of column cross sectional area), and M_W is the molar density of water (equal to 55.6 kmol/m³ or 3.47 lbmol/ft³). The number of transfer units in the air stripping column may be determined from the following relationship:

$$\text{NTU} = \left(\frac{R}{R-1}\right) \ln\left[\frac{(C_{in}/C_{out})(R-1)+1}{R}\right]$$

where C_{in} and C_{out} are the concentrations of the contaminant in the liquid phase in the influent and effluent, respectively, and R is the stripping factor, which may be expressed as

$$R = H'\left(\frac{Q_a}{Q_w}\right)$$

For the air stripping column, since H' is equal to Q_w/Q_a for a value of $R = 1$, this corresponds to the theoretical minimum amount of air required for stripping of a particular contaminant. Values of $R < 1$ would provide an insufficient flow of air to remove the target species. Due to the limitations on interphase mass transfer within the column, excess air is always required, and values for R in practice typically range from 1.5 to 5.0.

Membrane Filtration

Several advanced water treatment technologies are based on *membrane filtration* processes that have been widely used in specific industries that require water of high purity, and include *microfiltration, ultrafiltration, reverse osmosis*, and *electrodialysis*. Microfiltration makes use of membranes with pore sizes greater than 50–100 nm to remove turbidity-causing colloids as well as some biological contaminants that are greater than 0.1 μm in diameter. Particles that are 5–200 nm in size include nearly all colloids and biological agents, as well as high molecular weight organic compounds, and can be effectively removed by ultrafiltration membranes with pore sizes in the 2–50 nm range. Both of these processes use pressure to force water through the pores, while the undesired compounds are retained on the feedwater side of the membrane. The product water is often called the *permeate*, while the rejected water with the filtered particulate is called the *retentate*.

The volumetric flow rate through the membrane can be expressed as

$$Q_p = J_w A$$

where Q_p is the volumetric flow rate of the permeate (m³/s), A is the membrane area (m²), and J_w is the transmembrane water flux (m/s), which may be determined from the following expression:

$$J_w = \frac{\varepsilon r^2 \Delta P}{8 \mu \delta}$$

where ε is the membrane porosity (unitless), r is the pore size (m), ΔP is the net transmembrane pressure (Pa), μ is the fluid viscosity (Pa•s), and δ is the membrane thickness (m).

Historically used worldwide to desalinate seawater, reverse osmosis employs a semi-permeable membrane that restricts the flow of dissolved ions in an aqueous solution. Left to equilibrate, a vessel containing pure water on one side of a semi-permeable membrane and saline water on the other side will cause pure water to flow through the membrane due to *osmosis* in an attempt to equalize the ion concentration, and thus equalize the chemical potentials. If the vessel is maintained at a constant volume, the flow will cease when the increased pressure due to the accumulation of fluid on the saline side of the membrane equals the difference in the system chemical potential. The pressure difference is called the *osmotic pressure* (π), and can be determined by the following expression:

$$\pi = \phi v \frac{n}{V} RT$$

where ϕ is the osmotic coefficient, v is the number of ions formed from one molecule of electrolyte, n is the number of moles of electrolyte, V is the solvent volume, R is the universal gas constant, and T is the absolute temperature. For reference, the value of π for seawater at 35,000 mg/L TDS is 397 psi, or approximately 11.3 psi for every 1,000 mg/L of TDS.

If a pressure is applied to the saline side of the reservoir in excess of the osmotic pressure, pure water will flow against the osmotic pressure gradient through the membrane from the saline side to the pure water side. The flow of water through the membrane can be expressed as

$$Q_w = J_w A = W_p (\Delta P - \Delta \pi) A$$

where Q_w is the water production rate (quantity/time where the quantity may be volume, mass, or moles), J_w is the water flux (quantity per unit area per time), A is the membrane area, W_p is the membrane coefficient (quantity per unit area per time per unit pressure), ΔP is the difference in pressure between the feed water and product water, and $\Delta \pi$ is the difference in osmotic pressure between the feed water and product water.

Because diffusivity and viscosity are a function of temperature, the membrane coefficient is also temperature dependent. Correction is usually made by increasing or decreasing the required amount of membrane area required for water feeds of various temperatures. The area ratio is based on a reference temperature of 25°C and is generally expressed as the area required at temperature T versus the area required at a temperature of 25°C (A_T/A_{25}). Area ratios are 1.58 for 10°C, 1.34 for 15°C, 1.15 for 20°C, 1.00 for 25°C, and 0.84 for 30°C.

While the membrane is very selective, some ions are transported to the product water side during reverse osmosis. The rate of ion flux through the membrane can be expressed as

$$\dot{m}_s = Q_p C_{out} = J_s A = K_p (C_{in} - C_{out}) A = \left(\frac{D_s K_s}{\Delta Z}\right)(C_{in} - C_{out}) A$$

where $\dot{m}_s$ is the rate of salt passing through the membrane (mass/time or moles/time), Q_p is the volumetric flow rate of the permeate stream (volume/time), C_{in} and C_{out} are solute concentrations in the feed stream and permeate stream, respectively (mass/volume or moles/volume), J_s is the salt flux through the membrane (mass per area per time, or moles per area per time), A is the membrane area (length^2), K_p is the membrane solute mass transfer coefficient (length/time), D_s is the diffusivity of the solute in the membrane (area/time), K_s is the solute distribution coefficient (unitless), and ΔZ is the membrane thickness (length).

Another water purification technique that uses semi-permeable membranes is *electrodialysis*, where membranes selectively allow the passage of cations or anion through the pore openings. When an electric current is applied to a collection of these selective membranes stacked in alternating layers, cations are attracted to the cathode and pass through the cation selective membrane, but become trapped when they cannot pass through the anion selective membrane. Similarly, anions are attracted to the anode and pass through the anion selective membrane and become trapped when they cannot pass through the cation selective membrane. This creates alternating layers of pure water and water containing all of the ionic species, which are carried out of the system through separate flow pathways.

The current required for a stack of membranes is given by the expression

$$I = \frac{FQN}{n} \times \frac{E_1}{E_2}$$

where I is the current (amperes), F is Faraday's constant (96,487 ampere-seconds per gram equivalent weight removed), Q is the volumetric flow rate (L/s), N is the normality of the solution (gram-equivalents/L), n is the number of cells between the electrodes, E_1 is the electrolyte removal efficiency (fraction removed, typically 0.5), and E_2 is the current efficiency (fraction, typically 0.9). The voltage required for the electrodialysis process can be determined as

$$E = IR$$

where E is the voltage requirement (volts), I is the current determined above (amperes), and R is the resistance through the membrane stack (ohms, typically 4–8). Finally, the power (P) consumed during electrodialysis in units of watts can be determined as

$$P = I^2 R$$

Ion Exchange

Ion exchange is used widely in home water softeners and in commercial applications for removal of hardness but is less common in municipal drinking water treatment. The concept behind ion exchange is that opposite charges attract. An *ion exchange resin* has a charged surface, and undesirable ions in the water of the opposite charge (such as the divalent cations that cause water hardness) are attracted to the resin. Eventually, the resin surface becomes filled up (that is, its exchange capacity is *exhausted*), at which point the resin is washed with a regenerating solution. Once regenerated, the exchange resin can be reused for ion exchange.

Resins are either cationic (with a negative surface charge that attracts positive ions) or anionic (with a positive surface charge that attracts negative ions). Exchange capacity is expressed in terms of meq/mL or perhaps as kgr $CaCO_3/ft^3$ (kgr = kilograins, where 1 grain = 75 mg; 21.8 meq/mL in 1 kgr/ft^3).

Simplified ion exchange equations are provided below for the removal of carbonate hardness, noncarbonate hardness, and regeneration (U.S. EPA, 1996). In all cases, the ion exchange resin is denoted as X, and sodium is assumed to be the ion that exchanges from the resin.

Carbonate Hardness
$Ca(HCO_3)_2 + Na_2X \rightarrow CaX + 2NaHCO_3$
$Mg(HCO_3)_2 + Na_2X \rightarrow MgX + 2NaHCO_3$

Noncarbonate Hardness
$CaSO_4 + Na_2X \rightarrow CaX + Na_2SO_4$
$CaCl_2 + Na_2X \rightarrow CaX + 2NaCl$
$MgSO_4 + Na_2X \rightarrow MgX + Na_2SO_4$
$MgCl_2 + Na_2X \rightarrow MgX + 2NaCl$

Regeneration
$CaX + 2NaCl \rightarrow CaCl_2 + Na_2X$
$MgX + 2NaCl \rightarrow MgCl_2 + Na_2X$

Example 14.12

Ion exchange resin requirements

An ion exchange unit containing a resin with an exchange capacity of 1.1 meq/mL is to be used to remove nitrate from a groundwater source. The flow rate of the groundwater is 0.1 MGD, and the nitrate concentration is to be decreased from 30 mg/L to 0 mg/L. The goal is to regenerate the unit every other day. Estimate a volume of resin required.

Solution

The removal of nitrate will first be expressed in terms of eq/L:

$$(25 \text{ mg NO}_3^-/L) \cdot (1 \text{ mol}/62{,}000 \text{ mg}) \cdot (1 \text{ eq}/1 \text{ mol}) = 4 \cdot 10^{-4} \text{ eq/L}$$

The total equivalents of nitrate removed in two days is:

$$4 \cdot 10^{-4} \text{ eq/L} \cdot (0.1 \cdot 10^6 \text{ gal/day}) \cdot 2 \text{ days} \cdot (3.785 \text{ L}/1 \text{ gal}) = 303 \text{ eq}$$

The volume of resin required is:

$$303 \text{ eq} \cdot \frac{\text{mL}}{1.1 \text{ meq}} \cdot \frac{1000 \text{ meq}}{\text{eq}} \cdot \frac{1 \text{ m}^3}{10^6 \text{ mL}} = 0.28 \text{ m}^3$$

WATER TREATMENT RESIDUALS MANAGEMENT

There are four main categories of residuals produced from drinking water treatment plants:

1. *Sludge* is a mixture of solids and water. The solids in sludge consist of solids found in the source water and solids originating from chemicals added during treatment. Sludge solids originate from presedimentation, coagulation/flocculation, filter backwashing, lime softening, iron and manganese removal, etc.

2. *Concentrates* are wastewaters generated by membrane filtration and reverse osmosis reject water.

3. *Other solids* consist of spent GAC, ion exchange resins, and spent filter media.

4. *Air emissions* include off gases from air stripping, ozone contact tanks, etc.

This review will focus solely on the management of sludges. Sludges are produced by a variety of drinking water treatment processes, including lime softening and coagulation/flocculation. Solids in sludge are also generated during backwashing of filters and from PAC addition.

Sludge Properties

One of the most common sludge properties used by environmental engineers is the *total solids content,* or sometimes simply called the *solids content*. The total solids content of a sludge sample is defined as

$$TS = \frac{\text{mass of solids}}{\text{mass of sludge}}$$

Given that sludge consists of solids and water, the equation can be rewritten as

$$TS = \frac{\text{mass of solids}}{\text{mass of solids} + \text{mass of water}}$$

Total solids content is generally expressed as a percentage. Sludge exiting a sedimentation basin may have a total solids content of 1% or less. The goal of thickening and dewatering processes (described later in this section) is to increase the total solids content. As a result, thickening decreases the volume of sludge and decreases storage requirements, transportation costs, and so on. The characteristics of sludge vary greatly depending on the type of sludge and the solids content as shown in Table 14.7.

Table 14.7 Characteristics of sludge

Alum/Iron Coagulant Sludge Characteristics (ASCE/AWWA, 1990)

Solids Content	Sludge Characteristic
0–5%	Liquid
8–12%	Spongy, semisolid
18–25%	Soft clay
40–50%	Stiff clay

(continued)

(continued)

Chemical Softening Sludge Characteristics (ASCE/AWWA, 1990)

Solids Content	Sludge Characteristic
0–10%	Liquid
25–35%	Viscous liquid
40–50%	Semisolid
60–70%	Crumbly cake

Source: U.S. Environmental Protection Agency, *Management of Water Treatment Plant Residuals*, 625/R-95/008.

The dewaterability of sludge depends on the raw water turbidity, type of coagulant used, dose of coagulant, extent of lime use, and mechanism of coagulation. The pH also affects the solids content, as it controls the coagulation mechanism. Sludge produced by the charge neutralization mechanism (pH ~ 6.5) will produce a higher solids content than sludge produced by the sweep floc mechanism (pH ~ 8), as the former uses lower doses of alum. Moreover, alum sludges typically have a lower solids content than iron and lime sludges.

Specific resistance, r, measures the "dewaterability" of a sludge (that is, the sludge's ability to release water). The specific resistance depends on the permeability of the sludge, as well as how tightly the water is bound up within the sludge structure. Specific resistance is measured by removing water from a sample by applying a vacuum to a sample placed on a laboratory filtration apparatus. The volume of filtrate removed from the sample is measured as a function of time. The equation for specific resistance is

$$r = \frac{2PA^2 b}{\mu c}$$

where

r = specific resistance (L·M^{-1})

P = pressure drop across sludge cake (M·T^{-2}·L^{-3})

A = surface area of filter (L^2)

μ = filtrate viscosity (M·L·T^{-1})

c = weight of dry solids deposited per volume of filtrate (M·L^{-3})

b = slope of a plot of t/V vs. V (T·L^{-6})

t = time of filtrate (T)

V = filtrate volume (L^3)

Specific resistance varies widely among sludges (from $2 \cdot 10^{10}$ m/kg to $2 \cdot 10^{12}$ m/kg). Sludges for which r is less than $10 \cdot 10^{10}$ are said to dewater readily.

The density of sludge varies based on the water content and the nature of the flocs. Floc density is related to floc size, the type and dose of coagulant, the mechanism of coagulation, the suspended solids content. Floc specific gravities have been reported to be in the range of 1.1 to 1.3.

The specific gravity of sludge, SG_{sludge}, can be estimated based on its solids content and the specific gravity of the solids:

$$SG_{sludge} = SG_{solids} \cdot TS + SG_{water} \cdot (1 - TS)$$

where

SG_{solids} = specific gravity of sludge solids

TS = total solids content (expressed as a decimal)

SG_{water} = specific gravity of water

Sludge Quantity

The mass of alum sludge can be predicted by the following equation (Davis and Cornwell, 2008):

$$S = 8.34Q \cdot (0.44Al + SS + A)$$

where

S = sludge solids produced (lb/day)

Q = plant flow rate (MGD)

Al = alum dose (mg/L)

SS = raw water suspended solids (mg/L)

A = net solids from additional chemicals added such as polymer or PAC (mg/L)

The mass of iron coagulant sludge can be predicted by the following equation:

$$S = 8.34Q \cdot (2.9Fe + SS + A)$$

where

Fe = iron dose (mg/L, as Fe)

Alternatively, the mass of sludge produced due to the coagulant alone can be estimated from Table 14.8. The total mass of sludge produced is this mass plus the mass of suspended solids and any additional chemicals required.

The mass of softening sludge can be predicted by the following equation (Davis and Cornwell, 2008):

$$S = 8.34Q(2CaCH + 2.6MgCH + CaNCH + 1.6MgNCH + CO_2)$$

where

$CaCH$ = calcium carbonate hardness removed as $CaCO_3$ (mg/L)

$MgCH$ = magnesium carbonate hardness removed as $CaCO_3$ (mg/L)

$CaNCH$ = noncarbonated calcium hardness removed as $CaCO_3$ (mg/L)

$MgNCH$ = noncarbonated magnesium hardness removed as $CaCO_3$ (mg/L)

CO_2 = carbon dioxide removed by lime addition as $CaCO_3$ (mg/L)

For units of flow in m^3/s and sludge production in kg/day, the 8.34 factor is replaced with 86.4.

Table 14.8 Sludge generation rates

Process	Unit	Range	Typical Value
Coagulation			
Alum, $Al_2(SO_4)_3 \cdot 14H_2O$	kg dry sludge/kg coagulant	0.33–0.44	0.33
Ferric sulfate, $Fe_2(SO_4)_3$	kg dry sludge/kg coagulant	0.59–0.8	0.59
Ferric chloride, $FeCl_3$	kg dry sludge/kg coagulant	0.48–1.0	0.48
PACl	kg dry sludge/kg PACl	(0.0372–0.0489) × Al(%)	(0.0489) × (Al, %)
Polymer addition	kg dry sludge/kg coagulant	1.0	1.0
Turbidity removal	mg TSS/NTU removed	0.9–1.5	1.25
Softening			
Ca^{2+a}	kg dry sludge/kg Ca^{2+} removed	2.0	2.0
Mg^{2+b}	kg dry sludge/kg Mg^{2+} removed	2.6	2.6

[a] Sludge is expressed as $CaCO_3$.
[b] Sludge is expressed as $Mg(OH)_2$.

Source: MWH, *Water Treatment: Principles and Design*, 2nd ed., © 2005, John Wiley & Sons, Inc. Reprinted by permission.

The *volume* of sludge produced per day (V) is given by the following equation:

$$V = \frac{S}{\rho_{sludge} \cdot TS}$$

where

ρ_{sludge} = sludge density ($M \cdot L^{-3}$)

= $S.G._{sludge} \cdot \rho_{water}$

ρ_{water} = 998.2 kg/m³ at 20°C

TS = total solids content, expressed as decimal

Example 14.13

Sludge generation rates

Consider a 30,000 m³/day drinking water treatment plant. In the treatment process, 3.0 mg/L of PAC are utilized and 30 mg/L of ferric sulfate ($Fe_2(SO_4)_3$) are added. The total solids content of sludge leaving the settling basin is 1.5%. Estimate the daily volume of sludge exiting the settling basin. The specific gravity of the PAC is 0.55 and is 1.6 for all other solids.

The raw water has a turbidity of 5 NTU. A student worker has collected raw water data and created a graph (Exhibit 4) showing the correlation between suspended solids and turbidity.

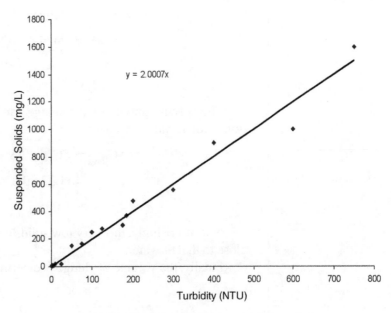

Exhibit 4 Suspended solids—turbidity correlation

Solution

First, express the ferric sulfate dose in terms of mg/L as Fe:

$$\left(\frac{30 \text{ mg Fe}_2(SO_4)_3}{L}\right)\left(\frac{1 \text{ mole Fe}_2(SO_4)_3}{400 \cdot 10^3 \text{mg}}\right)\left(\frac{2 \text{ mole Fe}}{\text{mole Fe}_2(SO_4)_3}\right)\left(\frac{55.8 \cdot 10^3 \text{ mg}}{\text{mole Fe}}\right)$$

$= 8.4$ mg/L as Fe

The solids produced each day are obtained as:

$$S = 8.34 Q \cdot (2.9\text{Fe} + SS + A)$$

$$Q = \left(\frac{30{,}000 \text{ m}^3}{d}\right)\left(\frac{1 \text{ gal}}{3.785 \cdot 10^{-3} \text{ m}^3}\right)$$

$= 7.9 \cdot 10^6$ gal $= 7.9$ MGD

Fe $= 8.4$ mg/L

$SS = 10$ mg/L (from Exhibit 4)

$A = 3$ mg/L

Therefore, the total mass of solids produced is 2561 lb/day, or 1162 kg/day.

To determine the volume of residuals produced, the density of the sludge needs to be known. The specific gravity of the solids is a weighted average of the PAC (SG = 0.55) and the remaining solids (SG = 1.6). The mass of PAC, M_{PAC}, used daily is

$$M_{PAC} = 8.34 \cdot 3 \text{ mg/L} \cdot 1.5 \text{ MGD}$$

$= 659$ lb/day, or 299 kg/day

the mass of the remaining solids, M_{other}, is

$$M_{other} = 1162 \text{ kg/day} - 299 \text{ kg/day}$$

$= 863$ kg/day

The specific gravity of the sludge solids is then calculated to be

$$SG_{solids} = \frac{(299 \text{ kg/day} \cdot 0.55) + (863 \text{ kg/day} \cdot 1.6)}{1162 \text{ kg/day}}$$

$$= 1.33$$

The specific gravity of the sludge can be determined by the following equation, with all values known:

$$SG_{sludge} = SG_{solids} \cdot TS + SG_{water} \cdot (1 - TS)$$

$$= (1.33 \cdot 0.015) + (1 \cdot 0.985)$$

$$= 1.005$$

Not surprisingly, this very low solids content sludge has a specific gravity very close to that of water.

Finally, the volume of sludge is determined as:

$$V = \frac{S}{\rho_{sludge} \cdot TS}$$

$$= \frac{1162 \text{ kg/day}}{1.005 \cdot 998 \text{ kg/m}^3 \cdot 0.015}$$

$$= 77 \text{ m}^3/\text{day}$$

Thickening

Thickening increases the solids content of sludges. It usually follows immediately after sedimentation, filtration, or water softening and is often used as "predewatering" step. Thickening is accomplished most commonly by gravity thickening, in which the solids enter the thickener in the center of the tank. The solids settle to the bottom where they are collected, while the effluent is collected by weirs at the periphery of the tank. Metal hydroxide residuals only thicken to between 1% and 6%, while softening sludges can thicken to between 15% and 30% total solids. Sludge can also be thickened by dissolved air flotation, lagoons, or drying beds.

Coagulant Recovery

The purpose of coagulant recovery is to conserve coagulant use and limit metal concentration in the sludge to be disposed. Coagulants are most often extracted by acidification, with the pH between 1.8 and 3 used for acid contact times between 10 and 20 minutes. Simplified reactions for coagulant recovery using sulfuric acid are provided as follows:

$$2Al(OH)_3 + 3H_2SO_4 \rightarrow Al_2(SO_4)_3 + 6H_2O$$

$$2Fe(OH)_3 + 3H_2SO_4 \rightarrow Fe_2(SO_4)_3 + 6H_2O$$

According to MWH (2005), coagulant recovery is limited in practice due to contaminants found in the recovered coagulant.

Conditioning

The purpose of sludge conditioning is to enhance the dewaterability of the sludge, typically by addition of chemicals. The most common chemical conditioners are ferric chloride, lime, or polymer.

Dewatering

Similar to thickening, dewatering processes increase the solids content; some have defined dewatering as increasing the solids content to greater than 8% (U.S. EPA, 1996). Dewatering is most often accomplished mechanically. Common means of mechanical dewatering include belt filter presses, centrifuges, and vacuum filters:

- **Centrifugation.** Centrifuges use the concept of centrifugal force to remove solids from the water. In effect, a centrifuge is a sedimentation device that increases the settling speed of sludge solids by centrifugal force. In operation, sludge is fed in through the center of the centrifuge. Solids are forced to the periphery, where they are removed by a rotating screw conveyer. Centrifugation is most effective for lime softening sludges, and solids contents between 35% and 50% are common. If properly conditioned, solids contents up to 25% may be obtained for alum sludges.

- **Vacuum filtration.** In vacuum filtration, a cylindrical drum, covered with a filter fabric, rotates through a partly submerged vat of sludge. A vacuum is applied that pulls water through the fabric and into the drum, where it is released. The solids adhere to the filter fabric, from which they are removed by scraping.

- **Plate and frame.** Pressure filters, such as a plate and frame press, use positive pressure (greater than 100 psi) to force water out of the sludge. In the plate and frame press, sludge enters the filter pack, which is a series of chambers held together by the press "skeleton." As the chamber is filled, water passes through a filter cloth medium that lines each chamber and is discharged. Once the chamber completely fills with solids, the chambers are opened and the filter cake is removed.

- **Belt filter press.** A belt filter press dewaters sludge by a combination of gravity drainage and mechanical application of pressure. Conditioned sludge is placed on a moving belt; initially, filtrate drains out. The sludge is then "sandwiched" between the original belt and a second belt, which moves across a series of rollers, effectively squeezing out the water.

The increase in solids content decreases the volume and mass of sludge produced, with concomitant decreases in storage volume required, transportation costs, disposal costs, and so on. The decrease in volume as a result of a thickening or dewatering process is provided in the following equation:

$$\frac{V_2}{V_1} = \frac{P_1}{P_2}$$

where

V_1 = volume of sludge before thickening (or dewatering) (L^3)

V_2 = volume of sludge after thickening (L^3)

P_1 = total solids concentration of sludge before thickening

P_2 = total solids concentration of sludge after thickening

It is important to note that this equation assumes that the density of the sludge does not change as the total solids content changes. Also, the equation assumes that 100% of the solids are captured by the thickening and dewatering devices.

Drying

The purpose of drying is to remove water by evaporation and is defined as increasing the solids content to greater than 35%.

The most common method of drying uses sand drying beds, on which drying takes place by evaporation and drainage. Sludge is placed on the bed, the bottom of which is composed of sand or some other filter medium. Depending on the dryness of the climate, an underdrain system is installed to remove water that has drained from the sludge. Decanting also aids in managing the operation of a drying bed. The design goal may be to minimize the number of application and removal cycles or minimize land area. The designer must balance the need for drying speed (which increases with decreasing thickness of the applied sludge), space availability, and operations concerns. Thin layers require more cycles of loading and unloading and/or increase the land area required for the sand beds. Davis and Cornwell (1998) recommend a total of three or four beds for flexibility of operations.

Operation of a sand drying bed entails applying between 6 and 12 inches of sludge at a time, allowing the sludge to dry, and then removing the dried sludge. Chemical conditioning hastens drying time. Design requires either small-scale testing, or knowledge of *net evaporation rates* (net evaporation rate is the difference between evaporation rate and precipitation rate). In cold climates, freezing effectively conditions the sludge; upon thawing, water drains rapidly from the sludge.

In certain parts of the country, drying occurs in lagoons. The lagoons are operated as gravity settlers for a period of time, followed by a period of drying. Lagoons are equipped with decant capabilities and may be designed to allow drainage. Long-term settling in a lagoon produces solids contents for metal hydroxide sludges on the order of 10%. Without underdrain capabilities, drying of the sludge could take years, depending on the climate and the depth of sludge. Drying time is strongly affected by the fact that many sludges form a crust that dramatically decreases the sludge drying rate. Conversely, if cracking of the surface occurs, evaporation rates will increase significantly. Dewatering lagoons, as compared to storage lagoons without underdrain systems, offer the benefits of a sand drying bed in addition to the ability to store large volumes of sludge periodically.

Example 14.14

Preliminary design of a dewatering train

A 5 MGD water treatment utility produces 1250 lb of solids per day at a concentration of 1.5% total solids. In the past, the sludge has been sent to the wastewater treatment facility. However, reduced capacity at the wastewater treatment facility necessitates the design and construction of a new dewatering train. The proposed dewatering train will consist of a gravity thickener followed by a belt filter press. The gravity thickener produces a solids content of 3.5%, while the filter press produces a solids content of 22%. The filter cake produced by the belt filter press will be sent to drying beds. The climate permits that each bed can be cycled twice per year, and sludge will be spread in 6-inch layers on the drying beds.

Estimate the number of acres of drying bed required to accommodate a total of four beds, with one bed to be used as a spare.

Solution

A sketch of the system is provided in Exhibit 5.

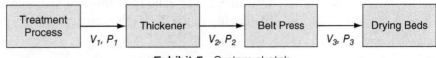

Exhibit 5 System sketch

For this problem, the density of the sludge will be assumed to remain constant at 8.34 lb/gal, irrespective of the solids content.

The daily volume of sludge entering the dewatering train is determined by first finding the daily mass of sludge produced and dividing this mass by the sludge density. The mass of sludge produced is determined using the definition of total solids:

$$\text{mass of sludge} = \frac{\text{mass of solids}}{\text{TS of sludge}} = \frac{1250 \text{ lb}}{0.015} = 83{,}000 \text{ lb of sludge per day}$$

$$\text{volume of sludge} = \frac{\text{mass of sludge}}{\text{sludge density}} = \frac{83{,}000 \text{ kg}}{8.34 \text{ lb/gal}}$$

$$= 10{,}000 \text{ gal of sludge per day}$$

The volume of sludge exiting the gravity thickener is determined as:

$$\frac{V_2}{V_1} = \frac{P_1}{P_2}$$

$V_1 = 10{,}000$ gal

$P_1 = 1.5\%$

$P_2 = 3.5\%$

Therefore, $V_2 = 4300$ gallons.

The volume of sludge exiting the belt filter press is determined using a similar analysis:

$$\frac{V_3}{V_2} = \frac{P_2}{P_3}$$

Given that $P_3 = 22\%$, the volume of sludge exiting the filter press and traveling to the drying beds, V_3, is 680 gallons. Thus, the sand drying beds must be sized to handle 680 gallons of sludge per day (or 250,000 gallons/year).

The total land area needed is estimated by dividing the volume of sludge generated per year by the depth of application:

$$\frac{\dfrac{250{,}000 \text{ gal}}{\text{yr}} \cdot \dfrac{\text{ft}^3}{7.5 \text{ gal}}}{0.5 \text{ ft}} = 67{,}000 \text{ ft}^2$$

Given that each bed can be filled twice per year, the land area required is 33,500 ft². This area must be distributed among three beds, so each bed is about 11,000 ft². The total land area that must be set aside for the four beds is:

$$\text{total area} = 11{,}000 \text{ ft}^2 \cdot 4 \text{ beds} \cdot 1.5$$

$$= 66{,}000 \text{ ft}^2$$

$$= 1.5 \text{ acres}$$

The factor of 1.5 used in the above equation takes into account roadways and berms that will need to be constructed.

Recall that the problem statement assumed that density did not vary with solids content. The validity of this assumption can be examined by first determining the density of the 22% solids content sludge, assuming that the solids have a specific gravity of 2.5.

$$SG_{sludge} = SG_{solids} \cdot TS + SG_{water} \cdot (1 - TS)$$
$$SG_{sludge} = 2.5 \cdot 0.22 + 1 \cdot (0.78)$$
$$= 1.33$$

Clearly, the sludge has a specific gravity significantly greater than 1. To determine how this impacts the volume, consider that the mass of solids is conserved. Thus, the mass of solids entering the drying beds is equal to 1250 lb. The mass of sludge entering the drying beds is

$$TS = \frac{\text{mass of solids}}{\text{mass of sludge}}; \quad \text{mass of sludge} = \frac{\text{mass of solids}}{TS} = \frac{1250 \text{ lb}}{0.22} = 5700 \text{ lb}$$

The volume of this sludge is then

$$\text{Vol} = \frac{\text{mass}}{\text{density}} = \frac{5700 \text{ lb/day}}{1.33 \cdot 8.34 \text{ lb/gal}} = 514 \text{ gal/day}$$

This is significantly less than the estimated 680 gal/day.

Disposal and Reuse

The following disposal options are commonly used in the United States (U.S. EPA, 1996):

- Landfilling
  - Daily cover
  - Codisposal with municipal solid waste (MSW)
- Land application on...
  - Agricultural land
  - Forest land
  - Other designated disposal site
- Sewer disposal
- Direct discharge to receiving body of water
- Reuse
  - Coagulant recovery
  - Turf farming
  - Building materials
  - Fill material

The most common disposal option in the United States is landfilling in a municipal solid waste or hazardous waste landfill. The latter option is rare, as drinking water treatment sludges typically pass the TCLP test (see Chapter 16 for a description of the TCLP test).

Land application has seen mixed results, as the benefits are much less obvious than the benefits of land application of wastewater treatment sludge. The spreading of coagulant sludges has raised concern about the increased concentrations of aluminum in the soil, and the ability of the sludges to bind plant-available phosphorus. Lime sludges can be beneficially used on agricultural fields in place of commercially available lime.

A growing concern with land application of water treatment sludges is increasing arsenic concentrations. As regulations require greater removal of arsenic from drinking water, the arsenic ends up in the sludge.

WASTEWATER TREATMENT

Wastewater is generally defined as *liquid and liquid-carried solid wastes*. Wastewater that comes from residential, commercial, and institutional sources is called *sanitary* or *domestic*, while *municipal* wastewater may have some fraction of permitted industrial sources. *Industrial* wastewaters that contain high levels of contaminants that would inhibit biological activity in the treatment plant must be pretreated to permitted discharge levels on-site prior to discharge.

Wastewater treatment processes are those steps taken to transform municipal wastewater resources into a sufficiently clean state so as to be able to be discharged into surface water receiving streams, or sometimes for unrestricted use (for example, irrigation water). Some of the fundamental concepts used in water treatment can be applied directly to wastewater, such as design of clarifiers and disinfection units; however, the focus in wastewater treatment is the removal of organic matter through *secondary treatment* processes, also called *biological aeration*. WWTPs are generally categorized into the following treatment categories: preliminary, primary settling, secondary (biological) aeration, secondary clarification, tertiary (advanced) treatment, disinfection, and sludge processing.

Wastewater Characteristics

The composition of wastewater entering the wastewater treatment plant (WWTP) is usually described with respect to the constituents of primary concern. The primary water quality indicators (WQIs) that have discharge limits are TSS, BOD, fecal coliforms, oil and grease, and pH. Typically, TSS enter the WWTP at approximately 220–250 mg/L, approximately 75% of which are volatile. BOD_5 enters with an average concentration of 200 mg/L, an indication of the amount of organic matter (OM) present, while fecal coliforms can be on the order of 10^6–10^8 CFU/100 mL. Oil and grease are difficult to determine in the influent but typically are present at concentrations near 90 mg/L and are readily skimmed during processing. Values for pH are generally in the range of 6–8.

Treatment plant effluent standards may vary based on geographic region, and, often, more stringent standards may apply for a particular WWTP than those listed below. However, as a minimum, effluent wastewater characteristics for TSS and BOD are limited to 30 mg/L (monthly average), 45 mg/L (weekly average), or a minimum of 85% removal for influents below 200 mg/L. Fecal coliform discharge limits vary on end use; however, for discharge to recreational waters, concentrations should be below 200 CFU/100 mL (monthly average) and 400 CFU/100 mL (weekly average). For unrestricted use, fecal coliforms need to be below 2 CFU/100 mL, which requires considerable tertiary treatment. Oil and grease need to be below 10 mg/L (monthly average) and 20 mg/L (weekly average), while pH is maintained in the 6–9 range. You will note that, while monthly and weekly averages must be met, discharges that exceed the average may still occur, albeit on an infrequent basis. A summary of wastewater characteristics entering and leaving the WWTP can be found in Table 14.9.

Table 14.9 Average values for municipal wastewater

WQI	Influent	Effluent
pH	6–8	6–9
TSS	220 mg/L	< 30 mg/L
BOD	200 mg/L	< 30 mg/L
Total N	35 mg/L	25 mg/L
Total P	8 mg/L	5 mg/L
Coliforms	10^6–10^8 CFU/100 mL	200 CFU/100 mL

Preliminary Treatment

Traditional preliminary treatment units serve to remove large particles prior to primary treatment. Historically, this is accomplished through the use of coarse or medium screens and grit removal. Additional processes that may occur prior to primary treatment include pumping, size reduction (shredding), and flow measurement. It is not uncommon to have these three combined through the use of metered grinding pumps that perform size reduction and flow logging while lifting raw wastewater to the grit chamber. Further, flow equalization may be required and either occurs just prior to or immediately following primary settling.

Screening

Screens are often referred to as bar racks, due to the physical configuration of the units. Often consisting of vertically oriented bars, classification of screens are based on the bar spacing.

Coarse screens are located just upstream of the main plant lift pumps and serve to protect the pumps from large debris. Since WWTP lift pumps can often accommodate 3-inch diameter solids, bar spacing is generally in the 2–2¾-inch range. Medium screens may be placed in open channels immediately following the pumps, often leading to the grit chamber. Medium screens typically have openings in the ½–1½-inch range and serve to remove additional solids prior to grit removal. Design considerations for screens typically require approach velocities of 2–2.5 fps and through velocities of 2.5–3 fps to minimize forcing solids through the screen and screen damage. Cleaning of screens is often by mechanical scraper; however, small facilities may still clean screens through manual raking.

Grit Removal

Grit is defined as particles (sand, gravel, cinders, bone chips, etc.) that have settling velocities much greater than the organic material typical of wastewater suspended solids. Its removal prior to primary settling is preferred due to the difference in nature of the solids and the potential for fouling (and increased maintenance) of downstream treatment units and pumps. Often, grit is washed and disposed of as solid waste, while the rinse water is returned to the WWTP flow stream. Grit removal is accomplished in horizontal-flow tanks with or without aeration, or in vortex-type units.

Often, raw wastewater is mixed in an *aerated grit chamber*, which provides enough mixing to keep organics suspended, while the reduced density of the bulk fluid allows heavier particles to settle. Sometimes the aerated grit chamber is used for the process of *preaeration*, where more dissolved oxygen is supplied to the biological community and, while little BOD is consumed, prepares the water for the subsequent biological aeration process.

Although covered in this chapter previously for water treatment, the fundamentals of gravity-induced settling are repeated here (with an example problem) as they also apply to wastewater treatment processes. Examinees confident with this topic are encouraged to continue on to the next section.

The rate at which particles move in the vertical direction is described by their *terminal settling velocity* (V_t), which can be expressed as

$$V_t = \left[\frac{4g(\rho_p - \rho_f)d}{3 C_D \rho_f}\right]^{1/2}$$

where g is the gravity constant; ρ_p and ρ_f are the densities of the particle and the fluid, respectively; d is the diameter of the particle; and C_D is the drag coefficient. C_D is dependent on Reynolds number (Re) and can be determined as

$$C_D = \frac{24}{\text{Re}} \qquad \text{Re} \leq 1$$

$$C_D = \frac{24}{\text{Re}} + \frac{3}{\sqrt{\text{Re}}} + 0.34 \qquad 1 \leq \text{Re} \leq 10^4$$

$$C_D = 0.4 \qquad \text{Re} > 10^4$$

where the Reynolds number is defined as

$$\text{Re} = \frac{\rho D V}{\mu} = \frac{DV}{\nu}$$

Since V_t and Re are dependent on each other, an iterative solution is required where a velocity is assumed and used to determine the value of Re. This number is then used to determine C_D, and this drag coefficient is used to calculate a corrected V_t. The Reynolds number is corrected using the new velocity, and the process is repeated until the error in V_t is small. To reduce the complexity of the solution method, it is often assumed that *Stoke's law* is applicable (Re ≤ 1), which is used to describe external flow around spherical bodies at low values for Re. Low Re values are possible for very small particles (on the order of micrometers in diameter) and low settling velocities (on the order of centimeters per second) and simplifies the expression for V_t by setting $C_D = 24/\text{Re}$. In these cases, V_t can be expressed as

$$V_t = \frac{g(\rho_p - \rho_f)d^2}{18\mu}$$

where g, ρ_p, ρ_f, and d are defined as previously, and μ is the absolute viscosity of the fluie—in this case, water.

Example 14.15

Settling velocity

Determine the distance a particle will settle in 1 hour if the particle has a SG = 1.01 and a diameter of 0.5 mm. Assume the water temperature is 15°C and the Re < 1.

Solution

Since we know Re is less than 1, Stoke's law applies, and we may use the Stoke's form of the settling equation as follows:

$$V_t = \frac{g(\rho_p - \rho_f)d^2}{18\mu}$$

Converting the diameter of the particle to meters and interpolating a value for the viscosity of water at 15°C, we may write

$$V_t = \frac{\left(9.81 \frac{m}{s^2}\right)\left[(1.01)(999) - (999) \frac{kg}{m^3}\right](0.0005\ m)^2}{(18)\left(1.1545 \times 10^{-3} \frac{kg}{m \cdot s}\right)} = 0.0012 \frac{m}{s}$$

The distance traveled by a particle can be found by multiplying the velocity by the time period, which may be calculated as follows:

$$d = V_t \times t = \left(0.0012 \frac{m}{s}\right)(1\ hr)\left(\frac{3600\ s}{hr}\right) = 4.32\ m$$

It is often a good idea to check the Stoke's law assumption by calculating Re as

$$Re = \frac{\rho D V}{\mu} = \frac{\left(999 \frac{kg}{m^3}\right)(0.0005\ m)\left(0.0012 \frac{m}{s}\right)}{1.1545 \times 10^{-3} \frac{kg}{m \cdot s}} = 0.52$$

Since this is less than 1, Stoke's law is valid.

Horizontal-flow grit chambers are often rectangular or square in geometry. Typical detention times are 1 minute with horizontal velocities of 1 fps. Detention time is sometimes called retention time or residence time (t_R) and has been defined previously as

$$t_R = \frac{V}{Q}$$

where V is the volume of the basin and Q is the volumetric flow through the basin. Note that the units on V and Q need to be equal, and the time units for t_r will be the same as the time units on Q.

Detention times in an aerated grit chamber range from 2–5 minutes at peak flow (typically 3 minutes) for mixing purposes to 20 minutes for a preaeration basin. Air is usually supplied at a rate of 3–8 ft³/min per foot of basin length. Width-to-depth (W:D) ratios are 1:1 to 2:1, and length-to-width (L:W) ratios are typically 3:1 to 5:1.

Flow Equalization

Due to the diurnal flow pattern of typical wastewater flows, it is anticipated that flow into the WWTP will vary by a factor of 4–10 or more. As treatment units are designed based on detention time, which is a function of this flow, variations in influent rates may have a negative impact on plant performance. One method to deal with this variation is to design all units for peak flow to ensure removal percentages at the highest flows. While this ensures all flows will be sufficiently treated, it does require all process units to be sized larger than they might have otherwise. Another method is to provide flow equalization, where some of the influent is diverted to a holding tank during the highest flows and later released back into the process stream at times when the flow is below the daily average. The equalization basin is placed after grit removal, most often prior to primary settling. This would require the basin to be mixed and aerated to maintain organic solids in suspension and to ensure the wastewater does not become anaerobic.

Required storage capacity is determined through the evaluation of inflows of a WWTP over the course of several seasons, if the data is available. To use the *graphical technique*, a plot is made of cumulative inflow as a function of time for a 24-hour period. Assuming the process flow is at a constant rate over the 24 hours, the average inflow curve will be a straight line. To determine the required size of the basin, a line parallel to the average inflow curve is drawn tangent to the points where the two curves are at a maximum and minimum difference, as shown in Figure 14.9. If the flow pattern is such that no maximum exists, the average flow curve would be used. The maximum vertical distance between the cumulative inflow curve and any of the parallel lines is the required storage.

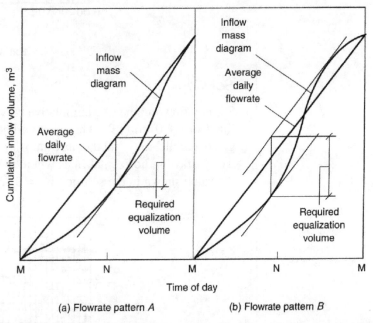

Figure 14.9 Graphical technique for estimation of equalization basin volume

Source: Metcalf and Eddy, *Wastewater Engineering: Treatment and Reuse*, 4th ed., © 2003, McGraw-Hill Education. Reprinted by permission of the McGraw-Hill Companies.

Two spreadsheet methods calculate both cumulative inflows and outflows and the disparity between the two for each date. In the first method, the difference is calculated as inflow minus outflow, which are summed over the time period to determine the cumulative difference. The required storage is determined by finding the maximum value from the cumulative difference column and subtracting the subsequent minimum value. This is called the *difference technique*. In the second method, the deficiency is calculated as outflow minus inflow. In the cumulative deficiency column, any negative values are set equal to zero, and the summation is continued. The required storage is taken directly from the maximum value in the cumulative deficiency column. This is called the *deficiency technique*.

Example 14.16

Flow equalization

Using the WWTP cumulative inflow curve provided in Exhibit 6, determine the volume of a flow equalization basin.

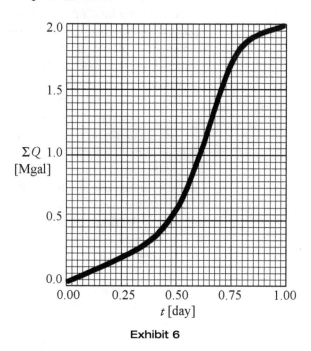

Exhibit 6

Solution

Start by drawing the Q_{ave} cumulative curve as shown in Exhibit 7, which is the 45° line from (0,0) to (1,2). Then draw a line tangent to the cumulative inflow curve given, parallel to Q_{ave} at the maximum and minimum points on the curve. Now, find Δy by drawing horizontal lines where the tangent lines cross any value of t. Note the value shown in the solution is $t = 0.75$.

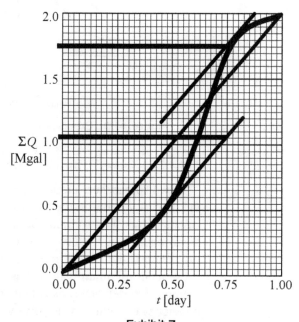

Exhibit 7

Δy = equalization basin volume = 1.75 − 1.05 = 0.7 Mgal

Primary Treatment

Primary treatment is accomplished through gravity-induced settling, usually called *sedimentation* or *clarification*, and is the process in which solid particles are allowed to fall out of suspension in relatively quiescent water. Tanks are designed to greatly reduce flow velocities such that particles are not being agitated by movement of fluid and are allowed to collect at the bottom of the tank. Sedimentation process fundamentals for wastewater treatment are identical to those in water treatment, where the presence of biological solids is a primary concern, and clarifiers are differentiated by task.

In addition to the detention time defined previously, the flow of water in the clarifier can be further characterized by the fluid velocity in the vertical direction, called the *surface settling rate*, and is expressed as

$$V_0 = \frac{Q}{A}$$

where V_0 has units ft/s or m/s, Q is the volumetric flow rate into the clarifier (cfs or m³/s), and A is the surface area of the clarifier (ft² or m²). Any particles possessing $V_t > V_0$ are collected in the clarifier, while particles with $V_t < V_0$ are collected only if they enter the clarifier at a height less than the water depth in the basin.

For design purposes, the surface settling rate is more commonly expressed in the common English units for Q, gallons per day (gpd), and is called the *overflow rate*, with units of gpd/ft². Another design parameter is the *side-water depth*, which usually describes the depth of water not including the sludge blanket that forms on the bottom of the tank. Because t_R, V_0, and side-water depth are not independent, care must be taken when using them in clarifier design to keep from overspecifying the design.

To ensure the wastewater does not cause mixing due to high velocity, another design consideration is the *horizontal* (or *flow-through*) *velocity* (V_h), which may be defined as

$$V_h = \frac{Q}{WH}$$

where Q is volumetric flowrate, W is the tank width, and H is the tank depth.

An additional design parameter is the *weir loading* (*wl*), sometimes called the *weir overflow rate* (*WOR*), which is a measure of the rate of effluent flow over the discharge weir and can be expressed as

$$wl = \frac{Q}{L_w}$$

where *wl* has common units of gpd/ft, L_w is the length of the outlet weir (ft), and Q is as above.

Clarifier shape is usually rectangular, although some circular tanks exist, with the size of the tanks determined through the use of empirical design criteria. These criteria are based on operating conditions that have been found to be most effective in practice. Rectangular tanks generally have a L:W ratio of from 3:1 to 5:1, and weirs are set inside the tank at one end in a box or finger arrangement. For circular tanks, influent is in the center of the tank, and water flows radially toward the weir, located at the tank perimeter, with weir length equal to the tank perimeter.

Primary clarifiers are designed to treat raw wastewater, and removal rates are approximately 50% for suspended solids and 30%–40% for BOD. Primary wastewater clarifiers also allow for the removal of grease and oil by means of skimming at the air-water interface. Design criteria for clarifiers are often listed in tables and are provided for different wastewater treatment applications in Table 14.10. In general, V_0 for primary clarifiers ranges from 800–1200 gpd/ft² (approximately equal to 32–48 m³ per m² per day, or equivalently 32–48 m/d), and *horizontal* (or *flow-through*) *velocities* (V_h), should be kept below 0.5 ft/min. The design criterion for detention time is typically 2 hours, side-water depths should be maintained greater than 10 ft, and weir loadings are typically 20,000 gpd/ft.

Table 14.10 Design criteria for wastewater clarifiers based on average flows

Type of Basin	V_0 (gpd/ft²)	m³/(m² · day)	wl gpd/ft	m³/(m · day)	t_R hr
Primary clarifiers	800–1200	32–48	20,000	250	2
Fixed film reactors	400–800	16–32	20,000	250	2
Air-activated sludge	400–800	16–32	20,000	250	2
Extended aeration	200–400	8–16	10,000	125	4

Example 14.17

Rectangular clarifier design

Determine the length of a rectangular clarifier for a flow of 3.1 mgd if the design detention time is 2 hours, the tank has a depth of 10 ft, and the L:W = 4:1.

Solution

Volume of a tank can be found given detention time and volumetric flow rate as

$$t_R = \frac{V}{Q} \Rightarrow V = Q\, t_R$$

$$V = (3.1 \times 10^6 \text{ gpd})(2 \text{ hr})\left(\frac{1 \text{ day}}{24 \text{ hr}}\right) = 258{,}333 \text{ gal} \times \left(\frac{\text{ft}^3}{7.48 \text{ gal}}\right) = 34{,}537 \text{ ft}^3$$

The area of the tank can be found from the volume and depth as follows:

$$A = \frac{V}{\text{depth}} = \frac{34{,}537 \text{ ft}^3}{10 \text{ ft}} = 3454 \text{ ft}^2$$

Further, for a L:W ratio of 4:1, we can find the tank dimensions as

$$A = LW = (4W)(W) = 4W^2 = 3454 \text{ ft}^2 \Rightarrow W = 29.4 \text{ ft} \approx 30 \text{ ft}$$

$$\text{and} \quad L = 4W = 120 \text{ ft}$$

Secondary Treatment

Secondary treatment is also called biological treatment or biological aeration due to the fact that microorganisms (primarily bacteria) are employed to reduce wastewater BOD to acceptable effluent levels. Optimization of this naturally occurring process requires the knowledge of biology covered in Chapter 7, coupled with process unit experience based on reactor characteristics. Most biological treatment process can be categorized as either suspended or attached growth processes. In suspended growth, individual organisms interact in a well-mixed environment that contains the food, oxygen, and nutrients required for removal of BOD. Attached growth processes use a biofilm surface and pass the food/nutrient/oxygen-containing wastewater over the film. Three types of biological treatment are discussed in this review: activated sludge, fixed film reactors, and treatment ponds and lagoons.

Activated Sludge

One method of encouraging interaction between wastewater organic matter, the bacteria that consume that organic matter as substrate, and the oxygen required to serve as terminal electron acceptor is a process known as *activated sludge*. As seen in Figure 14.10, a highly mixed reactor is designed to distribute the needed oxygen and keep the biological mass in suspension.

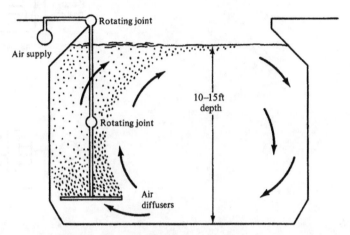

Figure 14.10 Cross section of a typical channel demonstrating mixing through aeration

Source: Warren Viessman and Mark J. Hammer, *Pollution Control*, 7th ed., © 2005. Reprinted by permission of Pearson Education, Inc., Upper Saddle River, N.J.

As bacteria consume the organic matter, the numbers increase significantly, converting BOD to TSS that still requires removal prior to final effluent discharge. However, these "biosolids" are rich with active bacteria that could be employed to remove additional organic matter from the wastewater. It is the return of the "activated" mass of organisms that provides the name for the activated sludge process. There are several typical process layouts as presented in Figure 14.11, but, in all cases, significant numbers of the bacteria removed in the final clarifier are subsequently returned to the aeration tank to continue the consumption of organic matter.

Continuous accumulation of biosolids is avoided through the natural process of cell lysis, through the wasting of a small amount of the returned sludge, and from the fraction that is discharged in the effluent as TSS. Modifications to the processes shown include the use of multiple feeds to the tank and direct aeration without the use of a primary clarifier.

System design considerations may be based on historical operation parameters or may be determined through kinetic evaluation of the biological process. The process of designing a system through operational parameters relies on a range of values that have been shown to be effective in the reduction of BOD through biological activity. Once the type of process is selected, aeration tank volume (V_A) may be calculated using a known wastewater flow rate (Q_0) and the *hydraulic residence time* (θ), also called the *aeration period*, which can be expressed as

$$\theta = \frac{V_A}{Q_0}$$

It is worthwhile to note that this is equivalent to the mean residence time (t_R) used previously, and that the volumetric flow rate used does not include any flow into the tank attributed to recirculated biosolids from the final clarifier. Typical values for θ are based on reactor configuration and are presented in Table 14.11, along with several additional activated sludge design parameters. It should be noted that definitions for each of the design criteria listed in Table 14.11 will be provided in the following portions of this section.

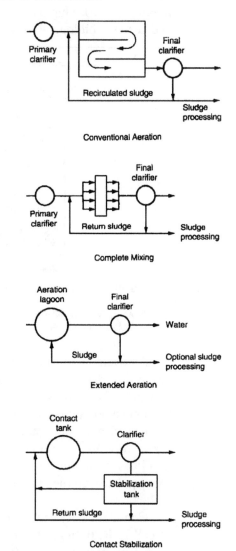

Figure 14.11 Variations of the activated sludge process

Table 14.11 Design criteria for activated sludge treatment systems

Process	SRT days	F/M (kg BOD/d)/ kg MLVSS	Volumetric Loading		MLSS mg/L	θ hr	RAS % of Q
			lb BOD/ 1000 ft³ · d	kg BOD/ m³ · d			
High-rate aeration	0.5–2	1.5–2.0	75–150	1.2–2.4	200–1000	1.5–3	100–150
Contact stabilization							
Contact basin	5–10	0.2–0.6	60–75	1.0–1.3	1000–3000	0.5–1	50–150
Stabilization basin					6000–10,000	2–4	

(continued)

	(continued)

High-purity oxygen	1–4	0.5–1.0	80–200	1.3–3.2	2000–5000	1–3	25–50
Conventional plug flow	3–15	0.2–0.4	20–40	0.3–0.7	1000–3000	4–8	25–75
Step feed	3–15	0.2–0.4	40–60	0.7–1.0	1500–4000	3–5	25–75
Complete mix	3–15	0.2–0.6	20–100	0.3–1.6	1500–4000	3–5	25–100
Extended aeration	20–40	0.04–0.10	5–15	0.1–0.3	2000–5000	20–30	50–150
Oxidation ditch	15–30	0.04–0.10	5–15	0.1–0.3	3000–5000	15–30	75–150
Sequencing batch reactor	10–30	0.04–0.10	5–15	0.1–0.3	2000–5000	15–40	NA

Source: adapted from Metcalf & Eddy, *Wastewater Engineering: Treatment and Reuse*, 4th ed. (McGraw-Hill, 2002), Table 8.16, p. 747.

Example 14.18

Activated sludge tank design

A conventional activated sludge system treats 6.4 mgd. Based on a minimum hydraulic residence time, what is the length of aeration channel required if the cross section has dimensions of 15 × 12 feet?

Solution

From Table 14.11, the hydraulic residence time (also called aeration period) for a conventional activated sludge system is from 4 to 8 hours. Using the minimum of 4 hours, tank volume can be calculated as follows:

$$\theta = \frac{V}{Q} \Rightarrow V = \theta Q = (4 \text{ hr})\left(\frac{1 \text{ day}}{24 \text{ hr}}\right)\left(6.4 \frac{\text{Mgal}}{\text{day}}\right)$$

$$= 1.07 \text{ Mgal} \times \frac{1 \text{ ft}^3}{7.485 \text{ gal}} = 142{,}500 \text{ ft}^3$$

Given the tank volume, the channel length can be determined as

$$V = L \cdot W \cdot H \Rightarrow L = \frac{V}{H \cdot W} = \frac{142{,}500 \text{ ft}^3}{(15 \text{ ft})(12 \text{ ft})} = 792 \text{ ft}$$

Tank volume calculations using hydraulic loadings must also satisfy the loading rate of the substrate (organic matter), which is approximated analytically using a value of BOD (or COD). The *volumetric substrate loading rate* (V_L) is defined as the *mass rate of BOD (or COD) fed per unit volume of aeration tank* and can be expressed as

$$V_L = \frac{Q_0 S_0}{V_A} = \frac{S_0}{\theta}$$

where Q_0 is the flow rate into the aeration tank and S_0 is the BOD (or COD) concentration fed to the aeration tank, not including any recirculated flow. Typical values for V_L are also based on reactor configuration and are given in Table 14.11. It should be noted that primary clarifiers remove 30%–40% of the BOD_5 that enters the plant, and activated sludge systems that are located after a primary clarifier should have their influent BOD_5 values reduced as such.

Example 14.19

BOD loading criteria

Determine the BOD loading in units of kg/m³/day to the system in Example 14.18 if the BOD concentration into the activated sludge tank is 140 mg/L.

Solution

The units for the answer are in the metric system. We should start by converting the tank volume to metric units as follows:

$$V = 142{,}500 \text{ ft}^3 \times \left(\frac{1 \text{ m}}{3.28 \text{ ft}}\right)^3 = 4038 \text{ m}^3$$

We can calculate the BOD load by multiplying BOD concentration and flow rate as

$$\text{BOD load} = \left(140 \frac{\text{mg}}{\text{L}}\right)\left(6.4 \times 10^6 \frac{\text{gal}}{\text{day}}\right)\left(3.78 \frac{\text{L}}{\text{gal}}\right)\left(10^{-6} \frac{\text{kg}}{\text{mg}}\right) = 3387 \frac{\text{kg}}{\text{day}}$$

Now the BOD loading to the tank can be determined as

$$\text{BOD loading} = 3387 \frac{\text{kg}}{\text{day}} \div 4038 \text{ m}^3 = 0.84 \frac{\text{kg}}{\text{m}^3 \cdot \text{day}}$$

Once tank volume is determined, geometric configuration can be determined depending on the type of activated sludge process unit. For conventional plug-flow systems, channels are assumed to have square or rectangular cross sections with depths of 10–15 feet. Total length of channel is determined by dividing tank volume by channel cross-sectional area. This channel length is divided into equal segments that will fit into the desired footprint, and overall tank dimensions are determined.

The concentration of bacteria in the aeration tank is called the *mixed liquor suspended solids* (MLSS) concentration. Typical values for MLSS are given in Table 14.11 for a variety of activated sludge processes. As a parameter in biological kinetic evaluations, MLSS is usually expressed by the variable X_A and is often expressed units of mg biomass (or COD) per L of tank volume. Another method of evaluating substrate loading is to regulate the food supply based on the amount of bacteria present in the aeration tank that consume the organic matter. This mass ratio is called the *food to microorganism (mass) ratio* (F/M) and can be calculated as

$$\frac{F}{M} = \frac{Q_0 S_0}{V_A X_A}$$

where F/M has units of mass of BOD per day per mass of biosolids. It should be noted that Table 14.11 uses the units of mass of BOD per day per unit mass of MLVSS, or mixed liquor volatile suspended solids. MLVSS is considered a better measure of the biomass fraction of the MLSS, as all bacterial would be considered volatile based on the definitions for solids analysis covered at the beginning of this chapter.

The F/M ratio is a powerful tool used to control the activated sludge process. As seen in Figure 14.12, high F/M ratios are good for high rates of consumption and, therefore, smaller aeration tanks, but the effluent from that tank exhibits poor settling characteristics. Further, the presence of large quantities of food to a relatively small bacteria population may result in the microbes preferentially consuming the simpler substrates, thus the potential for incomplete consumption of

BOD. In order to achieve good solids removal in the final clarifier, the F/M ratio is lowered, and, while the substrate consumption rate is decreased, the consumption percentage is increased.

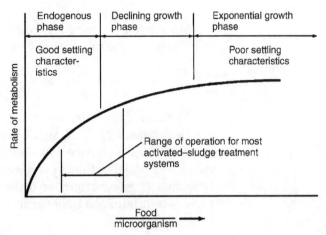

Figure 14.12 Effect of F/M ratio on degradation rate and settling characteristics

Source: Warren Viessman and Mark J. Hammer, *Pollution Control*, 7th ed., © 2005. Reprinted by permission of Pearson Education, Inc., Upper Saddle River, N.J.

Since the majority of the solids that were removed from the wastewater in the final clarifier are returned to the aeration tank, an individual cell has the potential of cycling many times through the system. The average length of time that a single cell remains in the system is called the *mean solids (cell) residence time* (θ_c), often called the *sludge age* or *solids retention time* (SRT), and can be expressed as

$$\theta_c = \text{SRT} = \frac{V_A X_A}{Q_e X_e + Q_w X_w} = \frac{V \times \text{MLSS}}{Q_e SS_e + Q_w SS_w}$$

where the subscript w indicates the waste stream, and the subscript e indicates the plant effluent stream. Notice that, while there may be substantial recirculation flow, the volumetric flow rate of the water in the effluent plus the wasted solids must equal the original flow into the aeration tank (that is, $Q_0 = Q_e + Q_w$) in order to avoid an accumulation of liquid volume in the system. Typical values for SRT are given in Table 14.11 for several different activated sludge treatment systems.

In the absence of other data, estimates for several terms in the sludge age equation can be made. Since there are regulatory limits on discharged suspended solids, X_e may be assumed to be < 30 mg/L and a value of 20 mg/L is typical. Further, the volumetric flow of the wasted solids is quite small compared to the effluent flow (Q_w is often ≤ 1% of Q_0); therefore, Q_e is often assumed to be equal to Q_0. Finally, the $Q_w X_w$ product represents the mass rate of wasted solids and can be estimated as a function of F/M ratio as

$$Q_w X_w = k Q_0 S_0$$

where $k \approx 0.43$ for F/M = 0.2 and $k \approx 0.5$ for F/M = 0.4, with k estimated for other values of F/M through interpolation or extrapolation.

For completely mixed activated sludge systems, the microbial kinetics expressions developed in Chapter 7 can be employed to design system parameters. Generally, laboratory data is generated that provides estimates for μ_{max}, K_s, k_d, and Y. Values for k_d range between 0.01 to 0.1 day^{-1} with a typical value of 0.05 day^{-1}, and Y ranges from 0.4 to 1.0 with a typical value of 0.7. The mean cell residence time, θ_c, as defined previously has units of mass of cells divided by the mass rate of change in the cells leaving the final clarifier, which may be expressed as

$$\theta_c = \frac{X_A}{(dX/dt)}$$

Substituting the biomass growth expression developed in Chapter 7 allows θ_c to be related to the substrate utilization rate as follows:

$$\frac{1}{\theta_c} = \frac{(dX/dt)}{X_A} = \frac{Y(dS/dt)}{X_A} - k_d = \frac{\mu_{max} S}{K_s + S} - k_d = \mu - k_d$$

From this expression, the mean cell residence time can be estimated using the kinetic parameters μ_{max}, K_s, and k_d, along with a value for the substrate concentration inside of the aeration tank. Often, this substrate concentration is described as the soluble BOD, and since completely mixed reactors have the same effluent concentration as the concentration in the tank, the effluent soluble BOD concentration (S_e) can be determined from the above expression as

$$S_e = \frac{K_s(1 + k_d \theta_c)}{\theta_c(\mu_{max} - k_d) - 1}$$

Therefore, for a set of laboratory determined kinetic parameters, the effluent soluble BOD concentration can be estimated for a fixed value for sludge age.

Since substrate utilization is the change in substrate concentration that takes place during one unit of time in the aeration tank, we can express substrate utilization as

$$\frac{dS}{dt} = \frac{S_0 - S_e}{\theta} = \frac{Q_0(S_0 - S_e)}{V_A}$$

This may now be substituted into the sludge age expression above and written as

$$\frac{1}{\theta_c} = \frac{Y Q_0 (S_0 - S_e)}{V_A X_A} - k_d = Y\left(\frac{F}{M}\right)\left(\frac{E}{100}\right) - k_d$$

where E is the BOD removal efficiency and may be expressed as

$$E = \frac{S_0 - S_e}{S_0} \times 100$$

Solving this expression for the tank volume (V_A) yields the following:

$$V_A = \frac{\theta_c Y Q_0 (S_0 - S_e)}{X_A (1 + k_d \theta_c)}$$

or, if tank volume is known, the concentration of cells in the aeration tank (MLSS) can be estimated as

$$X_A = \text{MLSS} = \frac{\theta_c Y Q_0 (S_0 - S_e)}{V_A (1 + k_d \theta_c)} = \frac{\theta_c Y (S_0 - S_e)}{\theta (1 + k_d \theta_c)}$$

Finally, the net mass of cell biomass produced (P_x) in g/d may be calculated as

$$P_x = \frac{Y Q_0 (S_0 - S_e)}{1 + k_d \theta_c}$$

Fixed Film Treatment Units

Another type of biological treatment system is modeled from the natural tendency of microorganisms to attach to a surface and feed off of dissolved substrates that are carried past them. Because this biomass forms a thin film on the stationary surface, usually referred to as a biofilm, the process is called *fixed film* or *attached growth* biological treatment.

The most common type of fixed film treatment unit is the *trickling filter*, where the term *filter* does not refer to solids removal but the consumption of organic matter through biological utilization. The first trickling filters used rock as the biofilm surface, due to its ready availability anywhere a treatment system would be constructed. The surface area and void volume were low, however, and the weight of the media limited the depth of the units to 5–7 ft. Today, tricking filters consists of a tank that holds a high specific surface area (>100 m²/m³), high void fraction media (90%–95% voids), such as engineered plastic in the form of spheres, cylinders, or sheets. Additional characteristics of good filter media include durability, low unit weight (<100 kg/m³), low cost, and uniform flow properties.

Operation of a trickling filter requires that clarified wastewater is sprayed uniformly over the surface of the filter bed and allowed to trickle down through the bed and to collect in the underdrains below. Sufficient void space is required to maintain both a continuous flow of wastewater and a continuous flow of air, which provides oxygen to the wastewater as DO is needed by the bacteria for aerobic metabolism. Nearly all systems recirculate a portion of the treated wastewater at a ratio of 0.5 to 3.0 times the influent flow rate (that is, $Q_R = 0.5$ to $3 \times Q_0$) in order to reduce the strength of the wastewater that contacts the biofilm and provide a sufficient quantity of wastewater, regardless of flow variations that occur throughout the day. Several possible recirculation schemes may be applied, and some of the possible layouts are presented in Figure 14.13.

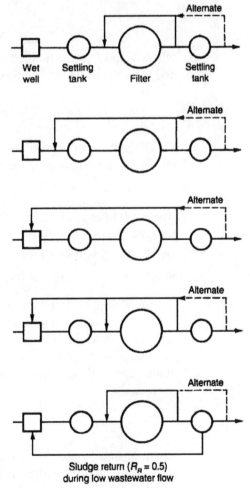

Figure 14.13 Process flow diagrams for some possible trickling filter recirculation scenarios

Design of a trickling filter unit is based on both hydraulic loading, which has units of volumetric flow rate per unit surface area of the tank in plan view (Q/A_s), and substrate loading, which has units of mass of BOD applied per day per unit volume of tank ($Q \cdot BOD/V$). While substrate loading is based upon the influent wastewater after primary sedimentation and ignores any BOD in the recirculation flow, hydraulic loading is based upon the combined flow (that is, $Q_0 + Q_R$).

The BOD removal efficiency for stone media trickling filters is calculated using the NRC equation, which can be expressed as

$$E_1 = \frac{100}{1 + 0.0561 \left(\frac{w}{VF}\right)^{1/2}}$$

where E_1 is the BOD removal efficiency at 20°C (%), w is the applied BOD load (lb/day), V is the volume of the filter (ft³ × 10⁻³), and F is the recirculation factor, which can be expressed as

$$F = \frac{1+R}{(1+0.1\,R)^2}$$

where R is the recirculation ratio, which is calculated as Q_R/Q_0. For the case where a high strength wastewater is being treated by trickling filter, it is common to require a second filter that is placed in series after an intermediate clarifier. The NRC efficiency equation for the second filter can be expressed as

$$E_2 = \frac{100}{1 + \left(\frac{0.0561}{1-E_1}\right)\left(\frac{w_2}{V_2 F_2}\right)^{1/2}}$$

where V and F are as defined previously, the subscript 2 refers to the second filter, and w_2 is the BOD load applied to the second filter (lb/day) and can be calculated as

$$w_2 = \frac{(100 - E_1)}{100} w_1$$

Since the NRC equation expresses efficiency at 20°C, correction for wastewater temperature must be made as follows:

$$E_T = E_{20} (1.035)^{T-20}$$

where T is in units of °C. The efficiency as expressed by the NRC equation accounts for both the trickling filter and the final clarifier; therefore, the overall plant efficiency can be expressed as

$$E_{TOT} = 100 - 100\,(1-E_P)(1-E_{1T})(1-E_{2T})$$

In the case of deep beds of plastic media, this system is often called a *biological tower*, or simply a *biotower*. Biotowers are usually evaluated using biological kinetics to describe substrate utilization and, therefore, effluent quality and BOD removal efficiency. Laboratory data is generated to determine wastewater treatability, and BOD removal for single pass systems (that is, no recirculation flow) is expressed as

$$\frac{S_e}{S_0} = \exp\left[\frac{-k\,D}{q^n}\right]$$

where S_e and S_0 are the BOD concentration in the tower effluent and influent, respectively; k is the treatability constant (min⁻ⁿ, often min⁻¹ᐟ²); D is the filter depth (m); q is the hydraulic loading (m³/min per m² of filter surface area); and n is the empirical flow constant, generally assumed to be 0.5 for modular plastic media.

The treatability constant can range from 0.01–0.1 min^{-1}, with a value of 0.06 min^{-1} generally used when specific kinetic data is unknown. This constant is often given at 20°C, and temperature compensation must be made using the equation given above for stone media trickling filters.

For high strength wastewaters, applied BOD concentration (S_a) can be reduced by recirculating treated water, and is calculated as follows:

$$S_a = \frac{S_0 + R S_e}{1 + R}$$

The effluent concentration can now be determined as follows:

$$\frac{S_e}{S_a} = \frac{\exp\left[\dfrac{-k D}{q^n}\right]}{(1+R) - R \exp\left[\dfrac{-k D}{q^n}\right]}$$

where q is the total hydraulic load to the tower that is, $(Q_0 + Q_R)$/tower surface area).

Another form of the fixed film treatment system is called a *rotating biological contactor* (RBC). The RBC is a series of closely spaced circular plastic disks attached to a slowly rotating shaft, which are 10–15 feet in diameter and serve as the attachment surface for the biofilm. The disks are partially (~40%) submerged in the wastewater, which allows for contact between the organic matter and the bacteria while a section of the disk is submerged, and for contact with the air for replenishing depleted DO when rotated out of the wastewater. Figure 14.14 offers a schematic of an RBC along with the BOD removal efficiency expected as a function of hydraulic loading.

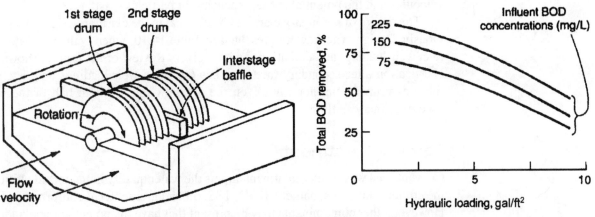

Figure 14.14 RBC schematic and BOD removal efficiency

Stabilization Ponds

Treatment of domestic wastewaters from small populations may be accomplished through the use of natural processes in shallow ponds. These ponds are often called *facultative ponds*, where symbiotic relationships between algae and aerobic bacteria cohabit with anaerobic activity in the sediment layer at the bottom of the pond. When stabilization ponds are used subsequent to secondary aeration, they are called *polishing* (*tertiary* or *maturation*) ponds. Because polishing ponds receive waters already treated, they generally possess detention times of only 10–15 days and maintain depths of 2–3 feet.

Volumetric BOD loads in facultative ponds generally range from 0.1–0.3 lb of BOD per 1000 ft^3 of pond volume per day. Normal operating depths are kept greater than 3 feet to discourage rooted plant growth but less than 6 feet to maintain an aerobic aqueous environment. Detention times of 50–150 days (usually between

90 and 120 days) require substantial land area, and area-based BOD mass loadings are kept below 35 lb BOD per acre per day. Warmer climates can generally handle slightly larger BOD loads. However, freezing conditions essentially eliminate biological activity, and on-site storage volume should be sufficient to hold all wastewaters until late spring for locations that experience subfreezing temperatures.

Operation of the pond generally relies on controlling the depth as flow rates vary with the seasons. It is not uncommon for these ponds to have no discharge, as the evaporation rate is sufficient to maintain appropriate water depths. Often ponds are constructed in series or parallel configurations to provide additional control over flow distribution and to discourage short circuiting, and ponds may be lined with a compacted clay liner if groundwater contamination is a concern. Since odor is a minor problem during parts of the year, ponds are often located a sufficient distance downwind of any residential areas.

Secondary Clarification

Clarification following biological aeration may occur in two forms: *intermediate* and *final* clarifiers. Intermediate clarifiers are designed to treat the effluent of the first stage of a two-stage trickling filter. Final clarifiers treat the effluent from the secondary treatment aeration tanks, and their design depends on the type of biological treatment process used. One additional consideration for settling of activated sludge is the potential for *zone settling*, which may occur in waters of high biological cell concentrations. In these cases, a blanket of solids forms and settles as a mass. However, due to the particle density, fluid movement in the upwards direction is inhibited, and thus settling velocity is reduced. As water seeks a way back to the surface from the edges or breaks in the sludge blanket, high local fluid velocities and the potential for resuspension of particles exist.

The design of secondary clarifiers is identical to the clarifier design review previously. Design criteria were presented in Table 14.10, where values for overflow rate, weir loading, and detention time are given. Review of the criteria shows that clarification after secondary treatment has the same weir loadings and detention times as primary clarifiers; however, the overflow rate is about half of the value used in primary settling.

Tertiary Treatment

Conventional wastewater treatment targets the water quality indicators identified in the effluent standards, namely, BOD, TSS, oil and grease, fecal coliforms, and pH. However, other contaminants may be present that have the potential to degrade the ecosystem upon discharge. Two chemical species, N and P, are required for secondary treatment to be successful and usually are available in sufficient quantities. However, often these "nutrients" are present in amounts that may have a negative impact to receiving streams. For example, excess phosphorus may cause eutrophication of surface water bodies through the growth of algae and other aquatic plants, and ammonia nitrogen is toxic to fish at relatively low concentrations. Another possibility is that excess suspended solids may require additional processing for their removal to levels below discharge limits. Further, the potential for other chemicals, such as heavy metals or nonbiodegradable organic compounds from the improper disposal of hazardous chemicals into the municipal sewer system, may also pose a toxic threat to aquatic life forms and require additional processing for their removal.

Nitrification

Nitrification is the process of the biological oxidation of ammonia to nitrate through the following reaction:

$$NH_3 + 2O_2 = NO_3^- + H^+ + H_2O$$

In biological treatment, the aerobic degradation converts most of the organic nitrogen to ammonia, so nitrification systems are based upon the conversion of both ammonia and organic nitrogen to nitrate. This sum is known as the *total Kjeldahl nitrogen* (TKN). Theoretical oxygen requirement is 4.57 g O_2 per g N oxidized, although the actual amount is slightly less due to biological synthesis. Note that this process does not remove nitrogen from the system but instead converts it into the nontoxic nitrate (NO_3^-) form.

Nitrification can be achieved in a single-stage system (combined with activated sludge) or in a separate stage. The single-stage system requires no additional facilities but has limited control and results in increased oxygen utilization. A solids retention time of greater than 5 days is required at 20°C and greater than 10 days at 10°C to have effective nitrification. Nitrification can also be achieved in trickling filters and rotating biological contactors if the organic loading is limited. Organic loadings of less than about 5 lb BOD_5/1000 ft³ · d (80 kg/m³ · d) and 8 lb BOD_5/1000 ft³ · d (130 kg/m³ · d) for rock and plastic media, respectively, will result in greater than 90% nitrification. The kinetic coefficients are listed in Table 14.12.

Table 14.12 Kinetic coefficients for nitrification reactors

Coefficient	Value for Ammonia-N, 20°C
k	5 mg/mg·d
K_s	1 mg/L
k_d	0.05 d⁻¹
Y	0.2 mg/mg · d

Temperature corrections for k are

$$k_T = k_{20} e^{1.053(T-20)}$$

where k_T and k_{20} are the maximum N utilization rates (mg/mg · d) at temperatures T and 20°C, respectively. Nitrification is also strongly affected by pH below 7.2. The overall reaction produces a hydrogen ion, so adequate alkalinity must be present or added to maintain pH above 7.2.

Biological Denitrification

Biological denitrification is used to reduce nitrate to nitric oxide, nitrous oxide, and nitrogen gas. Coupled with nitrification, it is an integral part of biological nitrogen removal. Denitrification occurs in biological treatment under the presence of an organic substrate (BOD) and nitrate and the near absence of oxygen. A wide range of bacteria can use nitrate in the place of oxygen and are ubiquitous to wastewaters and wastewater treatment processes. The important parameter in designing denitrification systems is the ratio of BOD needed as an electron donor per NO_3 removed. Typical values range from about 2 to 5 mg BOD/mg NO_3-N.

Several different designs have been used to achieve denitrification for wastewaters. The complexity of the design results from having to provide BOD to reactors or zones of reactors that have high nitrate concentration. The BOD is commonly obtained from BOD in the influent or from endogenous decay of cells. Anoxic conditions can be easily obtained by eliminating aeration. The design of the systems in Figure 14.15 is beyond the scope of this review.

Phosphorus Removal

Phosphorus is removed from wastewater by precipitation with calcium (lime), aluminum (alum), or iron (ferric salt). Alum is the preferred process for a variety of reasons, including minimum sludge volumes and a greater removal at neutral pH. The metal salts can be added before the primary clarifier, aeration basin, or secondary clarifier as seen in Figure 14.16; however, alum is most effective when used downstream of the secondary treatment where a majority of the phosphate has been converted to orthophosphorus. Required alum dosages (mole Al:mole P) range from 1.4:1 for 75% phosphorus reduction, to 1.7:1 for 85% reduction, to 2.3:1 for 95% reduction.

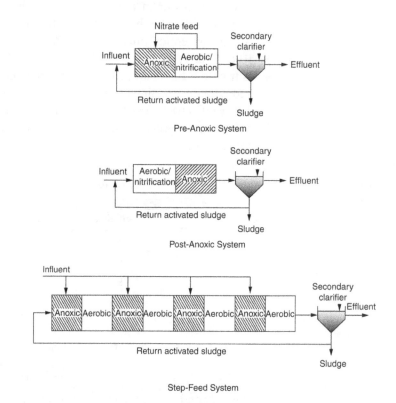

Figure 14.15 Examples of denitrification process systems

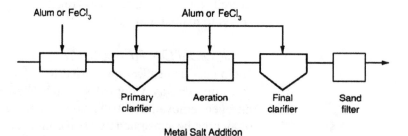

Figure 14.16 Phosphorus removal with alum or chloride

Adsorption

Removal of heavy metals and organic compounds that were not consumed in the secondary treatment unit may be removed through the transfer of mass from the aqueous phase to a suitable sorptive material through adsorption. For wastewater treatment, the adsorbent of choice is activated carbon (AC), either by passing the wastewater through a fixed bed of granular activated carbon (GAC) or through the mixing with powered activated carbon (PAC) in the aeration tank and its subsequent removal with the biosolids. The fine-grained nature of the PAC may require the addition of coagulants and the use of sand filters to achieve its complete removal from the process water prior to discharge.

Due to the common nature of the adsorption processes used throughout environmental engineering, the fundamentals of chemical sorption and the design of fixed-bed units will be covered in detail in Chapter 15.

Filtration

There may be several reasons to consider suspended solids removal using a granular media (sand filtration), such as to meet TSS effluent limits, to remove particulate prior to disinfection to increase chlorination efficiency, and to prevent the fouling of a GAC adsorption column. Although care must be taken to account for the different nature of the particulate and the potential for substantial variations in flow rates (in the absence of flow equalization), a typical dual-media gravity filter is the process unit of choice. It is also possible to use the dual-media filter under pressure to overcome the potential headloss issues due to the nature of the solids removed. Filtration rates are similar to water treatment units at 3–6 gpm/ft^2 of filter area. The design of filtration units was covered in detail in the water treatment section of this chapter.

Disinfection

The primary consideration for disinfection of WWTP effluents is the removal of pathogens, as quantified through the fecal coliform test, although fecal viruses, protozoal cysts, and helminth eggs may need to be addressed. Two popular choices are chlorination and ultraviolet radiation.

Chlorination

Traditional chlorine disinfection as covered previously in this chapter is the most common form of treatment for wastewater effluents. However, the major difference between chlorination of water supplies and wastewater is the reaction of chlorine with ammonia to produce chloramines as

$$NH_3 + HOCl = NH_2Cl + H_2O$$

$$NH_2Cl + HOCl = NHCl_2 + H_2O$$

$$NHCl_2 + HOCl = NCl_3 + H_2O$$

Chloramines can undergo further oxidation to nitrogen gas with their subsequent removal.

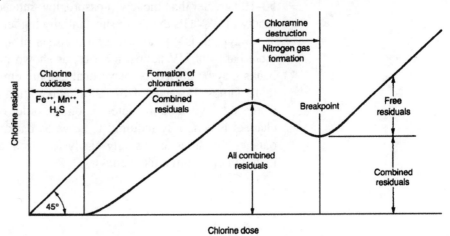

Figure 14.17 Chlorine dosage versus chlorine residual

This stepwise oxidation of various compounds is shown in Figure 14.17. As chlorine is initially added, it reacts with easily oxidized compounds such as reduced iron, sulfur, and manganese. Further addition results in the formation of chloramines, collectively termed combined-chlorine residual. This combined-chlorine residual results in the destruction of microorganisms, although much less effectively than with HOCl. Further addition of chlorine results in the destruction of the chloramines and ultimately leads to the formation of free-chlorine residuals such as HOCl and OCl$^-$. The development of a free-chlorine residual is termed *breakpoint chlorination*.

The more contaminants present in the wastewater, the greater the chlorine demand before free chlorine is formed. Chlorination requirements vary by state; a typical requirements is a chlorine residual of 0.5 mg/L after 15 minutes of contact time. Typical dosages required to meet such a requirement are listed in Table 14.13. In practice, chlorine is introduced in a rapid-mix tank, followed by a minimum of 30 minutes of contact (designed at peak flow). Contact basins often are serpentine and/or baffled tanks to ensure adequate contact and limit short circuiting.

Table 14.13 Chlorine dosages for wastewaters

Wastewater	Dosage, mg/L
Untreated	6–25
Primary effluent	5–20
Activated sludge effluent	2–8
Trickling filter effluent	3–15
Sand filtered effluent	2–6

Ultraviolet Radiation

Gaining in popularity is the use of ultraviolet (UV) radiation for pathogen control in WWTP effluents. Similar to the $C \cdot t$ product used in chlorination of drinking water, UV disinfection is a product of the radiation intensity and time of exposure. Treatment times are determined based on the lamp intensity as well as the lamp construction. Low-pressure, low-intensity lamps emit light at 254 nm, which is nearly optimal for disinfection effectiveness. However, lamp efficiency is temperature dependent, with a maximum efficiency at 40°C, and power output is limited to approximately 25 W. Low-pressure, high-intensity lamps provide 2–20 times the power output, more emission stability over a greater temperature range, and a 25% life-span increase. Medium-pressure, high-intensity lamps possess a power output 50–100 greater than the low-pressure, low-intensity lamps and are used more commonly in WWTPs that have substantially higher flow rate.

A typical UV treatment unit consists of an open-channel filled with two or more banks of UV lamps in series, as shown in Figure 14.18. Orientation of the banks may be horizontal or vertical and are comprised of modules containing up to 16 lamps. Because of the short penetration distance of UV light into water, distance between lamp centers is on the order of 3 inches. Further, an additional channel is generally required to serve as backup for emergency or maintenance outages. Modules are usually easily removed, as regular cleaning is required to maintain optimum effectiveness.

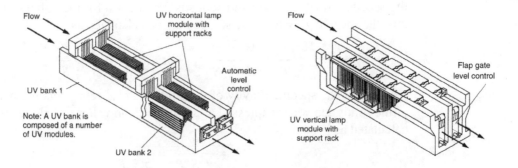

Figure 14.18 Open-channel UV disinfection systems
Source: Metcalf and Eddy, *Wastewater Engineering: Treatment and Reuse*, 4th ed., © 2003, McGraw-Hill Education. Reprinted by permission of the McGraw-Hill Companies.

WASTEWATER SLUDGE TREATMENT

Wastewater treatment processes are designed to remove contaminants from solution, either through direct physical separation (sedimentation and filtration) or after biological (secondary aeration) treatment. While some sludge treatment processes differ based on the origin of the solids, many are the same regardless of source. The general progression of processing is (1) storage, (2) thickening, (3) conditioning, (4) dewatering, and (5) disposal.

Sludge Generation

Wastewater treatment results in a variety of sludge types, the vast majority from primary and secondary sedimentation. Typical quantities can be estimated based upon the type of processing unit generating the solids, as presented in Table 14.14.

Table 14.14 Sludge generation characteristics and quantities

Source	Specific Gravity		Quantity of Dry Solids (lb/1000 gal)
	Solids	Sludge	
Primary sedimentation	1.4	1.02	1.25
Activated sludge	1.25	1.005	0.7
Trickling filter	1.45	1.025	0.6
Extended aeration	1.3	1.015	0.8
Aerated lagoon	1.3	1.01	0.8

Sludge is composed primarily of water. The specific gravity of a particular sludge is given as

$$\frac{W_s}{S_s \gamma_w} = \frac{W_f}{S_f \gamma_w} + \frac{W_v}{S_v \gamma_w} \quad \text{or} \quad \frac{M_s}{S_s \rho_w} = \frac{M_f}{S_f \rho_w} + \frac{M_v}{S_v \rho_w}$$

where W is the weight, M is the mass, S is the specific gravity, γ is the specific weight, ρ is the density, and the subscripts s, f, v, and w indicate the solids, fixed solids, volatile solids, and water fractions, respectively. Typical values assumed for the specific gravity of the fixed solids is 2.5 and 1.0 for the volatile solids. From this, the specific gravity of the sludge may be calculated as

$$\frac{M_{sl}}{S_{sl}} = \frac{M_s}{S_s} + \frac{M_w}{S_w} \quad \text{or} \quad \frac{1}{S_{sl}} = \frac{X_s}{S_s} + \frac{X_w}{S_w}$$

where S_{sl} is the specific gravity of the sludge, and X represents the mass fraction of the solids and water expressed as a decimal. The volume of sludge may be also calculated as

$$V = \frac{W_s}{\gamma_w S_{sl} P_{sl}} = \frac{M_s}{\rho_w S_{sl} P_{sl}}$$

where S_{sl} is the specific gravity of sludge, and P_{sl} is the percent solids of sludge expressed as a decimal. Solids content of several different sludges are provided in Table 14.15.

Table 14.15 Solids content of several sludge types

Sludge Type	Sludge Solids, % Dry Solids
Primary	5.0
Primary and activated sludge	4.0
Primary and trickling filter	5.0
Activated sludge	0.8
Pure oxygen-activated sludge	2.0
Trickling filter	1.5
Rotating biological contactor	1.5
Primary after gravity thickening	8.0
Activated sludge after air flotation	4.0
Activated sludge after centrifugation	5.0
Primary after anaerobic digestion	7.0
Primary and activated sludge after anaerobic digestion	3.5
Primary and trickling filter after anaerobic digestion	4.0
Primary after aerobic digestion	3.5
Primary and trickling filter after aerobic digestion	2.5
Activated sludge after aerobic digestion	1.3

Example 14.20

Sludge volume

Determine the volume of 1000 kg of primary sludge if 55% of the solids are volatile.

Solution

First determine the specific gravity of all of the solids in the primary sludge as

$$\frac{M_s}{S_s\rho_w} = \frac{M_f}{S_f\rho_w} + \frac{M_v}{S_v\rho_w} = \frac{(1000 \text{ kg})(0.45)}{(2.5)\left(1000 \frac{\text{kg}}{\text{m}^3}\right)} + \frac{(1000 \text{ kg})(0.55)}{(1.0)\left(1000 \frac{\text{kg}}{\text{m}^3}\right)}$$

$$= 0.73 = \frac{(1000 \text{ kg})}{S_s\left(1000 \frac{\text{kg}}{\text{m}^3}\right)}$$

or $S_s = 1.37$. Knowing that primary sludge is 5% solids (from Table 14.15), we can now determine the specific gravity of the sludge as

$$\frac{1}{S_{sl}} = \frac{X_s}{S_s} + \frac{X_w}{S_w} = \frac{0.05}{1.37} + \frac{0.95}{1.0} = 0.9865 \quad \text{or} \quad S_{sl} = 1.014$$

Now the volume of sludge may be calculated as

$$V = \frac{M_s}{\rho_w S_{sl} P_{sl}} = \frac{1000 \text{ kg}}{\left(1000 \frac{\text{kg}}{\text{m}^3}\right)(1.014)(0.05)} = 19.7 \text{ m}^3$$

That is equivalent to 5200 gallons of sludge.

Example 14.21

Sludge production

Determine the solids loading to the wastewater treatment plant sludge digesters in kg/day given the following plant operating parameters: Q_{ave} = 8.2 mgd; plant inlet SS = 240 mg/L; a primary clarifier removes 50% of inlet SS; waste rate from final clarifier = 30%; final clarifier sludge concentration = 1% solids; sludge return to aeration tank = 105,000 gpd.

Solution

First, we can determine the solids loading from the primary clarifiers from the information given in the problem statement as follows:

$$\left(240 \frac{\text{mg}}{\text{L}}\right)\left(8.2 \frac{\text{Mgal}}{\text{day}}\right)\left(8.34 \frac{\frac{\text{lb}}{\text{day}}}{\frac{\text{mg}}{\text{L}} \cdot \frac{\text{Mgal}}{\text{day}}}\right)(0.5) = 8207 \frac{\text{lb}}{\text{day}}$$

For the final clarifier, we need to determine the waste mass flow rate. Since we were given a waste rate of 30%, we know that the returned sludge is at a 70% rate. Since this flow rate is given, we can find the waste flow rate as

$$Q_R = 0.7 Q_T \quad \Rightarrow \quad Q_T = \frac{Q_R}{0.7} = \frac{105,000 \text{ gpd}}{0.7} = 150,000 \text{ gpd}$$

Now the waste flow rate can be determined as

$$Q_W = 0.3 Q_T \Rightarrow Q_W = (0.3)(150{,}000 \text{ gpd}) = 45{,}000 \text{ gpd}$$

Now the mass rate of solids can be found noting that a concentration of 1% solids is equal to 10,000 mg/L (i.e., 1,000,000 mg/L × 0.01 = 10,000 mg/L) as follows:

$$\text{loading} = \left(45{,}000 \frac{\text{gal}}{\text{day}}\right)\left(\frac{1 \text{ Mgal}}{10^6 \text{ gal}}\right)\left(10{,}000 \frac{\text{mg}}{\text{L}}\right)\left(8.34 \frac{\frac{\text{lb}}{\text{day}}}{\frac{\text{mg}}{\text{L}} \cdot \frac{\text{Mgal}}{\text{day}}}\right) = 3753 \frac{\text{lb}}{\text{day}}$$

Finally, total solids can be found by adding the two sources and converting to kg as follows:

$$\text{total} = 8207 + 3753 = 11{,}960 \frac{\text{lb}}{\text{day}} \times \frac{1 \text{ kg}}{2.205 \text{ lb}} = 5424 \frac{\text{kg}}{\text{day}}$$

Preconditioning

Storage often occurs in the processing units where the solid is collected (for example, at the bottom of the clarifier) and pumped daily to the sludge handling units. When necessary, a separate solids holding tank can be designed for temporary storage prior to processing. In some cases (as in aerobic digestion), sludges are pumped directly to the appropriate sludge processing unit as they are generated. *Thickening* is often accomplished using gravity settling, where the clarifier is designed with a greater depth, longer detention times, and lower overflow rates. Alternatives include *gravity belt thickening* for water treatments plant residues, where water drains from a conveyor belt that carries the sludge, to *air flotation thickening*, where fine bubbles lift the small diameter solids that are a result of activated sludge treatment to the liquid surface for subsequent removal via skimming.

Conditioning of sludges is dependent on the origin of the solid, which determines the composition and, as such, the treatment requirement. Water processing residues are basically inorganic slurries, and conditioning generally refers to additional lime or polymer additions to coagulate dissolved chemical species and increase overall solids removal in subsequent processes. Wastewater treatment sludges may be conditioned through *anaerobic digestion* or *aerobic digestion*, where biological activity is used to reduce sludge organic matter and produce a more stable product with no putrefaction.

Conditioning through Anaerobic Digestion

Anaerobic digestion has a long history of treating biomass due to the ability to recover an energy resource (methane) and biosolids possessing characteristics that are amenable to subsequent mechanical dewatering processes. This process occurs in two stages, the first being the hydrolysis of high molecular weight organic compounds to organic acids by specific bacterial strains. The second stage is the formation of methane and carbon dioxide by methanogenic bacteria. The volume of methane gas produced can be given by the following expression

$$V_{CH_4} = 0.35 \left[\frac{(S_0 - S_e)Q}{10^3} - 1.42 P_x \right]$$

where methane volume has units of m³/d measured at STP, S_0 and S_e are the influent and effluent BOD (or biodegradable COD) concentrations in mg/L, Q is the volumetric flow rate in m³/d, and P_x is the mass of cell production per day in kg/day. The expression for determination of P_x was presented previously in this chapter. Design parameters for anaerobic digesters are presented in Table 14.16.

Table 14.16 Design parameters for anaerobic sludge digestion

Parameter	Standard-Rate	High-Rate
Solids residence time [days]	30–90	10–20
Volatile solids loading [kg/m³/d]	0.5–1.6	1.6–6.4
Digested solids concentration [%]	4–6	4–6
Volatile solids reduction [%]	35–50	45–55
Gas production [m³/kg VSS added]	0.5–0.55	0.6–0.65
Methane content [%]	65	65

Source: *Fundamentals of Engineering Supplied-Reference Handbook*, 7th ed. (NCEES, 2005), p. 160.

The reactor volume required to anaerobically process wastewater sludge at the standard rate can be expressed as

$$V = \frac{V_1 + V_2}{2} t_r + V_2 t_s$$

where V_1 is the volumetric flow rate of influent raw sludge (volume/day), V_2 is the volumetric accumulation of digested sludge in the tank (volume/day), t_r is the digestion period (days), and t_s is the storage period (days). Digestion periods are generally 25–30 days at ~ 90°F, and storage periods range from 30–120 days, with 60 days a typical value.

For high-rate digestion, the process is typically split into two tanks, one for each stage of the digestion process. The volume of the first tank in high-rate anaerobic digestion can be determined as follows:

$$V_I = V_1 t_r$$

where V_1 and t_r are as defined previously. The second stage reactor volume can be calculated as

$$V_{II} = \frac{V_1 + V_2}{2} t_t + V_2 t_s$$

where V_1, V_2, and t_s are as defined previously, and t_t is the thickening period (days). While storage periods are similar for the two types of processes, the digestion period in the first stage of the high-rate system is reduced to 10–15 days.

Conditioning through Aerobic Digestion

Aerobic digestion offers a stable end product with low capital cost and relative ease of operation, although operating costs may be greater due to oxygenation requirements. The process is similar to activated sludge treatment, where the slurry is maintained in the endogenous phase. After consuming the small amount of remaining BOD, the only food source available is from the biomass as cell lysis provides the remaining bacteria with the energy needed to maintain cell function. System design parameters are provided in Table 14.17.

Table 14.17 Design parameters for aerobic sludge digestion

Parameter	Value	Units
Hydraulic retention time, 20°C		
Waste activated sludge only	10–15	days
AS without primary settling	12–18	days
Primary plus waste activated or trickling-filter sludge	15–20	days
Solids loading	0.1–0.3	lb TVS/ft^3 · d
Oxygen requirements		
Cell tissue	~2.3	lb O$_2$/lb solids
BOD$_5$ in primary sludge	1.6–1.9	lb O$_2$/lb solids
Energy requirements for mixing		
Mechanical aerators	0.7–1.5	hp/1000 ft^3
Diffused-air mixing	20–40	ft^3/1000 ft^3 · min
Dissolved-oxygen residual in liquid	1–2	mg/L
Reduction in volatile suspended solids	40–50	%

Source: Metcalf and Eddy, *Wastewater Engineering: Treatment and Reuse*, 3d ed., ©1991 McGraw-Hill Education. Reprinted by permission of the McGraw-Hill Companies.

Tank design is based primarily on the mean cell residence time (θ_c) required to reduce the volatile solids to the desired level. Tank volume (V) can be expressed as

$$V = \frac{Q_i (X_i + F S_i)}{X_d \left(K_d P_v + \dfrac{1}{\theta_c} \right)}$$

where Q_i is the volumetric flow rate to the digester, X_i is the suspended solids concentration in the influent, F is the fraction of BOD$_5$ from raw primary sludge, S_i is the influent BOD$_5$, X_d is the suspended solids concentration in the digester, K_d is the reaction rate constant, P_v is the fraction of the digester suspended solids that is volatile, and θ_c is the mean cell residence time in the digester. For sludge processing where primary sludge is not included, the FS_i term may be dropped from the above expression. A sludge age of 40 days is required for liquid temperatures of 20°C, increasing to 60 days as the temperature drops to 15°C. Oxygen requirements are approximately 2.3 pounds of oxygen per pound of cell mass, plus 1.6–1.9 pounds of oxygen per pound of BOD applied, where applicable.

Postconditioning

Dewatering has historically been completed in *sand beds*, where conditioned sludge was spread over a layer of sand containing drain pipe and allowed to dry until the desired solids content was obtained. However, more recent designs have attempted to reduce the manual labor with mechanized systems, including *vacuum filtration* and *pressure filtration* using belt filters or filter presses, or *centrifugation*, which is often used for large facilities. Final disposal of processed sludges is dependent on composition. Water treatment plant residues are often codisposed with municipal solid waste or directed into a waste monofill, whereas wastewater

Wastewater Sludge Treatment 655

treatment plant sludges are often land-applied for agricultural purposes. Wastewater biosolids may also be composted, often with other organic waste matter, prior to use as a soil amendment.

Example 14.22

Comprehensive wastewater plant design

A small midwestern city produces a wastewater flow of 5.5 mgd and has an influent BOD of 170 mg/L. The WWTP consists of an aerated grit chamber (AGC), primary clarifier, activated sludge (AS) tank, final clarifier, and disinfection through UV radiation, with biosolids treated through anaerobic digestion. The primary removes 60% of the solids and 35% of the BOD. Laboratory data for the AS process determined a biological yield of 0.6 and a microbial decay constant of 0.06 day^{-1} in the 155,000 ft^3 tank when there was an MLSS concentration of 2100 mg/L. The complete-mix high-rate anaerobic digester has a biological yield of 0.08, a microbial decay constant of 0.03 day^{-1}, and a solids retention time (SRT) of 14 days. Final plant effluent has a BOD of 10 mg/L. Determine the following:

a). The dimensions of the AGC for the flow above if the detention time is 3 minutes at peak flow, the width:depth ratio is 1.5:1, and the length:width ratio is 4:1

b). The depth of the circular primary clarifier if the overflow rate is 1000 gpd/ft^2 and the detention time is 2 hours at Q_{AVE}

c). The SRT in the AS tank

d). Estimate of the monthly energy costs for disinfection if the electricity costs 5.5 cents per kW · hr. The UV system utilizes modules of 16 vertical lamps placed in rectangular flow channels. Each module can treat 0.5 mgd, and each lamp is rated at 100 W.

e). Estimate of the daily volume of CH_4 produced at this WWTP. The anaerobic digesters were installed to provide CH_4 to produce power to run the WWTP operations.

Solutions

a). Assuming an average per capita contribution of 100 gpd, the city population is approximately 55,000, which yields a peaking factor of approximately 3, or

$$Q_{peak} = 3Q_{ave} = (3)(5.5 \text{ Mgd})\left(1.547 \frac{\text{cfs}}{\text{Mgd}}\right) = 25.53 \text{ cfs}$$

AGC tank volume can now be calculated from the design detention time as

$$t_R = \frac{V}{Q} \Rightarrow V = t_R \times Q = (3 \text{ min})\left(60 \frac{\text{s}}{\text{min}}\right)\left(25.53 \frac{\text{ft}^3}{\text{s}}\right) = 4595 \text{ ft}^3$$

Using the dimension ratios we know that

$$W = 1.5D \quad \text{and} \quad L = 4W = (4)(1.5D) = 6D$$

Therefore,

$$V = LWD = (6D)(1.5D)D = 9D^3 = 4595 \text{ ft}^3$$

Rearranging and solving for D gives us L and W also:

$$D = \left(\frac{4595 \text{ ft}^3}{9}\right)^{1/3} = 8.0 \text{ ft}$$

$$W = 1.5 D = (1.5)(8 \text{ ft}) = 12 \text{ ft}$$

$$\text{and } L = (4)(12 \text{ ft}) = 48 \text{ ft}$$

b). Using the design detention time to find clarifier volume we get

$$V = t_R \times Q = (2 \text{ hr})\left(\frac{1 \text{ day}}{24 \text{ hr}}\right)\left(5.5 \times 10^6 \frac{\text{gal}}{\text{day}}\right)\left(\frac{1 \text{ ft}^3}{7.48 \text{ gal}}\right) = 61,275 \text{ ft}^3$$

Using the expression for overflow rate, we can find tank surface area as

$$V_o = \frac{Q}{A_S} \Rightarrow A_S = \frac{Q}{V_o} = \frac{5.5 \times 10^6 \text{ gpd}}{1000 \frac{\text{gpd}}{\text{ft}^2}} = 5500 \text{ ft}^2$$

Now tank depth can be determined as

$$\text{depth} = \frac{V}{A_S} = \frac{61,275 \text{ ft}^3}{5500 \text{ ft}^2} = 11.14 \text{ ft}$$

c). With the information provided in the problem statement, we can use the following expressions to find SRT:

$$\frac{1}{SRT} = Y\left(\frac{F}{M}\right)\left(\frac{E}{100}\right) - k_d \quad \text{where} \quad E = \frac{S_0 - S}{S_0} \times 100 \quad \text{and} \quad \left(\frac{F}{M}\right) = \frac{Q S_0}{V X}$$

For the AS process, S_0 is defined as the BOD concentration entering the tank, after the 35% is removed in the primary, therefore:

$$S_0 = \left(170 \frac{\text{mg}}{\text{L}}\right)(1 - 0.35) = 110.5 \frac{\text{mg}}{\text{L}}$$

Knowing the effluent BOD, we can calculate efficiency (E) as

$$E = \frac{110.5 - 10}{110.5} \times 100 = 90.95\%$$

Noting X is the MLSS concentration in the AS tank, we can calculate the F/M ratio as

$$\left(\frac{F}{M}\right) = \frac{\left(5.5 \times 10^6 \frac{\text{gal}}{\text{day}}\right)\left(110.5 \frac{\text{mg}}{\text{L}}\right)}{(155,000 \text{ ft}^3)\left(7.48 \frac{\text{gal}}{\text{ft}^3}\right)\left(2100 \frac{\text{mg}}{\text{L}}\right)} = 0.25 \text{ day}^{-1}$$

Combining expression allows for the calculation of SRT as

$$\frac{1}{SRT} = Y\left(\frac{F}{M}\right)\left(\frac{E}{100}\right) - k_d = (0.6)(0.25 \text{ day}^{-1})\left(\frac{90.95}{100}\right) - 0.06 \text{ day}^{-1}$$

$$= 0.0764 \text{ day}^{-1}$$

By taking the reciprocal of 0.0764 day^{-1}, the SRT is found to be 13.1 days.

d). The number of modules required can be determined as

$$\frac{5.5 \times 10^6 \text{ gpd}}{500{,}000 \dfrac{\text{gpd}}{\text{module}}} = 11 \text{ modules}$$

Total monthly power usage can now be determined as

$$(11 \text{ modules})\left(16\frac{\text{lamps}}{\text{module}}\right)\left(100\frac{\text{W}}{\text{lamp}}\right)\left(\frac{1\text{ kW}}{1000\text{ W}}\right)\left(30\frac{\text{days}}{\text{mo}}\right)\left(24\frac{\text{hr}}{\text{day}}\right)$$

$$= 12{,}672 \frac{\text{kW}\cdot\text{hr}}{\text{mo}}$$

Power usage multiplied by the utility rate yield monthly cost as

$$\left(12{,}672 \frac{\text{kW}\cdot\text{hr}}{\text{mo}}\right)\left(\frac{\$0.055}{\text{kW}\cdot\text{hr}}\right) = \$697 \text{ per month}$$

e). With the information provided in the problem statement, we can use the following expressions to find the volume of methane produced as

$$V_{CH_4} = (0.35)\left[\frac{(S_0 - S)Q}{10^3} - 1.42\, P_X\right] \quad \text{where} \quad P_X = \frac{YQ(S_0 - S)}{10^3\left[1 + (k_d)(SRT)\right]}$$

In these expressions, Q has units of m³/day, S has units of g/m³ (note that this is the same as mg/L), k_d has units of day⁻¹, SRT has units of days, and Y is unitless. It is also important to note that the S_0 here is the BOD entering the WWTP, not the wastewater entering the AS unit. This is because the BOD removed in the primary is included in the sludge to be treated anaerobically. This yields P_X in units of kg/day, which is converted to m³/day by the constant 0.35. First, Q must be converted into the correct units as follows:

$$Q = \left(5.5 \times 10^6 \frac{\text{gal}}{\text{day}}\right)\left(3.785 \frac{\text{L}}{\text{gal}}\right)\left(\frac{1\text{ m}^3}{1000\text{ L}}\right) = 20{,}818 \frac{\text{m}^3}{\text{day}}$$

Solving for P_X yields:

$$P_X = \frac{YQ(S_0 - S)}{10^3\left[1+(k_d)(SRT)\right]} = \frac{(0.08)\left(20{,}818\,\dfrac{\text{m}^3}{\text{day}}\right)\left(170 - 10\,\dfrac{\text{g}}{\text{m}^3}\right)}{\left(10^3\,\dfrac{\text{g}}{\text{kg}}\right)\left[1+(0.03\text{ day}^{-1})(14\text{ days})\right]} = 188 \frac{\text{kg}}{\text{day}}$$

Solving for the volume of methane produced yields:

$$V_{CH_4} = \left(0.35\,\frac{\text{m}^3\text{ CH}_4}{\text{kg CH}_4}\right)\left[\frac{\left(170-10\,\dfrac{\text{g}}{\text{m}^3}\right)\left(20{,}818\,\dfrac{\text{m}^3}{\text{day}}\right)}{\left(10^3\,\dfrac{\text{g}}{\text{kg}}\right)} - (1.42)\left(188\,\frac{\text{kg}}{\text{day}}\right)\right]$$

$$= 1072 \frac{\text{m}^3}{\text{day}}$$

658 Chapter 14 Water and Wastewater

PROBLEMS

14.1 A flow of 0.3 m³/s enters a single 1000 m³ CSTR with a concentration of species A at 200 mg/L. If the first order reaction rate coefficient is 5.09 hr⁻¹, determine the concentration of species A in the reactor effluent.
   a. 25 mg/L
   b. 35 mg/L
   c. 45 mg/L
   d. 55 mg/L

14.2 The flow described in Problem 14.1 enters a series of 5 CSTRs, each with a volume of 200 m³. Determine the concentration of species A leaving the final reactor.
   a. 25.0 mg/L
   b. 18.7 mg/L
   c. 7.2 mg/L
   d. 2.9 mg/L

14.3 Determine the number of cylindrical rapid mix tanks (H:D = 1.5:1), running in a parallel flow configuration, necessary for a total plant flow of 6 mgd. Assume the design detention time is 30 seconds and the maximum tank dimension is 1.8 m.
   a. 1
   b. 2
   c. 3
   d. 4

14.4 A flocculator with a $G \cdot t$ product of 10^5 is 30 ft × 12 ft × 35 ft and treats a flow of 3.4 mgd. Two paddle boards are located on each arm at a radius of 2.5 and 5 ft from the shaft and each has a surface area of 9.33 ft². If the slip coefficient is 0.7 and there are six total paddle arms (two arms on each of three separate shafts), determine the rotational speed in revolutions per minute. You may use a value of 2.36×10^{-5} lb·s/ft² for the viscosity of water.
   a. 4.6 rpm
   b. 5.3 rpm
   c. 7.6 rpm
   d. 9.1 rpm

14.5 Determine the tons of lime (CaO) required per day (tpd) for excess lime softening of water with the following quality (from Example 14.2) if the plant flow rate is 6 mgd: Ca^{2+} = 150 mg/L as $CaCO_3$; Mg^{2+} = 100 mg/L as $CaCO_3$; Na^+ = 46 mg/L; K^+ = 39 mg/L; HCO_3^- = 225 mg/L as $CaCO_3$; SO_4^{2-} = 72 mg/L; Cl^- = 71 mg/L.
   a. 3.5 tpd
   b. 4.5 tpd
   c. 5.4 tpd
   d. 6.3 tpd

14.6 Determine the length of a rectangular clarifier for a flow of 3 mgd if the design detention time is 2 hours, the tank has a depth of 8 ft, and the $L:W = 4:1$.
  a. 130 ft
  b. 120 ft
  c. 110 ft
  d. 100 ft

14.7 Determine the minimum number of rapid sand filters necessary to treat a flow of 20 mgd if the maximum design hydraulic loading is 5 gpm/ft². Assume the maximum dimension is 24 feet and the $L:W = 1.2:1$.
  a. 6
  b. 7
  c. 8
  d. 9

14.8 Determine the initial head loss through a single media filter with a maximum hydraulic loading of 5 gpm/ft² if the average particle size is 1.5 mm. You may assume the bed depth is 2 m, the water temperature is 15°C, and the bed porosity is 32 percent.
  a. 0.8 m
  b. 1.0 m
  c. 1.2 m
  d. 1.4 m

14.9 Find the porosity of the filter described in Problem 14.8 during backwashing if the bed depth is 3 m during fluidization.
  a. 50 percent
  b. 55 percent
  c. 60 percent
  d. 65 percent

14.10 Find the length and width of a square basin for the virus disinfection of a WTP flow of 6 mgd. You may assume a depth of 5 ft, a $C \cdot t$ product for 1 log removal of 6 (mg/L)·min, a target free chlorine residual of 0.5 mg/L and a reactor safety factor of 3.
  a. 45 ft
  b. 55 ft
  c. 65 ft
  d. 75 ft

14.11 An air stripping column has a diameter of 5 m and volumetric flow rates of 4.6 m³ of air per second and 0.35 m³ of water per second (note that $Q_{water} = L/M_w$). It is desired to remove 99 percent of a volatile organic that has a unitless Henry's constant of 0.23. If the column has an overall mass transfer coefficient of 0.5 min⁻¹, determine the height of the column.
  a. 8.8 m
  b. 10.1 m
  c. 11.9 m
  d. 13.5 m

14.12 Brackish water with a salt concentration of 0.5 percent is to be treated by reverse osmosis at a volumetric flow rate of 20 L/s. Determine the pressure difference between the feed water and the product water if the membrane coefficient/ area product for the selected system is 1.2×10^{-5} s • m. You may assume an osmotic pressure of 11.3 psi is generated for each 0.1 percent increase in salt concentration.
a. 800 kPa
b. 1200 kPa
c. 1600 kPa
d. 2000 kPa

14.13 Determine the power required to operate an electrodialysis system designed to treat 0.15 m³/s of a 5 mN solution of NaCl. The system consists of a stack of 350 cells and has a total resistance of 6 ohms. You may assume typical values for removal efficiency of 50 percent and current efficiency of 90 percent.
a. 40 kW
b. 60 kW
c. 80 kW
d. 100 kW

14.14 Determine the depth of a cubical aerated grit chamber necessary to treat a flow of 6.2 mgd if the design detention time is 3 minutes at Q_{avg}.
a. 9 ft
b. 10 ft
c. 11 ft
d. 12 ft

14.15 Find the diameter of a circular clarifier designed to treat 3.2 mgd. You may assume a detention time of 2 hours and a depth of 8 feet.
a. 55 ft
b. 65 ft
c. 75 ft
d. 85 ft

14.16 A flow of 2.6 mgd leaves a primary clarifier with a BOD of 131 mg/L. Determine the aeration period of an activated sludge tank that has a BOD loading of 35 lbs of BOD per 1000 ft³ per day.
a. 6.8 hr
b. 6.4 hr
c. 6.0 hr
d. 5.6 hr

14.17 Determine the sludge age of the facility described in Problem 14.16 given the following data: MLSS concentration is 2100 mg/L, effluent suspended solids are equal to 20 mg/L, and the mass rate of wasted solids is equal to 44 percent of the BOD load.
a. 4 days
b. 6 days
c. 8 days
d. 10 days

14.18 For the flow described in Problem 14.16 and the activated sludge tank described in Problem 14.17, determine the biological yield given a bacterial decay rate constant of 0.02 day^{-1}.
 a. 0.8
 b. 0.7
 c. 0.6
 d. 0.5

14.19 The flow in Problem 14.16 is to be treated using a trickling filter that has the same BOD loading as the activated sludge tank. Find the efficiency of a stone media filter at 20°C if the recirculation flow is 5.2 mgd. The NRC equation is given below.

$$E_1 = \frac{100}{1 + 0.0561\left(\frac{w}{VF}\right)^{1/2}} \quad \text{and} \quad F = \frac{1+R}{(1+0.1R)^2}$$

 a. 87 percent
 b. 84 percent
 c. 81 percent
 d. 78 percent

14.20 Your boss asked you to redesign the trickling filter in Problem 14.19 using high efficiency plastic media to fit into a 20 foot diameter tower. If the treatability constant is 0.06 min^{-1} and the empirical flow constant is 0.5, determine the depth of filter needed when the recirculation flow is reduced to 2.6 mgd.
 a. 23 ft
 b. 26 ft
 c. 29 ft
 d. 32 ft

14.21 Determine the minimum number of acres needed for a stabilization pond to treat the flow described in Problem 14.16.
 a. 80 acres
 b. 65 acres
 c. 50 acres
 d. 35 acres

14.22 The wasted sludge from Problem 14.17 is mixed with 1500 lb/day of sludge from the primary clarifier. Determine the volume of the anaerobic digester given the following sludge characteristics and design parameters; the combined raw sludge is 4 percent solids, and 70 percent of the solids are volatile; the digester has a volatile solids reduction of 40 percent; the digestion period is 25 days and the storage period is 60 days; the digested sludge is 6 percent solids.
 a. 22,000 ft^3
 b. 32,000 ft^3
 c. 42,000 ft^3
 d. 52,000 ft^3

14.23 An aerobic digester is being considered to treat only the wasted sludge from Problem 14.17, which has a solids fraction of 1 percent. The volatile fraction of the sludge is 70 percent and the reaction rate constant is 0.05 day^{-1}. Determine the digester volume if the digester solids concentration is 6 percent and the mean cell residence time is 50 days.
  a. 4000 ft^3
  b. 6000 ft^3
  c. 8000 ft^3
  d. 10,000 ft^3

SOLUTIONS

14.1 b. From the chemical engineering section in the *FE Supplied-Reference Handbook*, we know that

$$\frac{\tau}{C_{A0}} = \frac{X_A}{-r_a}$$

Further, from the *FE Supplied-Reference Handbook*, we find that the reaction rate for a first order reaction can be expressed as

$$-r_a = kC_A$$

Combining these expressions and solving for retention time we may write

$$\frac{\tau}{C_{A0}} = \frac{X_A}{kC_A} \Rightarrow \tau = \frac{X_A C_{A0}}{kC_A}$$

Now, detention time may be determined as follows:

$$\tau = \frac{V}{Q} = \frac{1000 \text{ m}^3}{0.3 \frac{\text{m}^3}{\text{s}}} = 3333 \text{ s} \times \frac{1 \text{ hr}}{3600 \text{ s}} = 0.926 \text{ hr}$$

By definition from the *FE Supplied-Reference Handbook*, we may write X_A as

$$X_A = \frac{C_{A0} - C_A}{C_{A0}}$$

Substituting this into the expression above we get

$$\tau = \frac{\left(\frac{C_{A0}-C_A}{C_{A0}}\right)C_{A0}}{kC_A} = \frac{C_{A0}-C_A}{kC_A} \Rightarrow \tau kC_A = C_{A0} - C_A \Rightarrow C_A = \frac{C_{A0}}{1+\tau k}$$

Finally, substituting for the variables C_{A0} and k given in the problem statement and τ determined above, we can solve for C_A as

$$C_A = \frac{200 \frac{\text{mg}}{\text{L}}}{1+(0.926 \text{ hr})(5.09 \text{ hr}^{-1})} = 35.0 \frac{\text{mg}}{\text{L}}$$

14.2 c. Using the expression given in the chemical engineering section of the *FE Supplied-Reference Handbook* for CSTRs in series, and the data given in the problem statement for Problems 14.1 and 14.2, we can write

$$\tau = \frac{N}{k}\left[\left(\frac{C_{A0}}{C_{AN}}\right)^{1/N} - 1\right] \Rightarrow 0.926 = \frac{5}{5.09}\left[\left(\frac{200}{C_{AN}}\right)^{1/N} - 1\right]$$

Rearranging and solving for C_{AN} we get

$$\frac{(0.926)(5.09)}{5} + 1 = \left(\frac{200}{C_{AN}}\right)^{1/5} = 1.94 \Rightarrow C_{AN} = 7.3 \frac{\text{mg}}{\text{L}}$$

14.3 d. Given an H:D ratio of 1.5:1 and a maximum dimension of 1.8 m, we know that

$$H = 1.8 \text{ m} \quad \text{and} \quad D = \frac{H}{1.5} = \frac{1.8 \text{ m}}{1.5} = 1.2 \text{ m}$$

Therefore, we can determine the maximum volume of one reactor as follows:

$$V_1 = \frac{\pi}{4} D^2 H = \frac{\pi}{4}(1.2 \text{ m})^2(1.8 \text{ m}) = 2.036 \text{ m}^3$$

Next we need to know the total reactor volume required to treat the water. Using the design detention time, we may write

$$t_R = \frac{V_T}{Q} \Rightarrow V_T = t_R Q = (30 \text{ s})(6 \text{ mgd})\left(\frac{1.547 \text{ cfs}}{1 \text{ mgd}}\right)$$

$$= 278.5 \text{ ft}^3 \times \left(\frac{1 \text{ m}}{3.28 \text{ ft}}\right)^3 = 7.89 \text{ m}^3$$

Finally, the number of reactors can be calculated as follows:

$$\text{number of reactors} = \frac{V_T}{V_1} = \frac{7.89 \text{ m}^3}{2.036 \text{ m}^3} = 3.87 \approx 4 \text{ reactors}$$

14.4 a. Given a $G \cdot t$ product, we need to determine the detention time in order to calculate the velocity gradient. Detention time, in units of seconds, is calculated as

$$t_R = \frac{V}{Q} = \frac{(30 \text{ ft})(12 \text{ ft})(35 \text{ ft})}{(3.4 \text{ mgd})\left(1.547 \frac{\text{cfs}}{\text{mgd}}\right)} = 2396 \text{ s}$$

From this, we can calculate the velocity gradient as

$$G \cdot t = 10^5 = G(2396 \text{ s}) \Rightarrow G = 41.7 \text{ s}{-1}$$

Now, the power can be calculated using the velocity gradient equation given in the environmental engineering section of the *FE Supplied-Reference Handbook* and the viscosity of water given in the problem statement as follows:

$$G = \left(\frac{P}{\mu V}\right)^{1/2} \Rightarrow P = G^2 \mu V$$

$$P = G^2 \mu V = (41.7 \text{ s}^{-1})^2 \left(2.36 \times 10^{-5} \frac{\text{lb} \cdot \text{s}}{\text{ft}^2}\right)(30 \text{ ft})(12 \text{ ft})(35 \text{ ft})$$

$$= 517 \frac{\text{ft} \cdot \text{lb}}{\text{s}}$$

Given the equations in the environmental engineering section of the *FE Supplied-Reference Handbook*, the power delivered by one paddle at one distance from the shaft can be expressed as

$$P_{\text{BOARD}} = \frac{C_D A_P \rho_f v_P^3}{2} \quad \text{and} \quad v_P = K_S 2\pi r n$$

$$\Rightarrow P = \frac{C_D A_P \rho_f (K_S 2\pi r n)^3}{2}$$

Applying this equation using the data given in the problem statement, we find

$$P_{2.5} = \frac{C_D A_P \rho_f (K_S 2\pi rn)^3}{2} = \frac{(1.8)(9.33)(1.94)[(0.7)(2)(\pi)(2.5)n]^3}{2}$$

$$= 21{,}656 n^3 \frac{\text{ft} \cdot \text{lb}}{\text{s}}$$

$$P_5 = \frac{C_D A_P \rho_f (K_S 2\pi rn)^3}{2} = \frac{(1.8)(9.33)(1.94)[(0.7)(2)(\pi)(5)n]^3}{2}$$

$$= 173{,}249 n^3 \frac{\text{ft} \cdot \text{lb}}{\text{s}}$$

Further, the problem statement says that there are six of each of these boards, and we can then calculate total power as follows:

$$P_T = 6(21{,}656 n^3 + 173{,}249 n^3) = 1{,}169{,}430 n^3 = 517 \frac{\text{ft} \cdot \text{lb}}{\text{s}}$$

Finally, rotational speed may be determined from this relationship as follows:

$$n^3 = \frac{517}{1{,}169{,}430} = 0.000442 \quad \Rightarrow \quad n = 0.000442^{1/3}$$

$$= 0.0762 \text{ rps} \times \frac{60 \text{ s}}{\text{min}} = 4.57 \text{ rpm}$$

14.5 c. Taking the bar graph from Example 14.2 (see Exhibit 14.5), the following hardness species and compositions are able to be identified:

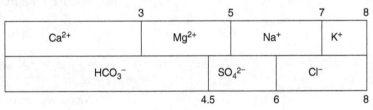

Exhibit 14.5

$$\text{Ca(HCO}_3)_2 = 3.0 \frac{\text{meq}}{\text{L}}; \quad \text{Mg(HCO}_3)_2 = 1.5 \frac{\text{meq}}{\text{L}}; \quad \text{MgSO}_4 = 0.5 \frac{\text{meq}}{\text{L}}$$

From the softening reactions provided in the *FE Supplied-Reference Handbook* we know the stoichiometric coefficient for lime in the Ca(HCO$_3$)$_2$ and MgSO$_4$ reactions is 1 and for Mg(HCO$_3$)$_2$ it is 2. Therefore the stoichiometric amount of lime needed may be determined as

$$(1)\left(3 \frac{\text{meq}}{\text{L}}\right) + (2)\left(1.5 \frac{\text{meq}}{\text{L}}\right) + (1)\left(0.5 \frac{\text{meq}}{\text{L}}\right) = 6.5 \frac{\text{meq}}{\text{L}}$$

However, for excess lime treatment, we need to add 1.25 meq/L, for a total of 7.75 meq/L. This quantity is converted to a mass concentration using the equivalent weight of lime as follows:

$$7.75 \frac{\text{meq}}{\text{L}} \times 28 \frac{\text{mg}}{\text{meq}} = 217 \frac{\text{mg}}{\text{L}}$$

Finally, daily mass required is found by multiplying concentration by daily flow rate as

$$\left(217\frac{mg}{L}\right)(6 \text{ mgd})\left(8.34\frac{lb \cdot L}{mg \cdot mgal}\right) = 10{,}859\frac{lb}{day} \times \frac{1 \text{ ton}}{2000 \text{ lb}} = 5.43 \text{ tpd}$$

14.6 a. Volume of a tank can be found given detention time and volumetric flow rate as

$$t_R = \frac{V}{Q} \quad \Rightarrow \quad V = Qt_R$$

$$V = (3 \times 10^6 \text{ gpd})(2 \text{ hr})\left(\frac{1 \text{ day}}{24 \text{ hr}}\right) = 250{,}000 \text{ gal} \times \left(\frac{ft^3}{7.48 \text{ gal}}\right)$$
$$= 33{,}422 \text{ ft}^3$$

The area of the tank can be found from the volume and depth as follows:

$$A = \frac{V}{depth} = \frac{33{,}422 \text{ ft}^3}{8 \text{ ft}} = 4178 \text{ ft}^2$$

Further, for a $L{:}W$ ratio of 4:1, we can find the tank dimensions as

$$A = L \times W = 4W \times W = 4W^2 = 4178 \text{ ft}^2$$
$$\Rightarrow \quad W = 32.3 \text{ ft} \quad \text{and} \quad L = 4W = 129.3 \text{ ft}$$

14.7 b. Given a $L{:}W$ ratio of 1.2:1 and a maximum dimension of 24 feet, we can calculate the maximum surface area as

$$L = 24 \text{ ft} \quad \text{and} \quad W = \frac{L}{1.2} = \frac{24 \text{ ft}}{1.2} = 20 \text{ ft} \quad \text{and}$$
$$A_1 = LW = (24 \text{ ft})(20 \text{ ft}) = 480 \text{ ft}^2$$

Next, we can convert volumetric flow rate to units of gpm as follows:

$$Q = 20 \text{ mgd} \times \frac{1 \text{ day}}{(24)(60) \text{ min}} = 13{,}889 \text{ gpm}$$

Now, applying the maximum filter rate, we can find the minimum required surface area as

$$A_T = \frac{13{,}889 \text{ gpm}}{5 \frac{gpm}{ft^2}} = 2778 \text{ ft}^2$$

The minimum number of filters can be found as follows:

$$n = \frac{A_T}{A_1} = \frac{2778 \text{ ft}^2}{480 \text{ ft}^2} = 5.78 \approx 6 \text{ filters}$$

Finally, we need to add one more to make sure the maximum flow rate is not exceeded when one filter is off-line during backwashing. Therefore

$$n = 7$$

14.8 a. Initial head loss through a single media filter can be estimated using the Carmen-Kozeny Equation, which is expressed in the environmental engineering section of the *FE Supplied-Reference Handbook* as

$$h_f = \frac{f'L(1-\eta)V_S^2}{\eta^3 g d_P} \quad \text{where} \quad f' = 150\left(\frac{1-\eta}{\text{Re}}\right) + 1.75 \quad \text{and} \quad \text{Re} = \frac{DV_S}{\nu}$$

First we need to determine f', which requires a value for Re. However, Re is dependent on V_s, which needs to be represented in units of m/s. This can be accomplished as follows:

$$V_S = 5\frac{\text{gpm}}{\text{ft}^2} \times \frac{\text{ft}^3}{7.48\text{ gal}} \times \frac{1\text{ min}}{60\text{ s}} = 0.01114\text{ fps} \times \frac{1\text{ m}}{3.28\text{ ft}} = 0.0034\frac{\text{m}}{\text{s}}$$

Using the value for kinematic viscosity from the fluid mechanics section of the *FE Supplied-Reference Handbook*, and converting the diameter into units of meters, Re is calculated as

$$\text{Re} = \frac{(0.0015\text{ m})\left(0.0034\frac{\text{m}}{\text{s}}\right)}{\left(1.139 \times 10^{-6}\frac{\text{m}^2}{\text{s}}\right)} = 4.48$$

Now, a value for f' can be calculated as follows:

$$f' = (150)\left[\frac{1-0.32}{4.48}\right] + 1.75 = 24.5$$

Finally, along with the data provided in the problem statement, these values can be substituted into the Carmen-Kozeny Equation to yield a head loss as follows:

$$h_f = \frac{(24.5)(2)(1-0.32)(0.0034)^2}{(0.32)^3(9.81)(0.0015)} = 0.799\text{ m}$$

14.9 b. Using the equation for the depth of the expanded bed during backwashing as

$$L_{fb} = \frac{L_0(1-\eta_0)}{1 - \left(\frac{V_B}{V_t}\right)^{0.22}} = \frac{L_0(1-\eta_0)}{1 - \eta_{fb}}$$

Substituting the values given in the problem statement, we can write

$$3\text{ m} = \frac{(2\text{ m})(1-0.32)}{1-\eta_{fb}} \Rightarrow 3 - 3\eta_{fb} = 1.36 \Rightarrow \eta_{fb} = 0.547 = 54.7\%$$

14.10 c. With the target free chlorine residual and $C \cdot t$ dose given in the problem statement, we can calculate contact time as follows:

$$C \cdot t = 6\frac{\text{mg}}{\text{L}} \cdot \text{min} = \left(0.5\frac{\text{mg}}{\text{L}}\right)t \Rightarrow t = 12\text{ min}$$

Using the safety factor of 3, the design detention time may be calculated as

$$t_R = (3)(12 \text{ min}) = 36 \text{ min}$$

With the volumetric flow rate given in the problem statement and the calculated detention time, we can size the chlorination basin volume as

$$t_R = \frac{V}{Q} \Rightarrow V = Qt_R = \frac{(36 \text{ min})(6 \text{ mgd})}{(60)(24)\frac{\min}{\text{day}}} = 0.15 \text{ mgal} \times \frac{1 \text{ ft}^3}{7.48 \text{ gal}}$$

$$= 20{,}053 \text{ ft}^2$$

Now, using the volume and depth, and the fact that the tank is square, we may calculate the dimensions as follows:

$$A = \frac{V}{\text{depth}} = \frac{20{,}053}{5} = 4{,}010.6 \text{ ft}^2 = L \times W \Rightarrow L = W = 63.3 \text{ ft}$$

14.11 d. The height of a stripping column can be determined by the equation provided in the environmental engineering section of the *FE Supplied-Reference Handbook* as

$$Z = (\text{HTU})(\text{NTU})$$

The height of a transfer unit can be estimated using the following relationship from the *Handbook*, making sure the units on the mass transfer coefficient and cross-sectional area of the column are compatible with the volumetric flow rate as follows:

$$\text{HTU} = \frac{Q_W}{K_L a} = \frac{\left(0.35 \frac{\text{m}^3}{\text{s}}\right)}{(0.5 \text{ min}^{-1})\left(\frac{1 \text{ min}}{60 \text{ s}}\right)\left(\frac{\pi}{4}\right)(5)^2 \text{m}^2} = 2.14 \text{ m}$$

The number of transfer units can be found using the following equation from the *Handbook*:

$$\text{NTU} = \left(\frac{R}{R-1}\right) \ln\left[\frac{(C_{in}/C_{out})(R-1)+1}{R}\right]$$

A value for the stripping factor (R) can be found from the unitless Henry's law constant and the relative volumetric flow rates as follows:

$$R = H'\left(\frac{Q_a}{Q_w}\right) = (0.23)\left(\frac{4.6}{0.35}\right) = 3.023$$

Further, using the information in the problem statement, the ratio of C_{in}/C_{out} can be estimated as

$$\frac{C_{in}}{C_{out}} = \frac{C_{in}}{(1-0.99)C_{in}} = \frac{C_{in}}{0.01 C_{in}} = 100$$

Substituting these values into the NTU expression, we get

$$\text{NTU} = \left(\frac{3.023}{2.023}\right) \ln\left[\frac{(100)(2.023)+1}{3.023}\right] = 6.29$$

Finally, the column height can be determined as

$$Z = (2.14)(6.29) = 13.46 \text{ m}$$

14.12 d. First, from the environmental engineering section of the *FE Supplied-Reference Handbook*, the mass flux through the membrane can be expressed as

$$\dot{m} = W_p A(\Delta P - \pi) \quad \Rightarrow \quad \dot{m} = \left(20\frac{L}{s}\right)\left(1\frac{kg}{L}\right) = 20\frac{kg}{s}$$

Assuming an osmotic pressure of 11.3 psi for every 0.1 percent salt, and converting pressure units into Pascals, the total osmotic pressure can be calculated as follows:

$$\pi = \frac{11.3 \text{ psi}}{0.1\% \text{ salt}} \times 0.5\% \text{ salt} = 56.5 \text{ psi} \times \frac{101,325 \text{ Pa}}{14.7 \text{ psi}} = 389,446 \text{ Pa}$$

Substituting this into the mass flux equation and solving for pressure difference yields

$$20\frac{kg}{s} = (1.2 \times 10^{-5} \text{ s} \cdot \text{m})(\Delta P - 389,446)$$

$$\Delta P = \frac{20}{1.2 \times 10^{-5}} + 389,446 = 2,056,113 \text{ Pa}$$
$$= 2056 \text{ kPa}$$

14.13 c. Starting with the power equation given in the environmental engineering section of the *FE Supplied-Reference Handbook* and substituting for the known resistance yields

$$P = I^2 R = I^2 \text{ (6 ohms)}$$

The current can be determined from the equation provided in the *FE Supplied-Reference Handbook* and the assumed efficiencies given in the problem statement as follows:

$$I = \frac{FQN}{n} \times \frac{E_1}{E_2} = \frac{(96,4878)(150)(0.005)(0.5)}{(350)(0.9)} = 114.86 \text{ amps}$$

Now power can be found as

$$P = I^2 R = (114.87)^2 (6) = 79,157 \text{ W} = 79.2 \text{ kW}$$

14.14 d. Starting with the expression for detention time and the data given in the problem statement, the tank volume can be determined as follows:

$$t_R = \frac{V}{Q} \quad \Rightarrow \quad V = Qt_R = \frac{(3 \text{ min})(6.2 \text{ mgd})}{(60)(24)\frac{\text{min}}{\text{day}}} = 12,917 \text{ gal} \times \frac{1 \text{ ft}^3}{7.48 \text{ gal}}$$
$$= 1727 \text{ ft}^3$$

With a known volume, a cubical tank can be dimensioned by taking the cube root of volume as

$$L = W = H = V^{1/3} = (1727 \text{ ft}^3)^{1/3} = 11.99 \text{ ft}$$

14.15 c. Again, tank volume can be found using the detention time as follows:

$$t_R = \frac{V}{Q} \Rightarrow V = Qt_R = (2 \text{ hr})\left(\frac{1 \text{ day}}{24 \text{ hr}}\right)(3.2 \text{ mgd})$$

$$= 266{,}667 \text{ gal} \times \frac{1 \text{ ft}^3}{7.48 \text{ gal}} = 35{,}651 \text{ ft}^3$$

Area is calculated by dividing volume by depth, and the diameter of a circular tank is found using the area as follows:

$$A = \frac{V}{\text{depth}} = \frac{35{,}651 \text{ ft}^3}{8 \text{ ft}} = 4456 \text{ ft}^2 = \frac{\pi}{4}D^2 \Rightarrow D = 75.3 \text{ ft}$$

14.16 d. The BOD loading to the primary can be determined using the BOD concentration and volumetric flow rate as follows:

$$\text{BOD load} = Q \times \text{BOD} = (2.6 \text{ mgd})\left(131 \frac{\text{mg}}{\text{L}}\right)(8.34) = 2841 \frac{\text{lb}}{\text{day}}$$

Remember, the factor 8.34 converts (mgd) × (mg/L) into the units of lb/day.

The volume of the tank can now be found by dividing the BOD load by the design loading rate as follows:

$$V = \frac{2841 \frac{\text{lb}}{\text{day}}}{35 \frac{\text{lb}}{1000 \text{ ft}^3 \cdot \text{day}}} = 81{,}171 \text{ ft}^3$$

Now, the aeration period can be calculated using the given flow rate and volume as follows:

$$\theta = \frac{V}{Q} = \frac{(81{,}171 \text{ ft}^3)\left(7.48 \frac{\text{gal}}{\text{ft}^3}\right)}{2.6 \times 10^6 \frac{\text{gal}}{\text{day}}} = 0.233 \text{ day} \times \frac{24 \text{ hr}}{\text{day}} = 5.6 \text{ hr}$$

14.17 b. The equation for sludge age is given in the environmental engineering section of the *FE Supplied-Reference Handbook* as

$$\theta_c = \frac{\text{MLSS} \times V}{Q_e SS_e + Q_w SS_w}$$

The mass rate of wasted solids is related to the BOD load determind in Solution 14.16 as described in the problem statement, and can be calculated as

$$Q_w SS_w = 44\% \times \text{BOD load} = (0.44)\left(2841 \frac{\text{lb}}{\text{day}}\right) = 1250 \frac{\text{lb}}{\text{day}}$$

We can simplify the sludge age expression if we assume that the effluent flow rate is approximately equal to the influent flow rate. Since Q_w is often ≤1 percent of Q_0, the assumption that Q_e = 99 percent Q_0 or $Q_e \approx Q_0$ is good enough for estimations. Therefore, if we assume $Q_e \approx$ 2.6 mgd, we can estimate the sludge age as follows:

$$\theta_c = \frac{\left(2100 \frac{\text{mg}}{\text{L}}\right)\left[(81{,}171 \text{ ft}^3)\left(7.48 \frac{\text{gal}}{\text{ft}^3}\right)\left(10^{-6} \frac{\text{Mgal}}{\text{gal}}\right)\right](8.34)}{1250 \frac{\text{lb}}{\text{day}} + (2.6 \text{ mgd})\left(20 \frac{\text{mg}}{\text{L}}\right)(8.34)}$$

$$= 6.32 \text{ days}$$

Notice the use of the 8.34 conversion factor in this problem as well. It should be noted that it works in the numerator with tank volume instead of volumetric flow rate, because it is really a conversion from mg to pounds and liters to million gallons (i.e., 8.34 = 2.205 lb/kg × 1 kg/10^6 mg × 3.785 L/gal × 10^6 gal/mgal). The time variable unit is not part of the conversion.

14.18 a. The expression for MLSS using the biological kinetic variables found in the environmental engineering section of the *FE Supplied-Reference Handbook* can be rearranged to express biological yield as a function of MLSS as follows:

$$\text{MLSS} = \frac{\theta_c Y(S_0 - S_e)}{\theta(1 + k_d \theta_c)} \Rightarrow Y = \frac{\text{MLSS}\,\theta(1 + k_d \theta_c)}{\theta_c(S_0 - S_e)}$$

Using the data given in the problem statements, biological yield can be calculated as

$$Y = \frac{\left(2100\,\frac{\text{mg biomass}}{\text{L}}\right)\left(\frac{5.6}{24}\,\text{days}\right)(1 + (0.02)(6.3))}{(6.3\,\text{days})\left(131 - 20\,\frac{\text{mg BOD}}{\text{L}}\right)} = 0.789\,\frac{\text{mg biomass}}{\text{mg BOD}}$$

14.19 c. The NRC equation is given in the problem statement. Remembering that the quantity w/V is the BOD loading in units of lb per 1000 ft³ per day, we write

$$\frac{w}{V} = 35\,\frac{\text{lb}}{1000\,\text{ft}^3 \cdot \text{day}}$$

Solving for the recirculation factor (F) requires the recirculation ratio (R) as follows:

$$R = \frac{Q_R}{Q_0} = \frac{5.2}{2.6} = 2$$

Now F may be calculated as

$$F = \frac{1+R}{(1+0.1R)^2} = \frac{3}{(1+0.2)^2} = 2.083$$

The efficiency of the trickling filter may now be determined as

$$E = \frac{100}{1 + 0.0561\left(\frac{w}{VF}\right)^{1/2}} = \frac{100}{1 + 0.0561\left(\frac{35}{2.083}\right)^{1/2}} = 81.3\%$$

14.20 d. The expression for biotower design is given in the environmental engineering section of the *FE Supplied-Reference Handbook*, however it is often useful to simplify the expression as follows:

$$\frac{S_e}{S_a} = \frac{\exp[-kD/q^n]}{(1+R) - R\exp[-kD/q^n]} \Rightarrow \frac{S_e}{S_a} = \frac{X}{(1+R) - RX}$$

Since S_e is known, S_a and R must be determined in order to solve for X. Since S_a is also a function of R, start by calculating the recirculation ratio as follows:

$$R = \frac{Q_R}{Q_0} = \frac{2.6}{2.6} = 1$$

It should be noted that the *FE Supplied-Reference Handbook* has a different equation for R as given in the biotower discussion when compared to the activated sludge section. The expression used above is the proper equation to determine R.

Using the data above and in the problem statement, S_a may be determined as

$$S_a = \frac{S_0 + RS_e}{1+R} = \frac{130\,\frac{\text{mg}}{\text{L}} + (1)\left(20\,\frac{\text{mg}}{\text{L}}\right)}{1+1} = 75\,\frac{\text{mg}}{\text{L}}$$

Combining this with the results above, we may solve for X in the modified expression above as

$$\frac{S_e}{S_a} = \frac{X}{(1+R)-RX} \Rightarrow \frac{20\,\frac{\text{mg}}{\text{L}}}{75\,\frac{\text{mg}}{\text{L}}} = \frac{X}{2-X} = 0.267 \Rightarrow X = 0.42$$

Now, X was a substitute quantity that may be expressed as

$$X = \exp\left[\frac{-kD}{q^n}\right] = 0.42 \Rightarrow 0.866 = \frac{kD}{q^n}$$

The quantity q is measured in m/min, but is actually expressed as m³ of flow per minute per m² of filter cross-section area, and may be calculated as

$$q = \frac{(5.2\text{ mgd})\left(1.547\,\frac{\text{cfs}}{\text{mgd}}\right)\left(\frac{60\text{ s}}{\text{min}}\right)}{\frac{\pi}{4}(20\text{ ft})^2} = 1.54\,\frac{\text{ft}}{\text{min}} \times \frac{1\text{ m}}{3.28\text{ ft}} = 0.47\,\frac{\text{m}}{\text{min}}$$

Notice the factor 1.547 used here is the same as used in the solutions to Problems 14.3 and 14.4. This is another great conversion factor to remember for use with wastewater problems where flows are given in mgd and more standard English units of ft³/s are desired.

Finally, tower diameter can be determined from q, and given values of k and n as follows:

$$0.866 = \frac{kD}{q^n} \Rightarrow (0.866)(0.47)^{0.5} = (0.06)D$$

$$\Rightarrow D = 9.9\text{ m} \times \frac{3.28\text{ ft}}{1\text{ m}} = 32.4\text{ ft}$$

14.21 a. Stabilization ponds must meet several criteria. The design process requires an evaluation of all possibilities and selects the most conservative (i.e., the one that meets all of the criteria). One criteria is that the maximum BOD load is 35 pounds of BOD per acre per day. Given the value for BOD load in solution 3.17, the minimum area required is

$$\text{Area} = \frac{2841\,\frac{\text{lb BOD}}{\text{day}}}{35\,\frac{\text{lb BOD}}{\text{ac·day}}} = 81\text{ acres}$$

Another design criteria is 0.1–0.3 pounds of BOD per day per 1000 ft³ of pond volume. Also, pond depth is limited to the range of

3–6 feet. Therefore the minimum area would occur at the maximum loading rate and the maximum pond depth, and can be calculated as

$$\text{Area} = \frac{2841 \frac{\text{lb BOD}}{\text{day}}}{0.3 \frac{\text{lb BOD}}{1000 \text{ft}^3 \cdot \text{day}}} = 9.47 \times 10^6 \text{ ft}^3 \div 6 \text{ ft} = 1.58 \times 10^6 \text{ ft}^2 \times \frac{1 \text{ acre}}{43,560 \text{ ft}^2}$$

$$= 36 \text{ acres}$$

The third design criteria is a minimum detention time of 50 days. Using the same maximum pond depth of 6 feet to determine the minimum area, we can calculate area as follows:

$$t_R = 50 d = \frac{V}{Q} \Rightarrow V = (50 \text{ days})(2.6 \text{ mgd}) = 130 \text{ Mgal} \times \frac{1 \text{ ft}^3}{7.48 \text{ gal}}$$

$$= 17.38 \times 10^6 \text{ ft}^3$$

$$\text{Area} = \frac{17.4 \times 10^6 \text{ ft}^3}{6 \text{ ft}} = 2.9 \times 10^6 \text{ ft}^2 \times \frac{1 \text{ acre}}{43,560 \text{ ft}^2} = 66.6 \text{ acres}$$

Therefore, to meet all of the criteria, the minimum pond size is 81 acres.

14.22 d. From Problem 14.17, we know that $Q_w SS_w = 1250$ lb/day. Therefore the total amount of sludge to be treated is

$$\text{total sludge} = 1250 + 1500 = 2750 \frac{\text{lb}}{\text{day}}$$

Given a solids concentration of 4 percent, we can calculate the mass rate of flow as follows:

$$\frac{2750 \frac{\text{lb}}{\text{day}}}{0.04} = 68,750 \frac{\text{lb}}{\text{day}}$$

If we assume this flow has a specific weight that is close to the specific weight of water, we may determine the volume of raw sludge entering the digester per day as follows:

$$V_1 = \frac{68,750 \frac{\text{lb}}{\text{day}}}{62.4 \frac{\text{lb}}{\text{ft}^3}} = 1102 \frac{\text{ft}^3}{\text{day}}$$

Note that the *FE Supplied-Reference Handbook* uses the variable *V* here for a volumetric flow rate. Next, we need to calculate the digested sludge accumulation. This is accomplished by noting that 30 percent of the solids are not volatile and will remain during treatment. The other 70 percent of the solids will degrade, but the reduction is only 40 percent, so 60 percent of the volatile solids remain. These solids are at a concentration of 6 percent solids, and are converted to a volumetric flow rate using the specific weight of water as in computing V_1 above. The expression for V_2 may be expressed as follows:

$$V_2 = \frac{\left(2750 \frac{\text{lb}}{\text{day}}\right)(0.3)}{(0.06)\left(62.4 \frac{\text{lb}}{\text{ft}^3}\right)} + \frac{\left(2750 \frac{\text{lb}}{\text{day}}\right)(0.7)(0.6)}{(0.06)\left(62.4 \frac{\text{lb}}{\text{ft}^3}\right)} = 529 \frac{\text{ft}^3}{\text{day}}$$

Finally, we can calculate reactor volume as follows:

$$V_r = \frac{V_1 + V_2}{2} t_r + V_2 t_s$$

$$= \left(\frac{1102 + 529 \frac{ft^3}{day}}{2}\right)(25 \text{ days}) + \left(529 \frac{ft^3}{day}\right)(60 \text{ days}) = 52{,}128 \text{ ft}^3$$

14.23 b. From Problem 14.17, we know that $Q_w SS_w = 1250$ lb/day and the problem statement gives a solids concentration of 1 percent, which is equivalent to 10,000 mg/L. We can therefore determine the volumetric flow rate as follows:

$$Q_w SS_w = 1250 \frac{\text{lb}}{\text{day}} = \left(10{,}000 \frac{\text{mg}}{\text{L}}\right) Q_w (8.34) \quad \Rightarrow \quad Q_w = 0.015 \text{ mgd}$$

$$Q_w = (0.015 \text{ mgd})\left(\frac{10^6 \text{ gal}}{\text{mgal}}\right)\left(\frac{1 \text{ ft}^3}{7.48 \text{ gal}}\right) = 2004 \frac{\text{ft}^3}{\text{day}}$$

Now, using the equation in the *FE Supplied-Reference Handbook* and data given in the problem statement, we may write

$$V = \frac{QX}{X_d\left(k_d P_v + \frac{1}{\theta_C}\right)} = \frac{\left(2004 \frac{\text{ft}^3}{\text{day}}\right)\left(10{,}000 \frac{\text{mg}}{\text{L}}\right)}{\left(60{,}000 \frac{\text{mg}}{\text{L}}\right)\left[(0.05 \text{ day}^{-1})(0.7) + \frac{1}{50 \text{ days}}\right]}$$

$$= 6073 \text{ ft}^3$$

CHAPTER 15

Air Quality and Atmospheric Pollution Control

OUTLINE

AIR QUALITY   675

METEOROLOGY   677

ATMOSPHERIC DISPERSION MODELING   678

VENTILATING   681
Ventilating Terms and Definitions ■ General Ventilation ■ Specific or Local Ventilation

INTRODUCTION TO ATMOSPHERIC POLLUTION CONTROL   684

REMOVAL OF PARTICULATE MATTER   684
Cyclones ■ Electrostatic Precipitation ■ Fabric Filters

ABSORPTION   689

ADSORPTION   690

THERMAL CONTROLS AND DESTRUCTION   694

PROBLEMS   696

SOLUTIONS   698

AIR QUALITY

Primary pollutants in air are those that are emitted by identifiable (albeit ubiquitous) sources, making it possible to address each specific discharge. *Secondary pollutants* are those that are formed in the atmosphere, often through reactions with primary pollutants. *Criteria pollutants* are those that have been identified as having a negative impact on the health of human populations and the health of ecosystems as a whole. The U.S. Environmental Protection Agency (U.S. EPA) has established criteria pollutant limits in ambient (outdoor) air called the *National Ambient Air Quality Standards* (NAAQS), as presented in Table 15.1. Large-scale (regional, national, and global) air pollution concerns include acid rain, global warming, and ozone depletion, each of which have identifiable causes with an opportunity for implementation of control technology.

Table 15.1 National Ambient Air Quality Standards for criteria pollutants

Criterion Pollutant	Concentration [µg/m³]	Concentration [ppm]	Averaging Period
CO	10,000	9	8-hour
CO	40,000	35	1-hour
Pb	1.5	—	3-month
NO_2	100	0.053	Annual
O_3	235	0.12	1-hour
O_3	157	0.08	8-hour
PM_{10}	150	—	24-hour
PM_{10}	50	—	Annual
$PM_{2.5}$	65	—	24-hour
$PM_{2.5}$	15	—	Annual
SO_2	80	0.03	Annual
SO_2	365	0.14	24-hour

Acid rain is primarily caused by the reaction of *sulfur dioxide* (SO_2) in the atmosphere, creating sulfuric acid, which is carried to the ground through precipitation events. SO_2 is a by-product from the combustion of fossil fuels containing sulfur compounds (which can be as high as 3%–4% by weight in some coals), and it is estimated that greater than 95% of all SO_2 is emitted from stationary sources such as power plants. Much has been done in the recent past to control SO_2 emissions, primarily through the implementation of scrubbing technologies, as covered in Example 15.5. Other techniques to reduce sulfur emissions include the removal of sulfur from the fuel (for example, coal washing) or switching to the use of low-sulfur fuels.

Although there are still some scientists and politicians who question the link to global warming, the fact remains that the concentration of greenhouse gases in the atmosphere is on a measurable increase. Historically, *volatile organic compounds* (VOCs) such as CFCs (chlorofluorocarbons) received the primary attention of scientists, first from the *ozone depletion potential* (ODP) and later from a *global warming potential* (GWP). With the replacement of VOCs in most commercial products, other species have come under scrutiny. The most abundant *hydrocarbon* (HC), a class of compounds comprised solely of hydrogen and carbon, is *methane* (CH_4), and although the majority is naturally released, it is considered a primary greenhouse gas. Two gases from combustion processes, *carbon dioxide* (CO_2) and *carbon monoxide* (CO), are also on the greenhouse gas list, and efforts to control CO_2 emissions on a global scale are being discussed in the regulatory and industrial communities. Carbon monoxide also can have serious negative health impacts when respired at sufficient concentrations.

Ozone (O_3) is interesting in the fact that its appearance in the lower atmosphere causes smog and irritation of the mucous membranes, while the depletion of ozone in the upper atmosphere has been associated with a reduction in the ability to filter cosmic radiation. Ozone is a criteria pollutant, although it is also a secondary pollutant because it is derived from atmospheric reactions with primary pollutants such as NO_X and HCs. Besides being a respiratory irritant, it may also damage materials through enhanced oxidation as well as inhibit plant growth.

Particulate matter (PM) is emitted in urban areas, primarily through combustion processes, and is classified by the aerodynamic diameter of the particle in μm (10^{-6} m). PM is suspected of having a negative impact on human health. While the criteria pollutant in the recent past has been PM_{10} (PM with diameters less than 10 μm), new standards target $PM_{2.5}$ as the required capture limit. PM in the 2.5–10-μm range may become trapped in the lungs, minimally causing irritation and potentially causing more serious chronic health problems. Control of PM is covered in detail in section 9.4.

Oxides of nitrogen (NO_x), such as NO, NO_2, and N_2O, are also produced in the combustion process. However, only a fraction are fuel derived (oxidation of N in the fuel), even though some coals can have nitrogen concentrations up to 1% by weight. The majority of the NO_x species are due to the oxidation of N_2 that is part of the combustion air. NO_x contributes to smog and is also a respiratory irritant. Control of thermal NO_x, the portion of the NO_x that is created from N_2 in the combustion air, generally examines operating conditions in the combustion zone and attempts to inhibit creation of NO_x by maintaining the flame temperature within specified limits. After generation, or for controlling fuel-based NO_x, facilities often employ *selective catalytic reduction* (SCR) of the NO_x species after the combustion zone to convert NO_x back to N_2.

The final criteria pollutants to be discussed here are heavy metals, particularly *mercury* (Hg) and *lead* (Pb). While still of interest, the concern of atmospheric lead has been addressed by the regulatory and scientific communities in the recent past (for example, the removal of lead from gasoline). Lead may still be present in some industrial settings or as airborne particulate, but substantial efforts have been made to remove it from most commercial products. Mercury still remains a problem, due primarily to its high toxicity at low concentrations, its tendency to bioconcentrate and bioaccumulate, and its presence in many natural materials such as coal. However, advancements in air pollution control technology has done much to reduce the Hg emissions from power plant flue gases, and many scientists argue that the natural sources of mercury emissions such as global volcanic activity far exceed those from anthropogenic (those derived from human activity) sources.

METEOROLOGY

The atmosphere is in constant motion, driven by *mechanical* and *thermal* forces such as the Earth's rotation and solar energy. Wind and weather patterns are produced by these effects and also contribute to the distribution of all compounds that are discharged into the atmosphere. Once in the atmosphere, they are subject to a variety of chemical and physical transformations and are redeposited on the Earth's surface during precipitation events. Prediction of the distribution of pollutants from a point source is a valuable tool in evaluating the potential health impact an atmospheric discharge may have on any downwind populations.

The point source Gaussian dispersion model is used to evaluate contaminant concentrations downwind of a discharge, generally considered to be an industrial stack. A primary consideration of the model is the *atmospheric stability*, which may be defined as the *tendency of the atmosphere to resist vertical motion*. Atmospheres are classified as *stable* when thermal effects *restrict* mechanical turbulence, *unstable* when thermal effects *enhance* mechanical turbulence, and *neutral* when thermal effects *neither enhance nor restrict* mechanical effects. An approximate stability classification was developed by Turner in which atmospheric stability increases from A to F, as defined in Table 15.2. These classifications were then related to prevailing environmental conditions that were based upon incident solar radiation and wind speed at an elevation of 10 m, as presented in the Environmental Engineering section of the *Fundamentals of Engineering Supplied Reference Handbook*.

Table 15.2 Atmospheric stability classifications

Degree of Stability	Classification
Extremely unstable	A
Unstable	B
Slightly unstable	C
Neutral	D
Slightly stable	E
Stable	F

The stability classification is used to predict the Gaussian (normal) dispersion distance as a function of distance downwind of a point source. As seen in the standard deviations of plumes graphs in the Environmental Engineering section of the *FE Supplied Reference Handbook*, dispersion in the vertical direction increases at variable rates with increased downstream distance, depending on the atmospheric stability classification. This is to be expected, because stability is defined as the tendency of the atmosphere to resist vertical motion. Horizontal dispersion still depends on stability classification; however, it does increase at a constant rate with downstream distance.

ATMOSPHERIC DISPERSION MODELING

The Gaussian dispersion model as presented below assumes that material is discharged from a single point at elevation H, and any material that reaches the ground is reflected back into the air and can be expressed as

$$C_{(x,y,z)} = \frac{Q}{2\pi \sigma_y \sigma_z U} \exp\left[-\frac{1}{2}\left(\frac{y}{\sigma_y}\right)^2\right] \left\{\exp\left[-\frac{1}{2}\left(\frac{z-H}{\sigma_z}\right)^2\right] + \exp\left[-\frac{1}{2}\left(\frac{z+H}{\sigma_z}\right)^2\right]\right\}$$

where $C_{(x,y,z)}$ is the species concentration at location (x,y,z) downwind of the discharge (g/m³), x is the downwind or centerline distance (m), y is the horizontal distance that is perpendicular to the centerline (m), z is the elevation from ground level (m), Q is the species mass flow rate (g/s), σ_y and σ_z are the horizontal and vertical standard deviations (m) obtained from the Environmental Engineering section of the *Fundamentals of Engineering Supplied Reference Handbook*, U is the wind speed at the point of discharge (m/s), and H is the *effective stack height* (m). Note that the variable used here for the emission rate (Q) has dimensions of mass per time, unlike in previous chapters where Q was defined as the volumetric flow rate (volume per time).

The effective stack height takes into account the elevation of the discharge, as well as the *plume rise*. Plume rise is a result of gas velocity exiting the stack, as well as the buoyancy effects due to the temperature differential between the emitted gas and the surrounding air, and can be estimated by the following relationships:

$$H = h + \Delta H$$

$$\Delta H = \frac{U_s d}{U}\left\{1.5 + \left[0.0268\, P\left(\frac{T_s - T_a}{T_s}\right) d\right]\right\}$$

where h is the stack height (m), ΔH is the plume rise (m), U_s is the stack gas velocity (m/s), d is the stack diameter (m), U is the wind speed at the point of discharge (m/s), P is the prevailing atmospheric pressure (kPa), and T_s and T_a are the temperatures (K) of the stack gas and ambient air, respectively. It should be noted that the temperatures are expressed in absolute units, where K = °C + 273.

Often, it is important to evaluate ground level concentrations, where impact to human health would most likely occur. In this case, the elevation term z is set equal to zero, and the Gaussian model is simplified to

$$C_{(x,y,0)} = \frac{Q}{\pi \sigma_y \sigma_z U} \exp\left[-\frac{1}{2}\left(\frac{y}{\sigma_y}\right)^2\right] \exp\left[-\frac{1}{2}\left(\frac{H}{\sigma_z}\right)^2\right]$$

Further, based on the assumption of normally distributed dispersion, the maximum concentration at any downwind distance x would occur on the center line. In this case, the horizontal distance term y is set equal to zero, and the Gaussian model is simplified to

$$C_{(x,0,0)} = \frac{Q}{\pi \sigma_y \sigma_z U} \exp\left[-\frac{1}{2}\left(\frac{H}{\sigma_z}\right)^2\right]$$

This expression may be rearranged to convey the maximum concentration of pollutant for a given wind speed and emission rate as a function of effective stack height and atmospheric dispersion as follows:

$$\left(\frac{CU}{Q}\right)_{max} = \frac{1}{\pi \sigma_y \sigma_z} \exp\left[-\frac{1}{2}\left(\frac{H}{\sigma_z}\right)^2\right]$$

The left-hand side of the expression is often plotted for various downwind distances and effective stack heights, and the curves for each atmospheric stability class have been regressed. The values for $(CU/Q)_{max}$ may be calculated from the following equation:

$$\left(\frac{CU}{Q}\right)_{max} = \exp\left[a + b \ln H + c\left(\ln H\right)^2 + d\left(\ln H\right)^3\right]$$

where C, U, Q, and H are as defined previously, and a, b, c, and d are constants based on atmospheric stability class as given in the Environmental Engineering section of the *Fundamentals of Engineering Supplied Reference Handbook*. The equation above is also solved for all stability classes as a function of effective stack height and presented graphically, also provided in the Environmental Engineering section of the *Fundamentals of Engineering Supplied Reference Handbook*.

Finally, for emissions that originate from ground level (for example, fires and fugitive emissions), the effective stack height H and elevation z may both be set equal to zero to yield an expression as follows:

$$C_{(x,y,0)} = \frac{Q}{\pi \sigma_y \sigma_z U} \exp\left[-\frac{1}{2}\left(\frac{y}{\sigma_y}\right)^2\right]$$

to determine the ground level concentrations at a downwind location off the centerline, or as

$$C_{(x,0,0)} = \frac{Q}{\pi \sigma_y \sigma_z U}$$

to determine the maximum downwind concentration at ground level, which would occur on the centerline.

Example 15.1

Plume dispersion modeling

A factory is on fire and is releasing smoke with a particle size less than 10 μm at a rate of 3 kg/s. What is a safe downwind distance if the NAAQS for PM_{10} is 150 μg/m³ for a 24-hour exposure? Assume a wind speed of 20 mph under neutral atmospheric conditions.

Solution

Noting that a factory fire would be considered a ground-level emission, the maximum downwind concentration can be estimated as

$$C_{(x,0,0)} = \frac{Q}{\pi \sigma_y \sigma_z U}$$

Plugging in the information given in the problem statement and converting units we get

$$C_{max} = 150 \frac{\mu g}{m^3} = 0.00015 \frac{g}{s} = \frac{\left(3000 \frac{g}{s}\right)}{\pi \sigma_y \sigma_z \left(20 \frac{mi}{hr}\right)\left(0.447 \frac{m/s}{mi/hr}\right)}$$

$$\Rightarrow \sigma_y \sigma_z = 712,100 \text{ m}^2$$

Now we can use the vertical and horizontal dispersion plots provided in the *Fundamentals of Engineering Supplied Reference Handbook* to estimate the required distance under neutral conditions (class D) to yield the value above for the product of σ_y and σ_z. Through trial and error we get

@ $X = 30$ km $\Rightarrow$ $\sigma_y = 1500$ m and $\sigma_z = 250$ m $\therefore$ $\sigma_y \sigma_z = 375,000$ m²

@ $X = 50$ km $\Rightarrow$ $\sigma_y = 2200$ m and $\sigma_z = 310$ m $\therefore$ $\sigma_y \sigma_z = 682,000$ m²

@ $X = 60$ km $\Rightarrow$ $\sigma_y = 2600$ m and $\sigma_z = 340$ m $\therefore$ $\sigma_y \sigma_z = 884,000$ m²

Since 50 km is too small and 60 km is too large, we can interpolate to get an approximation of

$$\left(\frac{712,100 - 682,000}{884,000 - 682,000}\right)(60 - 50) + 50 = 51.5 \text{ km} \approx 32 \text{ miles}$$

VENTILATING

Ventilating Terms and Definitions

The following terms are commonly used in ventilation work.

Cubical contents is the contents of the space to be ventilated expressed in cubic feet. Length × width × height = cubical contents. No deduction is made for equipment, tables, etc., within the space.

Cubic feet per minute (cfm) is the rate of air flow.

Capacity is the volume of air handled by a fan, group of fans, or by a ventilation system, usually expressed in cubic feet per minute.

Fan rating is a statement of fan performance for one condition of operation and includes fan size, speed, capacity, pressure, and horsepower.

Fan performance is a statement of capacity, pressure, speed, and horsepower input.

Fan characteristic is a graphical presentation of fan performance throughout the full range from free delivery to no delivery at constant speed for any given fan.

Resistance pressure (RP). The resistance pressure of any ventilating system is the total of the various resistance factors that oppose the flow of air in the system stated in inches water gauge (WG).

Static pressure (SP) is the force exerted by the fan to force air through the ventilating system. If exerted on the discharge side, it is said to be positive and if on the inlet side, it is negative or suction pressure. The total static pressure the fan exerts is the sum of these two pressure readings. In any system the SP exerted by the fan is equal to the RP of the system. The SP is measured by an inclined draft gauge in inches of water.

Velocity is the speed at which the air is traveling expressed in lineal feet per minute. It is measured by use of a velometer or anemometer. Average velocities may be calculated for a fan by dividing the cfm by the square-foot area of the fan discharge. Average velocities in any part of the system may be calculated by dividing the cfm flowing by the cross-sectional area of the duct.

Velocity pressure (VP) is a measure of the kinetic energy of horsepower in the moving air. It is measured directly by the use of a Pitot tube and a draft gauge. It can be calculated from the velocity from the formula

$$VP = (\text{velocity, fpm}/4005)^2.$$

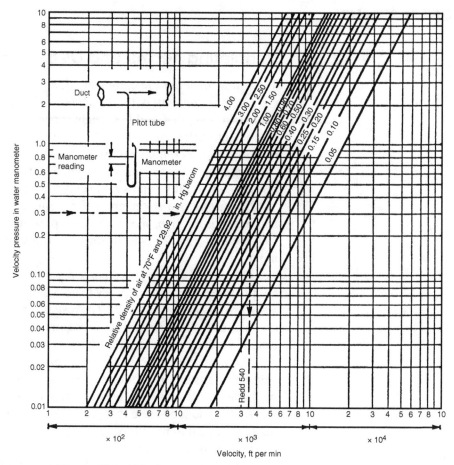

Figure 15.1 Velocity vs. velocity pressure, gases at various densities

Figure 15.1 shows a chart for determining the velocity of any gas within a duct from manometer readings in inches of water. First determine relative density of the flowing gas, remembering that density of standard air at 70°F and atmospheric pressure is 0.075 lb per cu ft. Then use manometer reading and read right to relative density line and down to actual velocity in feet per minute.

Example 15.2

Air at 100°F is flowing in a duct. Pitot tube reading for velocity pressure is 1/2 in. WG. What is the actual velocity of flow?

Solution

Air at 100°F has a density of 0.075 × (460 + 70)/(460 + 100), or 0.071 lb per cu ft. Relative density is

$$0.071/0.075 = 0.946$$

Now refer to Figure 15.1. From the left-hand side start at 1/2 in. and read right along this line to the sloping relative density line of 0.946. An interpolation must be made between 1.0 and 0.90. Read down to velocity of close to 2900 fpm. Normally for such temperatures no real deviation is made.

Horsepower output (air hp) of a fan, or air horsepower, is calculated from the formula

$$\text{Air hp} = \frac{\text{cfm} \times \text{TP}}{6356}$$

where cfm is capacity in cubic feet per minute; TP is total pressure of water or static pressure + velocity pressure.

Brake horsepower (bhp) is the horsepower required to drive the fan. Brake horsepower is the input to the fan shaft required to produce the output air hp.

Mechanical efficiency of a fan is the ratio of horsepower output to horsepower input. Therefore,

$$\text{Mechanical efficiency} = \frac{\text{Air hp}}{\text{bhp}}, \text{ expressed as a percentage.}$$

Fan discharge or outlet is the place provided for receiving a duct through which air leaves the fan.

Fan inlet is the place provided for receiving a duct through which air enters the fan.

General Ventilation

Where little or no ductwork is required to ventilate a space, the application is known as general ventilation. In most cases, exhaust fans high in the side walls or in the roof are used with general movement of air through windows or louvers across the space and out the fans. The air movement is caused to flow throughout the space to remove smoke, fumes, gases, excess moisture, heat, odors, or dust or simply to provide a constant inflow of fresh, outside air by the removal of foul, stale air.

To select the fans properly for the solution of general ventilation problems, the cubical contents of the space to be ventilated should be determined by multiplying the length by the width by the average height. All dimensions used must be in feet to give volume (cubage) in cubic feet.

The next step is to select the rate of air change required to give satisfactory ventilation. The ASHRAE *Guide* or manufacturers' catalogues contain much useful data and should be used to solve examination questions.

Specific or Local Ventilation

Throughout industry there is hardly an industrial plant that does not have at least one and usually many operations, machines, or processes that require special or, as it is known in industry, specific attention. By this we mean a system of ductwork and fans designed to prevent release or spread of smoke, fumes, odors, dusts, vapors, or excess heat into other working areas. This means the control of the problem at its source. This local ventilation requires well-designed hoods or special collecting systems and a thorough knowledge of air-flow principles and laws governing the behavior of gases and vapors. Because it is beyond the scope of this presentation to cover all the detailed data required by the infinite variety of local exhaust problems, we shall review with the help of fundamental charts and tables that which has been of paramount value.

The best ventilating system can be rendered almost useless unless there is a well-selected balance between fan and duct system. Care should also be exercised to be sure that the fumes, gases, or vapors are collected as close to the source of generation as possible. The closer the better, and the less will be the ventilation needs with attendant reduction in heating load for outside air make-up. Follow these simple rules and no real trouble will ensue.

a). Make all duct runs as short as possible with direct connections.

b). Make area of inlet and outlet ducts equal to outside diameter of fan for lowest friction loss.

c). Where hoods requiring a large area are required, use baffles to give higher edge velocities and reduce air volume required.

d). Do not forget that you cannot take more air out of a room than you are putting into it. Location and sizing of even gravity air inlets are frequently as important as the design of the exhaust system.

For exhaust systems the quantity of air to be moved should be determined by the selection of a suitable face velocity at hood entrance or by other considerations such as heat or moisture absorption, cubic feet per minute of liquid surface, by duct velocities required to convey the material, or combinations of the above. For supply systems the air flow usually is already determined by the volume being exhausted or by other considerations similar to those listed for the exhaust system. But let us digress for a moment to look into the matter of evaporation of water from tanks.

INTRODUCTION TO ATMOSPHERIC POLLUTION CONTROL

Due primarily to the negative impact on human health, as well as to the deterioration of the flora and fauna in natural environments, much effort has been extended to control the airborne emissions generated from anthropogenic (man-made) sources. While several control technologies exist to address primary and secondary pollutants in air, the majority may be classified into a few main categories, as addressed in this chapter. The control of particulate matter is accomplished through the use of cyclones, electrostatic precipitators, and baghouses. Certain chemical compounds may be removed through mass transfer operations such as absorption and adsorption or through the use of thermal controls such as manipulation of combustion zone properties and mass destruction through thermal oxidation processes.

REMOVAL OF PARTICULATE MATTER

As one of the criteria pollutants, particulate matter has been the focus of much research regarding the continuous capture and processing of airborne solids. Although it is estimated that all anthropogenic sources combined account for less than 7% of primary particulate emissions per year, primarily due to the combustion of fossil fuels in industry, these sources are an identifiable point source amenable to environmental control. The three primary particulate control devices used in industry today are the cyclone, the electrostatic precipitator (ESP), and the fabric filter (baghouse).

Cyclones

A *cyclone* relies upon the inertial forces generated when a particle-laden gas is forced to experience spiral motion. The *centrifugal force* causes the higher density particulate matter to contact the outer wall of the cyclone and fall out of the gas flow to a collection tube at the bottom of the cyclone. Standard dimensions for cyclones are based on the diameter of the body, as provided in the Environmental Engineering section of the *FE Supplied Reference Handbook*. A cyclone can be quite effective for removing particles that possess diameters ≥ 10 μm (PM_{10}), with efficiencies dropping off substantially for $PM_{2.5}$ or smaller, limiting their use to coarse dusts or as a pretreatment unit.

The estimation of particle capture efficiency is a function of the *approximate number of turns* (N_e) each particle will experience inside the cyclone and can be calculated as

$$N_e = \frac{1}{H}\left[L_b + \frac{L_c}{2}\right]$$

where H, L_b, and L_c are dimensions of the cyclone, as determined from the table of cyclone ratio of dimensions to body diameter as provided in the Environmental Engineering section of the *FE Supplied Reference Handbook*. The *cut diameter* (d_{pc}) can be defined as the *particle diameter that is collected at 50% efficiency* and may be calculated as

$$d_{pc} = \left[\frac{9\mu W}{2\pi N_e v_i (\rho_p - \rho_g)}\right]^{1/2}$$

where μ is the gas viscosity, W is the inlet width, N_e is as defined above, v_i is the inlet gas velocity (calculated as volumetric flow rate divided by the inlet cross-sectional area), and ρ_p and ρ_g are the particle and gas densities, respectively. Often, the gas density is assumed negligible as compared to the particle density and is dropped from the calculation.

Once the cut diameter is known for a given cyclone configuration and flow rate, the *particle capture efficiency* (η) may be calculated as

$$\eta = \frac{1}{1 + \left(\frac{d_{pc}}{d_p}\right)^2}$$

where d_p is the diameter of the particle of interest. Since the particle capture efficiency is only a function of the ratio of d_{pc} and d_p, it is common to find the solution of the above expression solved for a range of ratios and expressed graphically, as provided in the Environmental Engineering section of the *FE Supplied Reference Handbook*. Cyclone design is generally based on sizing a cyclone to achieve a desired removal efficiency for a specified particle size.

As seen in the table provided in the Environmental Engineering section of the *FE Supplied Reference Handbook*, the primary difference between high-efficiency and high-throughput cyclones is the inlet and outlet dimensions. Inlet dimensions establish the gas velocity, and increased inlet velocity increases collection efficiency (that is, decreases particle cut-diameter). Unfortunately, as gas volumes increase, particle collection at high efficiency increases pressure drop in the cyclone, and thus increases the power costs required to maintain flow. Generally, this is addressed through the use of *multi-cyclones* (several cyclones operating in a parallel configuration), or by using other particulate capture devices in series to serve as polishing devices with lower mass loads after a high-throughput cyclone.

Example 15.3

High-efficiency vs. high-throughput cyclone efficiency

A high-throughput cyclone captures a 10-μm particle at an efficiency of 40%. What would the capture efficiency be for a high-efficiency cyclone assuming the inlet width and velocity remain unchanged?

Solution

From the table provided in the Environmental Engineering section of the *FE Supplied Reference Handbook*, at an efficiency of 40% we see

$$\frac{d_p}{d_{pc}} = 0.8 \Rightarrow d_{pc} = \frac{d_p}{0.8} = \frac{10\,\mu m}{0.8} = 12.5\,\mu m$$

The particle cut-diameter may also be calculated using the following equation as

$$d_{pc} = \left[\frac{9\,W\mu}{2\pi N_e V_i (\rho_p - \rho_g)}\right]^{1/2}$$

Normally, values for W, N_e, and V_i are dependent upon the type of cyclone; however, the problem statement requires a constant value for W and V_i. Therefore, the difference in particle cut-diameter is only a function of N_e. Combining all other terms except N_e and representing them as a constant, we may write

$$d_{pc} = \left[\frac{C}{N_e}\right]^{1/2}$$

We can now determine the value of that constant term if we evaluate N_e for the high-throughput cyclone as follows:

$$N_e = \frac{1}{H}\left[L_b + \frac{L_C}{2}\right] = \frac{1}{0.8D}\left[1.7D + \frac{2D}{2}\right] = 3.4$$

Notice that the cyclone body diameter cancels out, so a value for diameter is not necessary. The result is that N_e is a constant that is only dependent on type of cyclone. With this value for N_e, we may now calculate the value for the lumped constant in the equation above as follows:

$$d_{pc} = \left[\frac{C}{N_e}\right]^{1/2} \Rightarrow C = d_{pc}^2 N_e = (12.5)^2 (3.4) = 531.25$$

To determine the particle cut-diameter for the high-efficiency cyclone, we will need to determine the value of N_e for the high-efficiency unit as follows:

$$N_e = \frac{1}{0.44D}\left[1.4D + \frac{2.5D}{2}\right] = 6.023$$

With the value for N_e and the constant determined previously, we may calculate the particle cut-diameter for the high-efficiency cyclone as follows:

$$d_{pc} = \left[\frac{531.25}{6.023}\right]^{1/2} = 9.4\,\mu m$$

From the particle cut-diameter, we can calculate the particle size ratio of d_p to d_{pc} as follows:

$$\frac{d_p}{d_{pc}} = \frac{10\,\mu m}{9.6\,\mu m} = 1.06$$

Now we can use this ratio with the cyclone collection efficiency plot as before to determine the efficiency of this unit. However, since the value of 1.06 is difficult to locate precisely on the plot, we can only estimate the efficiency as somewhere between 50% and 55%. An alternative method is to use the particle capture efficiency equation, noting that the ratio d_p to d_{pc} must be inverted as

$$\eta = \frac{1}{1+\left(\dfrac{d_{pc}}{d_p}\right)^2} = \frac{1}{1+\left(\dfrac{1}{1.06}\right)^2} = 0.529 = 52.9\%$$

Electrostatic Precipitation

The *electrostatic precipitator* (ESP) works by imparting a negative charge to particles entering the unit, which are then attracted to grounded collector plates. As particle mass is accumulated on the plates, they are tapped with hammers, causing the mass of solids to drop to a collection hopper below. Particle collection efficiency is dependent upon the velocity at which the particles travel toward the plates, called the *migration (drift) velocity* (W), and can be estimated as

$$W = \frac{qE_p C}{6\pi r \mu}$$

where q is the charge (C), E_p is the field intensity (V/m), r is the particle radius (m), μ is the gas viscosity (Pa · s), and C is the Cunningham correction factor, which may be calculated as

$$C = 1 + \frac{0.000621\, T}{d_p}$$

where T is the absolute temperature (K) and d_p is the particle diameter (μm). It should be noted that migration velocities are often provided for typical processes, with values of 0.015 to 0.018 m/s used for cement kilns and lime dust, and values of 0.08 to 0.17 m/s used for flyash from coal or solid waste incinerators.

Finally, the ESP collection efficiency (η) can be determined using the Deutsch-Anderson equation as follows:

$$\eta = 1 - \exp\left[\frac{-WA}{Q}\right]$$

where W is the migration velocity, A is the total plate collection area, and Q is the actual gas volumetric flow rate (at system temperature and pressure). Since many gas flows are given as *scfm* (*standard cubic feet per minute*), values of *acfm* (*actual cubic feet per minute*) must be obtained through the use of the ideal gas law. Standard conditions assume a pressure of 1 atmosphere and a temperature of 25°C, or equivalently 77°F. Since the pressure in the ESP is not significantly different from atmospheric pressure, usually only the temperature needs to be accounted for in ESP problems.

Example 15.4

ESPs

A small coal-fired power plant has an ESP unit that has four plates which are 4 m × 10 m each. If the efficiency must stay above 98%, what is the maximum actual gas flow rate allowed? You may assume an average migration velocity of 12 cm/s.

Solution

Starting with the ESP efficiency equation and substituting for the data given in the problem statement, converting cm/s to m/s, and recognizing that there are 4 total plates that are 4 × 10 meters with 2 sides each, we may write:

$$\eta = 1 - \exp\left[\frac{-WA}{Q}\right] = 0.98 \quad \Rightarrow \quad \ln(0.02) = \frac{-\left(0.12 \frac{m}{s}\right)(4)(4\ m)(10\ m)(2)}{Q \frac{m^3}{s}} = \frac{-38.4}{Q}$$

Solving for volumetric flow rate we get:

$$Q = \frac{-38.4}{\ln(0.02)} = 9.8 \frac{m^3}{s} \times \frac{35.3\ ft^3}{1\ m^3} \times \frac{60\ s}{1\ min} = 20,756\ \text{acfm}$$

Fabric Filters

Fabric filters are typified by the common filters used in most household air handling systems, where particle and gas flow rates are low. In industrial settings, the same principle is used on a large scale in a configuration often called a *baghouse*. A baghouse is a unit that holds several fabric tubes (bags), which are 8 to 12 inches in diameter and 10 to 20 feet long. The bags are open on one end, and flow is directed from inside the bag to the outside for particle collection in the bag (*shaker* or *reverse-flow* systems), or from outside the bag to inside for particle collection on the outside of the bag (*pulse-jet* systems).

Shaker systems have the advantage of using woven fabric, which generally has high tensile strength and therefore requires no additional support, and the flow from inside to outside keeps the bags inflated. Cleaning is dependent on mass loadings, which affect pressure drop and usually occur every 30 minutes to several hours. This is accomplished by isolating a compartment of bags by taking them offline and either shaking them or directing a flow of cleaned air in the reverse direction of flow, allowing the solids to drop into a hopper below. System design is based upon the *air-to-cloth* ratio to determine the total amount of fabric required for a specific application and are typically between 0.6 to 1.1 m³/min per m² of fabric for shaker units.

Pulse-jet systems use a felt fabric and require wire cages to maintain the cylindrical bag shape as flow goes from outside the bag to inside. Cleaning occurs quite frequently (every few minutes) through a pulse of high-pressure air inside the bag, which expands the bag slightly and causes accumulated dust on the outer surface to fall into the hopper below. Air-to-cloth ratios are typically larger for pulse-jet systems, usually in the range of 1.5 to 4.0 m³/min per m² of fabric.

ABSORPTION

In Chapter 14 of this review, *absorption* was described as a phenomenon through which contaminant mass is transferred from the liquid phase to the gas phase in a process called *air stripping,* and design equations for air stripping towers were discussed. For air pollution control, the mass transfer is from the gas phase to the liquid phase in a process called *gas scrubbing*. The mass transfer fundamentals are identical, and both rely upon *Henry's law,* as expressed previously. The magnitude of the dimensionless Henry's Law constant (H') is a good indicator of the direction of contaminant transfer, where values less than 0.01 indicate species that have a strong preference for the aqueous phase, and thus high potential for removal by gas scrubbing. Further, chemical reactions may increase removal efficiencies if a water soluble or particulate form of a contaminant species is generated in the scrubbing process and removed from the system by the liquid stream.

The design of a packed tower for gas scrubbing using a non-reactive solution (water) relies upon the concentration gradient as the mass transfer driving force, and, therefore, is best operated in the countercurrent flow mode. This operation contacts the pure water inlet stream with the gas exiting the system, offering the potential to remove air contaminants to a very low level. The overall mass balance at steady state can be expressed as

$$(Q_g)_{in} y_{in} - (Q_g)_{out} y_{out} = (Q_w)_{out} C_{out} - (Q_w)_{in} C_{in}$$

where Q_g and Q_w are the volumetric flow rates of gas and water, respectively, y and C are gas phase and liquid phase contaminant concentrations, respectively, and the subscripts in and out refer to material entering and exiting the tower, respectively. For the specific case where the inlet water is pure ($C_{in} = 0$) and the contaminant is present in low concentrations (i.e., $\Delta Q_g = 0$ and $\Delta Q_w = 0$), the mass balance may be written as

$$Q_g (y_{in} - y_{out}) = Q_w C_{out}$$

Design of the packed tower is based on the model presented previously as

$$Z = \text{HTU} \times \text{NTU}$$

where Z is the height of the stripping tower packing material (m), HTU is the *height of a transfer unit* (m), and NTU is the *number of transfer units* (unitless). For the gas scrubber, the height of a transfer unit is a function of the HTU for the gas (HTU_g) and water (HTU_w) phases, and may be expressed as

$$\text{HTU} = \text{HTU}_g + \left(\frac{H' Q_g}{Q_w}\right) \text{HTU}_w = \frac{Q_g}{(K_L a)_g A} + R\left(\frac{Q_w}{(K_L a)_w A}\right)$$

where $(K_L a)_g$ and $(K_L a)_w$ are the overall mass transfer coefficients in the gas and water phase, respectively (s^{-1}), A is the stripping column cross-sectional area (m^2), H', Q_g, and Q_w are as defined previously, and R is the scrubbing factor, which was expressed previously as

$$R = H' \left(\frac{Q_a}{Q_w}\right)$$

Expressions for the overall mass transfer coefficients are quite complex and beyond the scope of this review; however, they are often able to be calculated based upon characteristic data supplied by the manufacturers of tower packing materials.

The number of transfer units in the gas-scrubbing tower may be determined from the following relationship:

$$\text{NTU} = \frac{\ln\left[\left(\frac{y_{in} - H'C_{in}}{y_{out} - H'C_{in}}\right)(1 - R) + R\right]}{1 - R}$$

Often, the contaminant concentration of the inlet water is assumed to be zero, and the expression is simplified to

$$\text{NTU} = \frac{\ln[(y_{in}/y_{out})(1 - R) + R]}{1 - R}$$

Example 15.5

Gas scrubbing

A 20-m-tall packed scrubber has an HTU of 2.1 m and scrubbing factor (R) of 1.5. Calculate the percent reduction in the contaminant concentration of the exiting gas.

Solution

The percent reduction in vapor phase concentration can be found in the expression above as y_{in}/y_{out}. To obtain a value for this ratio, we need the scrubbing factor given in the problem statement and the value for the number of transfer units (NTU). This may be calculated as

$$Z = \text{HTU} \times \text{NTU} \quad \Rightarrow \quad \text{NTU} = \frac{Z}{\text{HTU}} = \frac{20 \text{ m}}{2.1 \text{ m}} = 9.5$$

Now we can rearrange the expression given to solve for y_{in}/y_{out} as follows:

$$\text{NTU} = \frac{\ln\left[\left(\frac{y_{in}}{y_{out}}\right)(1 - R) + R\right]}{1 - R} \quad \Rightarrow \quad \frac{y_{in}}{y_{out}} = \frac{\exp\left[(\text{NTU})(1 - R)\right] - R}{1 - R}$$

Substituting in the values for NTU and R, we can solve for y_{in}/y_{out} as follows:

$$\frac{y_{in}}{y_{out}} = \frac{\exp\left[(9.5)(1 - 1.5)\right] - 1.5}{1 - 1.5} = \frac{\exp\left[(9.5)(-0.5)\right] - 1.5}{-0.5} = 2.98$$

From this ratio, we may determine the percent reduction as follows:

$$\frac{y_{in}}{y_{out}} = 2.98 \quad \Rightarrow \quad y_{out} = \frac{y_{in}}{2.98} = 0.336 \, y_{in} = 33.6\% \text{ of } y_{in} \quad \Rightarrow \quad \text{reduction} = 66.4\%$$

ADSORPTION

Adsorption is a process in which contaminant molecules (called the *adsorbate* or *solute*) that are present either in the gas or liquid stream are removed from the fluid through their attachment to a solid surface (called the *adsorbent*). This technology is attractive for contaminants present in low concentrations that have a high monetary value or may impart a substantial negative impact to the environment upon release. The adsorbate preferentially attaches to the adsorbent due to chemical or physical attraction, and the attachment is generally stable as long as the fluid properties (temperature, pressure, constituent compositions) remain constant.

Regeneration of an adsorbent occurs when the majority of active sites have been saturated with adsorbate and a change in the fluid properties shifts the equilibrium of attachment, thereby releasing the previously adsorbed molecules. These released contaminant molecules are usually in much higher concentrations in the regeneration flow and may be further processed for recovery and reuse, or destruction and final, secure disposal.

Materials that make good adsorbents have large capacities due to extremely high specific surface areas that can be attributed to a highly porous internal structure. Commonly used adsorbents include *granular* or *powered activated carbon* (GAC or PAC), *activated alumina, silica gel,* and *zeolites* (aluminosilicate compounds often referred to as *molecular sieves*). Specific surface areas can range from 200–400 m^2/g for activated alumina, to 300–900 m^2/g for silica gel, to 700–1500 m^2/g for GAC and PAC.

The preference that a particular chemical contaminant has for either the fluid phase or the adsorbent is a function of the thermodynamic equilibrium of the system and is described by the *adsorption isotherm*. The isotherm data is obtained from laboratory tests. When testing a liquid medium, varying masses of GAC are placed in containers and each container is filled with the aqueous solution of interest. The containers are sealed and shaken for a period of time such that equilibrium conditions are achieved, which is typically on the order of days. The concentration of the contaminant in the liquid phase (C_e, where the subscript refers to the equilibrium conditions) is then measured. The mass of contaminant adsorbed, x, can be calculated using a mass balance.

These isotherms are often presented graphically as a plot of the ratio of mass of adsorbate adsorbed per mass of adsorbent versus the contaminant concentration in the fluid phase at equilibrium. Several researchers have mathematically described the curves that represent the data, the most common model suggested by Freundlich. The *Freundlich isotherm* is commonly applied to activated carbon adsorption in water and wastewater treatment processes and may be expressed as

$$\frac{x}{m} = X = KC_e^{1/n}$$

where x is the mass of adsorbate adsorbed, m is the mass of adsorbent, X is the mass ratio of the adsorbate adsorbed per mass of adsorbent, K is the Freundlich capacity factor, C_e is the equilibrium concentration of the solute in the aqueous phase after adsorption (mg/L), and $1/n$ is the Freundlich intensity factor. The constants K and $1/n$ in the Freundlich isotherm may be determined by plotting the adsorption data as ln (X) versus ln (C_e), and performing a regression using the linearized form of the equation, which may be expressed as

$$\ln X = \frac{1}{n} \ln C_e + \ln K$$

Values for K and $1/n$ for many compounds of industrial and environmental significance have been determined and may be found in tables in many reference texts.

Another model frequently used to describe the behavior of contaminants in the gas phase is the *Langmuir isotherm*, which is expressed in the environmental engineering section of the *FE Supplied-Reference Handbook* as

$$\frac{x}{m} = X = \frac{aKC_e}{1+KC_e}$$

where values for the empirical constants a and K may be regressed from the linearized form of the equation as follows:

$$\frac{m}{x} = \frac{1}{X} = \frac{1}{a} + \frac{1}{aK}\frac{1}{C_e}$$

It should be noted that the Langmuir constant K in the *FE Supplied-Reference Handbook* supplied equation is more commonly represented by the variable b in the literature and other texts on the subject.

Adsorption treatment units generally consist of a mass of adsorbent as a fixed bed in a column, sized to provide contaminant removal for a specified amount of time, with regularly scheduled regeneration cycles. A schematic representing the depth and travel of the adsorption zone in a column is presented in Figure 15.2. It would be useful to the student to refer to that figure during the discussion that follows. During the early stages of operation, contaminant mass is adsorbed onto the adsorbent closest to the fluid entrance region. As this adsorbent becomes saturated with the adsorbate, the contaminant travels further down the column until adsorbent with open active sites is encountered. In this way, the effluent fluid has a near-zero concentration of contaminant while this *adsorption zone* (Z_s) travels down the length of the column.

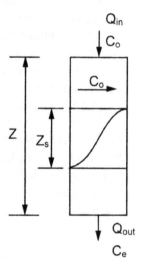

Figure 15.2 Schematic of adsorption column

When the majority of the column has been saturated with adsorbate (that is, the adsorption zone reaches the outlet of the column), low concentrations of contaminant begin to exit the adsorption column in a phenomenon called *breakthrough*. This concept can be visualized if we plot the effluent concentration of the contaminant species as a function of time (or throughput volume) as presented in Figure 15.3. The breakthrough point is typically defined as the time (or throughput volume) at which the contaminant concentration in the effluent stream reaches a value that is 5% of the inlet concentration. The column is considered *exhausted* when the effluent concentration of the contaminant reaches a value that is 95% of the influent concentration. Throughput volume is related to time by the volumetric flow rate of the fluid as $t = V/Q$.

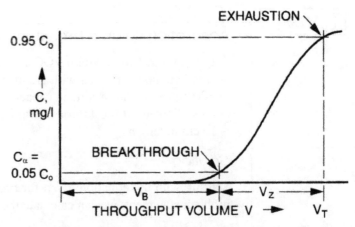

Figure 15.3 Effluent concentration profile (breakthrough curve) for adsorption column

Because it is undesirable to have any contaminant discharged from the column, it is common practice to run two or more columns in series. This allows for operation of the first column to the point of exhaustion, thereby taking advantage of all of the adsorbent capacity. After the first column is saturated with adsorbate, it is taken off-line, regenerated, and usually placed back in service as the last column in the series. Another configuration uses several columns in parallel, operated so as to stagger the regeneration cycles. The regeneration cycle may be triggered by the breakthrough of adsorbate or may be scheduled to precede breakthrough by some small time. In either configuration, it is beneficial to be able to predict the breakthrough time.

The breakthrough time t_B can be determined if the distance the adsorption zone travels is divided by the velocity of the adsorption zone V_S as it progresses down the column. Since the distance traveled is equal to the depth of the column Z minus the depth of the adsorption zone Z_S, this can be represented mathematically as

$$t_B = \frac{Z - Z_S}{V_S} \quad (15.4)$$

The depth of the adsorption zone may be calculated as

$$Z_S = Z \left[\frac{V_Z}{V_T - 0.5 V_Z} \right] \quad (15.5)$$

$$V_Z = V_T - V_B$$

where V_T and V_B are the volumes of fluid treated at exhaustion and breakthrough, respectively. The velocity of the adsorption zone can be expressed as

$$V_S = \frac{Q C_i}{X \rho_s A_c} = \frac{\dot{m}}{X \rho_s A_c} \quad (15.6)$$

where Q is the volumetric flow rate of the fluid (m³/s), C_i is the inlet concentration of the contaminant (g/m³), $\dot{m}$ is the mass flow rate of the contaminant species (g/s), X is the mass ratio of adsorbate to adsorbent (g/kg), ρ_s is the bulk density of the adsorbent as packed (kg/m³), and A_c is the cross-sectional area of the column (m²).

Example 15.6

Adsorption column breakthrough

A gas flows at 1 m³/s with a VOC contaminant at a concentration of 50 µg/m³. A 3-m-tall column is packed with granular activated carbon (GAC) at a bulk density of 300 kg/m³ and has a sorbent capacity of 442 mg VOC per kg AC. Determine the column diameter if breakthrough was observed in 11 days and the adsorption zone is 15 cm in length.

Solution

Starting with the equation given for breakthrough time and the data given in the problem statement, we can calculate the velocity of the sorption zone as follows:

$$t_B = \frac{Z - Z_S}{V_S} \Rightarrow V_S = \frac{Z - Z_S}{t_B} = \frac{3 - 0.15 \text{ m}}{11 \text{ days}} = 0.26 \frac{\text{m}}{\text{day}} \times \frac{1 \text{ day}}{(24)(60)(60) \text{ s}}$$

$$= 3.0 \times 10^{-6} \frac{\text{m}}{\text{s}}$$

Now we may use the definition for V_S to determine the diameter of the column by rearranging it and substituting $\pi D^2/4$ for cross-sectional area as

$$V_S = \frac{QC_i}{X\rho \frac{\pi}{4} D^2} \Rightarrow D = \left[\frac{4QC_i}{X\rho\pi V_S}\right]^{1/2}$$

Finally, the values given in the problem statement and calculated above may be substituted into this expression to determine column diameter as

$$D = \left[\frac{(4)\left(1 \frac{\text{m}^3 \text{ gas}}{\text{s}}\right)\left(50 \frac{\mu\text{g VOC}}{\text{m}^3 \text{ gas}}\right)}{\left(442 \frac{\text{mg VOC}}{\text{kg AC}}\right)\left(\frac{1000 \ \mu\text{g VOC}}{\text{mg VOC}}\right)\left(300 \frac{\text{kg AC}}{\text{m}^3}\right)\pi\left(3 \times 10^{-6} \frac{\text{m}}{\text{s}}\right)}\right]^{1/2} = 0.40 \text{ m}$$

Isotherm tests provide guidance on how well a contaminant will be removed by GAC, but care must be taken when using these results for the design of full-scale experiments. A better alternative is to conduct a series of column experiments on a pilot scale and use these results to design a full-scale GAC column.

THERMAL CONTROLS AND DESTRUCTION

As discussed previously, oxides of nitrogen (NO_x) are produced in combustion processes, primarily due to the oxidation of N_2 that is part of the combustion air. Control of NO_x can be accomplished by preventing their formation or through removal technologies employed downstream of the combustion zone. Several prevention technologies focus on the fact that N_2 is oxidized by O_2 during combustion processes where the temperature exceeds 1600 K. Decreased flame temperatures may be accomplished through lowering the amount of excess air fed, lean combustion with sufficient air to keep flame temperatures down, combustion gas recirculation (lowers O_2 concentration), or injection of water or steam. Facilities may also employ low NO_x burners, staged combustion, or secondary combustion technologies, which are practices where some of the fuel is combusted at a point downstream of the initial (primary) combustion zone.

Additional thermal technologies are used for the oxidation of unwanted species, as typified by the combustion of municipal solid waste, and are often classified under the broad category of *incineration* processes. For air pollutants, incineration is especially useful for CO and organic compounds but may be employed for any species that do not form postcombustion by-products that are more hazardous than the original material (for example, chlorinated hydrocarbons may produce free chlorine or HCl gas). A general form of the combustion reaction may be represented as

$$C_aH_bO_c + \left(\frac{4a+b-2c}{4}\right)(O_2 + 3.76N_2) \rightarrow$$

$$aCO_2 + \left(\frac{b}{2}\right)H_2O + \left(\frac{4a+b-2c}{4}\right)(3.76N_2)$$

While the above equation may allow the calculation of the theoretical amount of air required for complete combustion, the rate of air actually fed to the combustion chamber is in excess of the stoichiometric amount. This excess amount can range from 5% to 100% (that is, multiply the stoichiometric coefficient for air by a value between 1.05 and 2.00), depending on the amount of temperature control required for reduction of other flue-gas constituents. Thermal oxidation is generally classified as either *direct-flame incineration*, where compounds are combusted with or without supplemental fuels, or *catalytic oxidation*, where a catalytic compound on a ceramic substrate allows oxidation to take place at much lower temperatures (300–800°F).

Two common means to express the ability to remove a particular contaminant species using thermal oxidation systems are the *destruction and removal efficiency* (DRE) and the *combustion efficiency* (CE). The DRE is a measure of the mass loss of the principal hazardous compound as a result of oxidation and can be expressed as

$$DRE = \frac{\dot{m}_{in} - \dot{m}_{out}}{\dot{m}_{in}} \times 100\%$$

where $\dot{m}_{in}$ and $\dot{m}_{out}$ are the mass flow rates of the principal compound into and out of the unit, respectively. Federal performance standards require a DRE of 99.99% on one or more selected Principal Organic Hazardous Constituents during a supervised trial burn.

The CE is a measure of the degree of completion of the oxidation and can be expressed as a ratio of the amount of carbon dioxide emitted as a percentage of the total oxidized carbon (that is, the sum of carbon dioxide and carbon monoxide). This may be represented as

$$CE = \frac{[CO_2]}{[CO_2]+[CO]} \times 100\%$$

where bracketed quantities represent the dry volumetric concentrations (ppmv) of the indicated species.

PROBLEMS

15.1 An industrial stack with an effective stack height of 85 m emits SO_2 at a rate of 10 g/s. Estimate the maximum ground-level concentration at a distance 4 km downwind of the site if the wind is blowing at 5.5 m/s on a clear, sunny day.
    a. 6.1 µg/m³
    b. 19.5 µg/m³
    c. 35.2 µg/m³
    d. 48.6 µg/m³

15.2 Using the values for C, U, and H from Problem 15.1, estimate the maximum ground-level concentration of SO_2 that can be expected at any location downwind.
    a. 6.1 µg/m³
    b. 19.5 µg/m³
    c. 35.2 µg/m³
    d. 48.6 µg/m³

15.3 Acetylene (MW = 26) at close to atmospheric pressure is flowing in a pipeline at a temperature of 120°F and a manometer reading from a Pitot tube of 1 in. WG. What is the flow velocity?
    a. 1600 fpm
    b. 2600 fpm
    c. 3600 fpm
    d. 4600 fpm

15.4 Air flows through a duct system at the rate of 20,000 cfm. Resistance pressure (static and velocity pressure) is 1 in. WG. Estimate the brake horsepower of the fan assuming 70% fan efficiency.
    a. 2.21 bhp
    b. 3.15 bhp
    c. 4.50 bhp
    d. 5.36 bhp

15.5 An existing ESP at an MSW energy recovery facility operates at 156°C and has plate dimensions of 4 m × 10 m. How many plates are required to obtain a 95% capture efficiency for flyash particles (average migration velocity of 13 cm/s) that are 1 µm in diameter if the flow rate is 20,000 scfm?
    a. 8 plates
    b. 6 plates
    c. 4 plates
    d. 3 plates

15.6 A flue gas from a coal-fired power plant leaves the acid gas scrubbers at 55°C and at a flow rate of 50,000 scfm. Woven fabric filter bags are available from a manufacturer in units that contain 48 bags each, and each bag is 1 ft in diameter and 20 ft long. Determine the number of units required for the flow described.
   a. 5 units
   b. 6 units
   c. 7 units
   d. 8 units

15.7 A 60-ft-tall gas scrubber has a HTU of 1.6 ft and a scrubbing factor of 0.95. Determine the gas contaminant concentration in the inlet gas if the outlet gas has a concentration of 50 $\mu g/m^3$.
   a. 5600 $\mu g/m^3$
   b. 5000 $\mu g/m^3$
   c. 4400 $\mu g/m^3$
   d. 3800 $\mu g/m^3$

15.8 A serum bottle is filled with 100 mL of an aqueous solution containing 100 mg/L of toluene and an unknown quantity of activated carbon. After 48 hours, the equilibrium concentration of the aqueous phase is 20 mg/L of toluene. Assuming the Freundlich isotherm parameters for toluene and activated carbon are $K = 45$ mg/g and $1/n = 0.2$, respectively, determine the amount of activated carbon in the bottle.
   a. 50 mg
   b. 100 mg
   c. 150 mg
   d. 200 mg

15.9 A gas stream flowing at 1 m^3/s with a VOC contaminant at a concentration of 100 mg/m^3 is to be treated in an activated carbon adsorption column that is 0.5 m in diameter and 2 m long. Determine the breakthrough time (t_B) if the sorbent capacity is 120 g VOC per kg carbon. You may assume the packed density of the carbon is 300 kg/m^3 and the volumes treated at breakthrough and at capacity are 85 percent and 95 percent of the maximum sorbent capacity, respectively.
   a. 31 hr
   b. 35 hr
   c. 39 hr
   d. 43 hr

15.10 The destruction and removal efficiency for a particular VOC in an incinerator is 97.5 percent. Determine the mass flow rate of the VOC at the inlet if the volumetric flow rate is 15 m^3/s and the outlet concentration is 100 $\mu g/m^3$.
   a. 2.6 kg/day
   b. 3.6 kg/day
   c. 4.4 kg/day
   d. 5.2 kg/day

SOLUTIONS

15.1 a. First, use the "Atmospheric Stability under Various Conditions" table in the environmental engineering section of the *FE Supplied-Reference Handbook* to determine stability class given solar insolation and wind speed:

$U = 5.5$ m/s on a clear, sunny day; therefore assume stability class C

Using the plots in the environmental engineering section of the *FE Supplied-Reference Handbook* for vertical and horizontal standard deviation of a plume,

class C at $x = 4$ km $\Rightarrow$ $\sigma_y = 400$ m and $\sigma_z = 220$ m

Now, use the $C_{(x,0,0)}$ equation to determine the maximum ground level concentration as follows:

$$C_{(x,0,0)} = \frac{Q}{\pi \sigma_y \sigma_z U} \exp\left[-\frac{1}{2}\left(\frac{H}{\sigma_z}\right)^2\right]$$

$$= \frac{10 \frac{g}{s}}{\pi (400 \text{ m})(220 \text{ m})\left(5.5 \frac{m}{s}\right)} \exp\left[-\frac{1}{2}\left(\frac{85 \text{ m}}{220 \text{ m}}\right)^2\right]$$

$$C = 6.1 \times 10^{-6} \text{ g/m}^3 = 6.1 \text{ μg/m}^3$$

15.2 c. Since the stability class was determined in Solution 15.1, write the empirical expression and substitute the constants for atmospheric stability class C as

$$\left(\frac{CU}{Q}\right)_{max} = \exp\left[a + b \ln H + c (\ln H)^2 + d (\ln H)^3\right]$$

$$= \exp\left[-1.9748 - 1.998 \ln(85)\right] = 1.94 \times 10^{-5}$$

Now using the values for U and Q given in the problem statement, C may be estimated as

$$\left(\frac{CU}{Q}\right)_{max} = 1.94 \times 10^{-5} \Rightarrow C = \frac{\left(10 \frac{g}{s}\right)\left(1.94 \times 10^{-5} \text{ m}^{-2}\right)}{5.5 \frac{m}{s}}$$

$$= 3.524 \times 10^{-5} \frac{g}{m^3} = 35.2 \frac{\mu g}{m^3}$$

15.3 d. Density of the flowing gas is found to be as follows:

$$\frac{26}{379} \times \frac{460 + 60}{460 + 120} = 0.0615 \text{ lb per cu ft}$$

Relative density $= 0.0615/0.075 = 0.82$

From Figure 15.1 at a water pressure of 1 inch and a relative density of 0.82, we find the velocity to be 4600 fpm.

15.4 c. Air hp (output hp) = $\dfrac{\text{cfm} \times \text{TP}}{6356}$

$$\text{Air hp} = \dfrac{(20{,}000)(1 \text{ in } H_2O)}{6356} = 3.15 \text{ hp}$$

$$\text{Mechanical eff.} = \dfrac{\text{Air hp}}{\text{bhp}}$$

$$\eta_{\text{fan}} = \dfrac{3.15 \text{ hp}}{\text{bhp}} = 70\%$$

$$\text{bhp} = 4.5 \text{ hp}$$

15.5 b. Start with the ESP efficiency equation as

$$\eta = 1 - \exp\left[\dfrac{-WA}{Q}\right]$$

Since η and W are given in the problem statement, and since we are solving for A, we are left with calculating the actual gas flow rate. Given gas flow rate in scfm (standard ft³/min) assumes standard temperature (25°C) and standard pressure (1 atm). Assuming the pressure in the ESP is near atmospheric pressure, the only factor to correct for is operating temperature. However, gas phase calculations require the use of absolute temperature. This is accomplished as follows:

$$Q_{\text{actual}} = Q_{\text{scfm}} \dfrac{T_{\text{actual}}}{25°C} = (20{,}000 \text{ scfm}) \dfrac{(273.15 + 156)}{(298.15\ K)} = 28{,}789 \text{ acfm}$$

Next, we convert this into SI units as follows:

$$Q = 28{,}789 \dfrac{\text{ft}^3}{\text{min}} \times \dfrac{1 \text{ m}^3}{35.3 \text{ ft}^3} \times \dfrac{1 \text{ min}}{60 \text{ s}} = 13.6 \dfrac{\text{m}^3}{\text{s}}$$

Substituting all values into the efficiency equation, converting migration velocity to units of m/s and solving for A yields the required area as follows:

$$0.95 = 1 - \exp\left[\dfrac{-(0.13)(A)}{13.6}\right] \Rightarrow 0.05 = \exp\left[-0.00956\, A\right]$$

$$-2.996 = -0.00956\, A \Rightarrow A = 313.4 \text{ m}^2$$

Calculating the available surface area for each plate, we get

$$A = (4 \text{ m})(10 \text{ m})(2 \text{ sides}) = 80 \text{ m}^2$$

Therefore, the number of plates required is determined as

$$\dfrac{313.4 \text{ m}^2}{80 \text{ m}^2} = 3.92 \approx 4 \text{ plates}$$

15.6 d. Similar to Solution 5.5, we need to correct volumetric flow rate for gas temperature using the absolute temperature scale as follows:

$$Q_{actual} = Q_{scfm} \frac{T_{actual}}{0°C} = (50,000 \text{ scfm}) \frac{(273.15+55)}{(273.15)} = 60,068 \text{ acfm}$$

However, the tables in the environmental engineering section of the *FE Supplied-Reference Handbook* for estimating the amount of cloth needed for a baghouse have units of m³ per minute per m² of fabric. Converting flow rate to m³/min, we get

$$Q = 60,068 \text{ acfm} \times \left(\frac{1 \text{ m}^3}{35.3 \text{ ft}^3}\right) = 1702 \frac{\text{m}^3}{\text{min}}$$

Woven fabric requirements for flyash capture can be found in the table as 0.8 m³ per minute per m² of fabric. Therefore, the amount of fabric needed can be determined as

$$\frac{1702 \frac{\text{m}^3}{\text{min}}}{0.8 \frac{\text{m}^3}{\text{min} \cdot \text{m}^2}} = 2128 \text{ m}^2$$

Next, we need to determine the amount of fabric available in each unit as follows:

$$A_{unit} = \pi Dl \times 48 = \pi(1 \text{ ft})(20 \text{ ft})(48) = 3016 \text{ ft}^2 \times \left(\frac{1 \text{ m}}{3.28 \text{ ft}}\right)^2 = 280 \text{ m}^2$$

Finally, the number of units may be estimated as

$$\frac{2128 \text{ m}^2}{280 \text{ m}^2} = 7.6 \text{ units} \Rightarrow 8 \text{ units}$$

15.7 a. Using the data given for Z and HTU in the problem statement, we first need to calculate the number of transfer units as follows:

$$Z = (HTU)(NTU) = 60 = 1.6 \text{ NTU} \Rightarrow NTU = 37.5$$

Next, using the equation for NTU provided in the environmental engineering section of the *FE Supplied-Reference Handbook* and the data given for y_{out} and R in the problem statement, we can solve for y_{in} as follows:

$$NTU = \frac{\ln[(y_{in}/y_{out})(1-R)+R]}{1-R} = \frac{\ln[(y_{in}/50)(1-0.95)+0.95]}{1-0.95} = 37.5$$

$$1.875 = \ln[(y_{in}/50)(1-0.95)+0.95] \Rightarrow 6.52 = (y_{in}/50)(0.05)+0.95$$

$$\frac{6.52-0.95}{0.05} = (y_{in}/50) \Rightarrow y_{in} = 5570 \frac{\mu g}{\text{m}^3}$$

15.8 b. The Freundlich isotherm model can be found in the environmental engineering section of the *FE Supplied-Reference Handbook*. Substituting the values for K, C, and $1/n$ given in the problem statement allows us to solve for the concentration of toluene on the activated carbon as follows:

$$X = KC^{1/n} = \left(45\,\frac{mg}{g}\right)\left(20\,\frac{mg}{L}\right)^{0.2} = 81.9\,\frac{mg}{g}$$

To determine the amount of toluene associated with the carbon, we need to calculate how much toluene was in the aqueous phase initially and subtract the amount remaining in the aqueous phase at the end. This can be accomplished as follows:

$$\text{total toluene at beginning:} 100\,\frac{mg}{L} \times 0.1\,L = 10\,mg\,\text{toluene}$$

$$\text{total toluene at end:} 20\,\frac{mg}{L} \times 0.1\,L = 2\,mg\,\text{toluene}$$

$$\text{therefore, the carbon has } 10 - 2 = 8\,mg\,\text{toluene}$$

Now, we can estimate the amount of activated carbon in the bottle as follows:

$$\frac{8\,mg\,\text{toluene}}{81.9\left(\frac{mg\,\text{toluene}}{g\,\text{carbon}}\right)} = 0.0977\,g\,\text{carbon} = 97.7\,mg\,\text{carbon}$$

15.9 b. The expression for breakthrough time was given as

$$t_B = \frac{Z - Z_S}{V_S}$$

The variable Z was given in the problem statement. The variable Z_S needs to be evaluated and may be calculated using the following expressions found in the environmental engineering section of the *FE Supplied-Reference Handbook*:

$$Z_S = Z\left[\frac{V_Z}{V_T - 0.5V_Z}\right] \Rightarrow V_Z = V_T - V_B$$

The problem statement provided values for V_T and V_B as percentages of the system volume. Therefore, we may calculate V_T and V_B as follows:

$$V_{\text{column}} = \frac{\pi}{4}D^2H = \frac{\pi}{4}(0.5\,m)^2(2\,m) = 0.393\,m^3$$

$$V_T = (0.95)(0.393\,m^3) = 0.3734\,m^3$$

$$V_B = (0.85)(0.393\,m^3) = 0.334\,m^3$$

These can now be substituted into the expression for V_Z above to yield

$$V_Z = 0.3734\,m^3 - 0.334\,m^3 = 0.0394\,m^3$$

We can now solve for Z_S using these data and the expression above as follows:

$$Z_S = (2\,m)\left[\frac{0.0394\,m^3}{0.3734\,m^3 - (0.5)(0.0394\,m^3)}\right] = 0.223\,m$$

Now, using the expression for V_s given in the problem statement, we may write

$$V_s = \frac{QC_e}{X\rho_s A_c} = \frac{\left(1\,\frac{m^3}{s}\right)\left(100\,\frac{mg\,VOC}{m^3}\right)\left(\frac{1\,g\,VOC}{1000\,mg\,VOC}\right)}{\left(120\,\frac{g\,VOC}{kg\,AC}\right)\left(300\,\frac{kg\,AC}{m^3}\right)\frac{\pi}{4}(0.5)^2\,m^2} = 1.415\times 10^{-5}\,\frac{m}{s}$$

Finally, breakthrough time may be determined by substituting in the original expression as

$$t_B = \frac{2-0.223\,m}{1.415\times 10^{-5}\,\frac{m}{s}} = 125{,}583\,s \times \frac{1\,hr}{3600\,s} = 34.9\,hr$$

15.10 d. Starting with the equation for DRE given in the environmental engineering section of the *FE Supplied-Reference Handbook* and the value for DRE provided in the problem statement, we may write

$$\text{DRE} = \frac{\dot{m}_{in} - \dot{m}_{out}}{\dot{m}_{in}} = 0.975$$

Next, we can use the data in the problem statement to determine the mass flow rate out of the system as follows:

$$\dot{m}_{out} = QC_{out} = \left(15\,\frac{m^3}{s}\right)\left(100\,\frac{\mu g}{m^3}\right) = 1500\,\frac{\mu g}{s}$$

Substituting this into the DRE expression we can solve for mass flow rate into the system as

$$0.975 = \frac{\dot{m}_{in} - 1500}{\dot{m}_{in}} \Rightarrow 0.975\,\dot{m}_{in} = \dot{m}_{in} - 1500$$

$$\Rightarrow \dot{m}_{in} = 60{,}000\,\frac{\mu g}{s}$$

Now, convert this flow rate into units that are consistent with the answers given as follows:

$$\dot{m}_{in} = \left(60{,}000\,\frac{\mu g}{s}\right)\left(\frac{kg}{10^9\,\mu g}\right)\left(\frac{(60)(60)(24)\,s}{day}\right) = 5.184\,\frac{kg}{day}$$

CHAPTER 16

Solid and Hazardous Waste

OUTLINE

MUNICIPAL SOLID WASTE 704
Characterization

THERMAL TREATMENT OF MSW 705
Heating Value ■ Air Requirements and Combustion Efficiency ■ Combustion and Energy Recovery ■ Pyrolysis and Gasification

MSW COMPOSTING 714

LANDFILLS 717
Landfill Processes ■ Hydraulic Barriers and Leachate Collection ■ Landfill Biology and Gas Collection ■ Settlement

HAZARDOUS WASTE 730
Classification ■ TSD Facilities ■ Storage ■ Disposal ■ Biological Treatment ■ Physical-Chemical Treatment ■ Thermal Treatment

SITE REMEDIATION 740
Superfund Process ■ In-Situ vs. Ex-Situ ■ Biological Treatment ■ Physical-Chemical Treatment ■ Containment ■ Brownfields

RADIOACTIVE WASTE 748

PROBLEMS 750

SOLUTIONS 752

It is no accident that secure disposal is one of the least attractive options in the waste management hierarchy, just above direct release. The history of land disposal practices is replete with examples of negative environmental impacts due to everything from a lack of scientific understanding to outright criminal behavior. It is also a simple fact that nearly every human activity generates some quantity of *solid waste,* which can be defined as any unwanted and discarded materials that are solids or semisolids in their natural state. Often, other waste management options that are higher on the hierarchy may reduce the quantity of wastes through any of several processing methods. However, most of these still generate a final waste stream (usually solid) that requires further treatment and/or disposal.

MUNICIPAL SOLID WASTE

While the definition of solid waste above applies in all circumstances, the primary concern to most environmental engineers involves *municipal solid waste* (MSW). MSW is generally assumed to contain the solid wastes from residential, commercial, institutional, construction and demolition, and municipal services sources.

Characterization

There are two basic methods to determine total mass of MSW and waste generation rates. *Direct measurement* techniques weigh the vehicles that carry wastes as they enter and exit the dumping station, and are more accurate for the local community. However, extrapolation must be applied to very large, heterogeneous populations, which introduces the potential for substantial uncertainties. *Materials flow analyses* are based on mass balances performed at the generation source, prior to collection, and may offer good approximations for some sources, but fail to give a good total picture. Based on several studies, it is estimated that the average person in the United States generates 3.5 to 4.0 pounds of residential waste per day, which works out to be about 1400 pounds per year. If the other sources are added, total MSW generation in the United States is over 6 pounds per person per day, or approximately 2200 pounds per person per year.

Since the residential component of MSW accounts for 60 to 65 percent of the total, it is convenient to focus on this fraction when discussing composition. Typical components in residential MSW, with mass percentages in parentheses, include the following: paper (34%), yard wastes (18.5%), food wastes (9%), glass (8%), plastics (7%), cardboard (6%), steel cans (6%), other metals (3%), dirt and ash, etc. (3%), textiles (2%), wood (2%), aluminum (0.5%), rubber (0.5%), and leather (0.5%). Chemical properties of MSW are generally based on *proximate analysis,* which measures moisture content, volatile (combustible) matter, fixed carbon, and ash, or based on *ultimate analysis,* which provides values for percent C, H, O, N, S, and ash on a dry-weight basis. For a typical MSW stream, ultimate analysis indicates that the waste is 47% C, 6% H, 40% O, 1% N, 0.2% S, and 6% ash. Finally, *energy content* is a function of the types of combustible materials present in the waste stream, and will be discussed in more detail below.

There are several physical properties of MSW that are of interest to the environmental engineer, including the MSW *specific weight* (γ), which is often specified in units of lb/yd^3 and requires descriptive qualifiers such as *loose, as placed, uncompacted,* or *compacted.* Typical values for MSW under various scenarios are 200 to 300 lb/yd^3 loose, 500 lb/yd^3 in a compaction truck, 700 lb/yd^3 normally compacted in a landfill, and up to 1000 lb/yd^3 well-compacted in a landfill. Often it is desirable to know the *particle size distribution,* especially for material recovery facilities that process waste streams for the purpose of resource recovery. The amount of moisture is generally described by the *moisture content* (MC), and, while food and yard wastes can have a MC of 60 to 70 percent, the majority of the MSW mass has a MC of less than 10 percent, and the overall MC averages 20 to 25 percent for typical MSW. A related property is *field capacity,* which is a measure of the maximum amount of water that can be retained by the waste as placed in a landfill. While field capacity may be as high as 60 percent of the dry weight of the solids, this quantity is related to the overburden pressure (waste placed above), which reduces capacity. Water present in excess of the field capacity generally contributes to the discharge of landfill leachate.

Once generated, collection and transfer of the waste is generally accomplished through municipal or private *curb-side collection* from residences, or *container pick-up* from commercial establishments. The most common vehicle used for collection is the *compaction truck*, which allows for maximum time on route, minimizing lost time driving to and from the transfer station or landfill. *Inter-route transfer stations* can reduce the drive time of smaller collection trucks by consolidating MSW into larger carriers for transport to a distant processing facility or to the landfill. Much planning goes into the collection and inter-route transfer of MSW, as 70 to 85 percent of the total cost of MSW management is incurred in the collection phase. A *material recovery facility* (MRF) is used to segregate materials into fractions that can be recycled, composted, and/or combusted. Recycling efforts have been able to divert up to 20 percent of the U.S. waste stream, mostly paper and cardboard. While much has been done in the aluminum, steel, glass, and plastic recovery sectors, the fact that they comprise a small fraction of the total MSW stream means their removal does little to reduce the overall mass that requires final disposal.

THERMAL TREATMENT OF MSW

The majority of MSW is of a form that is amenable to certain transformation processes, due primarily to the high percentage of organic matter present in the waste. One of those technologies is *incineration,* or *municipal waste combustion* (MWC), which has developed from small, low-tech facilities in the 1970s to modern-day *waste-to-energy,* or *resource recovery,* power plants with stringent air pollution control devices. The attractiveness of MWC arises from the energy content of MSW, which averages 5000 Btu/lb as collected. Often, processing MSW removes inorganic (non-combustible) material and moisture, which can raise the energy content to 8000 to 9000 Btu/lb, while simultaneously reducing the potential for toxic substances to enter the atmosphere as flue-gas constituents. This value is comparable with many coals mined in the eastern United States.

Heating Value

The heat value of solid waste can be approximated using several methods. For example, the compositional analysis can be used to estimate the heat content as follows:

$$\text{Btu/lb} = 1238 + 15.6R + 4.4P + 2.7G \qquad (16.1)$$

where

R = plastics, % by weight on dry basis

P = paper, % by weight on dry basis

G = food waste, % by weight on dry basis

A second alternative is to estimate the heat value given the heat content of the various components of MSW. Sample data is provided in Table 16.1.

Table 16.1 MSW proximate analysis and energy content

Type of Waste	Proximate Analysis (% by weight)				Energy Content (Btu/lb)	
	Moisture	Volatiles	Fixed Carbon	Noncombustible	As Collected	Dry
Mixed food waste	70.0	21.4	3.6	5.0	1797	5983
Cardboard	5.2	77.5	12.3	5.0	7042	7428
Mixed paper	10.2	75.9	8.4	5.4	6799	7571
Newsprint	6.0	81.1	11.5	1.4	7975	8484
Mixed plastics	0.2	95.8	2.0	2.0	14,101	14,390
Textiles	10.0	66.0	17.5	6.5	7960	8844
Rubber	1.2	83.9	4.9	9.9	10,890	11,022
Yard wastes	60	30.0	9.5	0.5	2601	6503
Mixed wood	20.0	68.1	11.3	0.6	6640	8316
Residential MSW	21.0	52.0	7.0	20.0	5000	6250

Source: Adapted from Table 4-2, G. Tchobanoglous, H. Theisen, and S. A. Vigil, *Integrated Solid Waste Management* (McGraw-Hill, 1993).

Example 16.1

Calculating dry energy content

Confirm the value of the dry energy content of cardboard based on other information provided in Table 16.1.

Solution

The heat value of cardboard on an as-collected basis is 7042 Btu/lb. In other words, 1 lb of cardboard, including water and dry cardboard solids, is equivalent to 7042 Btu. From Table 16.1, cardboard has a moisture content of 5.2%; thus, every pound of as-collected waste contains 0.052 lb of water and 0.948 lb of dry cardboard. The heat value on a dry basis is simply:

$$\text{heat value}_{dry} = \frac{7042 \text{ Btu/lb}}{0.948 \text{ lb dry MSW}} = 7428 \text{ Btu/lb on a dry basis}$$

Example 16.2

Heat value of MSW

Consider a waste (20% moisture content) comprised of the following, with the percentages given on a dry weight basis:

Mixed paper	42%
Yard wastes	20%
Food wastes	9%
Glass	7%
Plastics	7%
Steel cans	5%
Other	10%

Use two methods to estimate the heat value of this waste per pound of as collected waste.

Solution

Determine the heat content for 1 lb of dry MSW by using a weighted average of each constituent's heat value.

heat content = (0.42 lb · 6799 Btu/lb) + (0.2lb · 2601 Btu/lb) + (0.09 lb · 1797 Btu/lb) + (0.07 lb · 14,101 Btu/lb)

= 4525 Btu/lb

The heat content per as collected pound of MSW is 4525 Btu/lb · (1 − 0.2), or 3620 Btu/lb.

Alternatively, the heat content can be determined from the Equation 16.1:

Btu/lb = 1238 + 15.6R + 4.4P + 2.7G

= 1238 + 15.6(7) + 4.4(42) + 2.7(9)

= 1556 Btu/lb

On a wet basis, the heat value is 1245 Btu/lb, which appears to be a severe underestimate.

If more precise estimates of the heat content of waste is desired, an *oxygen bomb calorimeter* (Figure 16.1) can be used. This calorimeter is used by placing a small sample of waste in a sealed sample holder and immersing the holder in a water bath. The sample is completely combusted within the sample holder, and the resulting release in energy causes the water bath temperature to increase.

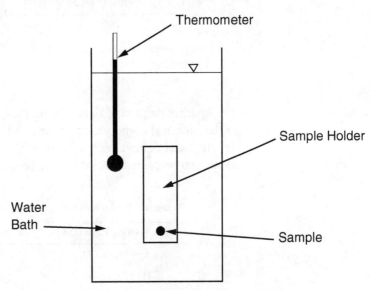

Figure 16.1 Oxygen bomb calorimeter

The change in temperature in the water bath can be related to the heat value (in units of cal/g) of the sample by Equation 16.2:

$$\text{heat value} = \frac{C_v \Delta T}{M} \quad (16.2)$$

where

C_v = heat capacity of the calorimeter (cal/°C)

ΔT = rise in temperature (°C)

M = mass of the unknown material (g)

Air Requirements and Combustion Efficiency

Incinerators require air to combust the organics, and the quantity of air required can be determined from stoichiometry, as illustrated in Example 16.3.

Example 16.3

Estimating air requirements for MSW combustion I

Assume that solid waste can be expressed by the chemical formula C_5H_{12}. Estimate the air requirements required to combust 10 tons per day of this waste.

Solution

Preliminary calculations include calculating the molecular weight of the waste (= 72 g/mol) and the total number of moles of waste:

$$\frac{10 \text{ ton}}{\text{day}} \frac{2000 \text{ lb}}{1 \text{ ton}} \frac{\text{kg}}{2.205 \text{ lb}} \frac{1000 \text{ g}}{\text{kg}} \frac{1 \text{ mol}}{72 \text{ g}} = 126{,}000 \text{ moles/day}$$

A balanced equation for the combustion of this MSW is:

$$C_5H_{12} + 8O_2 \rightarrow 5CO_2 + 6H_2O$$

Thus, every mole of waste requires eight moles of oxygen. The daily requirement of oxygen is $8 \times 126{,}000$ moles, or $1.0 \cdot 10^6$ moles/day.

Given that air is 21% oxygen, the daily requirements of air are

$$\frac{(1.0 \cdot 10^6 \text{ mol/day}) \cdot (32 \text{ g/mol})}{0.21} = 1.5 \cdot 10^8 \text{ g/day} = 339{,}000 \text{ lb/day}$$

Using a specific weight of air equal to 0.075 lb/ft³ yields a daily volume of $4.1 \cdot 10^6$ ft³/day.

Another method of determining the oxygen requirements requires knowledge of the chemical composition of the solid waste. By combining the data describing the chemical composition with stoichiometric requirements, the oxygen requirements can be estimated. Table 16.2 provides helpful information.

Table 16.2 Stoichiometric oxygen requirements

Governing Equation	1 gram of	...requires ___ g of O_2
$C + O_2 \rightarrow CO_2$	C	32/12
$2H_2 + O_2 \rightarrow 2H_2O$	H_2	32/4
$S + O_2 \rightarrow SO_2$	S	32/32.1

In practice, *excess air* is required to ensure complete combustion. Excess air also affects the composition and temperature of the flue gas.

Example 16.4

Estimating air requirements for MSW combustion II

A solid waste sample is characterized as follows:

Component	C	H	O	N	S
Weight percent	27	4	23	0.5	0.1

The remainder of the waste is comprised equally of water and inerts. Determine the air requirements, assuming 25% excess air.

Solution

Based on Table 16.2, the oxygen requirements, assuming 1 lb of waste, are:

Carbon: 0.27 lb · 32/12 = 0.72 lb O_2

Hydrogen: 0.04 lb · 32/4 = 0.32 lb O_2

Sulfur: 0.01 lb · 32/32.1 = 0.01 lb O_2

The total oxygen demand for combustion of these three elements is 1.05 lb. However, the pound of waste provides 0.23 lb of O_2, so the total requirement is 0.82 lb of O_2 per pound of waste.

Given that air is 21% oxygen, the air requirement per pound of waste is 0.82 lb/0.21, or 3.9 lb/lb of waste. At 25% excess air, the air requirements are 4.9 lb air/lb waste.

The efficiency of an MSW incinerator can be assessed by using an approach similar to the mass balance approach (Chapter 7). However, instead of auditing the mass entering and exiting the control volume, the energy entering and exiting the control volume will be tracked. The rate of energy entering the control volume must equal the rate of energy exiting the control volume (since energy cannot be created or destroyed). The energy leaving the control volume, which may be the physical boundaries of the incineration process, is either usable energy or wasted energy (for example, energy lost as heat to the surroundings).

The heat energy Q required to raise the temperature of a substance is related to the mass of the substance and the *specific heat* (sometimes referred to as the *heat capacity*):

$$Q = m \cdot c \cdot \Delta T \quad (16.3)$$

where

M = mass of the substance (kg)

c = heat capacity (e.g., kJ/kg · K)

ΔT = change in temperature (K)

The specific heat of water is 1.0 Btu/lbm · °F, or 4.19 kJ/kg · K.

The heat of vaporization is the energy required to transform a substance from its liquid form to its vapor form (or vice versa). The heat of fusion is the energy required to transform a substance from its solid form to its liquid form (or vice versa). The heat of fusion of water is 79.72 cal · g^{-1} or 334.5 kJ · kg^{-1}. The heat of vaporization of water is 2260 kJ · kg^{-1} or 539.8 cal · g^{-1}.

Example 16.5

Energy generated by MSW incineration

Estimate the energy available due to incinerating MSW after analyzing for losses due to heat loss in flue gas, heat loss in bottom ash, heat loss to radiation, and heat loss to vaporization of water. Assume that 20 tons of MSW are incinerated per day, and that the MSW has a composition similar to that for "residential MSW" in Table 16.1. Other information is provided as follows:

Flue gas	mass flow rate = 0.25 kg/s
	temperature = 300°C
	specific heat, C_{flue} = 1.0 kJ/kg/K
Ash	temperature = 800°C
	specific heat, C_{ash} = 0.84 kJ/kg/K
Radiation	assume 5% loss of energy
Water vaporization	heat of vaporization = 2575 kJ/kg

Solution

From Table 16.1, the moisture content is 21%, the noncombustible fraction is 20%, and the waste has an energy content of 6250 Btu/lb ($1.454 \cdot 10^7$ J/kg) on a dry weight basis (5000 Btu/lb on an as-collected basis). From this information, the total energy available is

$$(5000 \text{ Btu/lb}) \cdot (20 \text{ ton/day}) \cdot (2000 \text{ lb/ton}) = 2 \cdot 10^8 \text{ Btu/day}$$

$$= 2.4 \cdot 10^6 \text{ J/s}$$

The heat lost in the flue gas can be determined as

$$\frac{0.25 \text{ kg}}{\text{s}} \cdot \frac{1 \text{ kJ}}{\text{kg} \cdot {}^\circ\text{C}} \cdot 300^\circ\text{C} = 75{,}000 \text{ J/s}$$

To determine the energy lost in the ash, the generation rate of ash needs to be determined. Given 20 tons/day of MSW with 20% of the waste being noncombustible, the ash production is estimated to be 8000 lb/day, or 0.042 kg/s.

$$\frac{0.042 \text{ kg}}{\text{s}} \cdot \frac{0.84 \text{ kJ}}{\text{kg} \cdot {}^\circ\text{C}} \cdot 800^\circ\text{C} = 28{,}000 \text{ J/s}$$

The heat lost to radiation is

$$0.05 \cdot (2.4 \cdot 10^6 \text{ J/s}) = 1.2 \cdot 10^5 \text{ J/s}$$

The energy required to vaporize the water from the solids is the product of the heat of vaporization of the water and the mass of water. The daily mass of water in the MSW is 20% of 20 tons, or 8000 lb. The energy required is thus

$$(8000 \text{ lb/day}) \cdot (2575 \text{ kJ/kg}) \cdot (1 \text{ kg}/2.205 \text{ lb}) = 9.3 \cdot 10^6 \text{ kJ/day}$$

$$= 1.1 \cdot 10^5 \text{ J/sec}$$

These energy flows are shown in Exhibit 1. From this diagram, the energy available for extracting useful energy is

$$(2.4 \cdot 10^6 \text{ J/s}) - [(7.5 \cdot 10^4 \text{ J/s}) + (1.2 \cdot 10^5 \text{ J/s}) + (2.8 \cdot 10^5 \text{ J/s}) + (1.1 \cdot 10^5 \text{ J/s})]$$
$$= 2.1 \cdot 10^6 \text{ J/s.}$$

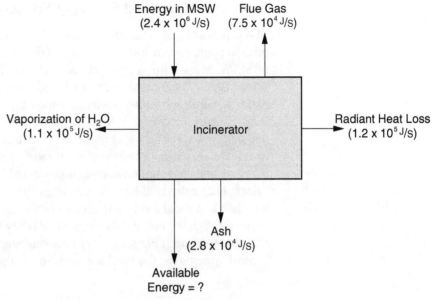

Exhibit 1 Energy flows in incinerator

Volume reduction is one significant benefit of MSW combustion. Typical values obtained are on the order of 90%.

Example 16.6

The effect of incineration on volume reduction

For the data of Example 16.2, estimate the volume reduction as a result of incineration, assuming an initial density of 300 lb/yd³ and a residue density of 950 lb/yd³.

Solution

This problem can be solved by referring to the noncombustible values of Table 16.1. The noncombustible fraction of the "other" portion of the MSW is assumed to be the same as for generic "residential MSW," and it is assumed that all water is removed. Thus, for every pound of waste, the mass of noncombustibles is:

$$\text{mass}_{\text{noncombustibles}} = (0.42 \text{ lb} \cdot 0.054) + (0.20 \text{ lb} \cdot 0.005) + (0.09 \text{ lb} \cdot 0.05) + (0.07 \text{ lb} \cdot 0.98) + (0.07 \text{ lb} \cdot 0.02) + (0.05 \cdot 0.98) + (0.1 \text{ lb} \cdot 0.2)$$

$$= 0.17 \text{ lb}$$

This represents an 83% reduction in mass.

The reduction in volume is estimated by considering the densities of the waste; that is, the volume of waste entering the facility corresponding to 1 lb of MSW is:

$$V_{\text{initial}} = 1 \text{ lb}/(350 \text{ lb/yd}^3)$$

$$= 2.9 \cdot 10^{-3} \text{ yd}^3$$

The final volume of ash is:

$$V_{\text{final}} = 0.17 \text{ lb}/(950 \text{ lb/yd}^3)$$

$$= 1.7 \cdot 10^{-4} \text{ yd}^3$$

The reduction in volume is $\dfrac{V_{\text{initial}} - V_{\text{final}}}{V_{\text{initial}}} = 94\%$.

Combustion and Energy Recovery

MWC processes are classified as mass-burn, where bulk waste is fed into the combustion zone, or refuse-derived fuel (RDF) fired, where the fuel is a processed form of MSW. Mass-burn systems allow MSW or RDF to travel on a sloped, *vibrating screen* (*grate*), and combustion air is fed from below, or overfed air is provided to material burning on *open hearths,* which are periodically cleared with hydraulic rams. Energy recovery systems generate steam for the production of electricity, and acid gas scrubbers and particulate control devices clean the gas stream prior to release. A cut-away view of a vibrating-grate mass-burn facility with energy recovery and air pollution control is presented in Figure 16.2. An alternative MWC system uses pelletized RDF in a *fluidized bed combustor* (FBC), which is a refractory-lined, vertical steel cylinder that is fed air from below. The velocity of the air rising through the column *fluidizes* the RDF pellets, similar to the action described in Chapter 14 on rapid sand filter backwashing. Energy recovery and air pollution control systems for the FBC are the same as for mass-burn MWC.

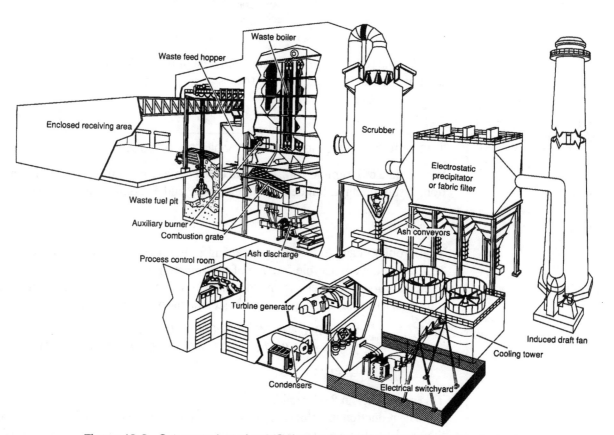

Figure 16.2 Cut-away view of an MSW mass-fired incinerator with energy recovery

MWC reduces the mass of MSW by approximately 80 percent and reduces the volume by up to 90 percent. However, the remaining 20 percent of the mass requires further consideration, usually land disposal. Particulate matter collected in the air pollution control (APC) system is comprised of *flyash* and scrubber byproduct, usually $CaSO_3$ or $CaSO_4$, and is often called *APC residue*. The partially-oxidized organic matter that exits the combustion chamber is quenched in a water pit and collected as *bottom ash*. Often, these two residuals are mixed together for disposal, and the mixture is given the name *combined ash*.

RDF is created by processing MSW before combustion to remove noncombustible materials and to create a smaller, more uniform particle size. RDF may be compressed to make densified RDF, which decreases transportation and storage costs. RDF offers the advantages of a more uniform heat source and requires

Thermal Treatment of MSW 713

less excess air. Excess air requirements are 50%, as compared to 100% or higher for mass burn plants. However, RDF use is limited in practice due to difficulties related to processing the RDF.

Example 16.7

RDF vs. mass burning

Consider the impact of using RDF rather than mass burning the MSW of Example 16.2 and Example 16.6; that is, determine the heat value and volume reduction of the waste, assuming that all noncombustible materials have been removed. Assume all other values remain the same.

Solution

The waste now consists of mixed paper, yard wastes, food wastes, and plastics. Thus, for one pound of the original waste, only 0.78 lb remains. The fraction of each of these materials in the RDF is:

Mixed paper	42% * 0.78 = 54%
Yard wastes	20% * 0.78 = 26%
Food wastes	9% * 0.78 = 12%
Plastics	7% * 0.78 = 9 %

The heat content of the RDF is

heat content = heat content of mixed paper + heat content of yard wastes + heat content of food wastes + heat content of plastic

$= (0.54 \cdot 6799 \text{ Btu/lb}) + (0.26 \cdot 2601 \text{ Btu/lb}) + (0.12 \cdot 1797 \text{ Btu/lb}) + (0.09 \cdot 14{,}101 \text{ Btu/lb})$

$= 5832 \text{ Btu/lb}$ (as compared to 4525 Btu/lb for Example 16.2)

To determine the change in volume, the mass of noncombustible materials needs to be determined.

$\text{mass}_{\text{noncombustibles}} = (0.54 \text{ lb} \cdot 0.054) + (0.26 \text{ lb} \cdot 0.005) + (0.12 \text{ lb} \cdot 0.05) + (0.09 \text{ lb} \cdot 0.02)$

$= 0.039 \text{ lb}$

This represents a 96% reduction in mass, which is much larger than the 83% reduction found in Example 16.6.

The volume of RDF entering the facility, corresponding to 1 lb of MSW is

$V_{\text{initial}} = 1 \text{ lb}/(350 \text{ lb/yd}^3)$

$= 2.9 \cdot 10^{-3} \text{ yd}^3$

The final volume of waste is:

$V_{\text{final}} = 0.039 \text{ lb}/(950 \text{ lb/yd}^3)$

$= 4.1 \cdot 10^{-5} \text{ yd}^3$

The reduction in volume is $\dfrac{V_{\text{initial}} - V_{\text{final}}}{V_{\text{initial}}} = 98.5\%$, which is greater than the volume reduction of 94% determined in Example 16.6.

Building an MSW incinerator has to overcome a number of hurdles:

- The regulatory requirements are extensive.
- Public resistance is often active and well organized.
- The cost of actual implementation can be very high. One infamous example is that of Detroit's MSW incinerator, which charges tipping fees more than five times the average cost of landfill tipping fees in Michigan.
- A variety of solid, liquid, and gaseous waste streams must be considered.
- Contaminants include waste heat, particulates, sulfur emissions, mercury, and dioxins, to name a few.

The contaminants created by MSW combustion can all be treated to acceptable levels, but such treatment comes at a cost. Contaminant concentrations can be reduced given aggressive source reduction and sorting practices. Proper operation of an incinerator can also decrease contaminant creation. For example, dioxin is thought to be created due to incomplete combustion of the waste, although the exact mechanisms of formation are not presently understood.

Pyrolysis and Gasification

Pyrolysis consists of heating the waste to high temperatures in the absence of oxygen. Thus it is not a combustion process. Pyrolysis is endothermic (that is, it requires an external heat source). The primary end products of pyrolysis are:

- Gases, including hydrogen, methane, carbon monoxide, carbon dioxide, and others
- A liquid fraction, which is a tar or oil stream that can be used as a synthetic fuel oil with further processing
- A solid fraction, or char

The oils have a heat value of about 9000 Btu/lb, and the gases can potentially have heat values of around 700 Btu/ft^3.

Pyrolysis is used widely for conversion of organic material to the useable end products and is used in the production of charcoal, but to date there has only been minimal application to the MSW field.

Gasification, unlike pyrolysis, is an exothermic reaction. Air or oxygen is used to partly combust the waste. The partial combustion, with stoichiometric proportions of air or slightly less (as compared to the excess air used in combustion), produces a combustible fuel gas that can be used to heat the remainder of the process to temperatures between 700°C and 900°C. Air emissions are typically lower of gasification processes than for combustion processes.

MSW COMPOSTING

Another technology that takes advantage of the high organic matter and moisture content of MSW is *composting*, where, depending on the feed composition, conditions are optimized for the biological transformation of MSW into a valuable soil amendment. In general, effective biological activity requires a moisture content in the range of 50 to 60 percent, a carbon-to-nitrogen ratio of 20–25:1, and a pH in the neutral range. The selection of the feed material often dictates the transformation percentage and the usefulness of the end product. Choices include yard waste only, the organic fraction of the segregated MSW, or commingled MSW.

As discussed in the review of sludge processing (Chapter 14), aerobic and anaerobic processes are available for the conversion of organic matter. Aerobic processes are simpler to operate; however, anaerobic processes produce methane, which may be used for energy production. This review will only cover the fundamentals of the more familiar aerobic process, which has the primary goal of volume reduction (~50%) and a secondary goal of compost production.

Aerobic biological processes require organisms, organic matter, oxygen, and water; and the complete biological transformation can be approximated by the following chemical reaction:

$$C_aH_bO_cN_d + \left(\frac{4a+b-2c+3d}{4}\right)O_2 \rightarrow aCO_2 + \left(\frac{b-3d}{2}\right)H_2O + dNH_3$$

This reaction assumes that the chemical composition given for the organic matter is comprised of the biodegradable fraction (i.e., plastics and other non-biodegradable compounds should not be included in the chemical formulation). Additionally, if sufficient time and oxygen are supplied, ammonia oxidizes to nitrate as follows:

$$NH_3 + 2O_2 \rightarrow H_2O + HNO_3$$

The source of the organisms and organic matter is the MSW, and, while some moisture is present, more is usually added during the processing. Oxygen is fed to the system by two possible methods. The *windrow* method utilizes large machinery to mix, or turn, a long pile with a triangular cross section to provide aeration throughout the mass. The *aerated static pile* employs an exhaust fan, connected to perforated pipe extending throughout the pile, to draw the necessary oxygen trough the pile from the surrounding air. In either case the processing time is 3 to 4 weeks; however, piles are allowed to cure for another 1 to 3 months without aeration or turning.

There are many benefits of composting, including keeping waste out of landfills. Compost is also an excellent soil conditioner with the ability to

- improve water drainage,
- increase water-holding capacity,
- improve nutrient-holding capacity,
- buffer pH,
- increase soil organic content, and
- reduce bulk density.

However, there are many challenges facing the implementation of commingled MSW recycling, including:

- The ability to find markets
- Odor problems
- The ability to create a uniformly high quality end product
- Inadequate design information

Composting is mediated by microorganisms, which need a carbon source, nutrients, moisture, oxygen, the proper particle size, and effective operating temperatures:

- *Carbon source.* Most MSW sources have adequate carbon sources, and when the MSW contains yard waste, the availability of carbon is not a problem. Most yard wastes (and food wastes) contain carbon that is highly bioavailable to the microorganisms, as compared to the carbon in lignins (from wood fibers), plastics, leather, and so on.

- *Nutrients.* Nitrogen is the most important nutrient of concern, and the ratio of carbon to nitrogen (C:N ratio) is considered critical to the success of the composting. C:N ratios of 20–30 are considered optimal. Nitrogen-rich feedstocks include food waste, yard trimmings, animal manures, and biosolids.

- *Moisture.* Moisture is also critical to the proper operation of a composting facility, and a moisture content between 50% and 60% is considered ideal. Lower moisture contents inhibit biological activity, while higher contents can lead to anaerobic conditions and the concomitant odors such as hydrogen sulfide.

- *Oxygen.* Oxygen is required for the aerobic degradation and, in practice, is supplied to the compost using a variety of means, as discussed below. The compost structure should have adequate void spaces to allow for proper transfer of oxygen through the pile.

- *Particle size.* the proper particle size allows for movement of oxygen through the compost. All other factors being equal, smaller particles tend to biodegrade much more rapidly than larger particles.

- *Temperature.* Figure 16.3 shows an idealized plot of temperature versus time in a large composting pile. As time progresses and the temperature varies, different types of microorganisms can flourish. In the mesophilic range, the most bioavailable organics are degraded and a corresponding rapid increase in temperature is found. The thermophilic range with temperatures between 40°C and 60°C provides near optimum conditions for a range of beneficial microorganisms. One important result of reaching the thermophilic range is that pathogen and weed seed destruction are achieved after three consecutive days at a temperature greater than 55°C. Following the thermophilic stage, the pile will enter the cooling/maturation phase. In reality, the actual temperature profile will vary significantly in response to turning of the pile.

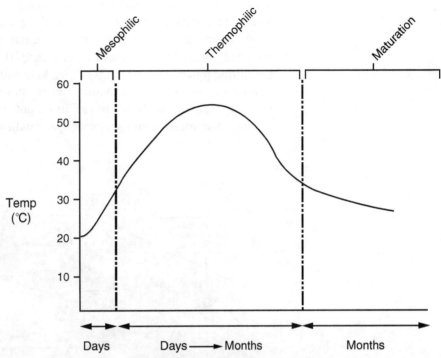

Figure 16.3 Temperature variation in a composting pile

Various means are available to create compost in practice and range from simple backyard composting piles to high-tech composting reactors. These composting methods vary one from another in terms of how the compost is mixed, how oxygen is provided, and how the proper moisture content and temperature are maintained.

The windrow method offers a relatively simple way of creating compost. The windrow width is usually twice the height, with widths between 14 feet and 16 feet common and heights between 4 feet and 8 feet. Optimal dimensions are large enough to allow for heat generation yet small enough for the transfer of oxygen. The piles are turned by specialized equipment. Turning adds oxygen to the pile and transfers material from the relatively cool outer portion of the pile to the relatively warm core. As a result of turning, the pile loses heat and may give off odors.

In the aerated static pile method, compost is placed in a large pile that is aerated. Aeration may be accomplished by a network of pipes underneath the pile hooked up to a blower or a vacuum. Air circulation provides relatively uniform oxygen concentrations throughout the pile, and the airflow can be controlled to maintain optimum temperatures. Since there is no turning, the outer layer of the pile may not reach proper temperatures to kill human pathogens and destroy weed seeds, and so a 6–12-inch "blanket" of finished compost may be added to help insulate the pile.

In-vessel composting systems are typically proprietary systems that allow for a high level of control of the composting system's temperature, moisture content, and oxygen level. Systems include drums, silos, digester bins, and tunnels. Some types of systems rotate to accomplish mixing, while other types are stationary with internal mixing mechanisms. Most in-vessel systems are continuous feed. Retention times are on the order of weeks. Minimal odors and, in some cases, no leachate are produced.

LANDFILLS

Final, secure disposal of solid and semisolid material generally occurs in a *landfill,* which is a repository that provides engineered control of the waste to minimize the potential negative environmental impact. Daily placement of waste is called a *cell* and is covered with a layer of soil, or similar material, to protect the waste from disturbance from water, wind, or animal, as well as to reduce problems with flies and odors. The development of a landfill begins with the site selection and permitting procedures, which is often a long and embattled process. Besides public opposition, major factors to address when considering a site include haul distance, location of major roadways or rail lines, potential for interaction with surface or groundwater sources, and adequate space with appropriate site geology.

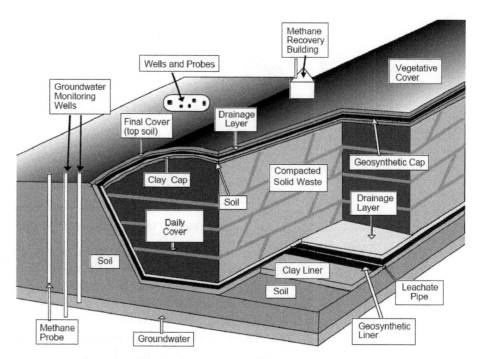

Figure 16.4 Schematic of a typical solid waste landfill

Source: P. Walsh and P. O'Leary, University of Wisconsin–Madison Solid and Hazardous Waste Education Center, reprinted from *Waste Age* Correspondence Course, 1991–1992.

The volume of a landfill may be approximated as that of a truncated pyramid (Figure 16.5). The volume of such a pyramid is shown in Equation 16.4:

$$V = \frac{h}{3}(a^2 + ab + b^2) \qquad (16.4)$$

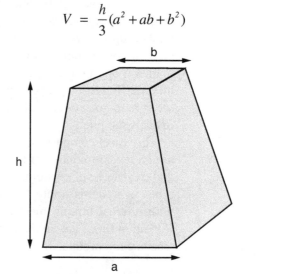

Figure 16.5 Truncated pyramid

Example 16.8

Preliminary landfill sizing

A landfill serves a single city of 100,000 residents. Waste is delivered five days per week to the landfill. The city has an effective recycling program that results in 30% of the waste being recycled. Waste is delivered at a density of 300 lb/yd³, and its density is expected to double due to compaction by equipment. Daily cells are to be approximately 3 ft thick. The area of the landfill that will accept waste is 12 acres. The landfill is able to contain a total depth of 36 ft of waste, including daily cover. Daily cover will consist of 6 in of soil. Final cover slopes will be 5:1.

a). Estimate the time for the landfill to be filled

b). Investigate the impact of using a removable textile product as daily cover

c). Investigate the impact of increasing recycling rates to 40%

Solution

a). Given a 30% recycling rate and assuming a MSW generation rate of 4 lb/day and that one-fifth of the town is serviced by waste pickup on each weekday, the daily volume of waste accepted at the landfill is

$$\frac{20{,}000 \text{ people}}{\text{day}} \cdot \frac{4 \text{ lb}}{\text{person}} (1 - 0.30) \frac{1 \text{ yd}^3}{300 \text{ lb}} = 187 \text{ yd}^3/\text{day}$$

Consequently, the volume of waste in one cell after compaction is 93.5 yd³/day, given the doubling of the waste's density.

The volume of daily cover added is the product of the area corresponding to the daily cell and the thickness of the daily cover layer. The area of the landfill corresponding to a 3-ft lift is

$$\frac{93.5 \text{ yd}^3/\text{day} \cdot 27 \text{ ft}^3/\text{yd}^3}{3 \text{ ft}} = \frac{842 \text{ ft}^2}{\text{day}}.$$

Thus, the volume of daily cover is 0.5 ft · 842 ft²/day = 421 ft³/day, or 15.6 yd³/day. Alternatively, this number could be arrived at by dividing the volume of compacted waste by six, since the daily cover amounts to one-sixth of the thickness of the daily lift.

The total volume of landfill airspace consumed in one day is 109 yd³ (93.5 yd³ + 15.6 yd³). The total volume of landfill available is found by determining the volume of a truncated pyramid. We will assume that the 12-acre site is a perfect square (723 feet on a side). Thus, the value a in Equation 16.4 is equal to 723 feet. Given the 5:1 side slopes, the value b in Equation 16.4 can be obtained by geometry to be equal to 468 ft. Thus, the volume of the truncated pyramid is

$$V = \frac{36 \text{ ft}}{3}(723 \text{ ft}^2 + 723 \text{ ft} \cdot 468 \text{ ft} + 468 \text{ ft}^2) = 480{,}000 \text{ yd}^3$$

Therefore, the landfill can accept waste for

$$\frac{480{,}000 \text{ yd}^3}{109 \text{ yd}^3/\text{day}} \cdot \frac{1 \text{ week}}{5 \text{ days}} \cdot \frac{1 \text{ year}}{52 \text{ weeks}} = 17 \text{ years}$$

b). Using the removable cover will decrease the daily consumption of air space from 109 yd³/day to 93.5 yd³/day. As a result, the landfill will be able to accept waste for about 20 years.

c). Using higher recycling rates will reduce the daily compacted cell volume to 80 yd³/day. In combination with 6 in of daily cover (or 13.3 yd³), the daily consumption of air space is 93.3 yd³/day, which is equivalent to the increase in available volume resulting from using a removable daily cover.

Landfill Processes

Stabilization of wastes in a landfill undergoes five distinct phases. These phases differ in terms of the type of physical, chemical, and biological processes taking place within the landfill. Consequently, the rate and quality of the leachate and landfill gas generated also varies with time. These five phases are described following, and the variation in leachate and landfill gas quantity and quality as a function of the phase are illustrated in Figure 16.6.

1. *Phase I* is the initial adjustment phase, or *lag phase*. It is during this phase that moisture begins to accumulate and the oxygen entrained in freshly deposited solid waste begins to be consumed.

2. *Phase II* is the transition phase in which the moisture content of the waste has increased such that the field capacity is exceeded. Microbial processes change from an aerobic to anaerobic environment as oxygen is depleted. Detectable levels of volatile organic acids (VOAs) and an increase in the chemical oxygen demand (COD) are noted in the leachate.

3. *Phase III* is the acid forming stage. Acidogenic bacteria convert the VOAs from Phase II, resulting in lower pH. The lower pH solubilizes metal species from the waste into the leachate. This phase is also characterized by peak COD levels in leachate.

4. *Phase IV* is the methane fermentation phase in which methanogenic bacteria convert acid compounds produced in earlier phases to methane and carbon dioxide gas. This phase marks a return to more neutral pH conditions and a corresponding reduction in the solubilization of metals in the leachate.

5. *Phase V* is the maturation phase, and biodegradable matter and nutrients become limiting. Landfill gas production drops, and the leachate strength is much lower and less variable than in previous stages. Degradation of organics continues at a much slower rate.

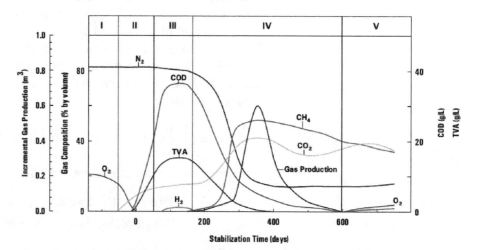

Figure 16.6 Progression of waste stabilization in landfills

Source: F. G. Pohland and S. R. Harper, *Critical Review and Summary of Leachate in Gas Production from Landfills*, EPA/600/2-86/073 (U.S. EPA, 1986).

Hydraulic Barriers and Leachate Collection

The construction of the landfill begins with the excavation of soil and the placement of *hydraulic barriers* (*liners*) and *leachate collection* systems. Liners may be constructed of natural materials, such as clays or soils with high clay content that are properly graded and compacted to a thickness of 2 to 4 feet, or through the use of synthetic materials, such as high density polyethylene (HDPE) and polypropylene (PP) plastic materials that have a thickness of 40 to 80 mils (1 mil = 0.001 inch). Liners that are composed of one layer of natural material on the bottom and one layer of synthetic material on top are called *composite* liners, and it is not uncommon to use a double-composite liner system, as seen in Figure 16.7. Leachate collection occurs through perforated pipe that is placed on top of the liner and covered by 1 to 2 feet of coarse sand or fine gravel. The layer of sand that is between the liners in a double-composite system is often called the *leakage detection layer*, as any water collected in this layer would indicate the probable failure of the synthetic liner above.

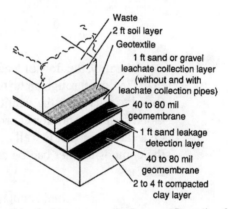

Figure 16.7 Possible double-composite liner configuration for an MSW landfill

Compacted clay liners act as hydraulic barriers due to very small permeability coefficients and effective porosities. Determination of the *breakthrough time* (t) for leachate through the liner, measured in years, may be expressed as

$$t = \frac{d^2 \eta}{K(d+h)}$$

where d is the thickness of the clay liner (ft), η is the effective porosity, K is the coefficient of permeability (ft/yr), and h is the hydraulic head (ft). Typical values for η range from 0.1 to 0.3, while K is generally in the range of 1.0 to 0.01 ft/yr (10^{-6} to 10^{-8} cm/s).

Geomembranes, or flexible membrane liners (FML), are required, in conjunction with clay, as a liner material for landfills. They have a very low permeability, but leaks are possible through seams, pinholes, and defects arising from the manufacturing process or due to accidental penetrations occurring during construction. Permeability rates can be estimated using Darcy's law (Chapter 13) or by applying the Bernoulli equation to a hole and treating the hole as an orifice. This equation is also known as Torricelli's equation:

$$Q = C_d \cdot A \cdot (2 \cdot g \cdot h)^{0.5} \tag{16.5}$$

where

C_d = discharge coefficient (typically 0.6)

A = area of hole, that is, of a defect in the membrane (L^2)

g = acceleration due to gravity (L · T^{-2})

h = head of water over hole (L)

Using this simple relationship, estimated flows through a composite liner as a function of hole size and hole density (holes/acre) can be calculated, and values are shown in Table 16.3. The number of holes per acre varies between 1 and 30, corresponding to high quality control and low quality control, respectively.

Table 16.3 Calculated leakage rate through a geomembrane

Size of Hole (cm²)	Number of Holes (hole/acre)	Flow Rate (gal/acre/day)
0.1	1	330
0.1	30	10,000
1	1	3300
1	30	100,000
10	1	33,000

In reality, the flow of leachate through a landfill liner will be more complicated than the flow through an orifice. The leakage through a composite system has been found to vary with the quality of contact between the geomembrane and the clay. Good contact conditions correspond to few waves or wrinkles in the geomembrane on a clay layer that is well compacted and has a smooth surface. The flow through the composite liner can be estimated from equations 16.6 and 16.7 (Quian et al., 2002):

$$\text{"Good" contact conditions: } Q = 0.21 \, A^{0.1} \cdot h^{0.9} \cdot k_s^{0.74} \quad (16.6)$$

$$\text{"Poor" contact conditions: } Q = 1.15 \, A^{0.1} \cdot h^{0.9} \cdot k_s^{0.74} \quad (16.7)$$

where

Q = leakage rate through membrane (m³/s)

A = area of hole in geomembrane (m²)

h = liquid head on top of the geomembrane (m)

k_s = hydraulic conductivity of the low-permeability soil component of the composite liner (m/s)

Typical values of k_s are 10^{-7} to 10^{-8} cm/s.

Example 16.9

Estimating landfill liner leakage rates

A 20-acre landfill liner is placed with very good quality control such that the FML and low permeability soil (10^{-8} cm/s) are in very good contact, and there are two 0.1-inch diameter holes per acre. Compare the leakage rate using Torricelli's equation to the leakage predicted by the equations above for composite liners.

Solution

From Torricelli's equation (Equation 16.5) and assuming a leachate head of 6 inches:

$$Q_{hole} = C_d \cdot A \cdot (2 \cdot g \cdot h)^{0.5}$$

$$Q_{hole} = 0.6 \frac{\pi (0.1 \text{ in}/12)^2}{4} \left[2 \cdot 32.2 \frac{\text{ft}}{\text{s}^2} \cdot 0.5 \text{ ft} \right]^{0.5} = 1.86 \cdot 10^{-4} \text{ cfs} = 120 \text{ gal/day}$$

$$Q_{total} = 120 \text{ gal/day/hole} \cdot 2 \text{ hole/acre} \cdot 20 \text{ acre}$$

$$= 4800 \text{ gal/day}$$

Considering the composite liner with good contact conditions, we will use Equation 16.6.

$$Q_{hole} = 0.21 \, A^{0.1} \cdot h^{0.9} \cdot k_s^{0.74}$$

This empirical relationship must use the units specified in the equation definition. Thus,

$A = \dfrac{\pi (0.1 \text{ in}/12)^2}{4} = 5.5 \cdot 10^{-5} \text{ ft}^2 = 5.1 \cdot 10^{-6} \text{ m}^2$

$h = 6 \text{ in} = 0.152 \text{ m}$

$k_s = 10^{-8} \text{ cm/s} = 10^{-10} \text{ m/s}$

Inserting these values into the above equation yields:

$$Q_{hole} = 4.5 \cdot 10^{-10} \text{ m}^3/\text{s}$$

$$= 0.01 \text{ gal/day}$$

$$Q_{total} = 0.01 \text{ gal/day/hole} \cdot 2 \text{ hole/acre} \cdot 20 \text{ acre}$$

$$= 0.4 \text{ gal/day}$$

The EPA defines leachate as the liquid that has passed through or emerged from solid waste and contains dissolved, suspended, or immiscible materials removed from the solid waste. The leachate generated in a landfill must be collected to minimize leakage from the landfill. Recall that the rate of leakage is directly related to the head of leachate over the landfill liner, and an effective leachate collection system minimizes this head. The goal of the system is to keep the leachate head less than 12 inches (30 cm).

A schematic of a typical leachate collection system is provided in Figure 16.8.

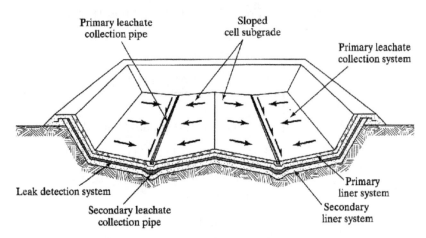

Figure 16.8 Leachate collection system

Source: X. Qian, R. M. Koerner, and D. H. Gray, *Geotechnical Aspects of Landfill Design and Construction*, © 2001. Reprinted by permission of Pearson Education, Inc., Upper Saddle River, N.J.

When designing a leachate collection system, the liner should be sloped at a minimum slope of 2%, and collection pipes should be sloped at a minimum slope of 1%. Increasing the slopes prevents accumulation in localized depression area but leads to more excavation and possibly loss of volume for placement of waste.

When leachate flows across a liner, it forms a *leachate mound*, as shown in Figure 16.12. The goal of the leachate collection system is to keep h_{max} below 12 inches (30 cm). Equation 16.8 estimates the pipe spacing based on the value of h_{max}. Equation 16.8 is most commonly used to find the pipe spacing L for a given leachate collection system configuration.

$$h_{max} = \frac{L}{2}\left(\frac{q}{K}\right)\left[\frac{K\tan^2\alpha}{q} + 1 - \frac{K\tan\alpha}{q}\left(\tan^2\alpha + \frac{q}{K}\right)^{1/2}\right] \qquad (16.8)$$

where

L = distance between collection pipes (L)

q = vertical inflow per unit horizontal area (L/T)

K = hydraulic conductivity of drainage layer (L/T)

α = slope of liner (L/L)

The flow into the drainage layer can be estimated using a model such as the HELP (Hydrologic Evaluation of Landfill Performance) model and can also be approximated using the water budget technique.

Landfills

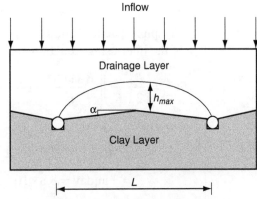

Figure 16.9 Leachate mound

Source: U.S. Environmental Protection Agency, *MSW Landfill Criteria Technical Manual*, 1993.

Example 16.10

Analysis of a leachate collection system

The preliminary layout for a leachate collection system is provided in Exhibit 2. (Exhibit 2 is simply a schematic, as the number of collection pipes are not known until the spacing is determined.)

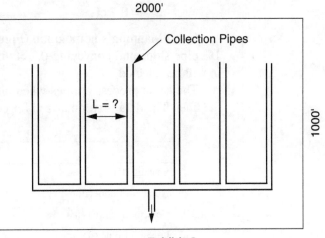

Exhibit 2

Compare the length and size of laterals required for a liner cross slope of 2% slope vs. a cross slope of 4%, given the following information:

Drainage layer hydraulic conductivity = 10^{-2} cm/s

Worst-case percolation rate = 3 in/day

Pipe slope = 1%

Solution

The pipe spacing, length, and diameter will first be estimated for a 2% cross slope of the liner using Equation 16.8.

$$h_{max} = \frac{L}{2}\left(\frac{q}{K}\right)\left[\frac{K\tan^2\alpha}{q} + 1 - \frac{K\tan\alpha}{q}\left(\tan^2\alpha + \frac{q}{K}\right)^{1/2}\right]$$

where

h_{max} = 30 cm

q = 3 in/day = $8.8 \cdot 10^{-5}$ cm/s

K = 10^{-2} cm/s

α = 0.02

These values yield a pipe spacing of 82 m, or 270 ft. Thus, a total of eight collection pipes will be required, separated by 250 ft spacing. The pipes on the ends will be placed 125 ft from the outer face of the landfill. A total pipe length of approximately 8000 ft will be required.

The flow in each pipe can be estimated based on the contributing land area. For each collection pipe, the contributing area is 1000 ft × 250 ft, or $2.5 \cdot 10^5$ ft². The flow is the product of the area and the percolation rate:

Q_{max} = 3 in/day · ($2.5 \cdot 10^5$ ft²) · (1 ft/12 in) · (1 day/86,400 s)

= 0.7 cfs

From Manning's nomograph (Figure 11.16), the diameter corresponding to a 1% pipe slope and conveying 0.7 cfs is between 6 in and 8 in, so a diameter of 8 in will be selected.

The same process applies when analyzing the impact of a 4% cross slope on the liner. The data is summarized for comparison's sake in the following table:

	2% cross slope	4% cross slope
Spacing (ft)	270 (use 250)	311 (use 300)
Number of collection pipes	8	7
Contributing area (ft²)	$2.5 \cdot 10^5$	$3 \cdot 10^5$
Length of pipe (ft)	8000	7000
Q_{max} (cfs)	0.7	0.87
Diameter (in)	8	8

Upon reaching placement capacity, the entire landfill is fitted with the *final cover* or *cap*, which is structured in a similar fashion to the liner and leachate collection system, as seen in Figure 16.10. This is usually covered with 2 to 5 feet of soil and seeded to encourage shallow-root vegetative cover, which decreases and slows surface water runoff, thereby minimizing erosion. While computer programs have been developed to assist in the estimation of water percolation through soil cover, a simple water balance may be performed for rough approximation, and may be expressed as

$$\Delta S_{LC} = P - R - ET - PER_{SW}$$

where ΔS_{LC} is the change in the storage of water in the soil cover (inches), P is the precipitation (inches), R is the runoff (inches), ET is the evapotranspiration loss (inches), and PER_{SW} is the amount of water that percolated through the cover and into the compacted waste (inches).

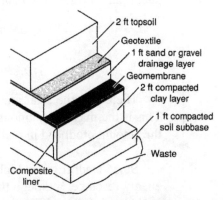

Figure 16.10 Example of possible final cap for a MSW landfill

Landfill Biology and Gas Collection

As with composting, biological activity within the landfill converts the organic matter through aerobic and anaerobic processes. Available oxygen is depleted quickly, and the majority of the biological activity is anaerobic, primarily through a process called *methanogenesis,* meaning methane creation. Landfill gas is roughly a 50-50 mix by volume of methane and carbon dioxide, with other trace gases such as non-methane organic compounds (NMOCs) and H_2S (responsible for the "rotten egg" odor). The transformation may be expressed chemically as

$$C_aH_bO_cN_d + \left(\frac{4a-b-2c+3d}{4}\right)H_2O$$

$$\rightarrow \left(\frac{4a+b-2c-3d}{8}\right)CH_4 + \left(\frac{4a-b+2c+3d}{8}\right)CO_2 + dNH_3$$

For unknown organic compositions, rapidly biodegradable material (e.g., food wastes, paper, cardboard, and yard wastes that do not include large stumps or branches) can be estimated to have the composition $C_{70}H_{110}O_{50}N$, while other biodegradable matter can be estimated to have the composition $C_{20}H_{30}O_{10}N$.

Example 16.11

Estimating gas production rates

Determine the gas production per pound of MSW, given an MSW composition of $C_{20}H_{30}O_{10}N$.

Solution

The governing equation is

$$C_{20}H_{30}O_{10}N + 33/4\,H_2O \rightarrow 87/8\,CH_4 + 73/8\,CO_2 + NH_3$$

The molecular weight of the MSW is 444 g/mol [(20 · 12) + (30 · 1) + (10 · 16) + (1 · 14)]. Consequently, 1 lb of MSW is equivalent to approximately 1 mole:

$$1\,lb\,\frac{1\,kg}{2.205\,lb}\,\frac{1000\,g}{1\,kg}\,\frac{1\,mol}{444\,g} = 1.02\,mole$$

Thus, every pound (or mole) of MSW produces 10.9 moles of methane and 9.1 moles of carbon dioxide. Assuming that methane and carbon dioxide act as ideal gases (22.4 L/mol), the volume of each gas can be determined:

$$10.9 \text{ mol CH}_4 \frac{22.4 \text{ L}}{\text{mol}} \frac{1 \text{ ft}^3}{28.3 \text{ L}} = 8.6 \text{ ft}^3$$

$$9.1 \text{ mol CO}_2 \frac{22.4 \text{ L}}{\text{mol}} \frac{1 \text{ ft}^3}{28.3 \text{ L}} = 7.2 \text{ ft}^3$$

Therefore, the total amount of gas generated per pound of MSW is 15.8 ft³. This is the maximum amount possible theoretically and is 4–16 times larger than the amount produced in reality.

Since landfill gas is high in methane content, it is now considered a potential energy resource, although this has not always been the case. While it was known in the past that landfill gas would be produced, it was usually allowed to diffuse through the mass and escape to the atmosphere. Current federal air regulations require any landfill producing greater than 50 mg/yr of NMOCs to collect gas from every area of the landfill where it can be produced. The movement of gas through the permeable soil cover may be estimated as

$$N_A = -\frac{D \eta^{4/3} \left(C_{\text{atm}}^A - C_{\text{fill}}^A\right)}{L}$$

where N_A is the gas flux of compound A (g/s per cm²), D is the diffusion coefficient (cm²/s), η is the dry soil porosity (unitless), C_{atm}^A and C_{fill}^A are the concentrations of compound A in the atmosphere (top of the cover) and in the landfill (bottom of the cover), respectively (g/cm³), and L is the depth of the soil cover (cm). Typical values of D are 0.20 cm²/s for CH_4 and 0.13 cm²/s for CO_2.

With the advent of modern landfill technology and low-permeability liner construction, it was discovered that extreme pressures may build up inside the landfill, eventually causing a blowout of the containment system. Prevention of this occurrence was usually addressed through the drilling of vent wells into the closed landfill, with disastrous consequences at times, where landfill gas was flared (open burning as discharged) to the atmosphere. In addition, the drilled vent wells had the unfortunate effect of placing breeches in the cover, which increases the potential for water infiltration and subsequent leachate generation if not sealed properly. A schematic of a possible gas collection system is presented in Figure 16.11.

One method of addressing the problem is the installation of perforated pipe in the horizontal plane along the periphery of the landfill during MSW placement, which is connected to a collection manifold that carries landfill gas out of the enclosed space through a single well. Gas collection may be passive, where the pressure developed inside the landfill finds its own way to the low-pressure outlet pipe, or through active means, where a small vacuum is drawn on the system to assist in the flow of gas.

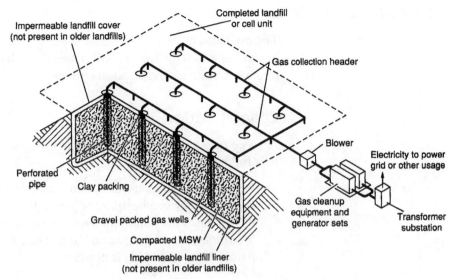

Figure 16.11 Gas recovery system for an MSW landfill

Settlement

One consideration for post-closure maintenance of landfills is the expectation of *settlement,* which arises from the conversion of solid mass to landfill gas that is allowed to exit the system. This loss of mass, coupled with the potential for additional compaction of placed MSW by the layers of material above, generally results in the settling of the landfill. This may cause damage to gas collection systems and/or generate breeches in the liner/cover system, and may subsequently interfere with the ability to use the property after closure. Studies have shown that the amount of settlement varies with the extent of compaction at placement, but in most cases, 90 percent of the settlement occurs in the first 5 years. The impact of *overburden pressure,* which is the additional compaction gained by placing MSW on top of previously placed material, may be estimated as

$$SW_p = SW_i + \frac{p}{a + bp}$$

where SW_p and SW_i are the specific weights of the MSW at pressure p and at initial placement, respectively (lb/yd³), p is the overburden pressure (psi), and a and b are empirical constants. Initial specific weight is often assumed to be in the range of 800 to 1000 lb/yd³, while the maximum specific weight is assumed to be in the 1750 to 2150 lb/yd³ range. This relationship may also be presented graphically for different values of SW_i, as seen in Figure 16.12.

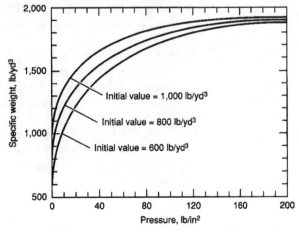

Figure 16.12 Effect of overburden pressure MSW specific weight

HAZARDOUS WASTE

The existence of hazardous waste in our environment has led to many deleterious health impacts. This section will define hazardous wastes and discuss a variety of treatment, storage, and disposal options that exist.

Classification

A waste may be defined as hazardous if it is a *listed* waste, or if it has one of the four following characteristics:

1. *Ignitability.* Ignitable wastes can create fires under certain conditions, are spontaneously combustible, or have a *flash point* less than 60°C (140°F). The flash point is the minimum temperature at which a liquid gives off a vapor in sufficient concentration to ignite when a source of ignition (sparks, open flames, cigarettes, etc.) is present.

2. *Corrosivity.* Corrosive wastes are acids or bases (pH less than or equal to 2 or greater than or equal to 12.5) that are capable of corroding metal containers, such as storage tanks, drums, and barrels.

3. *Reactivity.* Reactive wastes are unstable under normal conditions. They can cause explosions, toxic fumes, gases, or vapors when heated, compressed, or mixed with water.

4. *Toxicity.* Toxic wastes are harmful or fatal when ingested or absorbed (for example, containing mercury, lead, etc.). When toxic wastes are land disposed, contaminated liquid may leach from the waste and pollute groundwater. Toxicity is defined through a laboratory procedure called the Toxicity Characteristic Leaching Procedure (TCLP).

The TCLP test is a laboratory procedure designed to simulate the leaching a waste will undergo if disposed of in a sanitary landfill. In this procedure, a sample of a waste is extracted with an acetic acid solution. The extract obtained from the TCLP test (the *TCLP extract*) is analyzed to determine if any of the thresholds established for the 40 Toxicity Characteristic (TC) constituents have been exceeded.

Alternatively, a waste can be defined as a hazardous waste if it is *listed*. The U.S. Environmental Protection Agency (U.S. EPA) has defined four lists: F-list, K-list, P-list, and U-list.

- *F-list (nonspecific source wastes).* This list identifies wastes from common manufacturing and industrial processes, such as solvents that have been used in cleaning or degreasing operations. Because the processes producing these wastes can occur in different sectors of industry, the F-listed wastes are known as wastes from nonspecific sources.

- *K-list (source-specific wastes).* This list includes certain wastes from specific industries, such as petroleum refining or pesticide manufacturing. Certain sludges and wastewaters from treatment and production processes in these industries are examples of source-specific wastes.

- *P-list and U-list (discarded commercial chemical products).* These lists include specific commercial chemical products in an unused form. Some pesticides and some pharmaceutical products become hazardous waste when discarded.

The U.S. EPA allows for a waste to be *delisted* and considers the following categories from petitioners: cost savings and aggregate economic impacts, impacts of delisting on the environment, and impacts of delisting on the RCRA hazardous waste management program.

Universal wastes are wastes that do meet the regulatory definition of hazardous waste but are managed under special, tailored regulations. These wastes include:

- Batteries
- Pesticides
- Lamps/fluorescent bulbs
- Mercury-containing equipment/thermostats

Leftover household products that contain corrosive, toxic, ignitable, or reactive ingredients are considered to be *household hazardous waste,* or HHW. Americans generate 1.6 million tons of HHW per year. Products such as paints, cleaners, oils, batteries, and pesticides that contain potentially hazardous ingredients require special care for disposal. However, household hazardous wastes are exempt from RCRA Subtitle C regulations.

TSD Facilities[1]

RCRA Subtitle C established a *cradle-to-grave* concept, which controls hazardous waste from the time it is generated until its ultimate disposal. The latter stages of the cradle-to-grave time frame includes the *treatment, storage, and disposal (*TSD*)* of hazardous waste.

To track the hazardous waste from its cradle to grave, a *manifest* system is in place. The manifest system is a set of forms, reports, and procedures that tracks hazardous waste from the time it leaves the generator facility until it reaches the TSD facility. Manifests allow the waste generator to verify that its waste has been properly delivered and that no waste has been lost or unaccounted for in the process. A manifest provides information about the generator of the waste, the facility that will receive the waste, a description and quantity of the waste (including the number and type of containers), and how the waste will be routed to the receiving facility.

The EPA requires the following of TSD facilities (TSDFs):

- *Air emissions.* TSDFs must control the emissions of volatile organic compounds (VOCs) from process vents from certain hazardous waste treatment processes; hazardous waste management equipment (for example, valves, pumps, and compressors); and containers, tanks, and surface impoundments.

- *Closure.* When a TSDF ends operations and stops managing hazardous waste, it must be closed such that it will not pose a future threat to human health and the environment.

- *Corrective action/hazardous waste cleanup.* The RCRA Corrective Action Program allows TSDFs to address the investigation and cleanup of any hazardous releases themselves. Cleanup at closed or abandoned RCRA sites can also take place under the Superfund program.

---

1 U.S. EPA, *www.epa.gov/epaoswer/osw/tsds.htm.*

- *Financial assurance.* TSDFs must demonstrate that they will have the financial resources to properly close the facility or unit when its operational life is over, or provide the appropriate emergency response in the case of an accidental release.
- *Groundwater monitoring.* A TSDF must monitor the groundwater beneath its facility by installing groundwater monitoring wells and establishing a groundwater sampling regimen.

Building of TSDFs should be avoided in the following sensitive environments:

- 100-year floodplain (any land area that is subject to a 1 percent or greater chance of flooding in any given year from any source)
- Wetlands
- Recharge areas of high-quality groundwater
- Earthquake zones
- Karst terrain (rock such as limestone, dolomite, or gypsum that slowly dissolves when water passes through it and creates underground voids, tunnels, and caves, sometimes so large that their ceilings collapse forming large sinkholes[2])
- Unfavorable weather conditions (for example, siting an incinerator in an area prone to atmospheric inversion)
- Incompatible land use (for example, densely populated areas, areas near nursing homes, hospitals, day care facilities, etc.)

Community resistance to siting of a hazardous waste site can be extremely high, exacerbated by the NIMBY (not in my backyard), NIMTOO (not in my term of office), and NOPE (not on planet earth) mindsets. Citizens are concerned primarily over potential increases in risk due to increased levels of toxins in the environment and due to sudden, catastrophic failure of the TSDF. Other concerns include increased truck traffic, increased noise, objectionable odors, decrease in property values, decreased potential for economic development, and so on.

Storage

The EPA defines storage as the holding of waste for a temporary period of time prior to the waste being treated, disposed, or stored elsewhere. Hazardous waste is commonly stored prior to treatment or disposal and must be stored in one of the following, as defined by the U.S. EPA[3]:

- *Containers.* A hazardous waste container is any portable device in which a hazardous waste is stored, transported, treated, disposed, or otherwise handled. Examples of containers include 55-gallon drums, tanker trucks, railroad cars, buckets, bags, and even test tubes.
- *Tanks.* Tanks are stationary holding units constructed of nonearthen materials used to store or treat hazardous waste.

---

2   U.S. EPA, *Sensitive Environments and the Siting of Hazardous Waste Management Facilities,* www.epa.gov/epaoswer/hazwaste/tsds/site/sites.pdf.
3   www.epa.gov/epaoswer/osw/tsds.htm#store.

- *Drip pads.* A drip pad is a wood drying structure used by the pressure-treated wood industry to collect excess wood preservative drippage. Drip pads are constructed of nonearthen materials with a curbed, free-draining base that is designed to convey wood preservative drippage to a collection system for proper management.

- *Containment buildings.* Containment buildings are completely enclosed, self-supporting structures used to store or treat noncontainerized hazardous waste.

- *Waste piles.* A waste pile is an open, uncontained pile used for treating or storing waste. Hazardous waste piles must be placed on top of a double liner system to ensure leachate from the waste does not contaminate surface or groundwater supplies.

- *Surface impoundments.* A surface impoundment is a natural topographical depression, man-made excavation, or diked area such as a holding pond, storage pit, or settling lagoon. Surface impoundments are formed primarily of earthen materials and are lined with synthetic geomembranes to prevent liquids from escaping.

Disposal

Disposal may occur in one of the following: landfill, surface impoundment, waste pile, land treatment unit, injection well, salt dome formation, salt bed formation, underground mine, or underground cave. The latter four methods are geologic repositories. Such units vary greatly and are subject to environmental performance standards rather than prescribed technology-based standards.

Hazardous Waste Landfills

Hazardous waste landfills have many similarities to solid waste landfills; however, some additional considerations must be taken into account:

- A double composite liner is required, as is a leak detection system.

- Hazardous waste landfills must meet more stringent state requirements, which often include on-site state inspectors, additional groundwater monitoring wells, and restrictions on radioactive wastes.

- Liquid wastes are not allowed in hazardous waste landfills. Exceptions include very small containers (for example, laboratory ampules). Free liquids are eliminated by decanting or use of an absorbent material.

- All wastes delivered to a hazardous waste landfill must be manifested (to allow cradle-to-grave tracking).

- Care must be taken to ensure that incompatible wastes are not stored in close proximity to one another.

Operation of a hazardous waste landfill also differs from operation of a municipal solid waste landfill. Waste that is obtained as solid or semisolid form is spread in two- to three-foot layers and compacted. A 12-inch layer of daily cover must be used. Containers are placed upright in the cell. Space between the containers is filled with compatible bulk hazardous waste or soil.

Injection Wells

Injection wells have also been termed "deep-well disposal" or "subsurface injection." The wells must be deep enough to reach a porous, permeable, saline-water-bearing rock stratum that is confined by relatively impermeable layers above and beneath. Depths are on the order of several thousand feet. A schematic is shown in Figure 16.13. Some terms related to Figure 16.13 are defined as follows:

- *Surface casing.* This casing prevents contamination of any aquifers used for drinking water. It is cemented along its entire length.

- *Inner casing.* This casing is cemented along its entire length in order to seal off the injected wastes from any geologic formations above the injection zone.

- *Injection tubing.* Waste is injected into the injection zone through this tubing.

- *Annulus.* The area between the injection tubing and inner casing is filled with an inert, pressurized fluid and held in place by the *packer*. Thus, injected waste is prevented from entering the annulus.

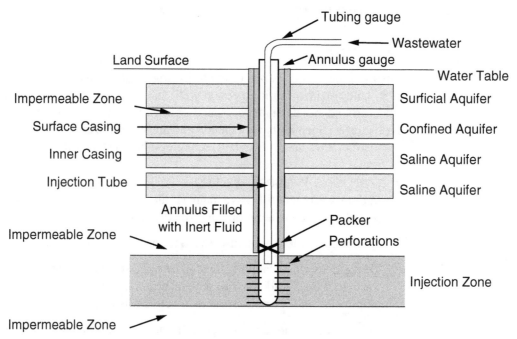

Figure 16.13 Deep-well injection

Source: *Remediation Technologies Screening Matrix and Reference Guide*, 4th ed., U.S. Army Environmental Command, SFIM-AEC-ET-CR-97053, January 2002.

The following limitations have been noted for deep-well injection by the Federal Remediation Technology Roundtable[4]:

- Injection is not feasible where seismic activity could occur.

- Injected wastes must be compatible with the mechanical components of the injection well system and the natural formation water. The waste generator may be required to perform pretreatment to ensure compatibility.

---

4 *FRTR Remediation Technologies Screening Matrix and Reference Guide, Version 4.0,* Federal Remediation Technology Roundtable, January 2002 (*www.frtr.gov/matrix2/section4/4-54.html*).

- High concentrations of suspended solids (typically > 2 ppm) can lead to plugging of the injection interval.

- Corrosive media may react with the injection well components, within the injection zone formation, or with confining strata with very undesirable results.

- High iron concentrations may result in fouling.

- Organic carbon may serve as an energy source for indigenous or injected bacteria resulting in rapid population growth and subsequent fouling.

- Extensive assessments are required before regulatory approval is obtained.

Biological Treatment

Biological treatment involves treatment of waste by bacteria, fungi, or algae to remove and degrade hazardous constituents. Examples include bioreactors (the principles of which are discussed in Chapter 14), constructed wetlands (Chapter 7), biopiles, landfarming, and slurry phase biological reactors.

Biopiles

Biopiles are a type of composting system in which contaminated soil is mixed with soil amendments (often proprietary) and aerated. The pile is placed on an impermeable membrane to protect groundwater quality and to allow leachate to be collected and recycled. VOCs may be released from a biopile, in which case further treatment of the VOCs may be necessary.

Land farming

Land farming is a common means of hazardous waste treatment. Although the land is not farmed in the sense that edible crops are produced (due to concerns with plant uptake of contaminants), the process follows many agricultural practices. In land farming, contaminated soil, sediment, or sludge is applied to lined beds, and periodically turned over or tilled to aerate the waste.

Slurry Phase Treatment

Bioremediation techniques have been successfully used to remediate soils, sludges, and sediments contaminated by explosives, petroleum hydrocarbons, solvents, pesticides, and a variety of other organic chemicals. The use of a bioreactor, such as slurry phase treatment, rather than *in situ* biological techniques is preferred for heterogeneous soils, low-permeability soils, areas where underlying groundwater would be difficult to capture, or when faster treatment times are required.

Slurry phase biological treatment involves the controlled treatment of contaminated soil in a bioreactor. Slurry phase treatment is similar to conventional suspended growth except for the high concentration of solids (10% to 30% solids by weight) and the fact that many of the solids may be inert. Although biological degradation is the primary removal process, removal can also occur via volatilization and desorption. A slurry phase bioreactor that is closed for cases when volatilization is a concern; the closed reactor allows the contaminated air to be treated and released in a controlled manner. Desorption can be achieved through pretreatment by adding surfactants.

Physical-Chemical Treatment

The following physical-chemical treatment processes will be discussed in this section: chemical oxidation, advanced oxidation, extraction, microencapsulation, neutralization, separation, soil washing, granular activated carbon, air stripping, and solidification/stabilization.

Chemical Oxidation/Advanced Oxidiation

Chemical oxidation uses strong oxidizing agents (for example, hypochlorite, peroxides, persulfates, percholorates, permanganates, etc.) to break down hazardous waste constituents to render them less toxic or mobile, while chemical reduction uses strong reducing agents (for example, sulfur dioxide, alkali salts, sulfides, iron salts, etc.). Oxidation may completely break down a contaminant, or at the least transform it into constituents that are more amenable to biodegradation.

Photolysis

Photolysis is the breaking of chemical bonds by the action of UV (ultraviolet) radiation or visible light. UV radiation has been demonstrated to degrade PCBs, dioxins, PAHs, BTEX, and so on. The energy of a photon E is governed by Planck's law:

$$E = h \cdot v$$

where

h = Planck's constant ($6.624 \cdot 10^{-27}$ erg $\cdot$ sec)

v = frequency (T^{-1})

The wavelength, frequency, and energy of various regions of the electromagnetic spectrum are shown in Table 16.4.

Table 16.4 Ranges of wavelengths, frequencies, and energies

	Wavelength (m)	Frequency (Hz)	Energy (J)
Radio	$> 1 \times 10^{-1}$	$< 3 \times 10^9$	$< 2 \times 10^{-24}$
Microwave	$1 \times 10^{-3} - 1 \times 10^{-1}$	$3 \times 10^9 - 3 \times 10^{11}$	$2 \times 10^{-24} - 2 \times 10^{-22}$
Infrared	$7 \times 10^{-7} - 1 \times 10^{-3}$	$3 \times 10^{11} - 4 \times 10^{14}$	$2 \times 10^{-22} - 3 \times 10^{-19}$
Optical (visible)	$4 \times 10^{-7} - 7 \times 10^{-7}$	$4 \times 10^{14} - 7.5 \times 10^{14}$	$3 \times 10^{-19} - 5 \times 10^{-19}$
UV	$1 \times 10^{-8} - 4 \times 10^{-7}$	$7.5 \times 10^{14} - 3 \times 10^{16}$	$5 \times 10^{-19} - 2 \times 10^{-17}$
X-ray	$1 \times 10^{-11} - 1 \times 10^{-8}$	$3 \times 10^{16} - 3 \times 10^{19}$	$2 \times 10^{-17} - 2 \times 10^{-14}$
Gamma-ray	$< 1 \times 10^{-11}$	$> 3 \times 10^{19}$	$> 2 \times 10^{-14}$

Source: M. D. LaGrega, P. L. Buckingham, and J. C. Evans, *Hazardous Waste Management*, 2nd ed., © 2001, McGraw-Hill Education. Reprinted by permission of McGraw-Hill Companies.

Extraction

Extraction is a process that separates hazardous constituents by means of settling, filtration, adsorption, absorption, or other means. Chemical extraction does not destroy wastes but is a means of separating hazardous contaminants from soils, sludges, and sediments. As a result, the volume of the hazardous waste to be treated is reduced. Physical separation steps may be required to separate the soil into coarse and fine fractions, as the fine constituents may contain most of the contamination.

Acid extraction may be used to extract heavy metals from the contaminated matrix. The residence time varies depending on the soil type, types of heavy metal, and heavy metal concentrations but generally ranges between 10 and 40 minutes. When extraction is complete, solids are separated and transferred to the rinse system. The heavy metals are concentrated in a form potentially suitable for recovery. Finally, the soils are dewatered and mixed with lime to neutralize any residual acid.

Solvent extraction uses an organic solvent to remove organic contaminants and organically bound metals. Solvent extraction has been shown to be effective in treating sediments, sludges, and soils containing primarily organic contaminants, such as polychlorinated biphenyls (PCBs), volatile organic compounds (VOCs), and petroleum wastes. Traces of solvent may remain within the treated soil matrix; thus, the toxicity of the solvent is an important consideration.

Microencapsulation
Microencapsulation is a process that coats the surface of the waste material with a chemical coating (such as a thin layer of plastic or resin) to prevent the material from leaching hazardous waste constituents.

Neutralization
Neutralization is a process that is used to treat corrosive hazardous waste streams. Low pH acidic corrosive waste streams are usually neutralized by bases. High pH corrosive waste streams are usually neutralized by adding acids.

Separation
Separation processes seek to remove the contaminant from its liquid or solid matrix, effectively concentrating the contaminant in the process of purifying the original waste matrix.

One means of separation is distillation, whereby a liquid waste is heated and contaminants of varying volatilities are removed. In simple distillation, heat is applied to a liquid mixture in a *still,* causing a portion of the liquid to vaporize. The vapors are cooled and condensed producing a liquid product called *distillate.* The relative concentration of volatile contaminants is higher in the distillate than in the mixture remaining in the still. In practice, distillation can be very complex and expensive and may require extensive land area and tall heights (possibly exceeding 200 ft).

Soil Washing
Soil washing differs from extraction techniques in that it uses a water-based solution to separate contaminants from contaminated soil. The process removes contaminants from soils by

- dissolving or suspending contaminants in the wash solution; or
- concentrating them into a smaller volume of soil through particle size separation, gravity separation, and attrition scrubbing.

The latter process (attrition scrubbing) is necessary because many contaminants preferentially bind to the clay, silt, and organic portions of soil. The clays, silts, and organics are attached to the larger inert solids (that is, sand and gravel). Soil washing can separate these smaller particles from the sand and gravel, and collecting these smaller particles effectively concentrates the contaminants. The sand and gravel can be returned to the site, perhaps without any further processing. Like the acid and solvent extraction processes, soil washing does not destroy contaminants, but transfers them from one media to another.

Air Stripping

Absorption is a phenomenon by which contaminant mass is transferred from the liquid phase to the gas phase (*air stripping*), or from the gas phase to the liquid phase (*gas scrubbing*). Both scenarios involve a column filled with a high surface area, high void volume packing material, often made of a plastic or ceramic compound, which is configured to provide uniform flow of both the liquid and gas phases. Mass transfer occurs at the interface between the liquid and gas and often a chemical is added to the liquid to further enhance the removal of the target species. While the governing equations were covered in detail in Chapter 14, Example 16.12 below will apply the concept to a contaminated groundwater source.

Example 16.12

Preliminary sizing of air stripper

Determine the stripping factor and the height of packing for removing ethylbenzene ($H' = 0.27$) from contaminated groundwater. The groundwater has an ethylbenzene concentration of 1.2 mg/L, and it must be decreased to 10 µg/L. Other data are

$K_L a = 0.020$ s^{-1}

$Q_w = 18$ m³/hr

Column diameter = 0.75 m

Air-to-water ratio (Q_0/Q_w) = 25

Solution

The following values are found from some simple preliminary calculations:

Cross-sectional area = 0.442 m²

$C_{in}/y_{out} = 120$

$Q_w = 5 * 10^{-3}$ m³/s

$Q_a = 1.25 * 10^{-4}$ m³/s

R, HTU, and NTU may be calculated from the equations provided in Chapter 14 as:

$$R = H' \frac{Q_a}{Q_w} = 0.27 * 25 = 6.75$$

$$HTU = \frac{Q_w}{K_L a A} = \frac{5 \cdot 10^{-3} \text{ m}^3/\text{s}}{0.02 \text{ s}^{-1} \cdot 0.442 \text{ m}^2} = 0.57 \text{m}$$

$$NTU = \frac{R}{R-1} \ln \left[\frac{(C_{in}/y_{out})(R-1)+1}{R} \right] = \frac{6.75}{5.75} \ln \left[\frac{(120)(5.75)+1}{6.75} \right]$$

$= 5.4$ transfer units

The height of the column packing is

$Z = HTU * NTU$

$= 0.57 \text{ m} \cdot 5.4 = 3.1 \text{ m}$

Adsorption

Adsorption is a process in which contaminant molecules (called the *adsorbate* or *solute*) that are present either in the gas or liquid stream are removed from the fluid through their attachment to a solid surface (called the *adsorbent*). This technology is attractive for contaminants present in low concentrations that have a high monetary value or may impart a substantial negative impact to the environment upon release. The adsorbate preferentially attaches to the adsorbent due to chemical or physical attraction, and the attachment is generally stable as long as the fluid properties (temperature, pressure, constituent compositions) remain constant.

Solidification/Stabilization

Stabilization and solidification processes reduce the mobility of the hazardous constituents of a waste or make the waste easier to handle. Wastes may be solidified *in situ* or *ex situ*. Solidification processes are useful on inorganic and organic wastes. When used with a waste containing heavy metals, the metals are bound in place and unable to enter the environment. Such decrease in contaminant mobility can be characterized by the TCLP test.

Solidification and stabilization are not synonyms and are defined as follows (U.S. EPA Document EPA/542-R-00-010):

- *Solidification* refers to processes that encapsulate a waste to form a solid material and to restrict contaminant migration by decreasing the surface area exposed to leaching and/or by coating the waste with low-permeability materials. Solidification can be accomplished by a chemical reaction between a waste and binding (solidifying) reagents or by mechanical processes. Solidification of fine waste particles is referred to as microencapsulation, while solidification of a large block or container of waste (e.g., a 55-gallon drum) is referred to as *macroencapsulation*.

- *Stabilization* refers to processes that involve chemical reactions that reduce the leachability of a waste. Stabilization chemically immobilizes hazardous materials or reduces their solubility through a chemical reaction. The physical nature of the waste may or may not be changed by this process.

Thermal Treatment

Similar to the thermal methods used for the treatment of MSW, hazardous wastes can also be treated by following thermal methods such as incineration and thermal desorption.

Incineration

Incineration of solid waste was discussed in detail previously. Incineration is the high temperature burning (rapid oxidation) of a waste, usually at 1400°F to 2500°F. This process, unlike many of the other processes described in this chapter, destroys contaminants rather than stabilizes them or transfers them to another medium. Often auxiliary fuels are employed to initiate and sustain combustion. The destruction and removal efficiency (DRE) for properly operated incinerators exceeds the 99.99% requirement for hazardous waste and can be operated to meet the 99.9999% requirement for PCBs and dioxins. However, off-gases and combustion residuals require further treatment (for example, bottom ash may require stabilization).

Three critical factors ensure the completeness of combustion in an incinerator, also known as the "three Ts":

1. The *temperature* in the combustion chamber
2. The length of *time* wastes are maintained at high temperatures
3. The *turbulence,* or degree of mixing, of the wastes and the air

Thermal Desorption

Thermal desorption is a physical separation process and as such does not destroy organics. Rather, it transfers the contaminants from the liquid phase to the vapor phase. Wastes are heated to volatilize water and organic contaminants. The process has been used extensively to treat nonhazardous wastes, such as those from petroleum production. Such wastes only require temperatures below 400°C (*low temperature desorption*). Conversely, many hazardous wastes require higher temperatures for desorption (540°C–650°C), still much lower than incineration temperatures.

SITE REMEDIATION

The Comprehensive Environmental Response, Compensation, and Liability Act (CERCLA), commonly known as Superfund, was enacted by Congress on December 11, 1980. CERCLA[5]

- established prohibitions and requirements concerning closed and abandoned hazardous waste sites,
- provided for liability of persons responsible for releases of hazardous waste at these sites, and
- established a trust fund to provide for cleanup when no responsible party could be identified.

The law authorizes two kinds of response actions:

1. Short-term removals to promptly address releases
2. Long-term remedial response actions that permanently and significantly reduce the dangers associated with releases or threats of releases of hazardous substances

Superfund Process

This section will focus on the cleanup process governed by Superfund. However, it is important to note that the steps followed to clean up a site under Superfund are similar to the steps required to clean up a site without being under the auspices of Superfund (that is, voluntary cleanup or cleanup regulated by individual states).

The Superfund cleanup process consists of the following steps.[6]

---

5   U.S. EPA, *www.epa.gov/superfund/action/law/cercla.htm*.
6   U.S. EPA, *www.epa.gov/superfund/action/process/sfproces.htm*.

Step 1: Site Discovery
Site discovery may be made by any number of parties, including local and state agencies, businesses, the U.S. Environmental Protection Agency (U.S. EPA), or by members of the public.

Step 2: Preliminary Assessment/Site Inspection (PA/SI)
Preliminary assessment evaluates the level of threat to human health and the environment. Sites requiring emergency response may be identified in the PA. If the PA recommends that additional investigations be made, a site inspection will be conducted. During the SI, soil, water, and air samples are obtained and analyzed. A vast amount of information is collected that serves as input to the Hazard Ranking System (HRS).

Step 3: HRS Scoring
The HRS is a numerical screening mechanism used to place sites on the National Priorities List (NPL). A high HRS score does not infer a higher priority for funding. The HRS score is the combination of scores from each of four pathways:

1. Groundwater migration
2. Surface water migration
3. Soil exposure
4. Air migration

Step 4: NPL Site Listing Process
A site is placed on the NPL by one of three mechanisms:

1. Magnitude of the HRS score
2. Designation by states or territories regardless of score
3. Ability to meet each of the three following requirements:
   - The Agency for Toxic Substances and Disease Registry (ATSDR) of the U.S. Public Health Service has issued a health advisory that recommends removing people from the site.
   - The U.S. EPA determines the site poses a significant threat to public health.
   - The U.S. EPA anticipates it will be more cost-effective to use its remedial authority (available only at NPL sites) than to use its emergency removal authority to respond to the site.

Step 5: Remedial Investigation/Feasibility Study (RI/FS)
The remedial investigation serves as the mechanism for collecting data in addition to that data collected for the HRS score in order to

- characterize site conditions,
- determine the nature of the waste,
- assess risk to human health and the environment, and
- conduct pilot- or laboratory-scale testing to evaluate the potential performance and cost of the treatment technologies that are being considered.

The feasibility study is the mechanism for the development, screening, and detailed evaluation of alternative remedial actions. The RI and FS are conducted concurrently. Data collected in the RI influence the development of remedial alternatives in the FS, which in turn affects the data needs and scope of treatability studies and additional field investigations.

Step 6: Record of Decision (ROD)

The Record of Decision (ROD) is a public document that explains which cleanup alternatives will be used to clean up a Superfund site.

Step 7: Remedial Design/Remedial Action (RD/RA)

Based on the specifications outlined in the ROD, full-scale remediation technologies are designed (remedial design) and constructed and implemented (remedial action).

Step 8: Construction Completion

Sites qualify for construction completion when

- all necessary physical construction is complete, regardless of whether final cleanup levels have been achieved; or
- the U.S. EPA has determined that the response action should be limited to measures that do not involve construction; or
- the site qualifies for deletion from the NPL.

Inclusion of a site on the Construction Completions List (CCL) has no legal significance.

Step 9: Postconstruction Completion

The goal of postconstruction tasks is to ensure the long-term protection of human health and the environment. Postconstruction completion steps may include:

- Operations and maintenance to ensure the effectiveness of the remedy
- Five-year reviews
- Assurance of proper institutional controls such as administrative and/or legal controls to minimize the potential for human exposure to contamination and/or protect the integrity of the remedy by limiting land or resource use
- Optimization of the remedy to decrease annual operating costs
- Deletion from the NPL if no further response action is appropriate
- Assurance that reuse activities do not adversely affect the implemented remedy

In-situ vs. Ex-Situ

Hazardous waste sites can be remediated using a wide array of technologies. One way to classify the treatment methods is *ex situ* or *in situ*. For *in situ* (or in place) treatment, the contaminant is not moved from the subsurface, while *ex situ* treatment involves the excavation of soil (or pumping of contaminated groundwater from the aquifer) for treatment.

Advantages of *in situ* treatment include the following:

- No costs for excavation or groundwater extraction
- Ability to treat soils under buildings and other structures without affecting the structure
- Avoidance of risks and costs associated with transportation
- Decreasing likelihood of spreading contaminants off-site

Advantages of *ex situ* treatment include:

- Shorter time periods for remediation
- Greater uniformity of treatment due to homogenization of solid phase (for example, soil or sludge) or the ability to monitor and continuously mix the groundwater

The *ex situ* treatment of groundwater is often termed "pump and treat." Pump-and-treat technologies are used at approximately three-quarters of all Superfund sites where groundwater is contaminated.

Biological Treatment

Bioremediation techniques destroy contaminants by creating a favorable environment for the microorganisms such that the microorganisms are able to grow and use the contaminants as a food and energy source.

Biological processes are implemented at a relatively low cost. Bioremediation has been used to remediate groundwater contaminated by petroleum hydrocarbons, solvents, pesticides, wood preservatives, and other organic chemicals. Care must be taken when contemplating a remediation alternative to consider that some compounds may be broken down into more toxic by-products during the bioremediation process (for example, TCE to vinyl chloride).

The following must be considered when designing an *in situ* biological treatment process:

- Oxygen must be supplied at rates sufficient to maintain aerobic conditions and may be accomplished by forced air, liquid oxygen injection, or hydrogen peroxide injection. Alternatively, anaerobic conditions can be used to degrade highly chlorinated contaminants. The dechlorinated contaminants can be biodegraded by subsequent aerobic treatment.

- Nutrients required for cell growth include nitrogen, phosphorus, potassium, sulfur, magnesium, calcium, manganese, iron, zinc, and copper. Some or all of these may be added to the subsurface (for example, as ammonium for nitrogen and as phosphate for phosphorus).

- Control of pH is critical, as it affects the solubility of constituents that can affect biological activity. For example, many metals that are potentially toxic to microorganisms are insoluble at elevated pH; therefore, elevating the pH of the treatment system can reduce this risk.

- Temperature affects microbial activity, with biological activity increasing as temperature increases. Provisions for heating the bioremediation site, such as use of warm-air injection, may speed up the remediation process. Increasing temperature promotes volatilization of VOCs, increases the solubility of most contaminants, and decreases the amount of oxygen that can be dissolved in the water.

- Bioaugmentation involves the use of microorganisms that have been specially bred for degradation of a specific contaminant or for survival under unusually severe environmental conditions. Bioaugmentation also takes the form of accelerating the growth of the natural microorganisms that preferentially feed on contaminants at the site.

- *Cometabolism* is the transformation of an organic compound by a microorganism that does not use the compound as a source of energy or as one of its constituent elements. For example, microorganisms growing on one compound can produce an enzyme that chemically transforms another compound. A practical example is that of certain microorganisms that degrade methane; in the process of degrading the methane, the microorganisms produce enzymes that can initiate the oxidation of a variety of carbon compounds.

Bioventing

Bioventing involves the supply of relatively low airflow rates to sustain microbiological activity in the vadose (unsaturated) zone. The air (or oxygen) is supplied to the unsaturated zone. Compounds volatilized in the process are further degraded biologically as they travel upwards through the soil matrix. The process is relatively inexpensive and simple to operate and creates minimal disturbance to the area. Care must be taken to ensure that the soil permeability is high enough to encourage air movement and to ensure that the seasonally high groundwater table does not inhibit the transfer of air in the treatment area.

Phytoremediation

Phytoremediation is a process that uses plants to remove, transfer, stabilize, and destroy contaminants in soil and sediment. Upon uptake of contaminants (*phyto-accumulation*), the contaminants may be stored in the roots, stems, or leaves; changed into less harmful chemicals within the plant (*phyto-degradation*); or changed into gases that are released into the air as the plant transpires. For example, it is thought that poplar trees can degrade trichloroethylene in which the carbon is used for tissue growth and the chloride is expelled through the roots. Phyto-accumulation by itself does not destroy contaminants but, in effect, extracts the contaminants from the soil; the resulting mass is much smaller than the mass of soil that would otherwise require disposal. Alternatively, in enhanced *rhizosphere biodegradation*, plant roots can create an environment that encourages biodegradation through release of nutrients and by attracting water from deeper levels of the subsurface. Microbial counts in rhizosphere soils can be one or two orders of magnitude greater than in nonrhizosphere soils.

Enhanced Bioremediation

Enhanced bioremediation stimulates the natural biodegradation process by providing nutrients, electron acceptors, and even additional microorganisms. The process destroys contaminants in the process. Oxygen is provided to the contaminated area typically by air sparging or addition of hydrogen peroxide. Air sparging is the process by which air is injected into the groundwater. Care must be taken when using hydrogen peroxide as high concentrations (above 1000 ppm) are toxic to microorganisms.

Monitored Natural Attenuation

Monitored natural attenuation is not a technology per se but a process whereby a number of fundamental processes are allowed to occur, thus reducing contaminant concentrations. These fundamental processes include dilution, dispersion, volatilization, biodegradation, radioactive decay, adsorption, and other chemical reactions.

Given the long-term time frame associated with natural attenuation, the U.S. EPA recommends this process only for those sites for which extensive off-site migration is not a concern. Also, the U.S. EPA encourages the use of natural attenuation in conjunction with more "active" techniques, perhaps as a follow-up step to more traditional active techniques.

Physical-Chemical Treatment

A variety of physical/chemical treatment processes are available that rely on such fundamental physical and chemical processes as volatilization, electrical separation, and a variety of chemical reactions.

Air Sparging

Air sparging was mentioned previously as a means of introducing air into the saturated portions of the aquifer for enhanced bioremediation. The incorporation of dissolved air into groundwater systems is common to many biological, physical, and chemical treatment processes, and the factors influencing the effectiveness of soil aeration systems are shown in Table 16.5.

Table 16.5 Factors affecting soil aeration systems

Soil Properties	Contaminant Properties	Environmenta Properties
Permeability	Henry's law constant	Temperature
Porosity	Solubility	Humidity
Grain size distribution	Adsorption coefficient	Wind speed
Moisture content	VOC concentration in soil	Solar radiation
pH	Polarity	Rainfall
Organic content	Vapor pressure	Terrain
Bulk density	Diffusion coefficient	Vegetation

Source: *Remediation Technologies Screening Matrix and Reference Guide*, 4th ed., U.S. Army Environmental Command, SFIM-AEC-ET-CR-97053, January 2002.

Passive/Reactive Treatment Walls

Treatment walls, or *permeable reactive barriers*, are trenches dug into the ground perpendicular to groundwater flow. The trench may be large enough to capture the entire contaminant plume; alternatively, a *funnel and gate* system may be used. A funnel and gate system relies on low-permeability walls formed using slurry walls or sheet piles. The groundwater plume cannot flow through these walls and is thus directed to gaps in the walls, or gates, that contain the reactive material.

Soil Flushing

Soil flushing involves the injection of an extraction fluid such as water or a mixture of water and chemicals into a contaminated soil mass and recovery of the fluid at a series of down-gradient wells. Contaminants in the soil partition into the extraction fluid, which is treated before being recycled or discharged from the site.

Soil Vapor Extraction

Soil vapor extraction (SVE) is used to remove contaminants from the unsaturated region (vadose zone) of the subsurface. Removal of VOCs from the vadose zone follows these two steps:

Step 1: Volatilization of VOCs to soil gas may occur directly from the residual contaminant adsorbed to the soil particle surface or via desorption from the soil particle surface; solubilization into soil water; and eventual volatilization into the soil gas. The transfer of the contaminant from the aqueous phase to the gas phase is governed by Henry's law, and the transfer from concentrated NAPL solutions to the gas phase is governed by Raoult's law.

Step 2: Once the contaminant has volatilized, movement of the contaminant through the soil media vapor phase occurs via advection (movement with bulk airflow) and diffusion (movement due to concentration gradient). The former controls in high permeability soil while the latter controls in low permeability soils.

Example 16.13

Soil vapor extraction of PCE

PCE has leaked from a dry cleaning facility such that 900 lb of PCE have contaminated the aquifer underlying the facility. Groundwater concentrations as high as 37 mg/L have been found. Treatment by SVE has commenced, using an extraction airflow of 300 ft³/min. The extracted air is being treated by a GAC bed. The average concentration of PCE in the air entering the GAC bed for each month following startup is shown in Exhibit 3.

Determine the mass of PCE that has been removed in the first 10 weeks and estimate the time by which 90% of the TCE will be removed.

Time (month)	1	2	3	4	5	6	7	8	9	10
Air concentration (mg/m³)	120	93	65	45	35	26	17	9	5	3

Exhibit 3

Solution

The mass of contaminant removed in the first month can be determined as follows:

$$\frac{120 \text{ mg}}{\text{m}^3} \cdot \frac{300 \text{ ft}^3}{\text{min}} \cdot \frac{1440 \text{ min}}{\text{day}} \cdot \frac{30 \text{ day}}{\text{month}} \cdot \frac{0.0283 \text{ m}^3}{1 \text{ ft}^3} \cdot \frac{2.205 \text{ lb}}{10^6 \text{ mg}} = 96 \text{ lb}$$

A similar procedure can be completed for each of the 10 months, and results are shown in Exhibit 4.

Time (month)	1	2	3	4	5	6	7	8	9	10
Mass (lb)	96.0	74.4	52.0	36.0	28.0	20.8	13.6	7.2	4.0	2.4

Exhibit 4 Mass of contaminant removed per month

The total mass removed is the sum of the monthly masses removed, or 334.5 lb. A first estimate for the time of cleanup would be to assume a linear model for removal; that is, if 334.5 lb were removed in the first 10-month period, an additional 334.5 lb could be removed in a second period of 10 months. However, a linear trend is clearly not the case, as shown in Exhibit 5. This decrease in the rate of removal occurs for many reasons, such as the decrease in diffusion rates as concentration gradients decrease. Moreover, removing the entire mass of PCE is not feasible, given the inability of extracted air to reach all of the contamination, the fact that some of the PCE is dissolved in the groundwater, and the fact that some of the PCE has most likely volatilized prior to implementation of SVE.

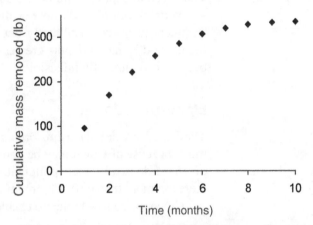

Exhibit 5 Cumulative mass of contaminant removed

Solidification/Stabilization

One of the most common means of in situ solidification and stabilization is vitrification. The concept behind in situ vitrification is the same as for ex situ vitrification where high temperatures (1600°C to 2000°C or 2900°F to 3650°F) are employed to melt the soil, producing a glasslike final product. In in situ vitrification, the high temperatures are obtained by passing an electric current through the soil matrix. The off-gases are collected in a large hood and treated before release. Organic compounds are destroyed by the process, while radionuclides and metals are retained within the vitrified mass in which they are resistant to leaching.

This process is not applicable to sites with large amounts of explosive or combustible materials. Also, if there are any buried electrical conduction paths (such as metal drums or cables), melting of the soil will not occur.

Containment

The purpose of containment is not to treat the waste but rather to isolate the waste. The isolated waste may be subjected to further treatment techniques or allowed to naturally attenuate.

Physical Barriers

Physical barriers such as *vertical cutoff walls* are used to contain or divert contaminated groundwater or divert uncontaminated groundwater flow from a contaminated region. These barriers can be installed quickly to provide additional time to design and implement a remediation process.

Physical barriers can be constructed using a variety of materials. *Sheet pile cutoff walls* can be constructed by driving steel or HDPE (high density polyethylene) piles into the soil and sealing the joints between the piles. Care must be taken when locating sheet pile walls to avoid underground structures and utilities.

Slurry walls are constructed by digging a 0.5–2-m-wide trench in the soil and backfilling with a slurry mixture of bentonite and native materials. Depths up to 50 m (160 ft) are possible. A third possibility is the use of *grout walls*, in which grouting compounds are forced under pressure into the ground. Grouting compounds include cement, bentonite, and silicate.

Cutoff walls can be designed as *hanging walls,* in which the bottom of the wall is "suspended" above the bottom of the aquifer; although water can pass under this wall, LNAPLs will be contained. Alternatively, the wall can be "keyed in" to the low permeable formation (for example, bedrock or clay) that serves as an aquitard. The keyed in depth should be 2–3 feet.

Vertical cutoff walls do not directly treat the contaminants, although such treatment may occur indirectly due to natural attenuation. Another limitation of the method is that at depths greater than 80 feet, sheet pile walls and slurry walls become economically infeasible.

Brownfields

The U.S. EPA defines a *brownfield* as "real property, the expansion, redevelopment, or reuse of which may be complicated by the presence or potential presence of a hazardous substance, pollutant, or contaminant." The U.S. EPA estimates that there are more than 450,000 brownfields in the United States.

Converting brownfields to usable properties has the following benefits:

- Increases local tax bases
- Facilitates job growth
- Utilizes existing infrastructure
- Takes development pressures off of undeveloped, open land
- Protects the environment

RADIOACTIVE WASTE

Radioactive wastes can be divided into six categories[7]:

1. Spent nuclear fuel from nuclear reactors
2. High-level radioactive waste from the reprocessing of spent nuclear fuel
3. Transuranic radioactive waste, resulting mainly from manufacture of nuclear weapons
4. Uranium mill tailings from the mining and milling of uranium ore
5. Low-level radioactive waste, generally in the form of radioactively contaminated industrial or research waste
6. Naturally occurring radioactive material

Remediation technologies include solidification (typically, cement ash or through vitrification) and stabilization. Stabilization involves the addition of chemical binders, such as cement, silicates, or pozzolans, which limit the solubility or mobility of radionuclides. Groundwater can be remediated by removal of radionuclides via coagulation and flocculation. Precipitation with hydroxides, carbonates, or sulfides is possible, as is ion exchange. Such technologies are generally applicable for low-level radioactive waste (LLW), transuranic waste (TRU), and/or uranium mill tailings. The technologies are not applicable to spent nuclear fuel and, for the most part, are not applicable for high-level radioactive waste.

---

7 U.S. EPA, *www.epa.gov/radiation/manage.htm.*

Some special considerations when remediating sites contaminated with radionuclides include the following[8]:

- Implementation of remediation technologies should consider the potential for radiological exposure to workers (internal and external). The degree of hazard is based on the radionuclide(s) present and the type and energy of radiation emitted (that is, alpha particles, beta particles, gamma radiation, and neutron radiation).

- Because radionuclides are not destroyed, ex situ techniques will require eventual disposal of residual radioactive wastes. These waste forms must meet disposal site waste acceptance criteria.

- Some remediation technologies result in the concentration of radionuclides. By concentrating radionuclides, it is possible to change the classification of the waste, which impacts requirements for disposal. For example, concentrating radionuclides could result in LLW becoming TRU waste (if TRU radionuclides were concentrated to greater than 100 nanocuries/gm with half-lives greater than 20 years per gram of waste). Also, LLW classifications (for example, Class A, B, or C for commercial LLW) could change due to the concentration of radionuclides. Waste classification requirements for disposal of residual waste (if applicable) should be considered when evaluating remediation technologies.

- Disposal capacity for radioactive and mixed waste is extremely limited. Mixed waste is waste comprised of radioactive waste and hazardous waste.

---

8 *Remediation Technologies Screening Matrix and Reference Guide, Version 4.0*, Federal Remediation Technology Roundtable, January 2002; *www.frtr.gov/matrix2/section2/2_9_1.html.*

PROBLEMS

16.1 The percentage of municipal solid wastes that are derived from wood, wood by-products, and plant matter is most nearly
a. 70 percent
b. 60 percent
c. 50 percent
d. 40 percent

16.2 The primary difference between a coal-fired power plant that burns high-sulfur coal and a mass-burn municipal waste combustion resource recovery facility is the
a. gas scrubbing system
b. particulate control system
c. combustion chamber
d. electric power generators

16.3 A composting windrow has a triangular cross section and is 10 feet wide at the base, 10 feet high at the peak, and 100 feet long. You may assume the waste has a specific weight of 400 lb/yd³ as placed in the windrow. The waste material is 40-percent biodegradable and has the ultimate composition $C_{60}H_{95}O_{40}N$. Determine the amount of oxygen required to stabilize the waste.
a. 53 tons
b. 37 tons
c. 21 tons
d. 5 tons

16.4 Determine the thickness of a clay liner that has a hydraulic conductivity of 1.55×10^{-8} cm/s and a porosity of 0.2 if there is a hydraulic head of 6 inches and the breakthrough time is 20 years.
a. 24 inches
b. 20 inches
c. 16 inches
d. 12 inches

16.5 The waste material described in Problem 16.3 is placed in a landfill instead of taken to the composting facility. Assuming complete decomposition of the biodegradable fraction, what would be the percentage of landfill gas, by volume, that is methane?
a. 46.7 percent
b. 49.9 percent
c. 51.6 percent
d. 53.4 percent

16.6 Waste initially placed at 800 lb/yd³ now has a specific weight of 1650 lb/yd³. Determine the depth of the material above if the average specific weight for all of the waste is 1200 lb/yd³.
a. 225 ft
b. 195 ft
c. 155 ft
d. 115 ft

16.7 Which of the following classes of compounds could be safely combined with trichloroethane?
   a. Non-oxidizing acids
   b. Aromatic hydrocarbons
   c. Aromatic amines
   d. Caustic alkalis

16.8 An old underground storage tank has leaked diesel fuel over a period of several years and has contaminated the subsurface soil and groundwater above action levels. Which of the following treatment methods is most appropriate for remediation of the site following tank removal?
   a. Natural attenuation
   b. Enhanced groundwater bioremediation
   c. Pump-and-treat with carbon adsorption
   d. Air sparging with soil vapor extraction

16.9 Which one of the following is not a necessary feature of a low-level waste storage facility?
   a. Storage capacity for 1000 years of material
   b. Ability to retrieve specific material
   c. Monitoring inside the facility
   d. Monitoring outside of the facility

SOLUTIONS

16.1 a. Paper and cardboard account for approximately 40 percent of MSW mass. Yard wastes are almost 20 percent of total mass and food waste almost 10 percent. Wood scraps add another 2 percent.

16.2 c. Electric power generation is essentially the same for all power plants that use steam turbines. Particulate control is required for all solid fuel combustion facilities, and acid gases will be scrubbed using similar wet/dry lime injection systems. The combustion chamber for mass-burn municipal solid waste facilities must be designed to deal with a highly heterogeneous fuel, with greater mean particle size, broader size distributions, and usually much higher bottom ash fractions.

16.3 c. First determine the volume of the windrow as follows:

$$V = \frac{1}{2}bhl = (0.5)(10)(10)(100) = 5000 \text{ ft}^3 \times \left(\frac{1 \text{ yd}}{3 \text{ ft}}\right)^3 = 185.2 \text{ yd}^3$$

Now, the mass of the waste in the windrow may be determined as

$$185.2 \text{ yd}^3 \times 400 \frac{\text{lb}}{\text{yd}^3} = 74{,}074 \text{ lb}$$

To use the reaction given, mass must be converted to moles. This requires the molecular weight of the biodegradable material, which is determined as follows:

$$MW = (12)(60) + (1)(95) + (16)(40) + (14)(1) = 1469 \frac{\text{lb}}{\text{lbmol}}$$

The number of moles reacted can now be calculated, remembering that only 40 percent of the waste is biodegradable:

$$n = \frac{74{,}074 \text{ lb}}{1469 \frac{\text{lb}}{\text{lbmol}}} = 50.4 \text{ lbmol} \times (0.4) = 20.17 \text{ lbmol}$$

The moles of oxygen required to complete this transformation is determined by multiplying the moles of waste consumed by the stoichiometric coefficient for oxygen in the reaction given, which may be calculated as follows:

$$n_{O_2} = \left(\frac{(4)(60) + 95 - (2)(40) + (3)(1)}{4}\right) 20.17 = 1301 \text{ lbmol}$$

Finally, the mass of oxygen required is calculated as

$$m_{O_2} = 1301 \text{ lbmol} \times 32 \frac{\text{lb}}{\text{lbmol}} = 41{,}632 \text{ lb} \, O_2 \times \frac{1 \text{ ton}}{2000 \text{ lb}} = 20.82 \text{ tons } O_2$$

16.4 a. Using the expression given in the *FE Supplied-Reference Handbook* and the data given in the problem statement, we can write

$$t = \frac{d^2 \eta}{k(d+h)} = \frac{d^2(0.2)}{k(d+0.5)} = 20 \text{ yrs}$$

The units for k in the expression are ft/yr, so converting units we find

$$1.55 \times 10^{-8} \frac{\text{cm}}{\text{s}} \times \frac{1 \text{ ft}}{30.48 \text{ cm}} \times (60)(60)(24)(365) \frac{\text{s}}{\text{yr}} = 0.016 \frac{\text{ft}}{\text{yr}}$$

Now, substituting this into the expression above and rearranging we get

$$(20)(0.016)(d+0.5) = 0.2d^2 = 0.32d + 0.16 \Rightarrow d^2 - 1.6d - 0.8 = 0$$

Solving this by using the quadratic formula yields two answers:

$$d = \frac{1.6 \pm (1.6^2 - (4)(1)(-0.8))^{1/2}}{(2)(1)} \Rightarrow d = 2, -0.4$$

Only the positive answer makes sense, therefore the thickness of the liner is

$$\text{depth} = 2 \text{ ft} = 24 \text{ in}$$

16.5 c. From Problem 16.3 we know that the waste has the composition $C_{60}H_{95}O_{40}N$. The amount of waste is not important in this problem because we only need the ratio of product gasses, which is only dependent on stoichiometry. From the stoichiometry given in the problem statement, we can determine the amount of gas generated per unit waste as

$$CH_4 = \frac{(4)(60) - (95) - (2)(40) - (3)}{8} = 31.5 \text{ lbmol}$$

$$CO_2 = \frac{(4)(60) - (95) + (2)(40) + (3)}{8} = 28.5 \text{ lbmol}$$

$$NH_3 = 1 \text{ lbmol}$$

The total amount of landfill gas produced per unit waste is then

$$\text{Total} = 31.5 + 28.5 + 1 = 61 \text{ lbmol}$$

Finally, the fraction of methane can be determined as

$$X_{CH_4} = \frac{31.5}{61} = 51.6\%$$

16.6 b. From Figure 16.12, waste initially placed at 800 lb/yd³ which is now at 1650 lb/yd³ requires a pressure of

$$P \approx 60 \text{ psi} \times \left(\frac{36 \text{ in}}{1 \text{ yd}}\right)^2 = 77{,}760 \frac{\text{lb}}{\text{yd}^2}$$

Since the average specific weight is 1200 lb/yd³, we can find the average depth as follows:

$$77{,}760 \frac{\text{lb}}{\text{yd}^2} \div 1200 \frac{\text{lb}}{\text{yd}^3} = 64.8 \text{ yd} \times \frac{3 \text{ ft}}{1 \text{ yd}} = 194.4 \text{ ft}$$

16.7 b. Trichloroethane is classified as a halogenated organic compound. The hazardous waste compatibility chart in the environmental engineering section of the *FE Supplied-Reference Handbook* indicates that it will form a toxic gas if mixed with acids or amines, and a flammable gas when mixed with caustics.

16.8 d. Natural attenuation is not acceptable for contamination above regulated action levels. Bioremediation and pump-and-treat technologies may successfully address the groundwater, but would do little to address soil contamination in the vadose zone. Air sparging will address volatile constituents in the groundwater while soil vapor extraction would address contamination in the unsaturated zone.

16.9 a. LLW storage facilities should have cataloged storage and the ability to retrieve specific wastes, as well as substantial monitoring of the environment both inside and outside of the facility. Storage capacity should be sufficient to hold material until natural decay reduces radiation to non-hazardous levels.